THIRD EDITION

PHYSICS

JOHN D. CUTNELL

KENNETH W. JOHNSON

Southern Illinois University at Carbondale

John Wiley & Sons, Inc.

New York Chichester

Brisbane Toronto Singapore

Acquisitions Editor: Cliff Mills
Developmental Editor: Kathleen Dolan
Marketing Manager: Catherine Faduska
Senior Production Editor: Katharine Rubin
Designer: Madelyn Lesure and HRS Electronic Text Management
Manufacturing Manager: Lori Bulwin
Photo Researchers: Hilary Newman, Ramon Rivera-Moret
Photo Editor: Charles Hamilton
Director of Photo Department: Stella Kupferberg
Senior Illustration Coordinator: Sigmund Malinowski
Production Service: HRS Electronic Text Management/Lorraine Burke
Cover/Frontispiece Photograph: Brian Bailey/Adventure Photo

This book was set in Palatino by Progressive Information Technologies and printed and bound by Von Hoffmann. The cover was printed by Phoenix.

Recognizing the importance of preserving what has been written, it is a policy of John Wiley & Sons, Inc. to have books of enduring value published in the United States printed on acid-free paper, and we exert our best efforts to that end.

The paper in this book was manufactured by a mill whose forest management programs include sustained yield harvesting of its timberlands. Sustained yield harvesting principles ensure that the number of trees cut each year does not exceed the amount of new growth.

Library of Congress Cataloging-in-Publication Data:

Cutnell, John D.
Physics / John D. Cutnell, Kenneth W. Johnson. — 3rd ed.
p. cm.
Includes index.
ISBN 0-471-59773-2 (alk. paper)
1. Physics. I. Johnson, Kenneth W.
QC23.C985 1995
530—dc20 94-14050
CIP

Printed in the United States of America

10 9 8 7 6 5 4 3 2

For my wife Joan Cutnell, whose patience, encouragement, and help have been beyond measure and generously given.

For my wife Anne Johnson and children, Lauri, Rick, and Rob, who have given their unqualified love, understanding, and support during this special venture.

CONTENTS

PART TWO: THERMAL PHYSICS

PART THREE : WAVE MOTION

PART FOUR : ELECTRICITY AND MAGNETISM

PART FIVE : LIGHT AND OPTICS

PART SIX : MODERN PHYSICS

FUNDAMENTAL CONSTANTS

Quantity	Symbol	Value*
Avogadro's number	N_A	$6.022\,1367 \times 10^{23}$ mol^{-1}
Boltzmann's constant	k	$1.380\,658 \times 10^{-23}$ J/K
Electron charge magnitude	e	$1.602\,177\,33 \times 10^{-19}$ C
Permeability of free space	μ_0	$4\pi \times 10^{-7}$ T·m/A
Permittivity of free space	ϵ_0	$8.854\,187\,817 \times 10^{-12}$ C^2/(N·m^2)
Planck's constant	h	$6.626\,0755 \times 10^{-34}$ J·s
Mass of electron	m_e	$9.109\,3897 \times 10^{-31}$ kg
Mass of neutron	m_n	$1.674\,9286 \times 10^{-27}$ kg
Mass of proton	m_p	$1.672\,6231 \times 10^{-27}$ kg
Speed of light in vacuum	c	$2.997\,924\,58 \times 10^{8}$ m/s
Universal gravitational constant	G	$6.672\,59 \times 10^{-11}$ N·m^2/kg^2
Universal gas constant	R	8.314 510 J/(mol·K)

* 1986 CODATA recommended values.

USEFUL PHYSICAL DATA

Acceleration due to earth's gravity	9.80 m/s^2 = 32.2 ft/s^2
Atmospheric pressure at sea level	1.013×10^5 Pa = 14.70 lb/in.^{2}
Density of air (0 °C, 1 atm pressure)	1.29 kg/m^3
Speed of sound in air (20 °C)	343 m/s
Water	
Density (4 °C)	1.000×10^3 kg/m^3
Latent heat of fusion	3.35×10^5 J/kg
Latent heat of vaporization	2.26×10^6 J/kg
Specific heat capacity	4186 J/(kg·C°)
Earth	
Mass	5.98×10^{24} kg
Radius (equatorial)	6.38×10^6 m
Mean distance from sun	1.50×10^{11} m
Moon	
Mass	7.35×10^{22} kg
Radius (mean)	1.74×10^6 m
Mean distance from earth	3.85×10^8 m
Sun	
Mass	1.99×10^{30} kg
Radius (mean)	6.96×10^8 m

CONVERSION FACTORS

Length

1 in. = 2.54 cm
1 ft = 0.3048 m
1 mi = 5280 ft = 1.609 km
1 m = 3.281 ft
1 km = 0.6214 mi
1 angstrom (Å) = 10^{-10} m

Mass

1 slug = 14.59 kg
1 kg = 1000 grams = 6.852×10^{-2} slug
1 atomic mass unit (u) = 1.6605×10^{-27} kg
(1 kg has a weight of 2.205 lb where the acceleration due to gravity is 32.174 ft/s^2)

Time

1 day = 24 h = 1.44×10^3 min = 8.64×10^4 s
1 yr = 365.24 days = 3.156×10^7 s

Speed

1 mi/h = 1.609 km/h = 1.467 ft/s = 0.4470 m/s
1 km/h = 0.6214 mi/h = 0.2778 m/s = 0.9113 ft/s

Force

1 lb = 4.448 N
1 N = 10^5 dynes = 0.2248 lb

Work and Energy

1 J = 0.7376 ft·lb = 10^7 ergs
1 kcal = 4186 J
1 Btu = 1055 J
1 kWh = 3.600×10^6 J
1 eV = 1.602×10^{-19} J

Power

1 hp = 550 ft·lb/s = 745.7 W
1 W = 0.7376 ft·lb/s

Pressure

1 Pa = 1 N/m^2 = 1.450×10^{-4} lb/in.^{2}
1 lb/in.^{2} = 6.895×10^3 Pa
1 atm = 1.013×10^5 Pa = 1.013 bar = 14.70 lb/in.^{2} = 760 torr

Volume

1 liter = 10^{-3} m^3 = 1000 cm^3 = 0.03531 ft^3
1 ft^3 = 0.02832 m^3 = 7.481 U.S. gallons
1 U.S. gallon = 3.785×10^{-3} m^3 = 0.1337 ft^3

Angle

1 radian = 57.30°
1° = 0.01745 radian

STANDARD PREFIXES USED TO DENOTE MULTIPLES OF TEN

Prefix	Symbol	Factor
Tera	T	10^{12}
Giga	G	10^9
Mega	M	10^6
Kilo	k	10^3
Hecto	h	10^2
Deka	da	10^1
Deci	d	10^{-1}
Centi	c	10^{-2}
Milli	m	10^{-3}
Micro	μ	10^{-6}
Nano	n	10^{-9}
Pico	p	10^{-12}
Femto	f	10^{-15}

BASIC MATHEMATICAL FORMULAE

Area of a circle = πr^2
Circumference of a circle = $2\pi r$
Surface area of a sphere = $4\pi r^2$
Volume of a sphere = $\frac{4}{3}\pi r^3$
Pythagorean theorem: $h^2 = h_o^2 + h_a^2$
Sine of an angle: $\sin\theta = h_o/h$
Cosine of an angle: $\cos\theta = h_a/h$
Tangent of an angle: $\tan\theta = h_o/h_a$
Law of cosines: $c^2 = a^2 + b^2 - 2ab\cos\gamma$
Law of sines: $a/\sin\alpha = b/\sin\beta = c/\sin\gamma$
Quadratic formula:
If $ax^2 + bx + c = 0$, then $x = (-b \pm \sqrt{b^2 - 4ac})/(2a)$

SI UNITS

Quantity	Name of Unit	Symbol	Expression in Terms of Other SI Units
Length	meter	m	Base unit
Mass	kilogram	kg	Base unit
Time	second	s	Base unit
Electric current	ampere	A	Base unit
Temperature	kelvin	K	Base unit
Amount of substance	mole	mol	Base unit
Velocity	—	—	m/s
Acceleration	—	—	m/s^2
Force	newton	N	$kg \cdot m/s^2$
Work, energy	joule	J	$N \cdot m$
Power	watt	W	J/s
Impulse, momentum	—	—	$kg \cdot m/s$
Plane angle	radian	rad	m/m
Angular velocity	—	—	rad/s
Angular acceleration	—	—	rad/s^2
Torque	—	—	$N \cdot m$
Frequency	hertz	Hz	s^{-1}
Density	—	—	kg/m^3
Pressure, stress	pascal	Pa	N/m^2
Viscosity	—	—	$Pa \cdot s$
Electric charge	coulomb	C	$A \cdot s$
Electric field	—	—	N/C
Electric potential	volt	V	J/C
Resistance	ohm	Ω	V/A
Capacitance	farad	F	C/V
Inductance	henry	H	$V \cdot s/A$
Magnetic field	tesla	T	$N \cdot s/(C \cdot m)$
Magnetic flux	weber	Wb	$T \cdot m^2$
Specific heat capacity	—	—	$J/(kg \cdot K)$ or $J/(kg \cdot C°)$
Thermal conductivity	—	—	$J/(s \cdot m \cdot K)$ or $J/(s \cdot m \cdot C°)$
Entropy	—	—	J/K
Radioactive activity	becquerel	Bq	s^{-1}
Absorbed dose	gray	Gy	J/kg
Exposure	—	—	C/kg

THE GREEK ALPHABET

Alpha	A	α	Iota	I	ι	Rho	P	ρ
Beta	B	β	Kappa	K	κ	Sigma	Σ	σ
Gamma	Γ	γ	Lambda	Λ	λ	Tau	T	τ
Delta	Δ	δ	Mu	M	μ	Upsilon	Y	υ
Epsilon	E	ϵ	Nu	N	ν	Phi	Φ	ϕ
Zeta	Z	ζ	Xi	Ξ	ξ	Chi	X	χ
Eta	H	η	Omicron	O	o	Psi	Ψ	ψ
Theta	Θ	θ	Pi	Π	π	Omega	Ω	ω

PERIODIC TABLE OF THE ELEMENTS

Symbol — Cl 17 — Atomic number
Atomic mass* — 35.453
$3p^5$ — Electron configuration

Group I	Group II	Transition elements										Group III	Group IV	Group V	Group VI	Group VII	Group 0
H 1 1.00794 $1s^1$																	**He** 2 4.00260 $1s^2$
Li 3 6.941 $2s^1$	**Be** 4 9.01218 $2s^2$											**B** 5 10.81 $2p^1$	**C** 6 12.011 $2p^2$	**N** 7 14.0067 $2p^3$	**O** 8 15.9994 $2p^4$	**F** 9 18.9984 $2p^5$	**Ne** 10 20.179 $2p^6$
Na 11 22.9898 $3s^1$	**Mg** 12 24.305 $3s^2$											**Al** 13 26.9815 $3p^1$	**Si** 14 28.0855 $3p^2$	**P** 15 30.9738 $3p^3$	**S** 16 32.06 $3p^4$	**Cl** 17 35.453 $3p^5$	**Ar** 18 39.948 $3p^6$
K 19 39.0983 $4s^1$	**Ca** 20 40.08 $4s^2$	**Sc** 21 44.9559 $3d^14s^2$	**Ti** 22 47.88 $3d^24s^2$	**V** 23 50.9415 $3d^34s^2$	**Cr** 24 51.996 $3d^54s^1$	**Mn** 25 54.9380 $3d^54s^2$	**Fe** 26 55.847 $3d^64s^2$	**Co** 27 58.9332 $3d^74s^2$	**Ni** 28 58.69 $3d^84s^2$	**Cu** 29 63.546 $3d^{10}4s^1$	**Zn** 30 65.39 $3d^{10}4s^2$	**Ga** 31 69.72 $4p^1$	**Ge** 32 72.59 $4p^2$	**As** 33 74.9216 $4p^3$	**Se** 34 78.96 $4p^4$	**Br** 35 79.904 $4p^5$	**Kr** 36 83.80 $4p^6$
Rb 37 85.4678 $5s^1$	**Sr** 38 87.62 $5s^2$	**Y** 39 88.9059 $4d^15s^2$	**Zr** 40 91.224 $4d^25s^2$	**Nb** 41 92.9064 $4d^45s^1$	**Mo** 42 95.94 $4d^55s^1$	**Tc** 43 (98) $4d^55s^2$	**Ru** 44 101.07 $4d^75s^1$	**Rh** 45 102.906 $4d^85s^1$	**Pd** 46 106.42 $4d^{10}5s^0$	**Ag** 47 107.868 $4d^{10}5s^1$	**Cd** 48 112.41 $4d^{10}5s^2$	**In** 49 114.82 $5p^1$	**Sn** 50 118.71 $5p^2$	**Sb** 51 121.75 $5p^3$	**Te** 52 127.60 $5p^4$	**I** 53 126.905 $5p^5$	**Xe** 54 131.29 $5p^6$
Cs 55 132.905 $6s^1$	**Ba** 56 137.33 $6s^2$	57–71	**Hf** 72 178.49 $5d^26s^2$	**Ta** 73 180.948 $5d^36s^2$	**W** 74 183.85 $5d^46s^2$	**Re** 75 186.207 $5d^56s^2$	**Os** 76 190.2 $5d^66s^2$	**Ir** 77 192.22 $5d^76s^2$	**Pt** 78 195.08 $5d^86s^2$	**Au** 79 196.967 $5d^{10}6s^1$	**Hg** 80 200.59 $5d^{10}6s^2$	**Tl** 81 204.383 $6p^1$	**Pb** 82 207.2 $6p^2$	**Bi** 83 208.980 $6p^3$	**Po** 84 (209) $6p^4$	**At** 85 (210) $6p^5$	**Rn** 86 (222) $6p^6$
Fr 87 (223) $7s^1$	**Ra** 88 226.025 $7s^2$	89–103	**Rf** 104 (261) $6d^27s^2$	**Ha** 105 (262) $6d^37s^2$	106 (263)	107 (262)	108 (265)	109 (266)									

Lanthanide series

La 57 138.906 $5d^16s^2$	**Ce** 58 140.12 $4f^26s^2$	**Pr** 59 140.908 $4f^36s^2$	**Nd** 60 144.24 $4f^46s^2$	**Pm** 61 (145) $4f^56s^2$	**Sm** 62 150.36 $4f^66s^2$	**Eu** 63 151.96 $4f^76s^2$	**Gd** 64 157.25 $5d^14f^76s^2$	**Tb** 65 158.925 $4f^96s^2$	**Dy** 66 162.50 $4f^{10}6s^2$	**Ho** 67 164.930 $4f^{11}6s^2$	**Er** 68 167.26 $4f^{12}6s^2$	**Tm** 69 168.934 $4f^{13}6s^2$	**Yb** 70 173.04 $4f^{14}6s^2$	**Lu** 71 174.967 $5d^14f^{14}6s^2$

Actinide series

Ac 89 227.028 $6d^17s^2$	**Th** 90 232.038 $6d^27s^2$	**Pa** 91 231.036 $5f^26d^17s^2$	**U** 92 238.029 $5f^36d^17s^2$	**Np** 93 237.048 $5f^46d^17s^2$	**Pu** 94 (244) $5f^66d^07s^2$	**Am** 95 (243) $5f^76d^07s^2$	**Cm** 96 (247) $5f^76d^17s^2$	**Bk** 97 (247) $5f^96d^07s^2$	**Cf** 98 (251) $5f^{10}6d^07s^2$	**Es** 99 (252) $5f^{11}6d^07s^2$	**Fm** 100 (257) $5f^{12}6d^07s^2$	**Md** 101 (258) $5f^{13}6d^07s^2$	**No** 102 (259) $6d^07s^2$	**Lr** 103 (260) $6d^17s^2$

* Atomic mass values are averaged over isotopes according to the percentages that occur on the earth's surface. For many unstable elements, the mass number of the most stable known isotope is given in parentheses. *Source: Handbook of Chemistry and Physics,* 68th ed., CRC Press, Boca Raton, FL. Reprinted by permission.

PART FOUR

ELECTRICITY AND MAGNETISM

In a very real sense, electricity and magnetism have made modern civilization what it is today. Take them away and you take away much of what fills our houses and our lives. For instance, electricity provides virtually all of our artificial lighting, much of our heating, and many time-saving household appliances. Magnetism is an integral part of the motors found in an enormous number of useful devices, including fans, hair driers, vacuum cleaners, pumps, and food processors. It is also an integral part of the transformers that play a vital role in the distribution of electric power throughout the country.

The widespread impact of electricity and magnetism on society has come about because they are closely related. One of the most exciting discoveries in physics was that electricity and magnetism are both manifestations of the same fundamental force, which is called the electromagnetic force. The close interplay between electricity and magnetism has led to a far greater number of practical applications than would otherwise have been possible. The entire magnetic tape recording industry is based on the relationship between them. Medical diagnostic techniques, such as magnetic resonance imaging, depend on that relationship. The commercial generation of electric power depends on it too. And so does radio, television, and communications technology, as well as the information storage technology that is the foundation of the computer industry.

In this part of the text, we will begin by discussing the electrically charged particles that are found in nature. We will see that these particles, whether moving or at rest, exert forces on one another. These forces are the electric part of the electromagnetic force. In addition, the magnetic part appears only when the charged particles are moving. Along with familiar concepts such as force and energy, we will encounter some new and useful concepts as we learn about the fundamental laws that govern electricity and magnetism.

Many of the applications of electricity and magnetism occur in the form of electric circuits, such as those in automobiles, airplanes, television sets, stereo systems, and computers. We will discuss circuits in two places. The first will occur in Chapter 20 after we study basic electricity. The second will be in Chapter 23 after we study magnetism and are in a position to understand some of the circuit components that depend on magnetism.

CHAPTER 18 ELECTRIC FORCES AND ELECTRIC FIELDS

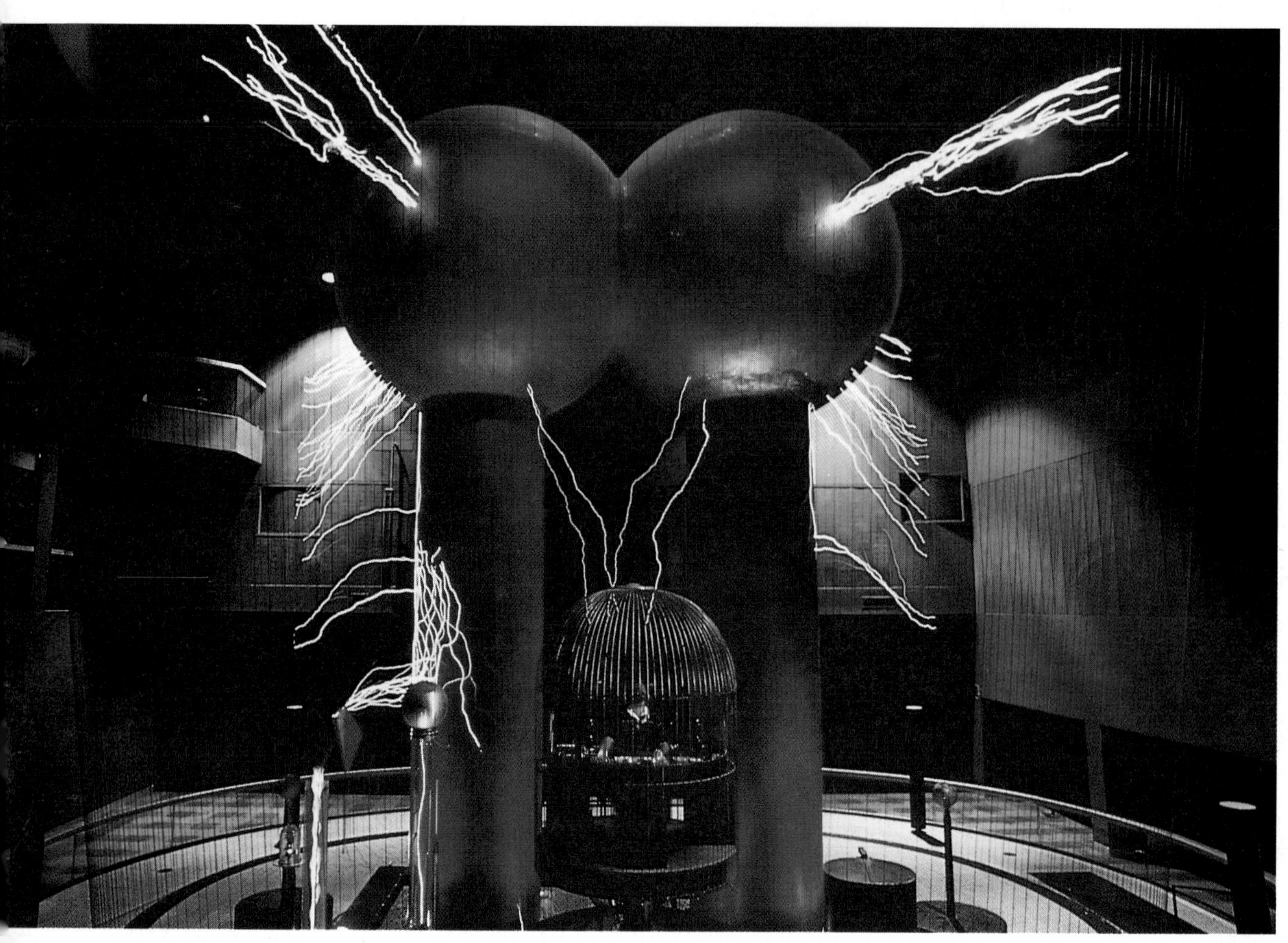

Lightning is nature's most spectacular display of electricity. The artificial lightning shown here is produced by (red-colored) Van de Graaff electric generators, while the wire frame of the cage protects the person inside it, just like an automobile shields its passengers from real lightning. In this chapter we will see that electricity consists of two kinds of electric charge, positive and negative. These charges exert a force, called the electric force, on each other. It is this force that holds together the positive and negative charges that make up atoms and molecules. Since our bodies are composed entirely of atoms and molecules, we owe our very existence to the electric force. In addition to exploring the properties of this force, we will show how it leads to the operation of many useful devices, such as photocopiers and laser printers.

18.1 THE ORIGIN OF ELECTRICITY

The electrical nature of matter is inherent in the atoms of all substances. An atom consists of a small, relatively massive nucleus that contains particles called protons and neutrons. Surrounding the nucleus is a diffuse cloud of orbiting particles known as electrons, as Figure 18.1 indicates. A proton has a mass of 1.67×10^{-27} kg, while an electron has a mass of 9.11×10^{-31} kg. Like mass, ***electric charge*** is an intrinsic property of protons and electrons, and only two types of charge have been discovered, positive and negative. A proton has a positive charge, and an electron has a negative charge.

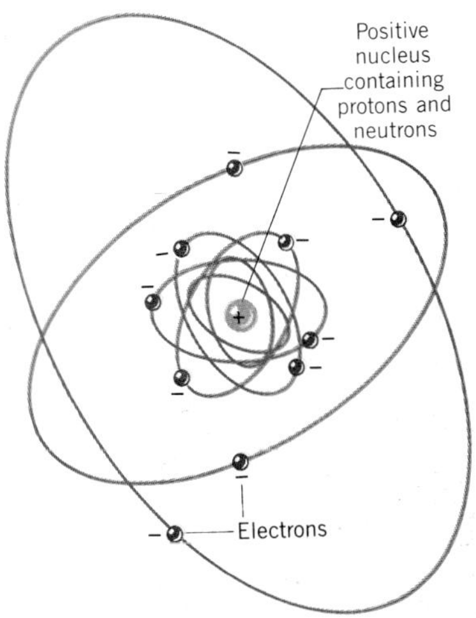

Figure 18.1 An atom contains a small, positively charged nucleus, about which the negatively charged electrons move. The closed-loop paths shown here are symbolic only. In reality, the electrons do not follow discreet paths, as Section 30.5 discusses.

Experiment reveals that the magnitude of the charge on the proton *exactly equals* the magnitude of the charge on the electron; the proton carries a charge $+e$, and the electron carries a charge $-e$. The SI unit for measuring the magnitude of an electric charge is the ***coulomb* (C)***, and e has been determined experimentally to have the value

$$e = 1.60 \times 10^{-19}\ \text{C}$$

It should be noted that the symbol e represents only the magnitude of the charge on a proton or an electron, and does not include an algebraic sign that indicates whether the charge is positive or negative. In nature, atoms are normally found with equal numbers of protons and electrons. Usually, then, an atom carries no net charge, because the algebraic sum of the positive charge of the nucleus and the negative charge of the electrons is zero. When an atom, or any object, carries no net charge, the object is said to be *electrically neutral*. The neutrons in the nucleus are electrically neutral particles.

The charge on an electron or a proton is the *smallest* amount of free charge that has been discovered. Charges of larger magnitude are built up on an object by adding or removing electrons. Thus, any charge of magnitude q is an integer multiple of e; that is, $q = Ne$, where N is an integer. Because any electric charge q occurs in integer multiples of elementary, indivisible charges e, electric charge is said to be *quantized*. Example 1 emphasizes the quantized nature of electrical charge.

Example 1 A Lot of Electrons

How many electrons are there in one coulomb of negative charge?

REASONING The negative charge is due to the presence of excess electrons, since they carry negative charge. Because an electron has a charge whose magnitude is $e = 1.60 \times 10^{-19}$ C, the number of electrons is equal to the charge q divided by the charge e on each electron.

SOLUTION The number N of electrons is

$$N = \frac{q}{e} = \frac{1.00\ \text{C}}{1.60 \times 10^{-19}\ \text{C}} = \boxed{6.25 \times 10^{18}}$$

Clearly, a negative charge of one coulomb contains an enormous number of electrons.

* At this time we omit a precise definition of the coulomb, since such a definition depends on electric currents and magnetic fields, concepts discussed in later chapters.

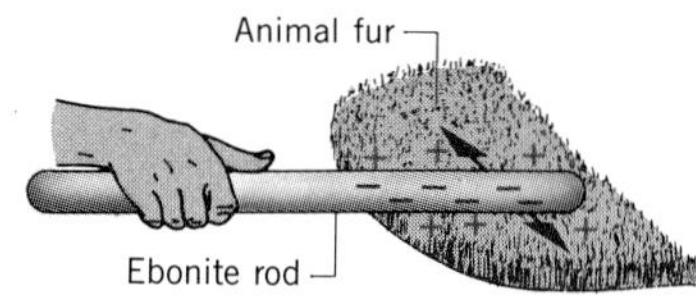

Figure 18.2 When an ebonite rod is rubbed against animal fur, electrons from the fur are transferred to the rod. This transfer gives the rod a negative charge (−) and leaves a positive charge (+) on the fur.

18.2 CHARGED OBJECTS AND THE ELECTRIC FORCES THAT THEY EXERT

THE SEPARATION OF CHARGES

Electricity has many useful applications, and they are related to the fact that it is possible to transfer electric charge from one object to another. Usually electrons are transferred, and the body that gains electrons acquires an excess of negative charge. The body that loses electrons has an excess of positive charge. Such separation of charge occurs often when two unlike materials are rubbed together. For example, when an ebonite (hard, black rubber) rod is rubbed against animal fur, some of the electrons from the fur are transferred to the rod. The ebonite becomes negatively charged, and the fur becomes positively charged, as Figure 18.2 indicates. Similarly, if a glass rod is rubbed with a silk cloth, some of the electrons are removed from the glass and deposited on the silk, leaving the silk negatively charged and the glass positively charged. There are many familiar examples of charge separation, as when you walk across a nylon rug, vigorously run a comb through dry hair, or remove a pullover sweater. In each case, objects become "electrified" as surfaces rub against one another and a transfer of electrons occurs.

In the operation of electrical equipment, charge separation plays a fundamental role. For instance, batteries, microphones, alternators in automobile electrical systems, and electric power generators all depend on the separation of electric charges.

THE CONSERVATION OF CHARGE

When an ebonite rod is rubbed with animal fur, the rubbing process serves only to separate electrons and protons already present in the materials. No electrons or protons are created or destroyed. Whenever an electron is transferred to the rod, a proton is left behind on the fur. Since the charges on the electron and proton have identical magnitudes but opposite signs, the algebraic sum of the two charges is zero, and the transfer does not change the net charge of the fur/rod system. If each material contains an equal number of protons and electrons to begin with, the net charge of the system is zero initially and remains zero at all times during the rubbing process.

Electric charges play a role in many situations other than rubbing two surfaces together. They are involved, for instance, in chemical reactions, electric circuits, and radioactive decay. A great number of experiments have verified that in any situation, the ***law of conservation of electric charge*** is obeyed.

Law of Conservation of Electric Charge
During any process, the net electric charge of an isolated system remains constant (is conserved).

THE ELECTRIC FORCE THAT CHARGES EXERT ON EACH OTHER

It is easy to demonstrate that two electrically charged objects exert a force on one another. Consider Figure 18.3*a*, which shows two small balls that have been

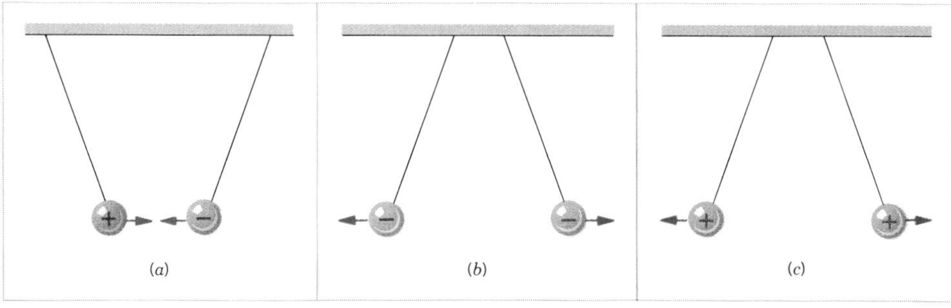

Figure 18.3 (*a*) A positive charge (+) and a negative charge (−) attract each other. (*b*) Two negative charges repel each other. (*c*) Two positive charges repel each other.

oppositely charged and are light and free to move. The balls are attracted toward each other. On the other hand, balls with the *same* type of charge, either both positive or both negative, repel each other, as parts *b* and *c* of the drawing indicate. It is a fundamental characteristic of electric charges that ***like charges repel and unlike charges attract each other.***

THE PHYSICS OF . . . an electrostatic air cleaner.

The electrostatic air cleaner is one device that puts the electric force to good use. This air cleaner is a great aid to people who suffer from respiratory problems, for it can remove up to 95% of all airborne particles, such as dust and pollen. Figure 18.4 shows how it works. A fan draws the contaminated air into the cleaner. The air first passes through an ordinary mesh filter that removes the larger particles, and then through a positively charged wire grid, known as the *charging electrode.* This electrode gives the contaminant particles a positive charge. The positively charged particles continue upward to the negatively charged grid, where they stick because of the electric force of attraction. Sometimes, after leaving the negative grid, the clean air passes through an activated charcoal filter before being discharged into the room. Activated charcoal is a good odor absorber and leaves the air smelling fresh.

THE PHYSICS OF . . . decontaminating soils.

The electric force that charges exert is also part of a technique for decontaminating soils deep underground. The technique, still in the experimental stage, also incorporates the use of sound waves. Soil particles typically have a negative charge, while contaminant particles, such as petroleum sludge, often have a

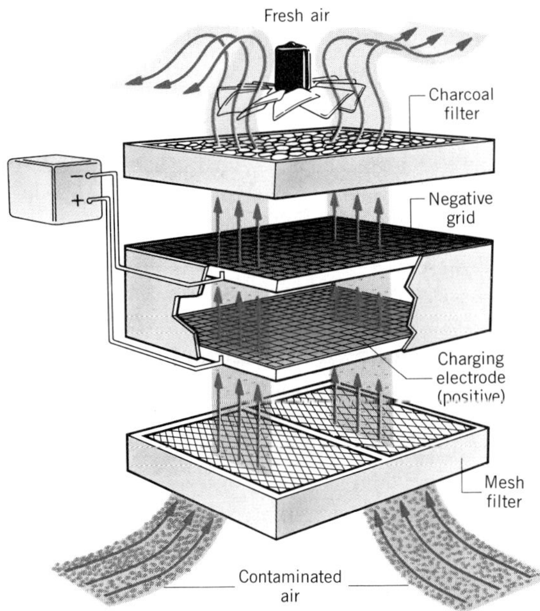

Figure 18.4 An electrostatic air cleaner. Unwanted airborne particles are given a positive charge as they pass the charging electrode. The positively charged particles are removed from the air stream, because they stick to the negative grid, which must be washed periodically.

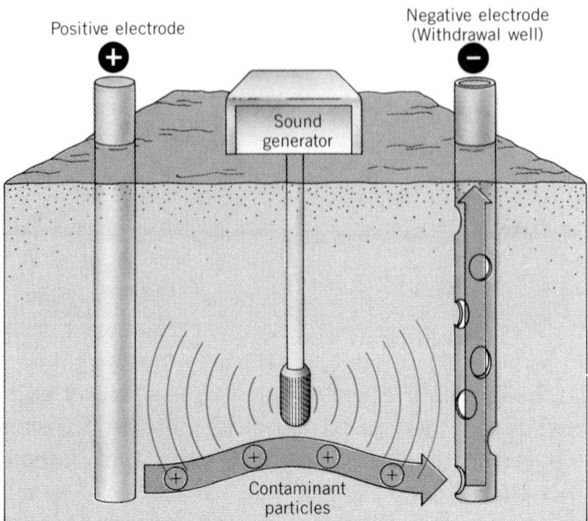

Figure 18.5 An experimental system for decontaminating underground soils.

positive charge. A sound generator driven into the ground (see Figure 18.5) produces a low-frequency sound wave that is thought to help loosen the positive contaminants from the negative soil. Positive and negative electrodes are also driven into the ground to provide an electric force that drives the positive contaminants toward the negative electrode. The negative electrode also serves as a withdrawal well for removing the contaminants.

18.3 CONDUCTORS AND INSULATORS

Not only can electric charge exist *on an object,* but it can also move *through an object.* However, materials differ vastly in their ability to allow electric charge to move or be conducted through them. To help illustrate such differences in conductivity, Figure 18.6*a* recalls the conduction of heat through a bar of material whose ends are maintained at different temperatures. As Section 13.2 discusses, metals conduct heat readily and, therefore, are known as thermal conductors. On the other hand, substances that conduct heat poorly are referred to as thermal insulators.

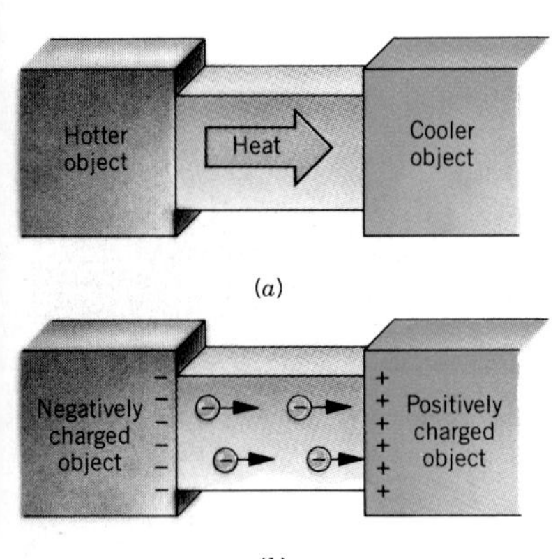

Figure 18.6 (*a*) Heat is conducted from the hotter end of the metal bar to the cooler end. (*b*) Electrons are conducted from the negatively charged end of the metal bar to the positively charged end.

A situation analogous to the conduction of heat arises when a metal bar is placed between two charged objects, as in Figure 18.6*b*. Electrons are conducted through the bar from the negatively charged object toward the positively charged object. Substances that readily conduct electric charge are called ***electrical conductors.*** Although there are exceptions, good thermal conductors are also good electrical conductors. Metals such as copper, aluminum, silver, and gold are excellent electrical conductors and, therefore, are used in electrical wiring. Materials that conduct electric charge *poorly* are known as ***electrical insulators.*** In many cases, thermal insulators are also electrical insulators. Common electrical insulators are rubber, many plastics, and wood. Insulators, such as the rubber or plastic that coats electrical wiring, prevent electric charge from going where it is not wanted.

The difference between electrical conductors and insulators is related to atomic structure. As electrons orbit the nucleus, those in the outer orbits experience a weaker force of attraction to the nucleus than do those in the inner orbits. Consequently, the outermost electrons (also called the valence electrons) can be dislodged more easily than the inner ones. In a good conductor, some valence electrons actually become detached from a parent atom and wander more or less freely throughout the material, belonging to no one atom in particular. The exact number of electrons detached from each atom depends on the nature of the material, but is usually between one and three. When one end of a conducting bar is placed in contact with a negatively charged object and the other end in contact with a positively charged object, as in Figure 18.6*b*, the "free" electrons are able to move readily away from the negative end and toward the positive end. The ready movement of electrons is the hallmark of a good conductor. In an insulator the situation is different, for there are very few electrons free to move throughout the material. Virtually every electron remains bound to its parent atom. Without the "free" electrons, there is very little flow of charge when the material is placed between two oppositely charged bodies, so the material is an electrical insulator.

18.4 CHARGING BY CONTACT AND BY INDUCTION

CHARGING BY CONTACT

When a negatively charged ebonite rod is rubbed on a metal object, such as the sphere in Figure 18.7*a*, some of the excess electrons from the rod are transferred to the object. Once the electrons are on the metal sphere, where they can move readily, their mutual repulsive forces cause them to spread out over the surface of the sphere. The insulated stand prevents the electrons from flowing to the earth, where they could spread out even more. When the rod is removed, as in part *b* of the picture, the sphere is left with a negative charge distributed over its surface. In a similar manner, the sphere would be left with a positive charge after being in contact with a positively charged rod. In this case, electrons from the sphere would be transferred to the rod. The process of giving one object a net electric charge by placing it in contact with a charged object is known as ***charging by contact.***

Ebonite rod
Metal sphere
Insulated stand
(*a*) (*b*)

Figure 18.7 (*a*) Electrons are transferred by rubbing the negatively charged rod on the metal sphere. (*b*) When the rod is removed, the electrons distribute themselves over the surface of the sphere.

CHARGING BY INDUCTION

It is possible to charge a conductor in a way that does not involve contact. In Figure 18.8, a negatively charged rod is brought close to, *but does not touch,* a metal sphere. In the sphere, the free electrons closest to the rod move to the other side, as part *a* of the drawing indicates. As a result, the part of the sphere nearest the rod becomes positively charged and the part farthest away becomes negatively charged. These positively and negatively charged regions have been "induced" or "persuaded" to form because of the repulsive force between the negative rod and the free electrons in the sphere. If the rod were removed, the free electrons would return to their original places, and the charged regions on either side of the sphere would disappear.

Under most conditions the earth is a good electrical conductor. So when a metal wire is attached between the sphere and the ground, as in Figure 18.8*b*, some of the free electrons leave the sphere and distribute themselves over the much larger earth. If the grounding wire is then removed, followed by the ebonite rod, the sphere is left with a positive net charge, as part *c* of the picture shows. The process of giving one object a net electric charge *without* touching the object to a second charged object is called ***charging by induction.*** The process could also be

In this demonstration, the wand is electrically charged and causes induced charges to appear on the girl's hair. The hair, in turn, is attracted to the wand because of the attraction between unlike charges.

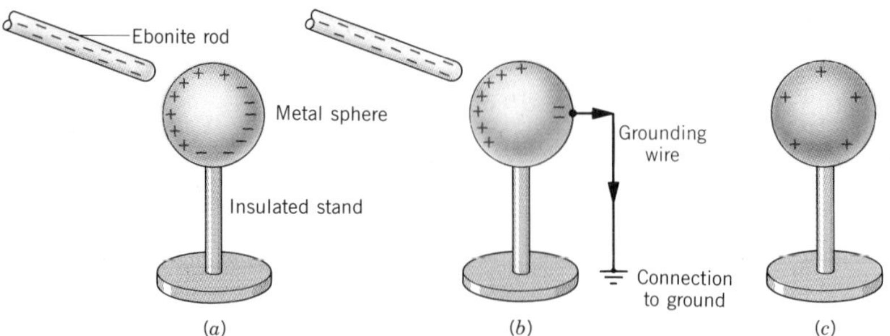

Figure 18.8 (*a*) When a charged rod is brought near the metal sphere without touching it, some of the positive and negative charges in the sphere are separated. (*b*) Some of the electrons leave the sphere through the grounding wire, with the result (*c*) that the sphere acquires a positive net charge.

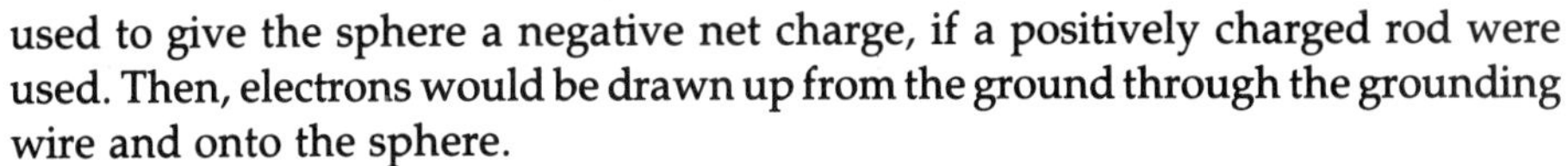

used to give the sphere a negative net charge, if a positively charged rod were used. Then, electrons would be drawn up from the ground through the grounding wire and onto the sphere.

If the sphere in Figure 18.8 were made from an insulating material like plastic, instead of metal, the method of producing a net charge by induction would not work, because very little charge would flow through the insulating material and down the grounding wire. However, the electric force of the charged rod would have some effect, as Figure 18.9 illustrates. The electric force would cause the positive and negative charges in the molecules of the insulating material to separate slightly, with the positive charges being "pulled" toward the rod. Although no net charge is created, the surface of the plastic does acquire a slight induced positive charge and is attracted to the negative rod. For a similar reason, one piece of cloth can stick to another in the phenomenon known as "static cling," which occurs when an article of clothing has acquired an electric charge while being tumbled about in a clothes drier.

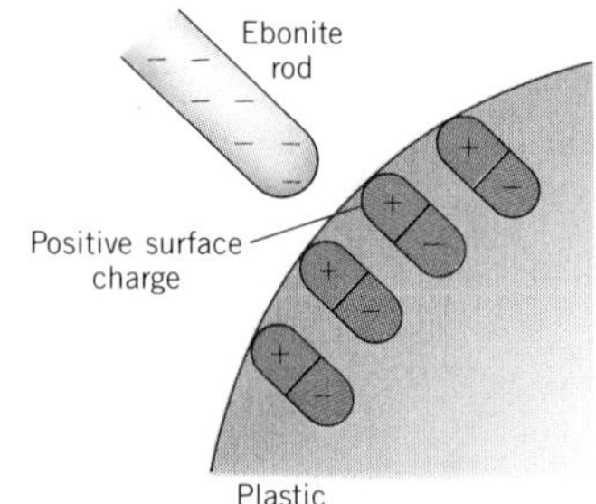

Figure 18.9 The negatively charged rod induces a slight positive surface charge on the plastic.

THE ELECTROSCOPE

The electroscope is a device for detecting small amounts of charge. The electroscope in Figure 18.10 consists of two thin strips, or leaves, of gold foil mounted at the bottom end of a metal rod. A metal knob caps the top of the rod, and glass windows enclose the leaves to prevent any effects due to air currents. An insulating rubber plug separates the metal rod from the metal case, so any charge on the rod does not leak away.

The electroscope can be used to determine if an insulator is charged and, if so, whether the charge is positive or negative. First, the electroscope is given a charge of known polarity, by touching the metal knob with a negatively charged ebonite rod, for example. As can be seen in Figure 18.11*a*, the negative charge spreads out over the leaves, and the repulsive force that the like charges exert on each other causes the leaves to spread apart. Then, the unknown charge is brought near the electroscope *without touching it.* If the unknown charge is positive, as in part *b* of the drawing, some of the electrons are drawn off the leaves and onto the metal knob. The loss of negative charge causes a reduction in the repulsive force between the leaves, and, as a result, they partially collapse. Conversely, if the unknown charge is negative, as in part *c*, it forces free electrons to leave the knob and increase the negative charge on the leaves. Consequently, the leaves spread apart even further. Conceptual Example 2 discusses some additional factors to consider when using an electroscope.

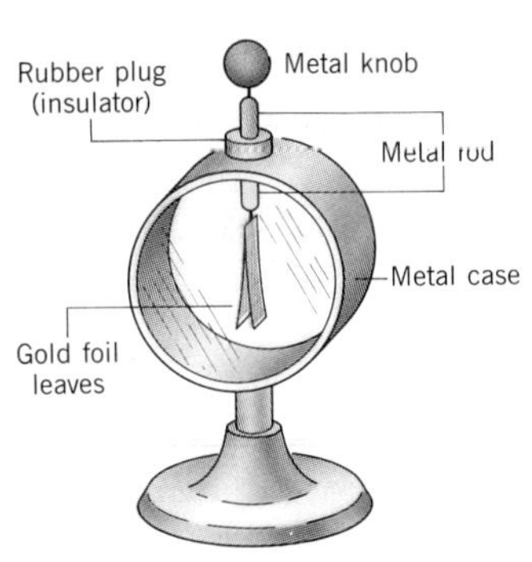

Figure 18.10 An electroscope.

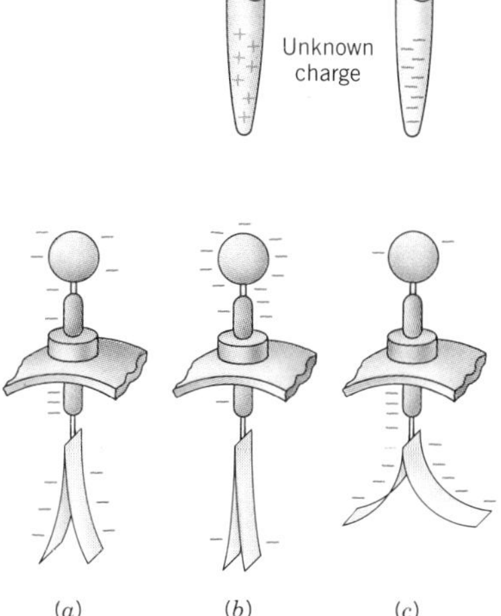

Figure 18.11 (*a*) A negative charge has been placed on the knob and leaves of an electroscope. (*b*) An unknown charge that is positive causes the leaves to collapse. (*c*) An unknown charge that is negative causes the leaves to diverge further.

Conceptual Example 2 The Behavior of the Leaves in an Electroscope

We have seen in Figure 18.11*b* that the leaves of the electroscope partially collapse as the positively charged rod is brought near the negatively charged electroscope. However, if the rod is brought very close, but still not touching the knob, the leaves are observed to collapse completely and then diverge again. What is the reason for this behavior?

REASONING The collapse of the leaves is a straightforward matter to understand. As the positive rod comes closer and closer to the knob, more electrons are drawn off the leaves and onto the knob. There comes a point when all the excess electrons have left the leaves, which are then electrically neutral and fully collapsed. But what does it mean that the leaves diverge again as the rod comes even closer? It means that the leaves somehow reacquire a like charge and repel one another. This is exactly what happens when the approaching rod draws more electrons from the neutral leaves. The leaves then become positively charged, as in Figure 18.12. The repulsive forces that the leaves exert on each other cause them to spread apart.

What happens to the leaves in Figure 18.11*c* as the negatively charged rod is brought closer to the knob of the electroscope? Do they subsequently collapse and then spread out? If not, what do they do? As always, provide a reason for your answers.

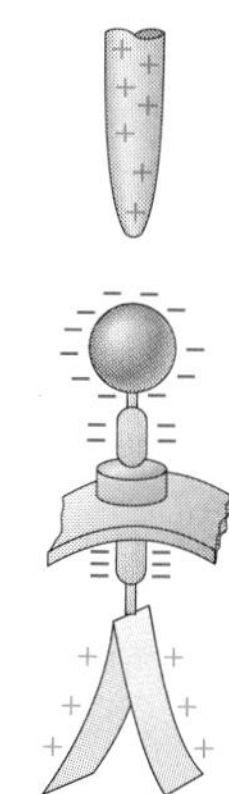

Figure 18.12 When the rod is brought very close to the knob, a sufficient number of electrons are drawn off the leaves to make them become positively charged. The repulsive force that each leaf exerts on the other causes them to diverge.

18.5 COULOMB'S LAW

THE FORCE THAT POINT CHARGES EXERT ON EACH OTHER

The electric force that stationary charged objects exert on each other is called the electrostatic force. This force depends on the amount of charge on and the distance between the objects. Experiments reveal that the greater the charges and the closer together they are, the greater is the force. To set the stage for explaining these features in more detail, Figure 18.13 shows two charged bodies. These objects are so small, compared to the distance r between them, that they can be regarded as mathematical points. The "point charges" have magnitudes q_1 and q_2. If the charges have *unlike* signs, as in part *a* of the picture, each charge is *attracted* to the other by a force that is directed along the line between them; $+\mathbf{F}$ is the electric force exerted on charge 1 by charge 2 and $-\mathbf{F}$ is the electric force

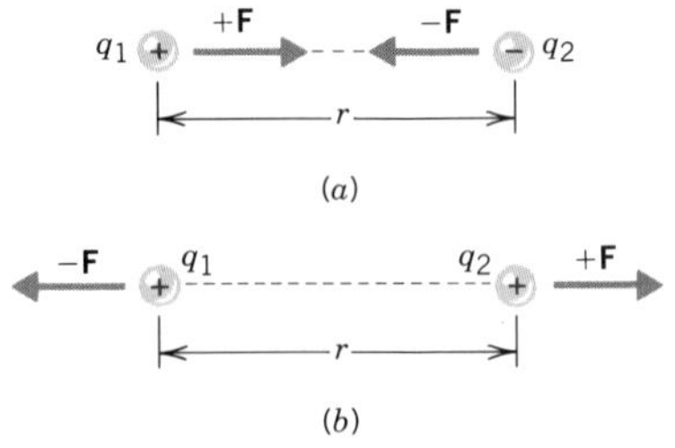

Figure 18.13 Each point charge exerts a force on the other. Regardless of whether the forces are (*a*) attractive or (*b*) repulsive, they are directed along the line between the charges and have equal magnitudes.

exerted on charge 2 by charge 1. If, as in part *b*, the charges have the *same* sign (both positive or both negative), each is repelled from the other. The repulsive forces, like the attractive forces, act along the line between the charges. Whether attractive or repulsive, the two forces are equal in magnitude but opposite in direction. These forces always exist as a pair, each one acting on a different object, in accord with Newton's action–reaction law.

The French physicist Charles-Augustin Coulomb (1736–1806) carried out a number of experiments to determine how the electric force that one point charge applies to another depends on the amount of each charge and the separation between them. His result, now known as ***Coulomb's law,*** is stated below.

Coulomb's Law

The magnitude F of the electrostatic force exerted by one point charge on another point charge is directly proportional to the magnitudes q_1 and q_2 of the charges and inversely proportional to the square of the distance r between the charges:

$$F = k\frac{q_1 q_2}{r^2} \tag{18.1}$$

where k is a proportionality constant whose value in SI units is $k = 8.99 \times 10^9$ $\text{N}\cdot\text{m}^2/\text{C}^2$. The electrostatic force is directed along the line joining the charges, and it is attractive if the charges have unlike signs and repulsive if the charges have like signs.

It is common practice to express k in terms of another constant ϵ_0, by writing $k = 1/(4\pi\epsilon_0)$; ϵ_0 is called the ***permittivity of free space*** and has a value of $\epsilon_0 = 1/(4\pi k) = 8.85 \times 10^{-12}\ \text{C}^2/(\text{N}\cdot\text{m}^2)$. Equation 18.1 gives the magnitude of the electrostatic force that each point charge exerts on the other. When using this equation, then, it is important to remember to substitute only the charge magnitudes (without algebraic signs) for q_1 and q_2, as Example 3 illustrates.

Electropainting is standard technology in automobile manufacturing. The body of the car and the paint are given opposite charges. The resulting electrical attraction draws the paint onto the body, covering it thoroughly.

Example 3 A Large Attractive Force

Two very small objects, whose charges are +1.0 and −1.0 C, are separated by 1.5 m. Find the magnitude of the attractive force that either charge exerts on the other.

REASONING Coulomb's law may be used to find the magnitude of the force that either charge exerts on the other, provided that only the *magnitudes of the charges* are used for the symbols q_1 and q_2 that appear in the law.

SOLUTION The magnitude F of the force is

$$F = k\frac{q_1 q_2}{r^2} = \frac{(8.99 \times 10^9\ \text{N}\cdot\text{m}^2/\text{C}^2)(1.0\ \text{C})(1.0\ \text{C})}{(1.5\ \text{m})^2} = \boxed{4.0 \times 10^9\ \text{N}} \tag{18.1}$$

The force calculated in Example 3 corresponds to 900 million pounds and is enormous. However, charges as large as one coulomb are usually encountered only in the most severe conditions, as in a lightning bolt, where as much as 25 coulombs can be transferred between the cloud and the ground. The typical charges produced in the laboratory are much smaller and are measured conveniently in microcoulombs (1 microcoulomb = 1 μC = 10^{-6} C).

Coulomb's law has a form remarkably similar to Newton's law of gravitation ($F = Gm_1m_2/r^2$). The force in both laws depends on the inverse square ($1/r^2$) of the distance between the two objects and is directed along the line between them. In addition, the force is proportional to the product of an intrinsic property of each of the objects, the magnitudes of the charges q_1 and q_2 in Coulomb's law and the masses m_1 and m_2 in the gravitation law. But there is a major difference between the two laws. The electrostatic force can be either repulsive or attractive, depending on whether or not the charges have the same sign; in contrast, the gravitational force is always an attractive force.

Section 5.5 discusses how the gravitational attraction between the earth and a satellite provides the centripetal force that keeps a satellite in orbit. Example 4 illustrates that the electrostatic force of attraction plays a similar role in a famous model of the atom created by the Danish physicist Niels Bohr (1885–1962).

Example 4 A Model of the Hydrogen Atom

In the Bohr model of the hydrogen atom, the electron ($-e$) is in orbit about the nuclear proton ($+e$) at a radius of $r = 5.29 \times 10^{-11}$ m, as Figure 18.14 shows. Determine the speed of the electron, assuming the orbit to be circular.

REASONING Recall from Section 5.2 that any object moving with speed v on a circular path of radius r has a centripetal acceleration of $a_c = v^2/r$. This acceleration is directed toward the center of the circle. Newton's second law, then, specifies that the net force ΣF needed to create this acceleration is $\Sigma F = ma_c = mv^2/r$, where m is the mass of the object. This equation can be solved for the speed: $v = \sqrt{(\Sigma F)r/m}$. Since the mass of the electron is $m = 9.11 \times 10^{-31}$ kg and the radius is given, we can calculate the speed, once a value for the net force is available. For the electron in the hydrogen atom, the net force is provided almost exclusively by the electrostatic force, as given by Coulomb's law. The electron is also pulled toward the proton by the gravitational force. However, the gravitational force is negligible in comparison to the electrostatic force.

SOLUTION The electron experiences an electrostatic force of attraction because of the proton, and the magnitude of this force is

$$F = k\frac{q_1q_2}{r^2} = \frac{(8.99 \times 10^9\ \text{N}\cdot\text{m}^2/\text{C}^2)(1.60 \times 10^{-19}\ \text{C})(1.60 \times 10^{-19}\ \text{C})}{(5.29 \times 10^{-11}\ \text{m})^2}$$

$$= 8.22 \times 10^{-8}\ \text{N}$$

Using this value for the net force, we find

$$v = \sqrt{\frac{(\Sigma F)r}{m}} = \sqrt{\frac{(8.22 \times 10^{-8}\ \text{N})(5.29 \times 10^{-11}\ \text{m})}{9.11 \times 10^{-31}\ \text{kg}}} = \boxed{2.18 \times 10^6\ \text{m/s}}$$

This orbital speed is almost five million miles per hour.

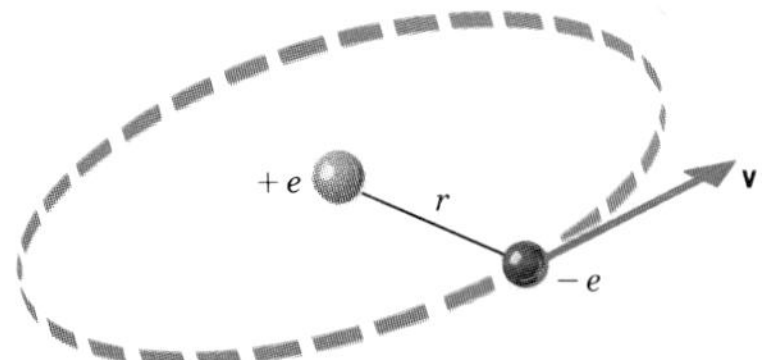

Figure 18.14 In the Bohr model of the hydrogen atom, the electron ($-e$) orbits the proton ($+e$) at a distance of $r = 5.29 \times 10^{-11}$ m. The velocity of the electron is **v**.

PROBLEM SOLVING INSIGHT

When using Coulomb's law ($F = kq_1q_2/r^2$), remember that the symbols q_1 and q_2 stand for the charge magnitudes. Do not substitute negative numbers for these symbols.

THE FORCE ON A POINT CHARGE DUE TO TWO OR MORE OTHER POINT CHARGES

Up to now, we have been discussing the electrostatic force on a point charge (magnitude q_1) due to another point charge (magnitude q_2). Suppose that a third point charge (magnitude q_3) were also present. What would be the net force on q_1 due to both q_2 and q_3? It is convenient to deal with such a problem in parts. First,

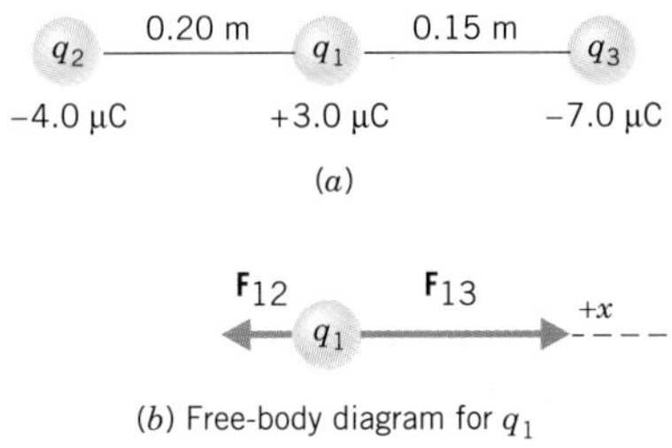

q_2 0.20 m q_1 0.15 m q_3

−4.0 μC +3.0 μC −7.0 μC

(a)

F_{12} q_1 F_{13} $+x$

(b) Free-body diagram for q_1

Figure 18.15 (a) Three charges lying along the x axis. (b) The force exerted on q_1 by q_2 is $\mathbf{F}_{12}$, while the force exerted on q_1 by q_3 is $\mathbf{F}_{13}$.

find the magnitude and direction of the force exerted on q_1 by q_2 (ignoring q_3). Then, determine the force exerted on q_1 by q_3 (ignoring q_2). The *net force* on q_1 is the *vector sum* of these two forces. Examples 5 and 6 illustrate this approach when the charges lie along a straight line and when they lie in a plane, respectively.

Example 5 Three Charges on a Line

Figure 18.15*a* shows three point charges that lie along the x axis. Determine the magnitude and direction of the net electrostatic force on q_1.

REASONING Part *b* of the drawing shows a free-body diagram of the forces that act on q_1. Since q_1 and q_2 have opposite signs, they attract one another. Thus, the force exerted on q_1 by q_2 is $\mathbf{F}_{12}$, and it points to the left. Similarly, the force exerted on q_1 by q_3 is $\mathbf{F}_{13}$ and is an attractive force. It points to the right in Figure 18.15*b*. The magnitudes of these forces can be obtained from Coulomb's law. The net force is the vector sum of $\mathbf{F}_{12}$ and $\mathbf{F}_{13}$.

SOLUTION The magnitudes of the forces are

$$F_{12} = k\frac{q_1 q_2}{r_{12}^2} = \frac{(8.99 \times 10^9\ \text{N}\cdot\text{m}^2/\text{C}^2)(3.0 \times 10^{-6}\ \text{C})(4.0 \times 10^{-6}\ \text{C})}{(0.20\ \text{m})^2} = 2.7\ \text{N}$$

$$F_{13} = k\frac{q_1 q_3}{r_{13}^2} = \frac{(8.99 \times 10^9\ \text{N}\cdot\text{m}^2/\text{C}^2)(3.0 \times 10^{-6}\ \text{C})(7.0 \times 10^{-6}\ \text{C})}{(0.15\ \text{m})^2} = 8.4\ \text{N}$$

Since $\mathbf{F}_{12}$ points in the negative x direction, and $\mathbf{F}_{13}$ points in the positive x direction, the net force $\mathbf{F}$ is

$$\mathbf{F} = \mathbf{F}_{12} + \mathbf{F}_{13} = (-2.7\ \text{N}) + (8.4\ \text{N}) = \boxed{+5.7\ \text{N}}$$

The plus sign in the answer indicates that the net force points to the right in the drawing.

PROBLEM SOLVING INSIGHT

Often charge magnitudes are specified in microcoulombs (μC). When using Coulomb's law, be sure to convert microcoulombs into coulombs (1 μC = 10^{-6} C) before substituting for the charge magnitudes q_1 and q_2.

Example 6 Three Charges in a Plane

Find the magnitude and direction of the net electrostatic force on q_1 in Figure 18.16*a*.

REASONING The force exerted on q_1 by q_2 is $\mathbf{F}_{12}$ and is an attractive force, because the two charges have opposite signs. It points along the line between the charges. The force exerted on q_1 by q_3 is $\mathbf{F}_{13}$ and is also an attractive force. It points along the line between q_1 and q_3. Coulomb's law allows us to calculate the magnitudes of these forces. Since the forces do not point in the same direction (see Figure 18.16*b*), we will use vector components to find the net force.

SOLUTION The magnitudes of the forces are

$$F_{12} = k\frac{q_1 q_2}{r_{12}^2} = \frac{(8.99 \times 10^9\ \text{N}\cdot\text{m}^2/\text{C}^2)(4.0 \times 10^{-6}\ \text{C})(6.0 \times 10^{-6}\ \text{C})}{(0.15\ \text{m})^2} = 9.6\ \text{N}$$

$$F_{13} = k\frac{q_1 q_3}{r_{13}^2} = \frac{(8.99 \times 10^9\ \text{N}\cdot\text{m}^2/\text{C}^2)(4.0 \times 10^{-6}\ \text{C})(5.0 \times 10^{-6}\ \text{C})}{(0.10\ \text{m})^2} = 18\ \text{N}$$

The net force $\mathbf{F}$ is the vector sum of $\mathbf{F}_{12}$ and $\mathbf{F}_{13}$, as part *b* of the drawing shows. The components of $\mathbf{F}$ that lie in the x and y directions are $\mathbf{F}_x$ and $\mathbf{F}_y$, respectively. Our approach to finding $\mathbf{F}$ is the same as that used in Chapters 1 and 4. The forces $\mathbf{F}_{12}$ and $\mathbf{F}_{13}$

are resolved into x and y components. Then, the x components are combined to give $\mathbf{F}_x$, and the y components are combined to give $\mathbf{F}_y$. Once $\mathbf{F}_x$ and $\mathbf{F}_y$ are known, the magnitude and direction of $\mathbf{F}$ can be determined using trigonometry.

Force	*x component*	*y component*
$\mathbf{F}_{12}$	$+(9.6 \text{ N}) \cos 73° = +2.8 \text{ N}$	$+(9.6 \text{ N}) \sin 73° = +9.2 \text{ N}$
$\mathbf{F}_{13}$	$+18 \text{ N}$	0
$\mathbf{F}$	$\mathbf{F}_x = +21 \text{ N}$	$\mathbf{F}_y = +9.2 \text{ N}$

The magnitude F and the angle θ of the net force are

$$F = \sqrt{F_x^2 + F_y^2} = \sqrt{(21 \text{ N})^2 + (9.2 \text{ N})^2} = \boxed{23 \text{ N}}$$

$$\theta = \tan^{-1}\left(\frac{F_y}{F_x}\right) = \tan^{-1}\left(\frac{9.2 \text{ N}}{21 \text{ N}}\right) = \boxed{24°}$$

PROBLEM SOLVING INSIGHT

The electrostatic force is a vector and has a direction as well as a magnitude. When adding electrostatic forces together, take into account the directions of all forces, using vector components as needed.

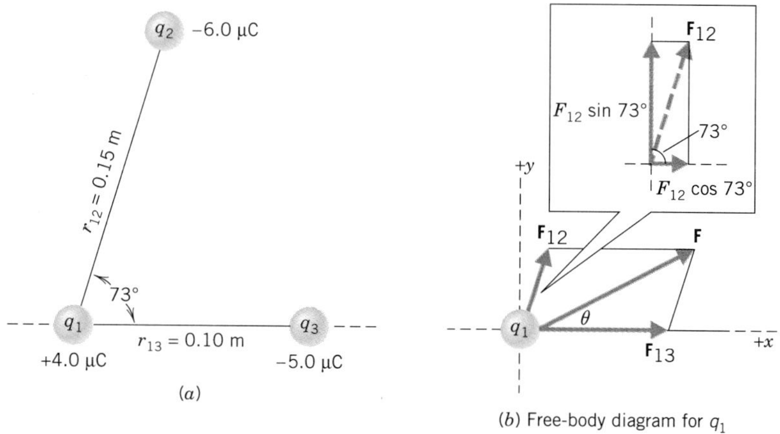

Figure 18.16 (*a*) Three charges lying in a plane. (*b*) The net force acting on q_1 is $\mathbf{F} = \mathbf{F}_{12} + \mathbf{F}_{13}$. The angle that $\mathbf{F}$ makes with the $+x$ axis is θ.

18.6 THE ELECTRIC FIELD

DEFINITION OF THE ELECTRIC FIELD

A charge can experience an electrostatic force due to the presence of other charges. For instance, the positive charge q_0 in Figure 18.17 experiences a force $\mathbf{F}$, which is the vector sum of the forces exerted by the charges on the rod and the two spheres. It is useful to think of q_0 as a ***test charge*** for determining the extent to which the surrounding charges generate a force. However, in using a test charge, we must be careful to select one with a very small magnitude, so that it does not alter the locations of the other charges. The next example illustrates how the concept of a test charge is applied.

Example 7 A Test Charge

The positive test charge in Figure 18.17 is $q_0 = +3.0 \times 10^{-8}$ C and experiences a force $F = 6.0 \times 10^{-8}$ N in the direction shown in the drawing. (a) Find the *force per coulomb* that the test charge experiences. (b) Using the result of part (a), predict the force that a charge of $+12 \times 10^{-8}$ C would experience if it replaced q_0.

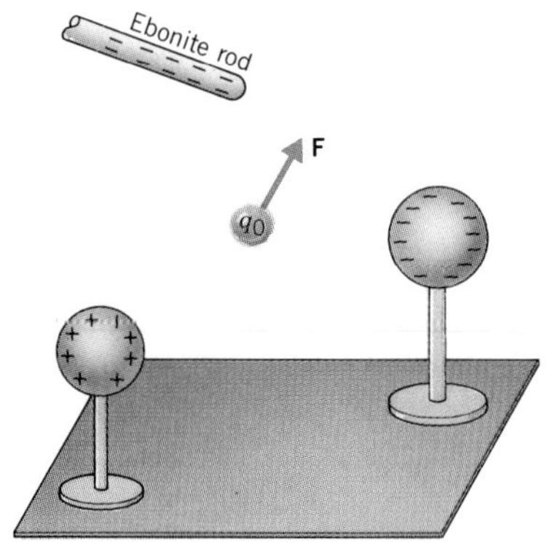

Figure 18.17 A positive charge q_0 experiences an electrostatic force **F** due to the surrounding charges on the ebonite rod and the two spheres.

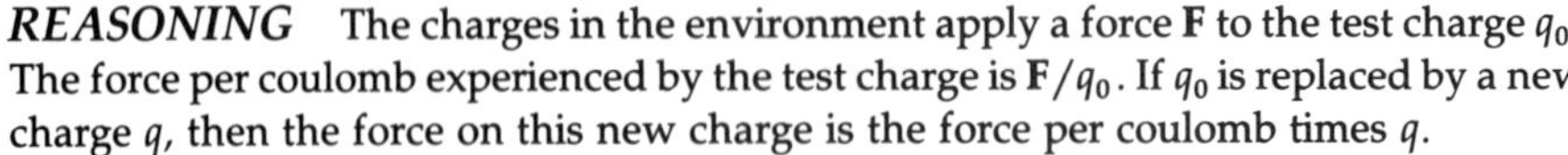

REASONING The charges in the environment apply a force **F** to the test charge q_0. The force per coulomb experienced by the test charge is $\mathbf{F}/q_0$. If q_0 is replaced by a new charge q, then the force on this new charge is the force per coulomb times q.

SOLUTION

(a) The force per coulomb of charge is

$$\frac{\mathbf{F}}{q_0} = \frac{6.0 \times 10^{-8}\ \text{N}}{3.0 \times 10^{-8}\ \text{C}} = \boxed{2.0\ \text{N/C}}$$

(b) The result from part (a) indicates that the surrounding charges can exert 2.0 newtons of force per coulomb of charge. Thus, a charge of $+12 \times 10^{-8}$ C would experience a force whose magnitude is

$$F = (2.0\ \text{N/C})(12 \times 10^{-8}\ \text{C}) = \boxed{24 \times 10^{-8}\ \text{N}}$$

The direction of this force would be the same as that experienced by the test charge, since both have the same positive sign.

The force per coulomb, $\mathbf{F}/q_0$, calculated in Example 7(a) is one illustration of a quantity that is very important in the study of electricity. This quantity is called the ***electric field.***

When the earth and an overhead cloud become oppositely charged, charges often flow between them because of the electrical force of attraction. In this instance, the flow of charge results in a spectacular lightning bolt that strikes a sycamore tree.

Definition of the Electric Field

The electric field **E** that exists at a point is the electrostatic force **F** experienced by a small test charge q_0 placed at that point divided by the charge itself:

$$\mathbf{E} = \frac{\mathbf{F}}{q_0} \tag{18.2}$$

The electric field is a vector, and its direction is the same as the direction of the force **F** on a positive test charge.

SI Unit of Electric Field: newton per coulomb (N/C)

Equation 18.2 indicates that the unit for the electric field is that of force divided by charge, which is a newton/coulomb (N/C) in SI units. The definition also emphasizes that the electric field is a vector with the same direction as the force on a *positive* test charge.

It is the surrounding charges that create an electric field at a given point. Any positive or negative charge placed at the point interacts with the field and, as a result, experiences a force, as the next example indicates.

Example 8 An Electric Field Leads to a Force

In Figure 18.18*a* the charges on the two metal spheres and the ebonite rod create an electric field **E** at the spot indicated. This field has a magnitude of 2.0 N/C and is directed as in the drawing. Determine the force on a charge placed at that spot, if the charge has a value of (a) $q_0 = +18 \times 10^{-8}$ C and (b) $q_0 = -24 \times 10^{-8}$ C.

REASONING The electric field at a given spot can exert a variety of forces, depending on the magnitude and sign of the charge placed there. The charge is assumed to be small enough so as not to alter the locations of the surrounding charges that create the field.

SOLUTION

(a) The magnitude of the force is the product of the magnitudes of q_0 and **E**:

$$F = q_0 E = (18 \times 10^{-8}\ \text{C})(2.0\ \text{N/C}) = \boxed{36 \times 10^{-8}\ \text{N}} \qquad (18.2)$$

Since q_0 is positive, the force points in the same direction as the electric field, as part *b* of the drawing indicates.

(b) In this case, the magnitude of the force is

$$F = q_0 E = (24 \times 10^{-8}\ \text{C})(2.0\ \text{N/C}) = \boxed{48 \times 10^{-8}\ \text{N}} \qquad (18.2)$$

The force on the negative charge points in the direction *opposite* to the force on the positive charge, that is, opposite to the electric field (see part *c* of the drawing).

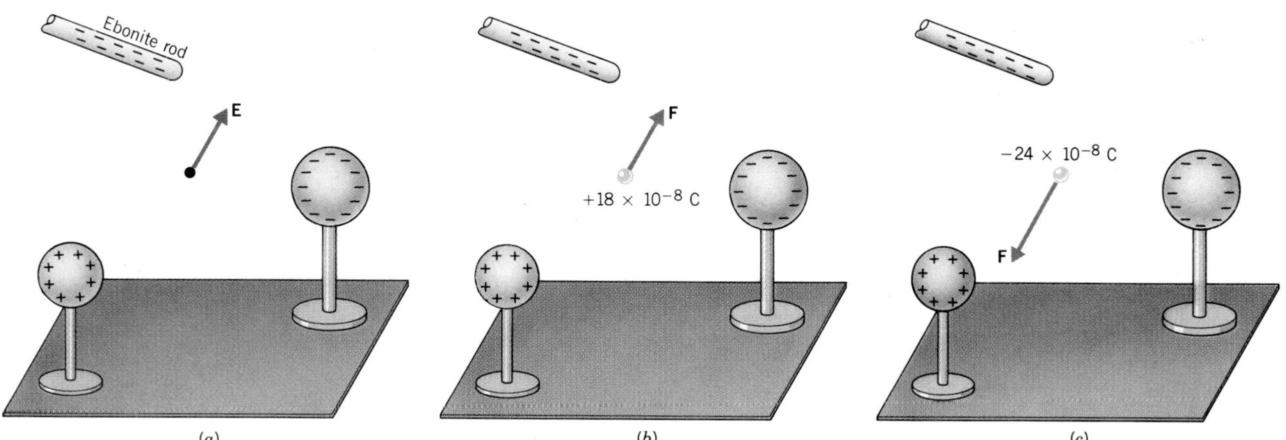

Figure 18.18 (*a*) The electric field **E** that exists at a given spot can exert a variety of forces. The force exerted depends on the magnitude and sign of the charge placed at that spot. (*b*) The force on a positive charge points in the same direction as **E**, while (*c*) the force on a negative charge points opposite to **E**.

At a particular point in space, each of the surrounding charges contributes to the net electric field that exists there. To determine the net field, it is necessary to obtain the various contributions separately and then find the vector sum of them all. Such an approach is an illustration of the principle of linear superposition, as applied to electric fields. (This principle is introduced in Section 17.1, in connection with waves.) Example 9 emphasizes the vector nature of the electric field.

Example 9 Electric Fields Add as Vectors Do

Figure 18.19 shows two charged objects, *A* and *B*. Each contributes as follows to the net electric field at point *P*: $\mathbf{E}_A = 3.00$ N/C directed to the right, and $\mathbf{E}_B = 2.00$ N/C directed downward. Thus, $\mathbf{E}_A$ and $\mathbf{E}_B$ are perpendicular. What is the net electric field at *P*?

REASONING The net electric field **E** is the vector sum of $\mathbf{E}_A$ and $\mathbf{E}_B$: $\mathbf{E} = \mathbf{E}_A + \mathbf{E}_B$. As illustrated in Figure 18.19, $\mathbf{E}_A$ and $\mathbf{E}_B$ are perpendicular, so **E** is the diagonal of the rectangle formed by $\mathbf{E}_A$ and $\mathbf{E}_B$. Thus, we can use the Pythagorean theorem to find the magnitude of **E** and trigonometry to find the directional angle θ.

SOLUTION The magnitude of the net electric field is

$$E = \sqrt{E_A{}^2 + E_B{}^2} = \sqrt{(3.00\ \text{N/C})^2 + (2.00\ \text{N/C})^2} = \boxed{3.61\ \text{N/C}}$$

The direction of **E** is given by the angle θ in the drawing:

$$\theta = \tan^{-1}\left(\frac{E_B}{E_A}\right) = \tan^{-1}\left(\frac{2.00\ \text{N/C}}{3.00\ \text{N/C}}\right) = \boxed{33.7^\circ}$$

PROBLEM SOLVING INSIGHT

The electric field is a vector and has a direction as well as a magnitude. When adding electric fields together, take into account the directions of all fields, using vector components as needed.

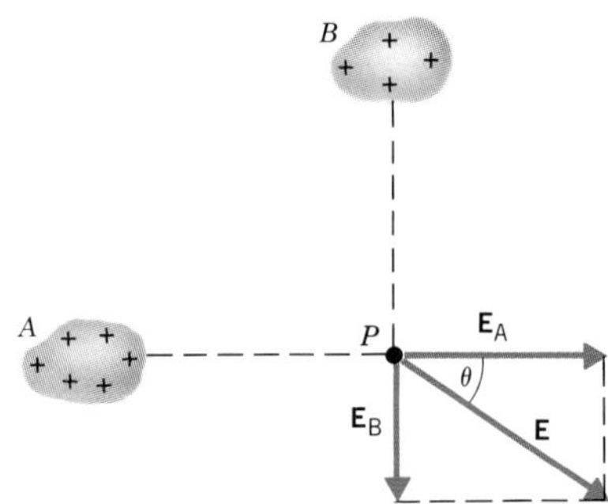

Figure 18.19 The electric field contributions $\mathbf{E}_A$ and $\mathbf{E}_B$, which come from the two charge distributions, are added vectorially to obtain the net field **E** at point *P*.

ELECTRIC FIELDS PRODUCED BY POINT CHARGES

A more complete understanding of the electric field concept can be gained by considering the field created by a point charge, as in the following example.

Example 10 The Electric Field of a Point Charge

There is an isolated point charge of $q = +15\ \mu\text{C}$ at the left in Figure 18.20*a*. Using a test charge of $q_0 = +0.80\ \mu\text{C}$, determine the electric field at point *P*, which is 0.20 m away.

REASONING Following the definition of the electric field, we place the test charge q_0 at point *P*, determine the force acting on the test charge, and then divide the force by the test charge.

SOLUTION The magnitude of the force can be found from Coulomb's law:

$$F = \frac{kq_0q}{r^2} = \frac{(8.99 \times 10^9\ \text{N}\cdot\text{m}^2/\text{C}^2)(0.80 \times 10^{-6}\ \text{C})(15 \times 10^{-6}\ \text{C})}{(0.20\ \text{m})^2} = 2.7\ \text{N} \quad (18.1)$$

The magnitude of the electric field is

$$E = \frac{F}{q_0} = \frac{2.7\ \text{N}}{0.80 \times 10^{-6}\ \text{C}} = \boxed{3.4 \times 10^6\ \text{N/C}} \quad (18.2)$$

The electric field **E** points in the *same direction* as the force **F** on the positive test charge. Since the test charge experiences a force of repulsion directed to the right, the electric field vector also points to the right, as Figure 18.20*b* shows.

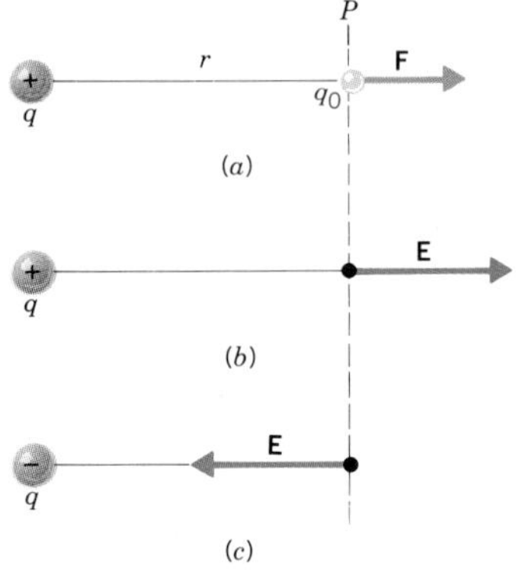

Figure 18.20 (*a*) At location *P*, a positive test charge q_0 experiences a repulsive force **F** due to the positive point charge *q*. (*b*) At *P* the electric field **E** is directed to the right. (*c*) If the charge *q* were negative rather than positive, the electric field would have the same magnitude as in (*b*) but would be directed to the left.

The electric field produced by a point charge q can be obtained in general terms from Coulomb's law. First, note that the magnitude of the force exerted by the charge q on a test charge q_0 is $F = kqq_0/r^2$. Then, divide this value by q_0 to obtain the magnitude of the field. Since q_0 is eliminated algebraically from the result, *the electric field does not depend on the test charge:*

[Point charge q] $$E = \frac{kq}{r^2} \qquad (18.3)$$

As in Coulomb's law, only the magnitude of q is used in Equation 18.3, without regard to whether q is positive or negative. If q is positive, then **E** is directed away from q, as in Figure 18.20*b*. On the other hand, if q is negative, then **E** is directed toward q, since a negative charge attracts a positive test charge. For instance, Figure 18.20*c* shows the electric field that would exist at P if there were a charge of $-q$ instead of $+q$ at the left of the drawing. Example 11 reemphasizes the fact that all the surrounding charges make a contribution to the electric field that exists at a given place.

Example 11 The Electric Fields from Separate Charges May Cancel

Two positive point charges, $q_1 = +16\ \mu C$ and $q_2 = +4.0\ \mu C$, are separated by a distance of 3.0 m, as Figure 18.21 illustrates. Find the spot on the line between the charges where the net electric field is zero.

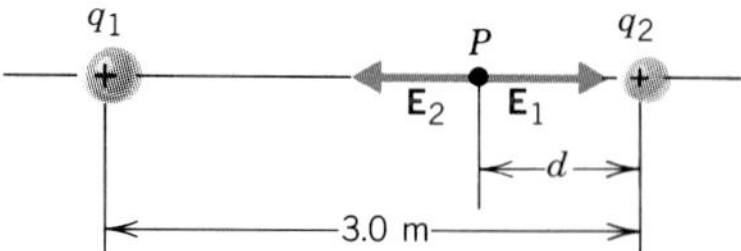

Figure 18.21 The two point charges q_1 and q_2 create electric fields E_1 and E_2 that cancel at a location P on the line between the charges.

REASONING Between the charges the two field contributions have opposite directions, and the net electric field is zero at the place where the magnitude of $\mathbf{E}_1$ equals that of $\mathbf{E}_2$. However, since q_2 is smaller than q_1, this location must be *closer* to q_2, in order that the field of the smaller charge can balance the field of the larger charge. In the drawing, the cancellation spot is labeled P, and its distance from q_2 is d.

SOLUTION At P, $E_1 = E_2$, and using the expression $E = kq/r^2$, we have

$$\frac{k(16 \times 10^{-6}\ \text{C})}{(3.0\ \text{m} - d)^2} = \frac{k(4.0 \times 10^{-6}\ \text{C})}{d^2}$$

Rearranging this expression shows that $4.0d^2 = (3.0\ \text{m} - d)^2$, and taking the square root of each side of this equation reveals that

$$\pm 2.0d = 3.0\ \text{m} - d$$

The plus and minus signs on the left occur because either the positive or negative root can be taken. Therefore, there are two possible values for d: $+1.0$ and -3.0 m. The negative value corresponds to a location off to the right of both charges, where the magnitudes of $\mathbf{E}_1$ and $\mathbf{E}_2$ are equal, but where the directions are the same. Thus, $\mathbf{E}_1$ and $\mathbf{E}_2$ do not cancel at this spot. The positive value corresponds to the location shown in the drawing and is the zero-field location: $\boxed{d = +1.0\ \text{m}}$.

When point charges are arranged in a symmetrical fashion, it is often possible to deduce useful information about the magnitude and direction of the electric field by taking advantage of the symmetry. Conceptual Example 12 illustrates the use of this technique.

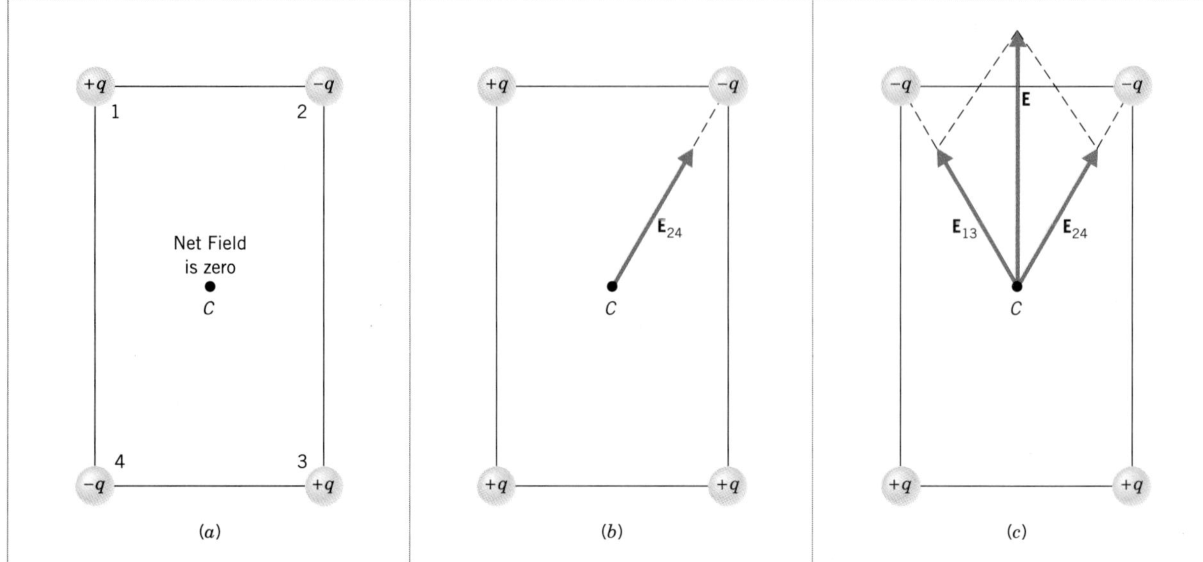

Figure 18.22 Charges of identical magnitude, but different signs, are placed at the corners of a rectangle. The charges give rise to different electric fields at the center C of the rectangle, depending on the signs of the charges.

Conceptual Example 12 Symmetry and the Electric Field

Figure 18.22 shows point charges fixed to the corners of a rectangle in three different ways. The charges all have the same magnitudes, but they have different signs. Consider the net electric field at the center C of the rectangle in each case. Rank the magnitudes of the net electric field in increasing order (smallest first).

REASONING We begin by noting that each charge in the drawing produces, by itself, an electric field at C that is directed along a diagonal of the rectangle. In Figure 18.22*a*, like charges occupy opposite corners. That is, the charges at corners 1 and 3 are both $+q$, while the charges at corners 2 and 4 are both $-q$. Let's first concentrate on the electric field produced at the center C of the rectangle by only the positive charges. The positive charge at corner 1 produces an electric field at C that points toward corner 3. In contrast, the positive charge at corner 3 produces an electric field at C that points toward corner 1. Thus, the two electric fields have opposite directions. The magnitudes of the fields are identical, because the charges have the same magnitude and are equally far from the center. Therefore, the fields from the two positive charges cancel. In a similar manner, the fields due to the negative charges on corners 2 and 4 also cancel. Consequently, the net electric field at the center of the rectangle due to all four charges is zero.

In part *b* of the drawing, we notice that there are identical positive charges on corners 1 and 3. This is just like part *a*, so the electric fields of these two charges cancel at the center. However, opposite charges are on corners 2 and 4. The electric field due to the negative charge at corner 2 points toward corner 2, and the field due to the positive charge at corner 4 points the same way. Furthermore, the magnitudes of these fields are equal, because each charge has the same magnitude and is located at the same distance from the center of the rectangle. As a result, the two fields combine to give the net electric field $\mathbf{E}_{24}$ shown in the drawing.

In Figure 18.22*c*, the charges on corners 2 and 4 are identical to those in part *b* of the drawing, so this pair gives rise to the electric field labeled $\mathbf{E}_{24}$. The charges on corners 1 and 3 are identical to those on corners 2 and 4, so they give rise to an electric field labeled $\mathbf{E}_{13}$, which has the same magnitude as $\mathbf{E}_{24}$. The net electric field $\mathbf{E}$ is the vector sum of $\mathbf{E}_{24}$ and $\mathbf{E}_{13}$, and is also shown in the drawing.

It is clear from the drawings that the ranking of the electric fields, smallest to largest, is as follows: *a*, *b*, and *c*.

Related Homework Material: *Problem 38*

THE ELECTRIC FIELD PRODUCED BY A PARALLEL PLATE CAPACITOR

Equation 18.3, which gives the electric field of a point charge, is a very useful result. With the aid of integral calculus, this equation can be applied in a variety of situations where point charges are distributed over a surface. One such example that has considerable practical importance is the ***parallel plate capacitor.*** As Figure 18.23 shows, this device consists of two parallel metal plates, each with area A. A charge $+q$ is spread uniformly over one plate, while a charge $-q$ is spread uniformly over the other plate. In the region between the plates and away from the edges, the electric field points from the positive plate toward the negative plate and is perpendicular to both. Using Gauss' law (see Section 18.9), it can be shown that the electric field has a magnitude of

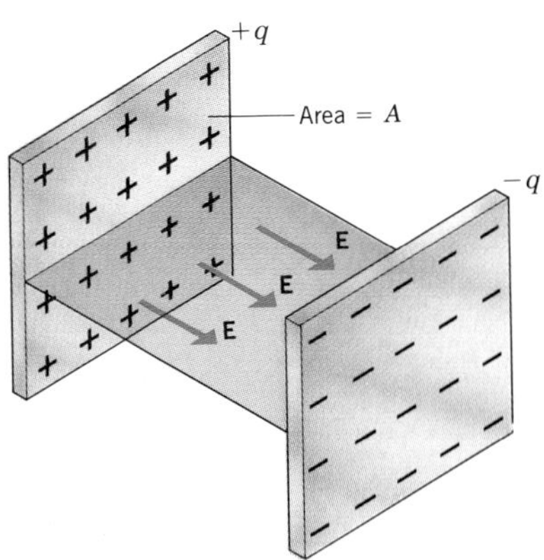

Figure 18.23 A parallel plate capacitor.

[Parallel plate capacitor]

$$E = \frac{q}{\epsilon_0 A} = \frac{\sigma}{\epsilon_0} \qquad (18.4)$$

In this expression the Greek symbol sigma (σ) denotes the charge per unit area ($\sigma = q/A$) and is sometimes called the charge density. Except in the region near the edges, the field has the same value at all places between the plates. The field does *not* depend on the distance from the charges, in distinct contrast to the field created by an isolated point charge.

18.7 ELECTRIC FIELD LINES

As we have seen, electric charges create an electric field in the space surrounding them. It is useful to have a kind of "map" that gives the direction and indicates the strength of the field at various places. The great English physicist Michael Faraday (1791–1867) proposed an idea that provides such a "map," the idea of ***electric field lines.*** Since the electric field is the electric force per unit charge, the electric field lines also provide information about electric forces and are sometimes called ***lines of force.***

To introduce the electric-field-line concept, Figure 18.24*a* shows a positive point charge $+q$. At the locations numbered 1–8, a positive test charge would experience a repulsive force, as the arrows in the drawing indicate. Therefore, the

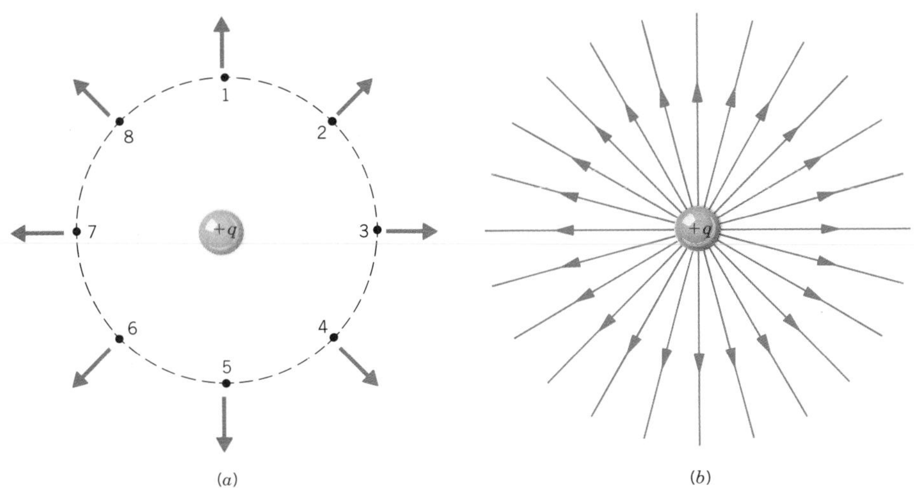

Figure 18.24 (*a*) At any of the eight marked spots around a positive point charge $+q$, a positive test charge would experience a repulsive force directed radially outward. (*b*) The electric field lines are directed radially outward from a positive point charge $+q$.

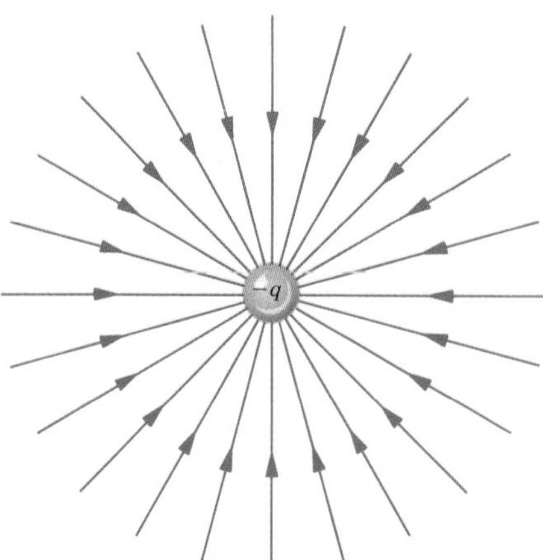

Figure 18.25 The electric field lines are directed radially inward toward a negative point charge $-q$.

electric field created by the charge $+q$ is directed radially outward. The electric field lines are lines drawn to show this direction, as part *b* of the drawing illustrates. The lines begin on the charge $+q$ and point radially *outward.* Figure 18.25 shows the electric field lines in the vicinity of a negative charge $-q$. In this case the lines are directed radially *inward,* because the force on a positive test charge is one of attraction, indicating that the electric field points inward. In general, ***electric field lines are always directed away from positive charges and toward negative charges.***

PROBLEM SOLVING INSIGHT

The electric field lines in Figures 18.24 and 18.25 are drawn in only two dimensions, as a matter of convenience. Electric field lines radiate from the charges in three dimensions, and an infinite number of lines could be drawn. However, for clarity only a small number is ever included in pictures. The number is chosen to be proportional to the magnitude of the charge; thus, five times as many lines would emerge from a $+5q$ charge as from a $+q$ charge.

The pattern of electric field lines also provides information about the magnitude or strength of the field. Notice in Figures 18.24 and 18.25 that near the charges, where the electric field is strongest, the field lines are close together. Conversely, at distances far from the charges, where the electric field is weaker, the electric field lines are more spread out. It is true in general that the electric field is strongest in regions where the field lines are closest together. In fact, no matter how many charges are present, the number of lines per unit area passing perpendicularly through a surface is proportional to the magnitude of the electric field.

In regions where the electric field lines are equally spaced, there is the same number of lines per unit area everywhere, and the electric field has the same strength at all points. For example, Figure 18.26 shows the field lines between the plates of a parallel plate capacitor. The lines are parallel and equally spaced, except near the edges where they bulge outward. The equally spaced, parallel lines indicate that the electric field has the same magnitude and direction at all points in the central region of the capacitor.

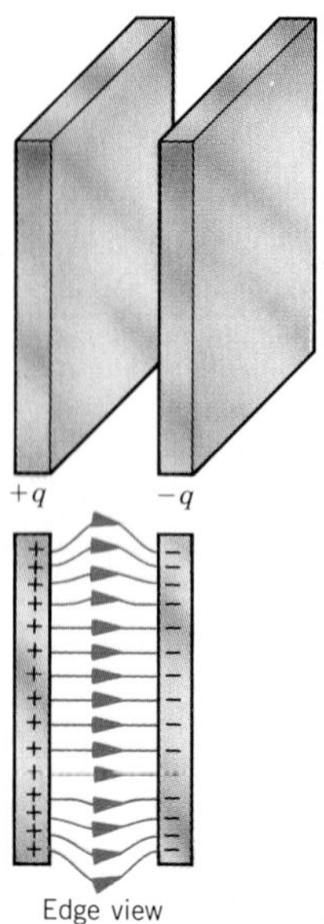

Figure 18.26 In the central region of a parallel plate capacitor the electric field lines are parallel and evenly spaced, indicating that the electric field there has the same magnitude and direction at all points.

Electric field lines are not always straight. More often they are curved, as in the case of an ***electric dipole.*** An electric dipole consists of two separated point charges that have the same magnitude but opposite signs. The electric field of a dipole is proportional to the product of the magnitude of one of the charges and the distance between the charges. This product is called the ***dipole moment.*** Many molecules, such as H_2O and HCl, have dipole moments. Figure 18.27 depicts the curved electric field lines in the vicinity of a dipole. For a curved field line, the electric field vector at a point is *tangent* to the line at that point (see points 1, 2, and 3 in the drawing). The pattern of the lines for the dipole indicates that the electric field is greatest in the region between and immediately surrounding the two charges, since the lines are closest together there.

Notice in Figure 18.27 that a given field line starts on the positive charge and ends on the negative charge. In general, ***electric field lines always begin on a positive charge and end on a negative charge and do not start or stop in mid-space. Furthermore, the number of field lines leaving a positive charge or entering a negative charge is proportional to the magnitude of the charge.*** This means, for example, that if 100 lines are drawn leaving a $+4\ \mu C$ charge, then 75 lines would have to end on a $-3\ \mu C$ charge and 25 lines on a $-1\ \mu C$ charge. Thus, 100 lines leave the charge of $+4\ \mu C$ and end on a *total charge* of $-4\ \mu C$, so the lines begin and end on equal amounts of total charge.

The electric field lines are also curved in the vicinity of two identical charges. Figure 18.28 shows the pattern associated with two positive point charges and reveals that there is an absence of lines in the region between the charges. The absence of lines indicates that the electric field is relatively weak between the charges.

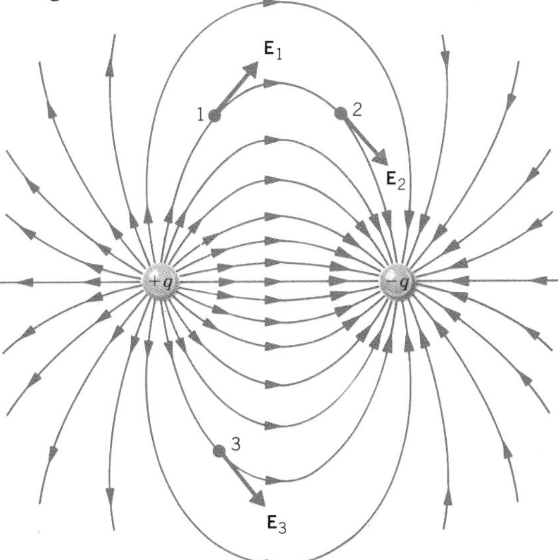

Figure 18.27 The electric field lines in the vicinity of an electric dipole are curved and extend from the positive charge to the negative charge. At any point, such as 1, 2, or 3, the electric field created by the dipole is tangent to the line that passes through the point.

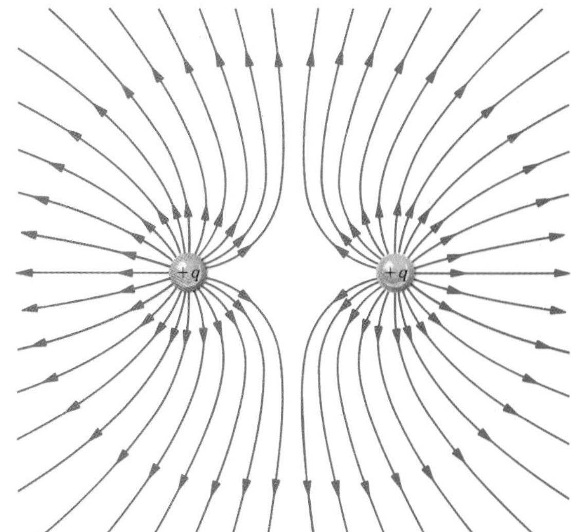

Figure 18.28 The electric field lines for two identical positive point charges. If the charges were both negative, the directions of the lines would be reversed.

Some of the important properties of electric field lines are reexamined in Conceptual Example 13.

Conceptual Example 13 Drawing Electric Field Lines

Figure 18.29*a* shows three negative point charges ($-q$, $-q$, and $-2q$) and one positive point charge ($+4q$), along with some electric field lines drawn between the charges. There are three things wrong with this drawing. What are they?

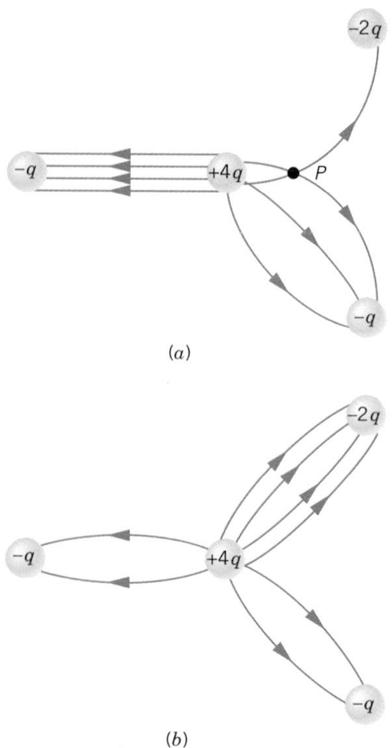

Figure 18.29 (*a*) Incorrectly and (*b*) correctly drawn electric field lines.

REASONING One aspect of Figure 18.29*a* that is incorrect is that electric field lines cross one another at point *P*. Electric field lines can never cross, because a charge placed at *P* experiences a single force due to the presence of other charges in its environment. Therefore, there is only one value for the electric field (which is the force per unit charge) at that point. If two electric field lines crossed one another, there would be two electric fields at the point where they crossed, one associated with each line. Since there can be only one value of the electric field at any point, there can be only one electric field line passing through that point, so electric field lines never cross each other.

Another mistake in Figure 18.29*a* is the number of electric field lines that end on the negative charges. Remember, the number of field lines leaving a positive charge or entering a negative charge is proportional to the magnitude of the charge. The $-2q$ charge has half the magnitude of the $+4q$ charge. Therefore, since 8 lines leave the $+4q$ charge, 4 of them (one-half of them) must enter the $-2q$ charge. Of the remaining 4 lines that leave the positive charge, 2 enter each of the $-q$ charges, according to a similar line of reasoning.

The third error in Figure 18.29*a* is the way in which the electric field lines are drawn between the $+4q$ charge and the $-q$ charge at the left of the drawing. As drawn, the lines are parallel and evenly spaced. This would indicate that the electric field everywhere in this region has a constant magnitude and direction, as is the case in the central region of a parallel plate capacitor. But the electric field between the $+4q$ and $-q$ charges is not constant everywhere. It certainly is stronger in places close to the $+4q$ or $-q$ charge than it is midway between them. The field lines, therefore, should be drawn with a curved nature, similar (but not identical) to those that surround a dipole. Figure 18.29*b* shows more nearly correct representations of the field lines for the four charges.

Related Homework Material: *Problems 29 and 58*

18.8 THE ELECTRIC FIELD INSIDE A CONDUCTOR: SHIELDING

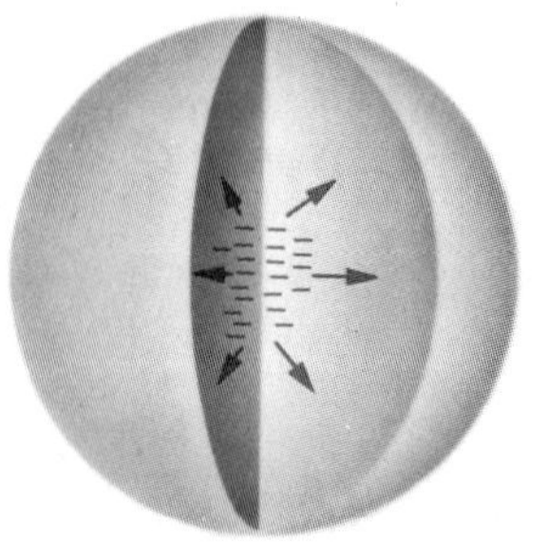

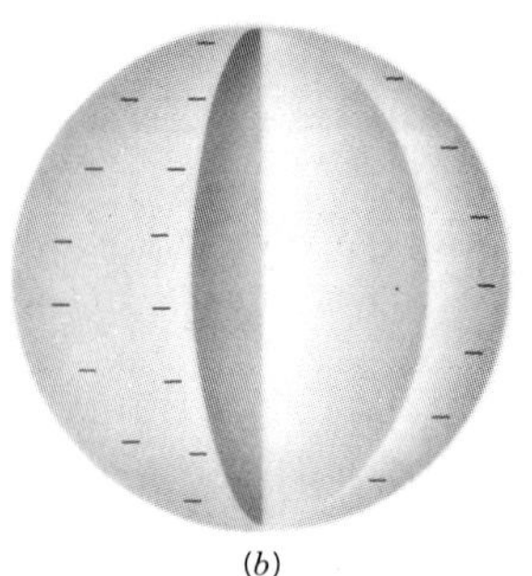

Figure 18.30 (*a*) Excess charge within a conductor moves quickly (*b*) to the surface.

In conducting materials such as copper, electric charges move readily in response to the forces that electric fields exert. This characteristic property of conducting materials has a major effect on the electric field that can exist within and around them. Suppose that a piece of copper carries a number of excess electrons somewhere within it, as in Figure 18.30*a*. Each electron would experience a force of repulsion because of the electric field of its neighbors. And, since copper is a conductor, the excess electrons move readily in response to that force. In fact, as a consequence of the $1/r^2$ dependence on distance in Coulomb's law, they rush to the surface of the copper. Once static equilibrium is established with all of the excess charge on the surface, no further movement of charge occurs, as part *b* of the drawing indicates. Similarly, excess positive charge also moves to the surface of a conductor. In general, ***at equilibrium under electrostatic conditions, any excess charge resides on the surface of a conductor.***

Now consider the interior of the copper in Figure 18.30*b*. The interior is electrically neutral, although there are still free electrons that can move under the influence of an electric field. The absence of a net movement of these free electrons indicates that there is no net electric field present within the conductor. In fact, the excess charges arrange themselves on the conductor surface precisely in the manner needed to make the total field zero within the material. Thus, ***at equilibrium under electrostatic conditions, the electric field at any point within a conducting material is zero.*** This fact has some fascinating implications.

Figure 18.31*a* shows an uncharged, solid, cylindrical conductor at equilibrium in the central region of a parallel plate capacitor. Induced charges on the surface of the cylinder alter the electric field lines of the capacitor. Since an electric field cannot exist within the conductor under these conditions, the electric field lines do not penetrate the cylinder and end or begin on the induced charges. Consequently, a test charge placed *inside* the conductor would feel no force due to the presence of the charges on the capacitor. In other words, ***the conductor shields any charge within it from electric fields created outside the conductor.*** The shielding results from the induced charges on the conductor surface.

Since the electric field is zero inside the conductor, nothing is disturbed if a cavity is cut from the interior of the material, as in part *b* of the drawing. Thus, the interior of the cavity is also shielded from external electric fields, a fact that has important applications, particularly for shielding electronic circuits. "Stray" electric fields are produced by various electrical appliances (e.g., hair driers, blenders, and vacuum cleaners), and these fields can interfere with the operation of sensitive electronic circuits, such as those in stereo amplifiers, televisions, and computers. To eliminate such interference, circuits are often enclosed within metal boxes that provide shielding from external fields.

THE PHYSICS OF . . .

shielding electronic circuits.

The blowup in Figure 18.31*a* shows another aspect of how conductors alter the electric field lines created by external charges. The lines are altered because ***the electric field just outside the surface of a conductor is perpendicular to the surface at equilibrium under electrostatic conditions.*** If the field were not perpendicular, there would be a component of the field parallel to the surface. Since the free electrons on the surface of the conductor can move, they would do so under the force exerted by the parallel component. But, in reality, no electron flow occurs at equilibrium. Therefore, there can be no parallel component, and the electric field is perpendicular to the surface.

The preceding discussion deals with aspects of the electric field within and around a conductor at equilibrium under electrostatic conditions. These features are related to the fact that conductors contain electrons that can move freely. Since insulators do not contain free electrons, these features *do not apply to insulators.* This section concludes with an example illustrating the behavior of a conducting material in the presence of an electric field.

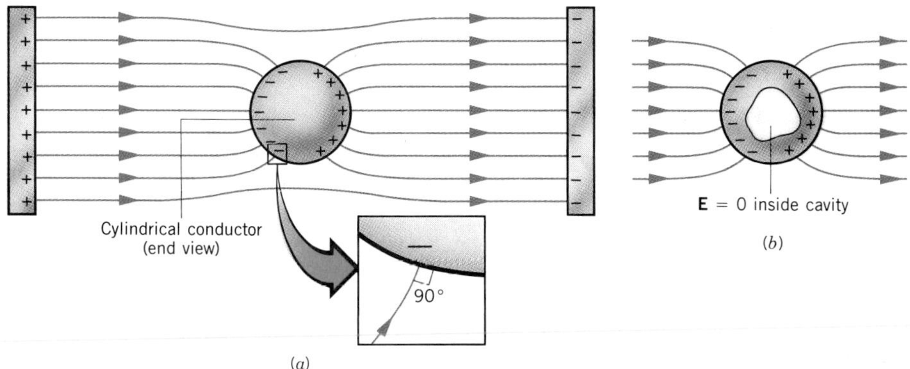

Figure 18.31 (*a*) A cylindrical conductor (shown in cross section) is placed between the oppositely charged plates of a capacitor. The electric field lines do not penetrate into the conductor. The blowup shows that, just outside the conductor, the electric field lines are perpendicular to its surface. (*b*) The electric field is zero in a cavity within the conductor.

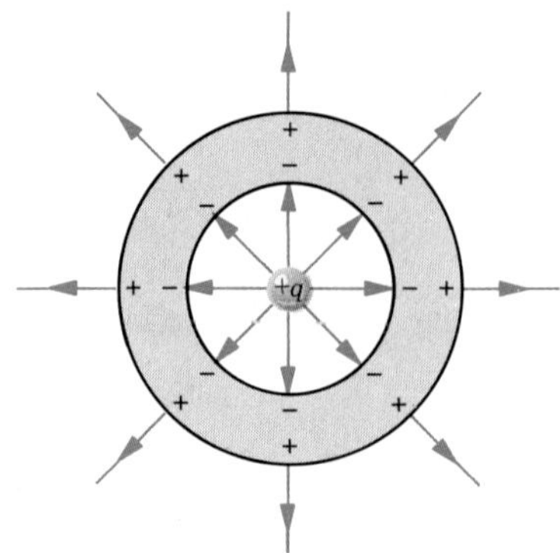

Figure 18.32 A positive charge $+q$ is suspended at the center of a hollow spherical conductor that is electrically neutral. Induced charges appear on the inner and outer surfaces of the conductor. The electric field within the conductor itself is zero.

Conceptual Example 14 A Conductor in an Electric Field

A charge $+q$ is suspended at the center of a hollow, electrically neutral, spherical conductor, as Figure 18.32 illustrates. Show that this charge induces (a) a charge of $-q$ on the interior surface and (b) a charge of $+q$ on the exterior surface of the conductor.

REASONING

(a) Electric field lines emanate from the positive charge $+q$. Since the electric field inside the metal conductor must be zero at equilibrium under electrostatic conditions, each field line ends when it reaches the conductor, as the picture shows. Consequently, there is an induced *negative* charge on the interior surface of the conductor, since field lines terminate only on negative charges. Furthermore, the lines begin and end on equal amounts of charge, so the magnitude of the total induced charge is the same as the magnitude of the charge at the center. Thus, the total induced charge on the interior surface is $-q$.

(b) Before the charge $+q$ is introduced, the conductor is electrically neutral. Therefore, it carries no net charge, and since an induced charge of $-q$ appears on the interior surface, a charge $+q$ must be induced on the outer surface. As we have seen, there can be no excess charge within the metal. The positive charge on the outer surface generates electric field lines that radiate outward (see the drawing) as if they originated from the central charge and the conductor were absent. The conductor does not shield the outside from the electric field produced by the charge on the inside.

Related Homework Material: *Problem 32*

18.9 GAUSS' LAW

Section 18.6 discusses how a point charge creates an electric field in the space around the charge. There are also many other situations in which an electric field is produced by charges that are spread out over a region, rather than by a single point charge. Such an extended collection of charges is called a charge distribution. For example, the electric field within the parallel plate capacitor in Figure 18.23 is produced by positive charges spread uniformly over one plate and an equal number of negative charges spread over the other plate. As we will see, Gauss' law describes the relationship between a charge distribution and the electric field it produces.

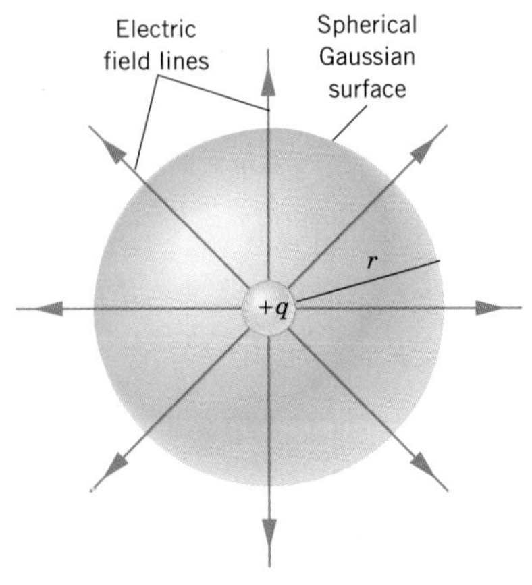

Figure 18.33 A positive point charge is located at the center of an imaginary spherical surface of radius r. Such a surface is one example of a Gaussian surface. Here the electric field is perpendicular to the surface and has the same magnitude everywhere on it.

GAUSS' LAW FOR A POSITIVE POINT CHARGE

For the moment we will postpone a discussion of the general form of Gauss' law and develop a version that is based on a familiar concept—the electric field produced by a positive point charge q. The electric field lines for such a charge radiate outward in all directions from the charge, as Figure 18.24*b* indicates. The magnitude E of the electric field at a distance r from the charge is $E = kq/r^2$, according to Equation 18.3. As mentioned in Section 18.5, the constant k can be expressed as $k = 1/(4\pi\epsilon_0)$, where ϵ_0 is the permittivity of free space. With this substitution, the magnitude of the electric field becomes $E = q/(4\pi\epsilon_0 r^2)$. We now place this point charge at the center of an imaginary spherical surface of radius r, as Figure 18.33 shows. Such a hypothetical closed surface is called a ***Gaussian surface,*** although in general it need not be spherical. The surface area A of a sphere is $A = 4\pi r^2$, and the magnitude of the electric field can be written in terms of this area as $E = q/(A\epsilon_0)$, or

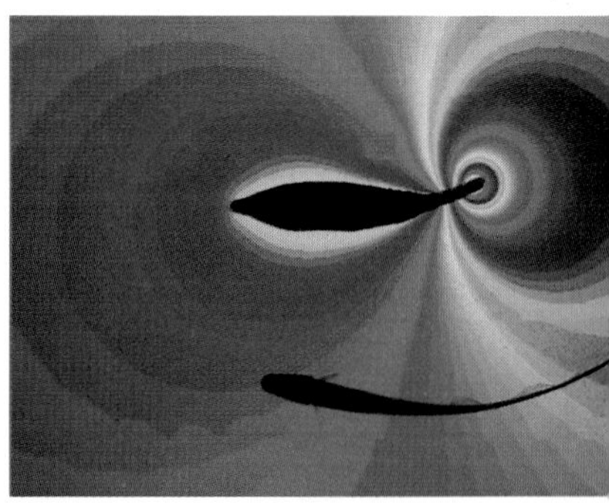

This computer image shows the electric field lines generated by the fish at the top of the picture. Through the electric field, the presence of other fish can be detected, such as the one silhouetted at the bottom.

Gauss' law for a point charge

$$\underbrace{EA}_{\substack{\text{Electric}\\ \text{flux, } \Phi_E}} = \frac{q}{\varepsilon_0} \qquad (18.5)$$

The left side of Equation 18.5 is the product of the magnitude E of the electric field at any point on the Gaussian surface and the area A of the surface. In Gauss' law this product is especially important and is called the ***electric flux, Φ_E***: $\Phi_E = EA$. (It will be necessary to modify this definition of the electric flux when we consider the more general case of a Gaussian surface with an arbitrary shape.)

Equation 18.5 is the result we have been seeking, for it is a form of Gauss' law that applies to a point charge. This result indicates, aside from the constant ϵ_0, that the electric flux Φ_E depends only on the charge q within the Gaussian surface and is independent of the radius r of the surface. We will now see how to generalize Equation 18.5 to account for situations where the electric field is produced by a distribution of charges and the Gaussian surface has an arbitrary shape.

GAUSS' LAW FOR A CHARGE DISTRIBUTION

Figure 18.34 shows a charge distribution whose net charge is labeled Q. The charge distribution is surrounded by a Gaussian surface, i.e., an imaginary closed surface. The surface can have *any arbitrary shape* (it need not be spherical), but *it must be closed* (an open surface would be like that of half an egg shell). The direction of the electric field is not necessarily perpendicular to the Gaussian

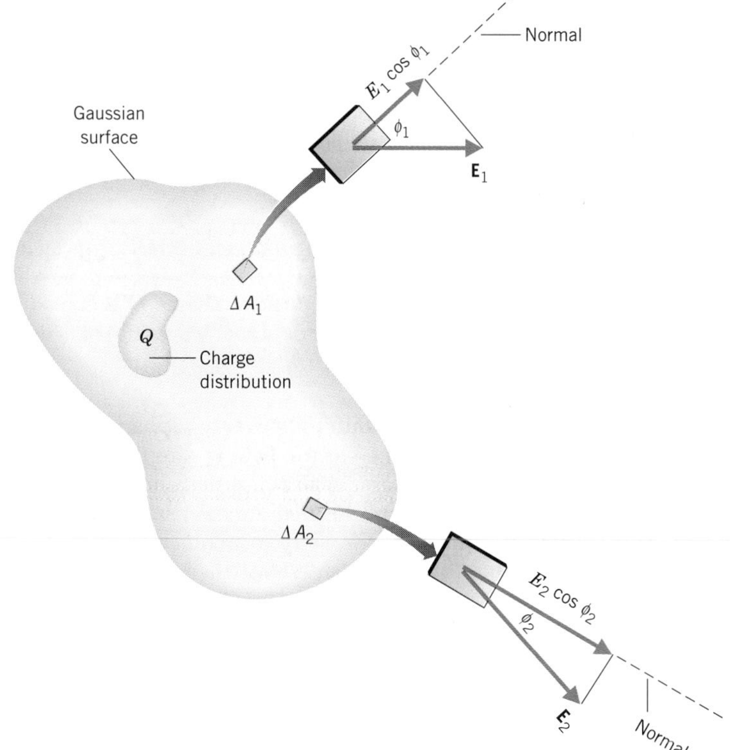

Figure 18.34 The charge distribution Q is surrounded by an arbitrarily shaped Gaussian surface. The electric flux Φ through any tiny segment of the surface is the product of $E \cos\phi$ and the area ΔA of the segment: $\Phi = (E \cos\phi)\,\Delta A$. The angle ϕ is the angle between the electric field and the normal to the surface.

surface. Furthermore, the magnitude of the electric field need not be constant on the surface, but can vary from point to point.

To determine the electric flux through such a surface, we divide the surface up into many tiny sections with areas, ΔA_1, ΔA_2, and so on. Each section is so small that it is essentially flat and the electric field **E** is a constant (both in magnitude and direction) over it. For reference, a dashed line called the "normal" is drawn perpendicular to each section on the outside of the surface. To determine the electric flux for each of the sections, we use only the component of **E** that is perpendicular to the surface, that is, the component of the electric field that passes through the surface. From the drawing it can be seen that this component has a magnitude of $E \cos \phi$, where ϕ is the angle between the electric field and the normal. The electric flux through any individual section is then $(E \cos \phi)\,\Delta A$. The electric flux Φ_E that passes through the entire Gaussian surface is the sum of all of these individual fluxes: $\Phi_E = (E_1 \cos \phi_1)\Delta A_1 + (E_2 \cos \phi_2)\Delta A_2 + \cdots$, or

$$\Phi_E = \Sigma(E \cos \phi)\,\Delta A \tag{18.6}$$

where, as usual, the symbol Σ means "the sum of." Gauss' law relates the electric flux Φ_E to the net charge Q enclosed by the arbitrarily shaped Gaussian surface.

Gauss' Law

The electric flux Φ_E through a Gaussian surface is equal to the net charge Q enclosed by the surface divided by ε_0, the permittivity of free space:

$$\underbrace{\sum (E \cos \phi)\Delta A}_{\text{Electric flux, } \Phi_E} = \frac{Q}{\varepsilon_0} \tag{18.7}$$

SI Unit of Electric Flux: $N \cdot m^2/C$

Gauss's law is often used to find the magnitude of the electric field produced by a distribution of charges. The law is most useful when the distribution is uniform and symmetrical. In the next three examples we will see how to apply Gauss' law in such situations.

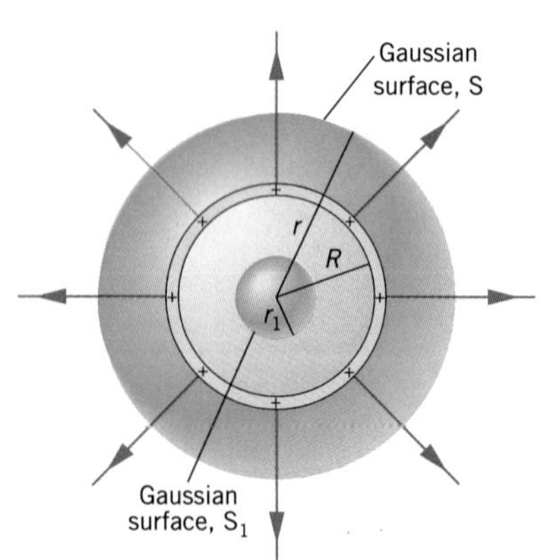

Figure 18.35 A uniform distribution of positive charge resides on a thin spherical shell of radius R. The spherical Gaussian surfaces S and S_1 are used in Example 15 to evaluate the electric flux outside and inside the shell, respectively.

Example 15 The Electric Field of a Charged Thin Spherical Shell

Figure 18.35 shows a thin spherical shell of radius R. A positive charge Q is spread uniformly over the shell. Find the magnitude of the electric field at any point (a) outside the shell and (b) inside the shell.

REASONING Because the charge is distributed uniformly over the spherical shell, the electric field is symmetric. This means that the electric field is directed radially outward in all directions, and its magnitude is the same at all points that are equidistant from the shell. All points that are equidistant from the shell lie on a sphere, so the symmetry is called *spherical symmetry*. With this symmetry in mind, we will use a spherical Gaussian surface to evaluate the electric flux Φ_E. We will then use Gauss' law to determine the magnitude of the electric field.

SOLUTION

(a) To find the magnitude of the electric field outside the charged shell, we construct a spherical Gaussian surface of radius r ($r > R$) that is concentric with the shell (see the surface labeled S in Figure 18.35). We now evaluate the electric flux $\Phi_E = \Sigma(E \cos \phi)$

ΔA. Since the electric field **E** is everywhere perpendicular to the Gaussian surface, $\phi = 0°$ and $\cos\phi = 1$. In addition, E has the same value at all points on the Gaussian surface, since they are equidistant from the charged shell. Being constant over the surface, E can be factored outside the summation, with the result that $\Phi_E = \Sigma(E\cos\phi)\Delta A = E(\Sigma\Delta A)$. The term $\Sigma\Delta A$ is just the sum of the tiny areas that make up the Gaussian surface. Since the area of a spherical surface is $4\pi r^2$, we have $\Sigma\Delta A = 4\pi r^2$. The electric flux is then $\Phi_E = E(4\pi r^2)$. Setting the electric flux equal to Q/ϵ_0, as specified by Gauss' law, yields $E(4\pi r^2) = Q/\epsilon_0$. Solving this equation for E gives

$$E = \frac{Q}{4\pi\epsilon_0 r^2} \qquad \text{(for } r > R\text{)}$$

This is a surprising result, for it is exactly the same as that for a point charge (see Equation 18.3). Thus, the electric field outside a uniformly charged spherical shell is the same as if all the charge Q were concentrated as a point charge at the center of the shell.

(b) To find the magnitude of the electric field inside the charged shell, we select a spherical Gaussian surface that lies inside the shell and is concentric with it (see the surface labeled S_1 in Figure 18.35). Inside the charged shell, the electric field (if it exists) must also have spherical symmetry. Therefore, using reasoning like that in part (a), the electric flux through the Gaussian surface is $\Phi_E = \Sigma(E\cos\phi)\,\Delta A = E(4\pi r_1{}^2)$. In accord with Gauss' law, the electric flux must be equal to Q/ϵ_0, where Q is the net charge *inside* the Gaussian surface. But now $Q = 0$, since all the charge lies on the shell that is *outside* the Gaussian surface. Consequently, we have $E(4\pi r_1{}^2) = Q/\epsilon_0 = 0$, or

$$E = 0 \qquad \text{(for } r < R\text{)}$$

Gauss' law allows us to deduce that there is no electric field inside a uniform spherical shell of charge. An electric field exists only on the outside.

Example 16 The Electric Field Inside a Parallel Plate Capacitor

Equation 18.4 states that the electric field inside a parallel plate capacitor, and away from the edges, is constant and has a magnitude of $E = \sigma/\epsilon_0$, where σ is the charge density (the charge per unit area) on a plate. Use Gauss' law to prove this result.

REASONING Figure 18.36*a* shows the electric field inside a parallel plate capacitor. Because the positive and negative charges are distributed uniformly over the surfaces of the plates, symmetry requires that the electric field be perpendicular to the plates. We will take advantage of this symmetry by choosing our Gaussian surface to be a small cylinder whose axis is perpendicular to the plates (see part *b* of the figure). With this choice, we will be able to evaluate the electric flux and then, with the aid of Gauss' law, to determine E.

SOLUTION Figure 18.36*b* shows that we have placed our Gaussian cylinder so that its left end is inside the positive metal plate, and the right end is in the space between the plates. To determine the electric flux through this Gaussian surface, we evaluate the flux through each of the three parts—labeled 1, 2, and 3 in the drawing—that make up the total surface of the cylinder and then add up the fluxes.

Surface 1—the left, flat end of the cylinder—is embedded inside the positive metal plate. As discussed in Section 18.8, the electric field is zero everywhere inside a conductor that is in equilibrium under electrostatic conditions. Since $E = 0$, the electric flux through this surface is also zero: $\Phi_1 = \Sigma(E\cos\phi)\,\Delta A = 0$.

Surface 2—the curved wall of the cylinder—is everywhere parallel to the electric

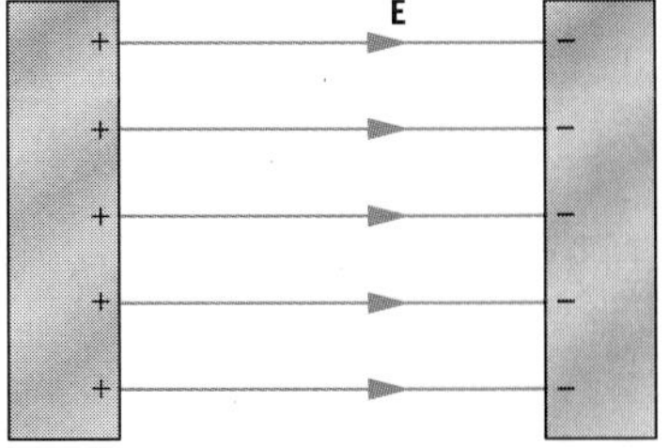

(*a*)

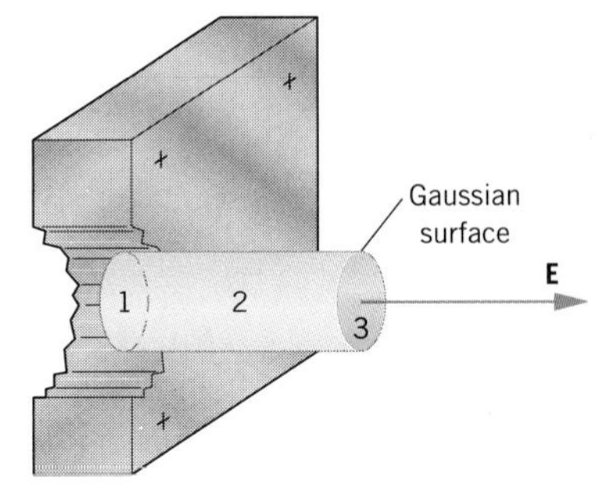

(*b*)

Figure 18.36 (*a*) A side view of a parallel plate capacitor, showing some of the electric field lines. (*b*) The Gaussian surface is a cylinder oriented so its axis is perpendicular to the positive plate and its left end is inside the plate.

field between the plates, so that $\cos \phi = \cos 90° = 0$. Therefore, the electric flux through this surface is also zero: $\Phi_2 = \Sigma(E \cos 90°)\,\Delta A = 0$.

Surface 3—the right, flat end of the cylinder—is perpendicular to the electric field between the plates, so $\cos \phi = \cos 0° = 1$. The electric field is constant over this surface, so E can be taken outside the summation in Equation 18.6. As a result, the electric flux through this surface is $\Phi_3 = \Sigma(E \cos 0°)\,\Delta A = E(\Sigma \Delta A) = EA$, where $\Sigma \Delta A = A$ is the area of surface 3.

The electric flux through the entire Gaussian cylinder is the sum of the three fluxes determined above:

$$\Phi_E = \Phi_1 + \Phi_2 + \Phi_3 = 0 + 0 + EA = EA$$

According to Gauss' law, we set the electric flux equal to Q/ϵ_0, where Q is the net charge *inside* the Gaussian cylinder: $EA = Q/\epsilon_0$. But Q/A is the charge per unit area, σ, on the plate. Therefore, we arrive at the value of the electric field inside a parallel plate capacitor: $\boxed{E = \sigma/\epsilon_0}$. Notice that the distance of the right end of the Gaussian cylinder from the positive plate does not appear in this result, indicating that the electric field has the same value everywhere between the plates.

Example 17 The Electric Field of a Long, Straight, Charged Wire

Figure 18.37 shows a long, straight wire with a positive charge distributed uniformly along it. Find the magnitude of the electric field at any point outside the wire.

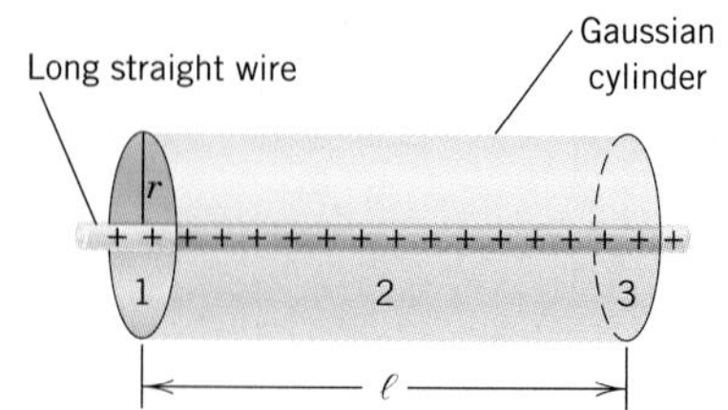

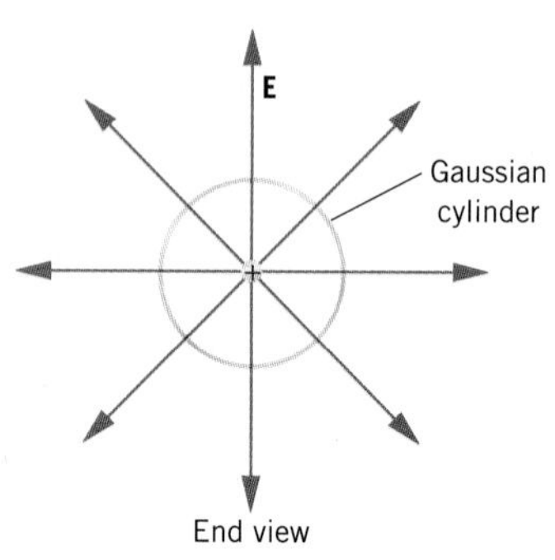

Figure 18.37 The Gaussian surface is a cylinder of radius r and length ℓ. The cylinder is concentric with the long straight wire. The electric field **E** is perpendicular to the curved part of the cylinder, or surface 2, and has the same magnitude everywhere on it.

REASONING Because the charge is distributed uniformly along the straight wire, the electric field is directed radially outward from the wire, as the end view of the wire illustrates. And because of symmetry, the magnitude of the electric field is the same at all points equidistant from the wire. In this situation we will use a Gaussian surface that is a cylinder concentric with the wire. The drawing shows that this cylinder is composed of three parts, the two flat ends and the curved wall. As in Example 16, we will evaluate the electric flux for this three-part surface and then set it equal to Q/ϵ_0 (Gauss' law) to find the magnitude of the electric field.

SOLUTION Surfaces 1 and 3—the flat ends of the cylinder—are parallel to the electric field, so $\cos \phi = \cos 90° = 0$. Thus, there is no flux through these two surfaces: $\Phi_1 = \Phi_3 = 0$.

Surface 2—the curved wall—is everywhere perpendicular to the electric field **E**, so $\cos \phi = \cos 0° = 1$. Furthermore, the magnitude E of the electric field is the same for all points on this surface, so it can be factored outside the summation in Equation 18.6: $\Phi_2 = \Sigma(E \cos 0°)\,\Delta A = E(\Sigma \Delta A)$. The area $\Sigma \Delta A$ of this surface is just the circumference $2\pi r$ of the cylinder times the length ℓ of the cylinder: $(\Sigma \Delta A) = (2\pi r)\ell$. The electric flux through the entire cylinder is, then,

$$\Phi_E = \Phi_1 + \Phi_2 + \Phi_3 = 0 + E(2\pi r\ell) + 0 = E(2\pi r\ell)$$

Following Gauss' law, we set Φ_E equal to Q/ϵ_0, where Q is the net charge inside the Gaussian cylinder: $E(2\pi r\ell) = Q/\epsilon_0$. The ratio Q/ℓ is the charge per unit length of the wire and is known as the linear charge density λ: $\lambda = Q/\ell$. Solving for E, we find that

$$\boxed{E = \frac{\lambda}{2\pi\epsilon_0 r}}$$

This result indicates that the electric field decreases as $1/r$, so E is weaker for points that are farther from the wire.

18.10 COPIERS AND COMPUTER PRINTERS

XEROGRAPHY

THE PHYSICS OF . . .

xerography.

The electrostatic force that charged particles exert on one another plays the central role in an office copier. The copying process is called *xerography,* from the Greek *xeros* and *graphos,* meaning "dry writing." The heart of a copier is the xerographic drum, an aluminum cylinder coated with a layer of selenium (see Figure 18.38*a*). Aluminum is an excellent electrical conductor. Selenium, on the

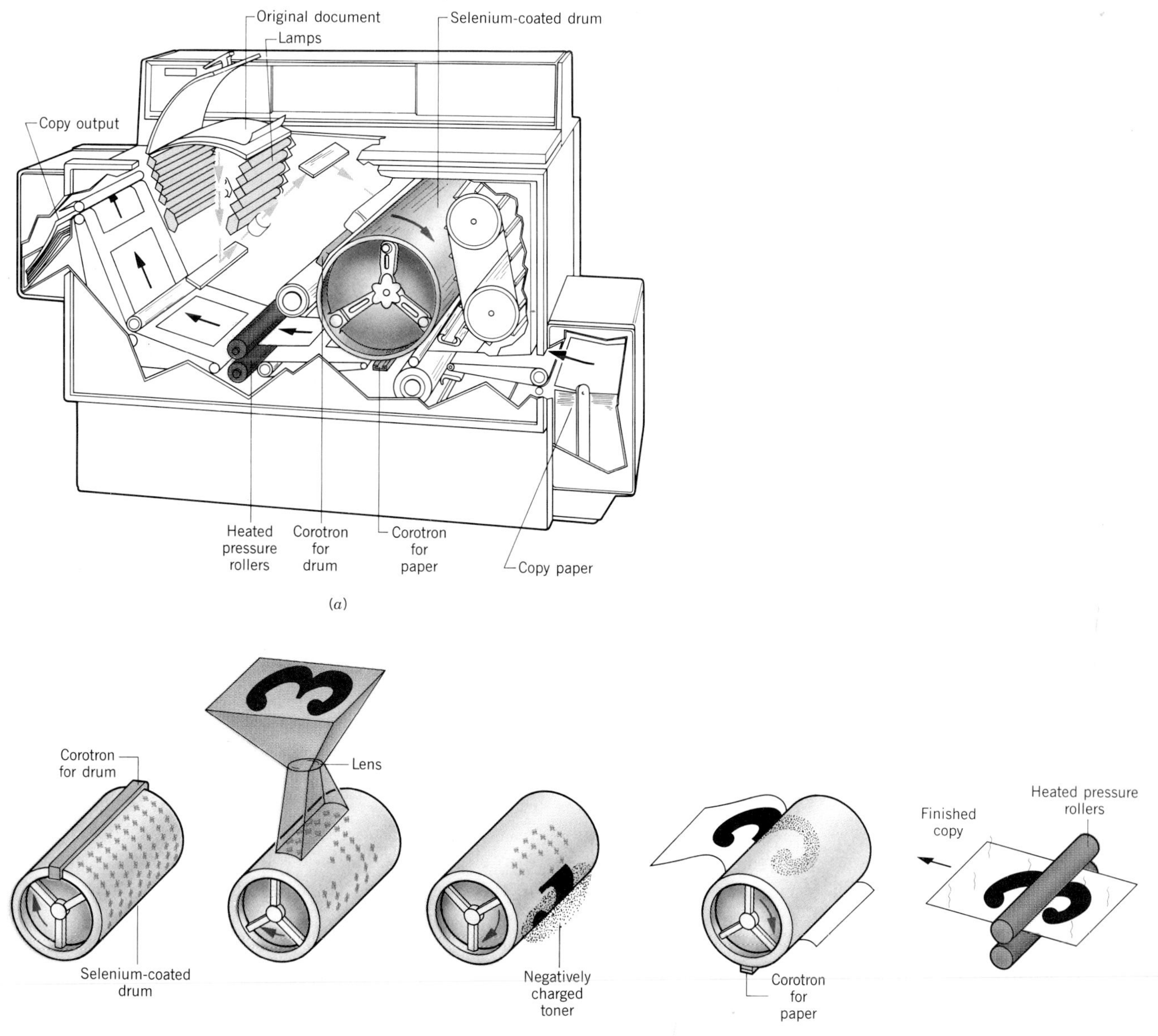

Figure 18.38 (*a*) This cutaway view shows the essential elements of a copying machine. (*b*) The five steps in the xerographic process.

other hand, is a photoconductor; it is an insulator in the dark but becomes a conductor when exposed to light. Consequently, a positive charge deposited on the selenium surface will remain there, provided the selenium is kept in the dark. When the drum is exposed to light however, electrons from the aluminum pass through the conducting selenium, and neutralize the positive charge.

The photoconductive property of selenium is critical to the xerographic process, as illustrated in Figure 18.38*b*. First, an electrode called a *corotron* gives the entire selenium surface a positive charge in the dark. Second, a series of lenses and mirrors focuses an image of a document onto the revolving drum. The dark and light areas of the document produce corresponding areas on the drum. The dark areas retain their positive charge, but the light areas become conducting and lose their positive charge, ending up neutralized. Thus, a positive-charge image of the document remains on the selenium surface. In the third step, a special dry, black powder, called the *toner,* is given a negative charge and then spread onto the drum, where it adheres selectively to the positively charged areas. The fourth step involves transferring the toner onto a blank piece of paper. However, the attraction of the positive-charge image holds the toner to the drum. To transfer the toner, the paper is given a *greater positive charge* than that of the image, with the aid of another corotron. Lastly, the paper and adhering toner pass through heated pressure rollers. As a result of the heat, the toner melts into the fibers of the paper and produces the finished copy.

THE PHYSICS OF . . .
a laser printer.

A LASER PRINTER

A laser printer is used with computers to provide high-quality copies of text and graphics. The laser printer is similar in operation to the xerographic machine, except the information to be reproduced is not on paper. Instead, the information comes from the computer's memory and is transferred to the printer through an electrical cable. Laser light is used to copy the information onto the selenium–aluminum drum. A laser beam, focused to a fine point, is scanned rapidly from side to side across the rotating drum, as Figure 18.39*a* indicates. While the light remains on, the positive charge on the drum is neutralized. As the laser beam moves, the computer turns the beam off at the right moments during each scan to produce the desired positive-charge image, which is the letter "A" in the picture.

Figure 18.39*b* shows the mechanism that turns the laser beam off and on and scans it across the xerographic drum. The light from the laser is sent through a device called a "modulator." The modulator also receives the information to be printed from the computer and, accordingly, allows the light to pass or blocks it. Thus, the laser output beam from the modulator is turned off and on, at rates often exceeding one million times per second. The output beam is then directed by a series of mirrors and lenses to a rotating polygonal mirror that causes the reflected beam to sweep from side to side. When the beam reflected from the rotating mirror is directed onto the xerographic drum, the beam scans across the drum and produces the positive-charge image.

THE PHYSICS OF . . .
an inkjet printer.

AN INKJET PRINTER

An inkjet printer is another type of printer that uses electric charges in its operation. While shuttling back and forth across the paper, the inkjet printhead ejects a thin stream of ink. Figure 18.40 illustrates the elements of one type of printhead. The ink is forced out of a small nozzle and breaks up into extremely small droplets, with diameters less than 1×10^{-4} m. About 150 000 droplets leave the nozzle each second and travel with a speed of approximately 18 m/s toward the

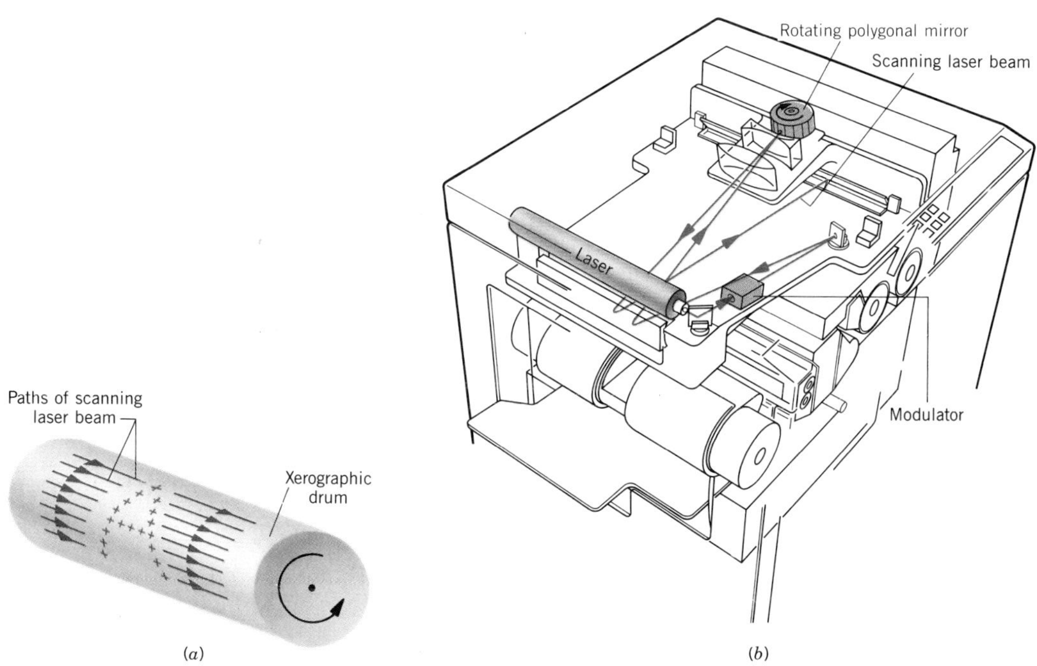

Figure 18.39 (*a*) As the laser beam scans back and forth across the surface of the xerographic drum, a positive-charge image of the letter "A" is created. (*b*) A laser printer showing the laser, modulator, and rotating mirror.

paper. During their flight, the droplets pass through two electrical components, a *charging electrode* and the *deflection plates* (a parallel plate capacitor). When the printhead moves over regions of the paper that are not to be inked, the charging electrode is turned on and gives the ink droplets a net charge. The deflection plates divert the charged droplets into a gutter and thus prevent them from reaching the paper. Whenever ink is to be placed on the paper, the charging control, responding to instructions from the computer, turns off the charging electrode. The uncharged droplets fly straight through the deflection plates and strike the paper.

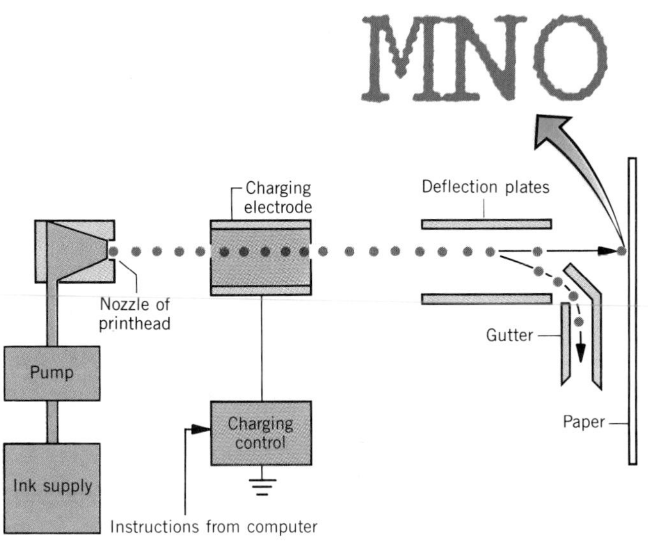

Figure 18.40 An inkjet printhead ejects a steady flow of ink droplets. The charging electrode is used to charge the droplets that are not needed on the paper. Charged droplets are deflected into a gutter by the deflection plates, while uncharged droplets fly straight onto the paper. The letters formed by an inkjet printer look normal, except when greatly enlarged, as they are here. Then the patterns from the drops become apparent.

INTEGRATION OF CONCEPTS

THE ELECTROSTATIC FORCE AND NEWTON'S LAWS OF MOTION

In this chapter we have seen that electrically charged particles exert a force on each other. This electrostatic force is part of one of nature's four fundamental forces, the one known as the electromagnetic force. Like any force, the electrostatic force can be used in Newton's laws of motion. These laws form the foundation of mechanics, the branch of physics that deals with the motion of objects and the forces that change it. The electrostatic force can cause an object to change its state of motion, that is, to accelerate, according to Newton's first and second laws. It can produce a change in the momentum of an object, in accordance with Newton's second law in the form known as the impulse–momentum theorem. It can do work and cause the kinetic energy of an object to change, according to the work–energy theorem. And the electrostatic force obeys Newton's third law, since the forces that two charged objects exert on each other are examples of action and reaction. The beauty of Newton's laws is that they apply to *all* types of forces, independent of the details that are specific to an individual kind of force, and independent of whether the force is one of nature's fundamental forces. Newton's laws relate to basic characteristics that forces of all types share, including the electrostatic force. It would be a complicated world indeed, if every type of force required a different type of Newton's laws to account for the effect of the force on the motion of an object. Fortunately, such is not the case.

SUMMARY

There are two kinds of **electric charge:** positive and negative; the SI unit of electric charge is the **coulomb** (C). The magnitude of the charge on an electron or a proton is $e = 1.60 \times 10^{-19}$ C. The electron carries a charge of $-e$, while the proton carries a charge of $+e$. The charge on any object, whether positive or negative, is **quantized,** in the sense that the charge consists of an integral number of protons or electrons. **The law of conservation of electric charge** states that the net electric charge of an isolated system remains constant during any process.

An **electrical conductor** is a material, such as copper, that conducts electric charge readily. An **electrical insulator** is a material, such as rubber, that conducts electric charge poorly. **Charging by contact** is the process of giving one object a net electric charge by placing it in contact with an object that is already charged. **Charging by induction** is the process of giving an object a net electric charge without touching it to a charged object.

One charge exerts an electric force on another charge. **For like charges the force is a repulsion, while for unlike charges the force is an attraction. Coulomb's law** gives the magnitude F of the electric force that two point charges exert on each other as $F = kq_1q_2/r^2$, where q_1 and q_2 are the magnitudes of the charges, r is the distance between the charges, and $k = 8.99 \times 10^9$ $N \cdot m^2/C^2$. The force acts along the line between the two point charges. The permittivity of free space ϵ_0 is defined by the relation $k = 1/(4\pi\epsilon_0)$.

The **electric field E** at a given spot is a vector and is the electrostatic force **F** experienced by a small test charge q_0 placed at that spot divided by the charge itself: $\mathbf{E} = \mathbf{F}/q_0$. The direction of the electric field is the same as the direction of the force on a positive test charge. The SI unit for the electric field is the newton per coulomb (N/C). The source of the electric field at any spot is the charged objects surrounding that spot. The magnitude of the electric field created by a point charge q is $E = kq/r^2$, where r is the distance from the charge. The magnitude of the electric field between the plates of a parallel plate capacitor is $E = \sigma/\epsilon_0$, where σ is the charge per unit area on either plate.

Electric field lines are lines that can be thought of as a "map" providing information about the direction and strength of the electric field. Electric field lines are directed away from positive charges and toward negative charges. The direction of the lines gives the direction of the electric field, since the electric field vector at a point is tangent to the line at that point. The electric field is strongest in regions where the number of lines per unit area passing perpendicularly through a surface is the greatest, that is, where the lines are closest together.

Excess negative or positive charge resides on the surface of a conductor at equilibrium under electrostatic conditions. In such a situation, the electric field at any point within the conducting material is zero, and the electric field just outside the surface of the conductor is perpendicular to the surface.

Gauss' law states that the electric flux Φ_E (see Equation 18.6) through a closed surface (a Gaussian surface) is equal to the net charge Q enclosed by the surface divided by ϵ_0, the permittivity of free space: $\Phi_E = Q/\epsilon_0$.

QUESTIONS

1. In Figure 18.8 the grounding wire is removed first, followed by the rod, and the sphere is left with a positive charge. If the rod were removed first, followed by the grounding wire, would the sphere be left with a charge? Account for your answer.

2. A metallic object is given a positive charge by the process of induction, as illustrated in Figure 18.8. (a) Does the mass of the object increase, decrease, or remain the same? Why? (b) What happens to the mass of the object if it is given a negative charge by induction? Explain.

3. A rod made from insulating material carries a net charge, while a copper sphere is neutral. The rod and the sphere do not touch. Is it possible for the rod and the sphere to (a) attract one another and (b) repel one another? Explain.

4. An electroscope is charged so that its leaves are spread apart. The spread between the leaves decreases slightly when an electrically neutral copper object is brought near the metal knob of the electroscope without touching it. Explain.

5. The leaves of an electroscope are given a positive charge. As an insulator is brought from far away and moved toward the knob of the electroscope, the leaves are observed to spread apart even further. Is the insulator charged positively or negatively? Give your reasoning.

6. Blow up a balloon and rub it against your shirt a number of times. In so doing you give the balloon a net electric charge. Now touch the balloon to the ceiling. Upon being released, the balloon will remain stuck to the ceiling. Why?

7. Suppose an electroscope is electrically neutral, rather than being negatively charged as in Figure 18.11*a*. Can this neutral electroscope be used to determine whether the net charge on an object is negative or positive? Justify your answer.

8. A particle is attached to a spring and is pushed so that the spring is compressed more and more. As a result, the spring exerts a greater and greater force on the particle. Similarly, a charged particle experiences a greater and greater force when pushed closer and closer to another particle that is fixed in position and has a charge of the same polarity. In spite of the similarity, the charged particle will *not* exhibit simple harmonic motion upon being released, as will the particle on the spring. Explain why not.

9. Identical point charges are fixed to opposite corners of a square. Where does a third point charge experience the greater force, at one of the empty corners or at the center of the square? Account for your answer.

10. On a thin, nonconducting rod, positive charges are spread evenly, so that there is the same amount of charge per unit length at every point. On another identical rod, positive charges are spread evenly over only the left half, and the same amount of negative charges are spread evenly over the right half. For each rod, deduce the *direction* of the electric field at a point that is located directly above the midpoint of the rod. Give your reasoning.

11. There is an electric field at point *P*. A very small charge is placed at this point and experiences a force. Another very small charge is then placed at this point and experiences a force that differs in both magnitude and direction from that experienced by the first charge. How can these two different forces result from the single electric field that exists at point *P*?

12. In Figure 18.27 there is no place on the line through the charges where the electric field is zero, neither to the left of the positive charge, nor between the charges, nor to the right of the negative charge. Now, suppose the magnitude of the negative charge were greater than the magnitude of the positive charge. Is there any place on the line through the charges where the electric field is zero? Justify your answer.

13. Drawings I and II show two examples of electric field lines. Decide which of the following statements are true and which are false, defending your choice in each case. (a) In both I and II the electric field is the same everywhere. (b) As you move from left to right in each case, the electric field becomes stronger. (c) The electric field in I is the same everywhere but becomes stronger in II as you move from left to right. (d) The electric fields in both I and II could be created by negative charges located somewhere on the left and positive charges somewhere on the right. (e) Both I and II arise from a single positive point charge located somewhere on the left.

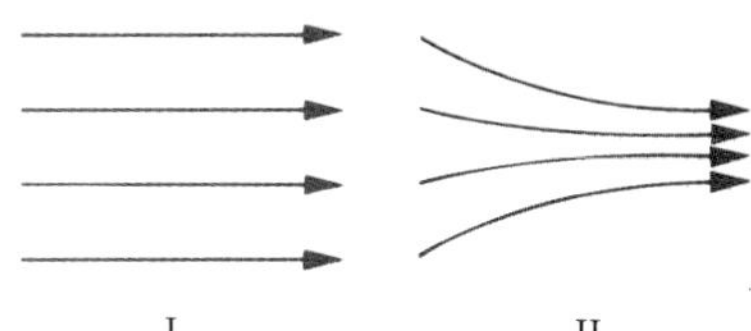

14. A positively charged particle is moving horizontally when it enters the region between the plates of a capacitor, as the drawing illustrates. (a) Draw the trajectory that the particle follows in moving through the capacitor. (b) When the particle is within the capacitor, which of the following four vectors, if any, are *parallel* to the electric field inside the capacitor: the particle's displacement, its velocity, its linear momentum, its acceleration? For each vector explain why the vector is, or is not, parallel to the electric field of the capacitor.

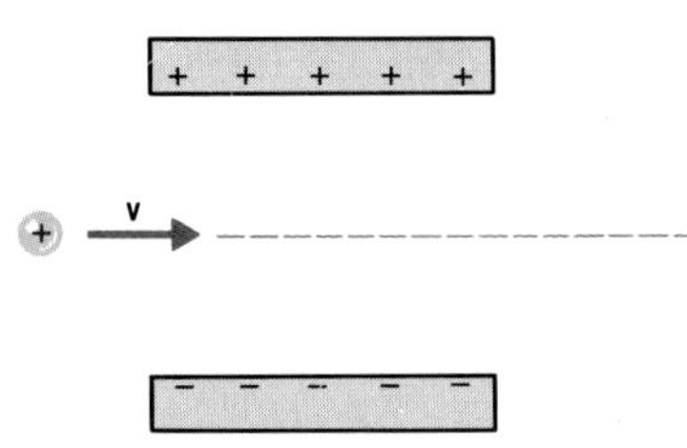

15. Refer to Figure 18.28. Imagine a plane that is perpendicular to the line between the charges, midway between them, and is half into and half out of the paper. The electric flux through this plane is zero. Explain why.

16. Two charges, $+q$ and $-q$, are inside a Gaussian surface. Since the net charge inside the Gaussian surface is zero, Gauss' law states that the electric flux through the surface is also zero, i.e., $\Phi_E = 0$. Does the fact that $\Phi_E = 0$ imply that the electric field **E** at any point on the Gaussian surface is also zero? Justify your answer. (*Hint: Imagine a Gaussian surface that encloses the two charges in Figure 18.27.*)

17. The drawing shows three charges, labeled q_1, q_2, and q_3. A Gaussian surface is drawn around q_1 and q_2. (a) Which charges determine the electric flux through the Gaussian surface? (b) Which charges produce the electric field at the point *P*? Justify your answers.

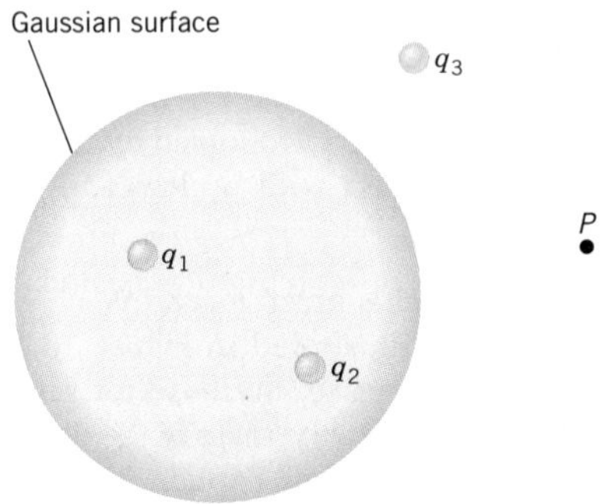

18. A charge $+q$ is placed inside a spherical Gaussian surface. The charge is *not* located at the center of the sphere. (a) Can Gauss' law tell us exactly where the charge is located inside the sphere? Justify your answer. (b) Can Gauss' law tell us about the magnitude of the electric flux through the Gaussian surface? Why?

PROBLEMS

Note: All charges are point charges, unless specified otherwise.

Section 18.1 The Origin of Electricity, Section 18.2 Charged Objects and the Electric Forces That They Exert, Section 18.3 Conductors and Insulators, Section 18.4 Charging by Contact and by Induction

1. How many electrons must be removed from an electrically neutral silver dollar to give it a charge of $+3.8\ \mu C$?

2. A metal sphere has a charge of $+8.0\ \mu C$. What is the net charge after 6.0×10^{13} electrons have been placed on it?

3. A rod has a charge of $-2.0\ \mu C$. How many electrons must be removed so that the charge becomes $+3.0\ \mu C$?

4. Consider three identical metal spheres, A, B, and C. Sphere A carries a charge of $+5q$. Sphere B carries a charge of $-q$. Sphere C carries no net charge. Spheres A and B are touched together and then separated. Sphere C is then touched to sphere A and separated from it. Lastly, sphere C is touched to sphere B and separated from it. How much charge ends up on sphere C?

5. (a) In problem 4, what is the net charge on the three spheres before they are allowed to touch each other? (b) What is the net charge on the three spheres after they have touched?

***6.** Object A is metallic and electrically neutral. It is charged by induction so that it acquires a charge of -3.0×10^{-6} C. Object B is identical to object A and is also electrically neutral. It is charged by induction so that it acquires a charge of $+3.0 \times 10^{-6}$ C. Find the *difference* in mass between the charged objects and state which has the greater mass.

Section 18.5 Coulomb's Law

7. The nucleus of the helium atom contains two protons that are separated by about 3.0×10^{-15} m. Find the magnitude of the electrostatic force that each proton exerts on the other. (The protons remain together in the nucleus because the repulsive electrostatic force is balanced by an attractive force called the strong nuclear force.)

8. The nucleus of a gold atom consists of 79 protons. If all but one of the orbiting electrons were removed from the neutral atom, what would be the magnitude of the electrostatic force exerted on the remaining electron when it is at a distance of 6.0×10^{-10} m from the nucleus?

9. Two very small spheres are initially neutral and separated by a distance of 0.50 m. Suppose that 3.0×10^{13} electrons are removed from one sphere and placed on the other. (a) What is the magnitude of the electrostatic force that acts on each sphere? (b) Is the force attractive or repulsive? Why?

10. At what separation distance do two point charges of $+1.0$ and -1.0 μC exert a force of attraction on each other that is 440 N?

11. Two charges attract each other with a force of 1.5 N. What will be the force if the distance between them is reduced to one-ninth of its original value?

12. An object has a mass of 215 kg and is located at the surface of the earth (radius $= 6.38 \times 10^6$ m). Suppose that this object and the earth each have an identical positive charge q. Assuming that the earth's charge is located at the center of the earth, determine q such that the electrostatic force exactly cancels the gravitational force.

13. Three charges are fixed to an xy coordinate system. A charge of $+18$ μC is on the y axis at $y = +3.0$ m. A charge of -12 μC is at the origin. Lastly, a charge of $+45$ μC is on the x axis at $x = +3.0$ m. Determine the magnitude and direction of the net electrostatic force on the charge at $x = +3.0$ m. Specify the direction relative to the $-x$ axis.

14. Two charges are placed on the x axis as follows: $q_1 = +11$ μC at $x = +0.85$ m, $q_2 = -4.6$ μC at $x = -0.22$ m. A third charge q is located at $x = 0$. The net electric force that acts on q has a magnitude of 6.0×10^{-4} N. Find the magnitude of q.

15. A charge of -3.00 μC is fixed at the center of a compass. Two additional charges are fixed on the circle of the compass (radius $= 0.100$ m). The charges on the circle are -4.00 μC at the position due north and $+5.00$ μC at the position due east. What is the magnitude and direction of the net electrostatic force acting on the charge at the center? Specify the direction relative to due east.

16. An equilateral triangle has sides of 0.15 m. Charges of -9.0, $+8.0$, and $+2.0$ μC are located at the corners of the triangle. Find the magnitude of the net electrostatic force exerted on the 2.0-μC charge.

17. Two particles, with identical positive charges and a separation of 2.60×10^{-2} m, are released from rest. Immediately after the release, particle 1 has an acceleration $\mathbf{a}_1$ whose magnitude is 4.60×10^3 m/s^2, while particle 2 has an acceleration $\mathbf{a}_2$ whose magnitude is 8.50×10^3 m/s^2. Particle 1 has a mass of 6.00×10^{-6} kg. Find (a) the charge on each particle and (b) the mass of particle 2.

***18.** A charge $+q$ is fixed to each of three corners of a square. On the empty corner a charge is placed, such that there is no net electrostatic force acting on the *diagonally opposite* charge. What charge (magnitude and sign) is placed on the empty corner? Express your answer in terms of q.

***19.** Two small objects, A and B, are fixed in place and separated by 2.00 cm. Object A has a charge of $+1.00$ μC, and object B has a charge of -1.00 μC. How many electrons must be removed from A and put onto B to make the electrostatic force that acts on each object an attractive force whose magnitude is 45.0 N?

***20.** A charge is placed on the x axis ($q = 7.00$ μC, $x = 0.600$ m), and another charge is placed on the y axis ($q = 9.00$ μC, $y = 0.400$ m). A third charge ($q = -6.00$ μC, $m = 5.00 \times 10^{-8}$ kg) is placed at the coordinate origin. If the charge at the origin were free to move, what would be (a) the magnitude of its acceleration and (b) the direction of its acceleration? Specify your answer in part (b) as an angle relative to the $+x$ axis.

***21.** Two small charged objects are attached to a horizontal spring, one at each end. The magnitudes of the charges are equal, and the spring constant is 180 N/m. The spring is observed to be stretched by 0.020 m relative to its unstrained length of 0.40 m. Determine (a) the possible polarities and (b) the magnitude of the charges.

***22.** Two positive charges, when combined, give a total charge of $+9.00$ μC. When the charges are separated by 3.00 m, the force exerted by one charge on the other has a magnitude of 8.00×10^{-3} N. Find the amount of each charge.

***23.** Suppose that all the electrons and protons in 2.00 grams of atomic hydrogen could be completely separated. Atomic hydrogen consists of one electron and one proton. If all the electrons were placed at the center of the earth and all the protons at the center of the sun, what would be the ratio of the electrostatic force of attraction to the gravitational force of attraction between the earth and sun?

****24.** There are four charges, each with a magnitude of 2.0 μC. Two are positive and two are negative. The charges are fixed to the corners of a 0.30-m square, one to a corner, in such a way that the force on any charge is directed toward the center of the square. Find the magnitude of the net electrostatic force experienced by any charge.

****25.** Two identical, small insulating balls are suspended by separate 0.25-m threads that are attached to a common point on the ceiling. Each ball has a mass of 8.0×10^{-4} kg. Initially the balls are uncharged and hang straight down. They are then given identical positive charges and, as a result, spread apart with an angle of 36° between the threads. Determine (a) the charge on each ball and (b) the tension in the threads.

****26.** Two objects are identical and small enough that their sizes can be ignored relative to the distance between them, which is 0.200 m. Each object carries a different charge, and they attract each other with a force of 1.20 N. The objects are brought into contact, so the net charge is shared equally, and then they are returned to their initial positions. Now it is found that the objects repel one another with a force whose magnitude is equal to that of the initial attractive force. What is the initial charge on each object? Note that there are two answers.

Section 18.6 The Electric Field, Section 18.7 Electric Field Lines, Section 18.8 The Electric Field Inside a Conductor: Shielding

27. An electric field of 280 000 N/C points due south at a certain spot. What are the magnitude and direction of the force that acts on a charge of $-4.0\ \mu C$ at this spot?

28. A charge of $+3.0 \times 10^{-5}$ C is located at a place where there is an electric field that points due east and has a magnitude of 15 000 N/C. What are the magnitude and direction of the force acting on the charge?

29. Review the important features of electric field lines discussed in Conceptual Example 13. Three point charges ($+q$, $+2q$, and $-3q$) are at the corners of an equilateral triangle. Sketch in six electric field lines between the three charges.

30. What electric field strength is needed to exert a 0.22-N force on a charge of 5.5×10^{-7} C?

31. The electric field at a distance of 0.50 m from a charge is 9.0×10^5 N/C, directed toward the charge. Find the magnitude and polarity of the charge.

32. Conceptual Example 14 in the text deals with the hollow spherical conductor in Figure 18.32. The conductor is initially electrically neutral, and then a charge $+q$ is placed at the center of the hollow space. Suppose the conductor initially has a net charge of $+2q$ instead of being neutral. What is the total charge on the interior and on the exterior surface when the $+q$ charge is placed at the center?

33. The helium atom contains two protons in its nucleus. When helium is singly ionized, one of its two electrons is completely removed. In the Bohr model of the singly ionized helium atom, the remaining electron orbits the nucleus at a radius of 2.65×10^{-11} m. (a) What is the electric field (magnitude and direction) created at the location of the electron by the positively charged nucleus? (b) What is the electrostatic force (magnitude and direction) acting on the electron? (c) Find the speed of the electron.

34. A charge of $q = +7.50\ \mu C$ is located in an electric field. The x and y components of the electric field are $E_x = 6.00 \times 10^3$ N/C and $E_y = 8.00 \times 10^3$ N/C, respectively. (a) What is the magnitude of the force on the charge? (b) Determine the angle that the force makes with the $+x$ axis.

35. The magnitude of the electric field between the plates of a parallel plate capacitor is 2.4×10^5 N/C. Each plate carries a charge whose magnitude is $0.15\ \mu C$. What is the area of each plate?

36. A charge of $+3.5\ \mu C$ is fixed on the x axis at $x = +0.55$ m, while a charge of $-15\ \mu C$ is fixed at the origin. (a) Determine the net electric field (magnitude and direction) on the x axis at $x = +0.80$ m. (b) What force (magnitude and direction) would act on a charge of $-8.0\ \mu C$ placed on the x axis at $x = +0.80$ m?

37. A small drop of water is suspended motionless in air by a uniform electric field that is directed upward and has a magnitude of 7530 N/C. The mass of the water drop is 4.25×10^{-9} kg. (a) Is the excess charge on the water drop positive or negative? Why? (b) How many excess electrons or protons reside on the drop?

38. Review Conceptual Example 12 before attempting this problem. The magnitude of each of the charges in parts *a* and *b* of Figure 18.22 is 8.60×10^{-12} C. The lengths of the sides of the rectangles are 3.00 cm and 5.00 cm. Find the magnitude of the electric field at the center of the rectangle in Figure 18.22*b* and *c*.

* **39.** A small object has a mass of 2.0×10^{-3} kg and a charge of $-25\ \mu C$. It is placed at a certain spot where there is an electric field. When released, the object experiences an acceleration of $3.5 \times 10^3\ m/s^2$ in the direction of the $+x$ axis. Determine the magnitude and direction of the electric field.

* **40.** A proton is moving parallel to a uniform electric field. The electric field accelerates the proton and increases its linear momentum to 5.0×10^{-23} kg·m/s from 1.5×10^{-23} kg·m/s in a time of 6.3×10^{-6} s. What is the magnitude of the electric field?

* **41.** A rectangle has a length of $2d$ and a height of d. Each of the following three charges is located at a corner of the rectangle: $+q_1$ (upper left corner), $+q_2$ (lower right corner), and $-q$ (lower left corner). The net electric field at the (empty) upper right corner is zero. Find the magnitude of q_1 and q_2. Express your answers in terms of q.

* **42.** A proton is released from rest at the positive plate of a parallel plate capacitor. The charge per unit area on each plate is $\sigma = 2.8 \times 10^{-7}\ C/m^2$, and the plates are separated by a distance of 1.5×10^{-2} m. How fast is the proton moving just before it reaches the negative plate?

* **43.** The drawing shows two positive charges q_1 and q_2 fixed to a circle. At the center of the circle they produce a net electric field that is directed upward along the vertical axis. Determine the ratio q_2/q_1.

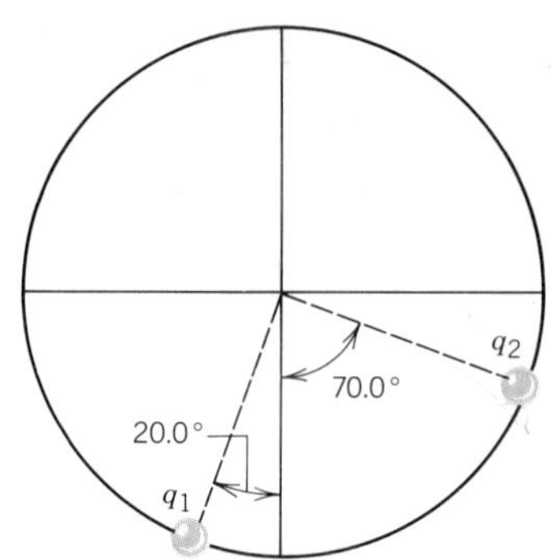

** **44.** The magnitude of the electric field between the plates of a parallel plate capacitor is 480 N/C. A silver dollar is placed between the plates and oriented parallel to the plates.

(a) Ignoring the edges of the coin, find the induced charge density σ on each face of the coin. (b) Assuming the coin has a radius of 1.9 cm, find the magnitude of the total charge on each face of the coin.

**45. A rectangle has a length of $3L$ and a height of L. A charge of magnitude q_1 is placed at the upper left corner, and another charge of magnitude q_2 is placed at the lower right corner. The electric field at the (empty) lower left corner is directed along the diagonal line that runs through the two empty corners. (a) Decide whether the charges have the same or different polarities. (b) Find the ratio q_2/q_1.

**46. The drawing shows an electron entering the lower left side of a parallel plate capacitor and exiting at the upper right side. The initial speed of the electron is 7.00×10^6 m/s. The capacitor is 2.00 cm long, and its plates are separated by 0.150 cm. Assume that the electric field between the plates is uniform everywhere and find its magnitude.

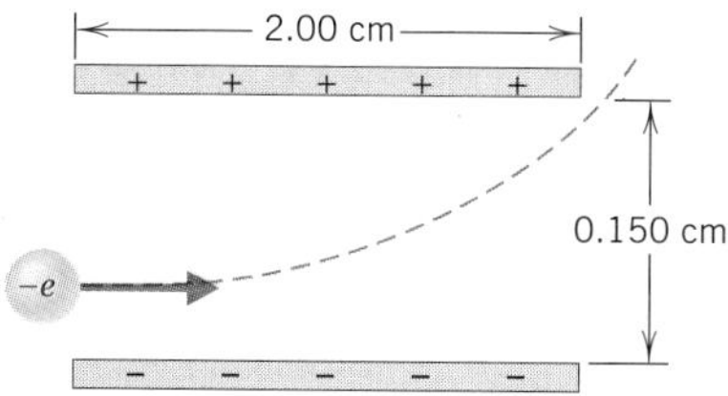

**47. A small plastic ball of mass 6.50×10^{-3} kg and charge $+0.150$ μC is suspended from an insulating thread and hangs between the plates of a capacitor (see the drawing). The ball is in equilibrium, with the thread making an angle of 30.0° with respect to the vertical. The area of each plate is 0.0150 m^2. What is the magnitude of the charge on each plate?

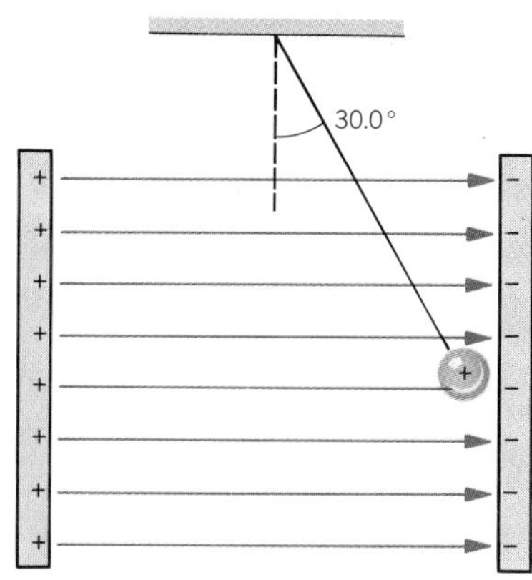

Section 18.9 Gauss' Law

48. A rectangular surface (0.16 m $\times$ 0.38 m) is oriented in a uniform electric field of 580 N/C. What is the maximum possible electric flux through the surface?

49. An electric flux of 160 N·m^2/C passes through a flat horizontal surface that has an area of 0.70 m^2. The flux is due to a uniform electric field. What is the magnitude of the field if the field points (a) vertically and (b) 15° above the horizontal?

50. A spherical surface completely surrounds a collection of charges. Find the electric flux through the surface if the collection consists of (a) a single $+3.5 \times 10^{-6}$ C charge, (b) a single -2.3×10^{-6} C charge, and (c) both of the charges in (a) and (b).

51. One day, the electric field in the atmosphere near ground level is 110 N/C. Assume that the magnitude of this field is the same everywhere around the earth and that the direction of the field is radially inward. The radius of the earth is 6.38×10^6 m. Calculate the net electric charge (magnitude and sign) on the earth.

52. A surface completely surrounds a $+2.0 \times 10^{-6}$ C charge. Find the electric flux through this surface when the surface is (a) a sphere with a radius of 0.50 m, (b) a sphere with a radius of 0.25 m, and (c) a cube with edges that are 0.25 m long.

*53. A cube is located with one corner at the origin of an x, y, z, coordinate system. One of the cube's faces lies in the x, y plane, another in the y, z plane, and another in the x, z plane. In other words, the cube is in the first octant of the coordinate system. The edges of the cube are 0.20 m long. A uniform electric field is parallel to the x, y plane and points in the direction of the $+y$ axis. The magnitude of the field is 1500 N/C. (a) Find the electric flux through each of the six faces of the cube. (b) Add the six values obtained in part (a) to show that the electric flux through the cubical surface is zero, as Gauss' law predicts, since there is no net charge within the cube.

*54. Two spherical shells have a common center. A -1.6×10^{-6} C charge is spread uniformly over the inner shell, which has a radius of 0.050 m. A $+5.1 \times 10^{-6}$ C charge is spread uniformly over the outer shell, which has a radius of 0.15 m. Find the magnitude and direction of the electric field at a distance (measured from the common center) of (a) 0.20 m, (b) 0.10 m, and (c) 0.025 m.

*55. A solid nonconducting sphere has a radius R and has a positive charge q spread uniformly throughout its volume. The charge density or charge per unit volume, therefore, is $q/(\frac{4}{3}\pi R^3)$. Use Gauss' law to show that the electric field at a point within the sphere at a radius r has a magnitude of $qr/(4\pi\epsilon_0 R^3)$. *(Hint: For a Gaussian surface, use a sphere of radius r centered within the solid sphere. Note that the net charge within any volume is the charge density times the volume.)*

**56. A flat sheet is infinitely large and has negligible thickness. This sheet is nonconducting and has positive charge spread uniformly over it. The charge per unit area is σ. Use Gauss' law to show that the electric field created by the sheet of charge has a magnitude of $\sigma/(2\epsilon_0)$. *(Hint: For a Gaussian surface, use a cylinder with its axis perpendicular to the sheet, one end on one side of the sheet, and one end on the other side of the sheet.)*

ADDITIONAL PROBLEMS

57. The force of repulsion that two like charges exert on each other is 3.5 N. What will the force be if the distance between the charges is increased to five times its original value?

58. Review Conceptual Example 13 as an aid in working this problem. Charges of $-4q$ are fixed to opposite corners of a square. A charge of $+5q$ is fixed to one of the remaining corners, and a charge of $+3q$ is fixed to the last corner. Assuming that ten electric field lines emerge from the $+5q$ charge, sketch the field lines in the vicinity of the four charges.

59. An electric field with a magnitude of 160 N/C exists at a spot that is 0.15 m away from a charge. At a place that is 0.45 m away from this charge, what is the electric field strength?

60. Three positive charges are fixed along a line. From left to right they are q_1, q_2, and q_3. The charge q_2 is situated one-fourth of the way between q_1 and q_3, with q_2 being nearer q_1, and experiences no net electrostatic force. Find the ratio q_3/q_1.

61. A tiny ball (mass $= 0.012$ kg) carries a charge of $-18\ \mu C$. What electric field (magnitude and direction) is needed to cause the ball to float above the ground?

62. A vertical wall (5.9 m $\times$ 2.5 m) in a house faces due east. A uniform electric field has a magnitude of 150 N/C. This field is parallel to the ground and points 35° north of east. What is the electric flux through the wall?

63. Three charges are located on the $+x$ axis as follows: $q_1 = +25\ \mu C$ at $x = 0$, $q_2 = +11\ \mu C$ at $x = +2.0$ m, and $q_3 = +45\ \mu C$ at $x = +3.5$ m. (a) Find the electrostatic force (magnitude and direction) acting on q_2. (b) Suppose q_2 were $-11\ \mu C$, rather than $+11\ \mu C$. Without performing any detailed calculations, specify the magnitude and direction of the force exerted on q_2. Give your reasoning.

64. Two charges, -16 and $+4.0\ \mu C$, are fixed in place and separated by 3.0 m. (a) At what spot along a line through the charges is the net electric field zero? Locate this spot relative to the positive charge. *(Hint: The spot does not necessarily lie between the two charges.)* (b) What would be the force on a charge of $+14\ \mu C$ placed at this spot?

65. Two tiny spheres have the same mass and carry charges of the same magnitude. The mass of each sphere is 2.0×10^{-6} kg. The gravitational force that each sphere exerts on the other is balanced by the electric force. (a) What polarities can the charges have? (b) Determine the charge magnitude.

***66.** A particle of mass 3.8×10^{-5} kg and charge $+12\ \mu C$ is released from rest in a region where there is a constant electric field of 470 N/C. (a) What is the acceleration of the particle due to the electric field? (b) How fast is the particle moving after traveling 0.020 m?

***67.** Two charges are placed between the plates of a parallel plate capacitor. One charge is $+q_1$ and the other is $q_2 = +5.00\ \mu C$. The charge per unit area on each plate has a magnitude of $\sigma = 1.30 \times 10^{-4}\ C/m^2$. The force on q_1 due to q_2 equals the force on q_1 due to the electric field of the parallel plate capacitor. What is the distance r between the two charges?

***68.** Two spheres are mounted on identical horizontal springs and rest on a frictionless table, as in the drawing. When the spheres are uncharged, the spacing between them is 0.0500 m, and the springs are unstrained. When each sphere has a charge of $+1.60\ \mu C$, the spacing doubles. Assuming that the spheres have a negligible diameter, determine the spring constant of the springs.

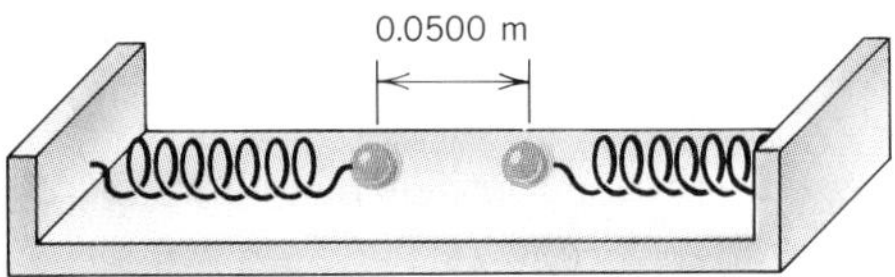

***69.** In the rectangle in the drawing, a charge is to be placed at the empty corner to make the net force on the charge at corner A point along the vertical direction. What charge (magnitude and sign) must be placed at the empty corner?

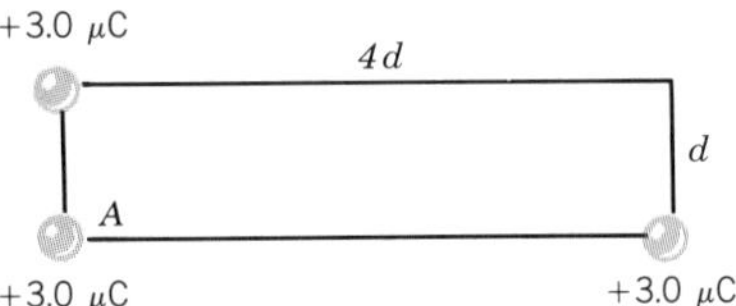

***70.** An electron is moving parallel to a uniform electric field whose magnitude is 2.5×10^5 N/C. The electron, being negatively charged, experiences an electrostatic force that is directed opposite to the electric field, and thus slows down. The initial kinetic energy of the electron is 2.0×10^{-17} J. How far does the electron go before it is stopped by the electric field?

****71.** A small spherical insulator of mass 8.00×10^{-2} kg and charge $+0.600\ \mu C$ is hung by a thin wire of negligible mass. A charge of $-0.900\ \mu C$ is held 0.150 m away from the sphere and directly to the right of it, so the wire makes an angle θ with the vertical (see the drawing). Find (a) the angle θ and (b) the tension in the wire.

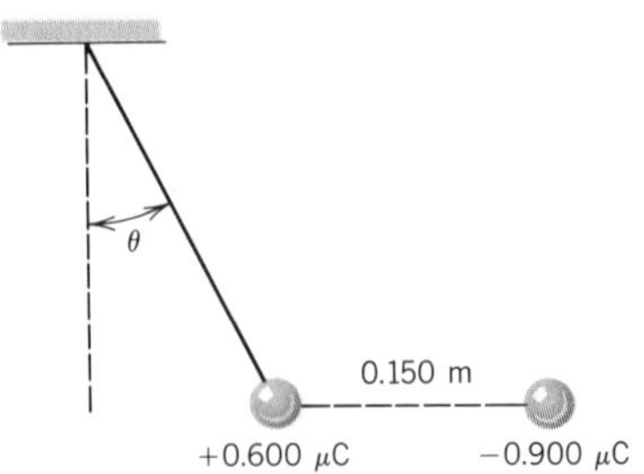

ELECTRIC POTENTIAL ENERGY AND THE ELECTRIC POTENTIAL

CHAPTER 19

Electrical generating plants send the power that they produce out along high-voltage transmission lines, which deliver it to electrical substations, such as the one shown here. These substations provide the first step in converting the high voltages transmitted into lower voltages that can be distributed to users. They help to make electricity so common that we hardly think about how it provides lighting, heating, air conditioning, and keeps the ice cream in the freezer from melting. It is remarkable that, with the flick of a switch, electricity can instantly provide a meager 60 watts for a light bulb or a hefty 10 000 watts for an air conditioner. In this chapter we will become acquainted with the concept of voltage and explore its relation to electrical energy and charge distributions. In addition, we will see that electrical energy and charge can be stored in an important device known as a capacitor.

19.1 POTENTIAL ENERGY

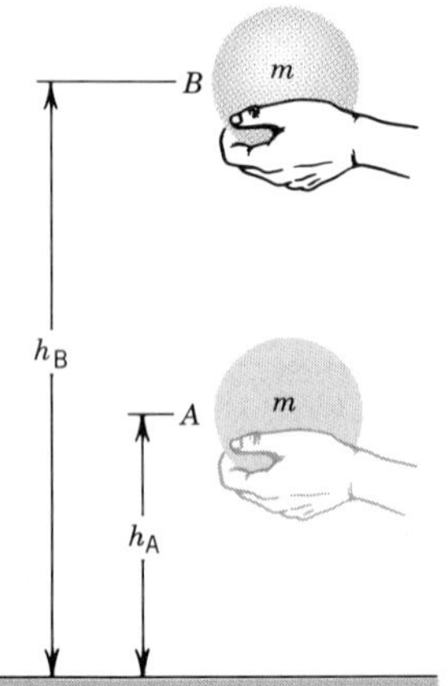

Figure 19.1 The gravitational potential energy of the ball at point *A* is $GPE_A = mgh_A$. Work W_{AB} (done by the hand) is required to raise the ball from *A* to *B*, where the gravitational potential energy is $GPE_B = mgh_B$.

In the last chapter we discussed the electrostatic force that two point charges exert on each other, $F = kq_1q_2/r^2$. The form of this equation is similar to that for the gravitational force that two particles exert on each other, which is $F = Gm_1m_2/r^2$, according to Newton's law of universal gravitation. Both of these forces are conservative and, as Section 6.4 explains, a potential energy can be associated with a conservative force. Thus, there is an electric potential energy that is analogous to the gravitational potential energy. To set the stage for a discussion of the electric potential energy, let's review some of the important aspects of the gravitational counterpart.

Figure 19.1 shows a ball of mass m located at point *A*, which is at a height h_A above the surface of the earth. Relative to the surface, the gravitational potential energy of the ball is given by Equation 6.5* as $GPE_A = mgh_A$, where g is the acceleration due to gravity. This expression assumes that the height is small compared to the radius of the earth, so g is essentially constant at all heights. If the ball is to be lifted to a higher point *B*, an external force must be applied to compensate for the gravitational force. The hand in the drawing provides the external force. When the ball is lifted upward, it starts from rest and ends at rest. Therefore, no change in the ball's kinetic energy KE occurs as a result of the motion, and the work W_{AB} done on the ball by the external force serves only to change the gravitational potential energy:

$$W_{AB} = \underbrace{(KE_B - KE_A)}_{=0} + (GPE_B - GPE_A) = mgh_B - mgh_A \tag{6.7a}$$

If the ball is released from rest at *B*, it is accelerated toward the earth by the gravitational force. As the ball falls, its potential energy decreases and its kinetic energy increases. In the absence of dissipative forces such as friction, the total mechanical energy is conserved; that is, the sum of the gravitational potential energy and the kinetic energy remains constant at all times during the fall.

Figure 19.2 helps to clarify the analogy between electric potential energy and gravitational potential energy. In this drawing a positive test charge $+q_0$ is situated between two oppositely charged plates. Because of the charges on the plates, the test charge experiences an electric force that is directed toward the lower plate (the gravitational force is being neglected here). The hand in the drawing provides the external force needed to compensate for the electric force and to move the test charge from *A* to *B*. The charge starts from rest and ends at rest, so that its kinetic energy does not change. Therefore, the work W_{AB} done on q_0 by the external force goes into changing the electric potential energy. The work equals the difference between the electric potential energy EPE at *B* and that at *A*:

$$W_{AB} = EPE_B - EPE_A \tag{19.1}$$

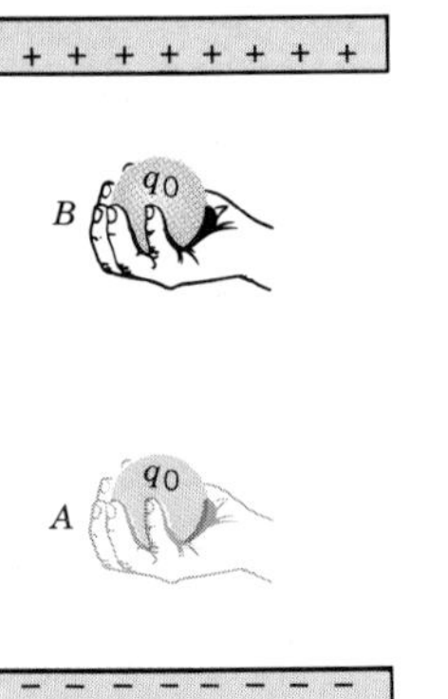

Figure 19.2 At locations *A* and *B*, the test charge $+q_0$ has electric potential energies of EPE_A and EPE_B. The work done to move the test charge from *A* to *B* at a constant speed is $W_{AB} = EPE_B - EPE_A$.

Because the electric force is a conservative force, the path along which the charge is moved from *A* to *B* is of no consequence, for the work W_{AB} is the same for all paths.

* The gravitational potential energy is denoted by GPE to distinguish it from the electric potential energy EPE.

19.2 THE ELECTRIC POTENTIAL DIFFERENCE

The work done to move the charge from A to B in Figure 19.2 depends on the magnitude of the charge, because the strength of the electric force opposing the motion depends on the magnitude of the charge. It is useful to express the work on a per-unit-charge basis by dividing both sides of Equation 19.1 by the charge q_0:

$$\frac{W_{\mathrm{AB}}}{q_0} = \frac{\mathrm{EPE_B} - \mathrm{EPE_A}}{q_0} \tag{19.2}$$

The ***electric potential energy per unit charge*** is an important concept in electricity and is known as the ***electric potential,*** or, simply, the ***potential.***

Definition of Electric Potential
The electric potential V at a given point is the electric potential energy EPE of a small test charge q_0 situated at that point divided by the charge itself: $$V = \frac{\mathrm{EPE}}{q_0} \tag{19.3}$$ ***SI Unit of Electric Potential:*** joule/coulomb = volt (V)

The SI unit of electric potential is a joule per coulomb, a quantity known as a *volt.* The name honors Alessandro Volta (1745–1827), who invented the voltaic pile, the forerunner of the battery. In spite of the similarity in names, the electric potential energy EPE and the electric potential V are *not* the same. The electric potential energy, as its name implies, is an *energy* and, therefore, is measured in joules. In contrast, the electric potential is an *energy per unit charge* and is measured in joules per coulomb, or volts.

According to Equations 19.2 and 19.3, the electric potential difference between two points A and B is related to the work per unit charge in the following way:

$$V_{\mathrm{B}} - V_{\mathrm{A}} = \frac{\mathrm{EPE_B} - \mathrm{EPE_A}}{q_0} = \frac{W_{\mathrm{AB}}}{q_0} \tag{19.4}$$

Often, the "delta" notation is used to express the difference in potentials and the difference in potential energies: $\Delta V = V_{\mathrm{B}} - V_{\mathrm{A}}$ and $\Delta(\mathrm{EPE}) = \mathrm{EPE_B} - \mathrm{EPE_A}$. In terms of this notation, Equation 19.4 takes the form

$$\Delta V = \frac{\Delta(\mathrm{EPE})}{q_0} = \frac{W_{\mathrm{AB}}}{q_0} \tag{19.4}$$

Neither the potential V nor the potential energy EPE can be determined in an absolute sense, for only the *differences* ΔV and Δ(EPE) are measurable in terms of the work W_{AB}. The gravitational potential energy has this same characteristic, since only the value at one height relative to that at some reference height has any significance. Example 1 emphasizes the relative nature of the electric potential.

The energy used to operate an electric car comes from the energy stored in the car's battery pack. When an electric charge passes through the battery pack, the voltage of the pack and the magnitude of the charge are the main factors that determine how much energy the charge obtains from the battery.

This torpedo ray contains a kidney-shaped organ that generates an electric potential difference, which can produce an electric discharge strong enough to liquefy tissue.

Example 1 Work, Electric Potential Energy, and Electric Potential

In Figure 19.2, the work done to move the test charge ($q_0 = +2.0 \times 10^{-6}$ C) at a steady speed from A to B is $+5.0 \times 10^{-5}$ J. (a) Find the difference in the electric potential energies of the charge between the two points. (b) Determine the potential difference between the two points.

REASONING Since work is required to move the test charge from A to B, this situation is like lifting an object uphill. At the top of the "hill" at B, the electric potential energy and the electric potential are greater than they are at the bottom at A.

SOLUTION

(a) The difference in potential energy between points A and B is equal to the work done in moving the charge from A to B, as expressed by Equation 19.1. Therefore, $\text{EPE}_B - \text{EPE}_A = W_{AB} = \boxed{+5.0 \times 10^{-5}\ \text{J}}$.

(b) The potential difference between A and B is the difference in potential energy divided by the charge

$$V_B - V_A = \frac{\text{EPE}_B - \text{EPE}_A}{q_0} = \frac{+5.0 \times 10^{-5}\ \text{J}}{+2.0 \times 10^{-6}\ \text{C}} = \boxed{+25\ \text{V}} \qquad (19.4)$$

The electric potential at B exceeds that at A by 25 V. It is not possible to determine separate values for V_B and V_A.

In Example 1, energy in the form of work is used to move the positive charge from the lower potential at A to the higher potential at B. According to the principle of the conservation of energy, this energy does not disappear. If the positive charge is released at B, it accelerates toward A because of the repulsion from the upper plate and the attraction to the lower plate. The speed of the charge increases as electric potential energy is converted into kinetic energy. Thus, ***a positive charge accelerates from a region of higher potential toward a region of lower potential.*** A negative charge behaves in the opposite fashion, since the electric force acting on it is directed opposite to that on a positive charge. ***A negative charge accelerates from a region of lower potential toward a region of higher potential.*** The next two examples illustrate the way positive and negative charges behave.

Higher potential

Lower potential

A B C

Figure 19.3 The electric potentials at points A, B, and C are different. Under the influence of these potentials, positive and negative charges accelerate in opposite directions.

Conceptual Example 2 The Accelerations of Positive and Negative Charges

Three points, A, B, and C, are located along a line, as Figure 19.3 illustrates. A positive test charge is released from rest at A and accelerates toward B. Upon reaching B, the test charge continues to accelerate toward C. Assuming that only motion along the line is possible, what will a negative test charge do when it is released from rest at B?

REASONING A negative charge will accelerate from a region of lower potential toward a region of higher potential. Therefore, before we can decide what the negative test charge will do here, it is necessary to know how the electric potentials compare at points A, B, and C. This information can be deduced from the behavior of the positive test charge, which accelerates from a region of higher potential toward a region of lower potential. Since the positive test charge accelerates from A to B, the potential at A must exceed that at B. And since the positive test charge accelerates from B to C, the potential at B must exceed that at C. We can see, then, that the potential at point B is between that at points A and C. When the negative test charge is released from rest at B, it will accelerate toward the region of higher potential. In other words, it will begin moving toward A.

Example 3 The Conservation of Energy

In Figure 19.2, point B has an electric potential that is 25 V greater than that at point A, so $V_B - V_A = 25$ V. A particle has a mass of 1.8×10^{-5} kg and a charge whose magnitude is 3.0×10^{-5} C. The effects of gravity and friction are negligible. (a) If the particle has a positive charge and is released from rest at B, what speed v_A does the particle have when it arrives at A? (b) If the particle has a negative charge and is released from rest at A, what speed v_B does the particle have at B?

REASONING The only force acting on the moving charge is the conservative electric force. Therefore, the total energy of the charge (the sum of the kinetic energy KE and the electric potential energy EPE) is the same at points A and B:

$$\text{KE}_A + \text{EPE}_A = \text{KE}_B + \text{EPE}_B$$

In the first part of this problem, the positive particle is at rest at point B and arrives at point A with a kinetic energy of $\text{KE}_A = \frac{1}{2}mv_A^2$. It follows, then, that $\frac{1}{2}mv_A^2 = \text{EPE}_B - \text{EPE}_A$. The difference in potential energies is related to the difference in potentials by Equation 19.4, $\text{EPE}_B - \text{EPE}_A = q_0(V_B - V_A)$. As a result, we find that $\frac{1}{2}mv_A^2 = q_0(V_B - V_A)$, which allows us to determine the speed v_A. In the second part of the problem, we need to take into account the negative charge of the particle.

PROBLEM SOLVING INSIGHT

A positive charge accelerates from a region of higher potential toward a region of lower potential. In contrast, a negative charge accelerates from a region of lower potential toward a region of higher potential.

SOLUTION

(a) Since $\frac{1}{2}mv_A^2 = q_0(V_B - V_A)$, we find that

$$v_A = \sqrt{\frac{2q_0(V_B - V_A)}{m}} = \sqrt{\frac{2(3.0 \times 10^{-5}\text{ C})(25\text{ V})}{1.8 \times 10^{-5}\text{ kg}}} = \boxed{9.1\text{ m/s}}$$

(b) A negative charge accelerates from a region of lower potential toward a region of higher potential. Therefore, when the negatively charged particle is released from rest at A, it accelerates toward B. The electric force causing this acceleration has a direction opposite to the force causing the acceleration in part (a). However, the magnitudes of the forces are the same. Thus, the speed v_B of the particle at B is the same as that calculated in part (a), $\boxed{v_B = 9.1\text{ m/s}}$.

As a familiar application of electric potential energy and electric potential, Figure 19.4 shows a 12-V automobile battery with a headlight connected between the battery terminals. The positive terminal has a potential that is 12 V higher than the potential at the negative terminal. Positive charges are repelled from the positive terminal and travel through the wires and headlight toward the negative terminal.* As the charges pass through the headlight, virtually all their potential energy is converted into heat, which causes the filament to glow "white hot" and emit light. When the charges reach the negative terminal, they no longer have any

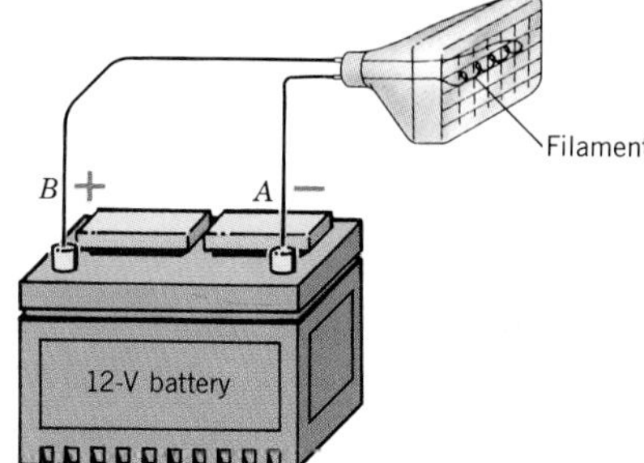

Figure 19.4 A headlight connected to a 12-V battery.

* Historically, it was believed that positive charges flow in the wires of an electric circuit. Today, it is known that negative charges flow in wires from the negative toward the positive terminal. Here, however, we follow the customary practice of describing the flow of negative charges by specifying the opposite but equivalent flow of positive charges. This hypothetical flow of positive charges is called the conventional electric current, as we will see in Section 20.1.

On a hot summer night insects can make it unbearable to sit outdoors. Bug-zappers provide a partial solution to the problem. These devices, with their characteristic bluish glow, attract certain insects and electrocute them with thousands of volts of electricity.

potential energy. The battery then gives the charges an additional "shot" of potential energy by moving them to the higher-potential positive terminal, and the cycle is repeated. In raising the potential energy of the charges, the battery does work W_{AB} on them, and draws from its reserve of chemical energy to do so. Example 4 illustrates the concept of the electric potential difference as applied to a battery.

Example 4 Operating a Headlight

Determine the number of particles, each carrying a charge of 1.60×10^{-19} C (the magnitude of the charge on an electron), that pass between the terminals of a 12-V car battery when a 60.0-W headlight burns for one hour.

REASONING To obtain the number of charged particles, we determine the total charge needed to provide the energy consumed by the headlight in one hour. Dividing the total charge by the charge on each particle gives the number of particles.

SOLUTION Using energy at a rate of 60.0 joules per second (60.0 watts) for one hour, the headlight consumes a total energy of

$$\text{Energy} = \text{Power} \times \text{Time} = (60.0\ \text{W})(3600\ \text{s}) = 2.2 \times 10^5\ \text{J} \qquad (6.10)$$

Equation 19.4 gives the total amount of charge that delivers this much energy upon passing through a 12-V potential difference:

$$q_0 = \frac{\Delta(\text{EPE})}{\Delta V} = \frac{2.2 \times 10^5\ \text{J}}{12\ \text{V}} = 1.8 \times 10^4\ \text{C}$$

The number of particles whose individual charges combine to provide this total charge is $(1.8 \times 10^4\ \text{C})/(1.60 \times 10^{-19}\ \text{C}) = \boxed{1.1 \times 10^{23}}$.

As used in connection with batteries, the volt is a familiar unit for measuring electric potential difference. The word "volt" also appears in another context, as part of a unit that is used to measure energy, particularly the energy of an atomic particle, such as an electron or a proton. This energy unit is called the *electron volt* (eV). ***One electron volt is the change in potential energy of an electron*** ($q_0 = 1.60 \times 10^{-19}$ C) ***when the electron moves through a potential difference of one volt.*** Since the change in potential energy equals $q_0\,\Delta V$, one electron volt is equal to $(1.60 \times 10^{-19}\ \text{C}) \times (1.00\ \text{V}) = 1.60 \times 10^{-19}$ J; thus,

$$1\ \text{eV} = 1.60 \times 10^{-19}\ \text{J}$$

One million (10^{+6}) electron volts of energy is referred to as one MeV, and one billion (10^{+9}) electron volts of energy is one GeV, where the "G" stands for the prefix "giga" (pronounced "jig'a").

19.3 THE ELECTRIC POTENTIAL DIFFERENCE CREATED BY POINT CHARGES

A positive point charge $+q$ creates an electric potential in a fashion that Figure 19.5 helps to explain. This picture shows two locations A and B, at distances r_A and r_B from the charge. At any position between A and B an electrostatic force of repulsion acts on a positive test charge $+q_0$. The magnitude of the force is given by Coulomb's law as $F = kq_0q/r^2$. Thus, to move the test charge from A to B at a constant speed, an external agent must apply a force $F = kq_0q/r^2$ to balance the electric force. Since r varies between r_A and r_B, F also varies, and the work done on q_0 is not simply the product of F and the distance between the points. However, the work can be found by using integral calculus. According to Equation 19.4, dividing the work W_{AB} by q_0 gives the potential difference between B and A. The result is

$$V_B - V_A = \frac{W_{AB}}{q_0} = \frac{kq}{r_B} - \frac{kq}{r_A} \quad (19.5)$$

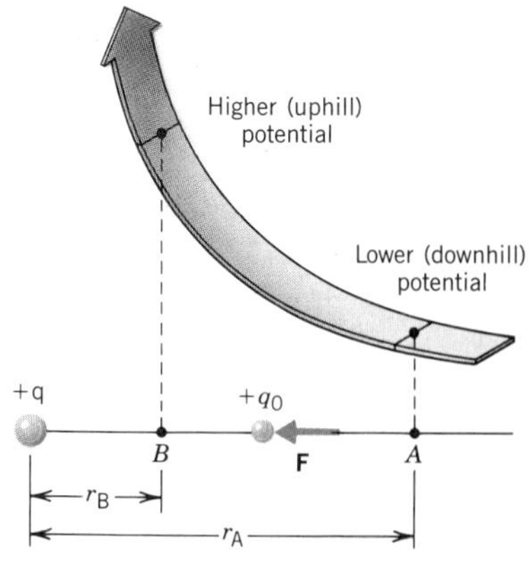

Figure 19.5 The positive test charge $+q_0$ experiences a repulsive force due to the positive point charge $+q$. As a result, work must be done by a force **F** to move the test charge from A to B. Consequently, the electric potential is higher (uphill) at B and lower (downhill) at A.

As point A is located farther and farther away from the charge q, r_A becomes larger and larger. In the limit that r_A becomes infinitely large, the term kq/r_A becomes zero, and it is customary to set V_A equal to zero and write $V_B = kq/r_B$. In this limit, then, we omit the subscripts and write the potential in the following form:

[Potential of a point charge]

$$V = \frac{kq}{r} \quad (19.6)$$

The symbol V in this equation does not refer to the potential in any absolute sense. Rather, $V = kq/r$ stands for the amount by which the potential at a distance r from a point charge differs from the potential at an infinite distance away. In other words, V refers to a potential difference with the arbitrary assumption that the potential at infinity is zero.

With the aid of Equation 19.6, we can describe the effect that a point charge q has on the surrounding space. When q is positive, the value of $V = kq/r$ is also positive, indicating that the positive charge has everywhere raised the potential above the zero reference value. Conversely, when q is negative, the potential V is also negative, indicating that the negative charge has everywhere decreased the potential below the zero reference value. The next example deals with these effects quantitatively.

Example 5 The Potential of a Point Charge

Using a zero reference potential at infinity, determine the amount by which a point charge of 4.0×10^{-8} C alters the electric potential at a spot 1.2 m away when the charge is (a) positive and (b) negative.

REASONING A point charge q alters the potential at every location in the surrounding space. In the expression $V = kq/r$, the effect of the charge in increasing or decreasing the potential is conveyed by the algebraic sign for the value of q.

SOLUTION

(a) Figure 19.6*a* shows the potential when the charge is positive:

$$V = \frac{kq}{r} = \frac{(8.99 \times 10^9 \text{ N}\cdot\text{m}^2/\text{C}^2)(+4.0 \times 10^{-8} \text{ C})}{1.2 \text{ m}} = \boxed{+300 \text{ V}} \quad (19.6)$$

(b) Part *b* of the drawing illustrates the results when the charge is negative. A calculation similar to that in part (a) shows the potential is now negative: $\boxed{-300 \text{ V}}$.

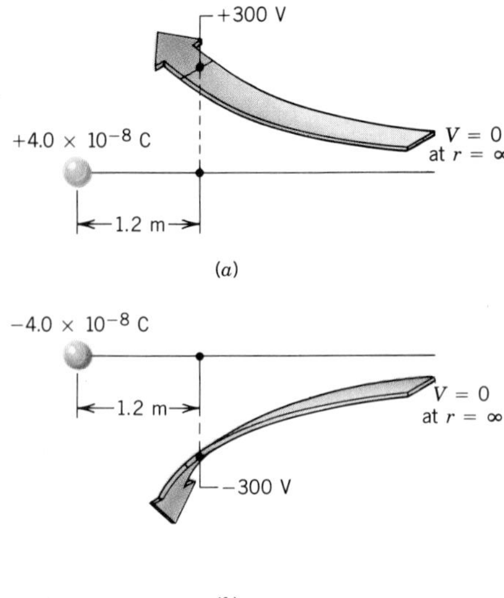

Figure 19.6 A point charge of 4.0×10^{-8} C alters the potential at a spot 1.2 m away. The potential is (*a*) increased by 300 V when the charge is positive and (*b*) decreased by 300 V when the charge is negative, relative to a zero reference potential at infinity.

The electric potential created by a point charge plays an important role in the structure of atoms, as Example 6 illustrates for the case of the hydrogen atom.

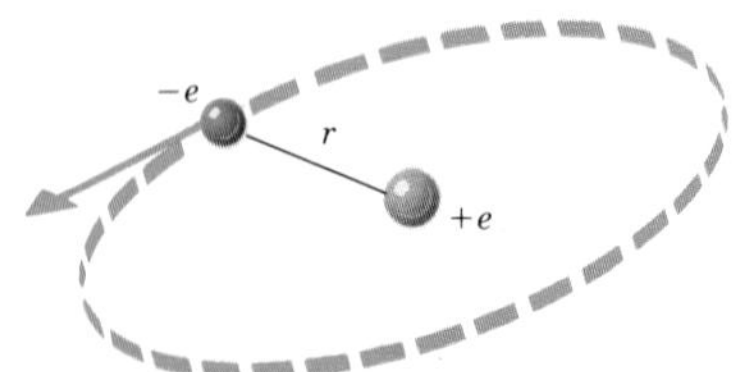

Figure 19.7 In the Bohr model of the hydrogen atom, the electron ($-e$) orbits the proton ($+e$) at a distance of $r = 5.29 \times 10^{-11}$ m.

Example 6 Ionization Energy

In the Bohr model of the hydrogen atom, the electron ($-e$) is in an orbit around the nuclear proton ($+e$) at a distance of $r = 5.29 \times 10^{-11}$ m, as Figure 19.7 shows. Find (a) the electric potential that the proton creates at this distance, (b) the total energy of the atom, and (c) the ionization energy for the atom. The ionization energy is the energy that must be put into the atom to remove the electron and place it at rest infinitely far from the proton. Express the answers to parts (b) and (c) in electron volts.

REASONING The total energy of the atom is the sum of the electric potential energy and the kinetic energy of the electron. The potential energy (relative to a zero reference value at infinity) can be obtained from Equation 19.4 as $\text{EPE} = qV$, where $q = -e$ is the charge on the electron and V is the electric potential created by the proton. The electric potential is the point charge potential given by Equation 19.6. The kinetic energy is $\text{KE} = \frac{1}{2}mv^2$, where m is the mass of the electron and v is its speed.

SOLUTION

(a) The electric potential created by the proton charge of $+e = +1.60 \times 10^{-19}$ C is

$$V = \frac{kq}{r} = \frac{(8.99 \times 10^9\ \text{N}\cdot\text{m}^2/\text{C}^2)(+1.60 \times 10^{-19}\ \text{C})}{5.29 \times 10^{-11}\ \text{m}} = \boxed{+27.2\ \text{V}} \quad (19.6)$$

(b) The total energy E of the atom is $E = \text{EPE} + \text{KE}$. Using the electric potential of +27.2 V obtained in part (a), we find that the electric potential energy is

$$\text{EPE} = qV = (-1.60 \times 10^{-19}\ \text{C})(+27.2\ \text{V}) = -4.35 \times 10^{-18}\ \text{J}$$

From Example 4 in Section 18.5, we know that the speed of the electron is $v = 2.18 \times 10^6$ m/s. Since the mass of the electron is 9.11×10^{-31} kg, its kinetic energy is $\text{KE} = \frac{1}{2}mv^2 = 2.17 \times 10^{-18}$ J. The total energy, then, is

$$E = \text{EPE} + \text{KE} = -4.35 \times 10^{-18}\ \text{J} + 2.17 \times 10^{-18}\ \text{J} = -2.18 \times 10^{-18}\ \text{J}$$

PROBLEM SOLVING INSIGHT

Be careful to distinguish between the concepts of potential V and electric potential energy EPE. Potential is electric potential energy per unit charge: $V = \text{EPE}/q$.

The total energy in electron volts (1 eV $= 1.60 \times 10^{-19}$ J) is $\boxed{E = -13.6 \text{ eV}}$.

(c) Removing the electron from its orbit and placing the electron at rest at infinity requires that +13.6 eV of energy be put into the atom. Then the total energy will be zero, the assumed reference value for the potential energy. Thus, the ionization energy of the hydrogen atom is $\boxed{+13.6 \text{ eV}}$, which agrees with the experimental value.

A single point charge raises or lowers the potential at a given location, depending on whether the charge is positive or negative. ***When two or more charges are present, the potential due to all the charges is obtained by adding together the individual potentials,*** as Examples 7 and 8 show.

PROBLEM SOLVING INSIGHT

Example 7 The Total Electric Potential

At locations *A* and *B* in Figure 19.8, find the total electric potential due to the two point charges.

REASONING At each location, each charge contributes to the total electric potential. We obtain the individual contributions by using $V = kq/r$ and find the total potential by adding the individual contributions algebraically. The two charges have the same magnitude. Thus, at *A* the total potential is positive, because this spot is closer to the positive charge, whose effect dominates over that of the more distant negative charge. At *B*, midway between the charges, the total potential is zero, since the potential of one charge exactly offsets that of the other.

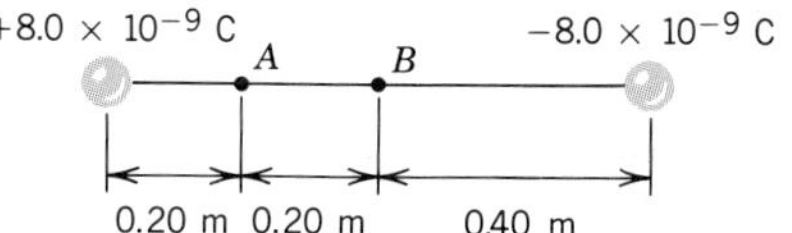

Figure 19.8 Both the positive and negative charges affect the electric potential at locations *A* and *B*.

SOLUTION

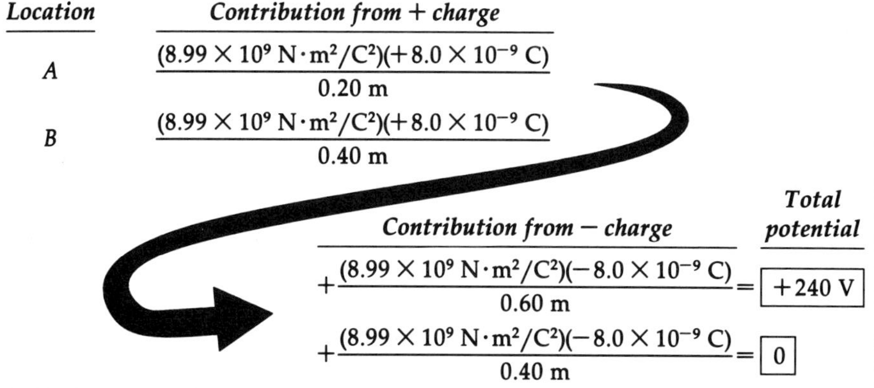

Location	*Contribution from + charge*	*Contribution from − charge*	*Total potential*
A	$\dfrac{(8.99 \times 10^9 \text{ N}\cdot\text{m}^2/\text{C}^2)(+8.0 \times 10^{-9} \text{ C})}{0.20 \text{ m}}$	$+\dfrac{(8.99 \times 10^9 \text{ N}\cdot\text{m}^2/\text{C}^2)(-8.0 \times 10^{-9} \text{ C})}{0.60 \text{ m}}$	$= \boxed{+240 \text{ V}}$
B	$\dfrac{(8.99 \times 10^9 \text{ N}\cdot\text{m}^2/\text{C}^2)(+8.0 \times 10^{-9} \text{ C})}{0.40 \text{ m}}$	$+\dfrac{(8.99 \times 10^9 \text{ N}\cdot\text{m}^2/\text{C}^2)(-8.0 \times 10^{-9} \text{ C})}{0.40 \text{ m}}$	$= \boxed{0}$

Conceptual Example 8 Where Is the Potential Zero?

Two point charges are fixed in place, as in Figure 19.9. The positive charge is $+2q$ and has twice the magnitude of the negative charge, which is $-q$. On the line that passes through the charges, how many places are there at which the total potential is zero?

REASONING At any location, the total potential is the algebraic sum of the individual potentials created by each charge. The total potential will be zero if the potential due to the positive charge is exactly offset by the potential due to the negative charge.

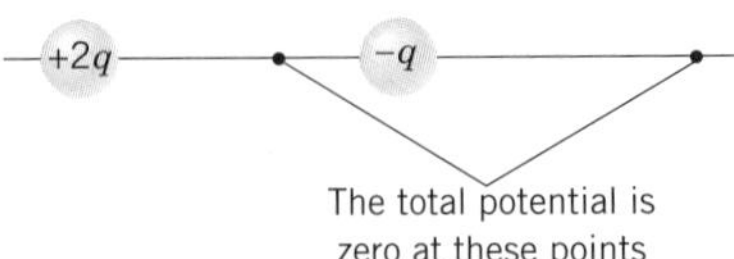

Figure 19.9 Two point charges, one positive and one negative. The positive charge, $+2q$, has twice the magnitude of the negative charge, $-q$.

We will consider the line in three sections: the part to the left of the positive charge, the part between the charges, and the part to the right of the negative charge. In the process, we must remember that the potential of a point charge is directly proportional to the charge and inversely proportional to the distance from the charge. Thus, the individual potentials cannot exactly offset one another anywhere to the left of the positive charge. The positive charge has the larger magnitude and is closer to any place in this region than the negative charge is. As a result, the potential of the positive charge in this region always dominates over that of the negative charge.

Between the charges there is a location where the individual potentials do balance. We saw a similar situation in Example 7, where the balance occurred at the midpoint between two charges that had equal magnitudes. Now the charges have unequal magnitudes, so the balance does not occur at the midpoint. Instead, it occurs at a location that must be closer to the charge with the smaller magnitude, namely, the negative charge. Then, since the potential of a point charge is inversely proportional to the distance from the charge, the effect of the smaller charge will be able to offset that of the more distant larger charge.

To the right of the negative charge, there is another location at which the individual potentials exactly offset one another. All places on this section of the line are closer to the negative than to the positive charge. In this region, therefore, a location exists at which the potential of the smaller negative charge again exactly offsets that of the more distant and larger positive charge.

In summary, there are two places on the line passing through the charges where the total potential is zero. One place is between the charges, but closer to the negative charge. The second place is to the right of the negative charge.

***Related Homework Material:** Problem 19*

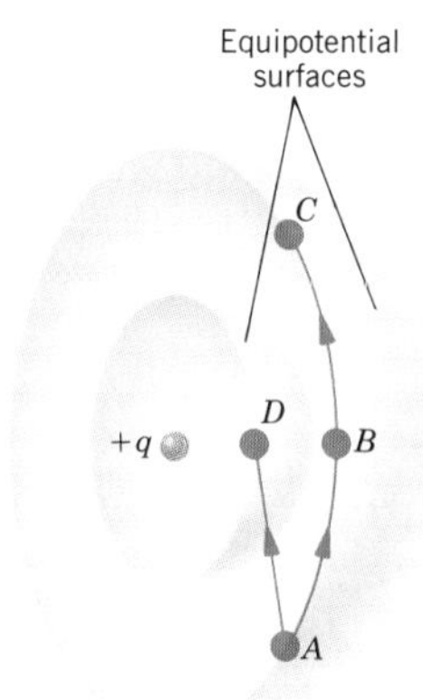

Figure 19.10 The equipotential surfaces that surround the point charge $+q$ are spherical. No work is required to move a charge at a constant speed on a path that lies on an equipotential surface, such as the path *ABC*. However, work is done when a charge moves between two equipotential surfaces, as along the path *AD*.

19.4 EQUIPOTENTIAL SURFACES AND THEIR RELATION TO THE ELECTRIC FIELD

EQUIPOTENTIAL SURFACES

An ***equipotential surface*** is a surface on which the electric potential is the same everywhere. The easiest equipotential surfaces to visualize are those that surround an isolated point charge. According to Equation 19.6, the potential at a distance r from a point charge q is $V = kq/r$. Thus, wherever r is the same, the potential is the same, and the equipotential surfaces are spherical surfaces centered on the charge. There are an infinite number of such surfaces, one for every value of r, and Figure 19.10 illustrates two of them. The larger the distance r, the smaller is the potential of the equipotential surface.

PROBLEM SOLVING INSIGHT

No work is required to move a charge at constant speed on an equipotential surface. This important characteristic arises because when work is done on an object, either the potential energy or the kinetic energy of the object changes. Since both types of energy remain unchanged when a charge is moved at constant speed on an equipotential surface, no work is needed to accomplish the motion. In Figure 19.10, for instance, no work is done in moving a test charge at constant speed along the circular arc *ABC*. The only force that must be applied to the test charge is that needed to counteract the electric force produced by q, and this applied force, being perpendicular to the path *ABC*, does no work. (See Section 6.1 for a review of work.) In contrast, work must be done to move a charge at a constant speed *between* equipotential surfaces, as from *A* to *D* in the picture. The work is the product of the charge and the difference between the potentials of the surfaces, in accord with Equation 19.4. Conceptual Example 9 further clarifies the idea of an equipotential surface.

Conceptual Example 9 Equipotential Surfaces and Zero Work

Figure 19.11 shows a positive point charge and two of its equipotential surfaces. A negative charge is moved at a constant speed along the path *ABC*, which lies on the inner surface. As we know, no work is required to move the charge at a constant speed on the equipotential surface. Now consider the path *AB′C*, where the point *B′* lies on the outer equipotential surface. The negative charge leaves the inner surface at *A* and returns at *C*, in each case with the same constant speed. Is any net work done on the negative charge along the path *AB′C*?

REASONING Since the speed along the path *AB′C* is constant, there is no change in the kinetic energy. There is also no change in the potential energy, since *A* and *C* lie on the same equipotential surface. Since neither kinetic nor potential energy has changed, no net work is done on the negative charge as it moves along path *AB′C*. There is no contradiction here of our earlier statement that work must be done to move a charge at a constant speed *between* equipotential surfaces. Positive work is required to move the negative charge from *A* to the outer surface. And negative work must be done on the charge to move it back to *C* on the inner surface. In each case, work is involved as the charge is moved *between* the equipotential surfaces. However, the positive and negative work are equal in magnitude, leading to a net work of zero for the path *AB′C*. This situation is analogous to that of an object lifted upward at a constant speed from the earth's surface and then lowered back down to the same height at the same constant speed.

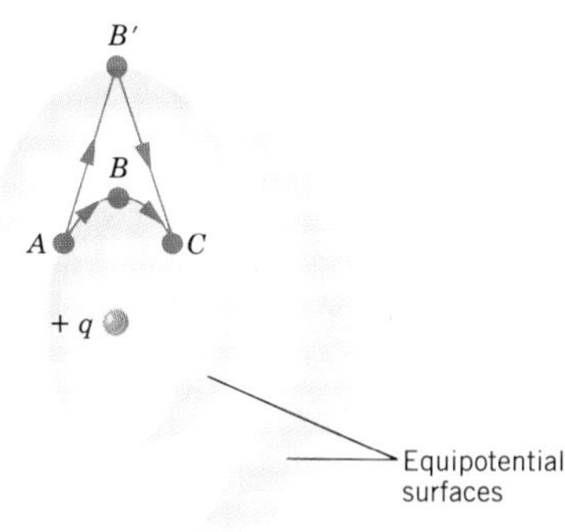

Figure 19.11 A negative charge (not shown) is moved along the path *ABC*, which lies entirely on the inner equipotential surface, and then along path *AB′C*, which does not lie entirely on one equipotential surface. In each case, the negative charge is moved at a constant speed. No net work is done on the charge for either path.

The spherical equipotential surfaces that surround an isolated point charge illustrate another characteristic of such surfaces. Figure 19.12 shows two equipotential surfaces around a positive point charge, along with some electric field lines. The electric field lines give the direction of the electric field, and for a positive point charge, the electric field is directed radially outward. Therefore, at each location on an equipotential sphere the electric field is perpendicular to the surface and points outward in the direction of decreasing potential, as the drawing emphasizes. This perpendicular relation is valid whether or not the equipotential surfaces result from a positive point charge or have a spherical shape; ***the electric field created by any group of charges is everywhere perpendicular to the associated equipotential surfaces and points in the direction of decreasing potential.*** For example, Figure 19.13 shows the electric field lines around an

PROBLEM SOLVING INSIGHT

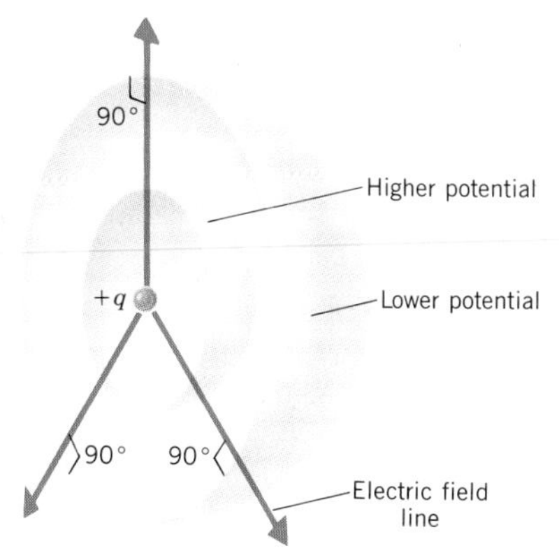

Figure 19.12 The radially directed electric field of a point charge is perpendicular to the spherical equipotential surfaces that surround the charge. The electric field points in the direction of *decreasing* potential.

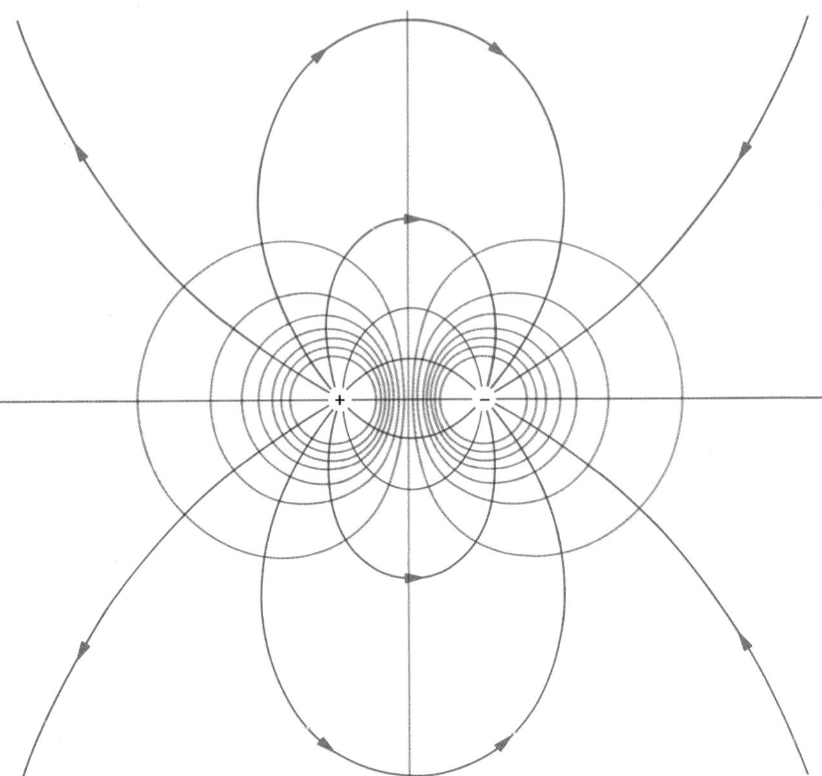

Figure 19.13 A cross-sectional view of the equipotential surfaces (in blue) of an electric dipole. The surfaces are drawn so that at every point they are perpendicular to the electric field lines (in red) of the dipole.

electric dipole, along with some equipotential surfaces. Since the field lines are not simply radial, the equipotential surfaces are no longer spherical, but, instead, have the necessary shape so as to be everywhere perpendicular to the field lines.

To see why an equipotential surface must be perpendicular to the electric field, consider Figure 19.14, which shows a hypothetical situation in which the perpendicular relation does *not* hold. If **E** were not perpendicular to the equipotential surface, there would be a component of **E** parallel to the surface. This field component would exert an electric force on a test charge placed on the surface. To move the test charge at a constant speed along the surface, work would have to be done to counteract the effect of this force. Thus, the surface could not be an equipotential surface as assumed. The only way out of the dilemma is for the electric field to be perpendicular to the surface, so there is no component of the field parallel to the surface.

We have already encountered one equipotential surface in the last chapter. In Section 18.8, we found that the direction of the electric field just outside an electrical conductor is perpendicular to the conductor's surface, when the conductor is at equilibrium under electrostatic conditions. Thus, the surface of any conductor is an equipotential surface under such conditions. In fact, since the electric field is zero everywhere inside a conductor whose charges are in equilibrium, the entire conductor can be regarded as an equipotential volume.

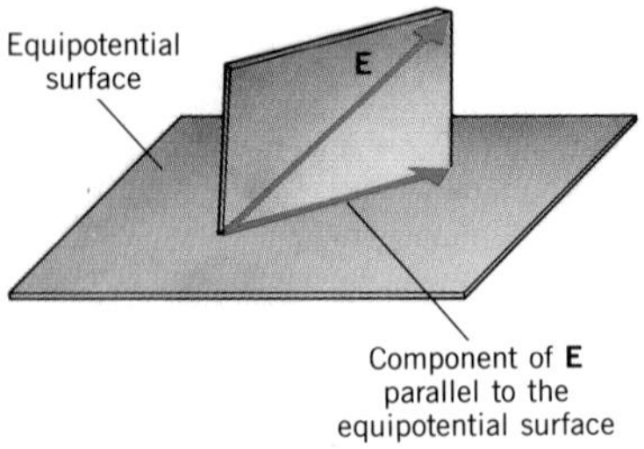

Figure 19.14 In this hypothetical situation, the electric field **E** is not perpendicular to the equipotential surface. As a result, there is a component of **E** parallel to the surface.

THE RELATION BETWEEN THE ELECTRIC FIELD AND THE ELECTRIC POTENTIAL

There is a quantitative relation between the electric field and the equipotential surfaces. One example that illustrates this relation is the parallel plate capacitor in

Figure 19.15. As Section 18.6 discusses, the electric field **E** between the metal plates is perpendicular to them and is the same everywhere, ignoring fringe fields at the edges. To be perpendicular to the electric field, the equipotential surfaces must be planes that are parallel to the plates, which themselves are equipotential surfaces. The potential difference between the plates is given by Equation 19.4 as $V_B - V_A = \Delta V = W_{AB}/q_0$, where A is a point on the negative plate and B is a point on the positive plate. The work required to move a positive test charge q_0 from A to B is $W_{AB} = F\,\Delta s$, where F is the magnitude of the applied force, directed to the left in opposition to the electric force, and Δs is the magnitude of the displacement along a line perpendicular to the plates. If the test charge is moved at a constant speed, the magnitude of the applied force equals the magnitude of the electric force, so $F = q_0 E$, and the work becomes $W_{AB} = F\,\Delta s = q_0 E\,\Delta s$. The potential difference between the capacitor plates is the work per unit charge, so

$$\Delta V = \frac{W_{AB}}{q_0} = \frac{q_0 E\,\Delta s}{q_0} \quad \text{or} \quad E = \frac{\Delta V}{\Delta s}$$

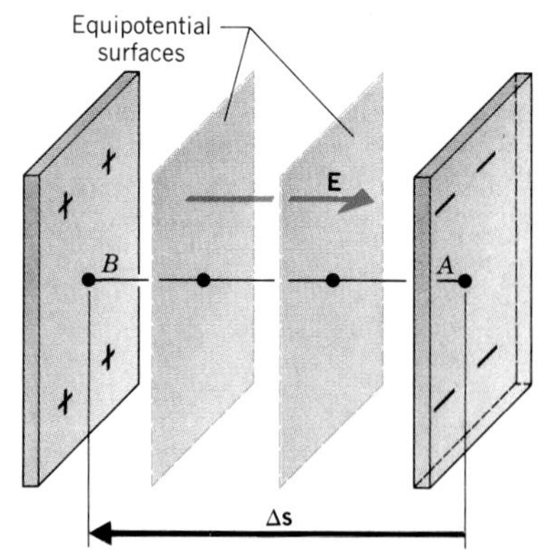

Figure 19.15 The metal plates of a parallel plate capacitor are equipotential surfaces. Two additional equipotential surfaces are shown between the plates. These two equipotential surfaces are parallel to the plates and are perpendicular to the electric field **E** between the plates.

As the test charge is moved from the negative to the positive plate, the potential difference increases with distance ($\Delta V/\Delta s$ is positive). However, the electric field is in the opposite direction, for it points from the positive toward the negative plate. To account for the fact that the electric field is directed opposite to the direction in which the potential increases, Equation 19.7 includes a minus sign:

$$E = -\frac{\Delta V}{\Delta s} \tag{19.7}$$

The quantity $\Delta V/\Delta s$ is referred to as the *potential gradient* and has units of volts per meter. The next example deals further with the equipotential surfaces between the plates of a capacitor.

Example 10 The Electric Field and Potential Are Related

The plates of the capacitor in Figure 19.15 are separated by a distance of 0.032 m, and the potential difference between them is 64 V. Between the two equipotential surfaces shown in color there is a potential difference of 3.0 V. Find the spacing between the two colored surfaces.

REASONING The magnitude of the electric field is $E = \Delta V/\Delta s$ (Equation 19.7 without the minus sign). To find the spacing between the two colored equipotential surfaces, we solve this equation for Δs, with $\Delta V = 3.0$ V and E equal to the magnitude of the electric field between the plates of the capacitor. A value for E can be obtained by using the values given for the distance and potential difference between the plates.

SOLUTION The magnitude of the electric field between the capacitor plates is

$$E = \frac{\Delta V}{\Delta s} = \frac{64\ \text{V}}{0.032\ \text{m}} = 2.0 \times 10^3\ \text{V/m}$$

The spacing between the colored equipotential surfaces can now be determined:

$$\Delta s = \frac{\Delta V}{E} = \frac{3.0\ \text{V}}{2.0 \times 10^3\ \text{V/m}} = \boxed{1.5 \times 10^{-3}\ \text{m}}$$

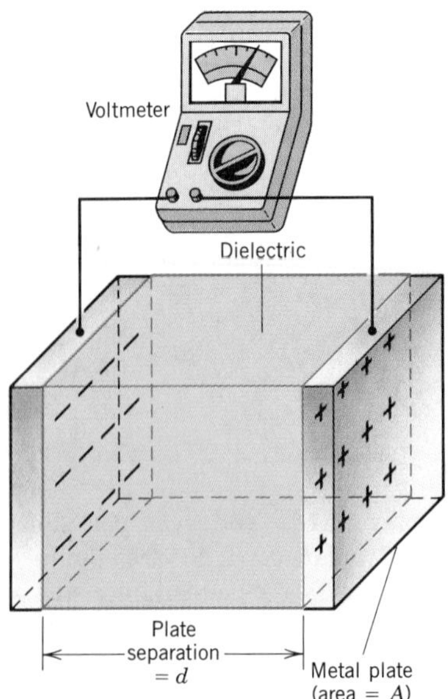

Figure 19.16 A parallel plate capacitor consists of two metal plates, one carrying a charge $+q$ and the other a charge $-q$. The potential of the positive plate exceeds that of the negative plate by an amount V. The region between the plates is filled with a dielectric.

19.5 CAPACITORS AND DIELECTRICS

THE CAPACITANCE OF A CAPACITOR

In Section 18.6 we saw that a parallel plate capacitor consists of two parallel metal plates placed near one another, but not touching. This type of capacitor is only one among many. In general, a ***capacitor*** consists of two conductors of any shape placed near one another without touching. For a reason that will become clear later on, it is common practice to fill the region between the conductors or plates with an electrically insulating material called a ***dielectric***, as Figure 19.16 illustrates.

A capacitor stores electric charge. Each capacitor plate carries a charge of the *same magnitude*, one positive and the other negative. Because of the charges, the electric potential of the positive plate exceeds that of the negative plate by an amount V, as Figure 19.16 indicates. Experiment shows that when the magnitude q of the charge on each plate is doubled, the electric potential difference V is also doubled, so q is proportional to V: $q \propto V$. Equation 19.8 expresses this proportionality with the aid of a proportionality constant C, which is the ***capacitance*** of the capacitor.

The Relation Between Charge and Potential Difference for a Capacitor

The magnitude q of the charge on each plate of a capacitor is directly proportional to the magnitude V of the potential difference between the plates:

$$q = CV \tag{19.8}$$

where C is the capacitance.

SI Unit of Capacitance: coulomb/volt = farad (F)

Equation 19.8 shows that the SI unit of capacitance is the coulomb per volt (C/V). This unit is called the *farad* (F), named after the English scientist Michael Faraday (1791–1867). One farad is an enormous capacitance. Usually smaller amounts, such as a microfarad (1 μF = 10^{-6} F) or a picofarad (1 pF = 10^{-12} F), are used in electric circuits. The capacitance reflects the ability of the capacitor to store charge, in the sense that a larger capacitance C allows more charge q to be put onto the plates for a given value of the potential difference V.

THE PHYSICS OF . . .

random-access memory (RAM) chips.

The ability of a capacitor to store charge lies at the heart of the random-access memory (RAM) chips used in computers, where information is stored in the form of the "ones" and "zeros" that comprise binary numbers. Figure 19.17 illustrates the role of a capacitor in a RAM chip. The capacitor is connected to a transistor switch, to which two lines are connected, an address line and a data line. A single RAM chip often contains hundreds of thousands of such transistor–capacitor units. The address line is used by the computer to locate a particular transistor-capacitor combination, and the data line carries the data to be stored. A pulse on the address line turns on the transistor switch. With the switch turned on, a pulse coming in on the data line can cause the capacitor to charge. A charged capacitor means that a "one" has been stored, while an uncharged capacitor means that a "zero" has been stored.

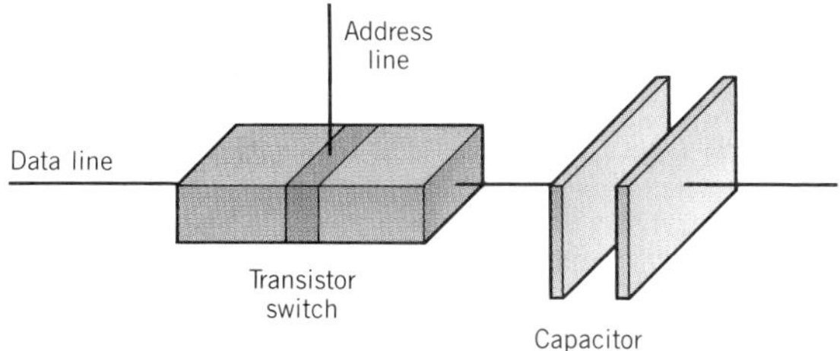

Figure 19.17 A transistor–capacitor combination is part of a RAM chip used in computer memories.

THE DIELECTRIC CONSTANT

If a dielectric is inserted between the plates of a capacitor, the capacitance can increase markedly, because of the way in which the dielectric alters the electric field between the plates. Figure 19.18 shows how this effect comes about. In part *a*, the region between the charged plates is empty. The electric field lines point from the positive toward the negative plate. In part *b*, a dielectric has been inserted between the plates. Since the capacitor is not connected to anything, the charge on the plates remains constant as the dielectric is inserted. In many materials (e.g., water) the molecules possess permanent dipole moments, even though the molecules are electrically neutral. The dipole moment exists, because one end of a molecule has a slight excess of negative charge while the other end has a slight excess of positive charge. When such molecules are placed between the charged plates of the capacitor, the negative ends are attracted to the positive plate and the positive ends are attracted to the negative plate. As a result, the dipolar molecules tend to orient themselves end-to-end, as in part *b*. Whether or not a molecule has a permanent dipole moment, the electric field can cause the electrons to shift position within a molecule, making one end slightly negative and the opposite end slightly positive. Once again, the result is similar to that in part *b*. Because of the end-to-end orientation, the extreme left surface of the dielectric becomes positively charged, and the extreme right surface becomes negatively charged. The surface charges are shown in red in the picture.

Because of the surface charges on the dielectric, not all the electric field lines generated by the charges on the plates pass through the dielectric. As Figure 19.18*c* shows, some of the field lines end on the negative surface charges and begin again on the positive surface charges. Thus, the electric field inside the dielectric is less than the electric field inside an empty capacitor, assuming the charge on the plates remains constant. This reduction in the electric field is described by the ***dielectric constant*** κ, which is the ratio of the field magnitude E_0 without the dielectric to the field magnitude E inside the dielectric:

$$\kappa = \frac{E_0}{E} \tag{19.9}$$

Being a ratio of two field strengths, the dielectric constant is a number without units. Moreover, since the field $\mathbf{E}_0$ without the dielectric is greater than the field $\mathbf{E}$ inside the dielectric, the dielectric constant is greater than unity. The value of κ depends on the nature of the dielectric material, as Table 19.1 indicates.

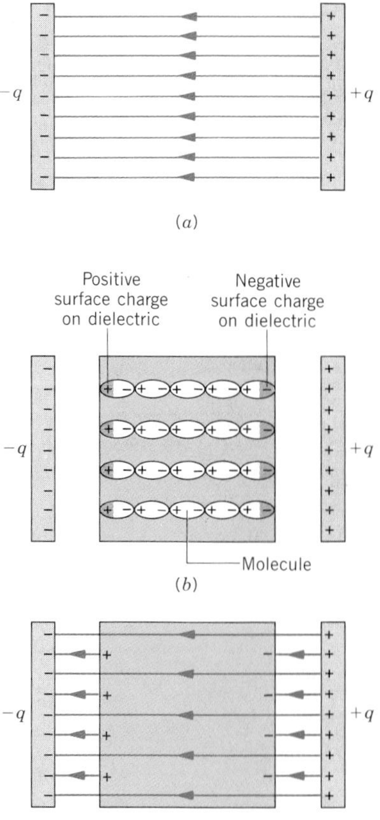

Figure 19.18 (*a*) The electric field lines inside an empty capacitor. (*b*) The electric field produced by the charges on the plates creates an end-to-end alignment of the molecular dipoles within the dielectric. (*c*) The resulting positive and negative surface charges on the dielectric cause a reduction in the electric field within the dielectric. The space between the dielectric and the plates is added only for clarity. In reality, the dielectric completely fills the region between the plates.

Table 19.1 Dielectric Constants of Some Common Substances[a]

Substance	Dielectric Constant, κ
Vacuum	1
Air	1.00054
Teflon	2.1
Benzene	2.28
Paper (royal gray)	3.3
Ruby mica	5.4
Neoprene rubber	6.7
Methyl alcohol	33.6
Water	80.4

[a] Near room temperature.

THE CAPACITANCE OF A PARALLEL PLATE CAPACITOR

The capacitance of a capacitor is affected by the geometry of the plates and the dielectric constant of the material between them. For example, Figure 19.16 shows a parallel plate capacitor in which the area of each plate is A and the separation between the plates is d. The magnitude of the electric field inside the dielectric is given by Equation 19.7 (without the minus sign) as $E = V/d$, where V is the potential difference between the plates. If the charge on each plate is kept fixed, the electric field inside the dielectric is related to the electric field in the absence of the dielectric via Equation 19.9. Therefore,

$$E = \frac{E_0}{\kappa} = \frac{V}{d}$$

Since the electric field within an empty capacitor is $E_0 = q/(\epsilon_0 A)$ (see Equation 18.4), it follows that $q/(\epsilon_0 A\kappa) = V/d$, which can be solved for q to give

$$q = \left(\frac{\kappa\epsilon_0 A}{d}\right) V$$

A comparison of this expression with $q = CV$ (Equation 19.8) reveals that the capacitance C is

[Parallel plate capacitor filled with a dielectric]

$$C = \frac{\kappa\epsilon_0 A}{d} \tag{19.10}$$

Notice that only the geometry of the plates (A and d) and the dielectric constant κ affect the capacitance. With C_0 representing the capacitance of the empty capacitor ($\kappa = 1$), Equation 19.10 shows that $C = \kappa C_0$. In other words, the capacitance with the dielectric present is increased by a factor of κ over the capacitance without the dielectric. It can be shown that the relation $C = \kappa C_0$ applies to any capacitor, not just to a parallel plate capacitor. One reason, then, that capacitors are filled with dielectric materials is to increase the capacitance. Example 11 illustrates the effect that increasing the capacitance can have on the amount of charge stored by a capacitor.

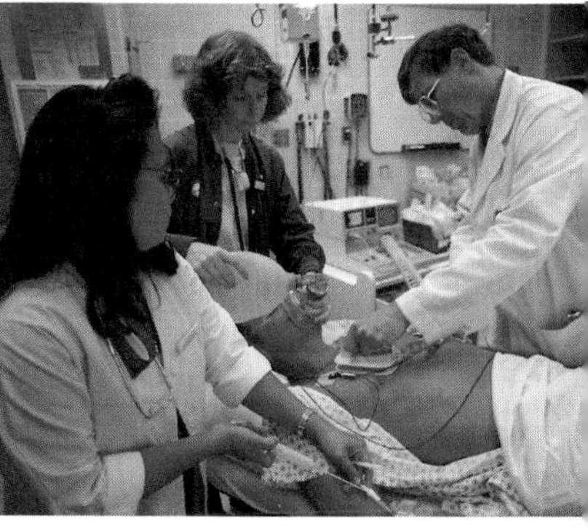

A defibrillator is used to revive a person who has suffered a heart attack. The device uses the charge stored in a capacitor to deliver a controlled electric shock that can restore normal heart rhythm.

Example 11 Storing Electric Charge

The capacitance of an empty capacitor is 1.2 μF. The capacitor is connected to a 12-V battery and charged up. With the capacitor connected to the battery, a slab of dielectric material is inserted between the plates. As a result, 2.6×10^{-5} C of *additional* charge flows from one plate, through the battery, and on to the other plate. What is the dielectric constant κ of the material?

REASONING The charge stored by a capacitor is $q = CV$, according to Equation 19.8. The battery maintains a constant potential difference of $V = 12$ volts between the plates of the capacitor, since the capacitor remains connected to the battery while the dielectric is inserted. Inserting the dielectric causes the capacitance C to increase, so that with V held constant, the charge q must increase. Thus, additional charge flows out of the battery and onto the capacitor plates. To find the dielectric constant, we apply Equation 19.8 to the empty capacitor and then to the capacitor filled with the dielectric material.

SOLUTION The empty capacitor has a capacitance $C_0 = 1.2\ \mu$F and stores an amount of charge $q_0 = C_0V$. With the dielectric material in place, the capacitor has a capacitance $C = \kappa C_0$ and stores an amount of charge $q = (\kappa C_0)V$. The additional charge that the battery supplies is $q - q_0 = (\kappa C_0)V - C_0V = 2.6 \times 10^{-5}$ C. Solving for the dielectric constant, we find that

$$\kappa = \frac{2.6 \times 10^{-5}\ \text{C}}{C_0V} + 1 = \frac{2.6 \times 10^{-5}\ \text{C}}{(1.2 \times 10^{-6}\ \text{F})(12\ \text{V})} + 1 = \boxed{2.8}$$

In Example 11 the capacitor remains connected to the battery while the dielectric is inserted between the plates. The next example discusses what happens if the capacitor is disconnected from the battery before the dielectric is inserted.

Conceptual Example 12 The Effect of a Dielectric When a Capacitor Has a Constant Charge

An empty capacitor is connected to a battery and charged up. The capacitor is then disconnected from the battery, and a slab of dielectric material is inserted between the plates. Does the voltage across the plates increase, remain the same, or decrease?

REASONING Our reasoning is guided by the following fact: Once the capacitor is disconnected from the battery, the charge on its plates remains constant, for there is no longer any way for charge to be added or removed. The charge q stored by the capacitor is $q = CV$. Inserting the dielectric causes the capacitance C to increase. Therefore, the voltage V across the plates must decrease in order for q to remain unchanged. The amount by which the voltage decreases from the value initially established by the battery depends on the dielectric constant of the slab.

Related Homework Material: *Problem 51*

Capacitors are used often in electronic devices, and Example 13 deals with one familiar application.

THE PHYSICS OF . . .

a computer keyboard.

Example 13 A Computer Keyboard

One common kind of computer keyboard is based on the idea of capacitance. Each key is mounted on one end of a plunger, the other end being attached to a movable metal plate (see Figure 19.19). The two plates of the key form a capacitor. When the key is pressed, the movable plate is pushed closer to the fixed plate, and the capacitance increases. Electronic circuitry enables the computer to detect the *change* in capacitance, thereby recognizing which key has been pressed. The separation of the plates is normally 5.00×10^{-3} m, but decreases to 0.150×10^{-3} m when a key is pressed. The plate area is 9.50×10^{-5} m^2, and the capacitor is filled with a material whose dielectric constant is 3.50. Determine the change in capacitance that is detected by the computer.

REASONING We can use Equation 19.10 directly to find the capacitance of the key, since the dielectric constant κ, the plate area A, and the plate separation d are known. We will use this relation twice, once to find the capacitance when the key is pressed and once when it is not pressed. The change in capacitance will be the difference between these two values.

SOLUTION When the key is pressed, the capacitance is

$$C = \frac{\kappa \epsilon_0 A}{d} = \frac{(3.50)[8.85 \times 10^{-12}\ \mathrm{C^2/(N \cdot m^2)}](9.50 \times 10^{-5}\ \mathrm{m^2})}{0.150 \times 10^{-3}\ \mathrm{m}}$$

$$= 19.6 \times 10^{-12}\ \mathrm{F} \quad (19.6\ \mathrm{pF}) \qquad (19.10)$$

A similar calculation shows that when the key is *not* pressed, the capacitance is 0.589×10^{-12} F (0.589 pF). The *change* in capacitance is $\boxed{19.0 \times 10^{-12}\ \mathrm{F}\ (19.0\ \mathrm{pF})}$.

The presence of the dielectric increases the *change* in the capacitance. The greater the change in capacitance, the easier it is for the circuitry within the computer to detect it.

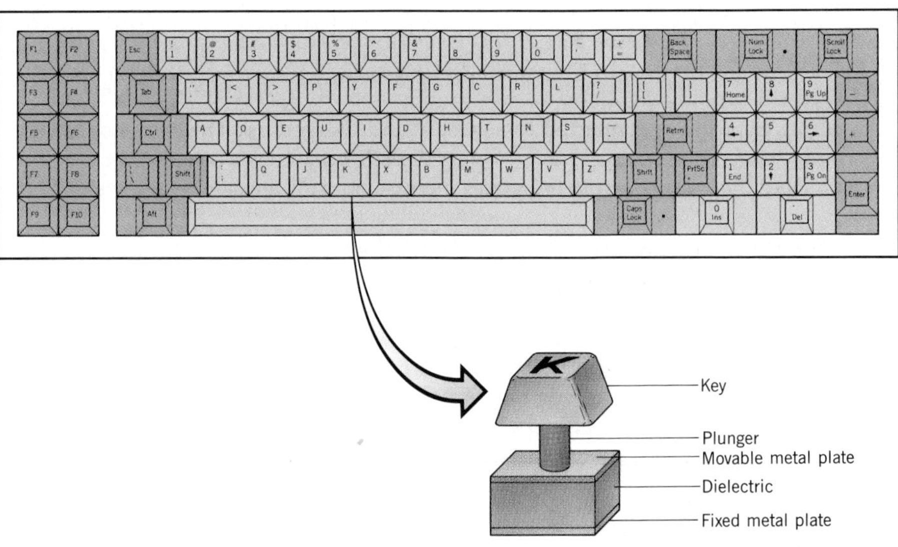

Figure 19.19 In one kind of computer keyboard, each key, when pressed, changes the separation between the plates of a capacitor.

ENERGY STORAGE IN A CAPACITOR

A capacitor is a device for storing charge. Alternatively, it is possible to view the capacitor as a device for storing energy. After all, the charge on the plates possesses electric potential energy, which arises because work was done to deposit the charge on the plates. In fact, as each small increment of charge is deposited during the charging process, the potential difference between the plates increases, and a larger amount of work is needed to bring up the next increment of charge. The total work W done in charging the capacitor, and hence the total electric potential energy EPE, can be obtained from Equation 19.4 by using an average potential difference $\overline{V}$; thus, $W = \text{EPE} = q\overline{V}$. As the capacitor becomes fully charged, the potential of the positive plate relative to the negative plate increases from 0 to V. The average potential difference is $\overline{V} = \frac{1}{2}V$, so the electric potential energy stored is $\text{EPE} = \frac{1}{2}qV$. Since $q = CV$, the energy stored becomes

$$\text{Energy} = \tfrac{1}{2}(CV)V = \tfrac{1}{2}CV^2 \tag{19.11}$$

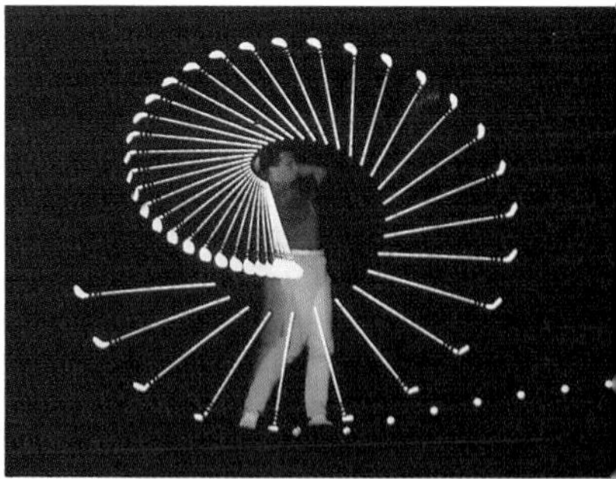

This time-lapse photo of a golfer teeing off was obtained by using an electronic flash attachment with the camera. The energy for each "flash" comes from the electrical energy stored in a capacitor.

It is also possible to regard the energy as being stored in the electric field between the plates. The relation between energy and field strength can be obtained for a parallel plate capacitor by substituting $V = Ed$ and $C = \kappa\epsilon_0 A/d$ into Equation 19.11:

$$\text{Energy} = \frac{1}{2}\left(\frac{\kappa\epsilon_0 A}{d}\right)(Ed)^2$$

Since the area A times the separation d is the volume between the plates, the energy per unit volume or ***energy density*** is

$$\text{Energy density} = \frac{\text{Energy}}{\text{Volume}} = \tfrac{1}{2}\kappa\epsilon_0 E^2 \tag{19.12}$$

It can be shown that this expression is valid for any electric field strength, not just that between the plates of a capacitor.

The energy-storing capability of a capacitor is often put to good use in electronic circuits. For example, in an electronic flash attachment for a camera, energy from the battery pack is stored in capacitors. The capacitors are then discharged between the electrodes of the flash tube, which converts the energy into light. Flash duration times range from 1/200 to 1/1 000 000 second or less, with the shortest flashes being used in high-speed photography. Some flash attachments automatically control the flash duration by monitoring the light reflected from the photographic subject and quickly stopping or quenching the capacitor discharge when the reflected light reaches a predetermined level.

THE PHYSICS OF . . .

an electronic flash attachment for a camera.

*19.6 MEDICAL APPLICATIONS OF ELECTRIC POTENTIAL DIFFERENCES

Several important medical diagnostic techniques depend on the fact that the surface of the human body is *not* an equipotential surface. Between various points on the body there are small potential differences (approximately 30–500 μV), which provide the basis for electrocardiography, electroencephalography, and electroretinography. The potential differences can be traced to the electrical characteristics of muscle cells and nerve cells. In carrying out their biological func-

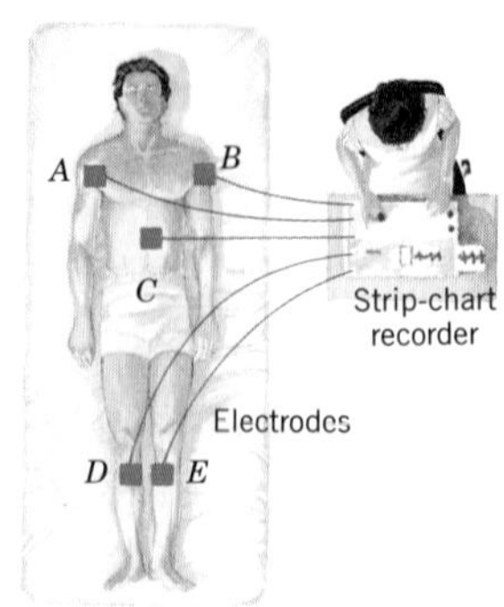

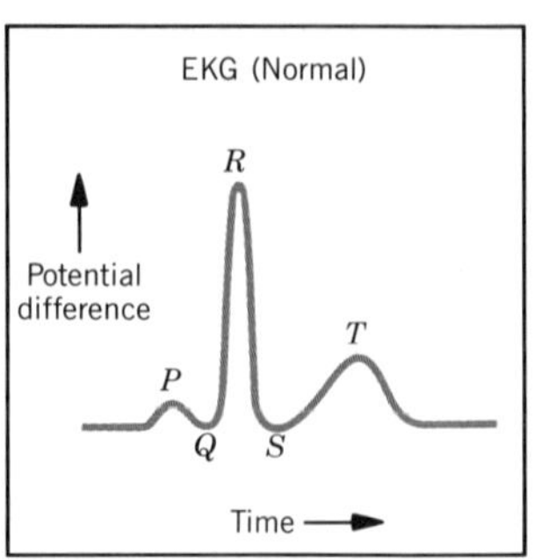

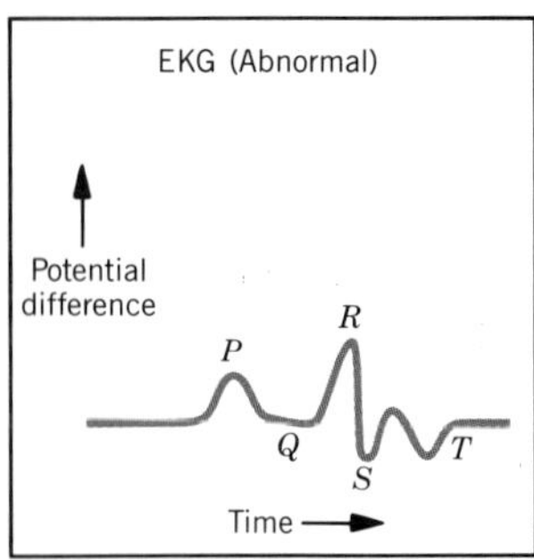

Figure 19.20 The potential differences generated by heart muscle activity provide the basis for electrocardiography. The normal and abnormal EKG patterns correspond to one heartbeat.

tions, these cells utilize positively charged sodium and potassium ions and negatively charged chlorine ions that exist within the cells and in the intercellular fluid outside the cells. As a result of such charged particles, electric fields are generated that extend to the surface of the body and lead to small potential differences.

THE PHYSICS OF . . . electrocardiography.

Figure 19.20 shows some locations on the body where electrodes are placed to measure potential differences in electrocardiography. The potential difference between two points changes as the heart beats and forms a repetitive pattern. The recorded pattern of potential difference versus time is called an electrocardiogram (ECG or EKG), and its shape depends on which pair of points in the picture (*A* and *B*, *B* and *C*, etc.) is used to locate the electrodes. The figure also shows some EKGs and indicates the regions (*P*, *Q*, *R*, *S*, and *T*) that can be associated with specific parts of the heart's beating cycle. The distinct differences between the EKGs of healthy and damaged hearts provide physicians with a valuable diagnostic tool.

THE PHYSICS OF . . . electroencephalography.

In electroencephalography the electrodes are placed at specific locations on the head, as Figure 19.21 indicates, and they record the potential differences that characterize brain behavior. The graph of potential difference versus time is known as an electroencephalogram (EEG). The various parts of the patterns in an EEG are often referred to as "waves" or "rhythms." The drawing shows an example of the main resting rhythm of the brain, the so-called alpha rhythm, and also illustrates the distinct differences that are found between the EEGs generated by healthy and diseased (abnormal) tissue.

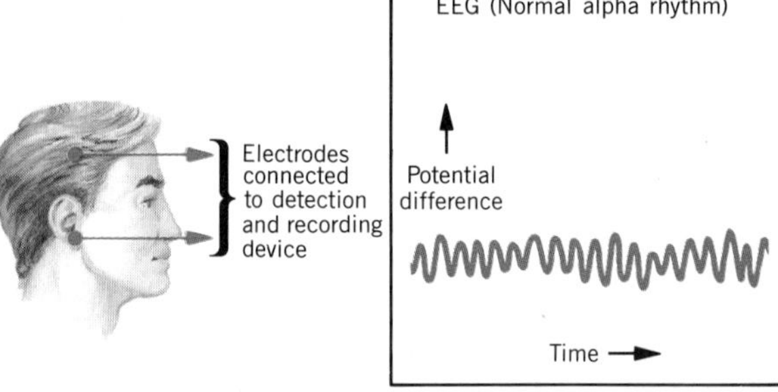

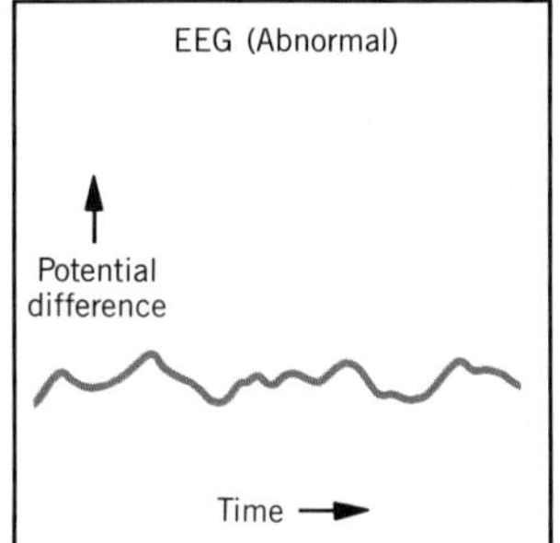

Figure 19.21 In electroencephalography the potential differences created by the electrical activity of the brain are used for diagnosing abnormal behavior.

THE PHYSICS OF . . . electroretinography.

The electrical characteristics of the retina of the eye lead to the potential differences measured in electroretinography. Figure 19.22 shows a typical electrode placement used to record the pattern of potential difference versus time that occurs when the eye is stimulated by a flash of light. One electrode is mounted on a contact lens, while the other is often placed on the forehead. The recorded pattern is called an electroretinogram (ERG), parts of the pattern being referred to as the "*A* wave" and the "*B* wave." As the graphs show, the ERGs of normal and diseased eyes can differ markedly.

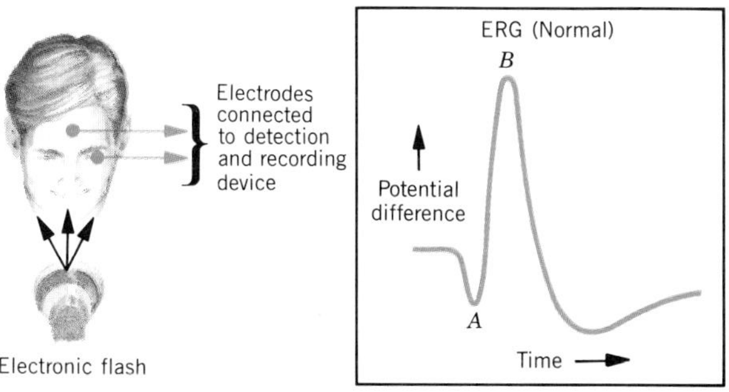

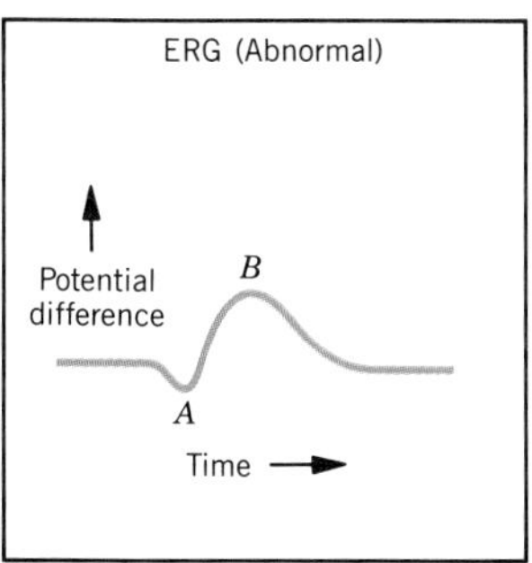

Figure 19.22 The electrical activity of the retina of the eye generates the potential differences used in electroretinography.

INTEGRATION OF CONCEPTS

ELECTRIC POTENTIAL ENERGY AND THE CONSERVATION OF ENERGY

A number of times in this text we have returned to the principle of conservation of energy. The concept of energy includes kinetic and potential energies, and the conservation principle states that energy can neither be created nor destroyed, but can only be converted from one form to another. There are different types of potential energy, and in previous chapters we have encountered the gravitational and elastic forms. To this list we now add the electric potential energy. According to the conservation principle, energy may be interconverted between each of these potential energies and kinetic energy as well. For example, homework problems 5, 6, 10, and 56 at the end of this chapter deal with the interconversion of electric potential energy and kinetic energy. Each type of potential energy is associated with a conservative force. It is the conservative nature of the electrostatic force, just as it is for the gravitational and the elastic forces, that allows us to introduce the concept of electric potential energy. Once the potential energies are known, the principle of conservation of energy provides a common framework that describes how the effects of conservative forces change the motion of a system. The individual mathematical expressions for the various kinds of potential energy are different, but each is a form of energy. And the great value of the conservation principle is that it treats all kinds of potential energy, as well as kinetic energy, in the same way.

SUMMARY

When work W_{AB} is done to move a positive test charge $+q_0$ at constant speed from point *A* to point *B*, the work equals the difference between the **electric potential energy** EPE at *B* and that at *A*: $W_{AB} = \text{EPE}_B - \text{EPE}_A$. The **electric potential** *V* is the electric potential energy per unit charge, so the electric potential difference between two points is $V_B - V_A = (\text{EPE}_B - \text{EPE}_A)/q_0 = W_{AB}/q_0$. A positive charge accelerates from a region of higher potential toward a region of lower potential. Conversely, a negative charge accelerates from a region of lower potential toward a region of higher potential. The electric potential at a distance *r* from a **point charge** *q* is $V = kq/r$. This expression for *V* assumes the potential is zero at an infinite distance away from the charge.

The **electron volt** (eV) is a unit of energy. One electron volt corresponds to 1.60×10^{-19} J and is the change in potential energy of an electron when the electron moves through a potential difference of one volt.

An **equipotential surface** is a surface on which the electric potential is the same everywhere. No work is needed to move a charge at a constant speed on an equipotential surface. The electric field created by any group of charges is always perpendicular to the associated equipotential surfaces and points in the direction of decreasing potential. The electric field is given by $E = -\Delta V/\Delta s$, where ΔV is the potential difference and Δs is the magnitude of the displacement perpendicular to the equipotential surfaces in the direction of increasing potential.

A **capacitor** is a device that can store charge and consists of two conductors that are near one another, but not touching. The magnitude q of the charge on each plate is given by $q = CV$, where V is the magnitude of the potential difference between the plates and C is the **capacitance.** The SI unit for capacitance is the farad (F), and one farad is one coulomb per volt (C/V). The insulating material included between the plates is called a **dielectric.** The dielectric constant κ of the material is $\kappa = E_0/E$, where E_0 and E are, respectively, the magnitudes of the electric fields between the plates without and with a dielectric, assuming the charge on the plates is kept fixed. The capacitance of a parallel plate capacitor is $C = \kappa\epsilon_0 A/d$, where $\epsilon_0 = 8.85 \times 10^{-12}\ \mathrm{C^2/(N\cdot m^2)}$ is the permittivity of free space, A is the area of each plate, and d is the distance between the plates. The **electric potential energy** stored in a capacitor is $\frac{1}{2}CV^2$. The **energy density** or energy stored per unit volume is $\frac{1}{2}\kappa\epsilon_0 E^2$.

SOLVED PROBLEMS

Solved Problem 1 The Potential Energy of a Group of Charges

Related Problems: *22 *24 **27

The drawing shows three point charges, initially very far apart, that are brought together and placed at the corners of an equilateral triangle. Each side of the triangle has a length of 0.50 m. Determine the electric potential energy of the group. In other words, determine the amount by which the electric potential energy of the triangular group differs from that of the three charges in their initial, widely separated locations.

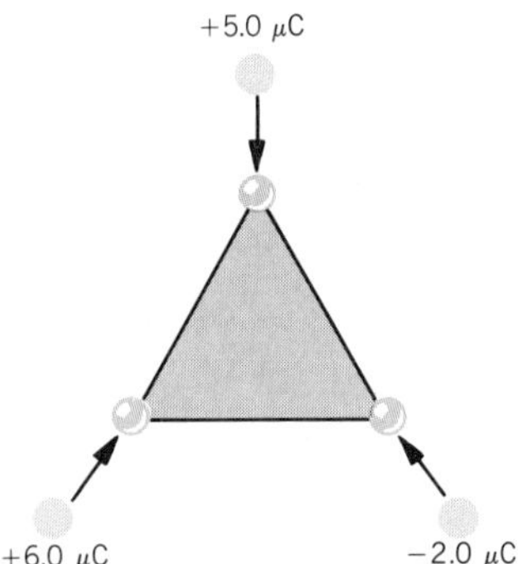

REASONING This problem is done by adding the charges to the triangle one at a time and calculating the energy needed to bring up each charge. According to Equation 19.4, the energy needed to bring up a charge is equal to the product of the charge and the potential at the spot where the charge is placed. The total potential energy of the triangular group is the sum of the energies needed to bring together the three charges.

SOLUTION The order in which the charges are put on the triangle does not matter; we begin with the charge of $+5.0\ \mu$C. No energy is required to place this charge at a corner of the triangle, since the two other charges are initially very far away. Once the charge is in place, the potential it creates at either empty corner ($r = 0.50$ m) is

$$V = \frac{kq}{r} = \frac{(8.99 \times 10^9\ \mathrm{N\cdot m^2/C^2})(+5.0 \times 10^{-6}\ \mathrm{C})}{0.50\ \mathrm{m}} = +9.0 \times 10^4\ \mathrm{V} \qquad (19.6)$$

The energy required to bring up the charge of $+6.0\ \mu$C and place it at one of the empty corners is

$$\mathrm{EPE} = qV = (+6.0 \times 10^{-6}\ \mathrm{C})(+9.0 \times 10^4\ \mathrm{V}) = +0.54\ \mathrm{J} \qquad (19.4)$$

The electric potential at the remaining empty corner is the sum of the potentials due to the two charges that are already in place:

$$V = \frac{(8.99 \times 10^9\ \mathrm{N\cdot m^2/C^2})(+5.0 \times 10^{-6}\ \mathrm{C})}{0.50\ \mathrm{m}} + \frac{(8.99 \times 10^9\ \mathrm{N\cdot m^2/C^2})(+6.0 \times 10^{-6}\ \mathrm{C})}{0.50\ \mathrm{m}} = +2.0 \times 10^5\ \mathrm{V}$$

The energy needed to bring up the charge of $-2.0\ \mu$C and place it at the third corner is

$$\mathrm{EPE} = qV = (-2.0 \times 10^{-6}\ \mathrm{C})(+2.0 \times 10^5\ \mathrm{V}) = -0.40\ \mathrm{J} \qquad (19.4)$$

The total potential energy of the triangular group differs from that of the widely separated charges by an amount that is the sum of the potential energies calculated above:

Total potential energy = 0.54 J − 0.40 J = $\boxed{+0.14\ \text{J}}$

This potential energy originates in the work that is done to bring the charges together.

SUMMARY OF IMPORTANT POINTS This problem shows how to find the electric potential energy of a group of point charges. Initially, the charges are widely separated. Then, each charge is brought up, one at a time, to form the group. The energy needed to add each charge to the group is determined. The total electric potential energy of the group is the sum of the individual energies. The value of the total energy does not depend on the order in which the individual charges are added to the group.

QUESTIONS

1. The drawing shows three possibilities for the potentials at two points, *A* and *B*. In each case, the same positive charge is moved from *B* to *A* at a constant speed. In which case, if any, is the most work done on the positive charge? Account for your answer.

A	*B*	*A*	*B*	*A*	*B*
150 V	100 V	25 V	−25 V	−10 V	−60 V
Case 1		Case 2		Case 3	

2. A positive point charge and a negative point charge have equal magnitudes. One charge is fixed to one corner of a square, and the other is fixed to another corner. On which corners should the charges be placed, so that the same potential exists at the empty corners? Give your reasoning.

3. Three point charges have identical magnitudes, but two of the charges are positive and one is negative. These charges are fixed to the corners of a square, one to a corner. No matter how the charges are arranged, the potential at the empty corner is positive. Explain why.

4. What point charges, all having the same magnitude, would you place at the corners of a square (one charge per corner), so that both the electric field and the electric potential (assuming a zero reference value at infinity) are zero at the center of the square? Account for the fact that the charge distribution gives rise to *both* a zero field and a zero potential.

5. Positive charge is spread uniformly around a circular ring, as the drawing illustrates. Equation 19.6 gives the correct potential at points along the line perpendicular to the plane of the ring at its center. However, the equation does not give the correct potential at points that do not lie on this line. In the equation, q represents the total charge on the ring. Why does Equation 19.6 apply for points on the line, but not for points off the line?

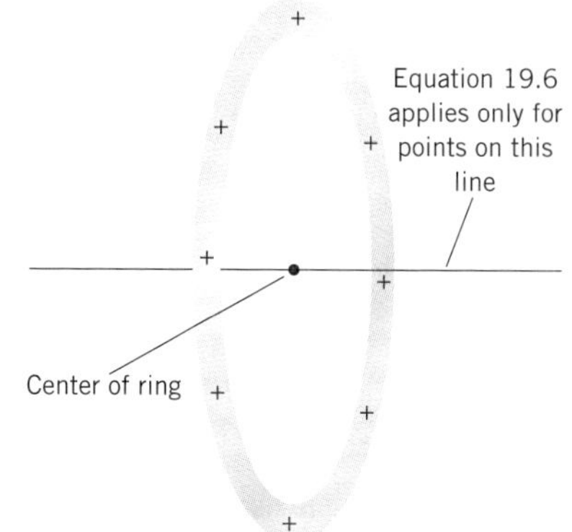
Equation 19.6
applies only for
points on this
line
Center of ring

Question 5

6. The electric field at a single location is zero. Does this fact necessarily mean that the electric potential at the same place is zero? Use a spot on the line between two identical point charges as an example to support your reasoning.

7. To measure the potential at the midpoint between the positive and negative charges of a dipole, a positive test charge is brought in from infinity at a constant speed. When the path followed is along the perpendicular bisector of the dipole, no work is required along any portion of the path. Why?

8. A proton is fixed in place. An electron is released from rest and allowed to collide with the proton. Then the roles of the proton and electron are interchanged, and the same experiment is repeated. Which is traveling faster when the collision occurs, the proton or the electron? Explain.

9. The potential is constant throughout a given region of space. Is the electric field zero or nonzero in this region? Justify your answer.

10. In a region of space where the electric field is constant everywhere, as it is inside a parallel-plate capacitor, is the potential constant everywhere? Account for your answer.

11. A positive test charge is placed in an electric field. In what direction should the charge be moved relative to the field, such that the charge experiences a constant electric potential? Account for your answer.

12. Is any work needed to make a charged particle accelerate from a lower to a higher speed on an equipotential surface? Justify your answer.

13. The location marked P in the drawing lies midway between the point charges $+q$ and $-q$. The blue lines labeled A, B, and C are edge-on views of three planes. Which one of these planes could be an equipotential surface? Why?

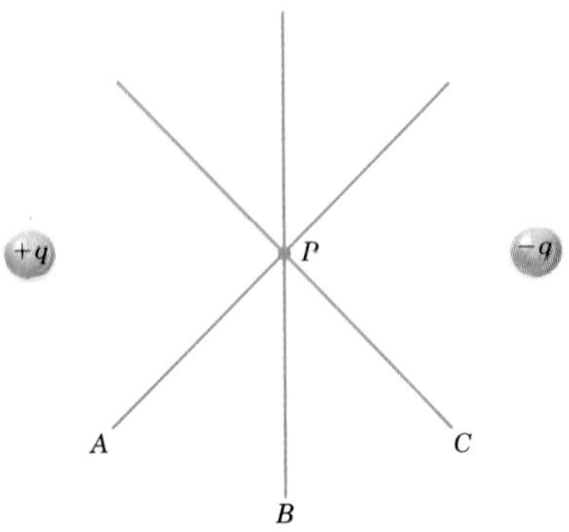

14. Imagine that you are moving a positive test charge along the line between two identical point charges. Is the midpoint on the line analogous to the top of a mountain or the bottom of a valley when the two point charges are (a) positive and (b) negative? In each case, explain your answer.

15. Repeat question 14, assuming that you are moving a negative instead of a positive test charge.

16. The electric field can be expressed in either of two units: newtons/coulomb or volts/meter. Show that these units for the electric field are equivalent.

17. A proton and an electron are released from rest at the midpoint between the plates of a charged parallel plate capacitor. Except for these particles, nothing else is between the plates. Ignore the attraction between the proton and the electron, and decide which particle strikes a capacitor plate first? Why?

PROBLEMS

Note: All charges are assumed to be point charges unless specified otherwise.

Section 19.1 Potential Energy, Section 19.2 The Electric Potential Difference

1. The electric potential inside a living cell is lower than the electric potential outside the cell. Suppose the electric potential difference between the inner and outer cell wall is 0.095 V, a typical value. To maintain the internal electrical balance, the cell "pumps" out sodium ions. How much work must be done to remove a single sodium ion (charge $+e$)?

2. An agent moves a charge of $+1.80 \times 10^{-4}$ C at a constant speed from point A to point B and performs 5.80×10^{-3} J of work on the charge. (a) What is the difference $EPE_B - EPE_A$ between the electric potential energies of the charge at the two points? (b) Determine the potential difference $V_B - V_A$ between the two points. (c) State which point is at the higher potential.

3. The potential of point A relative to point B is $V_A - V_B = +95$ V, while the potential of point C relative to point B is $V_C - V_B = +23$ V. How much work is needed to move a charge of $+45\ \mu$C from C to A?

4. The anode (positive terminal) of an X-ray tube is at a potential of $+125\ 000$ V with respect to the cathode (negative terminal). (a) How much work (in joules) is done on an electron that is accelerated from the cathode to the anode? (b) If the electron is initially at rest, what kinetic energy does the electron have when it arrives at the anode?

5. In a television picture tube, electrons strike the screen after being accelerated from rest through a potential difference of 25 000 V. The speeds of the electrons are quite large, and in accurate calculations of the speeds, the effects of special relativity must be taken into account. Ignoring such effects, find the electron speed just before the electron strikes the screen.

6. A particle with a charge of $-1.5\ \mu$C and a mass of 2.5×10^{-6} kg is released from rest at point X and accelerates toward point Y, arriving there with a speed of 42 m/s. (a) What is the potential difference $V_Y - V_X$ between X and Y? (b) Which point is at the higher potential? Give your reasoning.

7. An electric car accelerates for 8.0 s by drawing energy from its 320-V battery pack. During this time, 1300 C of charge pass through the battery pack. Find the minimum horsepower rating of the car.

***8.** The energy in a lightning bolt is enormous. Consider, for example, a lightning bolt in which 25 C of charge moves through a potential difference of 1.2×10^8 V. With the amount of energy in this bolt, how many kilograms of water at 100 °C could be boiled into steam at 100 °C?

***9.** The potential at location A is 452 V. A positively charged particle is released there from rest and arrives at location B with a speed v. The potential at location C is 791 V, and when released from this spot, the particle arrives at B with a speed of $2v$. Find the potential at B.

***10.** An electron and a proton, starting from rest, are accelerated through an electric potential difference of the same magnitude. In the process, the electron acquires a speed v_e, while the proton acquires a speed v_p. Find the ratio v_e/v_p.

****11.** A particle is uncharged and is thrown vertically upward from ground level with a speed of 25.0 m/s. As a result, it attains a maximum height h. The particle is then given a positive charge $+q$ and reaches the same maximum height h when thrown vertically upward with a speed of 30.0 m/s. The electric potential at the height h exceeds the electric potential at ground level. Finally, the particle is given a negative charge $-q$. Ignoring air resistance, determine the speed with which the negatively charged particle must be thrown vertically upward, so that it attains exactly the maximum height h.

Section 19.3 The Electric Potential Difference Created by Point Charges

12. To see how large a charge of one coulomb is, calculate the electric potential at a location that is 1.0 km away from a charge of +1.0 C.

13. There is an electric potential of +130 V at a spot that is 0.25 m away from a charge. Find the magnitude and sign of the charge.

14. Two charges of $+2.60 \times 10^{-8}$ and -5.50×10^{-8} C are separated by 1.40 m. What is the electric potential midway between them?

15. A charge of +125 μC is fixed at the center of a square that is 0.64 m on a side. How much work is required to move a charge of +7.0 μC from one corner of the square to any other empty corner? Explain.

16. Location A is 2.00 m from a charge of -3.00×10^{-8} C, while location B is 3.00 m from the charge. Find the potential difference $V_B - V_A$ between the two points, and state which point is at the higher potential.

17. Two identical point charges are fixed to diagonally opposite corners of a square that is 0.500 m on a side. Each charge is $+3.0 \times 10^{-6}$ C. How much work must be done to move one of the charges to an empty corner?

18. Three charges are located at the corners of a square whose sides are 2.0 m in length. The charges are +2.0, +14, and +5.0 μC. The empty corner of the square is opposite the 14-μC charge. How much work is required to bring up a fourth charge of +8.0 μC and place it at the empty corner?

19. Review Conceptual Example 8 as background for this problem. Two charges are fixed in place with a separation d. One charge is positive and has twice the magnitude of the other charge, which is negative. The positive charge lies to the left of the negative charge, as in Figure 19.9. Relative to the negative charge, locate the two spots on the line through the charges where the total potential is zero.

20. Charges of $+2q$ and $-q$ are fixed in place and separated by a distance of 2.0 m. A dashed line is drawn through the negative charge, perpendicular to the line between the charges. Relative to the negative charge, where on the dashed line is the total potential equal to zero? There are two places.

***21.** A charge of −3.00 μC is fixed in place. From a distance of 0.0450 m, a particle of mass 7.20×10^{-3} kg and charge −8.00 μC is fired with an initial speed of 65.0 m/s directly toward the fixed charge. How far does the particle travel before its speed is zero?

***22.** A square is 0.50 m on each side. How much work is done to bring in four identical charges (5.0 μC each) from infinity and place them on the square, one to a corner? *(See Solved Problem 1 for a related problem.)*

***23.** Identical point charges of +1.7 μC are fixed to diagonally opposite corners of a square. A third charge is then fixed at the center of the square, such that it causes the potentials at the empty corners to change signs without changing magnitudes. Find the sign and magnitude of the third charge.

***24.** Determine the electric potential energy for the array of three charges shown in the drawing, relative to its value when the charges are infinitely far away. *(See Solved Problem 1 for a related problem.)*

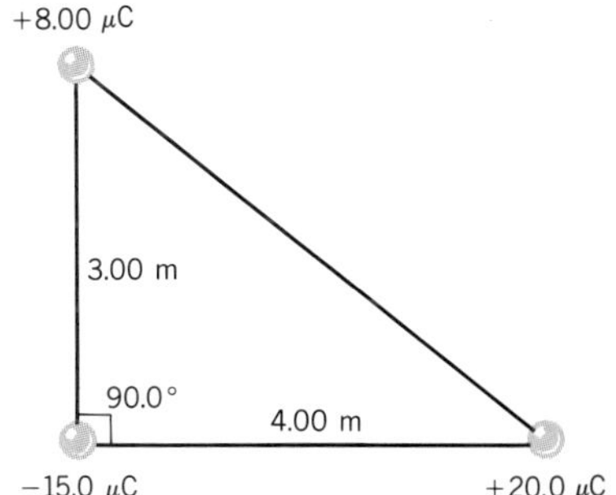

***25.** According to the Bohr model, a singly ionized helium atom contains one electron in orbit at a distance of 2.65×10^{-11} m from a nucleus that contains two protons and two neutrons. (The neutrons are electrically neutral and play no role in this problem.) (a) Calculate the ionization energy (in joules) for the atom. (b) Express the answer in electron volts.

****26.** A positive charge of $+q_1$ is located 3.00 m to the left of a negative charge $-q_2$. The charges have different magnitudes. On the line through the charges, the net *electric field* is zero at a spot 1.00 m to the right of the negative charge. On this line there are also two spots where the potential is zero. Locate these two spots relative to the negative charge.

****27.** Charges q_1 and q_2 are fixed in place, q_2 being located at a distance d to the right of q_1. A third charge q_3 is then fixed to the line joining q_1 and q_2 at a distance d to the right of q_2. The third charge is chosen so the potential energy of the group is zero; that is, the potential energy has the same value as that of the three charges when they are widely separated. Determine q_3, assuming that (a) $q_1 = q_2 = q$ and (b) $q_1 = q$ and $q_2 = -q$. Express your answers in terms of q. *(See Solved Problem 1 for a related problem.)*

****28.** A positive charge $+q_1$ is located to the left of a negative charge $-q_2$. On a line passing through the two charges, there are two places where the total potential is zero. The first place is between the charges and is 4.00 cm to the left of the negative charge. The second place is 7.00 cm to the right of the negative charge. (a) What is the distance between the charges? (b) Find q_1/q_2, the ratio of the magnitudes of the charges.

Section 19.4 Equipotential Surfaces and Their Relation to the Electric Field

29. An equipotential surface that surrounds a $+3.0 \times 10^{-7}$ C point charge has a radius of 0.15 m. What is the potential of this surface?

30. What is the radius of the +12-V equipotential surface that surrounds a charge of $+2.0 \times 10^{-10}$ C?

31. Consider the equipotential surfaces that surround a point charge of $+1.50 \times 10^{-8}$ C. How far from the 190-V surface is the 75-V surface?

32. The inner and outer surfaces of a cell membrane carry a negative and positive charge, respectively. Because of these charges, a potential difference of about 0.095 V exists across the membrane. The thickness of the membrane is 7.5×10^{-9} m. What is the magnitude of the electric field in the membrane?

33. When you walk across a rug on a dry day, your body can become electrified, and its electric potential can change. When the magnitude of the potential becomes large enough, a spark can jump between your hand and a metal surface. A spark occurs when the electric field strength created by the charges on your body reaches the dielectric strength of the air. The dielectric strength of the air is 3.0×10^6 N/C and is the electric field strength at which the air suffers electrical breakdown. Suppose a spark 3.0 mm long jumps between your hand and a metal doorknob. Assuming that the electric field is uniform, find the magnitude of the potential difference between your hand and the doorknob.

34. Two points, *A* and *B*, are separated by 0.016 m. The potential at *A* is +28 V, and that at *B* is +95 V. Determine the magnitude and direction of the electric field between the two points.

***35.** The drawing shows the potential at five points on a set of axes. Each of the four outer points is 6.0×10^{-3} m from the point at the origin. From the data shown, find the magnitude and direction of the electric field in the vicinity of the origin.

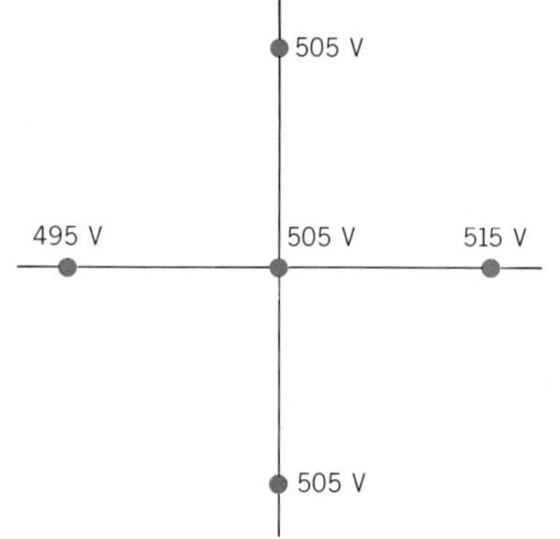

***36.** Equipotential surface *A* has a potential of 5650 V, while equipotential surface *B* has a potential of 7850 V. A particle has a mass of 5.00×10^{-2} kg and a charge of $+4.00 \times 10^{-5}$ C. The particle has a speed of 2.00 m/s on surface *A*. It is moved to surface *B* and arrives there with a speed of 3.00 m/s. How much work is done in moving the particle from *A* to *B*?

***37.** Two equipotential surfaces surround a point charge of $+1.00 \times 10^{-9}$ C. The surfaces are separated by a distance of 1.00 m and have a potential difference of 1.00 V between them. (a) What are the radii of the surfaces? (b) What is the potential of each surface?

****38.** At a distance of 1.60 m from a point charge of $+2.00 \times 10^{-6}$ C, there is an equipotential surface. At greater distances there are additional equipotential surfaces. The potential difference between any two successive surfaces is 1.00×10^3 V. Starting at a distance of 1.60 m and moving radially outward, how many of the additional equipotential surfaces are crossed by the time the electric field has shrunk to one-half its initial value? Do not include the starting surface.

Section 19.5 Capacitors and Dielectrics

39. What voltage is required to store 7.2×10^{-5} C of charge on the plates of a 6.0-μF capacitor?

40. One farad is a large capacitance. To see just how large, determine the area of each plate of an empty, one-farad parallel plate capacitor whose plate separation is one meter. Express your answer in square miles (1 square mile = 2.59×10^6 m^2).

41. A parallel plate capacitor has a capacitance of 7.0 μF when filled with a dielectric. The area of each plate is 1.5 m^2 and the separation between the plates is 1.0×10^{-5} m. What is the dielectric constant of the dielectric?

42. A defibrillator is a device used in emergency situations to stimulate the heart muscle and start the heart beating again. Two "paddles" are placed on the body near the heart, and the energy stored in a capacitor is discharged through them. The energy discharged is 510 J and the capacitance is 120 μF. What is the potential difference across the capacitor plates?

43. A capacitor has a capacitance of 2.5×10^{-8} F. In the charging process, electrons are removed from one plate and placed on the other plate. When the potential difference between the plates is 450 V, how many electrons have been transferred?

44. The membrane that surrounds a certain type of living cell has a surface area of about 5×10^{-9} m^2 and a thickness of about 1×10^{-8} m. Assume that the membrane behaves like a parallel plate capacitor and has a dielectric constant of 5. (a) If the potential on the outer surface of the membrane is +60 mV greater than that on the inside surface, how much charge resides on the outer surface? (b) If the charge in part (a) is due to K^+ ions (charge $+e$), how many such ions are present on the outer surface?

45. A capacitor stores 5.3×10^{-5} C of charge when connected to a 6.0-V battery. How much charge does the capacitor store when connected to a 9.0-V battery?

46. Each plate of a parallel plate capacitor has an area of 2.2×10^{-4} m^2 and stores a charge whose magnitude is 4.8×10^{-9} C. Determine the magnitude of the electric field between the plates when the capacitor is (a) empty and (b) filled with Teflon.

47. The electronic flash attachment for a camera contains a capacitor for storing the energy used to produce the flash. In one such unit, the potential difference between the plates of a 750-μF capacitor is 330 V. (a) Determine the energy that is used to produce the flash in this unit. (b) Assuming that the flash lasts for 5.0×10^{-3} s, find the effective "wattage" of the flash.

***48.** What is the potential difference between the plates of a 3.3-F capacitor that stores sufficient energy to operate a 75-W light bulb for one minute?

***49.** The equipotential surfaces between the plates of an empty parallel plate capacitor are such that two surfaces (not the metal plates), 2.00 mm apart, have a potential difference of 1.20×10^{-3} V. The area of each plate is 7.50×10^{-4} m^2. How much charge is on each plate?

***50.** An empty capacitor has a capacitance of 2.7 μF and is connected to a 12-V battery. A dielectric material ($\kappa = 4.0$) is inserted between the plates of this capacitor. What is the magnitude of the surface charge on the dielectric that is adjacent to either plate of the capacitor? *(Hint: The surface charge is equal to the difference in the charge on the plates with and without the dielectric.)*

***51.** Review Conceptual Example 12 before attempting this problem. An empty capacitor is connected to a 12.0-V battery and charged up. The capacitor is then disconnected from the battery, and a slab of dielectric material ($\kappa = 2.8$) is inserted between the plates. Find the amount by which the potential difference across the plates changes. Specify whether the change is an increase or a decrease.

****52.** The plate separation of a charged capacitor is 0.0800 m. A proton and an electron are released from rest at the midpoint between the plates. Ignore the attraction between the two particles, and determine how far the proton has traveled by the time the electron strikes the positive plate.

****53.** The drawing shows a parallel plate capacitor. One-half of the region between the plates is filled with a material that has a dielectric constant κ_1. The other half is filled with a material that has a dielectric constant κ_2. The area of each plate is A, and the plate separation is d. The potential difference across the plates is V. Note especially that the charge stored by the capacitor is $q_1 + q_2 = CV$, where q_1 and q_2 are the charges on the area of the plates in contact with materials 1 and 2, respectively. Show that $C = \epsilon_0 A(\kappa_1 + \kappa_2)/(2d)$.

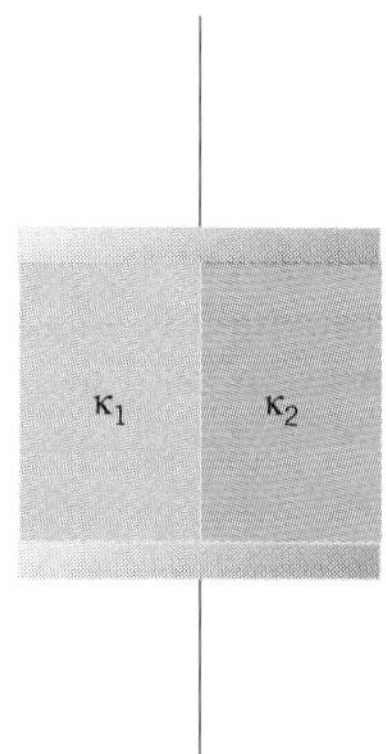

ADDITIONAL PROBLEMS

54. At a distance of 0.20 m from a charge, the electric potential is 164 V. What is the potential at a distance of 0.80 m?

55. Two positive point charges are held in place, 0.86 m apart. They are then moved so that their electric potential energy doubles. What is the new separation between the charges?

56. A proton, released from rest, accelerates from one point to another and gains kinetic energy, because there is a potential difference of 1.5 μV between the two points. To acquire the same amount of kinetic energy, through what height would the proton have to fall freely, after being released from rest in the presence of the earth's gravity?

57. An axon is the relatively long tail-like part of a neuron, or nerve cell. The outer surface of the axon membrane (dielectric constant = 5, thickness = 1×10^{-8} m) is charged positively, and the inner portion is charged negatively. Thus, the membrane is a kind of capacitor. Assuming that an axon can be treated like a parallel plate capacitor with a plate area of 5×10^{-6} m^2, what is its capacitance?

58. Point A is at a potential of +250 V, and point B is at a potential of −150 V. An α-particle is a helium nucleus that contains two protons and two neutrons; the neutrons are electrically neutral. An α-particle starts from rest at A and accelerates toward B. When the α-particle arrives at B, what kinetic energy (in electron volts) does it have?

59. A parallel plate capacitor is filled with ruby mica, and the area of each plate is 3.8 m^2. The capacitor stores 2.7 μC of charge when a 1.5-V flashlight battery provides the potential difference between the plates. What is the plate separation?

60. A charge of $+9q$ is fixed to one corner of a square, while a charge of $-8q$ is fixed to the opposite corner. Expressed in terms of q, what charge should be fixed to the center of the square, so the potential is zero at each of the two empty corners?

61. Surrounding a positive point charge there are two equipotential surfaces. Surface A has twice the area of surface B. Find the ratio V_A/V_B of the potentials of these surfaces.

***62.** The dielectric strength of an insulating material is the maximum electric field strength to which the material can be subjected without electrical breakdown occurring. Suppose a parallel plate capacitor is filled with a material whose dielectric constant is 3.5 and whose dielectric strength is 1.4×10^7 N/C. If this capacitor is to store 1.7×10^{-7} C of charge on each plate without suffering breakdown, what must be the radius of its circular plates?

***63.** The electric field has a constant value of 3.0×10^3 V/m and is directed downward. The field is the same everywhere. The potential at a point P within this region is 135 V. Find the potential at the following points: (a) 8.0×10^{-3} m directly above P, (b) 3.3×10^{-3} m directly below P, (c) 5.0×10^{-3} m directly to the right of P.

***64.** Two particles each have a mass of 6.0×10^{-3} kg. One has a charge of $+5.0 \times 10^{-6}$ C, and one has a charge of -5.0×10^{-6} C. They are initially held at rest at a distance of 0.80 m apart. Both are then released and accelerate toward each other. How fast is each particle moving when the separation between them is one-half its initial value?

****65.** The potential difference between the plates of a capacitor is 175 V. Midway between the plates, a proton and an electron are released. The electron is released from rest. The proton is projected perpendicularly toward the negative plate with an initial speed. The proton strikes the negative plate at the same instant the electron strikes the positive plate. Ignore the attraction between the two particles, and find the initial speed of the proton.

****66.** One particle has a mass of 3.00×10^{-3} kg and a charge of $+8.00$ μC. A second particle has a mass of 6.00×10^{-3} kg and the same charge. The two particles are initially held in place and then released. The particles fly apart, and when the separation between them is 0.100 m, the speed of the 3.00×10^{-3}-kg particle is 125 m/s. Find the initial separation between the particles. *(Hint: Both the energy and momentum of the two-particle system are conserved.)*

MYSTERIES SOLVED BY PHYSICS

Sharks: Electromagnetic Predators

Silent, powerful, and deadly, sharks are among the most feared predators on earth. Once they become aware that food is nearby, they are implacable in their efforts to locate and kill it. Their keen senses alone do not fully explain the mystery of their awesome ability to find their next meal. Consider the difficulties obstructing sharks' senses as they search for food. Sharks have sensitive noses, sometimes able to detect 1 part fish extract in 10^{10} parts water. But water currents often channel the scent of prey away, causing sharks to miss the smell even of nearby fish. Nonetheless, downstream fish are detected by sharks. How? By sound? The sounds in the 20- to 300-Hz range made by fish as they swim are heard clearly by sharks. But even fish that are absolutely motionless are detected. By sight? Some sharks do have eyes that are sensitive to the low light levels found deep in the oceans of the world. But even sharks that do not see well can locate fish in the darkest waters. In short, sharks can sense downstream, motionless fish in complete darkness. Sharks can even detect the presence of prey, such as stingrays, completely buried in the ocean bottom!

The mystery is solved by examining the physics of detecting hidden prey. Buried in the ocean bottom, a fish's scent is absorbed by the mud or sand around it. By not moving, a fish gives off no sound. Sharks must therefore have the ability to sense their prey *remotely*. One possibility is that fish create electric fields that pass through the ocean bottom and water. How could they do this? And how do sharks detect the fields?

Fish and other sea creatures create electric fields in a variety of ways. The surfaces of their bodies in contact with water act as miniature batteries, creating small electric currents. Their muscle contractions also create electric fields, as do the mucous linings of their mouths and gills. These fields vary as the creature moves. Unlike humans and most other animals, sharks have evolved specialized organs, called the *ampullae of Lorenzini,* that are extremely sensitive to electric fields. The tops of these jelly-filled capsules can be seen as dark spots embedded in the shark's snout. The ampullae are sensitive to electric potential differences in the water of 5×10^{-9} V created by the way in which the electric fields of their prey change with distance. Even a fish buried in the ocean bottom creates electric fields, as the water passes through its gills and as its heart beats, that are strong enough to be detected by nearby sharks.

The most dramatic example of a shark whose body has evolved to enhance its ability to detect electric fields is the hammerhead. Its wide head creates a large surface area with ampullae of Lorenzini spread throughout. By detecting different electric fields in different places throughout its snout, the hammerhead can home in on its prey very precisely.

For Further Reading See: Richard Ellis and John E. McCosker. *The Great White Shark.* New York: Harper Collins and Stanford University Press, 1991.

Sharks have specialized organs, called the ampullae of Lorenzini, *that can detect electric potential differences as small as 5×10^{-9} V. In this hammerhead shark the ampullae are spread throughout the large surface area of the animal's wide head.*

CHAPTER 20 ELECTRIC CIRCUITS

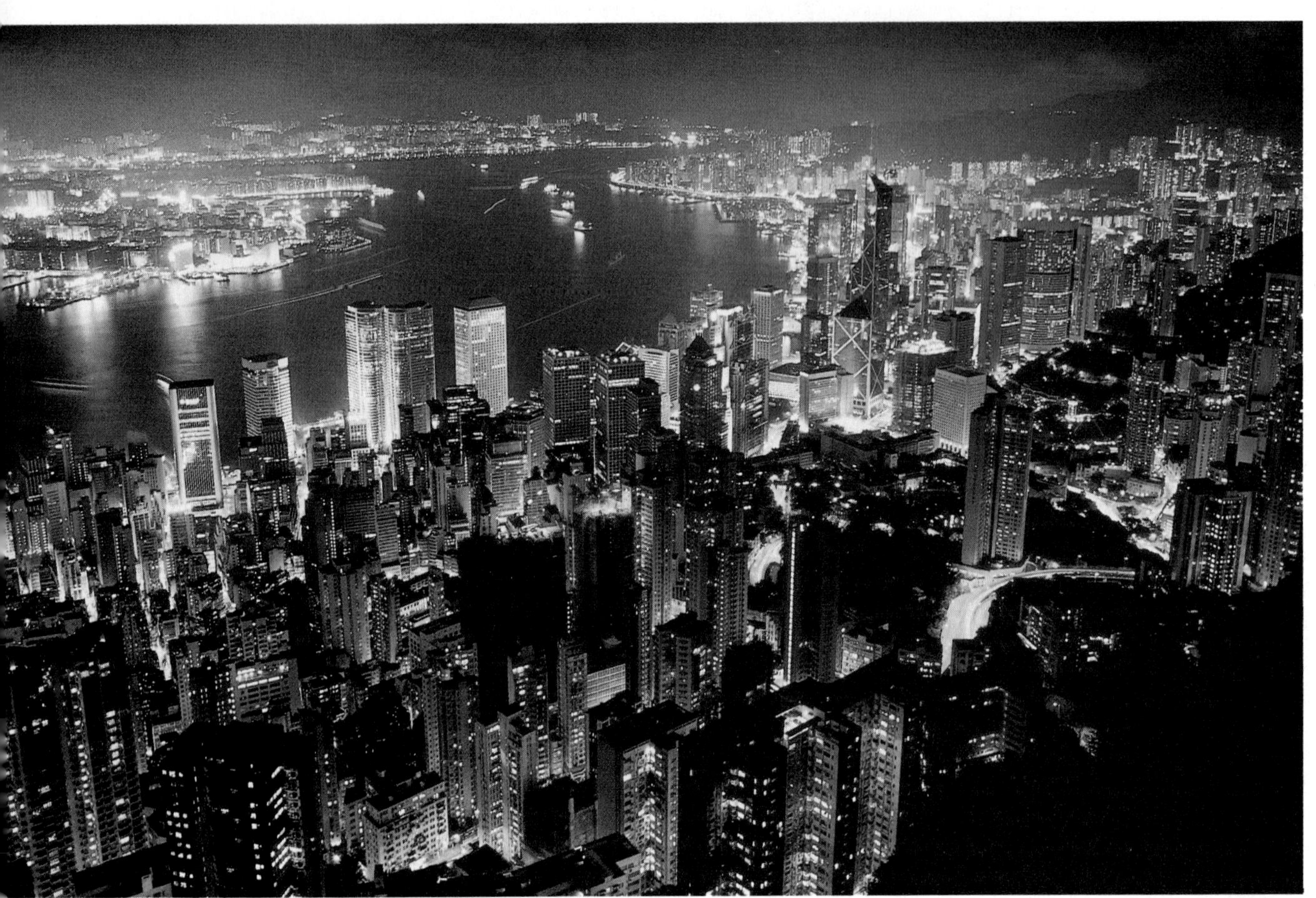

This night photograph of Hong Kong illustrates on a large scale the topic of this chapter, electric circuits. It is evident here that electrical energy is performing a wide variety of services in these buildings. The energy is used for lighting, heating and cooling offices, operating elevators, and cooking food in restaurants. At the core of all such services are electric circuits, which are "highways" that route electrical energy to where it is needed, much like the freeways that direct people into and out of a city. In this chapter we will examine the concepts and principles that apply to electric circuits. Many of these ideas will be familiar from earlier chapters, such as voltage, power, and the conservation of energy. Others, like current, resistance, and Ohm's law, will be introduced for the first time. We will discuss how electric circuits are assembled and the various ways in which appliances can be connected into them, with particular attention paid to series and parallel wiring. Since all homes and offices use alternating current and voltage as a means of supplying electrical energy, we will consider these concepts in some detail. The various types of meters used to measure currents, voltages, and resistances in a circuit will also be discussed.

20.1 ELECTROMOTIVE FORCE AND CURRENT

Look around you. Chances are that there is an electrical device nearby—a radio, a hair drier, a computer—something that needs electrical energy to operate. Such devices are common because electrical energy is convenient to use and because a variety of electrical energy sources are available. The battery is one of the more familiar sources of electrical energy. The energy needed to run a portable cassette player, for instance, comes from batteries, as Figure 20.1 illustrates. The transfer of energy takes place via an electric circuit, in which the source (the battery) and the energy-consuming device (the cassette player) are connected by conducting wires, through which electric charges move.

Within a battery, a chemical reaction occurs that transfers electrons from one terminal to the other, leaving one terminal negatively charged and the other positively charged. Figure 20.2 shows the two terminals of a car battery and of a flashlight battery. The drawing also illustrates the symbol (+|−) used to represent a battery in circuit drawings. Because of the positive and negative charges on the battery terminals, an electric potential difference exists between them. The maximum potential difference is called the ***electromotive force* (emf)*** of the battery, for which the symbol $\mathscr{E}$ is used. In a typical car battery, the chemical reaction maintains the potential of the positive terminal at a maximum of 12 volts (12 joules/coulomb) higher than the potential of the negative terminal, so the emf

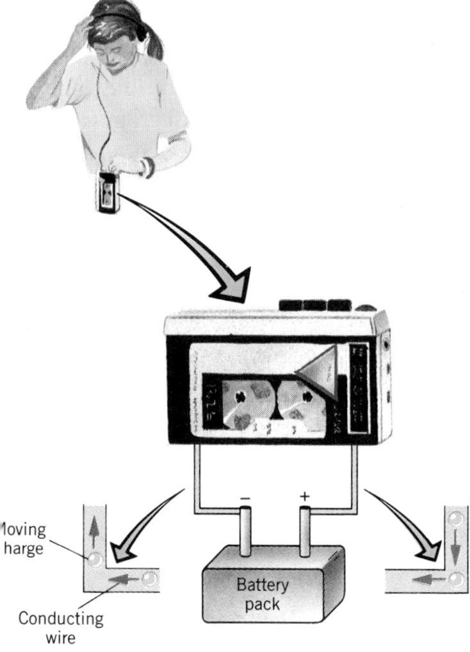

Figure 20.1 In an electric circuit, energy is transferred from a source (the battery pack) to a device (the cassette player) by charges that move through a conducting wire.

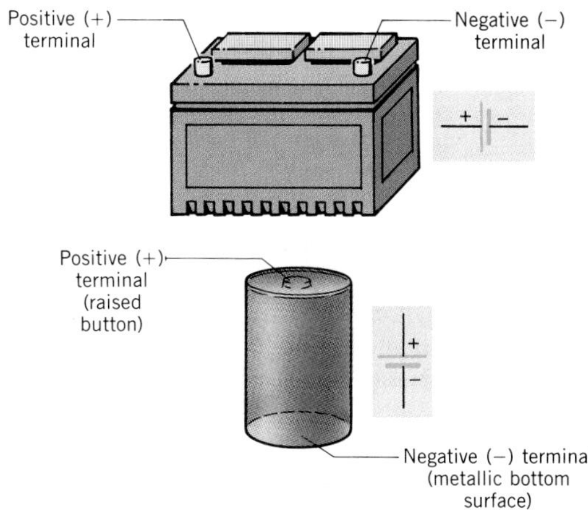

Figure 20.2 Typical batteries and the symbol (+|−) used to represent them in electric circuits.

* The word "force" appears in this context for historical reasons, even though it is incorrect; electric potential is energy per unit charge, which is not force.

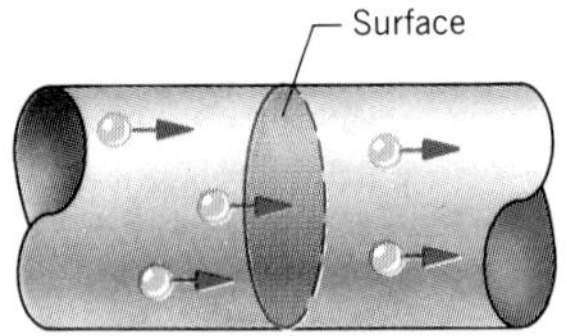

Figure 20.3 The electric current is the amount of charge per unit time that passes through a surface that is perpendicular to the motion of the charges.

is $\mathscr{E} = 12$ V. Thus, one coulomb of charge emerging from the battery and entering a circuit has at most 12 joules of energy. In a typical flashlight battery the emf is 1.5 V. In reality, the potential difference between the terminals of a battery is somewhat less than the maximum value indicated by the emf, for reasons that Section 20.9 discusses.

In a circuit such as that in Figure 20.1, the battery creates an electric field within† and parallel to the wire, directed from the positive toward the negative terminal. The field exerts a force on the free electrons in the wire, and they respond by moving. The resulting flow of charge is known as an ***electric current.*** Figure 20.3 shows charges moving inside a wire and crossing an imaginary surface that is perpendicular to their motion. The current is defined as the amount of charge per unit time that crosses this surface, in much the same sense that a river current is the amount of water flowing per unit of time. Suppose q is the amount of charge that passes through the surface in a time t. If the rate is constant, the current I is equal to the charge divided by the time:

$$I = \frac{q}{t} \tag{20.1}$$

If charge does not flow at a constant rate, then Equation 20.1 gives the average current. Since the units for charge and time are the coulomb (C) and the second (s), the SI unit for current is a coulomb per second (C/s). One coulomb per second is referred to as an ***ampere*** (A), after the French mathematician André-Marie Ampère (1775–1836).

If the charges move around a circuit in the same direction at all times, the current is said to be ***direct current (dc).*** Batteries, for example, produce direct current. In contrast, the current is said to be ***alternating current (ac)*** when the charges move first one way and then the opposite way, changing direction from moment to moment. Many energy sources produce alternating current, e.g., generators at power companies and microphones. Example 1 deals with the direct current produced by the battery in a pocket calculator.

† Here, an electric field exists inside a conductor, in contrast to the situation in electrostatics. The field exists here because the battery keeps the charges moving and prevents them from coming to equilibrium on the outer surface of the conductor, where they would cause the net electric field on the interior to be zero.

Example 1 A Pocket Calculator

The current from the 3.0-V battery of a pocket calculator is 0.17 mA (1 mA = 10^{-3} A). In one hour of operation, (a) how much charge flows in the circuit and (b) how much energy does the battery deliver to the calculator circuit?

REASONING Since current is defined as charge per unit time, the total charge that flows in one hour is the product of the current and the time (3600 s). The charge that leaves the 3.0-V battery has 3.0 joules of energy per coulomb of charge. Thus, the total energy delivered to the calculator circuit from the battery is the total charge (in coulombs) times the energy per unit charge (in volts or joules/coulomb).

SOLUTION

(a) The charge that flows in one hour can be determined from Equation 20.1:

$$q = It = (0.17 \times 10^{-3}\ \text{A})(3600\ \text{s}) = \boxed{0.61\ \text{C}}$$

(b) The energy delivered to the calculator circuit is

$$\text{Energy} = \text{Charge} \times \underbrace{\frac{\text{Energy}}{\text{Charge}}}_{\text{Battery emf}} = (0.61\ \text{C})(3.0\ \text{V}) = \boxed{1.8\ \text{J}}$$

Today, it is known that electrons flow in metal wires. Figure 20.4 shows the negative electrons emerging from the negative terminal of the battery and moving around the circuit toward the positive terminal. It is customary, however, *not* to use the flow of electrons when discussing circuits. Instead, a so-called ***conventional current*** is used, for reasons that date back to the time when it was believed that positive charges moved through metal wires. Conventional current is the hypothetical flow of positive charges that would have the same effect in the circuit as the movement of negative charges that actually does occur. In Figure 20.4, negative electrons *arrive* at the positive terminal of the battery. The same effect would have been achieved if an equivalent amount of positive charge had *left* the positive terminal. Therefore, the drawing shows the conventional current originating from the positive terminal. A conventional current of hypothetical positive charges is consistent with our earlier use of a positive test charge for defining electric fields and potentials. The direction of conventional current is always from a point of higher potential toward a point of lower potential, that is, from the positive terminal toward the negative terminal. In this text, the symbol I stands for conventional current.

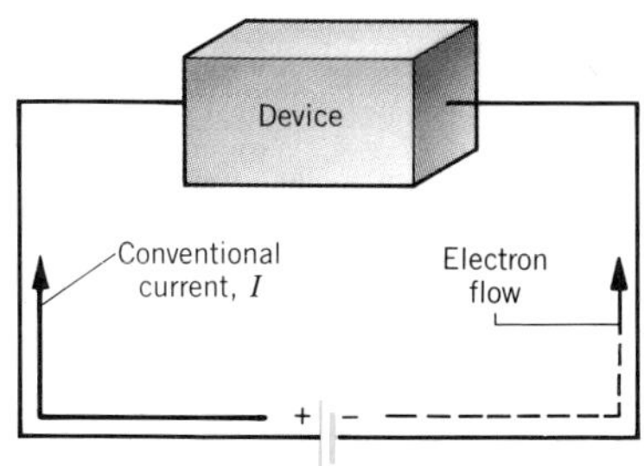

Figure 20.4 In a circuit, electrons actually flow through the metal wires. However, it is customary to use a conventional current I to describe the flow of charges.

20.2 OHM'S LAW

The current that a battery can push through a wire is analogous to the water flow that a pump can push through a pipe. Greater pump pressures lead to larger water flow rates, and, similarly, greater battery voltages* lead to larger electric currents. In the simplest case, the current I is directly proportional to the voltage V; that is, $I \propto V$. Thus, a 12-V battery leads to twice as much current as a 6-V battery, when each is connected to the same circuit.

In a water pipe, the flow rate is not only determined by the pump pressure but is also affected by the length and diameter of the pipe. Longer and narrower pipes offer higher resistance to the moving water and lead to smaller flow rates for a given pump pressure. A similar situation exists in electrical circuits. In the electrical case, the ***resistance*** (R) is defined as the ratio of the voltage V applied across a piece of material to the current I through the material, or $R = V/I$. When only a small current results from a large voltage, there is a high resistance to the moving charge. For many materials (e.g., metals), the ratio V/I is the same for a given piece of material over a wide range of voltages and currents. In such a case, the resistance is a constant, and the relation $R = V/I$ is referred to as ***Ohm's law,*** after the German physicist Georg Simon Ohm (1789–1854), who discovered it.

Thanks to miniaturized electric circuits, cassette players and CD players can be used wherever you are.

* The potential difference between two points, such as the terminals of a battery, is commonly called the voltage between the points.

Ohm's Law

The ratio V/I is a constant, where V is the voltage applied across a piece of material (such as a wire) and I is the current through the material:

$$\frac{V}{I} = R = \text{constant} \quad \text{or} \quad V = IR \tag{20.2}$$

R is the resistance of the piece of material.

SI Unit of Resistance: volt/ampere (V/A) = ohm (Ω)

The SI unit for resistance is a volt per ampere, which is called an *ohm* and is represented by the Greek capital letter omega (Ω). Ohm's law is not a fundamental law of nature like Newton's laws of motion. It is only a statement of the way certain materials behave in electric circuits.

To the extent that a wire or an electrical device offers resistance to the flow of charges, it is called a ***resistor.*** The resistance can have a wide range of values. The copper wires in a television set, for instance, have a negligibly small resistance. On the other hand, commercial resistors can have resistances up to many kilohms ($1\ \text{k}\Omega = 10^3\ \Omega$) or megohms ($1\ \text{M}\Omega = 10^6\ \Omega$). Such resistors play an important role in electric circuits, where they are used to limit the amount of current and establish certain voltage levels (see Figure 20.5).

In drawing electric circuits we follow the usual conventions: (1) a zigzag line (—ᴡᴡ—) represents a resistor and (2) a straight line (———) represents an ideal conducting wire, or one with a negligible resistance. Example 2 illustrates an application of Ohm's law to the circuit in a flashlight.

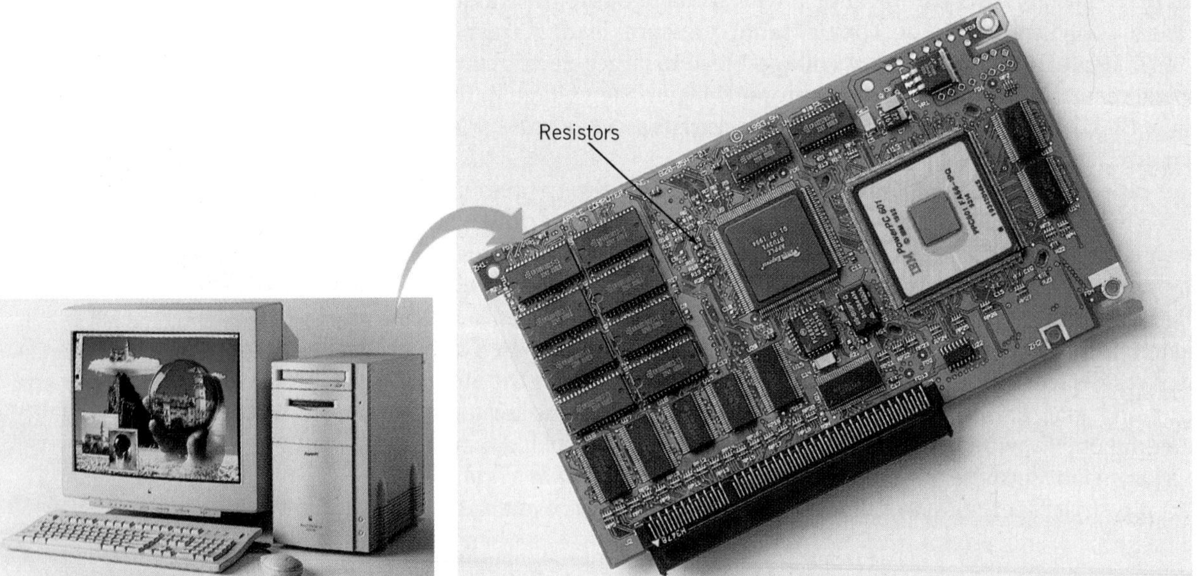

Figure 20.5 Resistors are found on circuit boards of computers.

Night skiers at the Quebec Winter Carnival in Canada use battery-operated light torches to produce this serpentine display. The ability of a battery to provide energy for light is indicated by the electric potential difference of the battery (expressed in volts).

Example 2 A Flashlight

The filament in a light bulb is a resistor in the form of a thin piece of wire. The wire becomes hot enough to emit light because of the current in it. Figure 20.6 shows a flashlight that uses two 1.5-V batteries (effectively a single 3.0-V battery) to provide a current of 0.40 A in the filament. Determine the resistance of the glowing filament.

REASONING The filament resistance is assumed to be the only resistance in the circuit. The potential difference applied across the filament is that of the 3.0-V battery. The resistance, given by Equation 20.2, is equal to this potential difference divided by the current.

SOLUTION The resistance of the filament is

$$R = \frac{V}{I} = \frac{3.0\ \text{V}}{0.40\ \text{A}} = \boxed{7.5\ \Omega} \tag{20.2}$$

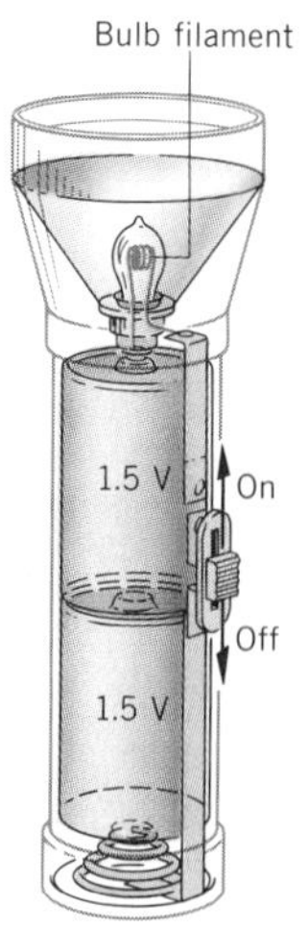

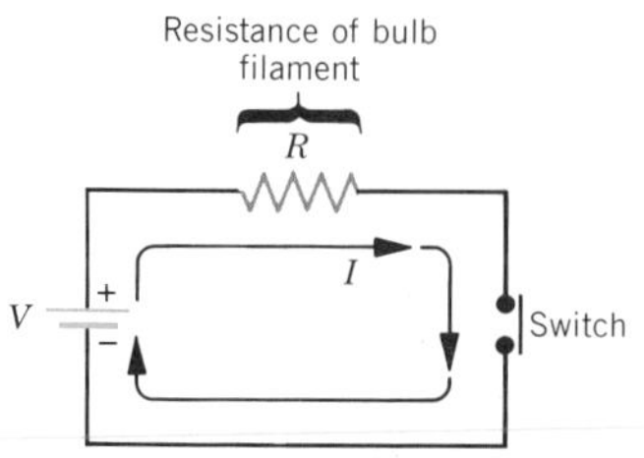

Figure 20.6 The circuit in this flashlight consists of a resistor (the filament of the light bulb) connected to a 3.0-V battery.

20.3 RESISTANCE AND RESISTIVITY

In a water pipe, the length and cross-sectional area of the pipe determine the resistance that the pipe offers to the flow of water. Longer pipes with smaller cross-sectional areas offer greater resistance. Analogous effects are found in the electrical case. For a wide range of materials, the resistance of a piece of material of length L and cross-sectional area A is

$$R = \rho \frac{L}{A} \tag{20.3}$$

Table 20.1 Resistivities[a] of Various Materials

Material	Resistivity ρ ($\Omega\cdot$m)	Material	Resistivity ρ ($\Omega\cdot$m)
Conductors		**Semiconductors**	
Aluminum	2.82×10^{-8}	Carbon	3.5×10^{-5}
Copper	1.72×10^{-8}	Germanium	0.5[b]
Gold	2.44×10^{-8}	Silicon	20–2300[b]
Iron	9.7×10^{-8}	**Insulators**	
Mercury	95.8×10^{-8}	Mica	10^{11}–10^{15}
Nichrome (alloy)	100×10^{-8}	Rubber (hard)	10^{13}–10^{16}
Silver	1.59×10^{-8}	Teflon	10^{16}
Tungsten	5.6×10^{-8}	Wood (maple)	3×10^{10}

[a] The values pertain to temperatures near 20 °C.

[b] Depending on purity.

where ρ is a proportionality constant known as the ***resistivity*** of the material. It can be seen from Equation 20.3 that the unit for resistivity is the ohm·meter ($\Omega\cdot$m), and Table 20.1 lists values for various materials. All the conductors in Table 20.1 are metals and have small resistivities. Insulators such as rubber have large resistivities. Materials like germanium and silicon have intermediate resistivity values and are, accordingly, called *semiconductors.*

Resistivity is an inherent property of a material, inherent in the same sense that the density of a material is an inherent property. Resistance, on the other hand, depends on both the resistivity and the geometry of the material. Thus, two wires can be made from copper, which has a resistivity of 1.72×10^{-8} $\Omega\cdot$m, but Equation 20.3 indicates that a short wire with a large cross-sectional area offers a smaller resistance to current than does a long, thin wire. Wires that carry large currents, such as main power cables, are thick rather than thin so that the resistance of the wires is kept as small as possible. Similarly, electric tools that are to be used far away from wall sockets require thicker extension cords, as Example 3 illustrates.

THE PHYSICS OF . . .

electrical extension cords.

Example 3 Longer Extension Cords

The instructions for an electric lawn mower suggest that a 20-gauge extension cord can be used for distances up to 35 m, but a thicker 16-gauge cord should be used for longer distances, to keep the resistance of the wire as small as possible. The cross-sectional area of 20-gauge wire is 5.2×10^{-7} m², while that of 16-gauge wire is 13×10^{-7} m². Determine the resistance of (a) 35 m of 20-gauge copper wire and (b) 75 m of 16-gauge copper wire.

REASONING AND SOLUTION We can use Equation 20.3, along with the resistivity of copper from Table 20.1, to find the resistance of the wires:

$$\textbf{[20-gauge wire]}\quad R=\frac{\rho L}{A}=\frac{(1.72\times10^{-8}\ \Omega\cdot\text{m})(35\ \text{m})}{5.2\times10^{-7}\ \text{m}^2}=\boxed{1.2\ \Omega}$$

$$\textbf{[16-gauge wire]}\quad R=\frac{\rho L}{A}=\frac{(1.72\times10^{-8}\ \Omega\cdot\text{m})(75\ \text{m})}{13\times10^{-7}\ \text{m}^2}=\boxed{0.99\ \Omega}$$

Even though it is more than twice as long, the thicker 16-gauge wire has less resistance than the thinner 20-gauge wire. It is necessary to keep the resistance as low as possible in order to minimize heating of the wire, thereby reducing the possibility of a fire, as Conceptual Example 7 in Section 20.5 emphasizes.

The resistivity of a material depends on temperature. In metals, the resistivity increases with increasing temperature, whereas in semiconductors the reverse is true. For many materials and limited temperature ranges, it is possible to express the temperature dependence of the resistivity as follows:

$$\rho = \rho_0[1 + \alpha(T - T_0)] \tag{20.4}$$

In this expression ρ and ρ_0 are the resistivities at temperatures T and T_0, respectively. The term α has the unit of reciprocal temperature and is the ***temperature coefficient of resistivity.*** Table 20.2 gives values of α for various materials. When the resistivity increases with increasing temperature, α is positive, as it is for metals. When the resistivity decreases with increasing temperature, α is negative, as it is for carbon, germanium, and silicon. Since resistance is given by $R = \rho L/A$, both sides of Equation 20.4 can be multiplied by L/A to show that resistance depends on temperature according to

$$R = R_0[1 + \alpha(T - T_0)] \tag{20.5}$$

Table 20.2 Temperature Coefficients[a] of Resistivity for Various Materials

Material	Temperature Coefficient of Resistivity α [$(C°)^{-1}$]
Aluminum	0.0039
Carbon	−0.0005
Copper	0.00393
Germanium	−0.05
Gold	0.0034
Iron	0.0050
Mercury	0.00089
Nichrome (alloy)	0.0004
Silicon	−0.07
Silver	0.0038
Tungsten	0.0045

[a] The values pertain to a temperature of 20 °C.

The next example illustrates the role of the resistivity and its temperature coefficient in determining the electrical resistance of a piece of material.

THE PHYSICS OF . . .

a heating element on an electric stove.

Example 4 The Heating Element of an Electric Stove

Figure 20.7 shows a heating element from an electric stove. The element contains a wire (length = 1.1 m, cross-sectional area = 3.1×10^{-6} m^2) through which electric charge flows. This wire is imbedded within an electrically insulating material that is contained within a metal casing. The wire becomes hot in response to the flowing charge and heats the casing. The material of the wire has a resistivity of $\rho_0 = 6.8 \times 10^{-5}\ \Omega\cdot$m at $T_0 = 320$ °C and a temperature coefficient of resistivity of $\alpha = 2.0 \times 10^{-3}\ (C°)^{-1}$. Determine the resistance of the heater wire at an operating temperature of 420 °C.

The cherry-red glow of a heating element on an electric stove indicates that the element becomes very hot as a result of the electric current.

REASONING We may use the relation $R = \rho L/A$ to find the resistance of the wire at 420 °C, provided the resistivity ρ can be determined at this temperature. Since the resistivity at 320 °C is given, Equation 20.4 can be employed to find the resistivity at 420 °C.

SOLUTION At the operating temperature of 420 °C, the material of the wire has a resistivity of

$$\rho = \rho_0[1 + \alpha(T - T_0)] \tag{20.4}$$

$$\rho = (6.8 \times 10^{-5}\ \Omega\cdot\text{m})[1 + (2.0 \times 10^{-3}\ (\text{C}°)^{-1})(420\ °\text{C} - 320\ °\text{C})]$$

$$= 8.2 \times 10^{-5}\ \Omega\cdot\text{m}$$

This value of the resistivity can be used along with the length and cross-sectional area to find the resistance of the heater wire:

$$R = \frac{\rho L}{A} = \frac{(8.2 \times 10^{-5}\ \Omega\cdot\text{m})(1.1\ \text{m})}{3.1 \times 10^{-6}\ \text{m}^2} = \boxed{29\ \Omega} \tag{20.3}$$

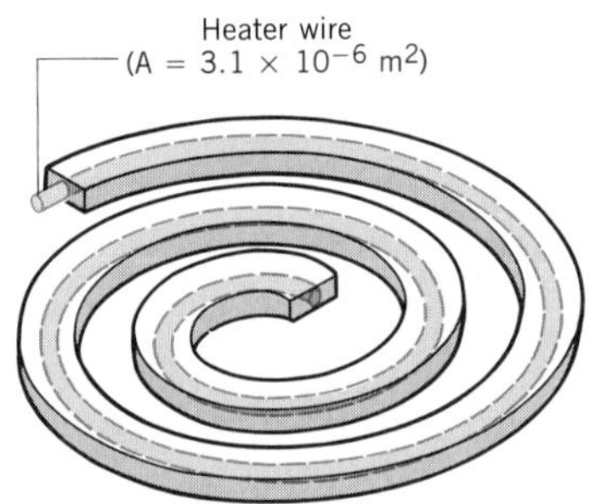

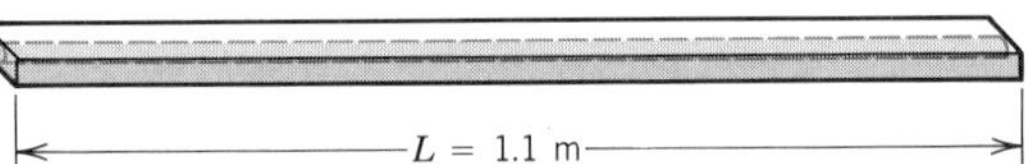

Figure 20.7 A heating element from an electric stove.

There is an important class of materials whose resistivity suddenly goes to zero below a certain temperature T_c called the *critical temperature,* commonly a few degrees above absolute zero. Below this temperature, such materials are called *superconductors.* The name derives from the fact that with zero resistivity, these materials offer no resistance to electric current and are, therefore, perfect conductors. One of the remarkable properties of zero resistivity is that once a current is established in a superconducting ring, it continues indefinitely without the need of an emf. Currents have persisted in superconductors for many years without measurable decay. In contrast, the current in a nonsuperconducting material drops to zero almost immediately after the emf is removed.

Many metals, such as aluminum ($T_c = 1.18$ K), tin ($T_c = 3.72$ K), lead ($T_c = 7.20$ K), and niobium ($T_c = 9.25$ K), become superconductors only at very low temperatures. Recently, ceramic materials have been developed that undergo the transition to the superconducting state at much higher temperatures. Superconducting transition temperatures as high as 125 K have been reported, and the ultimate hope is to discover materials that are superconducting at room temperature. Superconductors have many technological applications, including magnetic resonance imaging (Section 21.8), magnetic levitation of trains (Section 21.10), cheaper transmission of electrical power, powerful (yet small) electric motors, and faster computer chips.

20.4 ELECTRIC POWER

If the potential difference between the terminals of a battery is V, then a charge q emerging from the battery has an energy of qV, according to the definition of potential given in Equation 19.3. The product qV has units of energy (joules) since q is measured in coulombs and V is measured in joules/coulomb or volts. Since the amount of energy per second is the power, dividing qV by the time t gives the electric power qV/t provided to the circuit. The charge flowing per second q/t is the current I, so the electric power is IV, the product of current and voltage.

Electric Power

When there is a current I in a circuit as a result of a voltage V, the electric power P delivered to the circuit is

$$P = IV \tag{20.6}$$

SI Unit of Power: watt (W)

Power is measured in watts, and Equation 20.6 indicates, therefore, that the product of an ampere and a volt is equal to a watt.

Many electrical devices are essentially resistors that become hot when provided with sufficient electric power: toasters, irons, space heaters, heating elements on electric stoves, and incandescent light bulbs, to name a few. In such cases, it is possible to obtain two additional, but equivalent, expressions for the power. These two expressions follow directly upon substituting $V = IR$, or equivalently $I = V/R$, into the relation $P = IV$:

$$P = IV \tag{20.6a}$$

$$P = I(IR) = I^2R \tag{20.6b}$$

$$P = \left(\frac{V}{R}\right)V = \frac{V^2}{R} \tag{20.6c}$$

Example 5 deals with the electric power delivered to the bulb of a flashlight.

Example 5 The Power and Energy Used in a Flashlight

In the flashlight in Figure 20.6, the current is 0.40 A, and the voltage is 3.0 V. Find (a) the power delivered to the bulb and (b) the energy dissipated in the bulb in 5.5 minutes of operation.

REASONING The electrical power delivered to the bulb is the product of the current and voltage. Since power is energy per unit time, the energy delivered to the bulb is the product of the power and time.

SOLUTION

(a) The power is

$$P = IV = (0.40\text{ A})(3.0\text{ V}) = \boxed{1.2\text{ W}} \tag{20.6a}$$

The "wattage" rating of this bulb would therefore be 1.2 W.

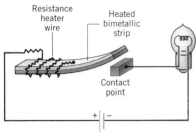

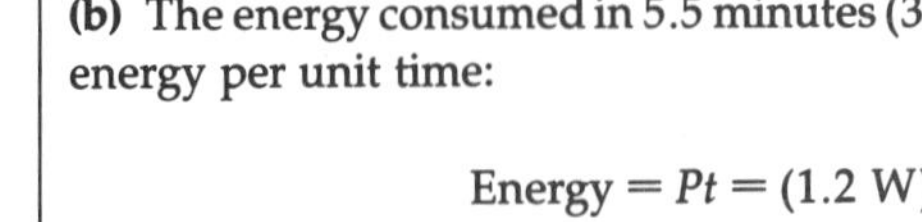

Figure 20.8 A bimetallic strip flasher.

(b) The energy consumed in 5.5 minutes (330 s) follows from the definition of power as energy per unit time:

$$\text{Energy} = Pt = (1.2\ \text{W})(330\ \text{s}) = \boxed{4.0 \times 10^2\ \text{J}}$$

Monthly electric bills specify the cost for the energy consumed during the month. Energy is the product of power and time, and electric companies compute your energy consumption by expressing power in kilowatts and time in hours. Therefore, a commonly used unit for energy is the *kilowatt-hour* (kWh). For instance, if you used an average power of 1440 watts (1.44 kW) for 30 days (720 h), your energy consumption would be (1.44 kW)(720 h) = 1040 kWh. At a cost of \$0.10 per kWh, your monthly bill would be \$104. One kilowatt-hour equals $(1000\ \text{J/s})(3600\ \text{s}) = 3.60 \times 10^6$ J of energy.

THE PHYSICS OF . . . a bimetallic strip flasher.

Figure 20.8 shows an interesting device that uses the heat generated when electric charge flows through a resistor. The device is called a bimetallic strip flasher. As Section 12.4 discusses, a bimetallic strip consists of two pieces of *dissimilar* metals fastened together. The metals expand by different amounts when heated, causing the strip to bend. The drawing shows a bimetallic strip with a resistance heater wire wrapped around it. While the strip is cool, its end touches the contact point. Charges from the battery pass directly through the strip and cause the light bulb to glow. However, as charges continue to flow, the resistance heater causes the bimetallic strip to become hot and bend away from the contact point, shutting off the current in the circuit. The light goes out and the heater shuts off. As the bimetallic strip cools, it bends back and touches the contact point again, turning the light back on. The on–off cycle repeats itself every second or so, and the result is a flashing light that is used as a warning device.

20.5 ALTERNATING CURRENT

Electric power companies send ac electricity over long distances by means of cables suspended from high towers.

Many electric circuits use batteries and involve direct current (dc). However, there are considerably more circuits that operate with alternating current (ac), in which the direction of charge flow reverses periodically. The common generators that create ac electricity depend on magnetic forces for their operation and are discussed in Chapter 22. In an ac circuit, these generators serve the same purpose as a battery serves in a dc circuit, that is, they give energy to the moving electric charges. This section deals with ac circuits that contain only resistance.

Since the electrical outlets in a house provide alternating current, we all use ac circuits routinely. Figure 20.9 shows the ac circuit that is formed when a toaster is plugged into a wall socket. The heating element of a toaster is essentially a thin wire of resistance R and becomes red hot when energy is dissipated in it. The circuit schematic in the picture introduces the symbol ⊝ that is used to represent the generator. In this case, the generator is located at the electric power company.

Figure 20.10 shows a graph that records the voltage produced between the terminals of the most common kind of ac generator at each moment of time. The voltage V fluctuates sinusoidally between positive and negative values as a function of time t;

$$V = V_0 \sin 2\pi f t \tag{20.7}$$

where V_0 is the maximum or peak value of the voltage, and f is the frequency at which the voltage oscillates. The angle $2\pi f t$ in Equation 20.7 is expressed in

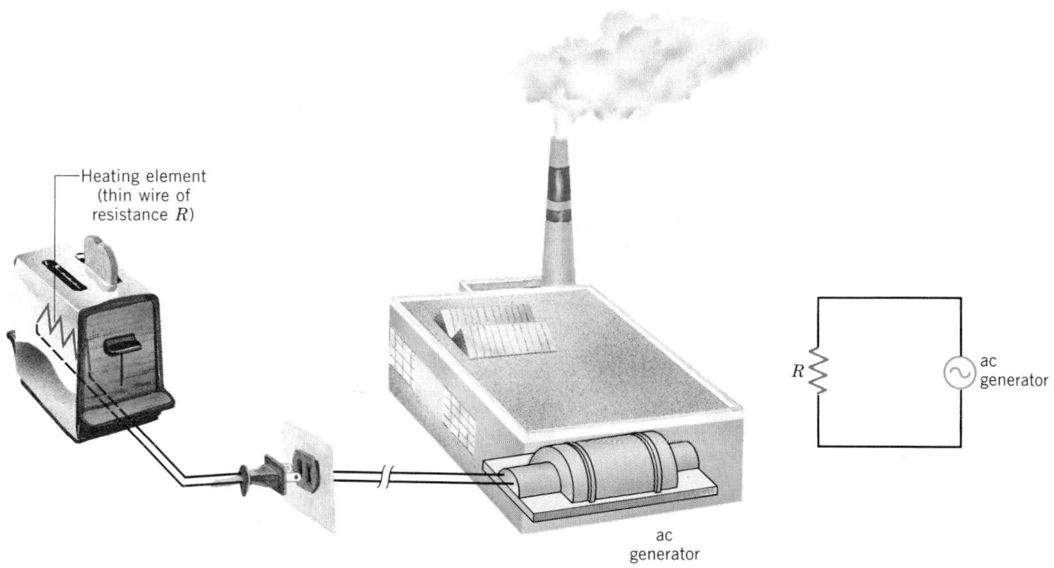

Figure 20.9 This circuit consists of a toaster (resistance $= R$) and an ac generator at the electric power company.

radians, so a calculator must be set to its radian mode before the sine of this angle can be evaluated. In the United States, the voltage at most home wall outlets has a *peak value* of approximately $V_0 = 170$ volts and oscillates with a frequency of $f = 60$ Hz. Thus, the period of each cycle is $\frac{1}{60}$ s, and the polarity of the generator terminals reverses twice during each cycle, as Figure 20.10 indicates.

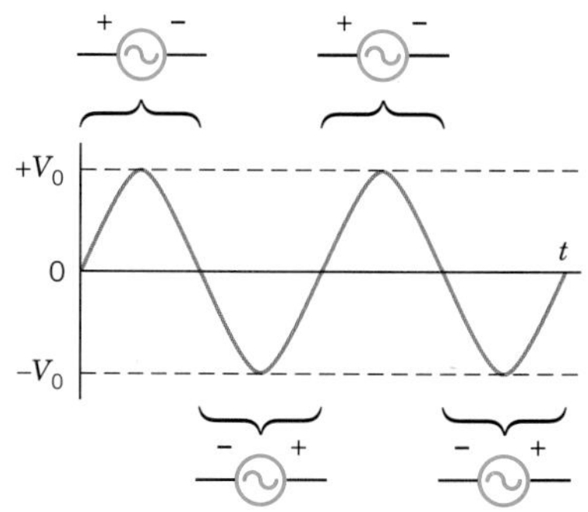

Figure 20.10 The voltage V produced between the terminals of an ac generator fluctuates sinusoidally in time, in the most common case. The circuits indicate the relative polarity of the generator terminals during the positive and negative parts of the sinusoidal graph.

The current in an ac circuit also oscillates. In circuits that contain only resistance, the current reverses direction each time the polarity of the generator terminals reverses. Thus, the current in a circuit like that in Figure 20.10 would have a frequency of 60 Hz and would change direction twice during each cycle. Substituting $V = V_0 \sin 2\pi ft$ into $V = IR$ shows that the current can be represented as

$$I = \left(\frac{V_0}{R}\right) \sin 2\pi ft = I_0 \sin 2\pi ft \tag{20.8}$$

The peak current is given by $I_0 = V_0/R$, so it can be determined if the peak voltage and the resistance are known.

The power delivered to an ac circuit by the generator is given by $P = IV$, just as it is in a dc circuit. However, since both I and V depend on time, the power fluctuates as time passes. Substituting Equations 20.7 and 20.8 for V and I into $P = IV$ gives

$$P = I_0 V_0 \sin^2 2\pi ft \tag{20.9}$$

This expression is plotted in Figure 20.11.

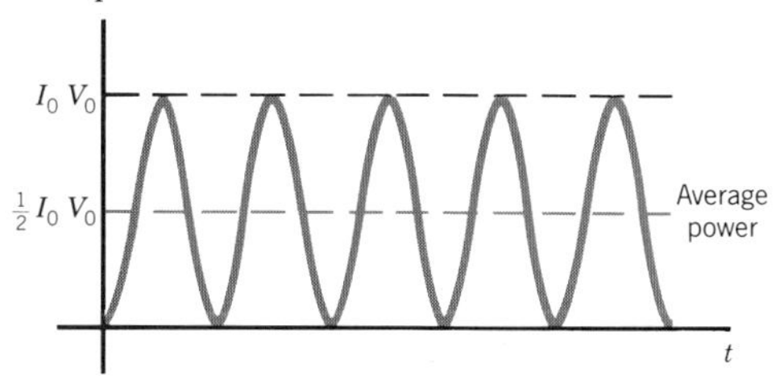

Figure 20.11 In an ac circuit, the power P delivered to a resistor oscillates between zero and a peak value of $I_0 V_0$. I_0 and V_0 are the peak current and the peak voltage, respectively.

Since the power fluctuates in an ac circuit, it is customary to consider the average power $\overline{P}$, which is one-half the peak power, as Figure 20.11 indicates:

$$\overline{P} = \tfrac{1}{2} I_0 V_0 \tag{20.10}$$

On the basis of this expression, a kind of average current and average voltage can be introduced that are very useful when discussing ac circuits. A slight rearrangement of Equation 20.10 reveals that

$$\overline{P} = \left(\frac{I_0}{\sqrt{2}}\right)\left(\frac{V_0}{\sqrt{2}}\right) = I_{rms} V_{rms} \tag{20.11}$$

I_{rms} and V_{rms} are called the ***root mean square*** (*rms*) current and voltage, respectively, and may be calculated from the peak values by dividing them by $\sqrt{2}$*:

$$I_{rms} = \frac{I_0}{\sqrt{2}} \tag{20.12}$$

$$V_{rms} = \frac{V_0}{\sqrt{2}} \tag{20.13}$$

Since the maximum ac voltage at a home wall socket in the United States is typically $V_0 = 170$ volts, the corresponding rms voltage is $V_{rms} = 170 \text{ volts}/\sqrt{2} = 120$ volts. Thus, when the instructions for an electric appliance specify 120 V, it is an rms voltage that is meant. Similarly, when we specify an ac voltage or current in this text, it is an rms value, unless indicated otherwise. Likewise, when we specify ac power, it is an average power, unless stated otherwise.

Except for dealing with average quantities, the relation $\overline{P} = I_{rms} V_{rms}$ has the same form as Equation 20.6a ($P = IV$). Moreover, Ohm's law can be written conveniently in terms of rms quantities:

$$V_{rms} = I_{rms} R \tag{20.14}$$

Substituting Equation 20.14 into $\overline{P} = I_{rms} V_{rms}$ shows that the average power can be expressed in the following ways:

$$\overline{P} = I_{rms} V_{rms} \tag{20.15a}$$

$$\overline{P} = I_{rms}^2 R \tag{20.15b}$$

$$\overline{P} = \frac{V_{rms}^2}{R} \tag{20.15c}$$

These expressions are completely analogous to $P = IV = I^2R = V^2/R$ for dc circuits. Example 6 deals with the average power in one familiar ac circuit.

* Equations 20.12 and 20.13 apply only for sinusoidal current and voltage.

† Other factors besides resistance can affect the current and voltage in ac circuits; they are discussed in Chapter 23.

Example 6 Electrical Power Sent to a Loudspeaker

A stereo receiver applies a peak ac voltage of 34 V to a speaker. The speaker is an 8.0-Ω speaker, for it behaves approximately† as an 8.0-Ω resistance. Figure 20.12 shows the circuit. Determine (a) the rms voltage, (b) the rms current, and (c) the average power for this circuit.

REASONING The rms voltage is, by definition, equal to the peak voltage divided by $\sqrt{2}$. We can find the rms current in the circuit from Ohm's law, as the rms voltage divided by the resistance of the speaker. The average power for the circuit is the product of the rms current and the rms voltage.

SOLUTION

(a) The peak value of the voltage is $V_0 = 34$ V, so the corresponding rms value is

$$V_{\text{rms}} = \frac{V_0}{\sqrt{2}} = \frac{34\text{ V}}{\sqrt{2}} = \boxed{24\text{ V}} \tag{20.13}$$

(b) The rms current can be obtained from Ohm's law:

$$I_{\text{rms}} = \frac{V_{\text{rms}}}{R} = \frac{24\text{ V}}{8.0\ \Omega} = \boxed{3.0\text{ A}} \tag{20.14}$$

(c) The average power is

$$\bar{P} = I_{\text{rms}} V_{\text{rms}} = (3.0\text{ A})(24\text{ V}) = \boxed{72\text{ W}} \tag{20.15a}$$

PROBLEM SOLVING INSIGHT

The rms values of the ac voltage and the ac current, V_{rms} and I_{rms}, respectively, are not the same as the peak values V_0 and I_0. The rms values are always smaller than the peak values by a factor of $\sqrt{2}$.

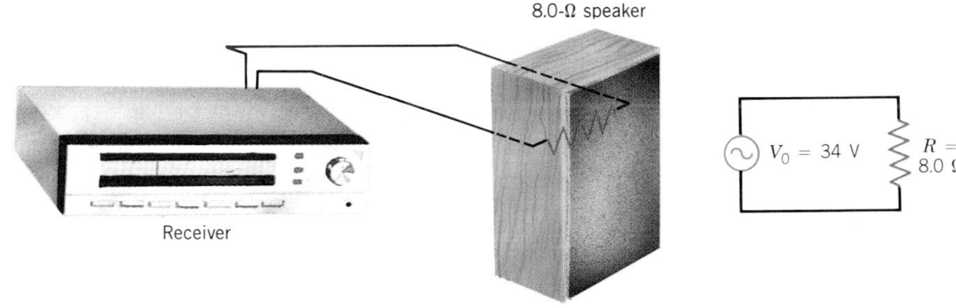

Figure 20.12 A receiver applies an ac voltage (peak value = 34 V) to an 8.0-Ω speaker.

The electrical power dissipated in a resistor causes it to heat up. Excessive power can lead to a potential fire hazard, as Conceptual Example 7 discusses.

Conceptual Example 7 Extension Cords and a Potential Fire Hazard

During the winter, many people use portable electric space heaters to keep warm. Sometimes, however, the heater must be located far from a 120-V wall receptacle, so an extension cord must be used (see Figure 20.13). However, manufacturers often warn against using an extension cord. Or, if one must be used, they recommend a certain wire gauge, or smaller. Why the warning, and why are smaller-gauge wires better than larger-gauge wires?

REASONING An electric space heater contains a heater element that is a piece of wire of resistance R, which is heated to a high temperature. The heating occurs because of the power $I_{\text{rms}}^2 R$ dissipated in the heater element. A typical heater uses a relatively large current I_{rms} of about 12 A. On its way to the heater, this current passes through the wires of any extension cord being used. Since these additional wires offer resistance to the current, the extension cord can also heat up, just as the heater element does. The extent of the heating depends on the resistance of the extension cord, with larger resistances producing greater amounts of heat. If the heating becomes excessive, the insulation around the wire can melt and possibly catch on fire. To keep the heating of the extension cord to a safe level, the resistance of the wire must be kept small. Recall

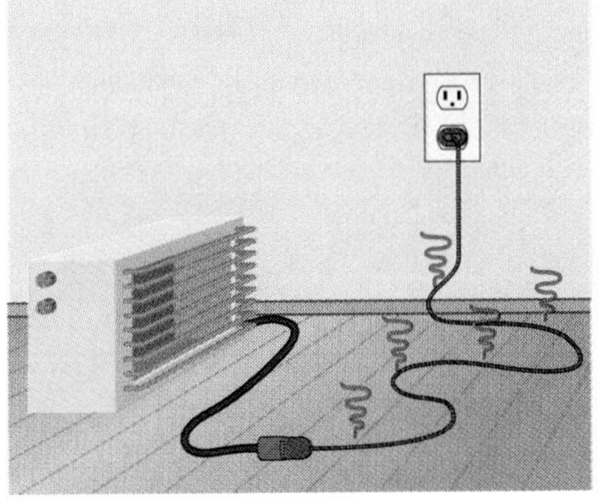

Figure 20.13 When an extension cord is used with a space heater, the cord must have a resistance that is sufficiently small to prevent overheating of the cord.

from Section 20.3 that the resistance of a wire depends inversely on its cross-sectional area. Thus, it is important to use an extension cord made from wire with a relatively large cross-sectional area. As mentioned in Example 3, the cross-sectional area of a wire is specified by its "gauge number." Smaller gauge numbers mean larger areas. Therefore, if an extension cord must be used with a space heater, the cord should have either the gauge number recommended by the manufacturer, or a smaller one.

Related Homework Material: *Problem 36*

20.6 SERIES WIRING

Thus far, we have dealt with circuits that include only a single device, such as a light bulb or a loudspeaker. There are, however, many circuits in which more than one device is connected to a voltage source. This section introduces one method by which such connections may be made, namely, series wiring. ***Series wiring means that the devices are connected in such a way that there is the same electric current through each device.*** Figure 20.14 shows a circuit in which two different devices, represented by resistors R_1 and R_2, are connected in series with a battery. Note that if the current in one resistor is interrupted, the current in the other is also interrupted. Such an interruption could occur, for example, if two light bulbs were connected in series, and the filament of one bulb broke. Because of the series wiring, the voltage V supplied by the battery is divided between the two resistors. The drawing indicates that the portion of the voltage across R_1 is V_1, while the portion across R_2 is V_2, so $V = V_1 + V_2$. Applying the definition of resistance to each resistor individually shows that

$$V = IR_1 + IR_2 = I(R_1 + R_2) = IR_S$$

where R_S is called the ***equivalent resistance*** of the series circuit. Thus, two resistors in series are equivalent to a single resistor whose resistance is $R_S = R_1 + R_2$, in the sense that there is the same current through R_S as there is through the series combination of R_1 and R_2. This line of reasoning can be extended to any number of resistors in series, with the result that

[Series resistors] $$R_S = R_1 + R_2 + R_3 + \cdots \qquad (20.16)$$

Example 8 illustrates the concept of the equivalent resistance in a series circuit.

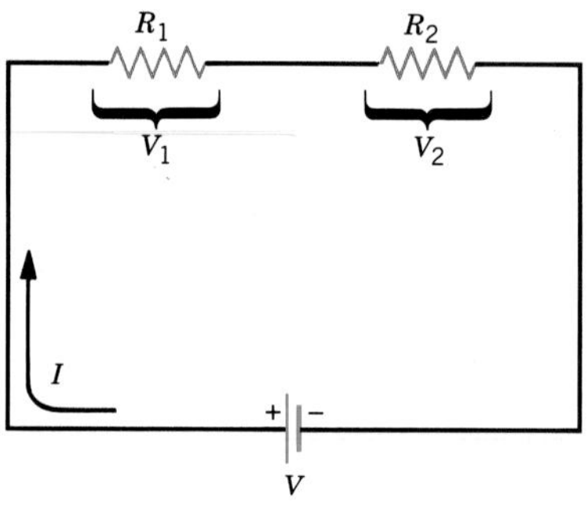

Figure 20.14 When two resistors are connected in series, the same current I is in both of them.

Example 8 Resistors in a Series Circuit

A 6.00-Ω resistor and a 3.00-Ω resistor are connected in series with a 12.0-V battery, as Figure 20.15 indicates. Assuming that the battery contributes no resistance to the circuit, find (a) the current, (b) the power dissipated in each resistor, and (c) the total power delivered to the resistors by the battery.

REASONING The current I can be determined from Ohm's law as $I = V/R_S$, where $R_S = R_1 + R_2$ is the equivalent resistance of the two resistors in series. The power delivered to each resistor is given by Equation 20.6b as $P = I^2R$, where R is the resistance of the resistor being considered and I is the current through it. The total power delivered by the battery is the power delivered to the 6.00-Ω resistor plus the power delivered to the 3.00-Ω resistor.

SOLUTION

(a) The equivalent resistance is

$$R_S = 6.00\ \Omega + 3.00\ \Omega = 9.00\ \Omega \tag{20.16}$$

Applying Ohm's law yields the current as

$$I = \frac{V}{R_S} = \frac{12.0\ \text{V}}{9.00\ \Omega} = \boxed{1.33\ \text{A}} \tag{20.2}$$

(b) Now that the current is known, the power dissipated in each resistor can be obtained from $P = I^2R$:

$$\begin{bmatrix}\textbf{6.00-}\boldsymbol{\Omega} \\ \textbf{resistor}\end{bmatrix} \qquad P = I^2R = (1.33\ \text{A})^2(6.00\ \Omega) = \boxed{10.6\ \text{W}} \tag{20.6b}$$

$$\begin{bmatrix}\textbf{3.00-}\boldsymbol{\Omega} \\ \textbf{resistor}\end{bmatrix} \qquad P = I^2R = (1.33\ \text{A})^2(3.00\ \Omega) = \boxed{5.31\ \text{W}}$$

(c) The total power delivered by the battery is the sum of the contributions in part (b): $P = 10.6\ \text{W} + 5.31\ \text{W} = 15.9\ \text{W}$. Alternatively, the total power can be obtained directly by using the equivalent resistance $R_S = 9.00\ \Omega$ and the current from part (a):

$$P = I^2R_S = (1.33\ \text{A})^2(9.00\ \Omega) = \boxed{15.9\ \text{W}}$$

In general, the total power delivered to any number of resistors in series is equal to the power delivered to the equivalent resistor.

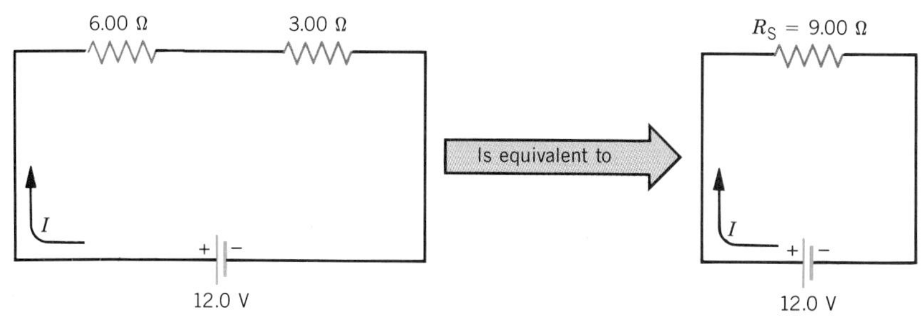

Figure 20.15 A 6.00-Ω and a 3.00-Ω resistor connected in series are equivalent to a single 9.00-Ω resistor.

20.7 PARALLEL WIRING

Parallel wiring is another method of connecting electrical devices together. ***Parallel wiring means that the devices are connected in such a way that the same voltage is applied across each device.*** Figure 20.16 shows two resistors connected in parallel between the terminals of a battery. Part *a* of the picture is drawn so as to emphasize that the entire voltage of the battery is applied across each resistor. Actually, parallel connections are rarely drawn in this manner; instead they are drawn as in part *b*, where the dots indicate the points where the wires for the two branches are joined together. Parts *a* and *b* are equivalent representations of the same circuit.

Parallel wiring is quite common. For example, when an electrical appliance is plugged into a wall socket, the appliance is connected in parallel with other appliances, as in Figure 20.17, where the entire voltage of 120 V is applied across the television, the stereo, and the light bulb. The presence of the unused socket or other devices that are turned off does not affect the operation of those devices that

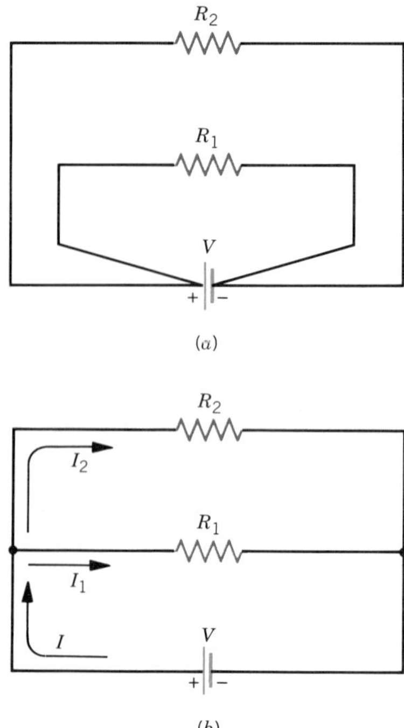

Figure 20.16 (*a*) When two resistors are connected in parallel, the same voltage V is applied across each resistor. (*b*) This circuit drawing is equivalent to that in part *a*. I_1 and I_2 are, respectively, the currents in R_1 and R_2.

are turned on. Moreover, if the current in one device is interrupted (perhaps by an opened switch or a broken wire), the current in the other devices is not interrupted. In contrast, if household appliances were connected in series, there would be no current through any appliance if the current at any point in the circuit were halted.

When two resistors R_1 and R_2 are connected as in Figure 20.16, each receives current from the battery as if the other were not present. Therefore, R_1 and R_2 together draw more current from the battery than does either resistor alone. According to the definition of resistance, a larger current arises from a smaller resistance. Thus, the two parallel resistors behave as a single equivalent resistance that is *smaller* than either R_1 or R_2. Figure 20.18 returns to the water flow analogy to provide additional insight into this important feature of parallel wiring. In part *a*, two sections of pipe that have the same length are connected in parallel with a pump. In part *b* these two sections have been replaced with a single pipe of the same length, whose cross-sectional area equals the combined cross-sectional areas of section 1 and section 2. The pump can push more water per second through the wider pipe in part *b* than it can through *either* of the narrower pipes in part *a*. In effect, the wider pipe (sections 1 and 2 acting together) offers less resistance to the flow than either of the narrower pipes offers individually.

As in a series circuit, it is possible to replace a parallel combination of resistors with an equivalent resistor that results in the same total current and power for a given voltage as the original combination. The equivalent resistance of two resistors connected in parallel can be determined in the following fashion. First, notice in Figure 20.16 that the total current I from the battery is the sum of I_1 and I_2, where I_1 is the current in resistor R_1 and I_2 is the current in resistor R_2: $I = I_1 + I_2$. Since the same voltage V is applied across each resistor, the definition of resistance indicates that $I_1 = V/R_1$ and $I_2 = V/R_2$. Therefore,

$$I = I_1 + I_2 = \frac{V}{R_1} + \frac{V}{R_2} = V\left(\frac{1}{R_1} + \frac{1}{R_2}\right) = V\left(\frac{1}{R_P}\right)$$

where R_P is the equivalent resistance. Hence, when two resistors are connected in parallel, they are equivalent to a single resistor whose resistance R_P is given by

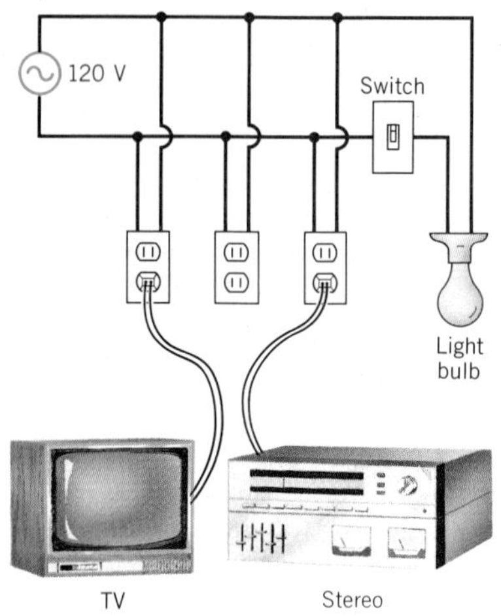

Figure 20.17 This drawing shows some of the parallel connections found in a typical home. Each wall socket provides 120 V to the appliance connected to it. In addition, 120 V is applied to the light bulb when the switch is turned on.

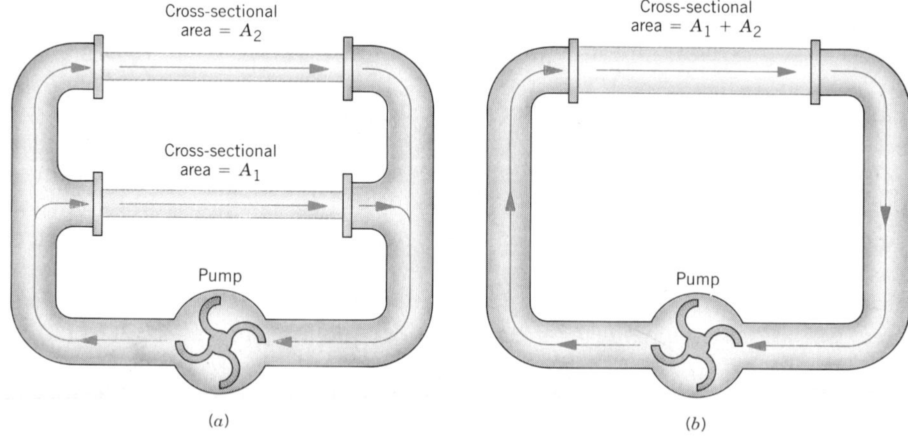

Figure 20.18 (*a*) Two equally long pipe sections, with cross-sectional areas A_1 and A_2, are connected in parallel to a water pump. (*b*) The two parallel pipe sections in part *a* are equivalent to a single pipe of the same length whose cross-sectional area is $A_1 + A_2$.

$1/R_P = 1/R_1 + 1/R_2$. For any number of resistors in parallel, a similar line of reasoning shows that

[Parallel resistors]
$$\frac{1}{R_P} = \frac{1}{R_1} + \frac{1}{R_2} + \frac{1}{R_3} + \cdots \qquad (20.17)$$

The next example deals with a parallel combination of resistors that occurs in a stereo system.

THE PHYSICS OF . . .

main and remote stereo speakers.

Example 9 Main and Remote Stereo Speakers

Most receivers allow the user to connect a pair of "remote" speakers (to play music in another room, for instance) in addition to the main speakers. Figure 20.19 shows that the remote speaker and the main speaker for the right stereo channel are connected to the receiver in parallel (for clarity, the speakers for the left channel are not shown). At the instant represented in the picture, the ac voltage across the speakers is 6.00 V. The main speaker has a resistance of 8.00 Ω, and the remote speaker has a resistance of 4.00 Ω.* Determine (a) the equivalent resistance of the two speakers, (b) the total current supplied by the receiver, (c) the current in each speaker, (d) the power dissipated in each speaker, and (e) the total power delivered by the receiver.

REASONING The total current supplied to the two speakers by the receiver can be obtained as $I_{rms} = V_{rms}/R_P$, where R_P is the equivalent resistance of the two speakers in parallel and is given by $1/R_P = 1/R_1 + 1/R_2$. The average power delivered to each speaker is the product of the current and the voltage. In the parallel connection the voltage applied to each speaker is the same, but the current in each speaker is different, a fact that we will have to take into account.

SOLUTION

(a) Since the speakers are in parallel, the equivalent resistance is

$$\frac{1}{R_P} = \frac{1}{8.00\ \Omega} + \frac{1}{4.00\ \Omega} = \frac{3}{8.00\ \Omega} \qquad (20.17)$$

$$R_P = \frac{8.00\ \Omega}{3} = \boxed{2.67\ \Omega}$$

This result is illustrated in part *b* of the drawing.

PROBLEM SOLVING INSIGHT

The equivalent resistance R_P of a number of resistors in parallel has a reciprocal given by $R_P^{-1} = R_1^{-1} + R_2^{-1} + R_3^{-1} + \cdots$, where R_1, R_2, and R_3 are the individual resistances. After adding together the reciprocals R_1^{-1}, R_2^{-1}, and R_3^{-1}, do not forget to take the reciprocal of the result to find R_P.

(b) The total current is

$$I_{rms} = \frac{V_{rms}}{R_P} = \frac{6.00\ \text{V}}{2.67\ \Omega} = \boxed{2.25\ \text{A}} \qquad (20.14)$$

(c) The current in each speaker can be determined from Ohm's law and the resistance of each speaker.

[8.00-Ω speaker]
$$I_{rms} = \frac{V_{rms}}{R} = \frac{6.00\ \text{V}}{8.00\ \Omega} = \boxed{0.750\ \text{A}}$$

[4.00-Ω speaker]
$$I_{rms} = \frac{V_{rms}}{R} = \frac{6.00\ \text{V}}{4.00\ \Omega} = \boxed{1.50\ \text{A}}$$

* In reality, frequency-dependent characteristics of the speaker (see Chapter 23) play a role in the operation of a loudspeaker. We assume here, however, that the frequency of the sound is low enough that the speakers behave as pure resistances.

The sum of these currents is equal to the total current from the receiver, as determined in part (b).

(d) The average power dissipated in each speaker can be calculated using $\overline{P} = I_{\text{rms}} V_{\text{rms}}$ with the individual currents obtained in part (c):

$$\begin{bmatrix}\textbf{8.00-}\boldsymbol{\Omega} \\ \textbf{speaker}\end{bmatrix} \qquad \overline{P} = (0.750 \text{ A})(6.00 \text{ V}) = \boxed{4.50 \text{ W}} \tag{20.15a}$$

$$\begin{bmatrix}\textbf{4.00-}\boldsymbol{\Omega} \\ \textbf{speaker}\end{bmatrix} \qquad \overline{P} = (1.50 \text{ A})(6.00 \text{ V}) = \boxed{9.00 \text{ W}} \tag{20.15a}$$

(e) The total power delivered by the receiver is the sum of the individual values found in part (d), $\overline{P} = 4.50 \text{ W} + 9.00 \text{ W} = 13.5 \text{ W}$. Alternatively, the total power can be obtained directly by using the equivalent resistance $R_P = 2.67\ \Omega$ and the current from part (b):

$$\overline{P} = I_{\text{rms}}^2 R_P = (2.25 \text{ A})^2(2.67\ \Omega) = \boxed{13.5 \text{ W}} \tag{20.15b}$$

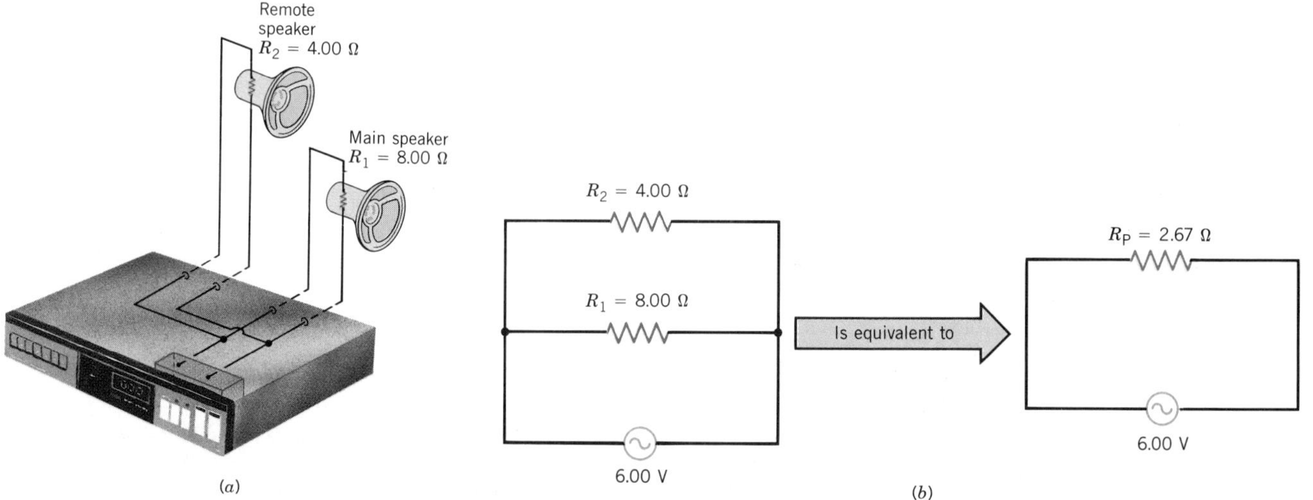

Figure 20.19 (*a*) The main and remote speakers in a stereo system are connected in parallel to the receiver. (*b*) The circuit schematic shows the situation when the ac voltage across the speakers is 6.00 V.

When a number of resistors are connected in parallel, the equivalent resistance is *less than* any of the individual resistances [see part (a) of Example 9]. In fact, it is the *smallest* resistance that has the largest impact in determining the equivalent resistance. If one resistance approaches zero, then according to Equation 20.17, the equivalent resistance also approaches zero. In such a case, the near-zero resistance is said to *short out* the other resistances, by providing a near-zero resistance path for the current to follow as a shortcut around the other resistances.

An interesting application of parallel wiring occurs in a 3-way light bulb, as discussed in Conceptual Example 10.

Conceptual Example 10 A 3-Way Light Bulb and Parallel Wiring

Three-way light bulbs are popular, because they can provide three levels of illumination (e.g., 50 W, 100 W, and 150 W) using a 120-V socket. The socket, however, must be equipped with a special 3-way switch that enables one to select the illumination level. This switch, however, does not select different voltages, for a 3-way bulb uses a single voltage of 120 V. Within the bulb are two separate filaments. When one of these burns out (i.e., vaporizes), the bulb can produce either its lowest or its intermediate illumination level, but not the highest level. How are the two filaments connected together, in series or in parallel? And how can two filaments be used to produce three different illumination levels?

THE PHYSICS OF . . .

a 3-way light bulb.

REASONING The filaments must be connected in parallel, as illustrated in Figure 20.20. If they were wired in series and one of them burned out, no current would be able to pass through the bulb and neither of the three illumination levels would be available, contrary to what is observed.

As to how two filaments can be used to produce three illumination levels, we note that the power dissipated in a resistance R is $\overline{P} = V_{\text{rms}}^2/R$, according to Equation 20.15c. With a single value of 120 V for the voltage V_{rms}, three different wattage ratings for the bulb can be obtained only if three different values for the resistance R can occur. In a 50 W/100 W/150 W three-way bulb, for example, one resistance R_{50} is provided by the 50-W filament, and the second resistance R_{100} comes from the 100-W filament. The third resistance R_{150} is the parallel combination of the other two and is given by $1/R_{150} = 1/R_{50} + 1/R_{100}$. Figure 20.20 illustrates how the 3-way switch operates to provide these three alternatives. In the first position of the switch, contact A is closed and contact B is open, so current passes only through the 50-W filament. In the second position, contact A is open and contact B is closed, so current passes only through the 100-W filament. In the third position, both contacts A and B are closed, and both filaments light up, giving 150-W of illumination.

Related Homework Material: Problem 52

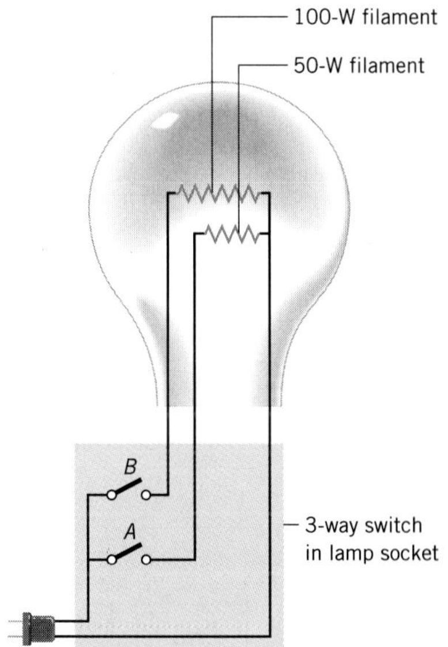

Figure 20.20 A 3-way light bulb uses two filaments connected in parallel. The filaments can be turned on one at a time, or both together.

20.8 CIRCUITS WIRED PARTIALLY IN SERIES AND PARTIALLY IN PARALLEL

Often an electric circuit is wired partially in series and partially in parallel. The key to determining the current, voltage, and power in such a case is to deal with the circuit in parts, with the resistances in each part being either in series or in parallel with each other. Example 11 shows how this analysis is carried out.

Example 11 A Four-Resistor Circuit

Figure 20.21 shows a circuit composed of a 24-V battery and four resistors, whose resistances are 110, 180, 220, and 250 Ω. Find (a) the total current supplied by the battery and (b) the voltage between points A and B in the circuit.

REASONING The total current supplied by the battery can be obtained from Ohm's law, $I = V/R$, where R is the equivalent resistance of the four resistors. The equivalent resistance can be calculated by dealing with the circuit in parts. The voltage V_{AB} between the points A and B is also given by Ohm's law, $V_{AB} = IR_{AB}$, where I is the current and R_{AB} is the equivalent resistance between the two points.

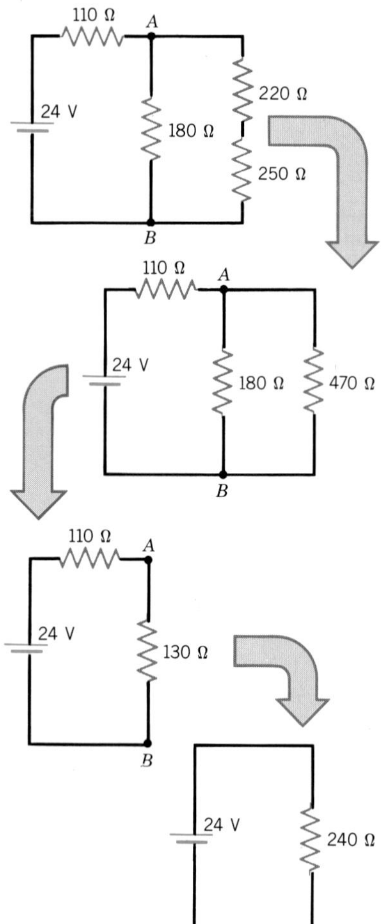

Figure 20.21 The four circuits shown in this picture are equivalent.

SOLUTION

(a) The 220-Ω resistor and the 250-Ω resistor are in series, so they are equivalent to a single resistor whose resistance is 220 Ω + 250 Ω = 470 Ω (see Figure 20.21). The 470-Ω resistor is in parallel with the 180-Ω resistor. Their equivalent resistance is given by Equation 20.17:

$$\frac{1}{R_{AB}} = \frac{1}{470\ \Omega} + \frac{1}{180\ \Omega} = 0.0077\ \Omega^{-1}$$

$$R_{AB} = \frac{1}{0.0077\ \Omega^{-1}} = 130\ \Omega$$

The circuit is now equivalent to a circuit containing a 110-Ω resistor in series with a 130-Ω resistor (see the drawing). This combination acts like a single resistor whose resistance is $R = 110\ \Omega + 130\ \Omega = 240\ \Omega$. The total current from the battery is, then,

$$I = \frac{V}{R} = \frac{24\ \text{V}}{240\ \Omega} = \boxed{0.10\ \text{A}}$$

(b) Ohm's law indicates that the voltage across the 130-Ω resistor between points A and B is

$$V_{AB} = IR_{AB} = (0.10\ \text{A})(130\ \Omega) = \boxed{13\ \text{V}}$$

Sometimes it is not necessary to carry out a complete quantitative analysis, such as that in Example 11. Conceptual Example 12 discusses how to determine qualitatively the relative amount of current drawn from a source when several appliances are connected to it.

Conceptual Example 12 More Appliances, More Current?

Figure 20.22 shows three different circuits. One circuit contains just a toaster (resistance = R_1), another contains the toaster and an iron (resistance = R_2), and the third contains the toaster, the iron, and a frying pan (resistance = R_3). Which circuit draws the least current from the 120-V source and which draws the most?

REASONING At first glance, one might think that the circuit with only the toaster would draw the least current and the circuit with the three appliances would draw the most. This would be true if all three appliances were wired in parallel with the source, as is typically done in home wiring. However, as we will now see, the circuit with the three appliances does not draw the most current.

As Figure 20.22*a* shows, a current I_1 passes through the toaster. When the iron is added in parallel with the toaster, as in part *b* of the drawing, an additional current I_2 passes through the iron, while the same current I_1 passes through the toaster. Thus, more current $(I_1 + I_2)$ is drawn from the source when the toaster and iron are both connected than when the toaster is alone. For a similar reason, circuit *c* draws more current than circuit *a*. Circuit *a*, then, draws the least current from the source.

Notice in part *c* of the drawing that the iron and frying pan are connected in series with each other. The equivalent resistance of these two appliances is $R_2 + R_3$, which is more than the resistance R_2 of the iron alone. According to Ohm's law, current is inversely proportional to resistance. Therefore, the current I_{23} through the two appliances must be *less* than the current I_2 through the iron when the frying pan is absent. Since I_{23} is less than I_2, we conclude that circuit *b* draws more current from the source than circuit *c*. Thus, of the three circuits, circuit *b* draws the most current from the source. We can also see that when a number of appliances are connected, the total current drawn from the source depends on how they are connected.

Related Homework Material: *Problem 60*

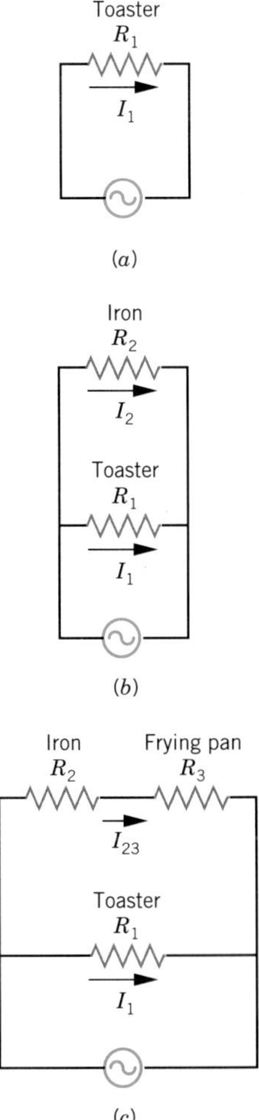

Figure 20.22 (*a*) Only the toaster is connected to the source. (*b*) A toaster and an iron are wired in parallel with each other. (*c*) A frying pan and a toaster are wired in series with each other, and the two are wired in parallel with the toaster.

20.9 INTERNAL RESISTANCE

So far, the circuits we have considered include batteries or generators that add only their emfs to a circuit. In reality, however, such devices also add some resistance. This resistance is called the ***internal resistance*** of the battery or generator, because it is located inside the device. In a battery, the internal resistance is the resistance encountered by the current due to the materials from which the battery is made. In a generator, the internal resistance is the resistance of wires and other components in the generator.

Figure 20.23 shows a schematic representation of the internal resistance r of a battery. The drawing emphasizes that when an external resistance R is connected to the battery, the resistance is connected *in series* with the internal resistance. The internal resistance of a functioning battery is typically small (several thousandths of an ohm for a new car battery). Nevertheless, the effect of the internal resistance may not be negligible. Example 13 illustrates that when current is drawn from a battery, the internal resistance causes the voltage between the terminals to drop below the maximum value specified by the battery's emf. The actual voltage between the terminals of a battery is known as the ***terminal voltage.***

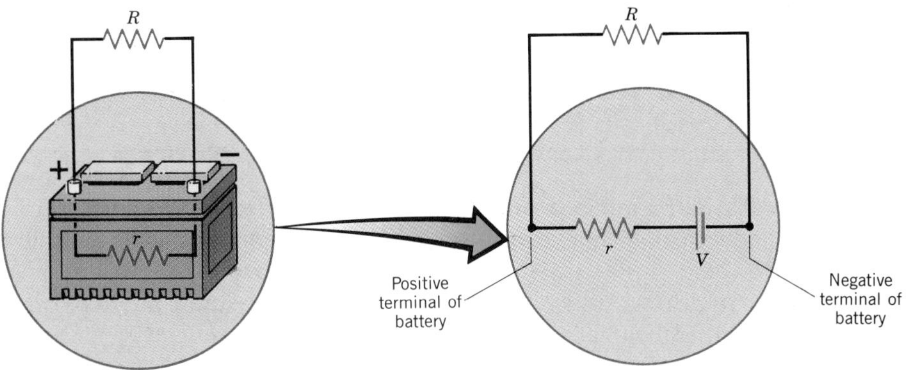

Figure 20.23 When an external resistance R is connected between the terminals of a battery, the resistance is connected in series with the internal resistance r of the battery.

Example 13 The Terminal Voltage of a Battery

Figure 20.24 shows a car battery whose emf is 12.0 V and whose internal resistance is 0.010 Ω. This resistance is relatively large because the battery is old and the terminals are corroded. What is the terminal voltage when the current I drawn from the battery is (a) 10.0 A and (b) 100.0 A?

REASONING The voltage between the terminals is not the entire 12.0-V emf, because part of the emf is needed to make the current go through the internal resistance.

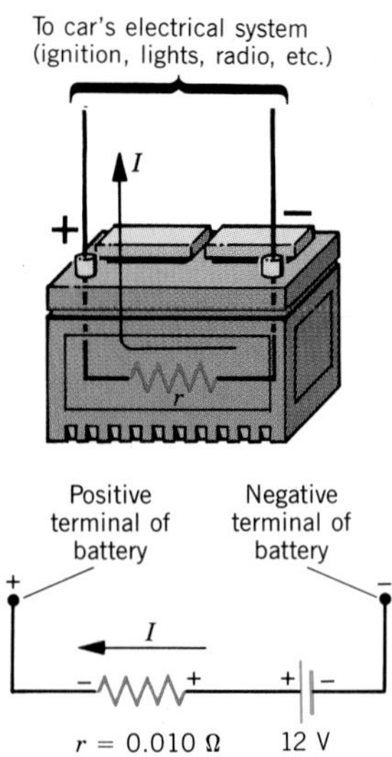

Figure 20.24 A car battery whose emf is 12 V and whose internal resistance is *r*.

The amount of voltage needed can be determined from Ohm's law as the current I through the battery times the internal resistance r. For larger currents, a larger amount of voltage is needed, leaving less of the emf between the terminals of the battery.

SOLUTION

(a) The amount of voltage needed to make a current of $I = 10.0$ A go through an internal resistance of $r = 0.010\ \Omega$ is $V = Ir = (10.0\ \text{A})(0.010\ \Omega) = 0.10$ V. To find the terminal voltage, remember that the direction of conventional current is always from a higher potential toward a lower potential. To emphasize this fact in the drawing, plus and minus signs have been included at the right and left ends, respectively, of the resistance r. The terminal voltage can be calculated by starting at the negative terminal of the battery and keeping track of how the voltage increases and decreases as we move toward the positive terminal. The voltage rises by 12.0 V due to the battery's emf. However, the voltage drops by 0.10 V because of the potential difference across the internal resistance. Therefore, the terminal voltage is $12.0\ \text{V} - 0.10\ \text{V} = \boxed{11.9\ \text{V}}$.

(b) When the current through the battery is 100.0 A, the amount of voltage needed to make the current go through the internal resistance is $V = (100.0\ \text{A})(0.010\ \Omega) = 1.0$ V. The terminal voltage of the battery now decreases to $12.0\ \text{V} - 1.0\ \text{V} = \boxed{11.0\ \text{V}}$.

Example 13 indicates that the terminal voltage of a battery is smaller when the current drawn from the battery is larger, an effect that any car owner can demonstrate. Turn the headlights on before starting your car, so the current through the battery is about 10 A, as in part (a) of Example 13. Then start the car. The starter motor draws a large amount of additional current from the battery, momentarily increasing the total current by an appreciable amount. Consequently, the terminal voltage of the battery decreases, causing the headlights to dim.

20.10 KIRCHHOFF'S RULES

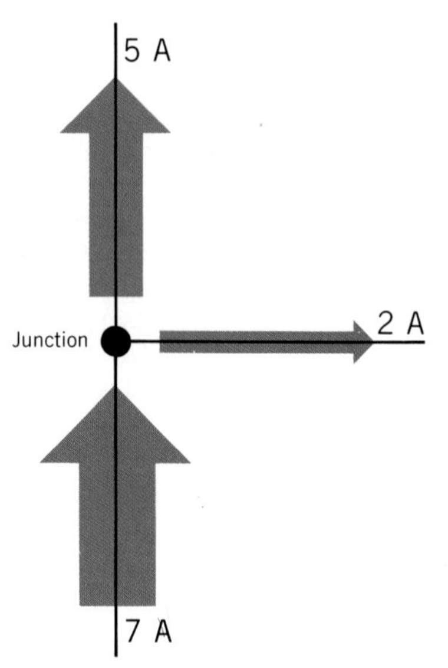

Figure 20.25 A junction is a point in a circuit where a number of wires are connected together. If 7 A of current is directed into the junction, then a total of 7 A of current must be directed out of the junction.

Electric circuits that contain a number of resistors can often be analyzed by combining individual groups of resistors in series and parallel, as Section 20.8 discusses. However, there are many circuits in which no two resistors are in series or in parallel. To deal with such circuits it is necessary to employ methods other than the series–parallel method. One alternative is to take advantage of Kirchhoff's rules, named after their developer Gustav Kirchhoff (1824–1887). There are two rules, one dealing with the currents in a circuit and the other dealing with the voltages.

Figure 20.25 illustrates the basic idea behind Kirchhoff's first rule, or ***junction rule,*** as it is called. The picture shows a junction where several wires are connected together. As Section 18.2 discusses, electric charge is conserved. Therefore, since there is no continual accumulation of charges at the junction itself, the total charge per second flowing into the junction must equal the total charge per second flowing out of the junction. In other words, the junction rule states that ***the total current directed into a junction must equal the total current directed out of the junction,*** or $7\ \text{A} = 5\ \text{A} + 2\ \text{A}$ for the specific case shown in the picture.

Figure 20.26 illustrates the basic idea behind Kirchhoff's second rule, or ***loop rule,*** as it is known. The drawing shows a circuit in which a 12-V battery is connected to a series combination of a 5-Ω and a 1-Ω resistor. The plus and minus signs associated with each resistor remind us that, outside a battery, conventional current is directed from a higher potential toward a lower potential. Thus, from left to right, there is a potential drop of 10 V across the first resistor and another drop of 2 V across the second resistor. Keeping in mind that potential is the electric potential energy per unit charge, let us follow a positive test charge clockwise* around the circuit. Starting at the negative terminal of the battery, we see that the test charge gains potential energy because of the 12-V rise in potential due to the battery. The test charge then loses potential energy because of the 10-V drop in potential across the first resistor and the 2-V drop across the second resistor, ultimately arriving back at the negative terminal. In traversing the closed circuit loop, the test charge is like a skier gaining gravitational potential energy in going up a hill on a chair lift and then losing it to friction in coming down and stopping. When the skier returns to the starting point, the gain equals the loss, so there is no net change in potential energy. Similarly, when the test charge arrives back at its starting point, there is no net change in electric potential energy, the gains matching the losses. This behavior of the test charge is an example of energy conservation, which the loop rule expresses in terms of the electric potential energy per unit charge. ***The loop rule states that for a closed circuit loop, the total of all the potential rises* (12 V) *is the same as the total of all the potential drops* (10 V + 2 V).**

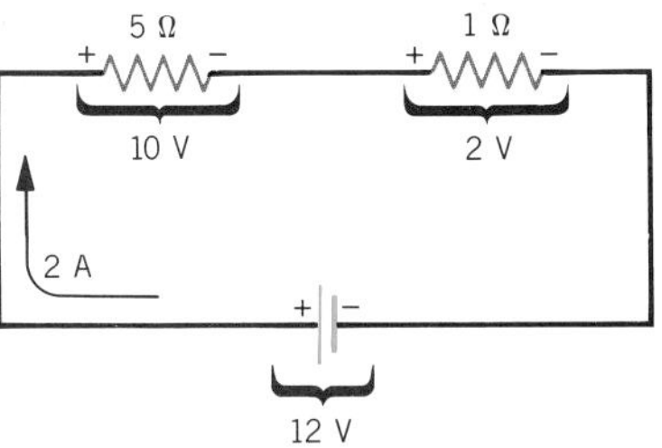

Figure 20.26 Following a positive test charge clockwise around the circuit, we see that the total voltage drop of 10 V + 2 V across the two resistors equals the voltage rise of 12 V provided by the battery. The plus and minus signs marking the ends of the resistors emphasize that, outside a battery, the conventional current of 2 A is directed from a higher potential (+) toward a lower potential (−).

Kirchhoff's rules can be applied to any circuit, even when the resistors are not in series or in parallel. The two rules are summarized below, and Examples 14 and 15 illustrate how to use them.

Kirchhoff's Rules

Junction Rule. The sum of the magnitudes of the currents directed into a junction equals the sum of the magnitudes of the currents directed out of the junction.

Loop Rule. Around any closed circuit loop, the sum of the potential drops equals the sum of the potential rises.

Example 14 Using Kirchhoff's Loop Rule

Figure 20.27 illustrates a circuit that contains two batteries and two resistors. Determine the current I in the circuit.

REASONING The first step is to draw the direction of the current, which we have chosen to be clockwise around the circuit. This direction is *arbitrary*, and if it is incorrect, we will find that the value for I is a negative number.

The second step in the solution is to mark the two resistors with plus and minus signs, which serve as an aid in identifying the potential drops and rises for Kirchhoff's loop rule. ***In marking the resistors, we remember that, outside a battery, conventional current is always directed from a higher potential* (+) *toward a lower potential* (−).** Thus, the plus and minus signs chosen for the two resistors *must* be those indicated in Figure 20.27, since the current is clockwise around the loop.

PROBLEM SOLVING INSIGHT

We may now apply Kirchhoff's loop rule to the circuit, starting at point A, proceeding clockwise around the loop, and identifying the potential drops and rises as we go. The potential across each resistor is given by Ohm's law as $V = IR$. The clockwise direction is arbitrary, and the same result is obtained with a counterclockwise path.

* The choice of the clockwise direction is arbitrary.

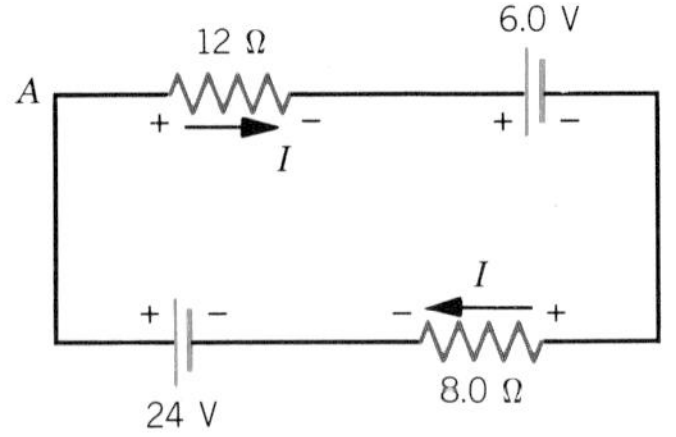

Figure 20.27 A single-loop circuit that contains two batteries and two resistors.

SOLUTION Starting at point *A*, and moving clockwise around the loop, there is:

1. A potential drop of $IR = I(12\ \Omega)$ across the 12-Ω resistor as we go from the + side to the − side of the resistor
2. A potential drop of 6.0 V as we proceed from the positive terminal to the negative terminal of the 6.0-V battery
3. A potential drop of $IR = I(8.0\ \Omega)$ across the 8.0-Ω resistor
4. A potential rise of 24 V across the 24-V battery, as we proceed from the negative to the positive terminal.

Setting the sum of the potential drops equal to the sum of the potential rises, as required by Kirchhoff's loop rule, gives

$$\underbrace{I(12\ \Omega) + 6.0\ \text{V} + I(8.0\ \Omega)}_{\text{Potential drops}} = \underbrace{24\ \text{V}}_{\text{Potential rises}}$$

Solving this equation for the current yields $\boxed{I = 0.90\ \text{A}}$. The current is a positive number, indicating that our initial choice for the direction (clockwise) of the current was correct.

THE PHYSICS OF . . .

an automobile electrical system.

Example 15 The Electrical System of a Car

In a car, the headlights are connected to the battery and would "run down" the battery if it were not for the alternator, which is run by the engine. Figure 20.28 indicates how the headlights and the alternator are connected to the battery. The picture also gives a circuit schematic in which the alternator is approximated as an additional 14.00-V battery for the sake of simplicity. The circuit includes an internal resistance of 0.0100 Ω for the battery and its leads, an internal resistance of 0.100 Ω for the alternator, and a resistance of 1.20 Ω for the headlights. Determine the current through the headlights, the battery, and the alternator.

REASONING We begin by labeling the currents in the headlights (I_H), the battery (I_B), and the alternator (I_A). The drawing shows the directions chosen for these currents. The directions are arbitrary, and if any of them is incorrect, then the analysis will show that the corresponding value for the current is negative.

Next, we mark the resistors with the plus and minus signs that serve as an aid in identifying the potential drops and rises for the loop rule, recalling that, outside a battery, conventional current is always directed from a higher potential (+) toward a lower potential (−). Thus, given the directions selected for I_H, I_B, and I_A, the plus and minus signs *must* be those indicated in Figure 20.28.

Kirchhoff's junction and loop rules can now be used.

SOLUTION The junction rule can be applied to junction *B* or junction *E*. In either case, the same equation results:

$$\begin{bmatrix}\textbf{Junction rule}\\ \textbf{applied}\\ \textbf{at } \boldsymbol{B}\end{bmatrix} \qquad \underbrace{I_A + I_B}_{\substack{\text{Into}\\ \text{junction}}} = \underbrace{I_H}_{\substack{\text{Out of}\\ \text{junction}}}$$

In applying the loop rule to the lower loop *BEFA*, we start at point *B*, move clockwise around the loop, and identify the potential drops and rises. The clockwise direction is arbitrary, and the same result is obtained with a counterclockwise path. There is a potential rise of $I_B(0.0100\ \Omega)$ across the 0.0100-Ω resistor. This rise is followed by a drop of 12.00 V as we proceed from the positive to the negative terminal of the battery. Continuing around the loop, we find a 14.00-V rise across the alternator, followed by a drop of $I_A(0.100\ \Omega)$ across the 0.100-Ω resistor. Setting the sum of the potential drops equal to the sum of the potential rises gives the following result:

$$\left[\begin{array}{l}\textbf{Loop rule}\\ \textbf{applied}\\ \textbf{clockwise}\\ \textbf{around } BEFA\end{array}\right] \quad \underbrace{I_A(0.100\ \Omega) + 12.00\ \text{V}}_{\text{Potential drops}} = \underbrace{I_B(0.0100\ \Omega) + 14.00\ \text{V}}_{\text{Potential rises}}$$

Since there are three unknown variables in this problem, I_A, I_B, and I_H, a third equation is needed for a solution. To obtain the third equation, we apply the loop rule to the upper loop *CDEB*, choosing a clockwise path for convenience. The result is

$$\left[\begin{array}{l}\textbf{Loop rule}\\ \textbf{applied}\\ \textbf{clockwise}\\ \textbf{around } CDEB\end{array}\right] \quad \underbrace{I_B(0.0100\ \Omega) + I_H(1.20\ \Omega)}_{\text{Potential drops}} = \underbrace{12.00\ \text{V}}_{\text{Potential rises}}$$

These three equations can be solved simultaneously to show that

$$\boxed{I_H = 10.1\ \text{A},\ I_B = -9.0\ \text{A},\ I_A = 19.1\ \text{A}}\ .$$

The negative answer for I_B indicates that the current through the battery is not directed from right to left, as drawn in Figure 20.28. Instead, the 9.0-A current is directed from left to right, opposite to the way current would be directed if the alternator were not connected. It is the left-to-right current created by the alternator that keeps the battery charged.

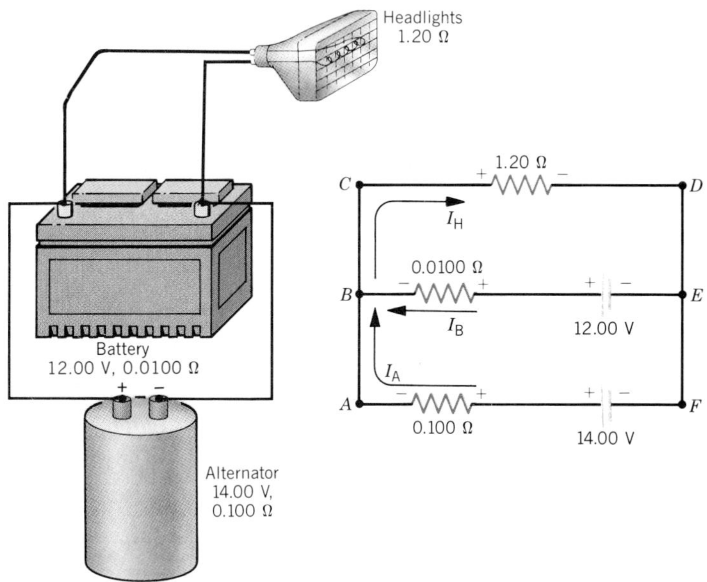

Figure 20.28 A circuit showing the headlight(s), battery, and alternator of a car.

20.11 THE MEASUREMENT OF CURRENT, VOLTAGE, AND RESISTANCE

THE GALVANOMETER

Current and voltage can be measured with devices known, respectively, as ammeters and voltmeters. There are two types of such devices: those that use digital electronics and those that do not. The essential feature of nondigital devices is the

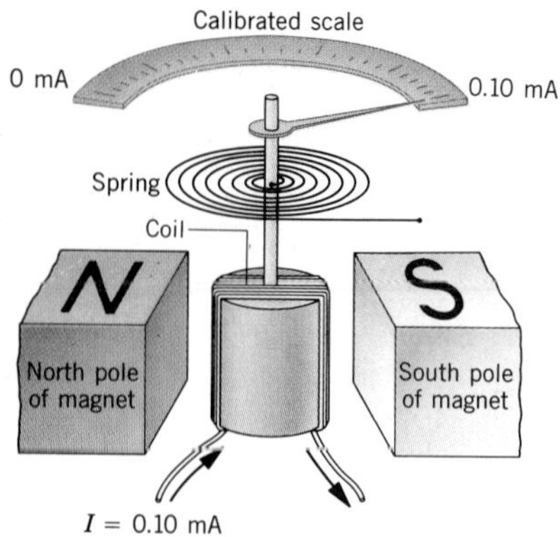

Figure 20.29 The essential parts of a dc galvanometer. The coil of wire and pointer rotate when there is a current in the coil.

dc *galvanometer.* As Figure 20.29 illustrates, a galvanometer consists of a magnet, a coil of wire, a spring, a pointer, and a calibrated scale. The coil is mounted in such a way that it can rotate, which causes the pointer to move in relation to the scale. As Section 21.7 will discuss, the coil rotates in response to the torque applied by the magnet when there is a current in the coil. The coil stops rotating when this torque is balanced by the torque of the spring.

A galvanometer has two important characteristics that must be considered when it is used as part of a measurement device. First, the amount of dc current that causes a full-scale deflection of the pointer indicates the sensitivity of the galvanometer. For instance, Figure 20.29 shows an instrument that deflects full scale when the current in the coil is 0.10 mA. The second important characteristic is the resistance R_C of the wire in the coil. Figure 20.30 shows how a galvanometer with a coil resistance of $R_C = 50\ \Omega$ is represented in a circuit diagram. Both the full-scale current and the coil resistance depend on the details of the galvanometer construction.

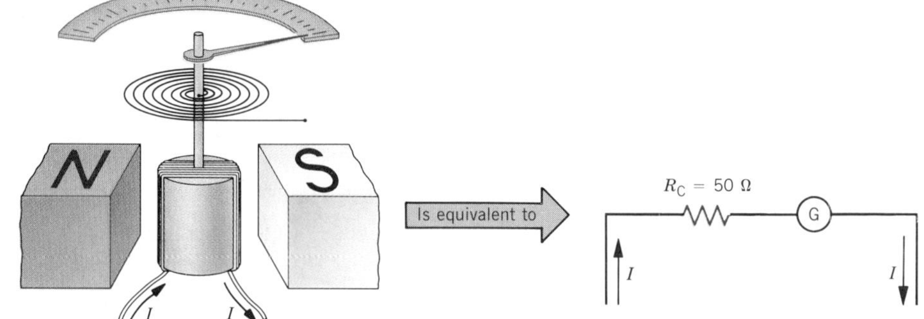

Figure 20.30 This picture shows how a galvanometer with a coil resistance of $R_C = 50\ \Omega$ is represented in a circuit diagram.

THE PHYSICS OF . . .

an ammeter.

THE AMMETER

An ***ammeter*** is a device for measuring current and must be inserted into the circuit so the current passes directly through the ammeter, as Figure 20.31 shows. An ammeter includes a galvanometer and one or more *shunt resistors.* The purpose of a shunt resistor is to allow excess current to bypass the galvanometer coil, so the ammeter can be used to measure a current that exceeds the full-scale amount of the galvanometer. A shunt resistor is connected in parallel with the galvanometer coil, and the next example illustrates how the value of the shunt resistance is selected.

Example 16 An Ammeter

A galvanometer has a full-scale current of 0.100 mA and a coil resistance of $R_C = 50.0\ \Omega$. As Figure 20.32 shows, this galvanometer is used with a shunt resistor to form an ammeter that will register a full-scale deflection for a current of 60.0 mA. Determine the shunt resistance R.

REASONING Since only 0.100 mA out of the available 60.0 mA is needed to cause a full-scale deflection of the galvanometer, the shunt resistor must allow the excess current of 59.9 mA to detour around the meter coil, as Figure 20.32 indicates. The value for the shunt resistance can be obtained by recognizing that the 50.0-Ω coil resistance and the shunt resistance are in parallel, both being connected between points A and B in the drawing. Thus, the voltage across each resistance is the same.

SOLUTION Expressing voltage as the product of current and resistance, we find that

$$\underbrace{(59.9 \times 10^{-3}\ \text{A})(R)}_{\substack{\text{Voltage across} \\ \text{shunt resistance}}} = \underbrace{(0.100 \times 10^{-3}\ \text{A})(50.0\ \Omega)}_{\substack{\text{Voltage across} \\ \text{coil resistance}}}$$

$$R = \frac{(0.100 \times 10^{-3}\ \text{A})(50.0\ \Omega)}{(59.9 \times 10^{-3}\ \text{A})} = \boxed{0.0835\ \Omega}$$

Typically, an ammeter includes a number of shunt resistors that provide several selectable current ranges.

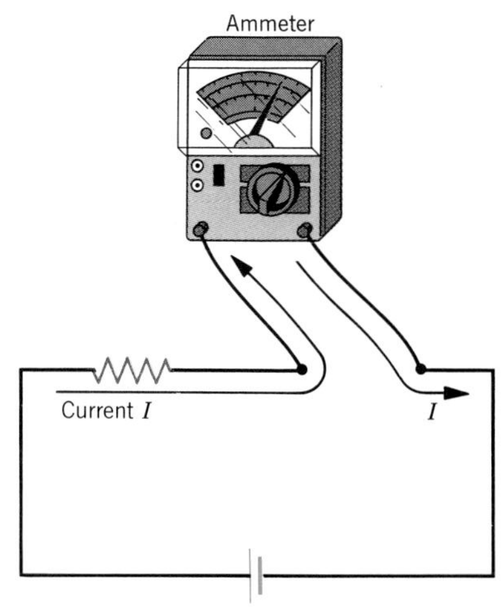

Figure 20.31 An ammeter measures the current I in a circuit. The ammeter is inserted into the circuit so that the current passes directly through the ammeter.

When an ammeter is inserted into a circuit, the equivalent resistance of the coil and the shunt resistor adds to the circuit resistance. Any increase in circuit resistance causes a reduction in current, and this is a problem, for an ammeter should only measure the current, not change it. Therefore, an *ideal* ammeter would have zero resistance. In practice, a good ammeter is designed with a sufficiently small equivalent resistance, so there is only a negligible reduction of the current in the circuit when the ammeter is inserted.

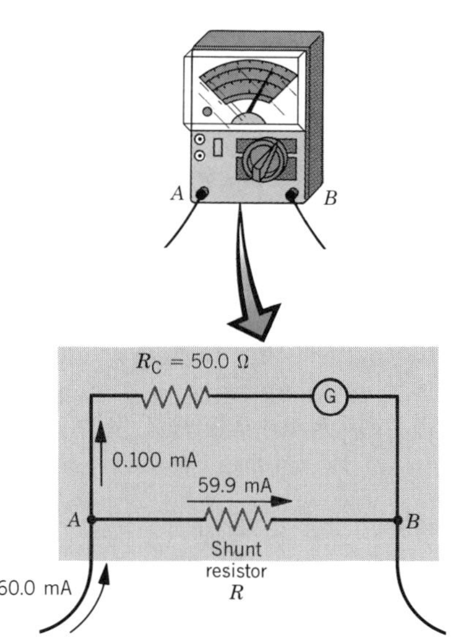

Figure 20.32 If a galvanometer with a full-scale deflection of 0.100 mA is to be used to measure a current of 60.0 mA, a shunt resistance R must be used, so the excess current of 59.9 mA can detour around the galvanometer coil.

THE VOLTMETER

THE PHYSICS OF . . .

a voltmeter.

A ***voltmeter*** is an instrument that measures the voltage between two points, A and B, in a circuit. Figure 20.33 shows that the voltmeter must be connected between the points and is *not* inserted into the circuit as an ammeter is. A voltmeter includes a galvanometer whose scale is calibrated in volts. Suppose, for instance, that the galvanometer in Figure 20.34 has a full-scale current of 0.1 mA and a coil resistance of 50 Ω. Under full-scale conditions, the voltage across the

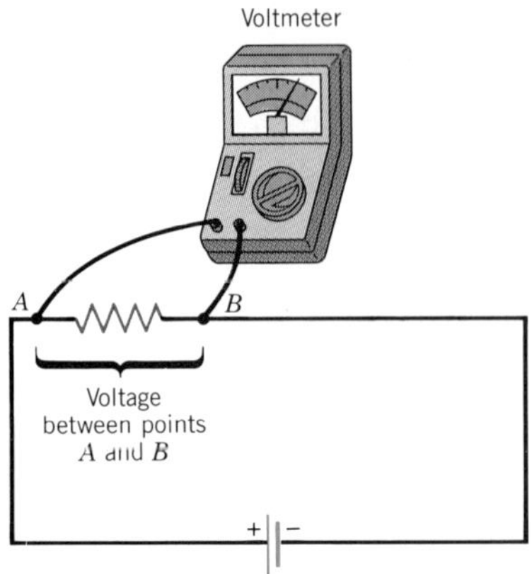

Figure 20.33 To measure the voltage between two points, *A* and *B*, in a circuit, a voltmeter is connected between the points.

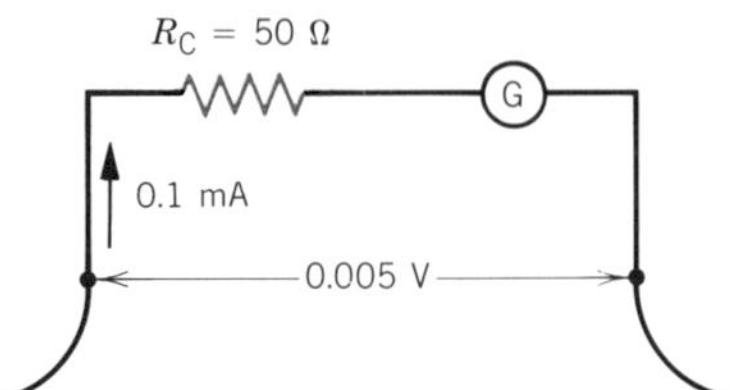

Figure 20.34 The galvanometer shown has a full-scale deflection of 0.1 mA and a coil resistance of 50 Ω.

coil would be $V = IR_C = (0.1 \times 10^{-3}\ \text{A})(50\ \Omega) = 0.005\ \text{V}$. Thus, this galvanometer could be used to register voltages in the range 0–0.005 V. A voltmeter, then, is a galvanometer used in this fashion, along with some provision for adjusting the range of voltages to be measured. Example 17 illustrates how the range of voltages can be extended by the simple expedient of connecting a resistor in series with the coil.

Example 17 A Voltmeter

A galvanometer has a full-scale current of 0.100 mA and a coil resistance of $R_C = 50\ \Omega$. Determine the resistance R that must be in series with the coil to produce a voltmeter that will register a full-scale voltage of 0.500 V.

REASONING Figure 20.35 shows the galvanometer in series with the resistance R. The resistance R is chosen so that when a voltage of 0.500 V is applied between points A and B in the drawing, the full-scale current of 0.100 mA will be in the galvanometer coil. The equivalent resistance of the series combination is $R + 50\ \Omega$. According to Ohm's law, the voltage across this combination is $V = I(R + 50\ \Omega)$, from which R can be determined.

SOLUTION Solving $V = I(R + 50\ \Omega)$ for R yields

$$R = \frac{V}{I} - 50\ \Omega = \frac{0.500\ \text{V}}{0.100 \times 10^{-3}\ \text{A}} - 50\ \Omega = \boxed{4950\ \Omega}$$

Usually a voltmeter includes a number of additional series resistors that provide a variety of selectable voltage ranges.

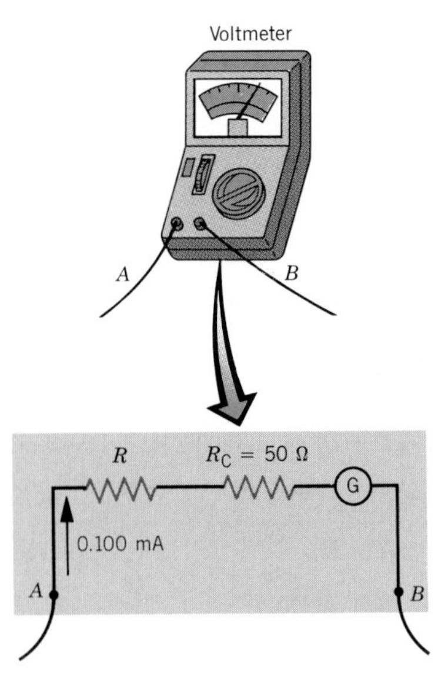

Figure 20.35 A voltmeter consists of a resistor R that is connected in series with the coil resistance R_C of the galvanometer.

Ideally, the voltage registered by a voltmeter should be the same as the voltage that exists when the voltmeter is not connected. However, a voltmeter takes some current from a circuit and, thus, alters the circuit voltage to some extent. An *ideal* voltmeter would have infinite resistance and draw away only an infinitesimal amount of current. In reality, a good voltmeter is designed with a resistance that is large enough so the unit does not appreciably alter the voltage in the circuit to which it is connected.

THE WHEATSTONE BRIDGE

THE PHYSICS OF . . .

a Wheatstone bridge.

One way to measure resistance is to use a circuit called a ***Wheatstone bridge,*** after Charles Wheatstone (1802–1875), the English physicist who established its usefulness. The circuit illustrates a method of measurement known as the ***null method.*** In addition to the unknown resistance R, a Wheatstone bridge includes three other resistances R_1, R_2, and R_v, as Figure 20.36 shows. The resistance R_v can be varied. A galvanometer records any current between points A and B. To measure the unknown resistance, the variable resistance R_v is adjusted until the galvanometer registers zero or null current, in which case the Wheatstone bridge is said to be "balanced." Example 18 shows how the unknown resistance can be obtained from the value of R_v and the ratio R_1/R_2 in a balanced Wheatstone bridge.

Example 18 Measuring a Resistance with a Wheatstone Bridge

When the Wheatstone bridge in Figure 20.36 is balanced, the variable resistance is $R_v = 173\ \Omega$, and the ratio R_1/R_2 is 0.100. Determine the value of the unknown resistance R.

REASONING The key to solving this problem lies in the information that the bridge is balanced. In other words, there is no current through the galvanometer. The direction of conventional current is from a higher toward a lower potential. But since there is no current through the galvanometer, points A and B in the circuit *must be at the same potential.* Furthermore, R_1 and R are connected at point C so the voltages across them must be the same. Similarly, since R_2 and R_v are connected at point D, the voltages across them must also be the same. We can use these observations to find the value R of the unknown resistance.

SOLUTION Since the voltage across R_1 is equal to the voltage across R and since voltage is the product of current and resistance (Ohm's law), it follows that

$$I_1 R_1 = IR$$

In a similar fashion, since the voltage across R_2 is the same as that across R_v, we find that

$$I_1 R_2 = IR_v$$

Dividing these two equations shows that $R_1/R_2 = R/R_v$ or

$$R = \frac{R_1}{R_2} R_v = (0.100)(173\ \Omega) = \boxed{17.3\ \Omega}$$

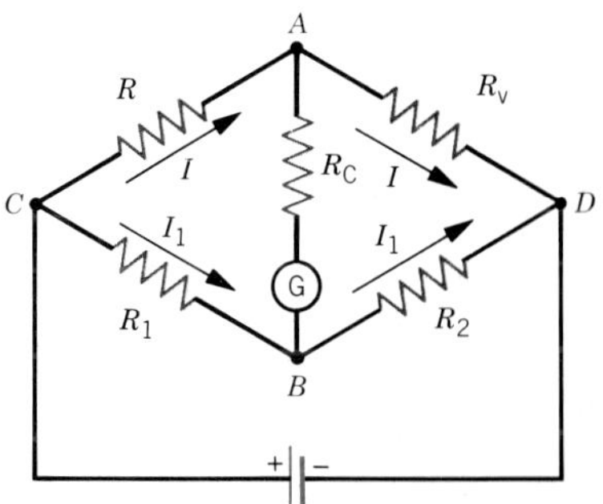

Figure 20.36 A Wheatstone bridge circuit. The resistor R_v is a variable resistor. The unknown resistance is R.

20.12 CAPACITORS IN SERIES AND PARALLEL

Capacitors, like resistors, can be connected in series and in parallel. Parallel capacitors are simpler to understand and will be considered first. Figure 20.37 shows two capacitors connected in parallel to a battery. Since the capacitors are in parallel, they have the same voltage V across their plates. However, the capacitors *contain different amounts of charge.* The charge stored by a capacitor is $q = CV$ (Equation 19.8), so $q_1 = C_1 V$ and $q_2 = C_2 V$.

As with resistors, it is always possible to replace a parallel combination of capacitors with an ***equivalent capacitor*** that stores the same charge and energy

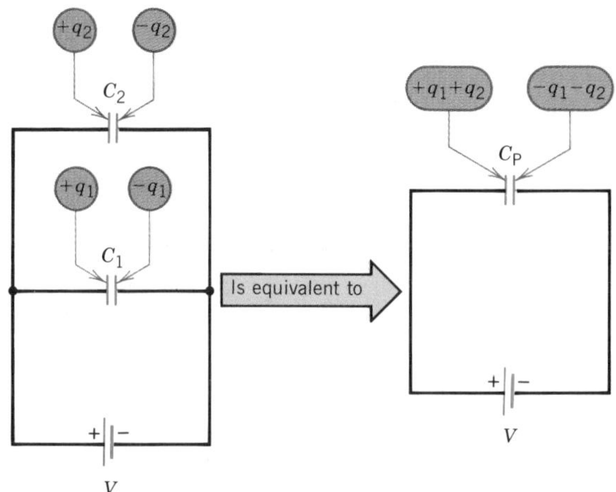

Figure 20.37 In a parallel combination of capacitances C_1 and C_2, the voltage V across each capacitor is the same, but the charges q_1 and q_2 on each capacitor are different.

for a given voltage as the combination does. To determine the equivalent capacitance C_P, note that the total charge q stored by the two capacitors is $q = q_1 + q_2$. Consequently,

$$q = q_1 + q_2 = C_1V + C_2V = (C_1 + C_2)V = C_PV$$

This result indicates that two capacitors in parallel can be replaced by an equivalent capacitor whose capacitance is $C_P = C_1 + C_2$. For any number of capacitors in parallel, the equivalent capacitance is

[Parallel capacitors]
$$C_P = C_1 + C_2 + C_3 + \cdots \quad (20.18)$$

Capacitances in parallel simply add together to give an equivalent capacitance. This behavior contrasts with that of resistors in parallel, which combine as reciprocals, according to Equation 20.17.

The equivalent capacitor not only stores the same amount of charge as the parallel combination of capacitors, but also stores the same amount of energy. For instance, the energy stored in a single capacitor is $\frac{1}{2}CV^2$ (Equation 19.11), so the total energy U stored by two capacitors in parallel is

$$U = \tfrac{1}{2}C_1V^2 + \tfrac{1}{2}C_2V^2 = \tfrac{1}{2}(C_1 + C_2)V^2 = \tfrac{1}{2}C_PV^2$$

which is equal to the energy stored in the equivalent capacitor C_P.

PROBLEM SOLVING INSIGHT

When capacitors are connected in series, the equivalent capacitance is different than when they are in parallel. As an example, Figure 20.38 shows two capacitors in series and reveals the following important fact: ***All capacitors in series, regardless of their capacitances, contain charges of the same magnitude, $+q$ and $-q$, on their plates.*** The battery places a charge of $+q$ on plate a of capacitor C_1, and this charge induces a charge of $+q$ to depart from the opposite plate a', leaving behind a charge $-q$. The $+q$ charge that leaves plate a' is deposited on plate b of capacitor C_2 (since these two plates are connected by a wire), where it induces a $+q$ charge to move away from the opposite plate b', leaving behind a charge of $-q$. Thus, all capacitors in series contain charges of the same magnitude on their plates.

PROBLEM SOLVING INSIGHT

Note the difference between charging capacitors in parallel and in series. ***In charging parallel capacitors, the battery moves a charge q that is the sum of the charges moved for each of the capacitors: $q = q_1 + q_2 + q_3 + \cdots$.*** In contrast,

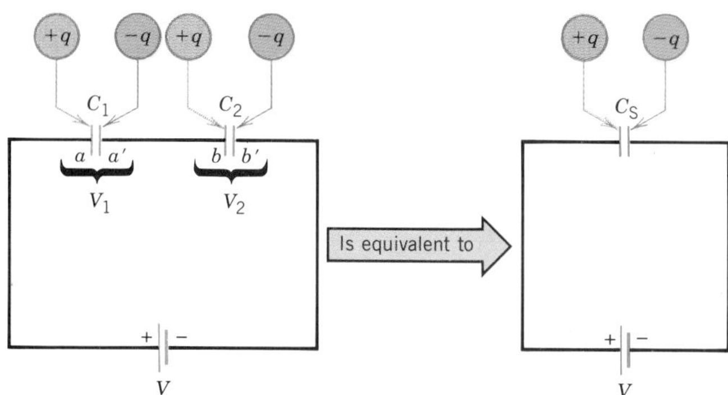

Figure 20.38 In a series combination of capacitances C_1 and C_2, the same amount of charge q is on the plates of each capacitor, but the voltages V_1 and V_2 across each capacitor are different.

in a series combination of n capacitors, the battery only moves a charge q, not nq, because the charge q passes by induction from one capacitor directly to the next one in line.

The equivalent capacitance C_S for the series connection in Figure 20.38 can be determined by observing that the battery voltage V is shared by the two capacitors, as it is in a series combination of resistors. The drawing indicates that the voltages across C_1 and C_2 are V_1 and V_2 and that $V = V_1 + V_2$. The voltage across each capacitor is related to the magnitude of the charge on each capacitor according to $V_1 = q/C_1$ and $V_2 = q/C_2$, the charge q being the same for each. Therefore,

$$V = \frac{q}{C_1} + \frac{q}{C_2} = q\left(\frac{1}{C_1} + \frac{1}{C_2}\right) = q\left(\frac{1}{C_S}\right)$$

where C_S is the equivalent capacitance. Thus, two capacitors in series can be replaced by a single capacitor whose capacitance C_S is given by $1/C_S = 1/C_1 + 1/C_2$. For any number of capacitors connected in series the result is

[Series capacitors]

$$\frac{1}{C_S} = \frac{1}{C_1} + \frac{1}{C_2} + \frac{1}{C_3} + \cdots \tag{20.19}$$

Equation 20.19 indicates that capacitances in series combine as reciprocals and do not simply add together as resistors in series do. It is left as an exercise (problem 96) to show that the equivalent series capacitance stores the same electrostatic energy as the sum of the energies of the individual capacitors.

It is possible to simplify circuits containing a number of capacitors in the same general fashion as that outlined for resistors in Example 11 and Figure 20.21. The capacitors in a parallel grouping can be combined according to Equation 20.18, and those in a series grouping can be combined according to Equation 20.19.

20.13 RC CIRCUITS

Many electric circuits contain both resistors and capacitors. Figure 20.39 illustrates an example of a resistor–capacitor or *RC* circuit. Part *a* of the drawing shows the circuit at a time t after the switch has been closed and the battery has begun to charge up the capacitor plates. The charge on the plates builds up gradually to its equilibrium value of $q_0 = CV_0$, where V_0 is the voltage of the battery. Assuming that the capacitor is uncharged at time $t = 0$ when the switch is closed, it can be shown that the magnitude q of the charge on the plates at time t is

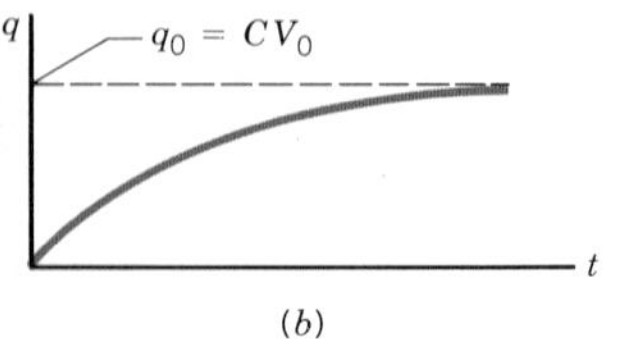

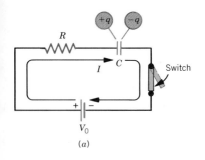

Figure 20.39 Charging a capacitor.

$$\left[\begin{array}{l}\textbf{Capacitor}\\\textbf{charging}\end{array}\right] \qquad q = q_0[1 - e^{-t/(RC)}] \tag{20.20}$$

where the exponential e has the value of 2.718. . . . Part b of the drawing shows a graph of this expression, which indicates that the charge is $q = 0$ when $t = 0$ and increases gradually toward the equilibrium value of $q_0 = CV_0$. The voltage V across the capacitor at any time can be obtained from Equation 20.20 by dividing the charges q and q_0 by the capacitance C, since $V = q/C$ and $V_0 = q_0/C$.

The term RC in the exponent in Equation 20.20 is called the ***time constant*** τ of the circuit:

$$\tau = RC \tag{20.21}$$

The time constant is measured in seconds; verification of the fact that an ohm times a farad is equivalent to a second is left as an exercise (see question 16). The time constant is the amount of time required for the capacitor to accumulate 63.2% of its equilibrium charge, as can be seen by substituting $t = \tau = RC$ in Equation 20.20; $q_0(1 - e^{-1}) = q_0(0.632)$. The charge approaches its equilibrium value rapidly when the time constant is small and slowly when the time constant is large.

Figure 20.40 shows a circuit at a time t after a switch is closed to allow a charged capacitor to begin discharging. There is no battery in this circuit, so the charge $+q$ on the left plate of the capacitor can flow counterclockwise through the resistor and neutralize the charge $-q$ on the right plate. Assuming that the capacitor has a charge q_0 at time $t = 0$ when the switch is closed, it can be shown that

$$\left[\begin{array}{l}\textbf{Capacitor}\\\textbf{discharging}\end{array}\right] \qquad q = q_0 e^{-t/(RC)} \tag{20.22}$$

where q is the amount of charge remaining on either plate at time t. The graph of this expression in part b of the drawing shows that the charge begins at q_0 when $t = 0$ and decreases gradually toward zero. Smaller values of the time constant RC lead to a more rapid discharge. Equation 20.22 indicates that when $t = \tau = RC$, the magnitude of the charge remaining on each plate is $q = q_0 e^{-1} = q_0(0.368)$. Therefore, the time constant is also the amount of time required for a charged capacitor to *lose* 63.2% of its charge.

The charging/discharging of a capacitor has many applications. For example, some automobiles come equipped with a feature that allows the windshield wipers to be used intermittently during a light drizzle. In this mode of operation

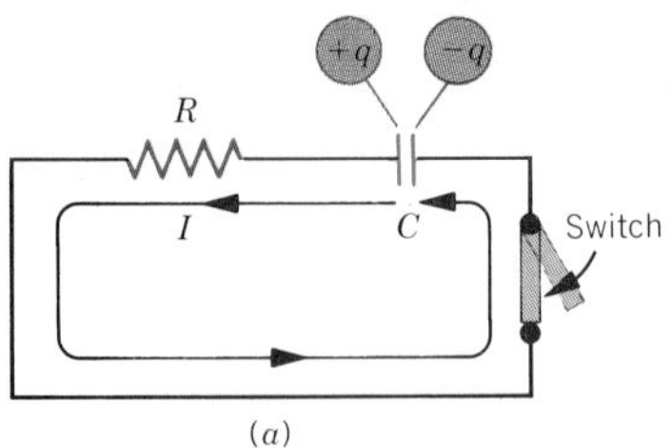

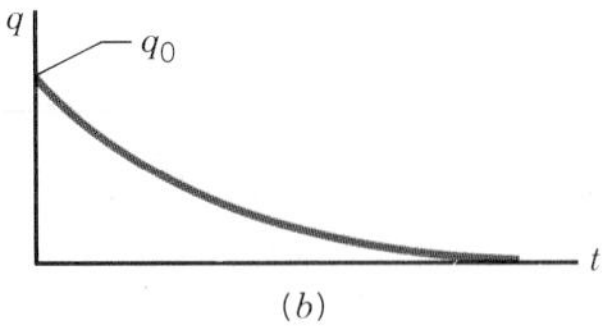

Figure 20.40 Discharging a capacitor.

the wipers remain off for a while and then turn on briefly. The timing of the on–off cycle is determined by the time constant of a resistor–capacitor combination.

*20.14 SAFETY AND THE PHYSIOLOGICAL EFFECTS OF CURRENT

THE PHYSICS OF . . .

electrical grounding.

Electric circuits, while very useful, can also be hazardous. To reduce the danger inherent in using circuits, proper ***electrical grounding*** is necessary. The next two figures help to illustrate what electrical grounding means and how it is achieved.

Figure 20.41*a* shows a clothes drier connected to a wall socket via an ordinary two-prong plug. The drier is operating normally; that is, the wires inside are insulated from the metal casing of the drier, so no charge flows through the casing itself. Notice that one terminal of the ac generator is customarily connected to ground (⏚) by the electric power company. Part *b* of the drawing shows the hazardous result that can occur if a wire comes loose and accidentally contacts the metal casing of the drier. A person touching the casing receives a shock, since electric charge flows from the generator, through the casing, the person's body, and the ground on the way back to the generator, as the picture illustrates.

Figure 20.42 shows the same clothes drier connected to a wall socket via a three-prong plug that provides safe electrical grounding. The third prong on the plug connects the metal casing of the drier directly to a copper rod driven into the ground or to a copper water pipe that is in the ground. This arrangement protects against electrical shock in the event that a broken wire touches the metal casing. The charge flows from the generator, through the casing, through the third prong of the plug, and into the ground, returning eventually to the generator. No charge flows into the person's body, since the copper rod connected to the third prong of the plug provides much less electrical resistance than does the body.

Serious and sometimes fatal injuries can result from electric shock. The severity of the injury depends on the magnitude of the current and the parts of the body through which the moving charges pass. The amount of current that the body senses as a mild tingling sensation is about 0.001 A. Currents on the order of 0.01–0.02 A can cause muscle spasms, in which a person "can't let go" of the

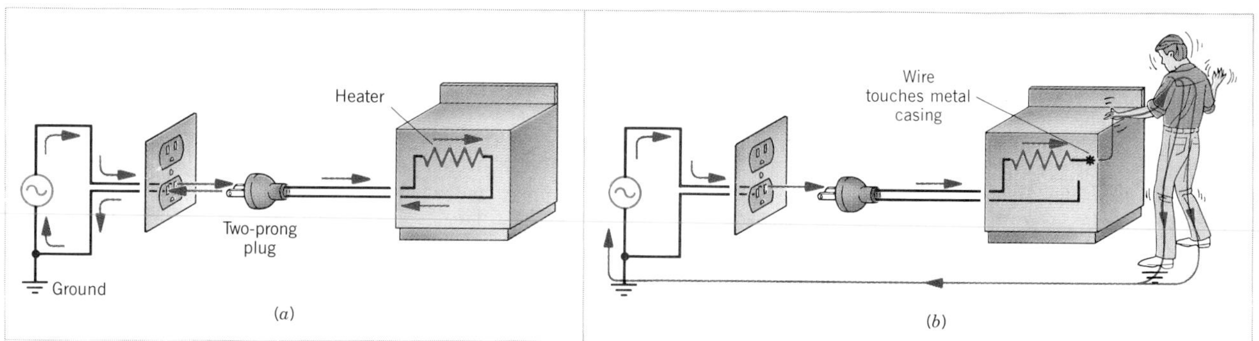

Figure 20.41 (*a*) A normally operating clothes drier that is connected to a wall socket via a two-prong plug. (*b*) The drier malfunctions, because an internal wire accidentally touches the metal casing. A person who touches the casing can receive an electrical shock.

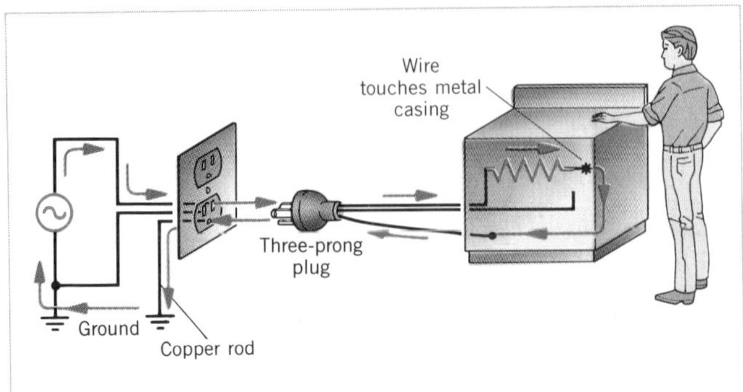

Figure 20.42 When the drier malfunctions, a person touching it receives no shock, since electric charge flows through the third prong and into the ground, rather than through the person's body.

object causing the shock. Currents of approximately 0.2 A are potentially fatal, because they can make the heart fibrillate, or beat in an uncontrolled manner. Substantially larger currents stop the heart completely. However, since the heart often begins beating normally again after the current ceases, the larger currents can be less dangerous than the smaller currents that cause fibrillation.

INTEGRATION OF CONCEPTS

KIRCHHOFF'S RULES AND CONSERVATION PRINCIPLES

Circuits are used to transfer electrical energy from sources to devices that use the energy. The currents and voltages in a circuit can be determined with the aid of Kirchhoff's two rules, the junction rule and the loop rule. Both of these rules are rooted in conservation principles.

The junction rule is based on the conservation of electric charge, which states that the net electric charge of an isolated system is conserved. When charge flows into a junction where two or more wires are connected and there is no buildup of charge at the junction, the net charge per second entering equals the net charge per second leaving the junction. Since the charge flowing per second is the current, the junction rule expresses the conservation of charge by stating that the magnitude of the total current directed into a junction equals the magnitude of the total current directed out of a junction. This idea is very similar to that expressed in Section 11.8 by the equation of continuity for a flowing fluid. The equation of continuity states that the mass per unit time that enters one end of a pipe must equal that leaving at the other end.

Kirchhoff's loop rule is based on the conservation of energy, which states that energy can neither be created nor destroyed, but can only be converted from one form to another. When electric charge moves in a circuit, it gains energy from sources such as batteries and generators. The electrical energy is converted into heat as the charge moves through resistors, the production of heat being accompanied by an equal loss in electrical energy. When the charge moves around a complete circuit, arriving back at its starting point, there is no overall change in energy, the gains having been matched by the losses, as energy is converted from one form to another. Kirchhoff's loop rule expresses the conservation of energy by stating that, around any closed loop, the sum of the potential drops (losses) equals the sum of the potential rises (gains).

SUMMARY

There must be at least one source or generator of electrical energy in an electric circuit. The **electromotive force (emf)** of a generator, such as a battery, is the maximum potential difference (in volts) that exists between the terminals of the generator.

The rate of flow of charge is called the **electric current.** If the rate is constant, the current I is given by $I = q/t$, where q is the magnitude of the charge crossing a surface in a time t, the surface being perpendicular to the motion of the charge. The SI unit for current is the coulomb per second (C/s), which is referred to as an ampere (A). When the charges flow only in one direction around a circuit, the current is called **direct current (dc).** When the direction of charge flow changes from moment to moment, the current is known as **alternating current (ac).**

The definition of **electrical resistance** is $R = V/I$, where V is the voltage applied across a piece of material and I is the current through the material. If the ratio V/I is constant for all values of V and I, the relation $R = V/I$ or $V = IR$ is referred to as **Ohm's law.** Resistance is measured in volts per ampere, a unit called an ohm (Ω).

The resistance of a piece of material of length L and cross-sectional area A is $R = \rho L/A$, where ρ is the **resistivity** of the material. The resistivity of a material depends on the temperature. For many materials and limited temperature ranges, the temperature dependence is given by $\rho = \rho_0[1 + \alpha(T - T_0)]$, where ρ and ρ_0 are the resistivities at temperatures T and T_0, respectively, and α is the **temperature coefficient of resistivity.**

In a circuit in which a current I results from a voltage V, the **electric power** delivered to the circuit is $P = IV$. Since $V = IR$, the power dissipated in a resistance R is also given by $P = I^2R$ or $P = V^2/R$.

The **alternating voltage between the terminals of an ac generator** can be represented by $V = V_0 \sin 2\pi ft$, where V_0 is the peak value of the voltage, t is the time, and f is the frequency at which the voltage oscillates. Correspondingly, in a circuit containing only resistance, the **ac current** is $I = I_0 \sin 2\pi ft$, where I_0 is the peak value of the current and is related to the peak voltage via $I_0 = V_0/R$. The **root mean square (rms)** voltage and current are related to the peak values according to $V_{rms} = V_0/\sqrt{2}$ and $I_{rms} = I_0/\sqrt{2}$. The power in an ac circuit is the product of the current and the voltage and oscillates in time. The **average power** is $\overline{P} = I_{rms} V_{rms}$. Since $V_{rms} = I_{rms} R$, the average power can also be written as $\overline{P} = I_{rms}^2 R$ or $\overline{P} = V_{rms}^2/R$.

When devices are connected **in series,** there is the same electric current through each device. The **equivalent resistance R_S of a series combination of resistances** (R_1, R_2, R_3, etc.) is $R_S = R_1 + R_2 + R_3 + \cdots$. The equivalent resistance dissipates the same total power as the series combination.

Connecting devices **in parallel** means that they are connected in such a way that the same voltage is applied across each device. In general, devices wired in parallel carry different currents. The **reciprocal of the equivalent resistance R_P of a parallel combination of resistances** is $1/R_P = 1/R_1 + 1/R_2 + 1/R_3 + \cdots$. The equivalent resistance dissipates the same total power as the parallel combination.

The **internal resistance** of a battery or generator is the electrical resistance encountered in the battery or generator by the current. The **terminal voltage** is the voltage between the terminals of a battery or generator and is equal to the emf only when there is no current through the device. When there is a current I, the internal resistance r causes the terminal voltage to be less than the emf by an amount Ir.

Kirchhoff's rules may be used to analyze the currents and potential differences in electric circuits. The **junction rule** states that the sum of the magnitudes of the currents directed into a junction equals the sum of the magnitudes of the currents directed out of the junction. The **loop rule** states that, around any closed circuit loop, the sum of the potential drops equals the sum of the potential rises.

A **galvanometer** is a device that responds to electric current and is used in nondigital ammeters and voltmeters. An **ammeter** is an instrument that measures current and must be inserted into a circuit so the current passes directly through the ammeter. A **voltmeter** is an instrument for measuring the voltage between two points in a circuit. For this measurement, a voltmeter must be connected between the two points and is not inserted into a circuit as an ammeter is. A **Wheatstone bridge** is a device that uses the **null method** for measuring resistance.

The **equivalent capacitance C_P for a parallel combination of capacitances** (C_1, C_2, C_3, etc.) is $C_P = C_1 + C_2 + C_3 + \cdots$. In general, each capacitor in a parallel combination carries a different amount of charge. The equivalent capacitor carries the same total charge and stores the same total energy as the parallel combination.

The **reciprocal of the equivalent capacitance C_S for a series combination of capacitances** is $1/C_S = 1/C_1 + 1/C_2 + 1/C_3 + \cdots$. Each capacitor in the combination carries the same amount of charge. The equivalent capacitor carries the same amount of charge as *any one* of the capacitors in the combination and stores the same total energy as the entire combination.

The **charging or discharging of a capacitor** in a dc series circuit (resistance R, capacitance C) does not occur instantaneously. Charge builds up gradually according to the relation $q = q_0[1 - e^{-t/(RC)}]$, where q is the charge on the capacitor at time t, q_0 is the equilibrium value of the charge, and RC is the **time constant** of the circuit. The discharging of a capacitor through a resistor is described by $q = q_0 e^{-t/(RC)}$, where q_0 is the charge on the capacitor at time $t = 0$.

QUESTIONS

1. The drawing shows a circuit in which a light bulb is connected to the household ac voltage via two switches S_1 and S_2. This is the kind of wiring, for example, that allows you to turn a carport light on and off from either inside the house or out in the carport. Explain which position A or B of S_2 turns the light on when S_1 is set to (a) position A and (b) position B.

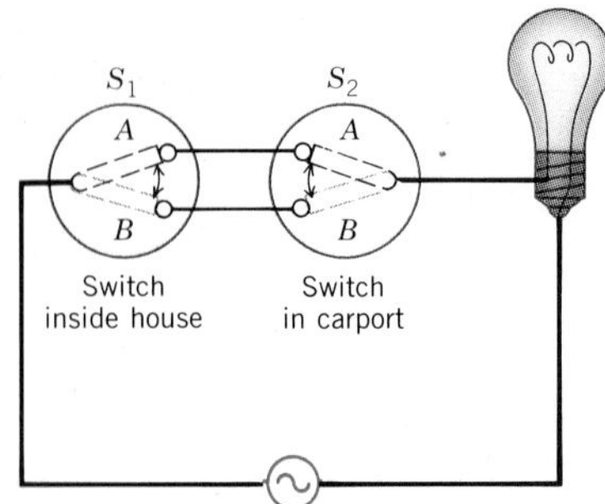

2. When an incandescent light bulb is turned on, the tungsten filament becomes white hot. What happens to the power delivered to the bulb as the filament heats up? Does the power increase, remain the same, or decrease? Justify your answer.

3. Two materials have different resistivities. Two wires of the same length are made, one from each of the materials. Is it possible for each wire to have the same resistance? Explain.

4. Does the resistance of a copper wire increase or decrease when both the length and the diameter of the wire are doubled? Justify your answer.

5. One electrical appliance operates with a voltage of 120 V, while another operates with 240 V. Based on this information alone, is it correct to say that the second appliance uses more power than the first? Give your reasoning.

6. Two light bulbs are designed for use at 120 V and are rated at 75 W and 150 W. Which light bulb has the greater filament resistance? Why?

7. Often, the instructions for an electrical appliance do not state how many watts of power the appliance uses. Instead, a statement such as "10 A, 120 V" is given. Explain why this statement is equivalent to telling you the power consumption.

8. A long extension cord is used to connect a light bulb to an electrical outlet. The current in the bulb is slightly less than that calculated using Ohm's law with the resistance of the light bulb and the voltage at the outlet. Why?

9. The power rating of a 1000-W heater specifies the power consumed when the heater is connected to an ac voltage of 120 V. Explain why the power consumed by two of these heaters connected in series with a voltage of 120 V is not 2000 W.

10. A number of light bulbs are to be connected to a single electrical outlet. Will the bulbs provide more brightness if they are connected in series or in parallel? Why?

11. A car has two headlights. The filament of one burns out, so charges can no longer flow out of the battery and through the headlight. However, the other headlight stays on. Draw a circuit diagram that shows how the headlights are connected to the battery.

12. One of the circuits in the drawing contains resistors that are neither in series nor in parallel. Which is it?

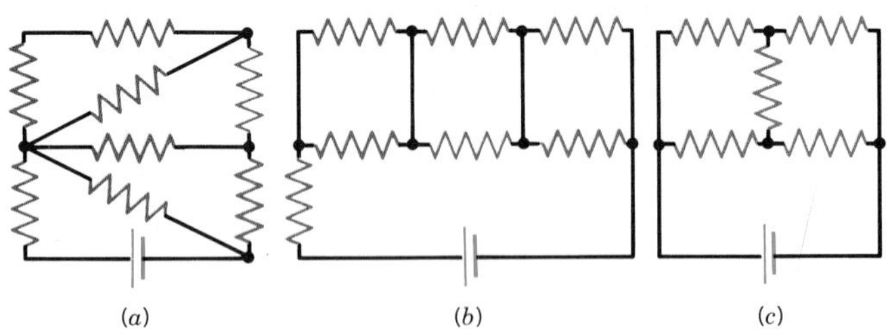

13. You have four identical resistors, each with a resistance of R. You are asked to connect these four together so that the equivalent resistance of the resulting combination is R. How many ways can you do it? Justify your answer(s).

14. Compare the resistance of an ideal ammeter with the resistance of an ideal voltmeter and explain why the resistances are so different.

15. Describe what would happen to the current in a circuit if a voltmeter, inadvertently mistaken for an ammeter, were inserted into the circuit.

16. The time constant of a series RC circuit is $\tau = RC$. Verify that an ohm times a farad is equivalent to a second.

PROBLEMS

Note: For problems that involve ac conditions, the current and voltage are rms values and the power is an average value, unless indicated otherwise.

Section 20.1 Electromotive Force and Current, Section 20.2 Ohm's Law

1. A lightning bolt delivers a charge of 35 C to the ground in a time of 1.0×10^{-3} s. What is the current?

2. Most of the wiring in a typical house can safely handle about 15 A of current. At this current level, how much charge flows through a wire in one hour?

3. A toaster has a resistance of 14 Ω and is plugged into a 120-V outlet. What is the current in the toaster?

4. The heating element of a clothes drier has a resistance of 11 Ω and is connected across a 240-V electrical outlet. What is the current in the heating element?

5. In the arctic, electric socks are useful. A pair of socks uses a 9.0-V battery pack for each sock. A current of 0.11 A is drawn from each battery pack by wire woven into the socks. Find the resistance of the wire in one sock.

6. A battery charger is connected to a dead battery and delivers a current of 8.0 A for 3.0 hours, keeping the voltage across the battery terminals at 12 V in the process. How much energy is delivered to the battery?

***7.** A car battery has a rating of 220 ampere·hours (A·h). This rating is one indication of the *total charge* that the battery can provide to a circuit before failing. (a) What is the total charge (in coulombs) that this battery can provide? (b) Determine the maximum current that the battery can provide for 38 minutes.

***8.** The filament of a light bulb has a resistance of 192 Ω, and the bulb is operating from a 120-V outlet. How much energy is delivered to the bulb in 45 minutes?

****9.** A beam of protons is moving toward a target in a particle accelerator. This beam constitutes a current whose value is 0.50 μA. (a) How many protons strike the target in 15 seconds? (b) Each proton has a kinetic energy of 4.9×10^{-12} J. Suppose the target is a 15-gram block of aluminum, and all the kinetic energy of the protons goes into heating it up. What is the change in temperature of the block at the end of 15 s?

Section 20.3 Resistance and Resistivity

10. High-voltage power lines are a familiar sight throughout the country. The aluminum wire used for some of these lines has a cross-sectional area of 4.9×10^{-4} m². What is the resistance of ten kilometers of this wire?

11. A coil of wire has a resistance of 38.0 Ω at 25 °C and 43.7 Ω at 55 °C. What is the temperature coefficient of resistivity?

12. Two wires have the same length and the same resistance. One is made from aluminum and the other from copper. Obtain the ratio of the cross-sectional area of the aluminum wire to that of the copper wire.

13. A copper wire has a cross-sectional area of 7.9×10^{-7} m². Find the resistance *per unit length* for this wire.

14. A cylindrical copper cable carries a current of 1200 A. There is a potential difference of 1.6×10^{-2} V between two points on the cable that are 0.24 m apart. What is the radius of the cable?

15. The filament in an incandescent light bulb is made from tungsten. The radius of the tungsten wire is 0.045 mm. If the bulb is to be plugged into a 120-V outlet and is to draw a current of 1.24 A, how long must the wire be?

16. A wire of unknown composition has a resistance of $R_0 = 35.0$ Ω when immersed in water at 20.0 °C. When the wire is placed in boiling water, its resistance rises to 47.6 Ω. What is the temperature of a hot summer day when the wire has a resistance of 37.8 Ω?

***17.** A wire has a resistance of 21.0 Ω. It is melted down, and from the same volume of metal a new wire is made that is three times longer than the original wire. What is the resistance of the new wire?

***18.** Liquid mercury in one rectangular container is poured into another rectangular container that has one-half the cross-sectional area. The resistance between the top and bottom surfaces of the mercury is measured in each case. Find the ratio of the resistance of the mercury in the new container to that of the mercury in the original container.

***19.** A toaster uses a Nichrome heating wire. When the toaster is turned on at 20 °C, the initial current is 1.50 A. A few seconds later, the toaster warms up and the current has a value of 1.30 A. The average temperature coefficient of resistivity for Nichrome wire is 4.5×10^{-4} $(C°)^{-1}$. What is the temperature of the heating wire?

*20. An iron wire has a resistance of 5.90 Ω at 20.0 °C, and a gold wire has a resistance of 6.70 Ω at the same temperature. At what temperature do the wires have the same resistance?

**21. A digital thermometer uses a thermistor as the temperature sensing element. A thermistor is a kind of semiconductor and has a large negative temperature coefficient of resistivity α. Suppose $\alpha = -0.060$ $(C°)^{-1}$ for the thermistor in a digital thermometer used to measure the temperature of a patient. The resistance of the thermistor decreases to 85% of its value at the normal body temperature of 37.0 °C. What is the patient's temperature?

Section 20.4 Electric Power

22. An automobile battery is being charged at a voltage of 12.0 V and a current of 19.0 A. How much power is being produced by the charger?

23. The heating element in a toaster has a resistance of 14 Ω. The toaster is plugged into a 120-V outlet. What is the power dissipated by the toaster?

24. A 240-V clothes drier draws 16 A of current for a period of 45 min. How much energy, in kilowatt-hours, does the drier consume?

25. An electric blanket is connected to a 120-V outlet and consumes 140 W of power. What is the current in the wire in the blanket?

26. An electric alarm clock uses a 5.0-W motor and runs all day, every day. If electricity costs $0.10 per kWh, determine the yearly cost of running the clock.

27. A commercial resistor can safely dissipate power only up to a certain rated value. Beyond this value, the resistor becomes excessively hot and often cracks apart. What is the largest voltage that can be applied across a 680-Ω resistor, when the resistor is rated at (a) 0.25 W and (b) 2.0 W?

*28. A stove is connected to a 240-V outlet and receives power P_{240}. When a restaurant owner uses the same stove with a 208-V outlet, the stove receives power P_{208}. Ignoring any change of resistance with temperature, find the ratio P_{240}/P_{208}.

*29. A power line between two towers has a length of 175 m. A current of 125 A exists in the aluminum wire of the line, and the potential difference between the ends of the wire is 0.300 V. Find the mass of the aluminum in this length of wire.

**30. An iron wire has a resistance of 12 Ω at 20.0 °C and a mass of 1.3×10^{-3} kg. A current of 0.10 A is sent through the wire for one minute and causes the wire to become hot. Assuming that all the electrical energy is dissipated in the wire and remains there, find the final temperature of the wire. *(Hint: Use the average resistance of the wire during the heating process, and see Table 12.3 for the specific heat capacity of iron.)*

Section 20.5 Alternating Current

31. The current in a circuit is ac and has a peak value of 2.50 A. Determine the rms current.

32. In the wire connecting an electric clock to a wall socket, how many times during a day does the current reverse its direction?

33. An ac voltage of $V = (65\text{ V}) \sin 2\pi(60\text{ Hz})t$ is applied across a 25-Ω resistor. What is the rms current in the resistor?

34. The heating element in an iron has a resistance of 16 Ω and is connected to a 120-V wall socket. (a) What is the average power consumed by the iron, and (b) the peak power?

35. A blow-drier and a vacuum cleaner each operate with an ac voltage of 120 V. The current rating of the blow-drier is 11 A, while that of the vacuum cleaner is 4.0 A. Determine the power consumed by (a) the blow-drier and (b) the vacuum cleaner. (c) Determine the ratio of the energy used by the blow-drier in 15 minutes to the energy used by the vacuum cleaner in one-half an hour.

36. Review Conceptual Example 7 as an aid in solving this problem. A portable electric heater uses 18 A of current. The manufacturer recommends that an extension cord attached to the heater produces no more than 2.0 W of heat per meter of length. What is the smallest radius of copper wire that can be used in the extension cord? *(Note: An extension cord contains two wires.)*

*37. The *recovery time* of a hot water heater is the time required to heat all the water in the unit to the desired temperature. Suppose that a 42-gal (1.00 gal $= 3.79 \times 10^{-3}$ m^3) unit starts with cold water at 11 °C and delivers hot water at 55 °C. The unit is electric and utilizes a resistance heater (120 V ac, 3.2 Ω) to heat the water. Assuming that no heat is lost to the environment, determine the recovery time (in hours) of the unit.

*38. On its highest setting, a heating element on an electric stove (see Figure 20.7) is connected to an ac voltage of 240 V. This element has a resistance of 29 Ω. (a) Find the power dissipated in the element. (b) Assuming that three-fourths of the heat produced by the element is used to heat a pot of water (the rest being wasted), find the time required to bring 1.9 kg of water (half a gallon) at 15 °C to a boil.

**39. To save on heating costs, the owner of a green house keeps 660 kg of water around in barrels. During a winter day, the water is heated by the sun to 10.0 °C. During the night the water freezes into ice at 0.0 °C in nine hours. What is the minimum ampere rating of an electric heating system (240 V) that would provide the same heating effect as the water does?

Section 20.6 Series Wiring

40. A 28-Ω resistor and a 62-Ω resistor are connected in series across a 48-V battery. What is the current in the circuit?

41. Three resistors, 25, 45, and 75 Ω, are connected in series, and a 0.51-A current passes through them. What is (a) the equivalent resistance and (b) the potential difference across the three resistors?

42. Three resistors, 9.0, 5.0, and 1.0 Ω, are connected in series across a 24-V battery. Find (a) the current in, (b) the voltage across, and (c) the power dissipated in each resistor.

43. A battery dissipates 2.50 W of power in each of two 47.0-Ω resistors connected in series. What is the voltage of the battery?

44. A 16.0-Ω resistor and an 8.0-Ω resistor are connected in series across a 12.0-V battery. What is the voltage across each resistor?

45. The current in a series circuit is 15.0 A. When an additional 8.00-Ω resistor is inserted in series, the current drops to 12.0 A. What is the resistance in the original circuit?

***46.** Two cylindrical rods, one copper and the other iron, are identical in lengths and cross-sectional areas. They are joined, end-to-end, to form one long rod. A 12-V battery is connected across the free ends of the copper–iron rod. What is the voltage between the ends of the copper rod?

***47.** A 47-Ω resistor can dissipate up to 0.25 W of power without burning up. What is the smallest number of such resistors that can be connected in series across a 9.0-V battery without any one of them burning up?

***48.** Three resistors are connected in series across a battery. The value of each resistance and its maximum power rating are as follows: 5.0 Ω and 20.0 W, 30.0 Ω and 10.0 W, and 15.0 Ω and 10.0 W. (a) What is the greatest voltage that the battery can have without one of the resistors burning up? (b) How much power does the battery deliver to the circuit in (a)?

Section 20.7 Parallel Wiring

49. A 16-Ω loudspeaker and an 8.0-Ω loudspeaker are connected in parallel across the terminals of an amplifier. Assuming the speakers behave as resistors, determine the equivalent resistance of the two speakers.

50. What resistance must be placed in parallel with a 155-Ω resistor to make the equivalent resistance 115 Ω?

51. Two resistors, 12.0 and 15.0 Ω, are connected in parallel. The current through the 15.0-Ω resistor is 4.0 A. (a) Determine the current in the other resistor. (b) What is the total power consumed by the two resistors?

52. For the 3-way bulb (50 W, 100 W, 150 W) discussed in Conceptual Example 10, find the resistance of each of the two filaments. Assume that the wattage ratings are not limited by significant figures and ignore any heating effects on the resistances.

53. How many 4.0-Ω resistors must be connected in parallel to create an equivalent resistance of one-sixteenth of an ohm?

54. A wire whose resistance is R is cut into three equally long pieces, which are then connected in parallel. In terms of R, what is the resistance of the parallel combination?

55. A 75-W lamp and a 15-W radio are connected in parallel to the same 120-V electrical outlet. What is the equivalent resistance of these two devices?

56. The drawing shows three resistors connected in parallel. At junction A, the current I divides equally. At junction B the current I_1 also divides equally. Find the values of R_1 and R_2.

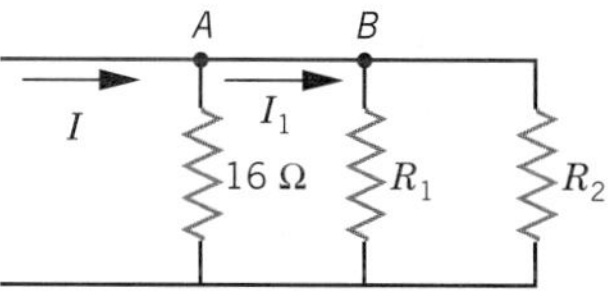

***57.** The total current delivered to a number of devices connected in parallel is the sum of the individual currents in each device. Circuit breakers are resettable automatic switches that protect against a dangerously large total current by "opening" to stop the current at a specified safe value. A 1650-W toaster, a 1090-W iron, and a 1250-W microwave oven are turned on in a kitchen. As the drawing shows, they are all connected through a 20-A circuit breaker to an ac voltage of 120 V. (a) Find the equivalent resistance of the three devices. (b) Obtain the total current delivered by the source and determine whether the breaker will "open" to prevent an accident.

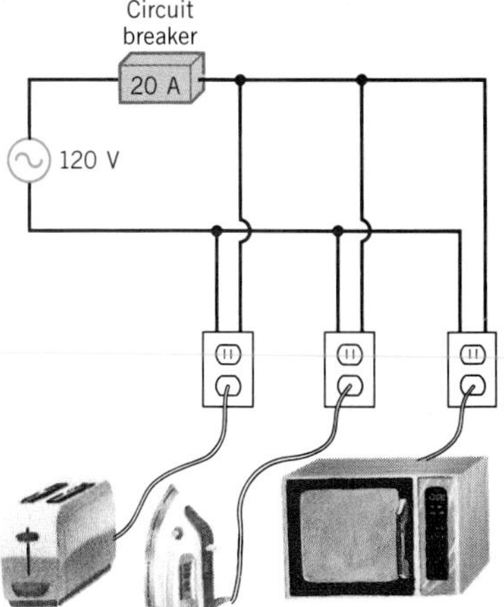

* **58.** A resistor (resistance = R) is connected first in parallel and then in series with a 2.00-Ω resistor. A battery delivers five times more current to the parallel combination than it does to the series combination. Determine the two possible values for R.

** **59.** The rear window defogger of a car consists of thirteen thin wires (resistivity = 88.0×10^{-8} Ω·m) embedded in the glass. The wires are connected in parallel to the 12.0-V battery, and each has a length of 1.30 m. The defogger can melt 2.10×10^{-2} kg of ice at 0 °C into water at 0 °C in two minutes. Assume that all the power dissipated in the wires is used immediately to melt the ice. Find the cross-sectional area of each wire.

Section 20.8 Circuits Wired Partially in Series and Partially in Parallel

60. Suppose in Conceptual Example 12 that the appliances have the following resistances: toaster ($R_1 = 14.0$ Ω), iron ($R_2 = 24.0$ Ω), frying pan ($R_3 = 16.0$ Ω). For each of the three circuits shown in Figure 20.22, find the total current provided by the source. Check to be sure that your results are consistent with the conclusions reached in the example.

61. For the combination of resistors shown in the drawing, determine the equivalent resistance between points A and B.

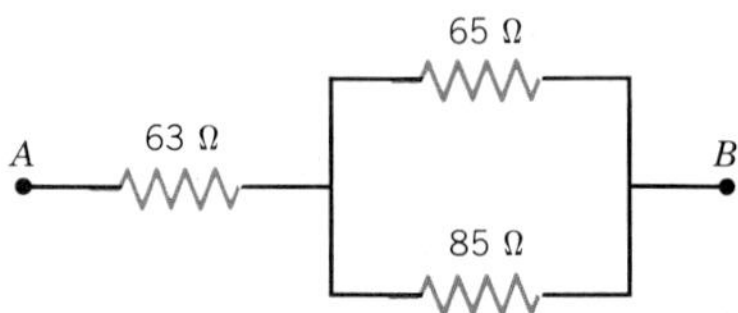

62. Circuit A has three resistors connected in series ($R_1 =$ 30 Ω, $R_2 = 70$ Ω, and $R_3 = 210$ Ω). Circuit B has three resistors (different from any of those in circuit A) connected in parallel. In circuit B each resistor has the same resistance. What is the resistance of each resistor in circuit B, such that the equivalent resistance of B equals the equivalent resistance of A?

63. Find the equivalent resistance between points A and B in the drawing.

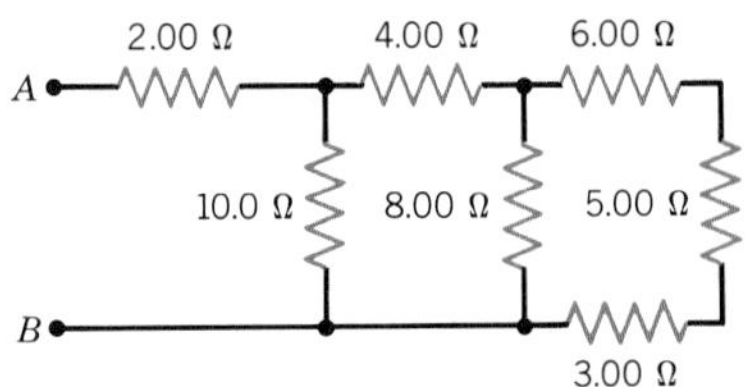

64. Two 25.0-Ω resistors are connected in series. This combination is connected between the terminals of a 75.0-V battery. A 50.0-Ω resistor is also connected across the battery, in parallel with the series combination. (a) How much current is supplied by the battery? (b) How much power is dissipated in one of the 25.0-Ω resistors?

65. Determine the equivalent resistance between the points A and B for the group of resistors in the drawing.

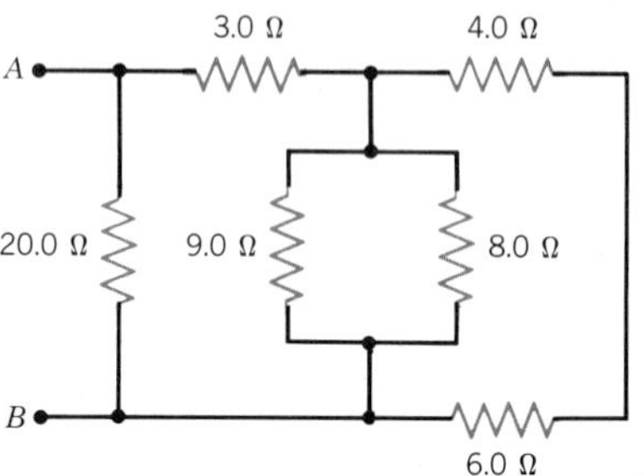

* **66.** A computer repairman needs a 200-Ω resistor to repair a faulty circuit. He is temporarily out of these resistors, but has a single 50-Ω resistor and three 450-Ω resistors. How can these four resistors be wired so as to have an equivalent resistance of 200 Ω?

* **67.** Determine the power dissipated in the 2.0-Ω resistor in the circuit shown in the drawing.

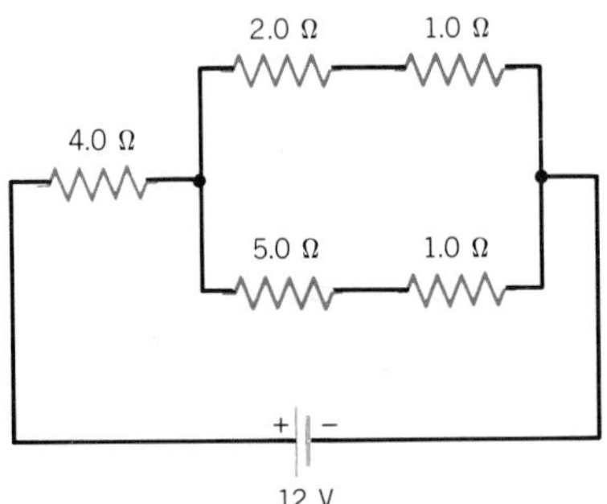

* **68.** Three identical resistors are connected in parallel. The equivalent resistance increases by 700 Ω when one resistor is removed and connected in series with the remaining two, which are still in parallel. Find the resistance of each resistor.

** **69.** The current in the 8.00-Ω resistor in the drawing is 0.500 A. Find the current in the 20.0-Ω resistor and in the 9.00-Ω resistor.

Section 20.9 Internal Resistance

70. A new "D" battery has an emf of 1.5 V. When a wire of negligible resistance is connected between the terminals of the battery, a current of 28 A is produced. Find the internal resistance of the battery.

Problem 69

71. A battery has an emf of 12.0 V and an internal resistance of 0.15 Ω. What is the terminal voltage when the battery is connected to a 1.50-Ω resistor?

72. A 2.00-Ω resistor is connected across a 6.00-V battery. The voltage between the terminals of the battery is observed to be only 4.90 V. Find the internal resistance of the battery.

73. A battery has an emf of 6.4 V and an internal resistance of 0.0048 Ω. An aluminum wire (length = 0.50 m, cross-sectional area = 2.0×10^{-6} m^2) is connected between the terminals of the battery. What is the current in the wire?

74. A battery has an internal resistance of 0.50 Ω. A number of identical light bulbs, each with a resistance of 15 Ω, are connected in parallel across the battery terminals. The terminal voltage of the battery is observed to be one-half the emf of the battery. How many bulbs are connected?

***75.** A 75.0-Ω and a 45.0-Ω resistor are connected in parallel. When this combination is connected across a battery, the current delivered by the battery is 0.294 A. When the 45.0-Ω resistor is disconnected, the current from the battery drops to 0.116 A. Determine (a) the emf and (b) the internal resistance of the battery.

Section 20.10 Kirchhoff's Rules

76. A current of 2.0 A exists in the partial circuit shown in the drawing. What is the magnitude of the potential difference between the points (a) *A* and *B*, and (b) *A* and *C*?

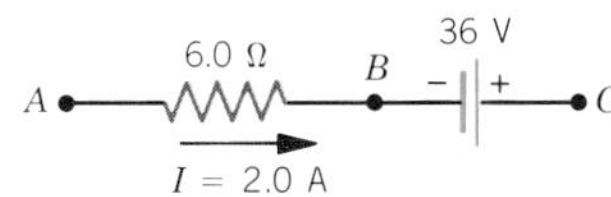

77. Consider the circuit in the drawing. Determine (a) the magnitude of the current in the circuit and (b) the magnitude of the voltage between the points labeled *A* and *B*. (c) State which point, *A* or *B*, is at the higher potential.

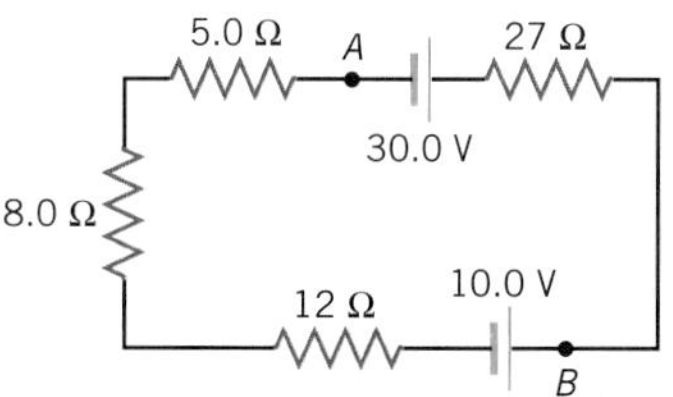

78. The drawing shows resistors that are partly in series and partly in parallel. (a) Find the current in the 4.0-Ω resistor without using Kirchhoff's rules. (b) Redetermine the current in the 4.0-Ω resistor, this time using Kirchhoff's rules. Verify that the answer obtained is the same as that in part (a).

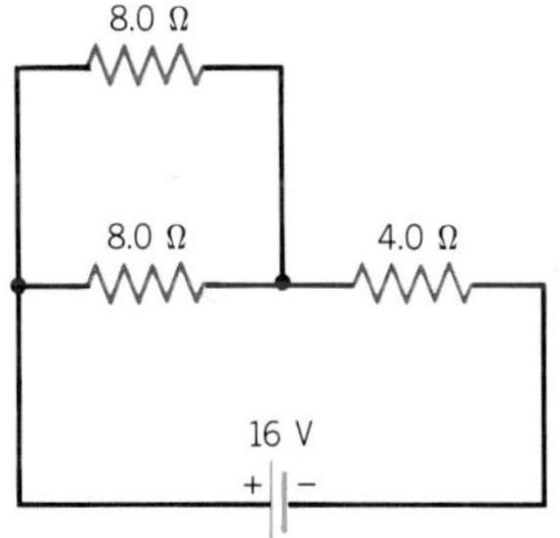

79. Two batteries, each with an internal resistance of 0.015 Ω, are connected as in the drawing. In effect, the 9.0-V battery is being used to charge the 8.0-V battery. What is the current in the circuit?

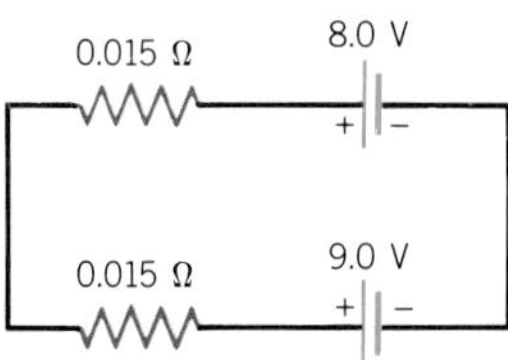

80. For the circuit shown in the drawing, find the current in the 3.00-Ω resistor. Be sure to specify the direction of the current.

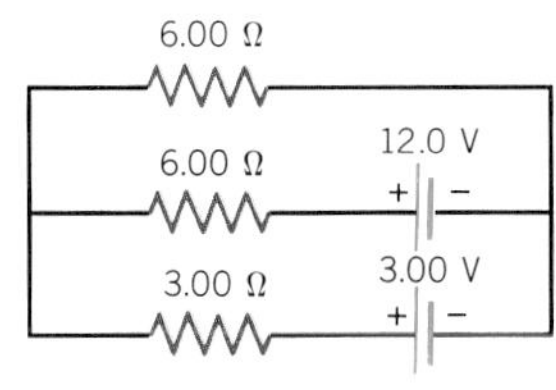

*81. Determine the voltage across the 5.0-Ω resistor in the drawing. Which end of the resistor is at the higher potential?

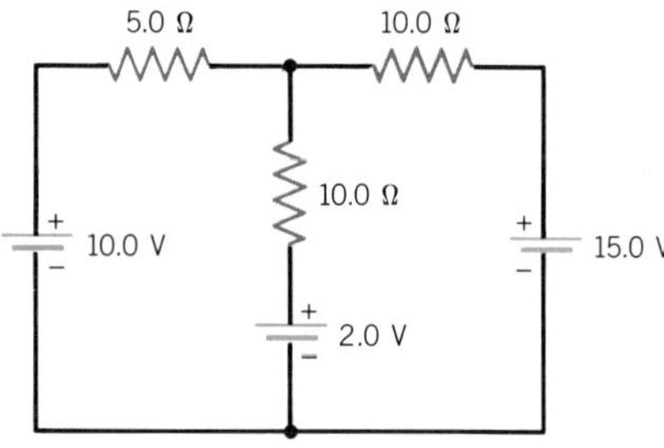

*82. For the circuit in the drawing, find the current in the 10.0-Ω resistor. Specify the direction of the current.

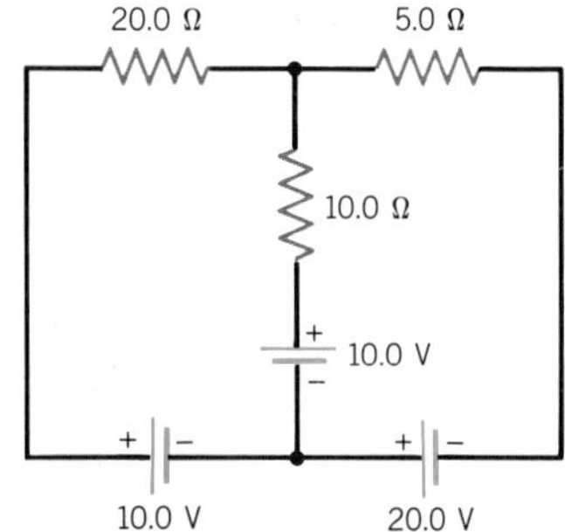

**83. Suppose the resistors in Figure 20.36 have the following resistances: $R = 10.0\ \Omega$, $R_1 = 20.0\ \Omega$, $R_2 = 30.0\ \Omega$, $R_v = 40.0\ \Omega$, and $R_C = 50.0\ \Omega$. If the battery is a 10.0-V battery, what is the voltage between points *A* and *B* in the circuit? State which point is at the higher potential.

Section 20.11 The Measurement of Current, Voltage, and Resistance

84. A galvanometer has a coil resistance of 250 Ω and requires a current of 1.5 mA for full-scale deflection. This device is to be used in an ammeter that has a full-scale current of 25.0 mA. What is the value of the shunt resistance?

85. A voltmeter utilizes a galvanometer that has a 180-Ω coil resistance and a full-scale current of 8.30 mA. The voltmeter measures voltages up to 30.0 V. Determine the resistance that is connected in series with the galvanometer.

86. A galvanometer has a coil resistance of 36 Ω. To make an ammeter, a 3.0-Ω shunt resistor is connected in parallel with the galvanometer. What percentage of the current entering the ammeter passes through the galvanometer?

87. A galvanometer with a coil resistance of 16.0 Ω and a full-scale current of 0.250 mA is used with a shunt resistor to make an ammeter. The ammeter registers a maximum current of 6.00 mA. Find the equivalent resistance of the ammeter.

*88. Two scales on a voltmeter measure voltages up to 20.0 and 30.0 V, respectively. The resistance connected in series with the galvanometer is 1680 Ω for the 20.0-V scale and 2930 Ω for the 30.0-V scale. Determine the coil resistance and the full-scale current of the galvanometer that is used in the voltmeter.

**89. In measuring a voltage, a voltmeter uses some current from the circuit. Consequently, the voltage measured is only an approximation to the voltage present when the voltmeter is not connected. Consider a circuit consisting of two 1550-Ω resistors connected in series across a 60.0-V battery. (a) Find the voltage across one of the resistors. (b) A voltmeter has a full-scale voltage of 60.0 V and uses a galvanometer with a full-scale deflection of 5.00 mA. Determine the voltage that this voltmeter registers when it is connected across the resistor used in part (a).

Section 20.12 Capacitors in Series and Parallel

90. Three capacitors (3.0, 7.0, and 9.0 μF) are connected in series. What is their equivalent capacitance?

91. A 4.0-μF and an 8.0-μF capacitor are connected in parallel across a 25-V battery. Find (a) the equivalent capacitance and (b) the total charge stored on the two capacitors.

92. Determine the equivalent capacitance between *A* and *B* for the group of capacitors in the drawing.

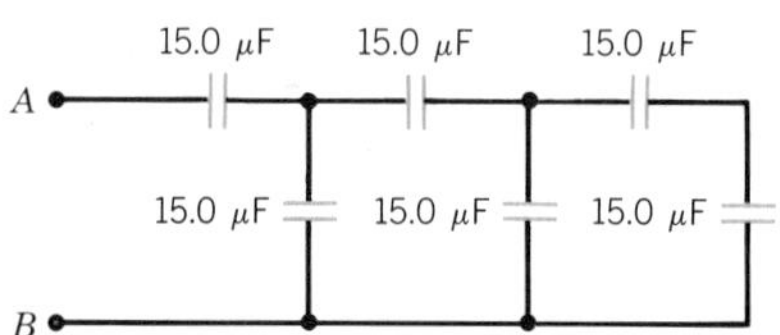

93. A 3.0-μF capacitor and a 4.0-μF capacitor are connected in series across a 40.0-V battery. A 10.0-μF capacitor is also connected directly across the battery terminals. Find the total charge that the battery delivers to the capacitors.

94. Three capacitors (4.0, 6.0, and 12.0 μF) are connected in series across a 50.0-V battery. Find the voltage across the 4.0-μF capacitor.

95. Three capacitors have identical geometries. One is filled with a material whose dielectric constant is 3.00. Another is filled with a material whose dielectric constant is 5.00. The third capacitor is filled with a material whose dielectric constant κ is such that this single capacitor has the same capacitance as the series combination of the two. Determine κ.

96. Suppose two capacitors (C_1 and C_2) are connected in series. Show that the sum of the energies stored in these capacitors is equal to the energy stored in the equivalent capacitor. [*Hint: The energy stored in a capacitor can be expressed as* $q^2/(2C)$.]

*97. A 16.0-μF and a 4.0-μF capacitor are connected in parallel and charged by a 22-V battery. What voltage is required to charge a series combination of the two capacitors to the same total energy?

*98. A 3.00-μF and a 5.00-μF capacitor are connected in series across a 30.0-V battery. A 7.00-μF capacitor is then connected in parallel across the 3.00-μF capacitor. Determine the voltage across the 7.00-μF capacitor.

**99. The drawing shows two fully charged capacitors ($C_1 = 2.00\ \mu$F, $q_1 = 6.00\ \mu$C; $C_2 = 8.00\ \mu$F, $q_2 = 12.0\ \mu$C). The switch is closed, and charge flows until equilibrium is reestablished (i.e., until both capacitors have the same voltage across their plates). Find the resulting voltage across either capacitor.

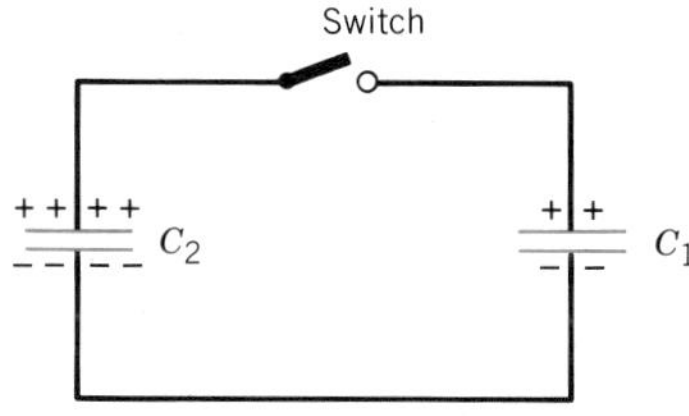

Section 20.13 *RC* Circuits

100. An electronic flash attachment for a camera produces a flash by using the energy stored in a 750-μF capacitor. Between flashes, the capacitor recharges through a resistor whose resistance is chosen so the capacitor recharges with a time constant of 3.0 s. Determine the value of the resistance.

101. The 150-μF capacitor in the drawing is fully charged. When the switch is opened, the capacitor begins to discharge. What is the time constant for the discharge?

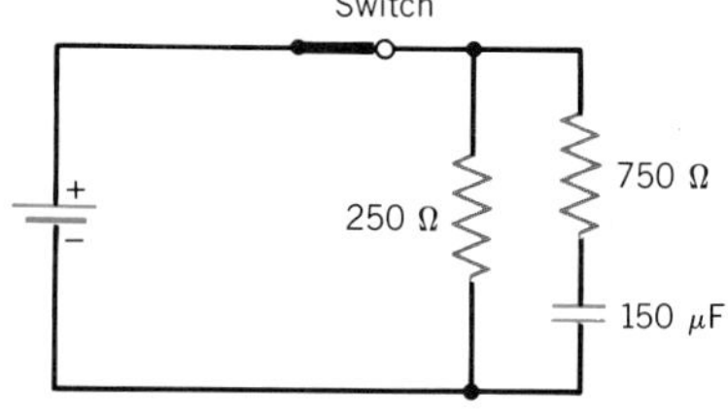

102. A charged capacitor is connected across a 9600-Ω resistor and allowed to discharge. The capacitor loses 63.2% of its original charge in a time of 8.3 s. What is the capacitance of the capacitor?

*103. Three identical capacitors are connected with a resistor in two different ways. When they are connected as in part *a* of the drawing, the time constant to charge up this circuit is 0.020 s. What is the time constant when they are connected with the same resistor as in part *b*?

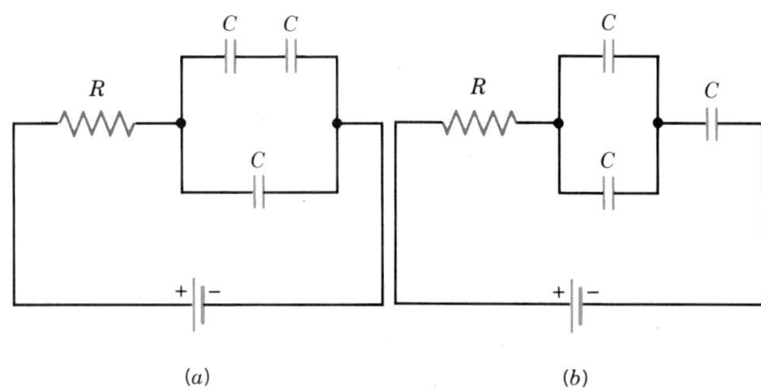

**104. How many time constants must elapse before a capacitor in a series *RC* circuit is charged to within 0.10% of its equilibrium charge?

ADDITIONAL PROBLEMS

105. A 2.00-μF and a 4.00-μF capacitor are connected to a 60.0-V battery. How much charge is supplied by the battery in charging the capacitors when the wiring is (a) in parallel and (b) in series?

106. A cigarette lighter in a car is a resistor that, when activated, is connected across the 12-V battery. Suppose a lighter dissipates 33 W of power. Find (a) the resistance of the lighter and (b) the current that the lighter draws from the battery.

107. A portable compact disc player is designed to play for 2.0 h on a fully charged battery pack. If the battery pack provides a total of 180 C of charge, how much current does the player use in operating?

108. Find the magnitude and direction of the current in the 2.0-Ω resistor in the drawing.

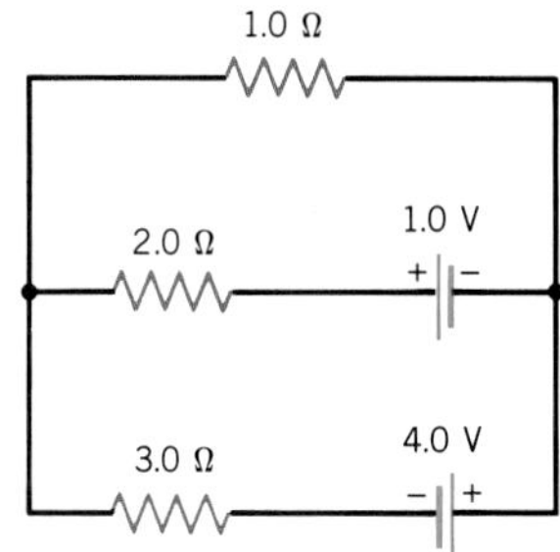

109. The filament of a light bulb has a resistance of 580 Ω. A voltage of 120 V is connected across the filament. How much current is in the filament?

110. In Section 12.3 it was mentioned that temperatures are often measured with electrical resistance thermometers made of platinum wire. Suppose that the resistance of a plati-

num resistance thermometer is 125 Ω when its temperature is 20.0 °C. The wire is then immersed in boiling chlorine, and the resistance drops to 99.6 Ω. The temperature coefficient of resistivity of platinum is $\alpha = 3.72 \times 10^{-3}\ (\text{C}°)^{-1}$. What is the temperature of the boiling chlorine?

111. The equivalent resistance of a voltmeter is 140 000 Ω. The voltmeter uses a galvanometer that has a full-scale deflection of 180 μA. What is the maximum voltage that can be measured by the voltmeter?

112. The two headlights of a car consume a total power of 120 W. A driver parks the car but leaves the lights on. The 12-V battery is rated at 95 A·h. (See problem 7 for an explanation of this rating.) How long does it take for the battery to lose its charge?

113. Eight different values of resistance can be obtained by connecting together three resistors (1.00, 2.00, and 3.00 Ω) in all possible ways. What are they?

114. An electric heater consumes 480 W of power when connected to a 120-V outlet. Two such heaters are connected in series, and the series combination is connected to a 120-V outlet. How much power does *each* heater now consume?

115. An electric furnace runs nine hours a day to heat a house during January (31 days). The heating element has a resistance of 5.3 Ω and carries a current of 25 A. The cost of electricity is $0.10 per kWh. Find the monthly cost of running the furnace.

***116.** A cylindrical aluminum pipe of length 1.50 m has an inner radius of 2.00×10^{-3} m and an outer radius of 3.00×10^{-3} m. The interior of the pipe is completely filled with copper. What is the resistance of this unit? *(Hint: Imagine that the pipe is connected between the terminals of a battery and decide whether the aluminum and copper parts of the pipe are in series or in parallel.)*

***117.** An extension cord is used with an electric weed trimmer that has a resistance of 15.0 Ω. The extension cord is made of copper wire that has a cross-sectional area of $1.3 \times 10^{-6}\ \text{m}^2$. The combined length of the two wires in the extension cord is 92 m. (a) Determine the resistance of the extension cord. (b) The extension cord is plugged into a 120-V socket. What voltage is applied to the trimmer itself?

***118.** A resistor has a resistance R, and a battery has an internal resistance r. When the resistor is connected across the battery, ten percent less power is dissipated in R than there would be if the battery had no internal resistance. Find the ratio r/R.

***119.** A piece of nichrome wire has a radius of 6.5×10^{-4} m. It is used in a laboratory to make a heater that dissipates 4.00×10^2 W of power when connected to a voltage source of 120 V. Ignoring the effect of temperature on resistance, estimate the necessary length of wire.

***120.** Three resistors are connected in series to a battery. From left to right, the resistances are R_1, $R_2 = 5.0\ \Omega$, and R_3. The voltage across R_1 and R_2 together is 8.0 V, while the voltage across R_2 and R_3 together is 4.0 V. The equivalent resistance of the three resistors is 22.0 Ω. Determine R_1, R_3, and the battery voltage.

****121.** A sheet of gold foil (negligible thickness) is placed between the plates of a capacitor and has the same area as each of the plates. The foil is parallel to the plates, at a position one-third of the way from one to the other. Before the foil is inserted, the capacitance is C_0. What is the capacitance after the foil is in place? Express your answer in terms of C_0.

****122.** Two wires have the same cross-sectional area and are joined end to end to form a single wire. One is tungsten and the other carbon. The total resistance of the composite wire is the sum of the resistances of the pieces. The total resistance of the composite does *not change with temperature.* What is the ratio of the lengths of the tungsten and carbon sections? Ignore any changes in length due to thermal expansion.

MAGNETIC FORCES AND MAGNETIC FIELDS

CHAPTER 21

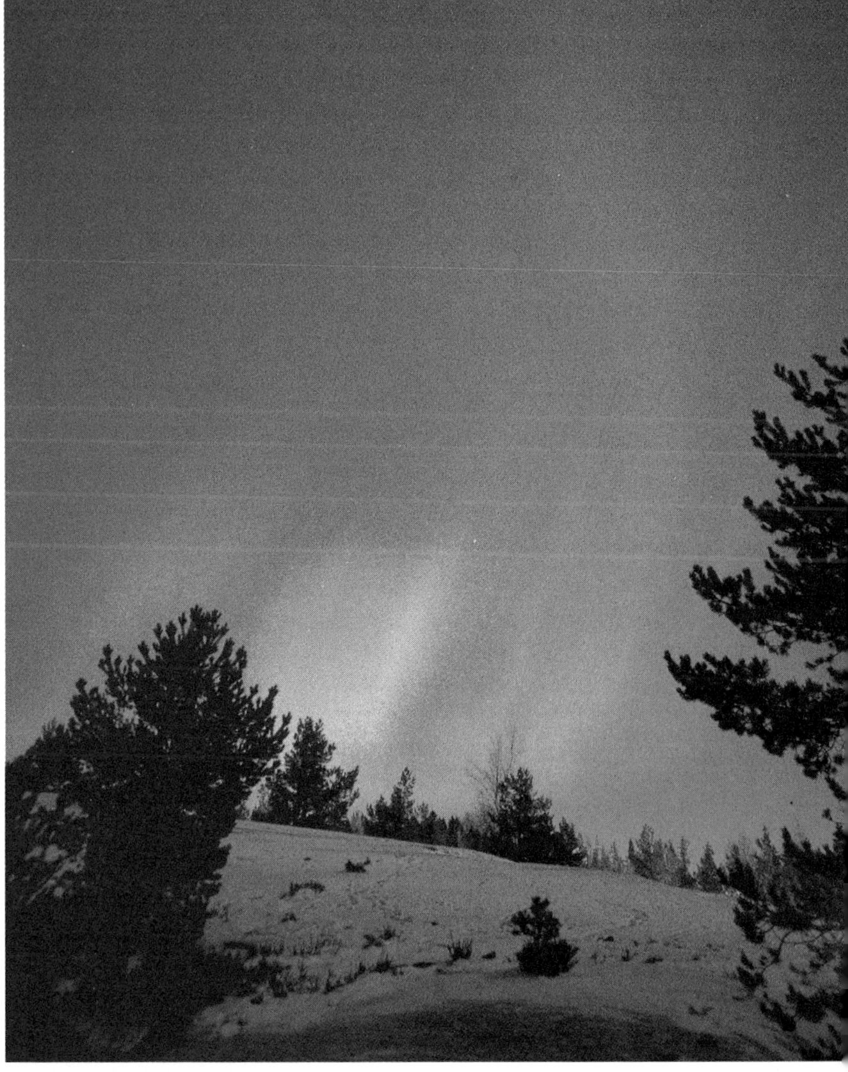

In the far north a dramatic display of colored lights sometimes appears in the night sky. This display is called the aurora borealis, which literally means "northern lights." In the far south the same phenomenon is known as the aurora australis or "southern lights." The lights depend on the earth's magnetic field, which captures electrons and protons streaming out from the sun. Some of the electrons collide with and give up energy to atoms and molecules in the upper atmosphere. In turn, these atoms and molecules radiate the energy in the form of the aurora. Energized oxygen atoms emit the green light in the display, while the pink color comes from nitrogen molecules. To capture the solar electrons and protons, the earth's magnetic field exerts a force on them. This chapter discusses magnetic forces and how they are produced by magnetic fields. We will see that a magnetic field can apply a magnetic force to an electric charge only when the charge is moving. Since moving charges form the current in an electric circuit, a magnetic field can exert a force on a current-carrying wire. We will find that when a wire is shaped into a coil, the magnetic force on the current can be used to produce a torque. Such a magnetic torque is used in electric motors to produce rotation. It is fortunate that magnetic fields can be used to produce forces and torques, but how do the fields arise in the first place? We will learn that they can be created by naturally occurring materials and also by electric currents themselves. An electric current, then, has a dual role to play in this chapter; it can experience a force from an external magnetic field and can also produce a magnetic field of its own. This interplay with electricity is one reason that magnetism is widely used, and we will discuss a number of its applications, including a magnetohydrodynamic propulsion system for ships, magnetic resonance imaging in medicine, and magnetically levitated trains.

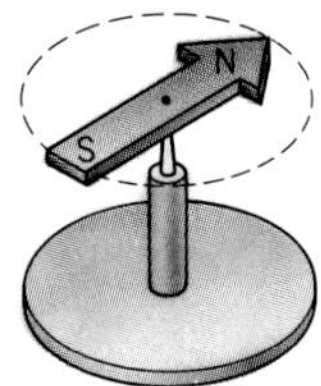

Figure 21.1 The needle of a compass is a permanent magnet that has a north magnetic pole (N) at one end and a south magnetic pole (S) at the other.

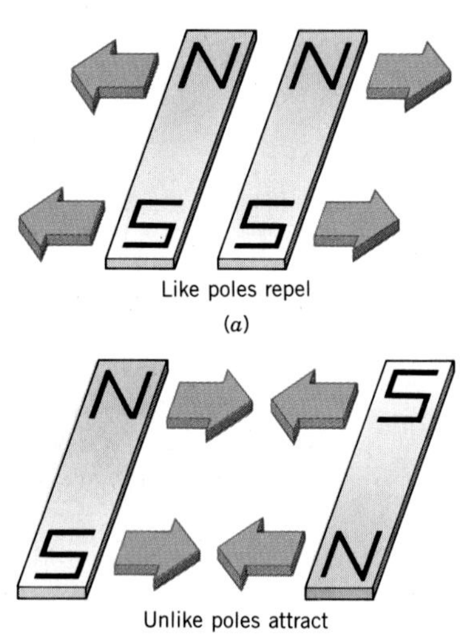

Figure 21.2 Bar magnets have a north magnetic pole at one end and a south magnetic pole at the other end. (*a*) Like poles repel each other, and (*b*) unlike poles attract.

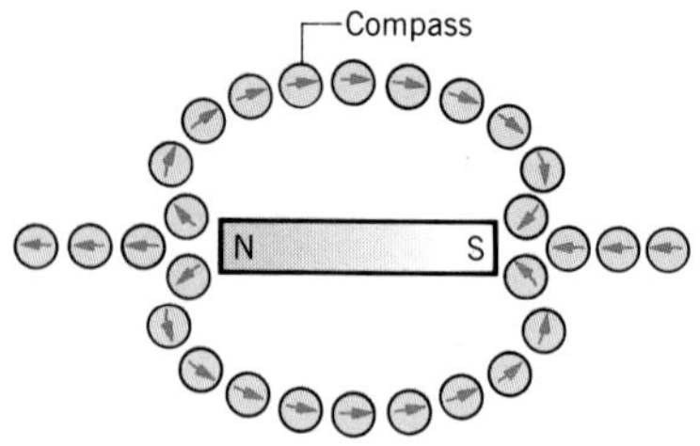

Figure 21.3 At any location in the vicinity of a magnet, the north pole (the arrowhead in this drawing) of a small compass needle points in the direction of the magnetic field at that location.

21.1 MAGNETIC FIELDS

PERMANENT MAGNETS

Permanent magnets have long been used in navigational compasses. As Figure 21.1 illustrates, the compass needle is a permanent magnet supported so it can rotate freely in a plane. When the compass is placed on a horizontal surface, the needle rotates until one end points approximately to the north. The end of the needle that points north is labeled the ***north magnetic pole;*** the opposite end is the ***south magnetic pole.***

Magnets can exert forces on each other. Figure 21.2 shows that the magnetic forces between north and south poles have the property that ***like poles repel each other, and unlike poles attract.*** This behavior is similar to that of like and unlike electric charges. However, there is a significant difference between magnetic poles and electric charges. It is possible to separate positive from negative electric charges and produce isolated charges of either kind. In contrast, no one has found a magnetic monopole (an isolated north or south pole). Any attempt to separate north and south poles by cutting a bar magnet in half fails, because each piece becomes a smaller magnet with its own north and south poles. Repeated cutting only produces more bar magnets, without yielding isolated magnetic poles.

Surrounding a magnet, there is a ***magnetic field.*** The magnetic field is analogous to the electric field that exists in the space around electric charges. Like the electric field, the magnetic field has both a magnitude and a direction. We postpone a discussion of the magnitude until Section 21.2, concentrating our attention here only on the direction of the field. ***The direction of the magnetic field at any point in space is the direction indicated by the north pole of a small compass needle placed at that point.*** In Figure 21.3 the compass needle is symbolized by an arrow, the head of the arrow being the north pole. The drawing shows how compasses can be used to map out the magnetic field in the space surrounding a bar magnet. Since like poles repel and unlike poles attract, the needle of each compass becomes aligned relative to the bar magnet in the manner shown in the picture. The compass needles provide a visual picture of the magnetic field that the bar magnet creates in the surrounding space.

As an aid in visualizing the electric field, we introduced the notion of electric field lines in Section 18.7. In a similar fashion, it is possible to draw magnetic field lines in the vicinity of a magnet. Figure 21.4*a* illustrates some magnetic field lines around a bar magnet. The lines appear to originate from the north pole and to end on the south pole; the lines do not start or stop in midspace. A visual image of the magnetic field lines in a plane can be created by sprinkling finely ground iron filings on a piece of paper that covers the magnet. Iron filings in a magnetic field behave like tiny compasses and align themselves along the magnetic field lines, as part *b* of the drawing shows.

As is the case with electric field lines, the magnetic field at any point is tangent to the magnetic field line at that point. Furthermore, the strength of the magnetic field is proportional to the number of lines per unit area that passes through a surface oriented perpendicular to the lines. Thus, the magnetic field is stronger in regions where the field lines are relatively close together and weaker where they are relatively far apart. For instance, in Figure 21.4*a* the lines are closest together near the north and south poles, reflecting the fact that the strength of the magnetic field is greatest in these regions. Away from the poles, the magnetic field becomes weaker. Notice in part *c* of the drawing that the magnetic field lines in the gap between the poles of the horseshoe magnet are nearly parallel and equally spaced, indicating that the magnetic field there is approximately constant.

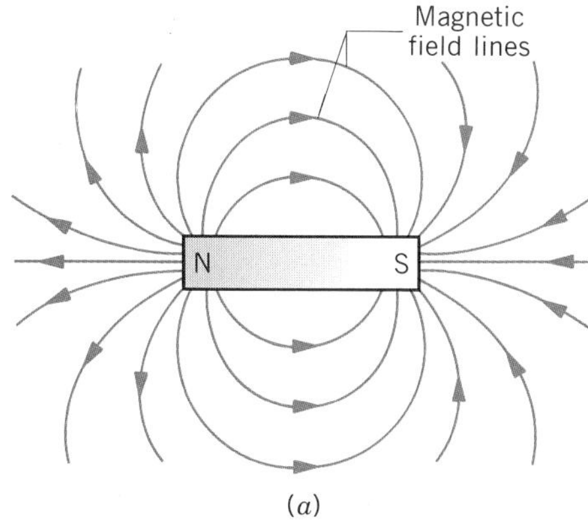

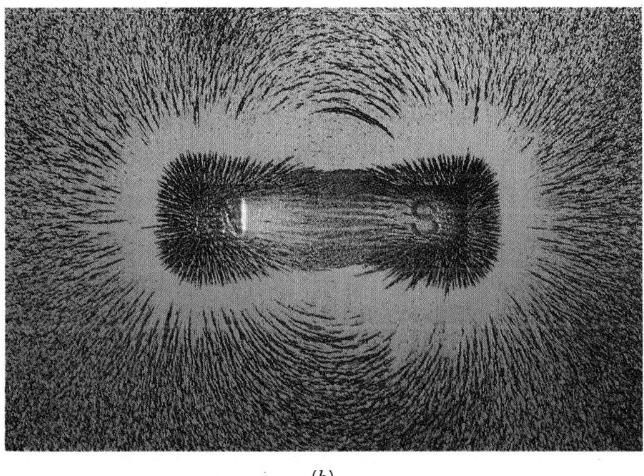

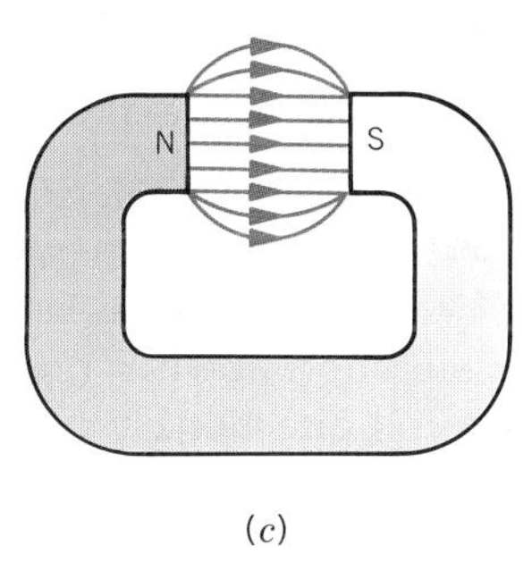

Figure 21.4 (*a*) The magnetic field lines and (*b*) the pattern of iron filings in the vicinity of a bar magnet. (*c*) The magnetic field lines in the gap of a horseshoe magnet.

GEOMAGNETISM

Although the north pole of a compass needle points northward, it does not point exactly at the north geographic pole of the earth. The north geographic pole is that point where the earth's axis of rotation crosses the surface in the northern hemisphere (see Figure 21.5). Measurements of the magnetic field surrounding the earth show that the earth behaves magnetically almost as if it were a bar magnet. As the drawing illustrates, the orientation of this fictitious bar magnet defines a magnetic axis for the earth. The location where the magnetic axis crosses the surface in the northern hemisphere is known as the north magnetic pole. The north magnetic pole of the earth is so named because it is the location toward which the north end of a compass needle points. Since unlike poles attract, the south pole of the earth's fictitious bar magnet lies beneath the north magnetic pole, as Figure 21.5 indicates.

The north magnetic pole does not coincide with the north geographic pole but, instead, lies in Hudson Bay, Canada, some 1300 km to the south. It is interesting to note that the position of the north magnetic pole is not fixed, but moves over the years. For example, the current location of the north magnetic pole is about 770 km northwest of its position in 1904. Pointing as it does at the north magnetic pole, a compass needle deviates from the north geographic pole. At any place on the surface of the earth, the angle that a compass needle deviates is called the *angle of declination* for that location. For New York City, the present angle of declination is about 12° west, meaning that a compass needle points 12° west of geographic north.

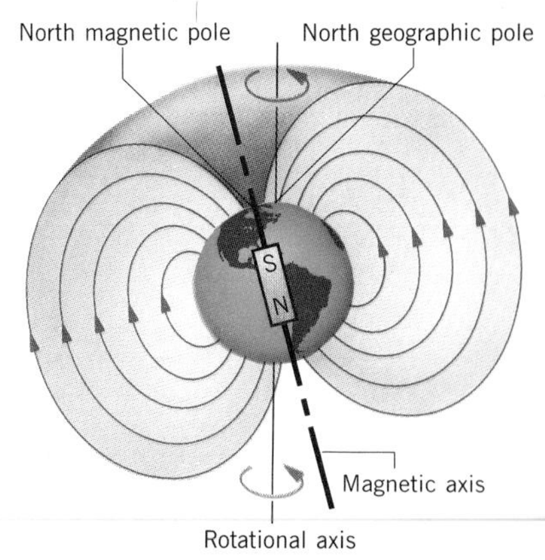

Figure 21.5 The earth behaves magnetically almost as if a bar magnet were located near its center. The axis of this fictitious bar magnet does not coincide with the earth's rotational axis; the two axes are currently about 11.5° apart.

Figure 21.5 shows that the earth's magnetic field lines are not parallel to the surface at all points. For instance, near the north magnetic pole the field lines are almost perpendicular to the surface of the earth. The angle that the magnetic field makes with respect to the surface at any point is known as the *angle of dip.*

21.2 THE FORCE THAT A MAGNETIC FIELD EXERTS ON A MOVING CHARGE

THE NATURE OF THE MAGNETIC FORCE

When a charge is placed in an electric field, the charge experiences an electric force. It is natural to ask, therefore, whether a charge placed in a magnetic field experiences a ***magnetic force.*** The answer is yes, provided two conditions are met:

1. The charge must be moving, for no magnetic force acts on a stationary charge.
2. The velocity of the moving charge must have a component that is perpendicular to the direction of the magnetic field.

To examine the second condition more closely, consider Figure 21.6, which shows a positive test charge $+q_0$ moving with a velocity **v** through a magnetic field labeled by the symbol **B**. The magnetic field is produced by an arrangement of magnets not shown in the drawing and is assumed to be constant in both magnitude and direction. If the charge moves *parallel or antiparallel* to the field, as in part *a* of the drawing, the charge experiences *no magnetic force.* If, on the other hand, the charge moves *perpendicular* to the field, as in part *b*, the charge experiences the *maximum possible force* **F**. In general, if a charge moves at an angle θ* with respect to the field (see part *c* of the drawing), only the velocity component $v \sin \theta$, which is perpendicular to the field, gives rise to a magnetic force. This force is smaller than the maximum possible force. The component of the velocity that is parallel to the magnetic field yields no force.

Figure 21.6 shows that the direction of the magnetic force **F** is perpendicular to both the velocity **v** and the magnetic field **B**; in other words, **F** is perpendicular to the plane defined by **v** and **B**. As an aid in remembering the direction of the force, it is convenient to use *Right-Hand Rule No. 1 (RHR-1),* as Figure 21.7 illustrates:

> ***Right-Hand Rule No. 1.*** Extend the right hand so the fingers point along the direction of the magnetic field **B** and the thumb points along the velocity **v** of the charge. The palm of the hand then faces in the direction of the magnetic force **F** that acts on a positive charge.

It is as if the open palm of the right hand pushes on the positive charge in the direction of the magnetic force. If the moving charge is *negative* instead of positive, the direction of the magnetic force is *opposite* to that predicted by RHR-1. Thus, there is an easy method for finding the force on a moving negative charge. First, assume that the charge is positive and use RHR-1 to find the direction of the force. Then, reverse this direction to find the direction of the force acting on the negative charge.

* The angle θ between the velocity of the charge and the magnetic field is chosen so that it lies in the range $0 \leq \theta \leq 180°$.

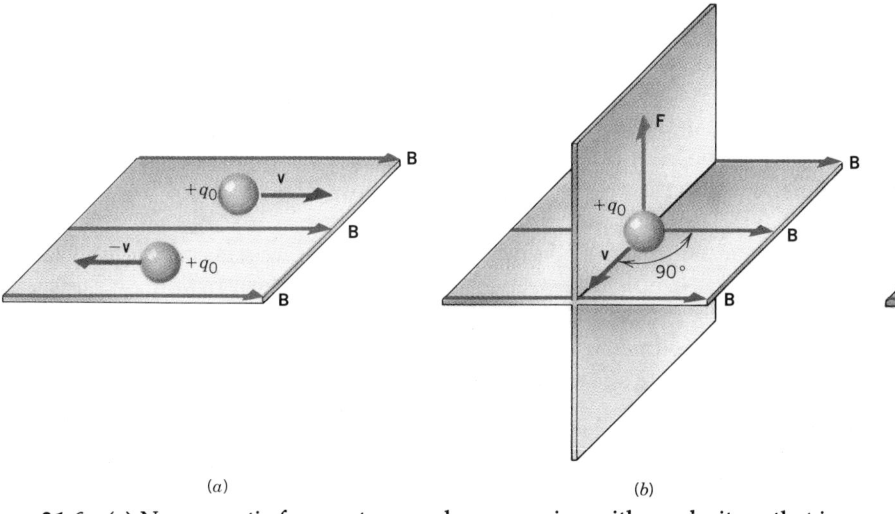

Figure 21.6 (*a*) No magnetic force acts on a charge moving with a velocity **v** that is parallel or antiparallel to a magnetic field **B**. (*b*) The charge experiences a maximum force **F** when the charge moves perpendicular to the field. (*c*) If the charge travels at an angle θ with respect to **B**, only the velocity component perpendicular to **B** gives rise to a magnetic force. This component is $v \sin \theta$.

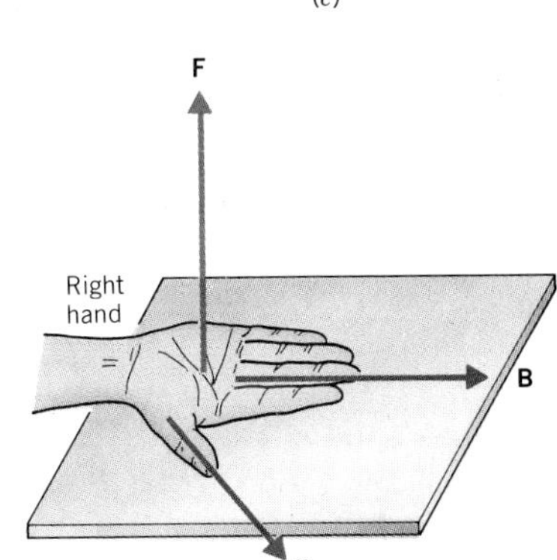

Figure 21.7 Right-Hand Rule No. 1 is illustrated. When the right hand is oriented so the fingers point along the magnetic field **B** and the thumb points along the velocity **v** of a positively charged particle, the palm faces in the direction of the magnetic force **F** applied to the particle.

DEFINITION OF THE MAGNETIC FIELD

The force that acts on a charge moving through a magnetic field has been studied experimentally in great detail. It has been found that the magnitude of this force is directly proportional to the magnitudes of (1) the charge and (2) the component of the velocity that is perpendicular to the magnetic field. Because of these facts, it is possible to define the magnitude of the magnetic field in a manner that is similar to that used for the electric field, although the details differ.

Recall that the electric field at any point in space is the force per unit charge that acts on a test charge q_0 placed at that point. In other words, to determine the electric field **E**, we divide the electric force **F** by the charge q_0: $\mathbf{E} = \mathbf{F}/q_0$. However, the magnetic force depends not only on the charge q_0, but also on the velocity component $v \sin \theta$ that is perpendicular to the magnetic field. Therefore, to determine the magnitude of the magnetic field, we divide the magnitude of the magnetic force not only by q_0, but also by $v \sin \theta$, according to the following definition:

Definition of the Magnetic Field

The magnitude B of the magnetic field at any point in space is defined as

$$B = \frac{F}{q_0(v \sin \theta)} \tag{21.1}$$

where F is the magnitude of the magnetic force on a positive test charge q_0 and **v** is the velocity of the charge and makes an angle $\theta (0 \le \theta \le 180°)$ with the direction of the magnetic field. The magnetic field **B** is a vector, and its direction can be determined by using a small compass needle.

SI Unit of Magnetic Field: $\dfrac{\text{newton} \cdot \text{second}}{\text{coulomb} \cdot \text{meter}} = 1$ tesla (T)

The unit of magnetic field strength that follows from Equation 21.1 is the $N \cdot s/(C \cdot m)$. This unit is called the *tesla* (T), a tribute to the Croatian-born, American engineer Nikola Tesla (1856–1943). Thus, one tesla is the strength of the magnetic field in which a unit test charge, traveling perpendicular to the magnetic field with a speed of one meter per second, experiences a force of one newton. Because a coulomb per second is an ampere (1 C/s = 1 A), the tesla is often written as $1\ T = 1\ N/(A \cdot m)$.

In many situations the magnetic field has a value that is considerably less than one tesla. For example, the strength of the magnetic field near the earth's surface is approximately 10^{-4} T. In such circumstances, a magnetic field unit called the *gauss* (G) is sometimes used. Although not an SI unit, the gauss is a convenient size for many applications involving magnetic fields. The relation between the gauss and the tesla is

$$1 \text{ gauss} = 10^{-4} \text{ tesla}$$

Example 1 deals with the magnetic force exerted on a moving proton and on a moving electron.

Example 1 Magnetic Forces on Charged Particles

A proton in a particle accelerator has a speed of 5.0×10^6 m/s. The proton encounters a magnetic field whose magnitude is 0.40 T and whose direction makes an angle of $\theta = 30.0°$ with respect to the proton's velocity (see Figure 21.6*c*). Find (a) the magnitude and direction of the magnetic force on the proton and (b) the acceleration of the proton. (c) What would be the force and acceleration if the particle were an electron instead of a proton?

REASONING For both the proton and the electron, the magnitude of the magnetic force is given by Equation 21.1. The magnetic forces that act on these particles have opposite directions, however, because the charges have opposite signs. In either case, the acceleration is given by Newton's second law, which applies to the magnetic force just as it does to any force. In using the second law, we must take into account the fact that the masses of a proton and an electron are not the same.

SOLUTION

(a) The positive charge on a proton is 1.60×10^{-19} C, and according to Equation 21.1, the magnitude of the magnetic force is $F = q_0 vB \sin \theta$. Therefore,

$$F = (1.60 \times 10^{-19}\ \text{C})(5.0 \times 10^6\ \text{m/s})(0.40\ \text{T})(\sin 30.0°) = \boxed{1.6 \times 10^{-13}\ \text{N}}$$

The direction of the magnetic force is given by RHR-1 and is directed upward in Figure 21.6*c*, with the magnetic field pointing to the right.

(b) The acceleration of the proton follows directly from Newton's second law as the magnetic force divided by the mass m_p of the proton:

PROBLEM SOLVING INSIGHT

The direction of the magnetic force exerted on a negative charge is opposite to that exerted on a positive charge, assuming both charges are moving in the same direction in the same magnetic field.

$$a = \frac{F}{m_p} = \frac{1.6 \times 10^{-13}\ \text{N}}{1.67 \times 10^{-27}\ \text{kg}} = \boxed{9.6 \times 10^{13}\ \text{m/s}^2} \quad (4.1)$$

(c) The magnitude of the magnetic force on the electron is the same as that on the proton, since both have the same speed and charge magnitude. However, the direction of the force on the electron is opposite to that on the proton, since the electron charge is negative. Furthermore, the electron has a smaller mass m_e and, therefore, experiences a significantly greater acceleration:

$$a = \frac{F}{m_e} = \frac{1.6 \times 10^{-13}\ \text{N}}{9.11 \times 10^{-31}\ \text{kg}} = \boxed{1.8 \times 10^{17}\ \text{m/s}^2}$$

21.3 THE MOTION OF A CHARGED PARTICLE IN A MAGNETIC FIELD

COMPARING PARTICLE MOTION IN ELECTRIC AND MAGNETIC FIELDS

The motion of a charged particle in an electric field is noticeably different from the motion in a magnetic field. For example, Figure 21.8*a* shows a positive charge moving between the plates of a parallel plate capacitor. Initially, the charge is moving perpendicular to the direction of the electric field. Since the direction of the electric force on a positive charge is in the same direction as the electric field, the particle is deflected sideways in the drawing. Part *b* of the drawing shows the same particle traveling initially at right angles to a magnetic field. An application of RHR-1 shows that when the charge enters the field, the charge is deflected upward (not sideways) by the magnetic force. As the charge moves upward, the direction of the magnetic force changes, always remaining perpendicular to both the magnetic field and the velocity. Conceptual Example 2 focuses on the difference in how electric and magnetic fields produce forces on a moving charge.

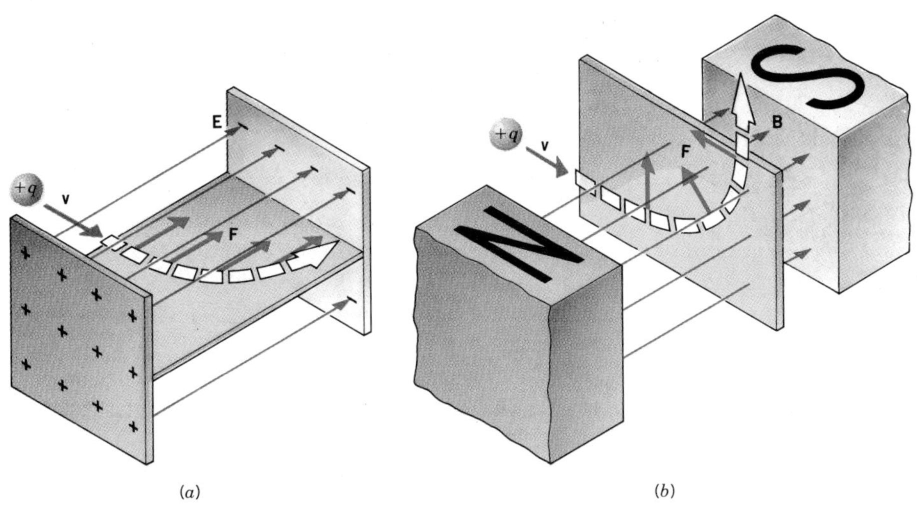

Figure 21.8 (*a*) The electric force **F** that acts on a positive charge is parallel to the electric field **E** and causes the particle's trajectory to bend in a horizontal plane. (*b*) The magnetic force **F** is perpendicular to both the magnetic field **B** and the velocity **v** and causes the particle's trajectory to bend in a vertical plane.

Conceptual Example 2 A Velocity Selector

A velocity selector is a device for measuring the velocity of a charged particle. The device operates by applying electric and magnetic forces to the particle in such a way that these forces balance. Figure 21.9*a* shows a particle with a positive charge $+q$ and a velocity v, which is perpendicular to a constant magnetic field* **B**. How should an electric field **E** be directed so that the force it applies to the particle can balance the magnetic force produced by **B**?

THE PHYSICS OF . . .

a velocity selector.

REASONING If the electric and magnetic forces are to balance, they must have opposite directions. By applying RHR-1, we find that the magnetic force acting on the positively charged particle in Figure 21.9*a* is directed upward. Therefore, the electric force must be directed downward. But the force applied to a positive charge by an electric field has the same direction as the field itself. Therefore, the electric field must point downward in Figure 21.9*a* if the electric force is to balance the magnetic force. In other words, the electric and magnetic fields are perpendicular. Figure 21.9*b* shows a velocity selector that uses perpendicular electric and magnetic fields. The device is a cylindrical tube located within the magnetic field **B**. Inside the tube is a parallel plate capacitor that produces the electric field **E**. The charged particle enters the left end of the tube perpendicular to both fields. If the strengths of the fields **E** and **B** are adjusted properly, the electric and magnetic forces acting on the particle will cancel each other. With no net force acting on the particle, the velocity remains unchanged, according to Newton's first law. As a result, the particle moves in a straight line at a constant speed and exits at the right end of the tube. The magnitude of the velocity "selected" can be determined from a knowledge of the strengths of the electric and magnetic fields. Particles with velocities different from the one "selected" are deflected and do not exit at the right end of the tube.

Related Homework Material: *Problems 20 and 21*

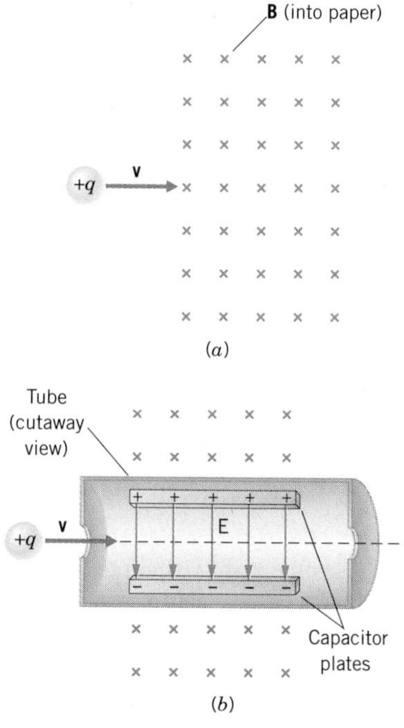

Figure 21.9 (*a*) A particle with a positive charge q and velocity **v** moves perpendicularly into a magnetic field **B**. (*b*) A velocity selector consists of a tube in which an electric field **E** is oriented perpendicular to a magnetic field, and the field magnitudes are adjusted so that the electric and magnetic forces acting on the particle balance.

We have seen that a charged particle traveling in a magnetic field experiences a magnetic force that is always perpendicular to the field. In contrast, the force applied by an electric field is always parallel (or antiparallel) to the field direction. Because of the difference in the way that electric and magnetic fields exert forces, the work done on a charged particle by each field is different, as we now discuss.

THE WORK DONE ON A CHARGED PARTICLE MOVING THROUGH ELECTRIC AND MAGNETIC FIELDS

In Figure 21.8*a* an electric field applies a force to a positively charged particle, and, consequently, the path of the particle bends in the direction of the force. Because there is a component of the particle's displacement in the direction of the electric force, the force does work on the particle. This work increases the kinetic energy and, hence, the speed of the particle, in accord with the work–energy theorem (see Section 6.2).

In contrast to an electric field, *a constant magnetic field does no work on the moving charged particle* in Figure 21.8*b*. This fact arises because the magnetic force

* In many instances it is convenient to orient the magnetic field **B** so its direction is perpendicular to the page. In these cases it is customary to use a dot to symbolize the magnetic field pointing out of the page (toward the reader); this dot symbolizes the tip of the arrow representing the **B** vector. A region where a constant magnetic field is directed *into the page* is drawn as a series of crosses that indicate the tail feathers of the arrows representing the **B** vectors. Therefore, regions where a magnetic field is directed out of the page or into the page are drawn as shown below:

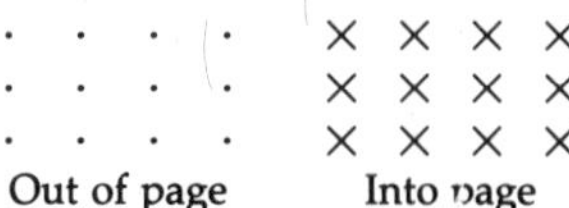

always acts in a direction that is perpendicular to the motion of the charge. Consequently, the displacement of the moving charge never has a component in the direction of the magnetic force. As a result, the magnetic force cannot do work and change the kinetic energy of the charge, although the force can alter the direction of the motion.

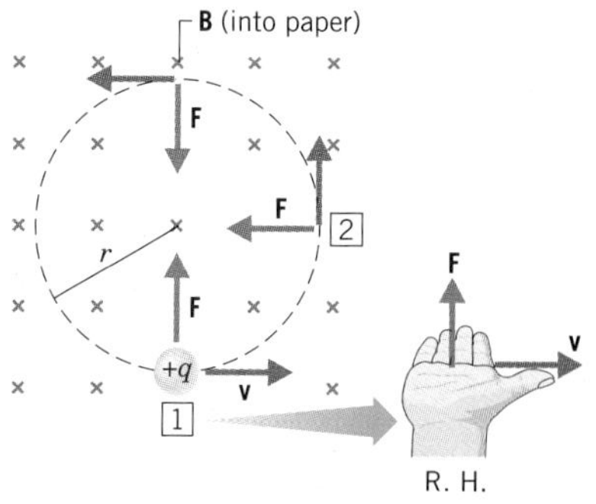

Figure 21.10 A positively charged particle is moving perpendicular to a constant magnetic field. The magnetic force **F** causes the particle to move on a circular path (R.H. = right hand).

THE CIRCULAR TRAJECTORY

To describe the motion of a charged particle in a magnetic field more completely, and to emphasize that the field does no work, we now discuss the special case in which the velocity of the particle is perpendicular to a uniform magnetic field. As Figure 21.10 illustrates, the magnetic force serves to move the particle in a circular path. To understand why the path is circular, consider two points on the circumference labeled 1 and 2. When the positively charged particle is at point 1, the magnetic force **F** is perpendicular to the velocity **v** and points directly upward in the drawing. This force causes the trajectory to bend upward. When the particle reaches point 2, the magnetic force still remains perpendicular to the velocity, but is now directed to the left in the drawing. ***The magnetic force always remains perpendicular to the velocity and is directed toward the center of the circular path.***

To find the radius of the path in Figure 21.10, we recall the concept of centripetal force from Section 5.3. The centripetal force is the net force, directed toward the center of the circle, that is needed to keep a particle moving along a circular path. The magnitude F_c of this force depends on the speed v and mass m of the particle, as well as the radius r of the circle:

$$F_c = \frac{mv^2}{r} \tag{5.3}$$

In the present situation, the magnetic force furnishes the centripetal force. Being perpendicular to the velocity, the magnetic force does no work in keeping the charge $+q$ on the circular path. According to Equation 21.1, the magnetic force is $qvB \sin 90°$, so $qvB = mv^2/r$ or

$$r = \frac{mv}{qB} \tag{21.2}$$

This result shows that the radius of the circle is inversely proportional to the magnitude of the magnetic field, with stronger fields producing "tighter" circular paths. Example 3 illustrates an application of Equation 21.2.

Example 3 The Motion of a Proton

A proton starts from rest at the positive plate of a parallel plate capacitor and is accelerated toward the negative plate by the electric force. The potential difference between the plates is $V = 2100$ volts. The high-speed proton leaves the capacitor through a small hole in the negative plate. Once outside the capacitor, the proton travels at a constant velocity until it enters a region of constant magnetic field of magnitude 0.10 T. The velocity and magnetic field are perpendicular, as in Figure 21.10. Find (a) the speed of the proton when it leaves the capacitor, (b) the change in the proton's kinetic energy due to the magnetic field, and (c) the radius of the circular path on which the proton moves in the magnetic field.

REASONING Initially, when the proton (charge $= +e$) is at the positive plate, the electric potential energy relative to the negative plate is eV. As the proton approaches the negative plate, all the potential energy is converted into kinetic energy, so $eV = \frac{1}{2}mv^2$ and the speed of the proton is $v = \sqrt{2eV/m}$. The proton enters the magnetic field with this speed. The magnetic field does no work on the proton, because the displacement of the proton is always perpendicular to the magnetic force. The magnetic field does, however, cause the proton to travel in a circular path.

SOLUTION

(a) The speed of the proton is

$$v = \sqrt{\frac{2eV}{m}} = \sqrt{\frac{2(1.60 \times 10^{-19}\ \text{C})(2100\ \text{V})}{1.67 \times 10^{-27}\ \text{kg}}} = \boxed{6.3 \times 10^5\ \text{m/s}}$$

(b) Since the magnetic field does no work on the moving proton, the kinetic energy of the proton does not change, according to the work–energy theorem.

(c) Since the kinetic energy remains constant, the speed of the proton does not change, and the radius of the circle can be found from Equation 21.2:

$$r = \frac{mv}{qB} = \frac{(1.67 \times 10^{-27}\ \text{kg})(6.3 \times 10^5\ \text{m/s})}{(1.60 \times 10^{-19}\ \text{C})(0.10\ \text{T})} = \boxed{6.6 \times 10^{-2}\ \text{m}}$$

One of the exciting areas in physics today is the study of elementary particles, which are the basic building blocks from which all matter is constructed. Important information about an elementary particle can be obtained from its motion in a magnetic field, with the aid of a device known as a bubble chamber. A bubble chamber contains a superheated liquid such as hydrogen, which will boil and form bubbles very readily. When an electrically charged particle passes through the chamber, a thin track of bubbles is left in its wake. This track can be photographed to show how a magnetic field affects the particle motion. Conceptual Example 4 illustrates how physicists deduce information from such photographs.

Conceptual Example 4 Particle Tracks in a Bubble Chamber

Figure 21.11*a* shows the bubble-chamber tracks resulting from an event that begins at point *A*. At this point a gamma ray, which is emitted by certain radioactive substances and travels in from the left, spontaneously transforms into two charged particles. There is no track from the gamma ray itself. These particles move away from point *A*, producing the two spiral tracks. A third charged particle is knocked out of a hydrogen atom and moves forward, producing the long track with the slight upward curvature. Each of the three particles has the same mass and carries a charge of the same magnitude. A uniform magnetic field is directed out of the paper toward the reader. Guided by RHR-1 and Equation 21.2, deduce the sign of the charge carried by each particle, identify which particle is moving most rapidly, and account for the two spiral paths.

REASONING To help in our reasoning, Figure 21.11*b* shows a positively charged particle traveling with a velocity **v** that is perpendicular to a magnetic field. The field is directed out of the paper. RHR-1 indicates that the magnetic force points downward. Therefore, downward-curving tracks in the photograph indicate a positive charge, while upward-curving tracks indicate a negative charge. Particles 1 and 3, then, must carry a negative charge. They are, in fact, electrons (e^-). In contrast, particle 2, must have a positive charge. It is called a positron (e^+), an elementary particle that has the same mass as an electron, but an opposite charge.

In Equation 21.2 the mass m, the charge magnitude q, and the magnetic field strength B are the same for each particle. Therefore, the radius r is proportional to the speed v, and a greater radius means a greater speed. The track for particle 3 has the greatest radius, so particle 3 has the greatest speed.

Each spiral indicates that the radius is decreasing as the particle moves. Since the radius is proportional to the speed, the speeds of particles 1 and 3 must be decreasing as these particles move. Correspondingly, the kinetic energies of these particles must be decreasing. They are losing energy each time they collide with a hydrogen atom in the bubble chamber.

Related Homework Material: *Questions 6 and 7, Problem 23*

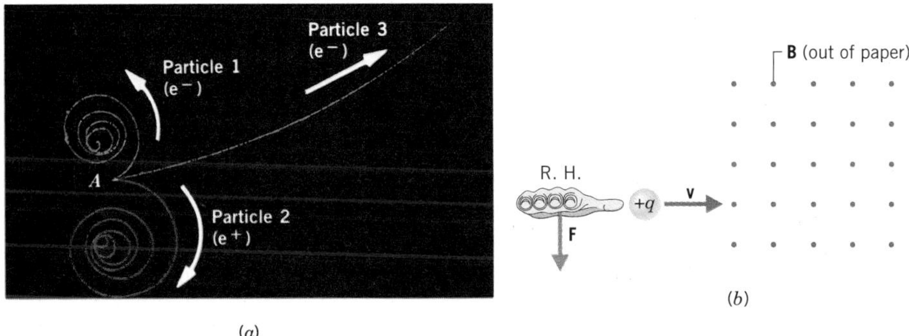

Figure 21.11 (*a*) A photograph of tracks in a bubble chamber. A magnetic field is directed out of the paper. At point A a gamma ray (not visible) spontaneously transforms into an electron (e^-) and a positron (e^+), which produce the spirals. In addition, an electron is knocked forward out of a hydrogen atom in the chamber. (*b*) In accord with RHR-1, the magnetic field applies a downward force to a positively charged particle that moves to the right.

21.4 THE MASS SPECTROMETER

Physicists use mass spectrometers for determining the relative masses and abundances of isotopes.† Chemists use these instruments to help identify unknown molecules produced in chemical reactions. Mass spectrometers are also used during surgery, where they give the anesthesiologist information on the gases, including the anesthetic, in the patient's lungs.

In the type of mass spectrometer illustrated in Figure 21.12, the atoms or molecules are first vaporized and then ionized by the ion source. The ionization process removes one electron from the particle, leaving it with a net positive charge of $+e$. The positive ions are then accelerated through the potential difference V, which is applied between the ion source and the metal plate. With a speed v, the ions pass through a hole in the plate and enter a region of constant magnetic field **B**, where they are deflected in semicircular paths. Only those ions following a path with the proper radius r strike the detector, which records the number of ions arriving per second.

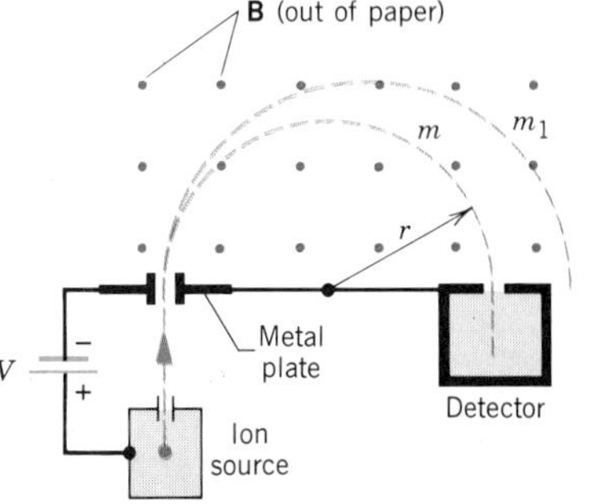

Figure 21.12 The basic features of a mass spectrometer. The dashed lines are the paths traveled by ions of different masses. Ions with mass m follow the path of radius r and enter the detector. Ions with the larger mass m_1 follow the outer path and miss the detector.

† Isotopes are atoms that have the same atomic number, but different atomic masses due to the presence of different numbers of neutrons in the nucleus. They are discussed in Section 31.1.

The mass m of the detected ions can be expressed in terms of r, B, and v by recalling that the radius of the path followed by a particle of charge $+e$ is $r = mv/(eB)$ (Equation 21.2). In addition, the Reasoning section in Example 3 shows that the ion speed v can be expressed in terms of the accelerating potential V as $v = \sqrt{2eV/m}$. Eliminating v from these two equations algebraically and solving for the mass gives

$$m = \left(\frac{er^2}{2V}\right) B^2$$

THE PHYSICS OF . . . a mass spectrometer.

This result shows that the mass of each ion reaching the detector is proportional to B^2. By experimentally changing the value of B and keeping the term in the parentheses constant, ions of different masses are allowed to enter the detector. A plot of the detector output as a function of B^2 then gives an indication of what masses are present and the abundance of each mass.

Figure 21.13 shows a record obtained by a mass spectrometer for naturally occurring neon gas. The results show that the element neon has three isotopes whose atomic mass numbers are 20, 21, and 22. These isotopes occur because neon atoms exist with different numbers of neutrons in the nucleus. Notice that the isotopes have different abundances, with neon-20 being the most abundant.

*21.5 THE HALL EFFECT

The current in a metal conductor is due to the motion of electrons, and electrons carry a negative charge. However, there are important materials in which the electric current is not necessarily caused by the motion of negative charge carriers. For example, semiconductors—most notably silicon and germanium—are important in the technology of integrated circuits. In contrast to the situation in metals, the charge carriers in semiconductors can be either negative or positive, depending on how the semiconductors are fabricated. (See Section 23.5.) When new types of semiconductors are developed, it is important to identify whether the charge carriers are negative or positive.

An experimental method for unambiguously determining the type of carrier was devised by Edwin H. Hall in 1879. Figure 21.14 illustrates Hall's method, which is widely used today. A thin, flat, conducting slab is placed in a constant magnetic field, such that the field is oriented perpendicular to the wide face of the slab. Suppose the current I in the drawing consists of moving positive charges. According to RHR-1, the charges are deflected upward by the magnetic force **F**. Thus, positive charges accumulate at the top edge of the slab, while corresponding negative charges accumulate at the bottom edge. Because of the buildup of positive and negative charges, an emf, called the *Hall emf* (or *Hall voltage*), appears across the slab, with the top of the slab being at a higher potential relative to the bottom. The Hall emf can be measured with a voltmeter, such as the one shown in the drawing. The emf builds up until the electric field produced by the separated positive and negative charges exerts an electric force on the current I that is equal and opposite to the magnetic force. Therefore, a current of positively charged carriers produces a situation in which the top of the slab becomes positively charged and the bottom becomes negatively charged.

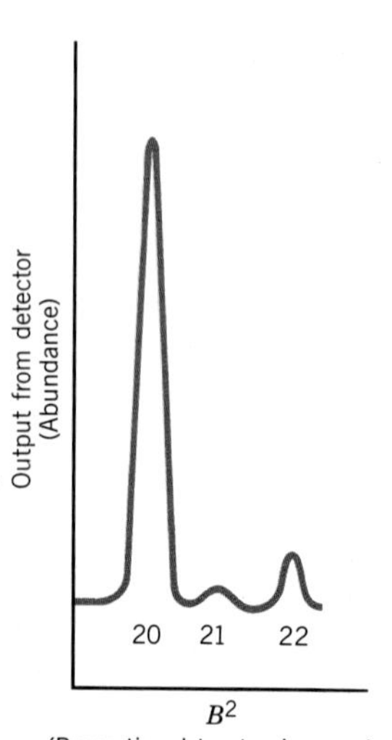

Figure 21.13 The mass spectrum of naturally occurring neon, showing three isotopes whose atomic mass numbers are 20, 21, and 22. The larger the peak, the more abundant the isotope.

On the other hand, the *same* current I could also have been caused by negative charge carriers moving to the *left* in Figure 21.14. An application of RHR-1 (with a reversal of the direction of the predicted force, since the moving charges are

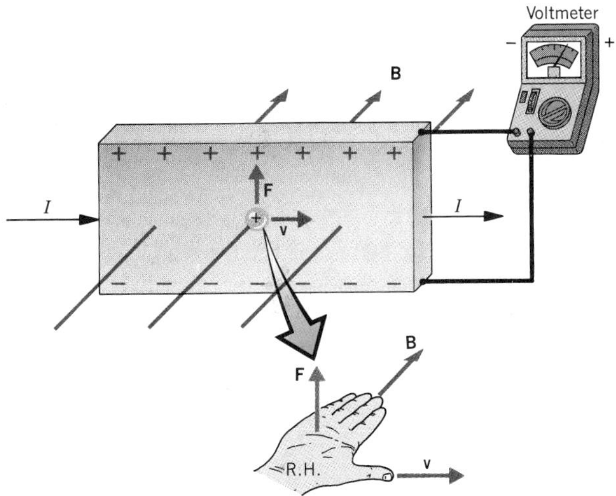

Figure 21.14 Positive charges moving to the right are deflected upward by the magnetic force **F**, giving the top surface of the slab a positive charge. The resulting potential difference between the top and bottom surfaces is called the Hall emf and is registered by the voltmeter.

negative) shows that the top of the slab now becomes negatively charged and the bottom becomes positively charged. Therefore, negative charges moving to the left generate a Hall emf of opposite polarity to that produced by positive charges moving to the right. Thus, the polarity of the Hall emf reveals whether the charge carriers are positive or negative.

THE PHYSICS OF . . .
a Hall probe.

Another use of the Hall effect is found in an instrument known as a Hall probe, which measures the strength of a magnetic field. It has been determined experimentally that the Hall emf is directly proportional to the strength of the magnetic field into which the conducting slab is placed. A Hall probe is a convenient, hand-held instrument that has been calibrated to register the strength of the magnetic field, rather than the Hall emf.

21.6 THE FORCE ON A CURRENT IN A MAGNETIC FIELD

As we have seen, a charge moving through a magnetic field can experience a magnetic force. Since an electric current is a collection of moving charges, a current in the presence of a magnetic field can also experience a magnetic force. In Figure 21.15, for instance, a current-carrying wire is placed between the poles of a magnet. When the direction of the current I is as shown, the moving charges experience a magnetic force that pushes the wire to the right in the drawing. The direction of the force is determined in the usual manner by using RHR-1, with the minor modification that the direction of the velocity of a positive charge is replaced by the direction of the conventional current I. If the current in the drawing were reversed by switching the leads to the battery, the direction of the force would be reversed, and the wire would be pushed to the left.

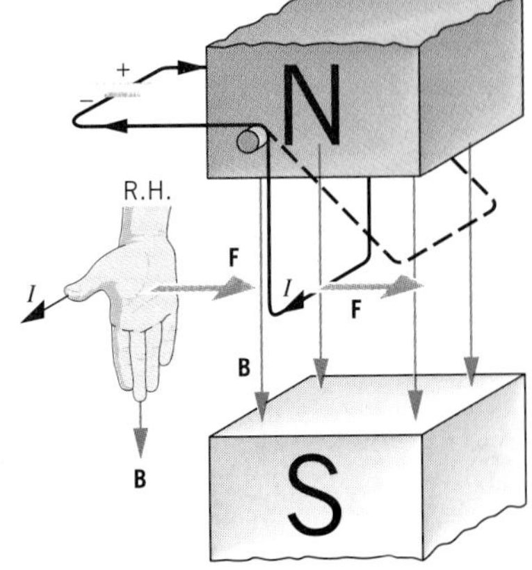

Figure 21.15 The wire carries a current I, and the bottom segment of the wire is oriented perpendicular to a magnetic field **B**. A magnetic force **F** deflects the wire to the right.

When a charge moves through a magnetic field, the magnitude of the force that acts on the charge is $F = qvB \sin \theta$. With the aid of Figure 21.16, this expression can be put into a form that is more suitable for use with an electric current. The drawing shows a wire of length L that carries a current I. The wire is oriented at an angle θ with respect to a magnetic field **B**. This picture is similar to Figure 21.6*c*, except that now the charges move in a wire. The magnetic force exerted on this

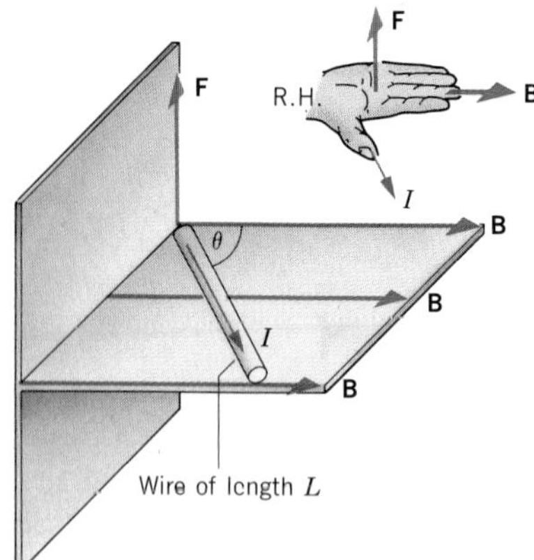

Figure 21.16 The current I in the wire, oriented at an angle θ with respect to a magnetic field **B**, is acted upon by a magnetic force **F**.

length of wire is the net force acting on the total charge q moving in the wire. Suppose that the time required for this charge to travel the length of the wire is t. Multiplying and dividing the right side of $F = qvB \sin\theta$ by t, we find that

$$F = \left(\frac{q}{t}\right)(vt)B \sin\theta$$

The term q/t is the current I in the wire, and the term vt is the length L of the wire. With these two substitutions, the expression for the magnetic force exerted on a current-carrying wire becomes

Magnetic force on a current-carrying wire of length L

$$F = ILB \sin\theta \tag{21.3}$$

As in the case of a single charge traveling in a magnetic field, the magnetic force on a current-carrying wire is a maximum when the wire is oriented perpendicular to the field ($\theta = 90°$) and vanishes when the current is parallel or antiparallel to the field ($\theta = 0°$ or $180°$). The direction of the magnetic force is given by RHR-1.

THE PHYSICS OF . . .

how a loudspeaker produces sound.

Most loudspeakers operate on the principle that a magnetic field exerts a force on a current-carrying wire. Figure 21.17*a* shows a speaker design that consists of three parts: a cone, a voice coil, and a permanent magnet. The cone is often made from specially treated, stiff paper and is mounted so it can vibrate back and forth. When vibrating, the cone pushes and pulls on the air in front of it, thereby creating sound waves. Attached to the apex of the cone is a hollow cardboard cylinder, around which many turns of wire are wound. This cylinder and its coils of wire are collectively called the "voice coil"; the voice coil is slipped over one pole of the permanent magnet, which is the north pole in the drawing. The permanent magnet itself does not move, but the voice coil is designed to move freely over the north pole. The two ends of the voice-coil wire are connected to the speaker terminals located on the back panel of a receiver.

The receiver acts as an ac generator, sending an alternating current to the voice coil. The alternating current interacts with the magnetic field to generate an alternating force that pushes and pulls on the voice coil and the attached cone. To see how the magnetic force arises, consider Figure 21.17*b*, which is a cross-sectional view of the voice coil and the magnet. In the cross-sectional view, the current is directed into the page in the upper half of the voice coil (⊗⊗⊗) and out of the page in the lower half (⊙⊙⊙). In both cases the magnetic field is perpendicular to the current, so the maximum possible force is exerted on the wire. An application of RHR-1 to both the upper and lower halves of the voice coil shows that the magnetic force **F** in the drawing is directed to the right, causing the cone to accelerate in that direction. One-half of a cycle later when the current is reversed, the direction of the magnetic force is also reversed, and the cone accelerates to the left. If, for example, the alternating current from the receiver has a frequency of 1000 Hz, the alternating magnetic force causes the cone to vibrate back and forth at the same frequency, and a 1000-Hz sound wave is produced. Thus, it is the magnetic force on a current-carrying wire that is responsible for converting an electrical signal into a sound wave. In Example 5 a typical force and acceleration in a loudspeaker are determined.

PROBLEM SOLVING INSIGHT

Whenever the current in a wire reverses direction, the force exerted on the wire by a given magnetic field also reverses direction.

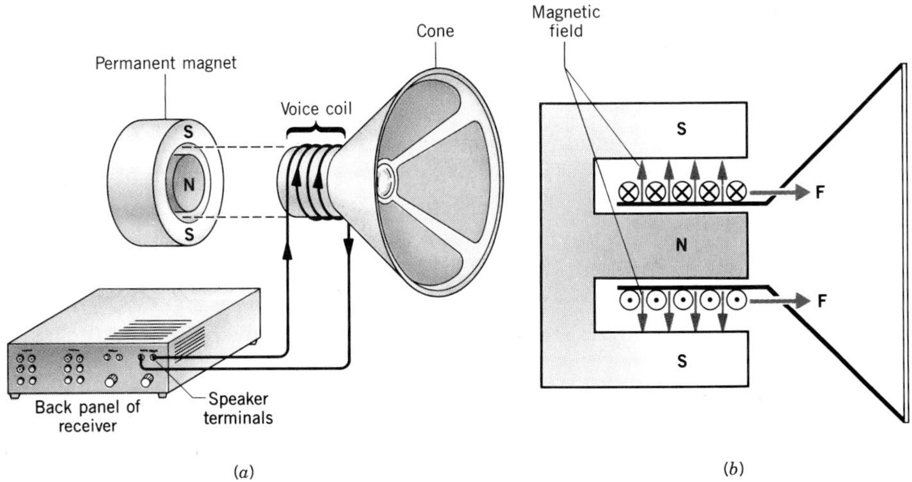

Figure 21.17 (*a*) An "exploded" view of one type of speaker design, which shows a cone, a voice coil, and a permanent magnet. (*b*) Because of the current in the voice coil (shown as ⊗ and ⊙), the magnetic field causes a force **F** to be exerted on the voice coil and cone.

Example 5 The Force and Acceleration in a Loudspeaker

The voice coil of a speaker has a diameter of $d = 0.025$ m, contains 55 turns of wire, and is placed in a 0.10-T magnetic field. The current in the voice coil is 2.0 A. (a) Determine the magnetic force that acts on the coil and cone. (b) If the voice coil and cone have a combined mass of 0.020 kg, find their acceleration.

REASONING The magnetic force that acts on the current-carrying voice coil is given by Equation 21.3 as $F = ILB \sin \theta$. The effective length L of the wire in the voice coil is very nearly the number of turns N times the circumference (πd) of one turn: $L = N\pi d$. The acceleration of the voice coil and cone is given by Newton's second law as the magnetic force divided by the combined mass.

SOLUTION

(a) Since the magnetic field acts perpendicular to all parts of the wire, $\theta = 90°$ and the force on the voice coil is

$$F = ILB \sin \theta \qquad (21.3)$$

$$F = (2.0 \text{ A})[55\pi(0.025 \text{ m})](0.10 \text{ T}) \sin 90° = \boxed{0.86 \text{ N}}$$

(b) According to Newton's second law, acceleration of the voice coil and cone is

$$a = \frac{F}{m} = \frac{0.86 \text{ N}}{0.020 \text{ kg}} = \boxed{43 \text{ m/s}^2} \qquad (4.1)$$

This acceleration is more than four times the acceleration due to gravity.

THE PHYSICS OF . . .

a voice-coil positioner for a hard disk drive.

The voice coil of a speaker moves when current is sent to it by a receiver. The same basic idea plays a role in some personal computer systems that incorporate a hard disk drive. In these drives the part that reads information from or writes information on the spinning disk is the read/write head (see Figure 21.18). Hard disk drives often use voice-coil positioners to move the read/write head to the proper location on the disk. In response to instructions from the user, current is sent to the voice-coil positioner, and a magnetic force causes the head to move across the surface of the disk to the appropriate location.

THE PHYSICS OF . . .

magnetohydrodynamic propulsion.

Magnetohydrodynamic (MHD) propulsion is a revolutionary type of propulsion system that uses a magnetic force on a current to power ships and submarines without propellers. The method eliminates motors, drive shafts, gears, as well as propellers, so it promises to be a low-noise system with great reliability at relatively low cost. Figure 21.19*a* shows the first ship to use MHD technology, the Yamato 1. In part *b* the schematic side view shows one of the two MHD propulsion units that are mounted underneath the vessel. Seawater enters the front of the unit and is expelled from the rear. This expulsion of water is analogous to how a jet engine uses air, taking it in the front of the engine and pushing it out the back to propel the plane forward.

Figure 21.19*c* presents an enlarged view of a propulsion unit. An electromagnet (see Section 21.8) within the vessel uses superconducting wire to produce a strong magnetic field. Electrodes (metal plates) mounted on either side of the unit are attached to a dc electrical generator. The generator sends an electric current through the seawater, from one electrode to the other, perpendicular to the magnetic field and the motion of the water. Consistent with RHR-1, the field applies a force to the current, as the drawing shows. The magnetic force pushes the seawater, similar to the way in which it pushes the wire in Figure 21.15. As a result, a water jet is expelled from the tube. Since the MHD unit exerts a magnetic force on the water, the water exerts a force on the unit and the vessel attached to it. According to Newton's third law, this "reaction" force is equal in magnitude, but opposite in direction to the magnetic force, and it is the reaction force that provides the thrust to drive the vessel.

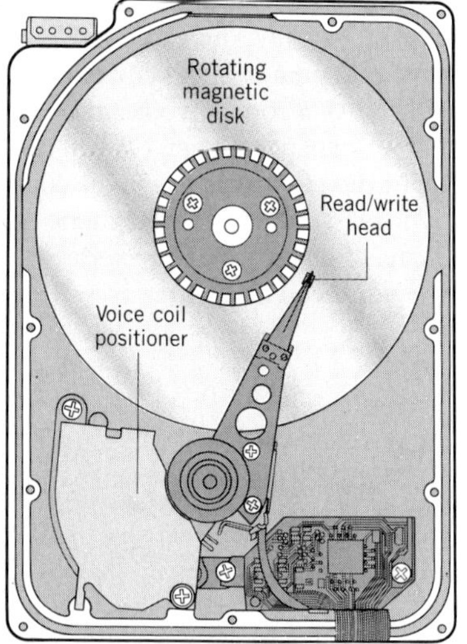

Figure 21.18 Many hard disk drives use a voice-coil positioner to move the read/write head to the appropriate location over the rotating disk.

(*a*)

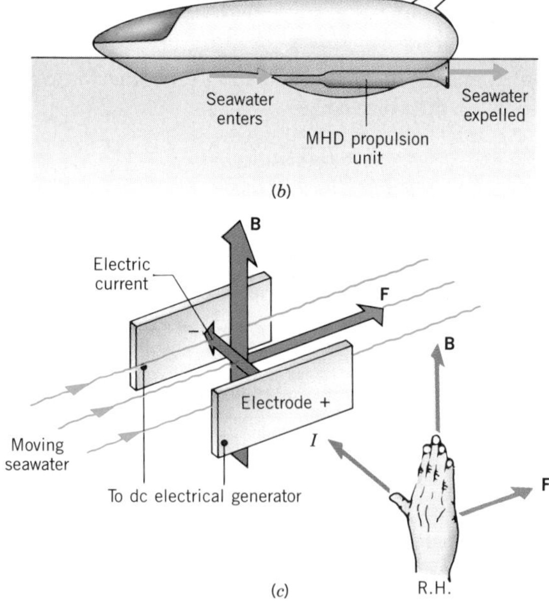

Figure 21.19 (*a*) The Yamato 1 is the first ship to use magnetohydrodynamic (MHD) propulsion. (*b*) A schematic side view of the Yamato 1. (*c*) In the MHD unit, the magnetic force exerted on the current forces water out the back.

21.7 THE TORQUE ON A CURRENT-CARRYING COIL

THE TORQUE

We have seen that a current-carrying wire can experience a force when placed in a magnetic field. If a loop of wire is suspended properly in a magnetic field, the magnetic force produces a torque that tends to rotate the loop. This torque is responsible for the operation of a number of useful devices, including galvanometers and electric motors.

Figure 21.20*a* shows a rectangular loop of wire attached to a vertical shaft. The shaft is mounted such that it is free to rotate in a uniform magnetic field. When there is a current in the loop, the loop rotates because magnetic forces act on the two vertical sides, labeled 1 and 2 in the drawing. Part *b* shows a top view of the loop and the magnetic forces **F** and −**F** on the two sides. These two forces have the same magnitude, but an application of RHR-1 shows that they point in opposite directions, so the loop experiences no net force. The loop does, however, experience a net torque that tends to rotate the loop in a clockwise fashion about the vertical shaft. Figure 21.21*a* shows that the torque is maximum when the normal to the plane of the loop is perpendicular to the field. In contrast, part *b* shows that the torque is zero when the normal is parallel to the field. ***When a current-carrying loop is placed in a magnetic field, the loop tends to rotate such that its normal becomes aligned with the magnetic field.*** In this respect, a current loop behaves like a magnet (e.g., a compass needle) suspended in a magnetic field, since a magnet also rotates to line itself up with the magnetic field.

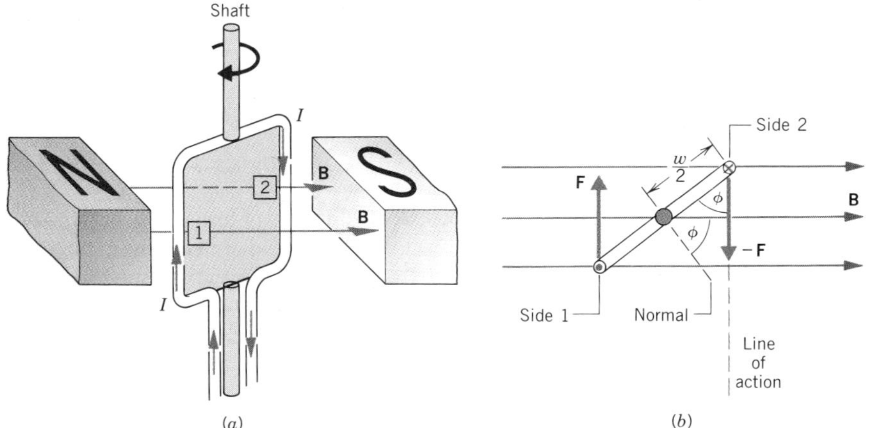

Figure 21.20 (*a*) A current-carrying loop of wire, which can rotate about a vertical shaft, is situated in a magnetic field. (*b*) A top view of the loop. The current in side 1 is directed out of the page (⊙), while that in side 2 is directed into the page (⊗). The current in side 1 experiences a force **F** that is opposite to the force exerted on side 2. The two forces produce a clockwise torque about the shaft.

It is possible to determine the magnitude of the torque on the loop. From Equation 21.3 the magnetic force on each vertical side has a magnitude of $F = ILB \sin 90°$, where L is the length of side 1 or side 2, and $\theta = 90°$ because the current I always remains perpendicular to the magnetic field as the loop rotates. As Section 9.1 discusses, the torque produced by a force is the product of the magnitude of the force and the lever arm. In Figure 21.20*b* the lever arm is the perpendicular distance from the line of action of the force to the shaft. This distance is given by $(w/2) \sin \phi$, where w is the width of the loop, and ϕ is the angle between the normal to the plane of the loop and the direction of the magnetic field. The net torque is the sum of the torques on the two sides, so

$$\text{Net torque} = \tau = ILB\,(\tfrac{1}{2}w \sin \phi) + ILB\,(\tfrac{1}{2}w \sin \phi) = IAB \sin \phi$$

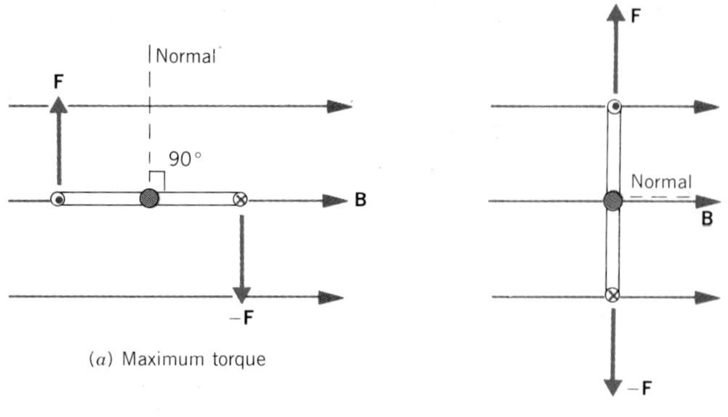

Figure 21.21 (*a*) Maximum torque occurs when the normal to the plane of the loop is perpendicular to the magnetic field, while (*b*) the torque is zero when the normal is parallel to the field.

where the product Lw has been replaced by the area A of the loop. If the wire is wrapped so as to form a coil containing N loops, each of area A, the force on each side is N times larger, and the torque becomes proportionally greater:

$$\tau = NIAB \sin \phi \tag{21.4}$$

Equation 21.4 has been derived for a rectangular coil, but it is valid for any shape of flat coil, such as a circular coil. It is apparent that the torque depends on (1) the geometric properties of the coil itself and the current in it (NIA), (2) the magnitude B of the magnetic field, and (3) the orientation of the normal to the coil with respect to the direction of the field ($\sin \phi$). The quantity NIA is known as the ***magnetic moment*** of the coil, and its units are ampere·meter^2. The greater the magnetic moment of a current-carrying coil, the greater the torque that the coil experiences when placed in a magnetic field. Example 6 discusses the torque that a magnetic field applies to such a coil.

Example 6 The Torque Exerted on a Current-Carrying Coil

A coil of wire has an area of 2.0×10^{-4} m^2, consists of 100 loops or turns, and contains a current of 0.045 A. The coil is placed in a uniform magnetic field of magnitude 0.15 T. (a) Determine the magnetic moment of the coil. (b) Find the maximum torque that the magnetic field can exert on the coil.

REASONING AND SOLUTION

(a) The magnetic moment of the coil is

$$\text{Magnetic moment} = NIA = (100)(0.045\text{ A})(2.0 \times 10^{-4}\text{ m}^2) = \boxed{9.0 \times 10^{-4}\text{ A}\cdot\text{m}^2}$$

(b) According to Equation 21.4, the torque is the product of the magnetic moment NIA and $B \sin \phi$. However, the maximum torque occurs when $\phi = 90°$, so

$$\tau = (\text{Magnetic moment})(B \sin 90°)$$

$$\tau = (9.0 \times 10^{-4}\text{ A}\cdot\text{m}^2)(0.15\text{ T}) = \boxed{1.4 \times 10^{-4}\text{ N}\cdot\text{m}}$$

THE PHYSICS OF . . . **a galvanometer.**

THE GALVANOMETER

As we have noted in Section 20.11, the galvanometer is the basic component of nondigital ammeters and voltmeters. In measuring the current, a galvanometer relies on the fact that a current-carrying coil can rotate when placed in a magnetic field. Figure 21.22 shows the coil (only one turn is shown) of a galvanometer suspended in a magnetic field and free to rotate about a vertical shaft. Attached to the shaft is a pointer and a spring. When there is a current in the coil, the magnetic torque causes the coil to rotate. As the coil rotates, the spring winds up and produces a countertorque. The coil comes to rest when the magnetic torque is counterbalanced by the spring torque. The greater the current, the greater the torque, and the further the coil rotates. In a properly designed instrument, the deflection of the coil and pointer is directly proportional to the current, so the measurement scale can be calibrated to indicate the magnitude of the current.

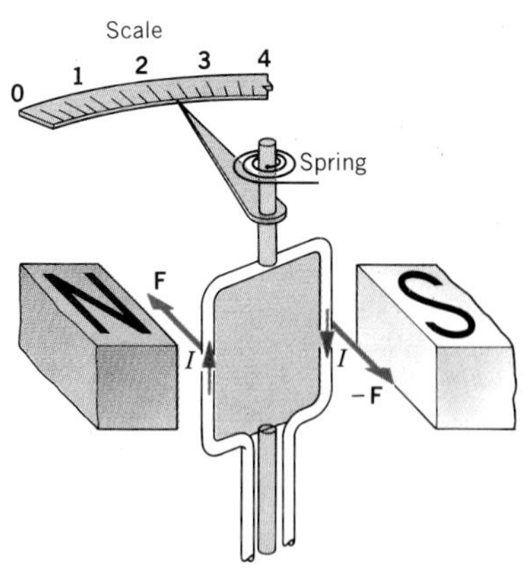

Figure 21.22 The basic elements of a galvanometer.

THE DIRECT-CURRENT ELECTRIC MOTOR

The electric motor is found in many devices, such as tape decks, turntables, automobiles, washing machines, and air conditioners. Figure 21.23 shows the essential parts of a direct-current (dc) motor. The elements of a motor are similar to those of a galvanometer, except the spring is removed so the coil can rotate continuously in one direction. The coil of wire contains many turns and is wrapped around a movable iron cylinder, although these features have been omitted to simplify the drawing. The coil and iron cylinder assembly is known as the armature. Each end of the wire coil is attached to a metallic half-ring. Rubbing against each of the half-rings is a graphite contact called a brush. While the half-rings rotate with the coil, the graphite brushes remain stationary. The two half-rings and the associated brushes are referred to as a split-ring commutator, the purpose of which will be explained shortly.

The operation of a motor can be understood by considering Figure 21.24. In part *a* the current from the battery enters the coil through the left brush and half-ring, goes around the coil, and then leaves through the right half-ring and brush. According to RHR-1, the directions of the forces on the two sides of the coil are as shown in the drawing, and these forces produce the torque that turns the coil. Eventually the coil reaches the position shown in part *b* of the drawing. In this position the half-rings momentarily lose electrical contact with the brushes, so that there is no current in the coil and no applied torque. However, like any moving object, the rotating coil does not stop immediately, for its inertia carries it onward. When the half-rings reestablish contact with the brushes, there again is a current in the coil, and a magnetic torque again rotates the coil in the same direction. The split-ring commutator ensures that the current is always in the proper direction to yield a torque that produces a continuous rotation of the coil. Conceptual Example 7 deals with one of the design considerations that must be faced when building a dc motor.

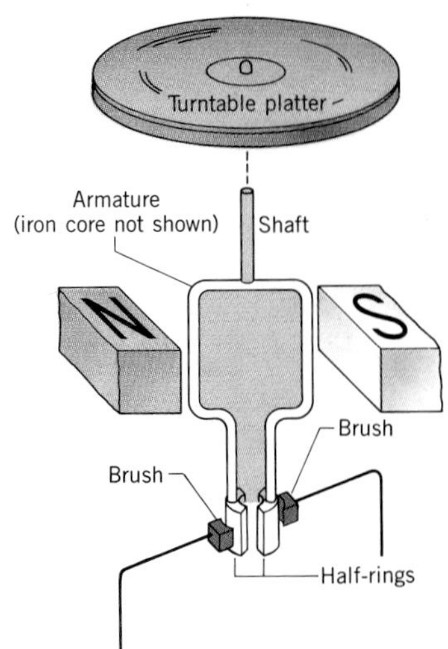

Figure 21.23 The basic components of a dc motor. The platter of a turntable is shown as it might be attached to the motor.

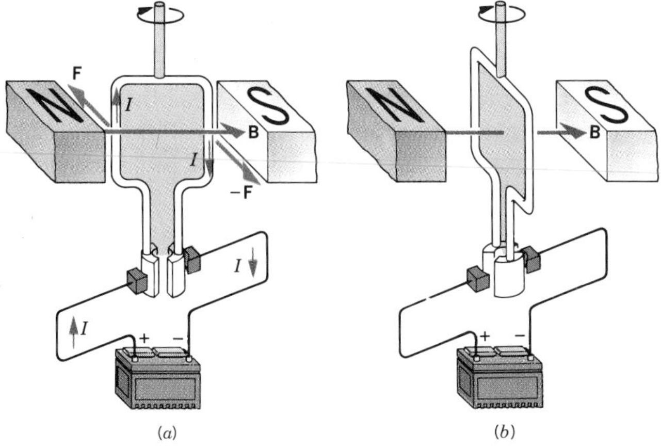

Figure 21.24 (*a*) When a current exists in the coil, the coil experiences a torque. (*b*) Because of its inertia, the coil continues to rotate when there is no current.

THE PHYSICS OF . . .

a direct-current electric motor.

Conceptual Example 7 Building a DC Motor

Suppose you are building a dc motor and use a wire of length L to make the coil. It is necessary to decide whether the coil should contain a smaller or a larger number of turns. To evaluate your options, consider a square coil containing one turn versus two turns. From the point of view of Equation 21.4 ($\tau = NIAB \sin \phi$), is more torque produced by using the wire to make one larger turn or two smaller turns?

REASONING For a given current I and magnetic field B, Equation 21.4 indicates that a larger torque τ results when the total effective area of the coil is larger. The total effective area is the number of turns N times the area A of one turn. Thus, the given length of wire L should be used to produce the largest total effective area. Figure 21.25 compares the total effective area obtainable from one turn versus two turns. With one turn, the length of each side of the square would be $\frac{1}{4} L$, whereas with two turns, the length of a side would be half as long, or $\frac{1}{8} L$. It is clear from the drawings that the total effective area of two turns is one-half that with one turn. Extending this line of reasoning leads to the conclusion that the total effective area NA available from a fixed length of wire is inversely proportional to N. This conclusion is valid, in fact, for any shape of flat coil, such as a circular coil. More torque is produced, then, by using the wire for one turn than for two turns. In general, you get more torque by using the wire for the fewest possible turns, consistent with any other design considerations that may apply.

Related Homework Material: Problem 45

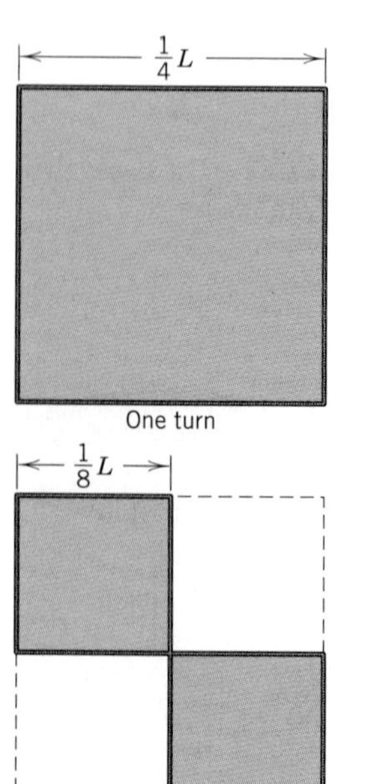

Figure 21.25 A wire of length L is used here to produce one turn or two turns of a square coil.

21.8 MAGNETIC FIELDS PRODUCED BY CURRENTS

We have seen that a current-carrying wire can experience a magnetic force when placed in a magnetic field. The magnetic field is assumed to be produced by some external source, such as a permanent magnet. In this section we consider the phenomenon in which ***a current-carrying wire produces a magnetic field.*** Hans Christian Oersted (1777–1851) first discovered this effect in 1820 when he observed that a current-carrying wire influenced the orientation of a nearby compass needle. The compass needle aligns itself with the net magnetic field produced by the current and the magnetic field of the earth. Oersted's discovery, which linked the motion of electric charges with the creation of a magnetic field, marked the beginning of an important discipline called ***electromagnetism.***

THE MAGNETIC FIELD PRODUCED BY A LONG, STRAIGHT, CURRENT-CARRYING WIRE

Figure 21.26*a* illustrates the essence of Oersted's discovery with a very long, straight wire. When a current is present, the compass needles are observed to point in a circular pattern about the wire. The pattern indicates that the magnetic field lines produced by the current are circles centered on the wire. If the direction of the current is reversed, the needles also reverse their directions, indicating that the direction of the magnetic field has reversed. The direction of the magnetic field can be obtained by using Right-Hand Rule No. 2 (RHR-2), as part *b* of the drawing indicates:

> ***Right-Hand Rule No. 2.*** Curl the fingers of the right hand into the shape of a half-circle. Point the thumb in the direction of the conventional current I, and the tips of the fingers will point in the direction of the magnetic field **B**.

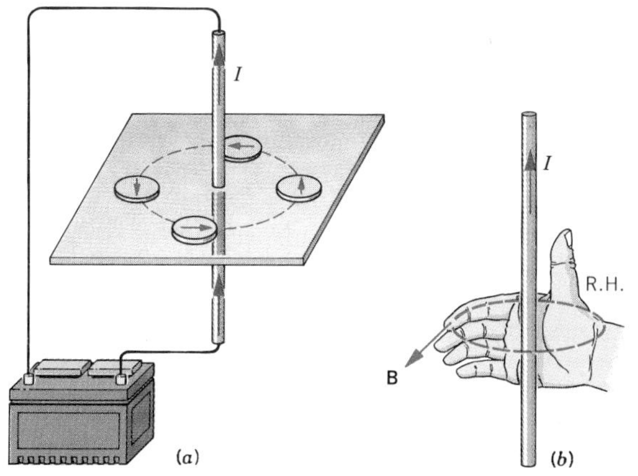

Figure 21.26 (*a*) A long, straight, current-carrying wire produces magnetic field lines that are circular about the wire. One such circular line is indicated by the compass needles. (*b*) If the thumb of the right hand (R.H.) is pointed in the direction of the current I, the curled fingers point in the direction of the magnetic field, according to RHR-2.

With a Hall-effect probe, or some other device that measures the magnetic field, the magnitude of **B** can be measured as a function of the current I in the wire and the radial distance r from the wire. It is found that the magnitude of the field is directly proportional to the current and inversely proportional to the radial distance: $B \propto I/r$. The proportionality constant is written as $\mu_0/2\pi$. Thus, the magnitude of the magnetic field created by the current in a very long, straight wire is

[Long, straight wire]
$$B = \frac{\mu_0 I}{2\pi r} \tag{21.5}$$

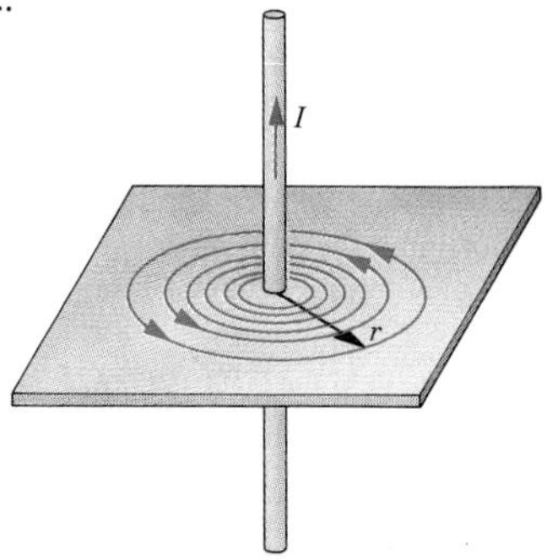

Figure 21.27 The magnetic field becomes stronger as the radial distance r decreases, so the field lines are closer together near the wire.

The constant μ_0 is known as the ***permeability of free space,*** and its value is $\mu_0 = 4\pi \times 10^{-7}$ T·m/A. The magnetic field becomes stronger nearer the wire, where r is smaller. Therefore, the field lines near the wire are closer together than those located farther away, where the field is weaker. Figure 21.27 shows the pattern of field lines.

THE PHYSICS OF . . .
an industrial robot.

Many factories use industrial robots to carry materials or parts from one place to another. One type of robot follows a current-carrying cable buried in the floor. As Figure 21.28 suggests, the robot follows the cable by using special sensors to detect the magnetic field around the cable.

The magnetic field that surrounds a current-carrying wire can exert a force on a moving charge, as the next example illustrates.

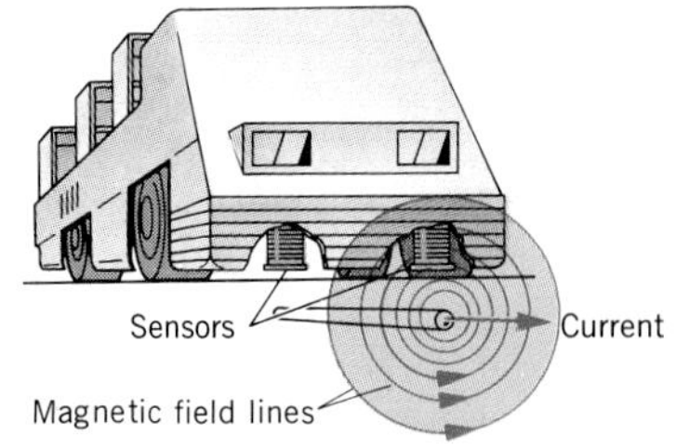

Figure 21.28 The robot follows the buried current-carrying cable by sensing the magnetic field that surrounds it.

Example 8 A Current Exerts a Magnetic Force on a Moving Charge

Figure 21.29 shows a long, straight wire carrying a current of $I = 3.0$ A. A particle of charge $q_0 = +6.5 \times 10^{-6}$ C is moving parallel to the wire at a distance of $r = 0.050$ m; the speed of the particle is $v = 280$ m/s. Determine the magnitude and direction of the magnetic force exerted on the moving charge by the current in the wire.

REASONING The current generates a magnetic field in the space around the wire. A charge moving through this field experiences a magnetic force **F** whose magnitude is given by Equation 21.1 as $F = q_0 vB \sin\theta$, where θ is the angle between the magnetic field and the velocity of the charge. The magnitude of the magnetic field follows from Equation 21.5 as $B = \mu_0 I/(2\pi r)$. Thus, the magnitude of the magnetic force can be expressed as

$$F = q_0 vB \sin\theta = q_0 v \left(\frac{\mu_0 I}{2\pi r}\right) \sin\theta$$

The direction of the magnetic force is predicted by RHR-1.

SOLUTION The drawing shows that the magnetic field **B** lies in the plane that is perpendicular to both the wire and velocity **v** of the particle. Thus, the angle between **B** and **v** is $\theta = 90°$, and the magnitude of the magnetic force is

$$F = q_0 v \left(\frac{\mu_0 I}{2\pi r}\right) \sin 90°$$

$$F = (6.5 \times 10^{-6}\ \text{C})(280\ \text{m/s}) \left[\frac{(4\pi \times 10^{-7}\ \text{T}\cdot\text{m/A})(3.0\ \text{A})}{2\pi(0.050\ \text{m})}\right] = \boxed{2.2 \times 10^{-8}\ \text{N}}$$

The direction of the magnetic force is predicted by RHR-1 and, as the drawing shows, is radially inward toward the wire.

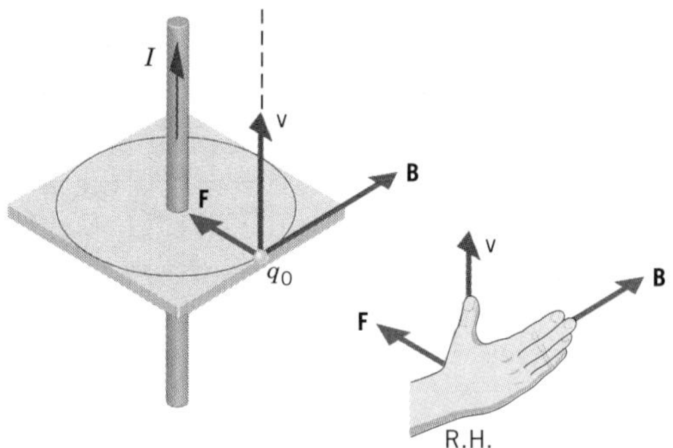

Figure 21.29 The moving charge experiences a magnetic force **F** because of the magnetic field **B** produced by the current in the wire.

We have now seen that an electric current can create a magnetic field of its own. We have also seen earlier that an electric current can experience a force created by another magnetic field. Therefore, the magnetic field that a current creates can exert a force on another nearby current. Examples 9 and 10 deal with the magnetic interaction between currents.

Example 9 *Two Current-Carrying Wires Exert Magnetic Forces on One Another*

Figure 21.30 shows two parallel straight wires. The wires are separated by a distance of $r = 0.065$ m and carry currents of $I_1 = 15$ A and $I_2 = 7.0$ A. Find the magnitude and direction of the force that the magnetic field of wire 1 applies to a 1.5-m length of wire 2 when the currents are (a) in opposite directions and (b) in the same direction.

REASONING The current I_2 in wire 2 is situated in the magnetic field produced by the current in wire 1. The magnitude F of the magnetic force experienced by a length L of wire 2 is given by Equation 21.3 as $F = I_2 LB \sin\theta$. Here B is the magnitude of the magnetic field produced by wire 1 and is given by Equation 21.5 as $B = \mu_0 I_1/(2\pi r)$. The direction of the magnetic force can be determined by using RHR-1.

SOLUTION

(a) At wire 2, the magnitude of the magnetic field created by wire 1 is

$$B = \frac{\mu_0 I_1}{2\pi r} = \frac{(4\pi \times 10^{-7}\ \text{T}\cdot\text{m/A})(15\ \text{A})}{2\pi(0.065\ \text{m})} = 4.6 \times 10^{-5}\ \text{T} \qquad (21.5)$$

The direction of this field is upward at the location of wire 2, as part *a* of the figure shows. The direction can be obtained using RHR-2 (thumb of right hand along I_1, curled fingers point upward at wire 2 and indicate the direction of **B**). The magnetic field is perpendicular to wire 2 ($\theta = 90°$), so the magnitude of the force on a 1.5-m length of wire 2 is

$$F = I_2 LB \sin \theta \tag{21.3}$$

$$F = (7.0 \text{ A})(1.5 \text{ m})(4.6 \times 10^{-5} \text{ T}) \sin 90° = \boxed{4.8 \times 10^{-4} \text{ N}}$$

The direction of the magnetic force on wire 2 is away from wire 1, as part *a* of the drawing indicates; the force direction is found by using RHR-1 (fingers of the right hand extended upward along **B**, thumb points along I_2, palm pushes in the direction of the force **F**).

In a like manner, the current in wire 2 also creates a magnetic field that produces a force on wire 1. Reasoning similar to that above shows that wire 1 is repelled from wire 2 with a force that also has a magnitude of 4.8×10^{-4} N. Thus, each wire generates a force on the other, and, if the currents are in *opposite* directions, the wires *repel* each other. The fact that the two wires exert equal, but oppositely directed forces on each other is consistent with Newton's third law, the action–reaction law.

(b) If the current in wire 2 is reversed, as part *b* of the drawing indicates, wire 2 is attracted to wire 1, because the direction of the magnetic force is reversed. However, the magnitude of the force is the same as that calculated in part *a* above. Likewise, wire 1 is attracted to wire 2. Two parallel wires carrying currents in the *same* direction *attract* each other.

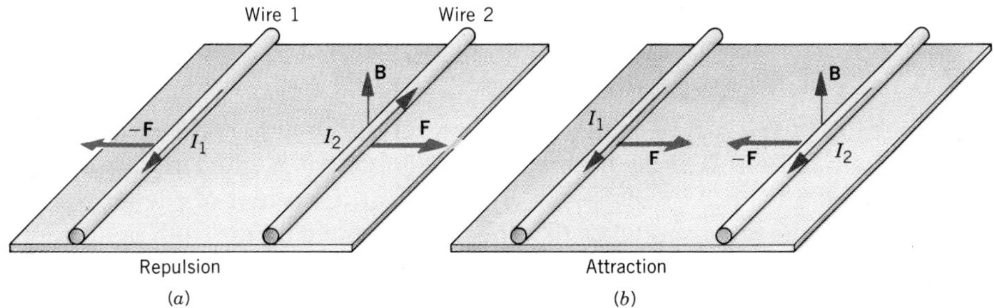

Figure 21.30 (*a*) Two long, parallel wires carrying currents I_1 and I_2 in opposite directions repel each other. (*b*) The wires attract each other when the currents are in the same direction.

Conceptual Example 10 *The Net Force That a Current-Carrying Wire Exerts on a Current-Carrying Coil*

Figure 21.31 shows a straight wire carrying a current I_1 and a rectangular coil carrying a current I_2. The wire and the coil lie in the same plane, with the wire parallel to the long sides of the rectangle. Is the coil attracted to or repelled from the wire?

REASONING The current in the straight wire exerts a force on each of the four sides of the coil. The net force acting on the coil is the vector sum of these four individual forces. To decide whether the net force is one of attraction or repulsion, we need to consider the directions and magnitudes of the individual forces. In the long side of the coil near the wire, the current I_2 has the same direction as the current I_1, and we have just seen in Example 9 that two such currents attract each other. In the long side of the coil farthest from the wire, I_2 has a direction opposite to that of I_1, and according to Example 9, they repel one another. However, the attractive force is stronger than the repulsive force, because the magnetic field produced by the current I_1 is stronger at shorter distances than it is at greater distances. Consequently, we reach the preliminary conclusion that the coil is attracted to the wire.

But what about the forces that act on the two short sides? Consider a small segment of each of the short sides, located at the same distance from the straight wire, as

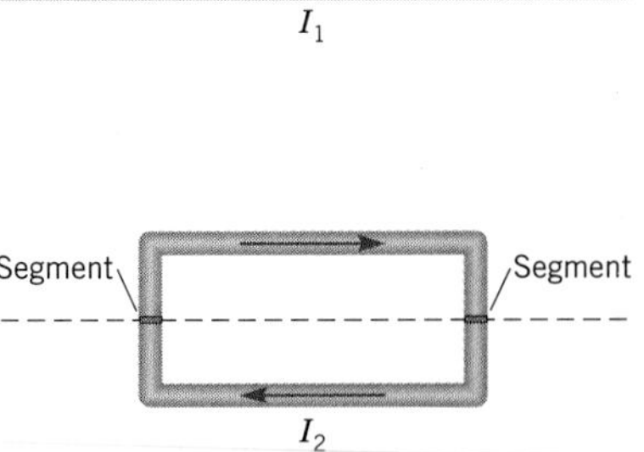

Figure 21.31 A straight wire carries a current I_1, while a rectangular coil carries a current I_2. The dashed line is parallel to the wire and locates a small segment on each short side of the coil.

indicated by the dashed line in Figure 21.31. Each of these segments experiences the same magnetic field from the current I_1. RHR-2 shows that this field is directed downward into the plane of the paper, so that it is perpendicular to the current I_2 in each segment. But the directions of I_2 in the segments are opposite. As a result, RHR-1 reveals that the magnetic field from the straight wire applies a force to one segment that is opposite to the force it applies to the other segment. Thus, the forces acting on the two short sides of the coil cancel, and our preliminary conclusion remains valid; the coil is attracted to the wire.

Related Homework Material: *Problem 59*

THE MAGNETIC FIELD PRODUCED BY A LOOP OF WIRE

If a current-carrying wire is bent into a circular loop, the magnetic field lines around the loop have the pattern shown in Figure 21.32*a*. At the *center* of a loop of radius R, the magnetic field is perpendicular to the plane of the loop and has the value $B = \mu_0 I/(2R)$, where I is the current in the loop. Often, the loop consists of N turns of wire that are wound sufficiently close together that they form a flat coil with a single radius. In this case, the magnetic fields of the individual turns add together to give a net field that is N times greater than that of a single loop. For such a coil the magnetic field at the center is

[Center of a circular loop]

$$B = N\frac{\mu_0 I}{2R} \tag{21.6}$$

The direction of the magnetic field at the center of the loop can be determined with the help of RHR-2. If the thumb of the right hand is pointed in the direction of the current and the curled fingers are placed at the center of the loop, as in Figure 21.32*b*, the fingers indicate that the magnetic field points from right to left.

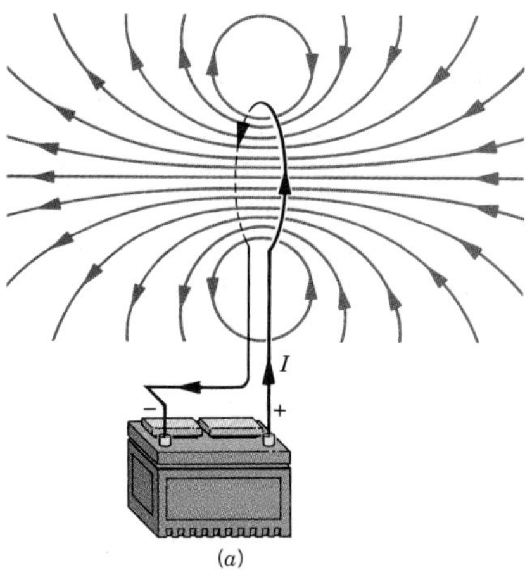

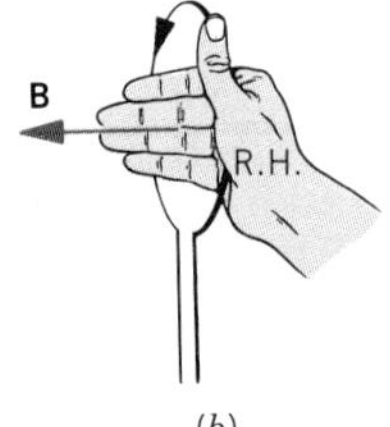

Figure 21.32 (*a*) The magnetic field lines in the vicinity of a current-carrying circular loop. (*b*) The direction of the magnetic field at the center of the loop is given by RHR-2.

Example 11 shows how the magnetic fields produced by the current in a loop of wire and the current in a long, straight wire combine to form a net magnetic field.

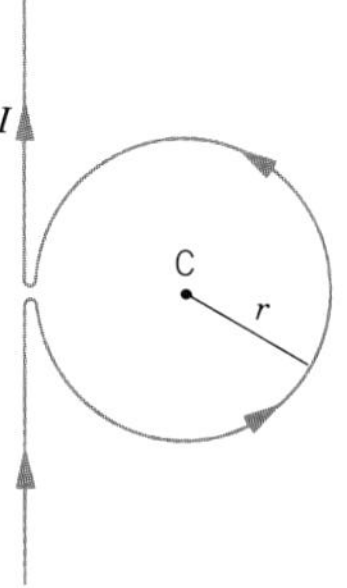

Figure 21.33 Part of a long, straight wire is bent into a circular loop that carries a current I.

Example 11 Finding the Net Magnetic Field

A long, straight wire carries a current of 8.0 A. A portion of the wire is then bent into a circular loop (one turn) of radius 0.020 m, as Figure 21.33 illustrates. Find the magnitude and direction of the net magnetic field at the center C of the loop.

REASONING The net magnetic field at the point C is the sum of two contributions: (1) the field that the circular loop produces, and (2) the field that the long, straight wire generates. An application of RHR-2 shows that the magnetic field generated by the circular loop at C is directed out of the plane of the paper, toward the reader. Similarly, RHR-2 shows that the magnetic field created at C by the long, straight wire is directed into the plane of the paper. Therefore, the directions of the two magnetic field contributions are opposite.

SOLUTION Taking the direction out of the paper as positive, the net magnetic field is

$$B = \underbrace{\frac{\mu_0 I}{2r}}_{\text{Center of loop}} - \underbrace{\frac{\mu_0 I}{2\pi r}}_{\text{Long wire}} = \frac{\mu_0 I}{2r}\left(1 - \frac{1}{\pi}\right)$$

$$B = \frac{(4\pi \times 10^{-7}\ \mathrm{T \cdot m/A})(8.0\ \mathrm{A})}{2\ (0.020\ \mathrm{m})}\left(1 - \frac{1}{\pi}\right) = \boxed{1.7 \times 10^{-4}\ \mathrm{T}}$$

The net field is directed perpendicularly out of the plane of the paper.

PROBLEM SOLVING INSIGHT

Do not confuse the formula for the magnetic field produced at the center of a circular loop with that of a long, straight wire. The formulas are similar, differing only by a factor of π in the denominator.

A comparison of the magnetic field lines around the current loop in Figure 21.32*a* with those in the vicinity of the short bar magnet in Figure 21.34*a* shows that the two patterns are quite similar. Not only are the patterns similar, but the loop itself behaves as a bar magnet with a "north pole" on one side and a "south pole" on the other side. To emphasize that the loop may be imagined to be a bar magnet, Figure 21.34*b* includes a "phantom" bar magnet at the center of the loop. The side of the loop that acts like a north pole can be determined with the aid of

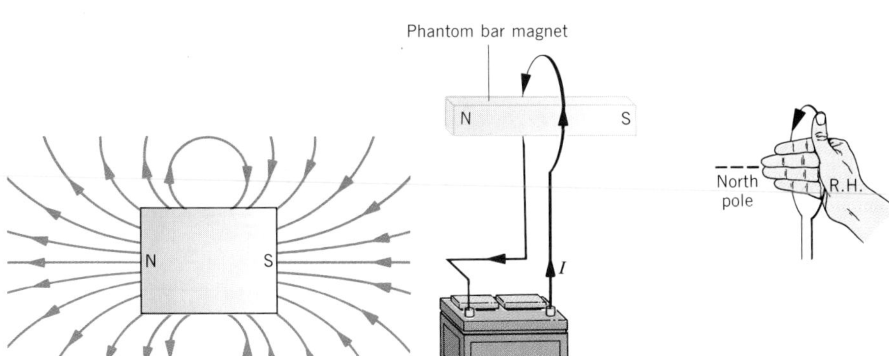

Figure 21.34 (*a*) The field lines around the bar magnet resemble those around the loop in Figure 21.32*a*. (*b*) The current loop can be imagined to be a phantom bar magnet with a north pole and a south pole.

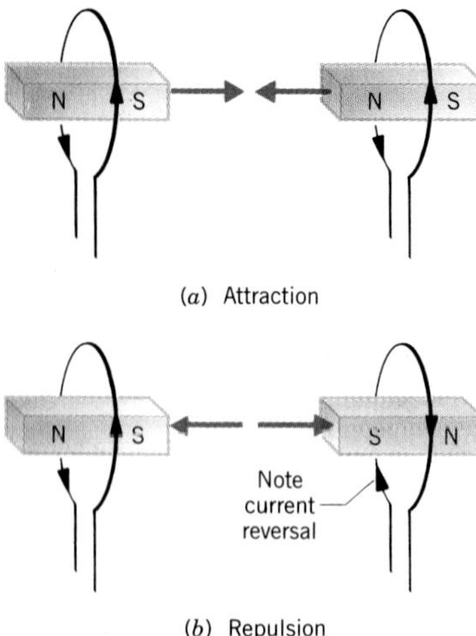

Figure 21.35 (*a*) The two current loops attract each other if the directions of the currents are the same and (*b*) repel each other if the directions of the currents are opposite. The "phantom" magnet included for each loop helps explain the attraction and repulsion between the loops.

RHR-2; the fingers of the right hand not only point in the direction of **B**, but they also point toward the north pole.

Because a current-carrying loop acts like a bar magnet, two adjacent loops can be either attracted to or repelled from each other, depending on the relative directions of the currents. Figure 21.35 includes a "phantom" magnet for each loop and shows that the loops are attracted to each other when the currents are in the same direction and repelled from each other when the currents are in opposite directions. This behavior is analogous to that of the two long, straight wires discussed in Example 9.

THE SOLENOID

A solenoid is a long coil of wire in the shape of a helix (see Figure 21.36). If the wire is wound so the turns are packed close to each other and the solenoid is long compared to its diameter, the magnetic field lines have the appearance shown in the drawing. Notice that the field inside the solenoid and away from its ends is nearly constant in magnitude and directed parallel to the axis. The direction of the magnetic field inside the solenoid is given by RHR-2, just as it is for a circular current loop. The magnitude of the magnetic field in the interior of a long solenoid is

[Interior of a long solenoid] $$B = \mu_0 n I \tag{21.7}$$

where n is the number of turns per unit length of the solenoid and I is the current. If, for example, the solenoid contains 100 turns and has a length of 0.05 m, the number of turns per unit length is $n = (100 \text{ turns})/(0.05 \text{ m}) = 2000 \text{ turns/m}$.

As with a single loop of wire, a solenoid can also be imagined to be a bar magnet, for the solenoid is just an array of connected current loops. And, as with a circular current loop, the location of the north pole can be determined with RHR-2. Figure 21.36 shows that the left end of the solenoid acts as a north pole, and the right end behaves as a south pole. Solenoids are often referred to as *electromagnets,* and they have several advantages over permanent magnets. For one thing, the strength of the magnetic field can be altered by changing the

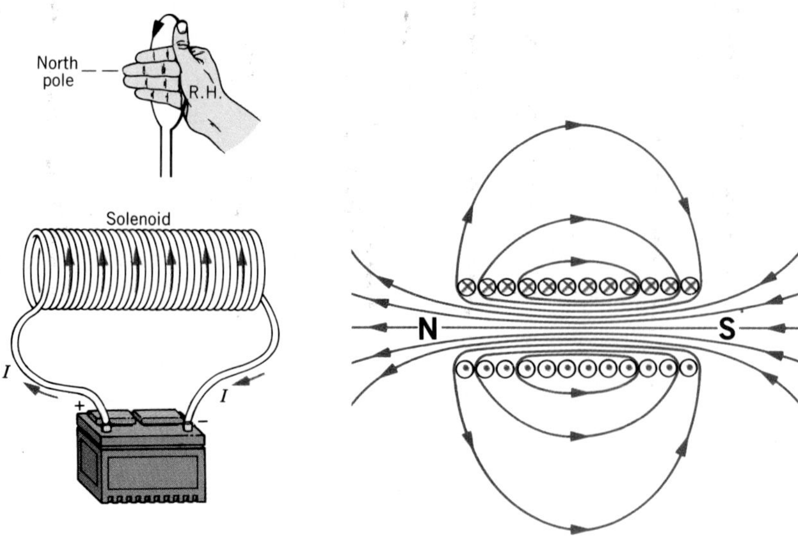

Figure 21.36 A solenoid and a cross-sectional view of it, showing the magnetic field lines and the north and south poles.

current and/or the number of turns per unit length. Furthermore, the north and south poles of an electromagnet can be readily switched by reversing the current.

THE PHYSICS OF . . .

magnetic resonance imaging (MRI).

Applications of the magnetic field produced by a current-carrying solenoid are widespread. An exciting medical application is in the technique of magnetic resonance imaging (MRI). With this technique, detailed pictures of the internal parts of the body can be obtained in a noninvasive way that involves none of the risks inherent in the use of X-rays. Figure 21.37 shows a patient inside a magnetic resonance imaging machine. The circular opening into which the patient is inserted is one end of a solenoid, which is typically made from superconducting wire. The superconducting wire facilitates the use of large currents to produce a strong magnetic field. In the presence of this field, the nuclei of certain atoms can be made to behave as tiny radio transmitters and emit radio waves similar to those used by FM stations. The hydrogen atom, which is so prevalent in the human body, can be made to behave in this fashion. The strength of the magnetic field determines where a given collection of hydrogen atoms will broadcast on an imaginary "FM dial." With a magnetic field that has a slightly different strength at different places, it is possible to associate the location on this imaginary FM dial with a physical location within the body. Computer processing of these locations produces the magnetic resonance image. When hydrogen atoms are used in this way, the image is essentially a map showing the distribution of hydrogen atoms within the body. Remarkably detailed magnetic resonance images can now be obtained, such as those shown in Figure 21.38. They provide doctors with a powerful diagnostic tool that complements those available from X-ray and other techniques.

THE PHYSICS OF . . .

television screens and computer display monitors.

Television sets and computer display monitors use electromagnets (solenoids) to produce images by exerting magnetic forces on moving electrons. An evacuated glass tube, called a cathode-ray tube (CRT), contains an electron gun that sends a narrow beam of high-speed electrons toward the screen of the tube, as illustrated in Figure 21.39*a*. The inner surface of the screen is covered with a phosphor coating, and when the electrons strike it, they generate a spot of visible light. This spot is called a pixel (a contraction of "picture element").

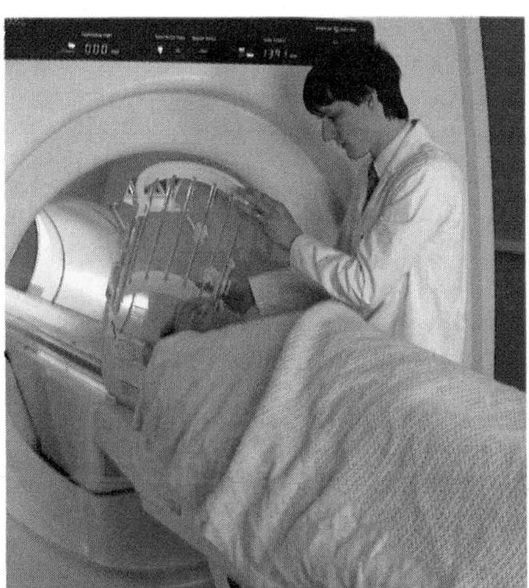

Figure 21.37 A magnetic resonance imaging machine. The patient is inserted into a circular opening at one end of a solenoid.

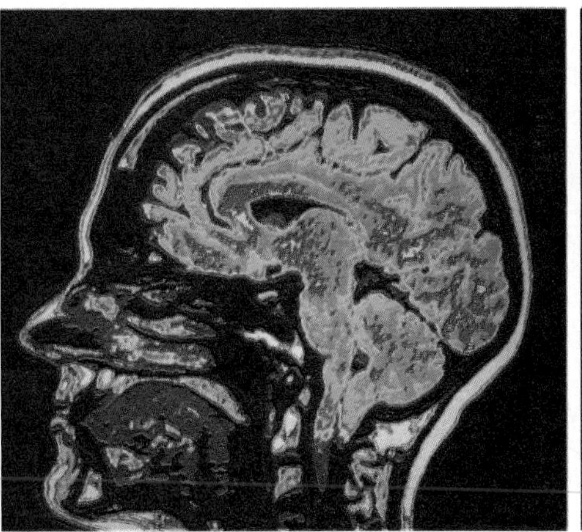

Healthy brain.

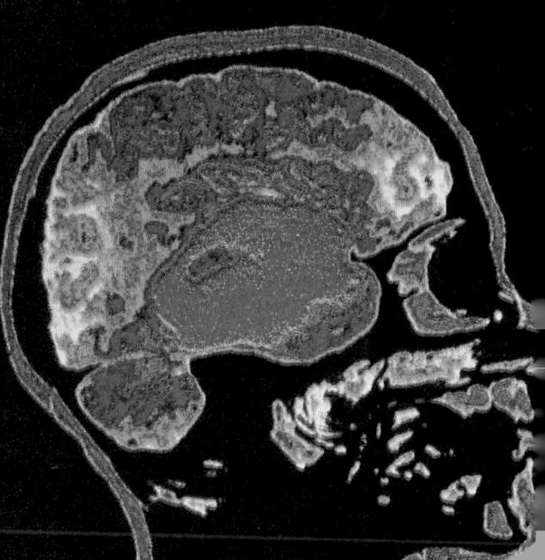

Diseased brain.

Figure 21.38 Magnetic resonance imaging provides one way to diagnose brain disorders. The MRI of the diseased brain shown here reveals the presence of a tumor (large red area). The colors are computer generated as an aid in distinguishing between different types of tissue.

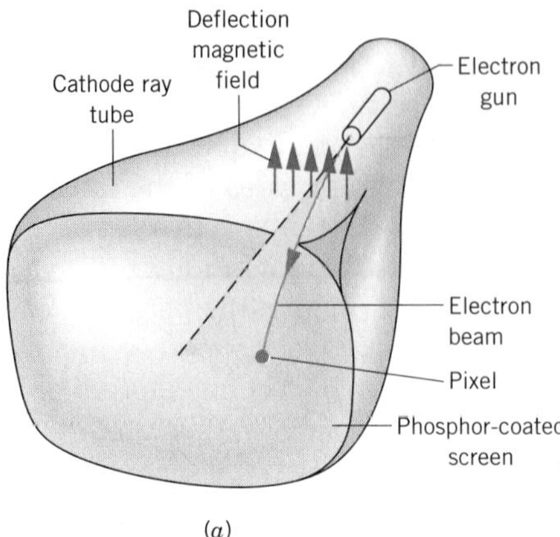

(*a*)

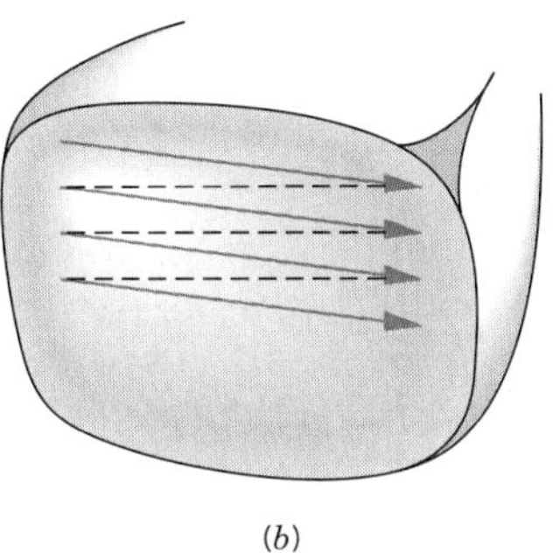

(*b*)

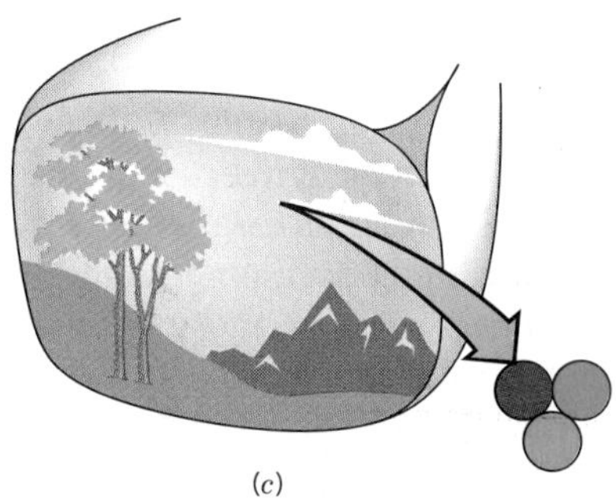

(*c*)

Figure 21.39 (*a*) A cathode-ray tube contains an electron gun, a magnetic field for deflecting the electron beam, and a phosphor-coated screen. (*b*) The image is formed by scanning the electron beam across the screen. (*c*) The red, green, and blue phosphors of a color TV.

To create a black-and-white picture, the electron beam is scanned rapidly from left to right across the screen. As the beam makes each horizontal scan, the intensity of the electrons striking the screen is changed by electronics controlling the electron gun, making the scan line brighter in some places and darker in others. When the beam reaches the right side of the screen, it is turned off and returned to the left side slightly below where it started (see part *b* of the figure). The beam is then scanned across the next line, and so on. In current TV sets, a complete picture consists of 525 scan lines from top to bottom and is formed in $\frac{1}{30}$ of a second. High-definition TV sets are being developed that use a greater number of scan lines.

The electron beam is deflected by a pair of electromagnets placed around the neck of the tube, between the electron gun and the screen. One electromagnet is responsible for producing the horizontal deflection of the beam and the other for the vertical deflection. For clarity, Figure 21.39*a* shows the net magnetic field at one instant generated by the electromagnets, and not the electromagnets themselves. The electric current in the electromagnets produces a net magnetic field that exerts a force on the moving electrons, causing their trajectories to bend and reach different points on the screen. Changing the current changes the field, so the electrons can be deflected to any point on the screen.

A color TV operates with three electron guns instead of one. And the single phosphor of a black-and-white TV is replaced by a large number of three-dot clusters of phosphors that glow red, green, and blue when struck by an electron beam, as indicated in Figure 21.39*c*. Each red, green, and blue color in a cluster is produced when electrons from one of the three guns strike the corresponding phosphor dot. The three dots are so close together that, from a normal viewing distance, they cannot be separately distinguished. Red, green, and blue are primary colors, so all other colors can be created by varying the intensities of the three beams focused on a cluster.

21.9 AMPERE'S LAW

We have seen that an electric current creates a magnetic field. However, the magnitude and direction of the field at any point in space depends on the specific geometry of the wire carrying the current. For instance, distinctly different magnetic fields surround a long, straight wire, a circular loop of wire, and a solenoid. Although different, each of these fields can be obtained from a general law known as ***Ampere's law.*** This law is valid for a wire with any geometrical shape and specifies the relationship between a current and its associated magnetic field.

To see how Ampere's law is stated, consider Figure 21.40, which shows two wires carrying currents I_1 and I_2. In general, there may be any number of currents. Around the wires we construct an arbitrarily shaped but closed path. Note that this path encloses or bounds a surface. The path is constructed from a large number of short segments, each of length $\Delta\ell$. Ampere's law deals with the product of $\Delta\ell$ and $B_{\parallel}$ for each segment, where $B_{\parallel}$ is the component of the magnetic field that is *parallel* to $\Delta\ell$ (see the enlarged cutaway view in the drawing). For magnetic fields that do not change as time passes, the law states that the sum of all the $B_{\parallel}\Delta\ell$ terms is proportional to the net current $I = I_1 + I_2$ passing through the surface bounded by the path. Ampere's law is stated below in equation form.

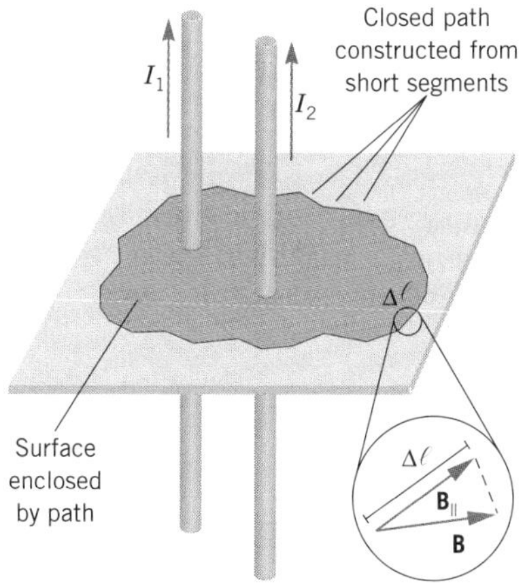

Figure 21.40 This setup is used in the text to explain Ampere's law.

Ampere's Law for Static Magnetic Fields

For any current geometry that produces a magnetic field that does not change in time,

$$\Sigma B_{\parallel}\Delta\ell = \mu_0 I \tag{21.8}$$

where $\Delta\ell$ is a small segment of length along a closed path of arbitrary shape around the current, $B_{\parallel}$ is the component of the magnetic field parallel to $\Delta\ell$, I is the net current passing through the surface bounded by the path, and μ_0 is the permeability of free space. The symbol Σ indicates that the sum of all $B_{\parallel}\Delta\ell$ terms must be taken around the closed path.

Ampere's law is valid for any current configuration, such as a long, straight wire, a single loop, or a solenoid. To illustrate the use of Ampere's law, we apply it in Example 12 to the special case of the current in a long, straight wire and show that it leads to the proper expression for the magnetic field.

Example 12 A Long, Straight Current-Carrying Wire

Use Ampere's law to obtain the magnetic field produced by the current in a long, straight wire.

REASONING Figure 21.26*a* shows that compass needles point in a circular pattern around the current-carrying wire, so we know that the magnetic field lines are circular.

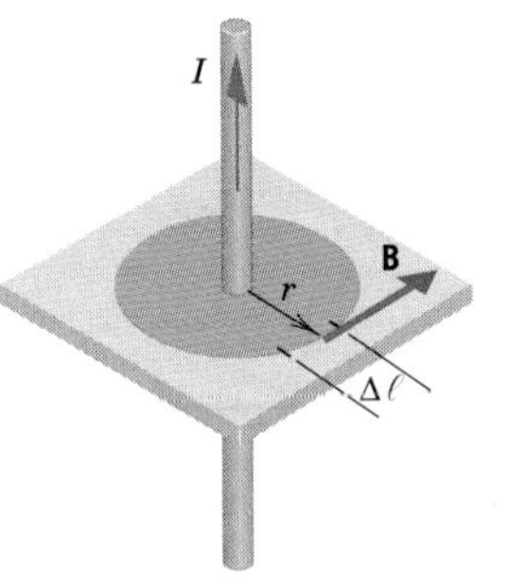

Figure 21.41 Example 12 uses Ampere's law to find the magnetic field in the vicinity of this long, straight, current-carrying wire.

Therefore, it is convenient to use a circular path of radius r when applying Ampere's law, as Figure 21.41 indicates.

SOLUTION Along the circular path in Figure 21.41, the magnetic field is everywhere parallel to $\Delta \ell$ and has a constant magnitude, since each point is at the same distance from the wire. Thus, $B_{\parallel} = B$ and

$$\Sigma B_{\parallel} \Delta \ell = B(\Sigma \Delta \ell) = \mu_0 I$$

But $\Sigma \Delta \ell$ is just the circumference $2\pi r$ of the circle, so Ampere's law reduces to

$$B(\Sigma \Delta \ell) = B(2\pi r) = \mu_0 I$$

Dividing both sides by $2\pi r$ shows that $\boxed{B = \mu_0 I/(2\pi r)}$, as given earlier in Equation 21.5.

21.10 MAGNETIC MATERIALS

FERROMAGNETISM

The similarity between the magnetic field lines in the neighborhood of a bar magnet and those around a current loop suggests that the magnetism in each case arises from a common cause. The field that surrounds the loop is created by the charges moving in the wire. The magnetic field around a bar magnet is also due to the motion of charges, but the motion is not that of a bulk current through the magnetic material. Instead, the motion responsible for the magnetism is that of the electrons within the atoms of the material.

The magnetism produced by electrons within an atom can arise from two motions. First, each electron orbiting the nucleus behaves like an atomic-sized loop of current that generates a small magnetic field; this situation is similar to the field created by the current loop in Figure 21.32. Second, each electron possesses a spin that also gives rise to a magnetic field. The net magnetic field created by the electrons within an atom is due to the combined fields created by their orbital and spin motions.

In most substances the magnetism produced at the atomic level tends to cancel out, with the result that the substance is nonmagnetic overall. However, there are some materials, known as ***ferromagnetic materials,*** in which the cancellation does not occur for groups of approximately $10^{16}-10^{19}$ neighboring atoms, because they have electron spins that are naturally aligned parallel to each other. This alignment results from a special type of quantum mechanical* interaction between the spins. The result of the interaction is a small but highly magnetized region of about 0.01 to 0.1 mm in size, depending on the nature of the material; this region is called a ***magnetic domain.*** Each domain behaves as a small magnet with its own north and south poles. Common ferromagnetic materials are iron, nickel, cobalt, chromium dioxide, and alnico (an *al*uminum–*ni*ckel–*co*balt alloy).

* The branch of physics called quantum mechanics is mentioned in Section 29.5, although a detailed discussion of quantum mechanics is beyond the scope of this book.

INDUCED MAGNETISM

Often the magnetic domains in a ferromagnetic material are arranged randomly, as Figure 21.42*a* illustrates for a piece of iron. In such a situation, the magnetic fields of the domains cancel each other, so the iron displays little, if any, overall magnetism. However, an unmagnetized piece of iron can be magnetized by placing it in an external magnetic field provided by a permanent magnet or an electromagnet. The external magnetic field penetrates the unmagnetized iron and *induces* (or "brings about") a state of magnetism in the iron by causing two effects on the domains. Those domains whose magnetism is parallel or nearly parallel to the external magnetic field grow in size at the expense of other domains that are not so oriented. Part *b* of the drawing shows the growing domains in gold. In addition, the magnetic alignment of some domains may rotate and become more oriented in the direction of the external field. The resulting preferred alignment of the domains gives the iron an overall magnetism, so the iron behaves like a magnet with associated north and south poles. In some types of ferromagnetic materials, such as the chromium dioxide used in cassette tapes, the domains remain aligned for the most part when the external magnetic field is removed, and the material thus becomes permanently magnetized.

The magnetism induced in a ferromagnetic material can be surprisingly large, even in the presence of a weak external field. For instance, it is not unusual for the induced magnetic field to be a hundred to a thousand times stronger than the external field that causes the alignment. For this reason, high-field electromagnets are constructed by wrapping the current-carrying wire around a solid core made from iron or other ferromagnetic material.

Induced magnetism explains why a permanent magnet sticks to a refrigerator door and why an electromagnet can pick up scrap iron at a junkyard. Notice in Figure 21.42*b* that there is a north pole at the end of the iron that is closest to the south pole of the permanent magnet. The net result is that the two opposite poles give rise to an attraction between the iron and the permanent magnet. Conversely, the north pole of the permanent magnet would also attract the piece of iron by inducing a south pole in the nearest side of the iron. In nonferromagnetic materials, such as aluminum and copper, the formation of magnetic domains does not occur, so magnetism cannot be induced into these substances. Consequently, magnets do not stick to aluminum cans or to copper pennies.

The fact that the soft drink truck in this photograph is hanging from an electromagnet indicates that the truck is made from an appreciable amount of ferromagnetic material. The field of the electromagnet acts on the magnetic domains in the ferromagnetic material, causing induced magnetism. Because of the induced magnetism, the electromagnet and the truck are attracted to one another with a force sufficiently strong to support the truck's weight.

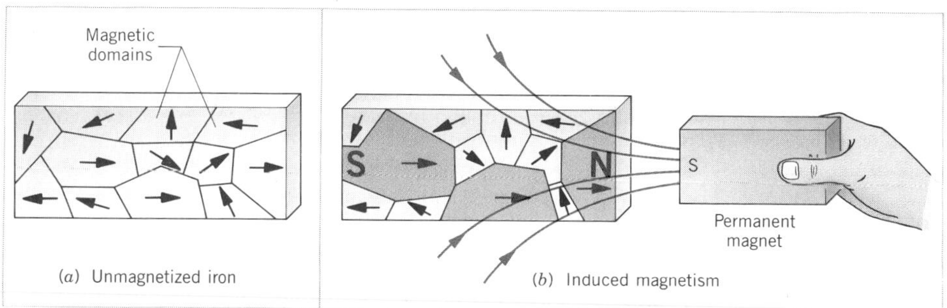

Figure 21.42 (*a*) Each magnetic domain is a highly magnetized region that behaves like a small magnet (represented by an arrow whose head indicates a north pole). An unmagnetized piece of iron consists of many domains that are randomly aligned. The size of each domain is exaggerated for clarity. (*b*) The external magnetic field of the permanent magnet causes those domains that are parallel or nearly parallel to the field to grow in size (shown in gold).

THE PHYSICS OF . . .

magnetic tape recording.

MAGNETIC TAPE RECORDING

The process of magnetic tape recording uses induced magnetism, as Figure 21.43 illustrates. The weak electrical signal from a microphone is routed to an amplifier where it is amplified. The current from the output of the amplifier is then sent to the recording head, which is a coil of wire wrapped around an iron core. The iron core has the approximate shape of a horseshoe with a small gap between the two ends. The ferromagnetic iron substantially enhances the magnetic field produced by the current in the wire.

When there is a current in the coil, the recording head becomes an electromagnet with a north pole at one end and a south pole at the other end. The magnetic field lines pass through the iron core and cross the gap. Within the gap, the lines are directed from the north pole to the south pole. Some of the field lines in the gap "bow outward," as Figure 21.43 indicates, the bowed region of magnetic field being called the *fringe field.* The fringe field penetrates the magnetic coating on the tape and induces magnetism in the coating. This induced magnetism is retained when the tape leaves the vicinity of the recording head and, thus, provides a means for storing audio information. Audio information is retained, because at any instant in time the way in which the tape is magnetized depends on the amount and direction of current in the recording head. The current, in turn, depends on the sound intensity picked up by the microphone, so that changes in the sound intensity that occur from moment to moment are preserved as changes in the tape's induced magnetization.

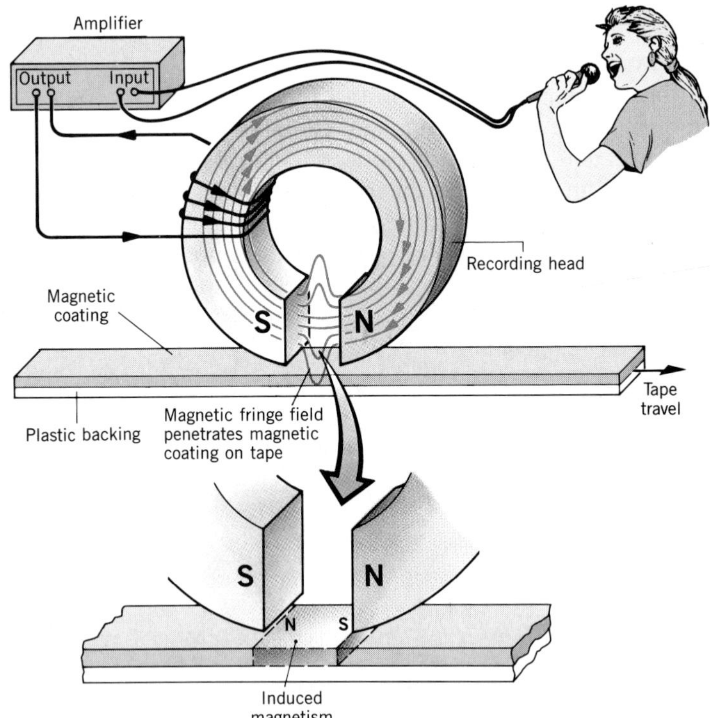

Figure 21.43 The magnetic fringe field of the recording head penetrates the magnetic coating on the tape and causes the coating to become magnetized.

MAGLEV TRAINS

THE PHYSICS OF . . .

a magnetically levitated train.

A magnetically levitated train—or maglev, for short—uses forces that arise from induced magnetism to levitate or float above a guideway. Since it rides a few centimeters above the guideway, a maglev does not need wheels. Freed from friction with the guideway, the train can achieve significantly greater speeds than do conventional trains. For example, the Transrapid maglev in Figure 21.44*a* has achieved speeds of 110 m/s.

Figure 21.44*a* shows that the Transrapid maglev achieves levitation with electromagnets mounted on arms that extend around and under the guideway. When a current is sent to an electromagnet, the resulting magnetic field creates induced magnetism in a rail mounted in the guideway. The upward attractive force from the induced magnetism is balanced by the weight of the train, so the train moves without touching the rail or the guideway.

Magnetic levitation only lifts the train and does not move it forward. Figure 21.44*b* illustrates how magnetic propulsion is achieved. In addition to the levitation electromagnets, propulsion electromagnets are also placed along the guideway. By controlling the direction of the currents in the train and guideway electromagnets, it is possible to create an unlike pole in the guideway just ahead of each electromagnet on the train and a like pole just behind. Each electromagnet on the train is thus both pulled and pushed forward by electromagnets in the guideway. By adjusting the timing of the like and unlike poles in the guideway, the speed of the train can be adjusted. Reversing the poles in the guideway electromagnets relative to those in the train serves to brake the train.

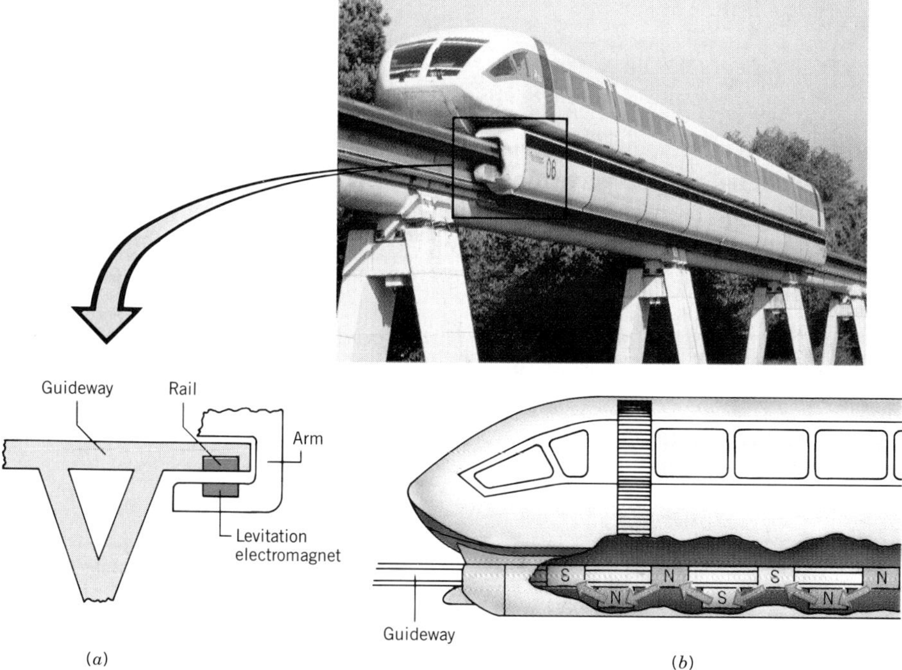

Figure 21.44 (*a*) The Transrapid maglev (a German train) has achieved speeds of 110 m/s (250 mph). The levitation electromagnets are drawn up toward the rail in the guideway, levitating the train. (*b*) The magnetic propulsion system.

21.11 OPERATIONAL DEFINITIONS OF THE AMPERE AND THE COULOMB

In Section 20.1 we defined current as the rate at which charge flows, or $I = q/t$ when the current is constant. Therefore, one way of measuring current is to determine the amount of charge q that flows in a time t. According to this procedure, one ampere of current exists when one coulomb of charge flows for one second. In practice, however, it is difficult to measure an ampere precisely by measuring the amount of charge flowing in a known time interval. It would be far superior if the ampere could be measured in terms of force and distance, quantities that can be measured with a high degree of precision. Such a measurement is possible in terms of the magnetic force that two current-carrying wires exert on each other.

Suppose the same current I is sent through two long, straight, parallel wires that are separated by a distance r. According to Equation 21.5, the magnetic field **B** produced at the location of one wire by the other wire has a magnitude of $B = \mu_0 I/(2\pi r)$. This magnetic field is perpendicular to the wire (see Figure 21.30) and exerts a force **F** on the wire:

$$F = ILB \sin 90° = \frac{\mu_0 I^2 L}{2\pi r} \tag{21.3}$$

With special instruments, this force can be measured accurately. The wire length L and the separation r can also be determined accurately, and μ_0 has been assigned the value of $4\pi \times 10^{-7}$ T·m/A. With these values, the equation above can be solved for the current I. For instance, suppose $F = 2.000 \times 10^{-7}$ N, $r = 1.000$ m, and $L = 1.000$ m. The current is

$$I = \sqrt{\frac{2\pi r F}{\mu_0 L}} = \sqrt{\frac{2\pi(1.000 \text{ m})(2.000 \times 10^{-7} \text{ N})}{(4\pi \times 10^{-7} \text{ T}\cdot\text{m/A})(1.000 \text{ m})}} = 1.000 \text{ A}$$

Therefore, one ampere of current is defined as the amount of electric current in each of two long, parallel wires that gives rise to a magnetic force per unit length of 2×10^{-7} N/m on each wire when the wires are separated by one meter. This definition provides a means for measuring current in terms of force and distance and obviates the need to define the ampere in terms of the amount of moving charge per unit time.

With the ampere defined in terms of force and distance, the coulomb can now be defined as the quantity of electrical charge that passes a given point in one second when the current is one ampere, or 1 C = 1 A·s. This definition is preferred, since it is possible to measure electric current and time more accurately than the amount of moving charge.

INTEGRATION OF CONCEPTS

FIELDS AND FORCES

The concept of an electric field is introduced in Chapter 18. An electric field is produced by one or more charged objects and exists in the region around them. Electric field lines are often drawn as an aid in visualizing the magnitude and direction of the electric field within the region. At any given location, the electric field exerts an electric force on a charged object placed there, the force being the product of the charge and the electric field at that point. The direction of the force is either parallel or antiparallel to the electric field, depending on whether the charge is positive or negative, respectively. In the present chapter, we see that a magnetic field is produced by permanent magnets or moving charges, such as an electric current, and exists in the region around them. Magnetic field lines are also drawn as an aid in visualizing the magnitude and direction of the magnetic field. A magnetic field, like an electric field, can exert a force on a charged object within it, but only if the object is moving and has a velocity component that is perpendicular to the magnetic field. The direction of the magnetic force is perpendicular to the plane defined by the velocity of the object and the magnetic field. Thus, the concept of a field is very useful, for it can be used to describe the electric and magnetic forces that are exerted on charged objects.

THE MAGNETIC FORCE AND NATURE'S FUNDAMENTAL FORCES

There are four fundamental forces in nature, fundamental in the sense that all other forces can be understood as manifestations of one or more of the four. Tension, friction, and the elastic force of a spring, for example, are not fundamental forces, but the gravitational force is. Another force that we have encountered is the force that one electrically charged particle exerts on another charged particle. This force is one part of a fundamental force called the electromagnetic force. The electromagnetic force contains two parts, an electric part and a magnetic part. Both parts, however, derive from the same source, the electric charge carried by the particles. Whether or not the particles are moving, they exert on each other the electric force specified in Chapter 18 by Coulomb's law. When the particles move, the other part of the electromagnetic force appears, the part that we have called the magnetic force in the present chapter.

SUMMARY

A magnet has a north pole and a south pole. The north pole is the end that points toward the north magnetic pole of the earth when the magnet is freely suspended. **Like poles repel each other, and unlike poles attract each other.**

A **magnetic field** exists in the space around a magnet. The magnetic field is a vector whose direction at any point is the direction indicated by the north pole of a small compass needle placed at that point. The magnitude B of the magnetic field at any point in space is defined as $B = F/(q_0 v \sin \theta)$, where F is the magnitude of the magnetic force that acts on a charge q_0 whose velocity **v** makes an angle θ with respect to the magnetic field. The SI unit for the magnetic field is the tesla (T). The direction of the magnetic force is perpendicular to both **v** and **B**, and for a positive charge the direction can be determined with the aid of Right-Hand Rule No. 1 (RHR-1, see Section 21.2). The magnetic force on a moving negative charge is opposite to the force on a moving positive charge.

If a particle of charge q and mass m moves with speed v perpendicular to a uniform magnetic field **B**, the magnetic force causes the charge to move on a circular path of radius $r = mv/(qB)$. A constant magnetic force does no work on the charged particle, because the direction of the force is always perpendicular to the motion of the particle. Being unable to do work, the magnetic force cannot change the kinetic energy of the particle; however, the magnetic force does change the direction in which the particle moves.

A **mass spectrometer** is an instrument that can determine the masses of atoms and molecules. The **Hall emf** develops across a current-carrying metal or semiconductor that has been placed in a magnetic field, because the moving charges are deflected by the magnetic force. The polarity of the Hall emf indicates whether the charge carriers are positive or negative.

An electric current, being composed of moving charges, can experience a magnetic force when placed in a magnetic field. For a straight wire that has a length L and carries a current I, the magnetic force has a magnitude of $F = ILB \sin \theta$, where θ is the angle between the directions of I and **B**. The direction of the force is perpendicular to both I and **B** and is given by RHR-1.

Magnetic forces can exert a **torque** on a current-carrying loop of wire and thus cause the loop to rotate. If a current I exists in a coil of wire with N turns, each of area A, in the presence of a magnetic field **B**, the coil experiences a torque of magnitude $\tau = NIAB \sin \phi$, where ϕ is the angle between the direction of the magnetic field and the normal to the plane of the coil.

An electric current produces a magnetic field, with different current geometries giving rise to different field patterns. For a **long, straight wire,** the magnetic field lines are circles centered on the wire, and their direction is given by RHR-2 (see Section 21.8). The magnitude of the field at a radial distance r from the wire is $B = \mu_0 I/(2\pi r)$, where I is the current and μ_0 is the permeability of free space ($\mu_0 = 4\pi \times 10^{-7}$ T·m/A). The magnetic field at the center of a **flat circular coil** consisting of N turns, each of radius R, is $B = N\mu_0 I/(2R)$. The coil has associated with it a north pole on one side and a south pole on the other side. The side of the coil that behaves like a north pole can be predicted by using RHR-2. A **solenoid** is a coil of wire wound in the shape of a helix. Inside a long solenoid the magnetic field is nearly constant and has the value $B = \mu_0 nI$, where n is the number of turns per unit length of the solenoid. One end of a solenoid behaves like a north pole, and the other end like a south pole, as can be predicted by using RHR-2.

Ampere's law, as presented in Equation 21.8, specifies the relationship between a current and its associated magnetic field.

Ferromagnetic materials, such as iron, are made up of tiny regions called domains, each of which behaves as a small magnet. In an unmagnetized ferromagnetic material, the domains are randomly aligned. In a permanent magnet, many of the domains are aligned, and a high degree of magnetism results. An unmagnetized ferromagnetic material can be induced into becoming magnetized by placing it in an external magnetic field.

QUESTIONS

1. Magnetic field lines, like electric field lines, never intersect. Suppose it were possible for two magnetic field lines to intersect at a point in space. Discuss what this would imply about the force(s) that act on a charge moving through such a point, thereby ruling out the possibility of field lines crossing.

2. Suppose you accidentally use your left hand, instead of your right hand, to determine the direction of the magnetic force on a positive charge moving in a magnetic field. Do you get the correct answer? If not, what direction do you get?

3. A charged particle, passing through a certain region of space, has a velocity whose magnitude and direction remain constant. (a) If it is known that the external magnetic field is zero everywhere in this region, can you conclude that the external electric field is also zero? (b) If it is known that the external electric field is zero everywhere, can you conclude that the external magnetic field is also zero? Explain.

4. Suppose that the positive charge in Figure 21.8*a* were launched from the negative plate toward the positive plate, directly opposite to the electric field. A sufficiently strong electric field would prevent the charge from striking the positive plate. Suppose the positive charge in Figure 21.8*b* were launched from the south pole toward the north pole, directly opposite to the magnetic field. Would a sufficiently strong magnetic field prevent the charge from reaching the north pole? Account for your answer.

5. A stationary charge is located between the poles of a horseshoe magnet. Is a magnetic force exerted on the charge? Why?

6. Review Conceptual Example 4 as background for this question. Three particles move through a constant magnetic field and follow the paths shown in the drawing. Determine whether each particle is positively charged, negatively charged, or neutral. Give a reason for each answer.

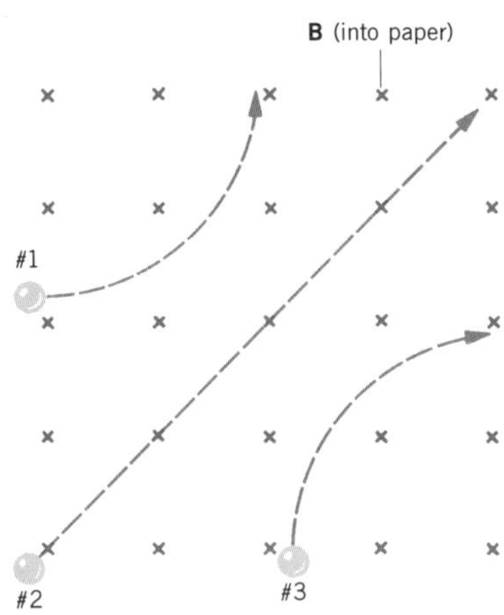

7. Review Conceptual Example 4 as background for this question. Three particles have identical charges and masses. They enter a constant magnetic field and follow the paths shown in the picture. Which particle is moving the fastest, and which is moving the slowest? Justify your answers.

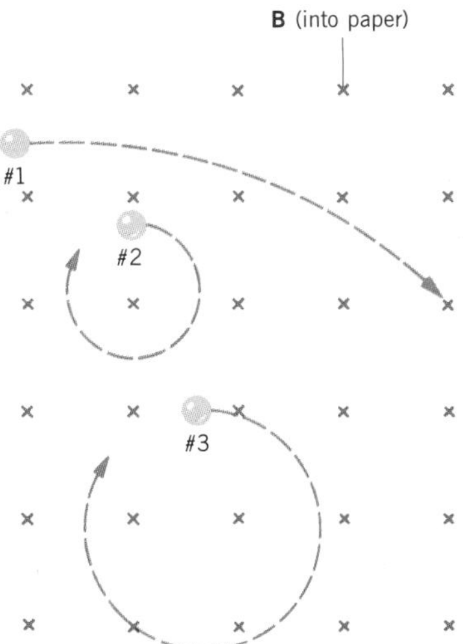

8. Refer to Figure 21.10. Assume that the particle in the picture is a proton. If an electron is projected at point 1 with the same velocity **v**, it will not follow the same path as the proton, unless the magnetic field is adjusted. Explain how the magnitude and direction of the field must be changed.

9. The drawing shows a top view of four interconnected chambers. A negative charge is fired into chamber 1. By turning on separate magnetic fields in each chamber, the charge can be made to exit from chamber 4, as shown. (a) Describe how the magnetic field in each chamber should be directed. (b) If the speed of the charge is v when it enters chamber 1, what is the speed of the charge when it exits chamber 4? Why?

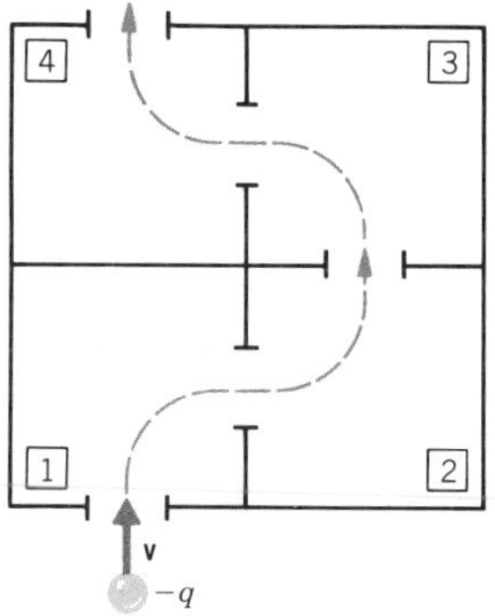

10. A positive charge moves along a circular path under the influence of a magnetic field. The magnetic field is perpendicular to the plane of the circle, as in Figure 21.10. If the velocity of the particle is reversed at some point along the path, will the particle retrace its path? If not, draw the new path. Explain.

11. A positively charged particle travels on a circular path in the presence of a magnetic field, as in Figure 21.10. A uniform electric field is then turned on. Draw the path of the particle when the electric field is directed parallel to the magnetic field.

12. The drawing shows a particle carrying a positive charge $+q$ at the coordinate origin, as well as a target located in the third quadrant. A uniform magnetic field is directed perpendicularly into the plane of the paper. The charge can be projected in the plane of the paper only, along the positive or negative x or y axis. Thus, there are four possible directions for the initial velocity of the particle. The particle can be made to hit the target for only two of the four directions. Which two are they? Give your reasoning, along with the two paths that the particle can follow on its way to the target.

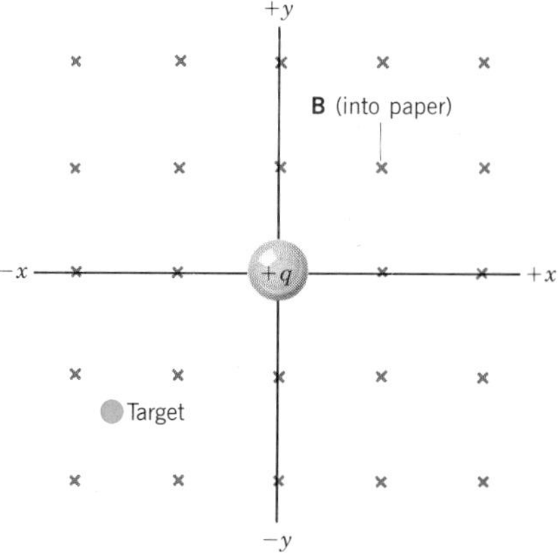

13. Refer to Figure 21.15. (a) What happens to the direction of the magnetic force if the current is reversed? (b) What happens to the direction of the force if *both* the current and the magnetic poles are reversed? Explain your answers.

14. Suppose that the magnet in a galvanometer has lost some of its magnetism due to age. Would the reading on the scale of the galvanometer be greater than or less than the value when the galvanometer was new? Account for your answer.

15. In Figure 21.30, assume that the current I_1 is larger than the current I_2. In parts *a* and *b*, decide whether there are places where the total magnetic field is zero. State whether they are located to the left of both wires, between the wires, or to the right of both wires. Give your reasoning.

16. The drawing shows an end-on view of three parallel wires that are perpendicular to the plane of the paper. In two of the wires the current is directed into the paper, while in the remaining wire the current is directed out of the paper. The two outermost wires are held rigidly in place. Which way will the middle wire move? Explain.

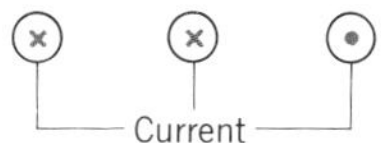

17. For each electromagnet at the left of the drawing, explain whether it will be attracted to or repelled from the adjacent magnet at the right.

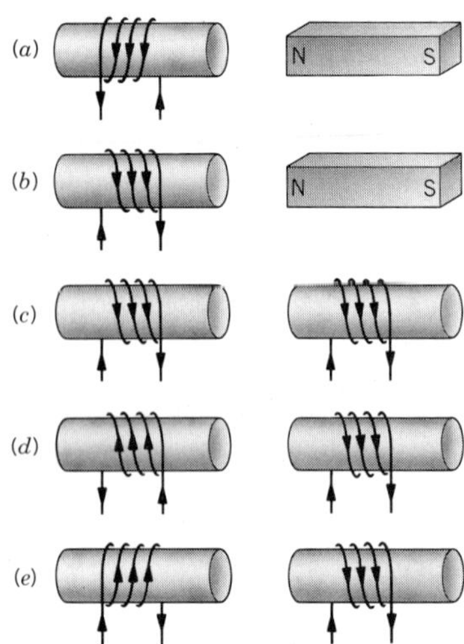

18. Refer to Figure 21.5. If the earth's magnetism is assumed to originate from a large circular loop of current within the earth, how is the plane of this current loop oriented relative to the magnetic axis, and what is the direction of the current around the loop?

19. There are four wires viewed end-on in the drawing. They are long, straight, and perpendicular to the plane of the paper. Their cross sections lie at the corners of a square. Currents of the same magnitude are in each of these wires. Choose the direction of the current for each wire, so that when any single current is turned off, the total magnetic field at point *P* (the center of the square) is directed toward a corner of the square. Account for your answer.

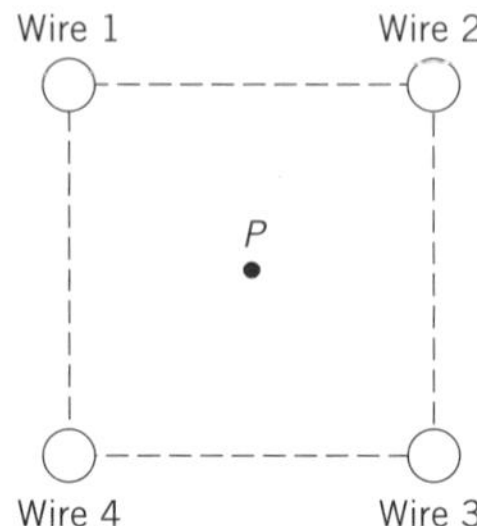

20. Suppose you have two bars, one of which is a permanent magnet and the other of which is not a magnet, but is made from a ferromagnetic material like iron. The two bars look exactly alike. (a) Using a third bar, which is known to be a magnet, how can you determine which of the look-alike bars is the permanent magnet and which is not? (b) Can you determine the identities of the look-alike bars with the aid of a third bar that is not a magnet, but is made from a ferromagnetic material? Give a reason for your answers.

21. In a TV commercial that advertises a soda pop, a strong electromagnet picks up a delivery truck carrying cans of the soft drink. The picture switches to the interior of the truck, where cans are seen to fly upward and stick to the roof just beneath the electromagnet. Are these cans made entirely of aluminum? Justify your answer.

PROBLEMS

Section 21.1 Magnetic Fields, Section 21.2 The Force That a Magnetic Field Exerts on a Moving Charge

1. A charge of 12 μC, traveling with a speed of 9.0×10^6 m/s in a direction perpendicular to a magnetic field, experiences a magnetic force of 8.7×10^{-3} N. What is the magnitude of the field?

2. A particle with a charge of $+6.0$ μC and a speed of 25 m/s enters a uniform magnetic field whose magnitude is 0.15 T. For each of the cases in the drawing, find the magnitude and direction of the magnetic force on the charge.

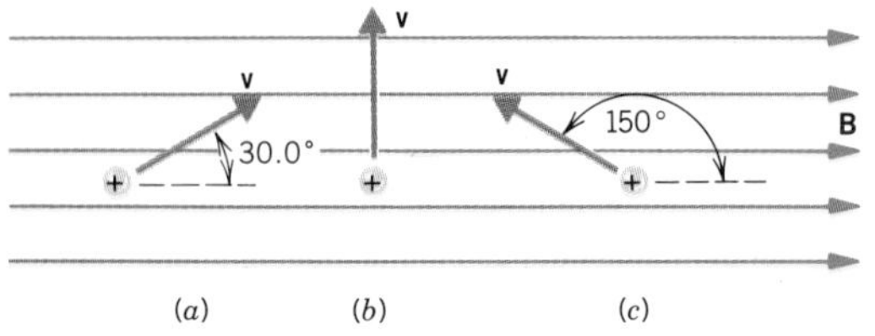

3. A proton, traveling with a velocity of 4.5×10^6 m/s due east, experiences a maximum magnetic force of 8.0×10^{-14} N due south. (a) What is the magnitude and direction of the magnetic field? (b) Answer part (a), assuming the proton is replaced by an electron.

4. A charged body, moving with a velocity of 8.0×10^4 m/s at an angle of 30.0° with respect to a magnetic field of strength 5.6×10^{-5} T, experiences a force of 2.0×10^{-4} N. What is the magnitude of the charge?

5. At a certain location, the horizontal component of the earth's magnetic field is 2.5×10^{-5} T, due north. A proton moves eastward with just the right speed, so the magnetic force on it balances its weight. Find the speed of the proton.

6. In New England, the horizontal component of the earth's magnetic field has a magnitude of 1.6×10^{-5} T. An electron is shot straight up from the ground with a speed of 2.1×10^6 m/s. What is the magnitude of the acceleration caused by the magnetic force?

7. A charged particle is projected perpendicularly into a magnetic field at a speed of 1400 m/s and experiences a force of magnitude F. If the speed of the particle were 1800 m/s, at what angle θ (less than 90°) with respect to the field would the particle experience the same force magnitude F?

* **8.** Draw an x, y axes system on a sheet of paper. A 0.048-T magnetic field points in the direction of the $+x$ axis. A 0.065-T field points in the direction of the $-y$ axis. A particle carrying a charge of $+2.0 \times 10^{-5}$ C is fired perpendicularly out of the plane of the paper, toward you, with a speed of 4.2×10^3 m/s. (a) Find the magnitude of the net magnetic force on the particle. (b) Find the angle with respect to the $+x$ axis that gives the direction of the force.

* **9.** In a television set, electrons are accelerated from rest through a potential difference of 15 kV. The electrons then pass through a 0.35-T magnetic field that deflects them to the appropriate spot on the screen. Find the magnitude of the maximum force that an electron can experience.

* **10.** There is a 0.200-T magnetic field directed along the $+x$ axis and a field of unknown magnitude along the $+y$ axis. A particle carrying a charge of 6.50×10^{-5} C experiences a maximum force of 0.455 N when traveling at a speed of 2.00×10^4 m/s through the region where the fields are. Find the magnitude of the unknown field.

Section 21.3 The Motion of a Charged Particle in a Magnetic Field, Section 21.4 The Mass Spectrometer, Section 21.5 The Hall Effect

11. An electron moves at a speed of 6.0×10^6 m/s perpendicular to a constant magnetic field. The path is a circle of radius 1.3×10^{-3} m. (a) Draw a sketch showing the magnetic field and the electron's path. (b) What is the magnitude of the field? (c) Find the magnitude of the electron's acceleration.

12. An ionized helium atom has a mass of 6.6×10^{-27} kg and a speed of 4.4×10^5 m/s. The atom moves perpendicular to a 0.75-T magnetic field on a circular path of radius 0.012 m. What is the magnitude of the charge of the helium atom?

13. A beam of protons moves in a circle of radius 0.25 m. The protons move perpendicular to a 0.30-T magnetic field. (a) What is the speed of each proton? (b) Determine the magnitude of the centripetal force that acts on each proton.

14. The solar wind is a thin, hot gas given off by the sun. Charged particles in this gas enter the magnetic field of the earth and can experience a magnetic force. Suppose a charged particle traveling with a speed of 9.0×10^6 m/s encounters the earth's magnetic field at an altitude where the field has a magnitude of 1.2×10^{-7} T. Assuming that the particle's velocity is perpendicular to the magnetic field, find the radius of the circular path on which the particle would move if it were (a) an electron and (b) a proton.

15. A mass spectrometer uses a potential difference of 2.00 kV to accelerate a singly charged ion ($+e$) to the proper speed. A 0.400-T magnetic field then bends the ion into a circular path of radius 0.226 m. What is the mass of the ion?

16. An ion source in a mass spectrometer produces deuterons. (A deuteron is a particle that has twice the mass of a proton, but the same charge.) Each deuteron is accelerated from rest through a potential difference of 2.00×10^3 V, after which it enters a 0.600-T magnetic field. Find the radius of its circular path.

17. In a given magnetic field, a proton and an electron move on circular paths that have the same radii. Find the ratio of their speeds, $v_{\text{electron}}/v_{\text{proton}}$.

18. Suppose that an ion source in a mass spectrometer produces *doubly* ionized gold ions (Au^{2+}), each with a mass of 3.27×10^{-25} kg. The ions are accelerated from rest through a potential difference of 1.00 kV. Then, a 0.500-T magnetic field causes the ions to follow a circular path. Determine the radius of the path.

19. Two isotopes of carbon, carbon-12 and carbon-13, have masses of 19.92×10^{-27} kg and 21.59×10^{-27} kg, respectively. These two isotopes are singly ionized ($+e$) and each is given a speed of 6.667×10^5 m/s. The ions then enter the bending region of a mass spectrometer where the magnetic field is 0.8500 T. Determine the spatial separation between the two isotopes after they have traveled through a half-circle. (*Hint: The spatial separation is the difference between the diameters of the trajectories.*)

20. Review Conceptual Example 2 before attempting this problem. Derive an expression for the magnitude v of the velocity "selected" by the velocity selector. This expression should give v in terms of the strengths E and B of the electric and magnetic fields, respectively.

* **21.** Review Conceptual Example 2 as background for this problem. A charged particle moves through a velocity selector at a constant speed in a straight line. The electric field of a velocity selector is 5.65×10^3 N/C, while the magnetic field is 0.114 T. When the electric field is turned off, the charged particle travels on a circular path whose radius is 2.90 cm. Find the charge-to-mass ratio of the particle.

* **22.** An electron moves in a circular orbit of radius 1.7 m in a magnetic field of 2.2×10^{-5} T. The electron moves perpendicular to the magnetic field. Determine the kinetic energy of the electron.

* **23.** As preparation for this problem, review Conceptual Example 4. The radius of the track for particle 3 (kinetic energy $= KE_3$) is 16 times larger than the initial radius of the track for particle 1 (initial kinetic energy $= KE_1$). Determine the ratio KE_3/KE_1.

* **24.** A proton with a speed of 2.2×10^6 m/s is shot into a region between two plates that are separated by a distance of 0.18 m. As the drawing shows, a magnetic field exists between the plates, and it is perpendicular to the velocity of the proton.

What must be the magnitude of the magnetic field, so the proton just misses colliding with the opposite plate?

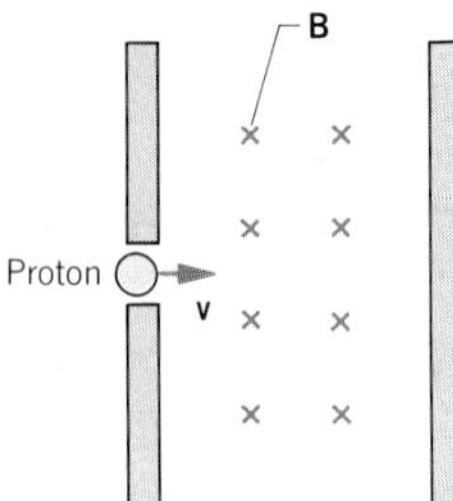

*25. A positively charged particle of mass 7.2×10^{-8} kg is traveling due east with a speed of 85 m/s. The particle enters a 0.31-T uniform magnetic field, and 2.2×10^{-3} s later the particle leaves the field one-quarter of a turn later, heading due south with a speed of 85 m/s. All during the motion the particle moves perpendicular to the magnetic field. (a) What is the magnitude of the magnetic force acting on the particle? (b) Determine the charge of the particle.

*26. An α-particle is the nucleus of a helium atom; the orbiting electrons are missing. The α-particle contains two protons and two neutrons, and has a mass of 6.64×10^{-27} kg. Suppose an α-particle is accelerated from rest through a potential difference and then enters a region where its velocity is perpendicular to a 0.0210-T magnetic field. With what angular speed ω does the α-particle move on its circular path?

**27. Refer to question 12 (not problem 12) before starting this problem. Suppose that the target discussed there is located at the coordinates $x = -0.10$ m and $y = -0.10$ m. In addition, suppose that the particle is a proton and the magnetic field has a magnitude of 0.010 T. The speed at which the particle is projected is the same for either of the two paths leading to the target. Find the speed.

Section 21.6 The Force on a Current in a Magnetic Field

28. The drawing shows wires of length L and current I, lying in a plane that is perpendicular to a magnetic field **B**. In all cases $B = 0.25$ T, $L = 0.60$ m, and $I = 15$ A. Find the magnitude and direction of the magnetic force on each wire.

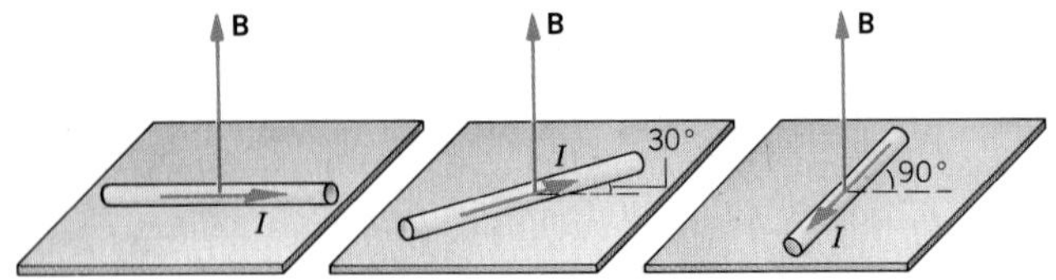

29. An electric power line carries a current of 1400 A in a location where the earth's magnetic field is 5.0×10^{-5} T. The line makes an angle of 75° with respect to the field. Determine the magnitude of the magnetic force on a 120-m length of line.

30. A straight wire in a magnetic field experiences a force of 0.030 N when the current in the wire is 2.7 A. What is the current in the wire when it experiences a force of 0.047 N?

31. A square coil of wire containing a single turn is placed in a uniform 0.25-T magnetic field, as the drawing shows. Each side has a length of 0.32 m, and the current in the coil is 12 A. Determine the magnitude of the magnetic force on each of the four sides.

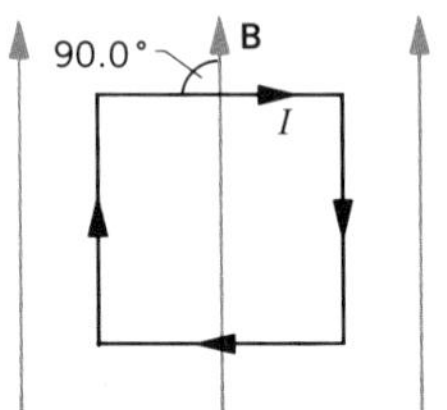

32. A 45-m length of wire is stretched horizontally between two vertical posts. The wire carries a current of 75 A and experiences a magnetic force of 0.15 N. Find the magnitude of the earth's magnetic field at the location of the wire, assuming the field makes an angle of 60.0° with respect to the wire.

33. A wire of length 0.655 m carries a current of 21.0 A. In the presence of a 0.470-T magnetic field, the wire experiences a force of 5.46 N. What is the angle (less than 90°) between the wire and the magnetic field?

34. At New York City, the earth's magnetic field has a vertical (downward) component of 5.2×10^{-5} T and a horizontal component of 1.8×10^{-5} T that is directed toward geographic north. What is the magnitude of the magnetic force on a long, straight wire, 8.0 m in length, that carries a 35-A current due east?

*35. A 125-turn rectangular coil of wire is hung from one arm of a balance, as the drawing shows. With the magnetic field turned off, a mass M is added to the pan on the other arm to balance the mass of the coil. When a constant magnetic field of magnitude 0.200 T is turned on and there is a current of 8.50 A in the coil, how much *additional* mass m must be added to regain the balance?

*36. The drawing shows a thin, uniform rod, which has a length of 0.40 m and a mass of 0.080 kg. This rod lies in the plane of the paper and is attached to the floor by a hinge at point P. A uniform magnetic field of 0.31 T is directed perpendicularly into the plane of the paper. There is a current $I = 3.8$ A in the rod, which does not rotate clockwise or counterclockwise. Find the angle θ. *(Hint: The magnetic force may be taken to act at the center of gravity.)*

*37. A copper rod of length 0.85 m is lying on a frictionless table (see the drawing). Each end of the rod is attached to a fixed wire by an unstretched spring whose spring constant is $k = 75$ N/m. A magnetic field with a strength of 0.16 T is

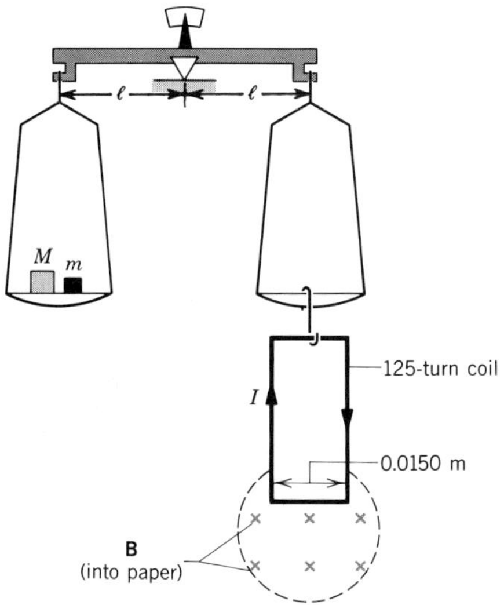

Problem 35

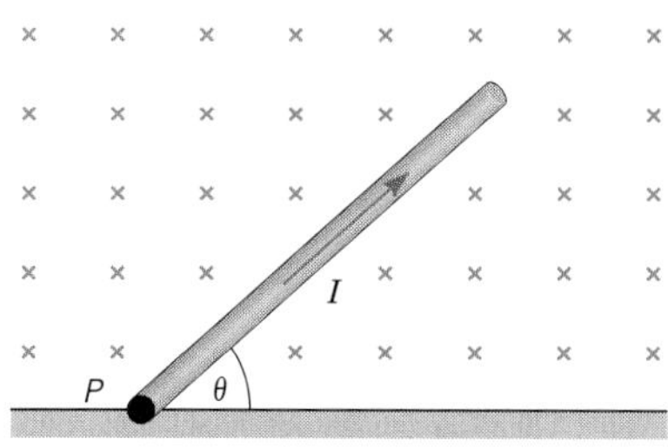

Problem 36

oriented perpendicular to the surface of the table. (a) What must be the direction of the current in the copper rod that causes the springs to stretch? (b) If the current is 12 A, by how much does each spring stretch?

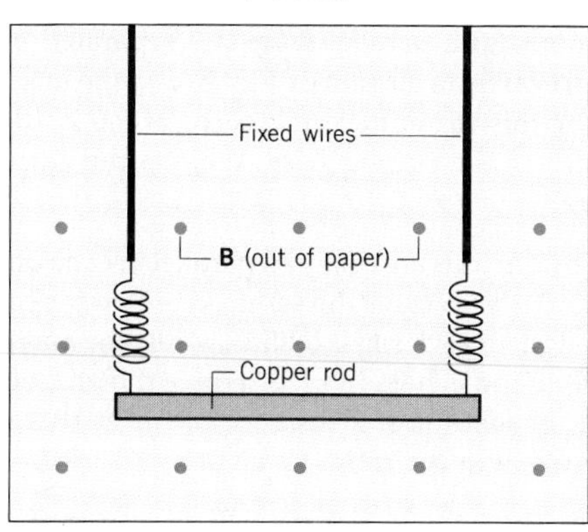

**38. A 0.20-kg aluminum rod is lying on top of two conducting rails that are separated by 1.6 m. A 0.050-T magnetic field has the direction shown in the drawing. The coefficient of static friction between the rod and a rail is $\mu_s = 0.45$. (a) How much current must be sent through the rod before the rod begins to move? (b) In what direction will the rod move, toward the battery or away from it? Explain.

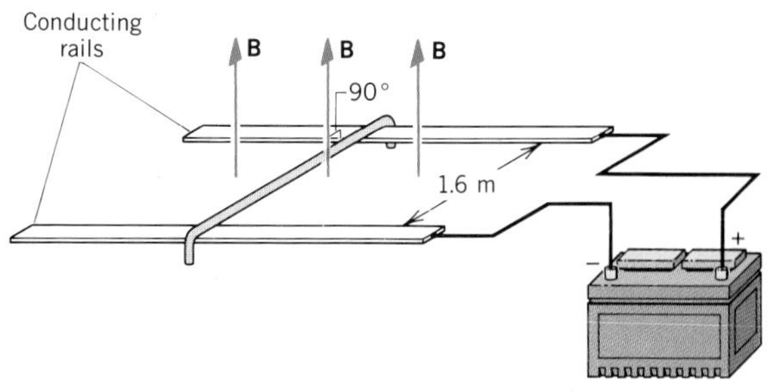

**39. A horizontal wire of length 0.20 m and mass 0.080 kg is hung from the ceiling of a room by two massless strings. A 0.070-T magnetic field is directed from the ceiling to the floor. When a current of 42 A passes through the wire, the wire swings upward through an angle ϕ, as the drawing shows. Find (a) the angle ϕ and (b) the tension in each of the two strings.

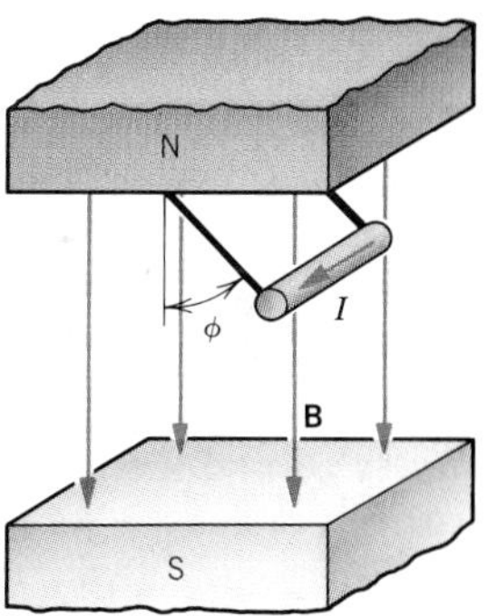

Section 21.7 The Torque on a Current-Carrying Coil

40. A circular coil of one turn is made from a wire of length 5.00×10^{-2} m. There is a current of 2.00 A in the wire. In the presence of a 1.50-T magnetic field, what is the largest torque that this loop can experience?

41. The proton has an intrinsic magnetic moment of 1.4×10^{-26} A·m². If the magnetic moment makes an angle of $\phi = 64°$ with respect to a 0.65-T magnetic field, what is the torque exerted on the proton?

42. The 1200-turn coil in a dc motor has an area per turn of 1.1×10^{-2} m². The design for the motor specifies that the magnitude of the maximum torque is 5.8 N·m when the coil is placed in a 0.20-T magnetic field. What is the current in the coil?

43. A 0.50-m length of wire is formed into a single-turn, square loop in which there is a current of 12 A. The loop is placed in a magnetic field of 0.12 T, as in Figure 21.21*a*. What is the maximum torque that the loop can experience?

44. A coil carries a current and experiences a torque due to a magnetic field. The value of the torque is 80.0% of the maximum possible torque. (a) What is the smallest angle between the magnetic field and the normal to the plane of the coil? (b) Make a drawing, showing how this coil would be oriented relative to the magnetic field. Be sure to include the angle in the drawing.

45. Review Conceptual Example 7 as background for this problem. A square coil of N turns carries a current I in a magnetic field B. The coil is made from a length L of wire. Derive an expression for the maximum torque that this coil experiences. Aside from numerical factors, the expression contains the variables N, I, B, and L.

46. Two pieces of the same wire have the same length. From one piece, a square coil containing a single turn is made. From the other, a circular coil containing a single turn is made. These coils, when carrying the same current in the same magnetic field, experience different maximum torques τ_{square} and τ_{circle}. Find the ratio $\tau_{\text{circle}}/\tau_{\text{square}}$.

* **47.** The coil in Figure 21.21*a* contains 380 turns and has an area per turn of 2.5×10^{-3} m^2. The magnetic field is 0.12 T, and the current in the coil is 0.16 A. A brake shoe is pressed perpendicularly against the shaft to keep the coil from turning. The coefficient of static friction between the shaft and the brake shoe is 0.70. If the radius of the shaft is 0.010 m, what is the minimum magnitude of the force pressing the brake shoe against the shaft?

* **48.** In the model of the hydrogen atom due to Niels Bohr, the electron moves around the proton at a speed of 2.2×10^6 m/s in a circle of radius 5.3×10^{-11} m. Considering the orbiting electron to be a small current loop, determine the magnetic moment associated with this motion. *(Hint: The electron travels around the circle in a time equal to the period of the motion.)*

* **49.** A charge of 4.0×10^{-6} C is placed on a small conducting sphere that is located at the end of a thin insulating rod whose length is 0.20 m. The rod rotates with an angular speed of $\omega = 150$ rad/s about an axis that passes perpendicularly through its other end. Find the magnetic moment of the rotating charge. *(Hint: The charge travels around a circle in a time equal to the period of the motion.)*

Section 21.8 Magnetic Fields Produced by Currents

50. A long, straight wire produces a 4.5×10^{-5}-T magnetic field at a distance of 0.16 m from the wire. What is the current in the wire?

51. In a lightning bolt, 15 C of charge flows in a time of 1.5×10^{-3} s. Assuming that the lightning bolt can be represented as a long, straight line of current, what is the magnitude of the magnetic field at a distance of 25 m from the bolt?

52. What must be the radius of a circular loop of wire so the magnetic field at its center is 1.4×10^{-4} T when the loop carries a current of 6.0 A?

53. A long solenoid consists of 1400 turns of wire and has a length of 0.65 m. There is a current of 4.7 A in the wire. What is the magnitude of the magnetic field within the solenoid?

54. A $+6.00$ μC charge is moving with a speed of 7.50×10^6 m/s parallel to a long, straight wire. The wire carries a current of 67.0 A in a direction opposite to that of the moving charge, and is 5.00 cm from the charge. Find the magnitude and direction of the force on the charge.

55. Suppose in Figure 21.30*a* that $I_1 = I_2 = 25$ A and that the separation between the wires is 0.016 m. By applying an external magnetic field (created by a source other than the wires) it is possible to cancel the mutual repulsion of the wires. This external field must point along the vertical direction. (a) Does the external field point up or down? Explain. (b) What is the magnitude of the external field?

56. Two long, straight wires are separated by 0.120 m. The wires carry currents of 8.0 A in opposite directions, as the drawing indicates. Find the magnitude of the net magnetic field at the points labeled A and B.

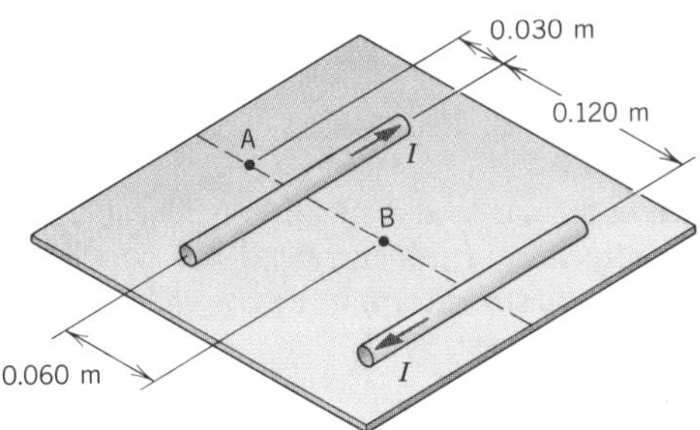

57. Two straight, parallel wires are separated by 0.15 m. The first wire carries a current of 125 A, and the magnetic field produced by this current exerts a force of 3.0×10^{-3} N on a 2.1-m length of the second wire. What is the current in the second wire?

58. Two rigid rods are oriented parallel to each other and to the ground. The rods carry the same current in the same direction. The length of each rod is 0.85 m, while the mass of each is 0.073 kg. One rod is held in place above the ground, and the other floats beneath it at a distance of 8.2×10^{-3} m. Determine the current in the rods.

* **59.** As background for this problem, review Conceptual Example 10. A rectangular current loop is located near a long, straight wire that carries a current of 12 A (see the drawing). The current in the loop is 25 A. Determine the net magnetic force that acts on the loop.

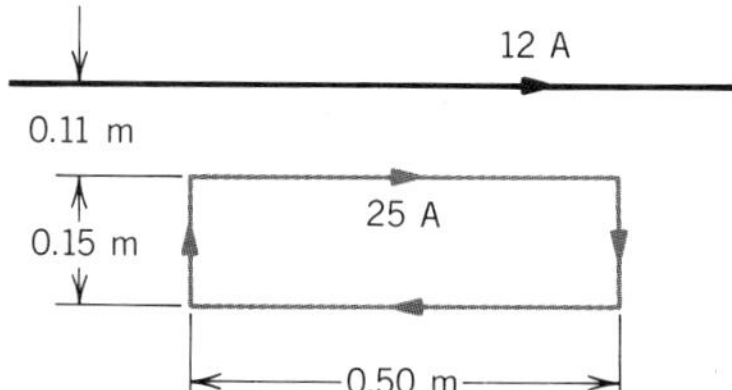

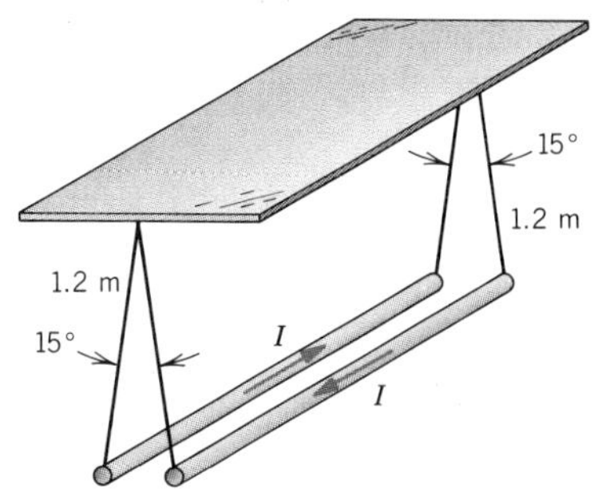

Problem 64

this wire. The current in the second wire is 25 A. Use Ampere's law to find the magnitude of the magnetic field at a distance of $r = 0.50$ m from the wires when the currents are (a) in the same direction and (b) in opposite directions.

***60.** Two circular coils are concentric and lie in the same plane. The inner coil contains 120 turns of wire, has a radius of 0.012 m, and carries a current of 6.0 A. The outer coil contains 150 turns and has a radius of 0.017 m. What must be the magnitude and direction (relative to the current in the inner coil) of the current in the outer coil, such that the net magnetic field at the common center of the two coils is zero?

***61.** A piece of copper wire has a resistance per unit length of 5.90×10^{-3} Ω/m. The wire is wound into a thin, flat coil of many turns that has a radius of 0.140 m. The ends of the wire are connected to a 12.0-V battery. Find the magnetic field strength at the center of the coil.

***62.** Two long, straight, parallel wires *A* and *B* are separated by a distance of one meter. They carry currents in opposite directions, and the current in wire *A* is one-third of that in wire *B*. On a line drawn perpendicular to the wires and passing through them, find the point where the net magnetic field is zero. Determine this point relative to wire *A*.

****63.** The drawing shows an end-on view of three wires. They are long, straight, and perpendicular to the plane of the paper. Their cross sections lie at the corners of a square. The currents in wires 1 and 2 are $I_1 = I_2 = I$ and are directed into the paper. What is the direction of the current in wire 3, and what is the ratio I_3/I, such that the total magnetic field at the empty corner is zero?

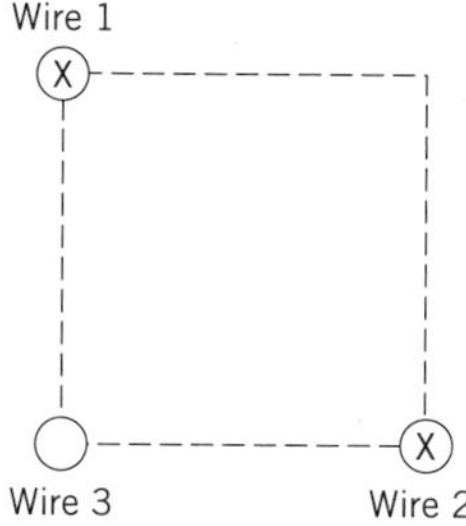

****64.** The drawing shows two long, straight wires that are suspended from a ceiling. Each of the four strings suspending the wires has a length of 1.2 m. The mass per unit length of each wire is 0.050 kg/m. When the wires carry identical currents in opposite directions, the angle between the strings holding the two wires is 15°. What is the current in each wire?

Section 21.9 Ampere's Law

65. The wire in Figure 21.41 carries a current of 15 A. Suppose that a second long, straight wire is placed right next to this wire. The current in the second wire is 25 A. Use Ampere's law to find the magnitude of the magnetic field at a distance of $r = 0.50$ m from the wires when the currents are (a) in the same direction and (b) in opposite directions.

66. Suppose a uniform magnetic field is everywhere perpendicular to this page. The field points directly upward toward you. A circular path is drawn on the paper. Use Ampere's law to show that there can be no net current passing through the circular surface.

***67.** A very long, hollow cylinder is formed by rolling up a thin sheet of copper. Electric charges flow along the copper sheet parallel to the axis of the cylinder. The arrangement is, in effect, a hollow tube of current *I*. Use Ampere's law to show that the magnetic field (a) is $\mu_0 I/(2\pi r)$ outside the cylinder at a distance *r* from the axis and (b) is zero at any point within the hollow interior of the cylinder. *(Hint: For closed paths, use circles perpendicular to and centered on the axis of the cylinder.)*

****68.** A long, cylindrical conductor is solid throughout and has a radius *R*. Electric charges flow parallel to the axis of the cylinder and pass uniformly through the entire cross section. The arrangement is, in effect, a solid tube of current I_0. The current per unit cross-sectional area (i.e., the current density) is $I_0/(\pi R^2)$. Use Ampere's law to show that the magnetic field inside the conductor at a distance *r* from the axis is $\mu_0 I_0 r/(2\pi R^2)$. *(Hint: For a closed path, use a circle of radius r perpendicular to and centered on the axis. Note that the current through any surface is the area of the surface times the current density.)*

ADDITIONAL PROBLEMS

69. Due to friction with the air, an airplane has acquired a net charge of 1.70×10^{-5} C. The plane moves with a speed of 2.80×10^{2} m/s at an angle θ with respect to the earth's magnetic field, the magnitude of which is 5.00×10^{-5} T. The magnetic force on the airplane has a magnitude of 2.30×10^{-7} N. Find the angle θ. (There are two possible angles.)

70. A charge $q_1 = 25.0$ μC moves with a speed of 4.50×10^{3} m/s perpendicular to a uniform magnetic field. The charge experiences a magnetic force of 7.31×10^{-3} N. A second charge $q_2 = 5.00$ μC travels at an angle of 40.0° with respect to the same magnetic field and experiences a $1.90 \times$

10^{-3}-N force. Determine (a) the magnitude of the magnetic field and (b) the speed of q_2.

71. A circular coil of wire has a radius of 0.10 m. The coil has 50 turns and a current of 15 A, and is placed in a magnetic field whose magnitude is 0.20 T. (a) Determine the magnetic moment of the coil. (b) What is the maximum torque the coil can experience in this field?

72. A long, straight wire carrying a current of 305 A is placed in a uniform magnetic field whose magnitude is 7.00×10^{-3} T. The wire is perpendicular to the field. Find a point in space where the net magnetic field is zero. Locate this point by specifying its perpendicular distance from the wire.

73. A long solenoid has 1400 turns per meter of length, and it carries a current of 3.5 A. A small circular coil of wire is placed inside the solenoid with the normal to the coil oriented at an angle of 90.0° with respect to the axis of the solenoid. The coil consists of 50 turns, has an area of 1.2×10^{-3} m^2, and carries a current of 0.50 A. Find the torque exerted on the coil.

74. Near the equator in South America the earth's magnetic field has a strength of 3.0×10^{-5} T; the field is parallel to the surface of the earth points due north. A straight wire, 25 m in length, has an east-west orientation and experiences a magnetic force of 0.041 N, directed vertically down (toward the earth). What is the magnitude and direction of the current in the wire?

75. A charged particle with a charge-to-mass ratio of 5.7×10^8 C/kg travels on a circular path that is perpendicular to a magnetic field whose magnitude is 0.72 T. How much time does it take for the particle to complete one revolution?

76. A circular loop of wire and a long, straight wire each carry the same current, as the drawing shows. The loop and wire lie in the same plane. The net magnetic field at the center of the loop is zero. Find the distance H, expressing your answer in terms of R, the radius of the loop.

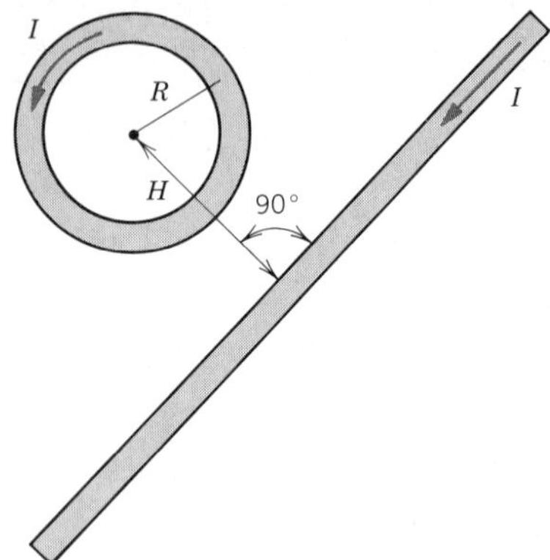

***77.** The drawing shows a charge entering a 0.52-T magnetic field. The charge has a speed of 270 m/s and moves perpendicular to the magnetic field. Just as the particle enters the magnetic field, an electric field is turned on. What must be the magnitude and direction of the electric field such that the *net* force on the particle is twice the magnetic force?

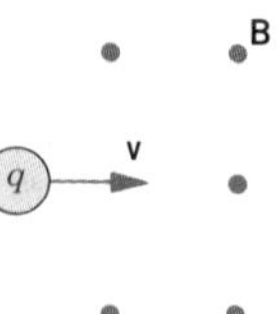

***78.** The electrons in the beam of a television tube have a kinetic energy of 2.40×10^{-15} J. Initially, the electrons move horizontally from west to east. The vertical component of the earth's magnetic field points down, toward the surface of the earth, and has a magnitude of 2.00×10^{-5} T. (a) In what direction are the electrons deflected by this field component? (b) What is the acceleration of an electron in part (a)?

***79.** A square coil and a rectangular coil are each made from the same length of wire. Each contains a single turn. The long sides of the rectangle are twice as long as the short sides. Find the ratio $\tau_{\text{square}}/\tau_{\text{rectangle}}$ of the maximum torques that these coils experience in the same magnetic field when they contain the same current.

***80.** A charge of $+3.5 \times 10^{-5}$ C is distributed uniformly around a thin ring of insulating material. The ring has a radius of 0.25 m and rotates with an angular speed of $\omega = 6500$ rad/s about an axis perpendicular to the plane of the ring and passing through its center. Determine the magnitude of the magnetic field produced at the center of the ring.

***81.** Two charged particles have *linear momenta* with the same magnitude, but particle 1 has three times the charge of particle 2. Both travel perpendicular to a uniform magnetic field. What is the ratio r_1/r_2 of the radii of the circles on which they move?

****82.** Singly ionized atoms of neon-20 and neon-22 follow circular paths in the bending region of a mass spectrometer. (a) What is the ratio of the neon-22 radius to that of neon-20? (b) What is the ratio of the radius of a *doubly* ionized atom of neon-22 to that of a singly ionized atom of neon-20?

****83.** The drawing shows two wires that carry the same current of $I = 85.0$ A and are oriented perpendicular to the plane of the paper. The current in one wire is directed out of the paper, while the current in the other is directed into the paper. Find the magnitude and direction of the net magnetic field at point P.

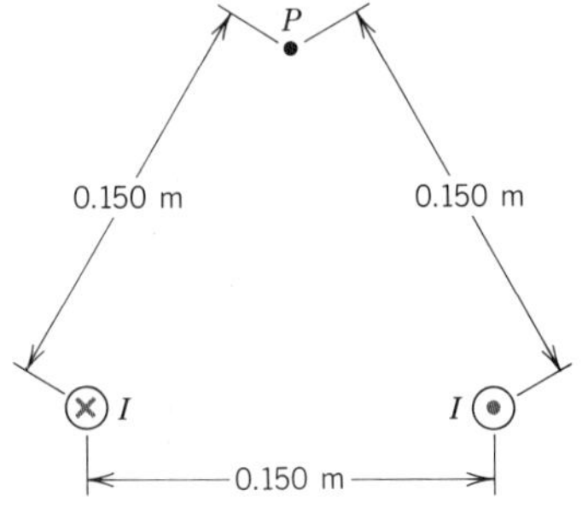

MYSTERIES SOLVED BY PHYSICS

Finding the Nuestra Señora de Atocha

Locating sunken wrecks is a long, painstaking, expensive, and dangerous job. However, the rewards can be considerable for the archaeologist, who can gain insights into past civilizations, or the treasure hunter, who can gain a fortune in precious metals and jewels. One of the greatest treasures ever lost was on board the *Nuestra Señora de Atocha,* a Spanish galleon that sank in 1622 off the Florida keys. Treasure hunter Mel Fisher set out to find the *Atocha* in 1964. He was not alone in this effort. Other treasure hunters were after the gold, silver, and jewels on board. After failing to find the *Atocha* where she was alleged to be, Fisher had the archives in Seville, Spain, searched for records of the ship. His perseverance in finding records of the *Atocha*'s last voyage paid off when his researcher found another location for the wreck. However, the *Atocha* was destroyed in a hurricane. Being made of wood, parts of it floated away after it broke up. As Fisher later discovered, the *Atocha* and its contents were actually strewn over eighteen square miles of ocean bottom.

How does one find traces of a ship tens, hundreds, or even thousands of meters underwater? Physics provides several alternatives. If any of the ship is resting on the ocean bottom, rather than buried in it, sonar can help locate it. If there is sufficient metal from the ship's hull or cargo, then a magnetometer can be used, which measures changes in the earth's magnetic field caused by the presence of the metal. Or, cameras and lights can be employed that are designed using basic physics and high technology to withstand pressures greater than 400 times atmospheric pressure. (The *Titanic* was discovered in this way.)

The *Atocha* was sunk on a sandy ocean bottom. After three and a half centuries, it was unlikely that the ship would be visible in the sand. The wood of the ship's hull would not create a magnetic distortion to register on a magnetometer, but its cargo of silver and gold and its bronze cannons would. Trailing a magnetometer sensitive to metal within a distance of twenty meters, Fisher's team began a long, systematic search covering 120 000 square miles, finding innumerable metal items, including a baby carriage, live torpedoes, toxic chemical dumps, airplanes, and meteorites. Then, in 1973 while investigating a strong magnetometer reading, Fisher's team discovered the *Atocha.*

Archaeologists shudder at the thought of quickly stripping the ocean floor bare of artifacts, but treasure hunters want their booty quickly. While this conflict may never be resolved, at least some of the *Atocha*'s legacy was studied by archaeologists as it was uncovered. To locate pieces from the wreck without disturbing them (so that archaeologists could study the relationships between different parts of the wreck), physics played a role in the form of hand-held magnetometers. Just like the ones used to comb dry land for dropped coins, the underwater magnetometers can locate metal without moving it. This is in contrast to the more drastic vacuuming or blowing operations that move both sand and artifacts.

The artifacts recovered from the *Atocha* provided archaeologists with valuable information about craftsmanship and geopolitical dynamics during the early 17th century. To Fisher and his investors, the *Atocha*'s treasures were worth more than 600 million dollars.

This diver is holding a small part of the treasure recovered from the Spanish galleon Nuestra Señora de Atocha, *which sank in 1622 off the coast of southern Florida.*

CHAPTER 22 ELECTROMAGNETIC INDUCTION

We have seen in Chapter 21 that an electric current produces a magnetic field. The interrelationship between electricity and magnetism goes even further than this, for we will now see that a magnetic field can be used to produce an emf, called an induced emf, which can generate a current in an electric circuit. In the photo, for example, the pickups in Lenny Kravits' electric guitar use magnetic fields to generate an induced emf. This emf is then routed to amplifiers and speakers, where it is converted into sound. The playback heads of a cassette deck, as well as many types of microphones, work on the same principle. Bicycle computers that display speed and other information for cyclists also use a magnetic field to produce an induced emf. And on a much grander scale, generators at power plants utilize magnetic fields in providing the electricity for our homes and industries. Furthermore, the distribution of electricity to homes and industries depends on a device called a transformer, the operation of which is based on induced emf. In this chapter we will examine the conditions under which an induced emf can be generated. These conditions are summarized in one of the most important laws in physics, Faraday's law of electromagnetic induction. We will show that the polarity of the induced emf in a circuit is such that the electrical energy being generated is consistent with the law of conservation of energy. The relationship between the polarity of the induced emf and the conservation of energy is known as Lenz' law. Finally, we will consider how electrical energy can be transferred from one circuit to another by means of self-induction and mutual-induction.

22.1 INDUCED EMF AND INDUCED CURRENT

There are a number of ways a magnetic field can be used to generate an electric current, and Figure 22.1 illustrates one of them. This drawing shows a bar magnet and a helical coil of wire to which an ammeter is connected. When there is no *relative* motion between the bar magnet and the coil, as in part *a* of the drawing, the ammeter reads zero, indicating there is no current. However, when the magnet moves toward the coil, as in part *b*, a current appears in the coil. As the magnet approaches, the magnetic field that it creates at the location of the coil becomes stronger and stronger, and it is this *changing* magnetic field that produces the current. When the magnet moves away from the coil, as in part *c*, a current also exists, but the direction of the current is reversed. Now the magnetic field at the coil becomes weaker as the magnet moves away. Once again, it is the *changing* magnetic field at the coil that generates the current.

A current would also be created in Figure 22.1 if the magnet were held stationary and the coil were moved, because the magnetic field at the coil would be changing as the coil approached or receded from the magnet. Only relative motion between the magnet and the coil is needed to generate a current; it does not matter which one moves.

The current in the coil is called an ***induced current,*** because the current is brought about (or "induced") by a changing magnetic field. Since a source of emf is always needed to produce a current, the coil itself behaves as if it were a source of emf. The emf is known as an ***induced emf.*** Thus, a changing magnetic field induces an emf in the coil, and the emf leads to an induced current.

THE PHYSICS OF . . .

bicycle computers.

Induced emf and induced current are used in bicycle computers, which are popular among cycling enthusiasts. This type of device consists of a small computer that attaches to the handlebar, a sensor that connects to the bike frame, and a magnet that attaches to the spokes (see Figure 22.2). The sensor contains a coil of wire in which an induced emf and an induced current appear each time the magnet passes by as the wheel turns. The computer counts the pulses of current. Since the computer also includes an internal clock, it can determine the number of pulses per second. From this information and the radius of the wheel, the computer determines the speed and presents the result on a liquid crystal display (LCD).

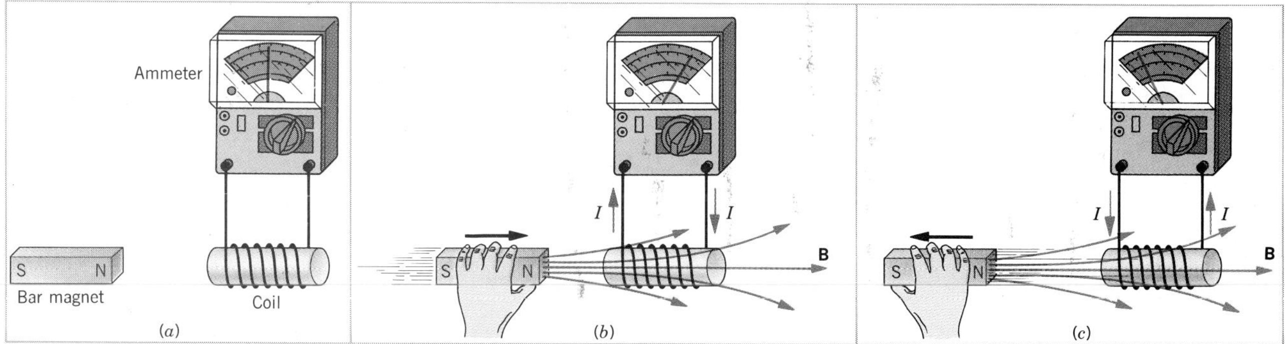

Figure 22.1 (*a*) When there is no relative motion between the coil of wire and the bar magnet, there is no current in the coil. (*b*) A current is created in the coil when the magnet moves toward the coil. (*c*) A current also exists when the magnet moves away from the coil, but the direction of the current is opposite to that in (*b*).

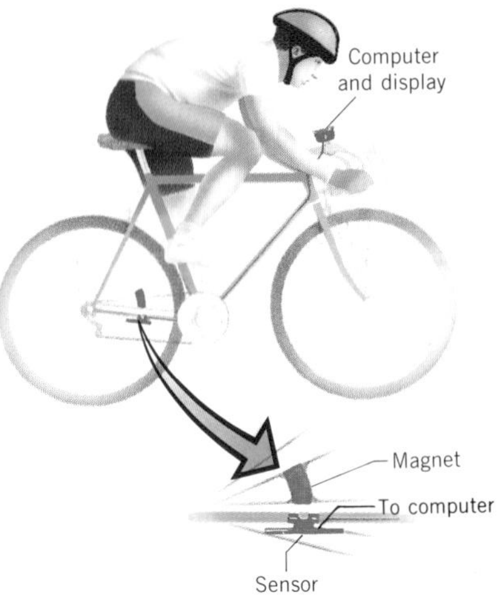

Figure 22.2 A bicycle computer uses the emf and current induced in a sensor coil to determine the speed of the bike.

Figure 22.3 shows other ways to induce an emf and a current in a coil. Parts *a* and *b* of the drawing illustrate that an emf can be induced by *changing the area* of a coil in a constant magnetic field. Here the shape of the coil is being distorted so as to reduce the area. As long as the area is changing, an induced emf and current exist; they vanish when the area is no longer changing. If the distorted coil is returned to its original circular shape, thereby increasing the area, an oppositely directed current is generated while the area is changing.

Part *c* of Figure 22.3 indicates that an induced emf is also generated when a coil of constant area is rotated in a constant magnetic field and the *orientation* of the coil *changes* with respect to the field. When the rotation stops, the emf, and hence the current, vanishes.

In each of the examples above, both an emf and a current are induced in the coil because the coil is part of a complete, or closed, circuit. If the circuit were open—perhaps because of an open switch—there would be no induced current. However, an emf would still be induced in the coil, whether the current exists or not.

Changing the magnetic field, changing the area of a coil, and changing the orientation of a coil are all methods that can be used to create an induced emf. The phenomenon of producing an induced emf with the aid of a magnetic field is called ***electromagnetic induction.*** The next section discusses yet another method by which an induced emf can be created, namely, by moving a conducting rod through a magnetic field.

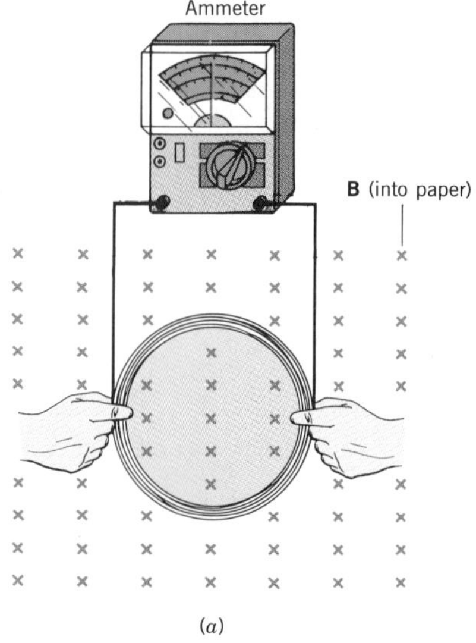

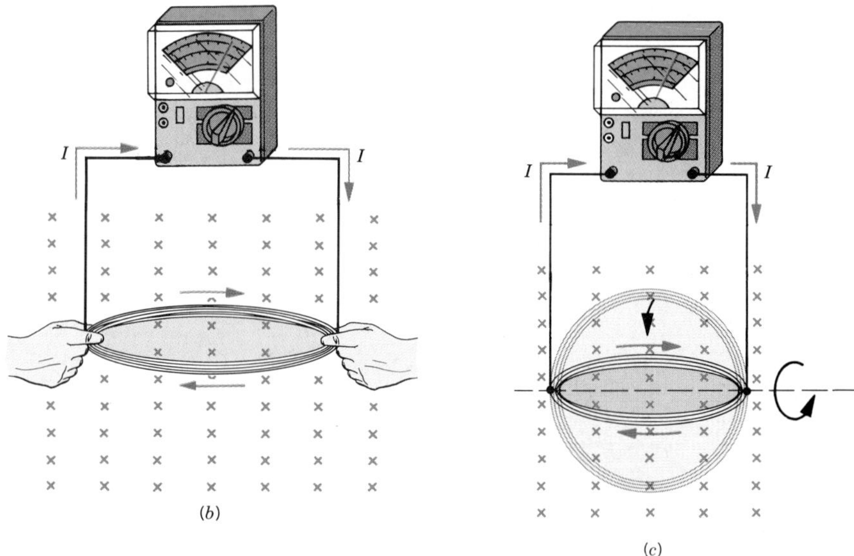

Figure 22.3 (*a*) No current exists in a coil of constant area (the shaded region) that is located in a constant magnetic field. (*b*) While the area of the coil is changing, an induced emf and current are generated. (*c*) An induced emf and current are also produced while the coil is rotating about an axis perpendicular to the magnetic field.

22.2 MOTIONAL EMF

THE EMF INDUCED IN A MOVING CONDUCTOR

When a conducting rod moves through a constant magnetic field, an emf is induced in the rod. This special case of electromagnetic induction arises as a result of the magnetic force (see Section 21.2) that acts on a moving charge. Consider the metal rod of length L moving to the right in Figure 22.4*a*. The velocity **v** of the rod is constant and is perpendicular to a uniform magnetic field ***B***. Each charge q within the rod also moves with a velocity **v** and experiences a magnetic force of magnitude $F = qvB$, according to Equation 21.1. By using RHR-1, it can be seen

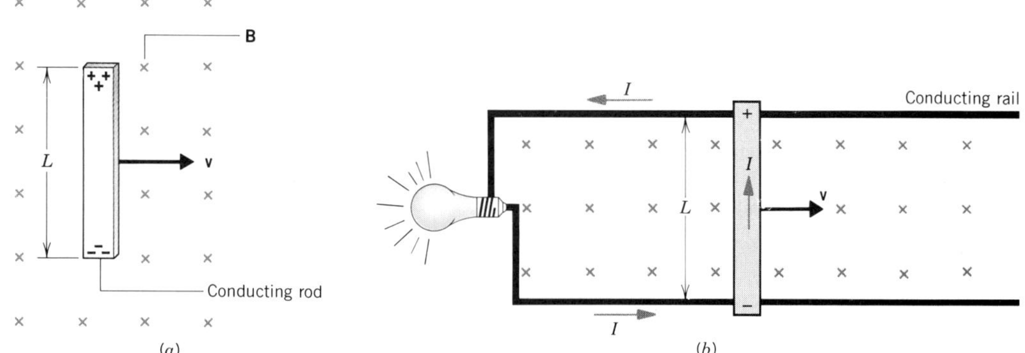

Figure 22.4 (*a*) When a conducting rod moves at right angles to a constant magnetic field, the magnetic force causes opposite charges to appear at the ends of the rod, giving rise to an induced emf. (*b*) The induced emf causes an induced current I to appear in the circuit.

that the mobile, free electrons are driven to the bottom of the rod, leaving behind an equal amount of positive charge at the top. (Remember to reverse the direction of the force that RHR-1 predicts, since the electrons have a negative charge.) The positive and negative charges continue to accumulate, until the attractive electric force that the positive and negative charge groups exert on each other becomes equal in magnitude to the magnetic force. When the electric force balances the magnetic force, equilibrium is reached and no further charge separation occurs.

The separated charges on the ends of the moving conductor give rise to an induced emf, called a ***motional emf,*** because it originates from the motion of charges through a magnetic field. The motional emf exists as long as the rod moves. If the rod is brought to a halt, the magnetic force vanishes, with the result that the attractive electric force reunites the positive and negative charges and the emf disappears. The emf of the moving rod is similar to that between the terminals of a battery. The difference between a battery and the rod is that the emf of a battery is produced by chemical reactions, whereas the emf of the rod is created by the agent that moves the rod through the magnetic field.

The fact that the electric and magnetic forces balance at equilibrium in Figure 22.4*a* can be used to determine the magnitude of the motional emf $\mathcal{E}$. The electric force acting on the positive charge q at the top of the rod is Eq, where E is the magnitude of the electric field due to the separated charges. According to Equation 19.7 (without the minus sign), the electric field magnitude is given by the voltage between the ends of the rod (the emf $\mathcal{E}$) divided by the length L of the rod. Thus, the electric force is $Eq = (\mathcal{E}/L)q$. The magnetic force is qvB, according to Equation 21.1, since the charge q moves perpendicular to the magnetic field. Since these two forces balance, it follows that $(\mathcal{E}/L)q = qvB$. The emf, then, is

[Motional emf when v, B, and L are mutually perpendicular]

$$\mathcal{E} = vBL \tag{22.1}$$

As expected, $\mathcal{E} = 0$ when $v = 0$, for no motional emf is developed in a stationary rod. Greater speeds and stronger fields lead to greater emfs for a given length L. As with batteries, $\mathcal{E}$ is expressed in volts. In Figure 22.4*b* the rod is sliding on conducting rails that form part of a closed circuit, and L is the length of the rod between the rails. Due to the emf, electrons flow in a clockwise direction around the circuit as long as the rod continues to move. Positive charge would flow in the direction opposite to the electron flow, so the conventional current I is drawn counterclockwise in the picture. Example 1 illustrates how to determine the electrical energy that the motional emf delivers to a device such as the light bulb in the drawing.

Example 1 Operating a Light Bulb with Motional Emf

Suppose the rod in Figure 22.4*b* is moving at a speed of 5.0 m/s in a direction perpendicular to a 0.80-T magnetic field. The rod has a length of 1.6 m and has negligible electrical resistance. The rails also have negligible resistance, and the light bulb has a resistance of 96 Ω. Find (a) the emf produced by the rod, (b) the induced current in the circuit, (c) the electrical power delivered to the bulb, and (d) the energy consumed by the bulb in 60.0 s.

REASONING The moving rod acts like an imaginary battery that supplies an emf to the circuit, the motional emf being given by vBL. The induced current can be determined from Ohm's law as the motional emf divided by the resistance of the light bulb. The electric power delivered to the bulb is the product of the induced current and the potential difference across the bulb (which, in this case, is the motional emf). The energy consumed is the product of the electric power and the time.

SOLUTION

(a) The motional emf is given by Equation 22.1 as

$$\mathscr{E} = vBL = (5.0\ \text{m/s})(0.80\ \text{T})(1.6\ \text{m}) = \boxed{6.4\ \text{V}}$$

(b) According to Ohm's law, the induced current is equal to the motional emf divided by the resistance of the circuit:

$$I = \frac{\mathscr{E}}{R} = \frac{6.4\ \text{V}}{96\ \Omega} = \boxed{0.067\ \text{A}} \qquad (20.2)$$

(c) The electrical power P delivered to the light bulb is the product of the current I and the potential difference across the bulb:

$$P = I\mathscr{E} = (0.067\ \text{A})(6.4\ \text{V}) = \boxed{0.43\ \text{W}} \qquad (20.6)$$

(d) Since power is energy per unit time, the energy E consumed in 60.0 s is the product of the power and the time:

$$E = Pt = (0.43\ \text{W})(60.0\ \text{s}) = \boxed{26\ \text{J}}$$

MOTIONAL EMF AND ELECTRICAL ENERGY

Motional emf arises because a magnetic force acts on the charges in a conductor that is moving through a magnetic field. Whenever this emf causes a current, a second magnetic force enters the picture. In Figure 22.4*b*, for instance, the second force arises because the current I in the rod is perpendicular to the magnetic field. The current, and hence the rod, experiences a magnetic force **F** whose magnitude is given by Equation 21.3 as $F = ILB \sin 90°$. Using the values of I, L, and B from Example 1, we see that $F = (0.067\ \text{A})(1.6\ \text{m})(0.80\ \text{T}) = 0.086\ \text{N}$. The direction of **F** is specified by RHR-1 and is *opposite* to the velocity **v** of the rod, as Figure 22.5 shows. Hence, **F** tends to *slow down* the rod, and here lies the crux of the matter. To keep the rod moving to the right with a constant velocity, a counterbalancing force must be applied to the rod by an external agent, such as somebody pushing it. The counterbalancing force must have a magnitude of 0.086 N and must be directed to the right in the drawing. If the counterbalancing force were removed, the rod would decelerate under the influence of the magnetic force **F** and eventu-

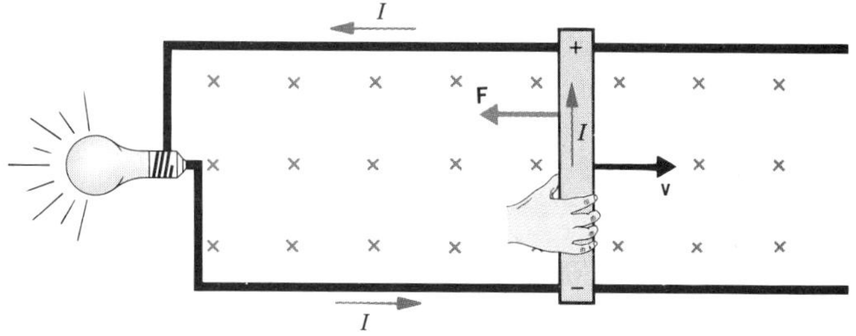

Figure 22.5 A magnetic force **F** is exerted on the current I in the moving rod and is opposite to the velocity **v**. The rod will slow down unless a counterbalancing force is applied by the hand.

ally come to rest. During the deceleration, the motional emf would decrease and the light bulb would eventually go out.

We can now answer an important question—Who or what provides the 26 J of electrical energy that the light bulb in Example 1 consumes in sixty seconds? The provider is the external agent that applies the 0.086-N counterbalancing force needed to keep the rod moving. This agent does work, and Example 2 shows that the work done is equal to the electrical energy consumed by the bulb.

Example 2 The Work Needed to Keep the Light Bulb Burning

In Example 1, an external agent (the hand in Figure 22.5) supplies the 0.086-N force that keeps the rod moving at a constant speed of 5.0 m/s. Determine the work done in 60.0 s by the external agent.

REASONING The work W done by the hand in Figure 22.5 is given by Equation 6.1 as $W = (F \cos \theta)x$. In this equation F is the magnitude of the external force, θ is the angle between the external force and the displacement of the rod ($\theta = 0°$), and x is the distance that the rod moves in a time t. The distance is the product of the speed of the rod and the time, $x = vt$, so the work can be expressed as $W = (F \cos 0°)\, vt$.

SOLUTION The work done by the external agent is

$$W = F\,vt = (0.086\ \text{N})(5.0\ \text{m/s})(60.0\ \text{s}) = \boxed{26\ \text{J}}$$

The 26 J of work done on the rod by the external agent is the same as the 26 J of energy consumed by the light bulb. Hence, the moving rod converts mechanical work into electrical energy, much as a battery converts chemical energy into electrical energy.

The direction of the current in Figure 22.5 reflects a fundamental law of physics. To focus on this law, Conceptual Example 3 speculates on what would happen if the direction of this current were *reversed.*

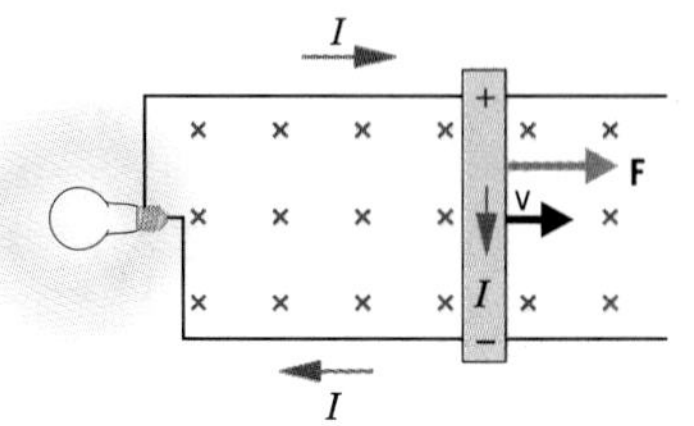

Figure 22.6 The current cannot be directed clockwise in this circuit, because the magnetic force **F** exerted on the rod would then be in the same direction as the velocity **v**. The rod would accelerate to the right and create energy on its own, violating the principle of conservation of energy.

Conceptual Example 3 Induced Current and the Conservation of Energy

If the direction of the current in Figure 22.5 were reversed, so that it was clockwise around the circuit rather than counterclockwise, would any law of physics be violated?

REASONING Figure 22.6 shows the reverse-current situation. Since the direction of the current in the rod is reversed, the magnetic force **F** is also reversed and now points in the same direction as the velocity **v** of the rod. As a result, the force would cause the rod to accelerate rather than decelerate. This would mean that the rod accelerates without the need of an external force (like that provided by the hand in Figure 22.5). The moving rod, however, creates a motional emf that supplies energy to the light bulb. Thus, this hypothetical generator would produce energy out of nothing, since there is no external agent. Such a device cannot exist, because it would violate the principle of conservation of energy, which states that energy cannot be created or destroyed, but can only be converted from one form to another. Therefore, the current cannot be directed clockwise around the circuit. In situations such at that in Figure 22.5, the current always has a direction that is consistent with the principle of conservation of energy.

We have seen that when a motional emf leads to an induced current, a magnetic force always appears that opposes the motion, in accord with the principle of conservation of energy. Conceptual Example 4 deals further with this important issue.

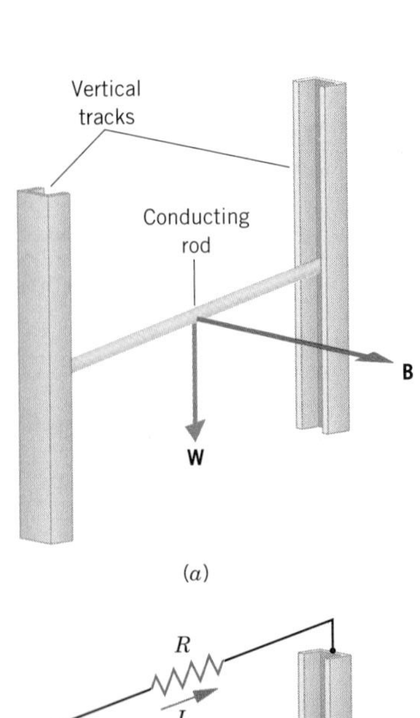

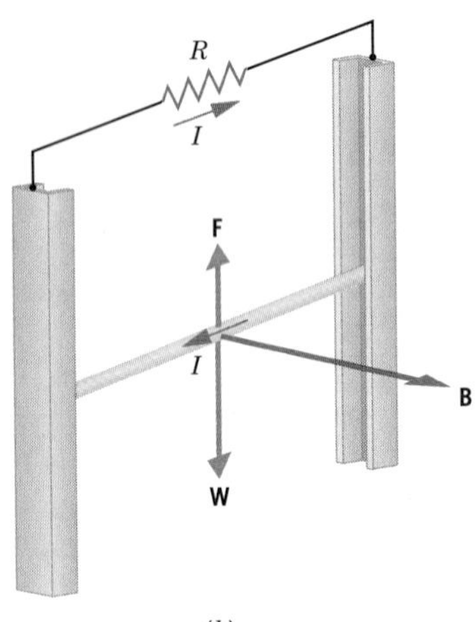

Figure 22.7 (*a*) Because there is no kinetic friction between the falling rod and the tracks, the only force acting on the rod is its weight **W**. (*b*) When an induced current I exists in the circuit, a magnetic force **F** also acts on the rod.

Conceptual Example 4 Conservation of Energy

Figure 22.7*a* illustrates a conducting rod that is free to slide down between two vertical copper tracks. There is no kinetic friction between the rod and the tracks, although the rod maintains electrical contact with the tracks during its fall. A constant magnetic field **B** is directed perpendicular to the motion of the rod, as the drawing shows. Because there is no friction, the only force that acts on the rod is its weight **W**, so the rod falls with an acceleration equal to the acceleration due to gravity, $a = 9.8\ \text{m/s}^2$. Suppose that a resistance R is connected between the tops of the tracks, as in part *b* of the drawing. (a) Does the rod now fall downward with the acceleration due to gravity? (b) How does energy conservation apply to what happens in part *b* of the drawing?

REASONING

(a) As the rod moves perpendicular to the magnetic field, a motional emf is induced between the ends of the rod, just as in Example 1. This emf is induced whether or not the resistance R is attached between the tracks. However, when R is present, a complete circuit is formed, and the emf produces an induced current I that is perpendicular to the magnetic field. The direction of this current is such that the rod experiences a magnetic force **F** that is directed upward, opposite to the weight of the rod (see part *b* of the drawing). The net force acting downward on the rod is **W** − **F**, which is less than the weight of the rod. In accord with Newton's second law of motion, the downward acceleration of the rod is proportional to the net force, so the rod falls with an acceleration that is *less than* the acceleration due to gravity. In fact, as the rod in Figure 22.7*b* speeds up during its descent, the magnetic force becomes larger and larger. There will come a time when the magnetic force becomes equal in magnitude to the weight of the rod. When this occurs, the net force on the rod is zero, and so its acceleration is zero. From then on, the rod falls at a constant velocity.

(b) For the freely falling rod in Figure 22.7*a*, gravitational potential energy (GPE) is converted into kinetic energy (KE) as the rod picks up speed on the way down. In Figure 22.7*b*, however, the rod always has a smaller speed than that of a freely falling rod at the same place. This is because the acceleration is less than the acceleration due to gravity. In terms of energy conservation, the smaller speed arises because only part of the GPE is converted into KE, with part also being dissipated as heat in the resistance R. In fact, when the rod reaches the point when it falls with a constant velocity, none of the GPE is converted into KE. During the constant velocity motion, all of the GPE is dissipated as heat in the resistance.

Related Homework Material: *Problem 10*

22.3 MAGNETIC FLUX

MOTIONAL EMF AND MAGNETIC FLUX

According to Equation 22.1, the emf induced in a rod moving perpendicular to a magnetic field is $\mathscr{E} = vBL$. This motional emf, as well as any other induced emf, can be described in terms of a concept called magnetic flux. To see how the expression for motional emf can be written in terms of this concept, look at Figure 22.8*a*, where a rod is shown moving through a magnetic field between a time $t = 0$ and a later time t_0. During this time interval, the rod moves a distance x_0 to the right. At an even later time t, the rod has moved an even greater distance x to the right, as part *b* of the drawing indicates. The speed v of the rod is the distance traveled divided by the elapsed time: $v = (x - x_0)/(t - t_0)$. Substituting this expression for v into $\mathscr{E} = vBL$ gives

$$\mathscr{E} = \left(\frac{x - x_0}{t - t_0}\right) BL = \left(\frac{xL - x_0 L}{t - t_0}\right) B$$

As the drawing indicates, the term $x_0 L$ is the area A_0 swept out by the rod in moving a distance x_0, while xL is the area A swept out in moving a distance x. In terms of these areas, the emf is

$$\mathscr{E} = \left(\frac{A - A_0}{t - t_0}\right) B = \frac{(BA) - (BA)_0}{t - t_0}$$

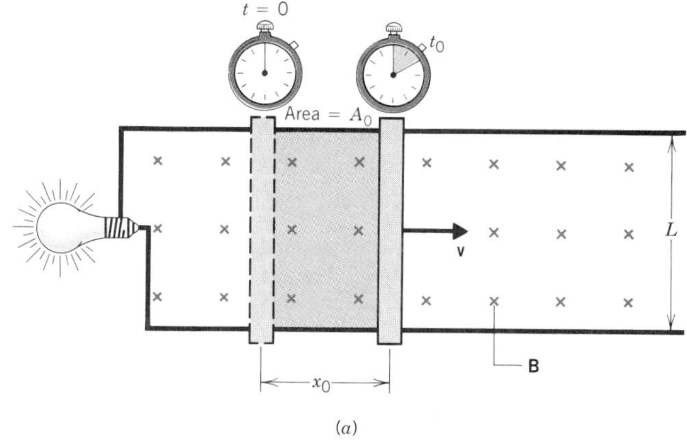

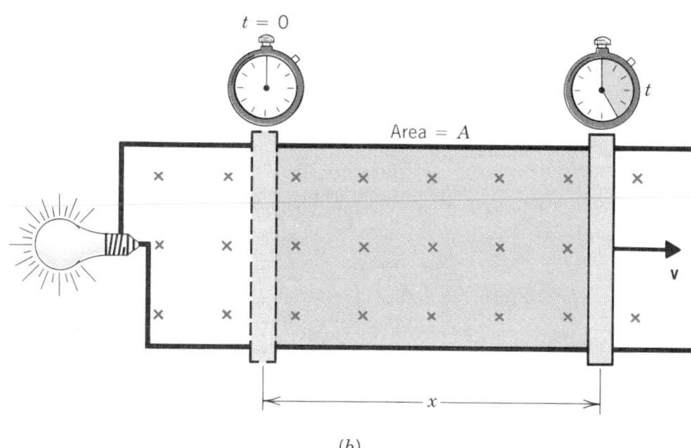

Figure 22.8 (*a*) In a time t_0, the moving rod sweeps out an area $A_0 = x_0 L$. (*b*) The area swept out in a time t is $A = xL$. In both parts of the figure the areas are shaded in color.

Notice how the product BA of the magnetic field strength and the area appears in the numerator of this expression. The quantity BA is given the name ***magnetic flux*** and is represented by the symbol Φ (Greek capital letter *phi*); thus, $\Phi = BA$. The magnitude of the induced emf is the *change* in flux $\Delta\Phi = \Phi - \Phi_0$ divided by the time interval $\Delta t = t - t_0$ during which the change occurs:

$$\mathscr{E} = \frac{\Phi - \Phi_0}{t - t_0} = \frac{\Delta\Phi}{\Delta t}$$

In other words, the induced emf equals the time rate of change of the magnetic flux.

You will almost always see the equation above written with a minus sign, namely, $\mathscr{E} = -\Delta\Phi/\Delta t$. The minus sign is introduced for the following reason: The direction of the current induced in the circuit is such that the magnetic force **F** acts on the rod to *oppose* its motion, thereby tending to slow down the rod (see Figure 22.5). The presence of the minus sign reminds us that the polarity of the induced emf sends the induced current in the proper direction so as to give rise to this opposing magnetic force.

The advantage of writing the induced emf as $\mathscr{E} = -\Delta\Phi/\Delta t$ is that this relation is far more general than our present discussion suggests. In Section 22.4 we will see that $\mathscr{E} = -\Delta\Phi/\Delta t$ can be applied to *all possible ways of generating induced emfs.*

A GENERAL EXPRESSION FOR MAGNETIC FLUX

In Figure 22.8 the direction of the magnetic field **B** is perpendicular to the surface swept out by the moving rod. In general, however, **B** may not be perpendicular to the surface. For instance, in Figure 22.9 the direction perpendicular to the surface is indicated by the normal to the surface, but the magnetic field is inclined at an angle ϕ with respect to this direction. In such a case the flux is computed using only the component of the field that is perpendicular to the surface, $B \cos \phi$. The general expression for magnetic flux is

$$\Phi = (B \cos \phi)A = BA \cos \phi \tag{22.2}$$

If either the magnitude B of the magnetic field or the angle ϕ is not constant over the surface, an average value of the product $B \cos \phi$ must be used to compute the flux. Equation 22.2 shows that the unit of magnetic flux is the tesla·meter^2 ($\mathrm{T \cdot m^2}$). This unit is called a *weber* (Wb), after the German physicist Wilhelm Weber (1804–1891): 1 Wb = 1 $\mathrm{T \cdot m^2}$. Example 5 illustrates how to determine the magnetic flux for three different orientations of the surface of a coil relative to the magnetic field.

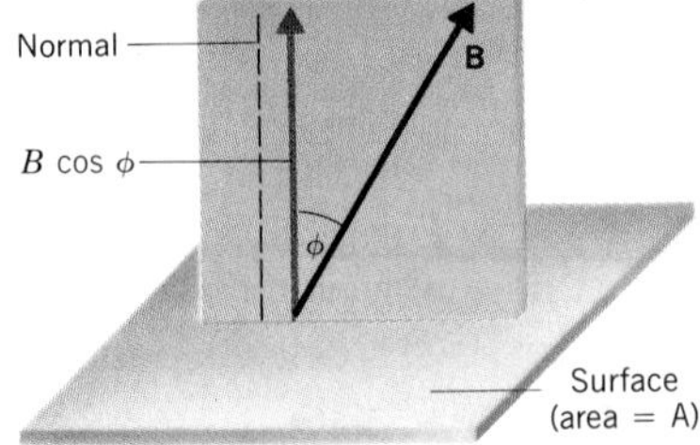

Figure 22.9 When computing the magnetic flux, the component of the magnetic field that is parallel to the normal to the surface must be used; this component is $B \cos \phi$.

Example 5 Magnetic Flux

A rectangular coil of wire is situated in a constant magnetic field whose magnitude is 0.50 T. The coil has an area of 2.0 m². Determine the magnetic flux for the three orientations, $\phi = 0°$, 60.0°, and 90.0°, shown in Figure 22.10.

REASONING AND SOLUTION The flux in the three cases can be computed using $\Phi = BA \cos \phi$:

[$\phi = 0°$] $\Phi = (0.50\ \text{T})(2.0\ \text{m}^2) \cos 0° = \boxed{1.0\ \text{Wb}}$

[$\phi = 60.0°$] $\Phi = (0.50\ \text{T})(2.0\ \text{m}^2) \cos 60.0° = \boxed{0.50\ \text{Wb}}$

[$\phi = 90.0°$] $\Phi = (0.50\ \text{T})(2.0\ \text{m}^2) \cos 90.0° = \boxed{0}$

PROBLEM SOLVING INSIGHT

The magnetic flux Φ is determined by more than just the magnitude *B* of the magnetic field and the area *A*. It also depends on the angle ϕ (see Figure 22.9 and Equation 22.2).

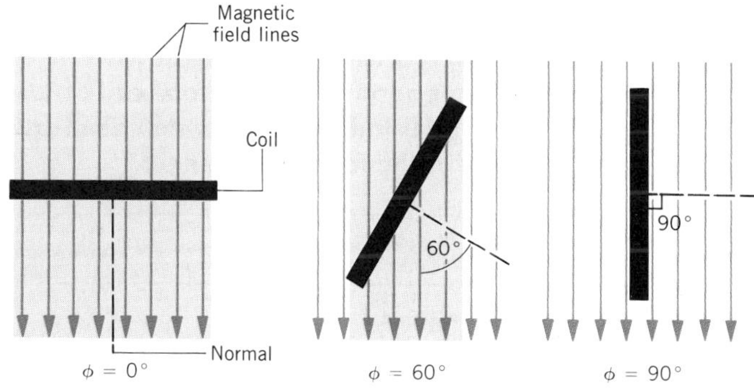

Figure 22.10 This picture shows three orientations of a rectangular coil (drawn as an edge view), relative to the magnetic field lines. The magnetic field lines that pass through the coil are those in the regions shaded in blue.

GRAPHICAL INTERPRETATION OF MAGNETIC FLUX

It is possible to interpret the magnetic flux graphically by noting that ***the flux is proportional to the number of magnetic field lines that passes through a surface.*** This useful interpretation stems from the fact that the magnitude of the magnetic field **B** in any region of space is proportional to the number of magnetic field lines per unit area that passes through a surface perpendicular to the field lines (see Section 21.1). For instance, the magnitude of **B** in Figure 22.11*a* is three times larger than it is in part *b* of the drawing, since the number of field lines drawn through the identical surfaces is in the ratio of 3:1. Because Φ is directly proportional to *B* for a given area, the flux in part *a* is also three times larger than that in part *b*. Therefore, we can say that the magnetic flux is proportional to the number of magnetic field lines that passes through a surface.

The graphical interpretation of flux also applies when the surface is oriented at an angle with respect to **B**. For example, as the coil in Figure 22.10 is rotated from $\phi = 0°$ to 60° to 90°, the number of magnetic field lines passing through the surface (see the field lines in the regions shaded in blue) changes in the ratio of 8:4:0 or 2:1:0. The results of Example 5 show that the flux in the three orientations changes by the same ratio. Because the magnetic flux is proportional to the number of field lines passing through a surface, one often encounters phrases like "the flux that passes through a surface bounded by a loop of wire."

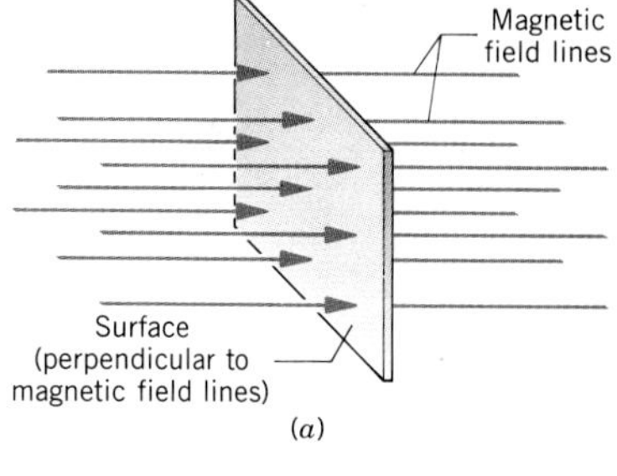

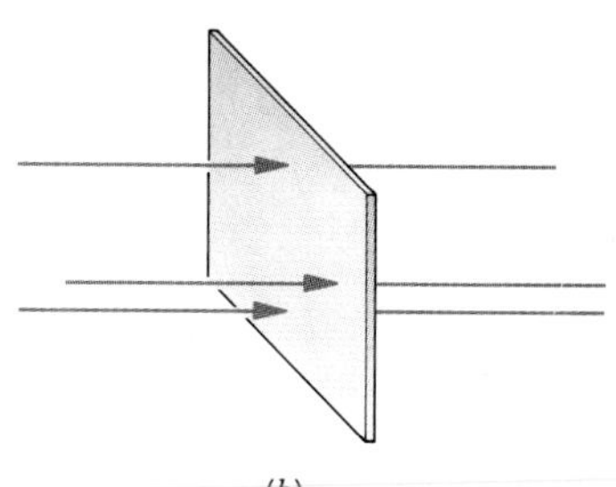

Figure 22.11 The magnitude of the magnetic field in (*a*) is three times greater than that in (*b*), because the number of magnetic field lines crossing the surfaces is in the ratio of 3:1.

22.4 FARADAY'S LAW OF ELECTROMAGNETIC INDUCTION

Two scientists are given credit for the discovery of electromagnetic induction: the Englishman Michael Faraday (1791–1867) and the American Joseph Henry (1797–1878). Although Henry was the first to observe electromagnetic induction, Faraday investigated it in more detail and published his findings first. Consequently, the law that describes the phenomenon bears his name.

Faraday discovered that whenever there is a *change in flux* through a loop of wire, an emf is induced in the loop. A constant flux creates no emf. In fact, Faraday found that the magnitude of the induced emf is equal to the time rate of change of the magnetic flux. This is consistent with the relation we obtained in Section 22.3 for the specific case of motional emf: $\mathscr{E} = -\Delta\Phi/\Delta t$.

Often the magnetic flux passes through a coil of wire containing more than one loop (or turn). If the coil consists of N loops, and if the same flux passes through each loop, it is found experimentally that the total induced emf is N times that induced in a single loop. An analogous situation occurs in a flashlight when two 1.5-V batteries are stacked in series on top of one another to give a total emf of 3.0 volts. For the general case of N loops, the total induced emf is described by ***Faraday's law of electromagnetic induction*** in the following manner.

Faraday's Law of Electromagnetic Induction

The average emf $\mathscr{E}$ induced in a coil of N loops is

$$\mathscr{E} = -N\left(\frac{\Phi - \Phi_0}{t - t_0}\right) = -N\frac{\Delta\Phi}{\Delta t} \qquad (22.3)$$

where $\Delta\Phi$ is the change in magnetic flux through one loop and Δt is the time interval during which the change occurs. The term $\Delta\Phi/\Delta t$ is the average time rate of change of the flux that passes through one loop.

SI Unit of Induced Emf: volt (V)

Faraday's law states that an emf is generated if the flux changes for any reason. Since the flux is given by Equation 22.2 as $\Phi = BA\cos\phi$, the flux depends on three factors, B, A, and ϕ, any of which may change. The examples in the remainder of this section illustrate how such changes can lead to an induced emf. Example 6 considers a changing magnetic field.

Example 6 The Emf Induced by a Changing Magnetic Field

A coil of wire consists of 20 turns, each of which has an area of 1.5×10^{-3} m^2. A magnetic field is perpendicular to the surface of each loop at all times, so that $\phi = \phi_0 = 0°$. At time $t_0 = 0$, the magnitude of the magnetic field at the location of the coil is $B_0 = 0.050$ T. At a later time of $t = 0.10$ s, the magnitude of the field at the coil has increased to $B = 0.060$ T. (a) Find the average emf induced in the coil during this time. (b) What would be the value of the induced emf if the magnitude of the magnetic field decreased from 0.060 T to 0.050 T in 0.10 s?

REASONING To find the induced emf, we use Faraday's law of electromagnetic induction (Equation 22.3), combining it with the definition of magnetic flux from

Equation 22.2. We note that only the magnitude of the magnetic field changes in time and that all other factors remain constant.

SOLUTION

(a) Since $\phi = \phi_0$, the induced emf is

$$\mathscr{E} = -N\left(\frac{\Phi - \Phi_0}{t - t_0}\right) = -N\left(\frac{BA\cos\phi - B_0A\cos\phi_0}{t - t_0}\right) = -NA\cos\phi\left(\frac{B - B_0}{t - t_0}\right)$$

We find that

$$\mathscr{E} = -(20)(1.5 \times 10^{-3}\ \text{m}^2)(\cos 0°)\left(\frac{0.060\ \text{T} - 0.050\ \text{T}}{0.10\ \text{s} - 0}\right) = \boxed{-3.0 \times 10^{-3}\ \text{V}}$$

(b) The reasoning here is similar to that in part (a), except the initial and final values of B are interchanged. This interchange reverses the sign of the emf, so $\boxed{\mathscr{E} = +3.0 \times 10^{-3}\ \text{V}}$. Because the polarity of the emf is reversed, the direction of the induced current is opposite to that in part (a).

PROBLEM SOLVING INSIGHT

The change in any quantity is the final value minus the initial value: e.g., the change in flux is $\Delta\Phi = \Phi - \Phi_0$ and the change in time is $\Delta t = t - t_0$.

The next example demonstrates that an emf can be created when a coil is rotated in a magnetic field.

Example 7 The Emf Induced in a Rotating Coil

A flat coil of wire has an area of 0.020 m^2 and consists of 50 turns. At $t_0 = 0$ the coil is oriented so the normal to its surface is parallel ($\phi_0 = 0°$) to a constant magnetic field of magnitude 0.18 T. The coil is then rotated through an angle of $\phi = 30.0°$ in a time of 0.10 s (see Figure 22.3*c*). (a) Determine the average induced emf. (b) What would be the induced emf if the coil were returned to its initial orientation in the same time of 0.10 s?

REASONING As in Example 6, we can determine the induced emf by using Faraday's law of electromagnetic induction, along with the definition of magnetic flux. In the present case, however, only the angle ϕ between the normal to the surface of the coil and the magnetic field changes in time, and all other factors remain constant.

SOLUTION

(a) Faraday's law yields

$$\mathscr{E} = -N\left(\frac{\Phi - \Phi_0}{t - t_0}\right) = -N\left(\frac{BA\cos\phi - BA\cos\phi_0}{t - t_0}\right) = -NBA\left(\frac{\cos\phi - \cos\phi_0}{t - t_0}\right)$$

$$\mathscr{E} = -(50)(0.18\ \text{T})(0.020\ \text{m}^2)\left(\frac{\cos 30.0° - \cos 0°}{0.10\ \text{s} - 0}\right) = \boxed{+0.24\ \text{V}}$$

(b) When the coil is rotated back to its initial orientation in a time of 0.10 s, the initial and final values of ϕ are interchanged. As a result, the induced emf has the same magnitude, but opposite polarity, so $\boxed{\mathscr{E} = -0.24\ \text{V}}$.

THE PHYSICS OF . . .

a ground fault interrupter.

One application of Faraday's law that is found in the home is a safety device called a ground fault interrupter. This device protects against electric shock from an appliance, such as a clothes drier, and plugs directly into (or sometimes replaces) a wall socket, as in Figure 22.12. The interrupter consists of a circuit breaker that can be triggered to stop the current to the drier, depending on whether an induced voltage appears across a sensing coil. The sensing coil is

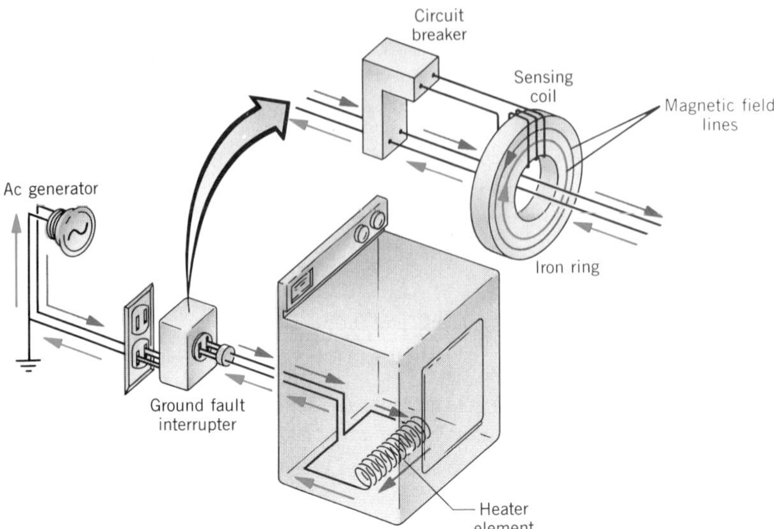

Figure 22.12 The clothes drier is connected to the wall socket through a ground fault interrupter. The drier is operating normally.

wrapped around an iron ring, and the two wires carrying current to and from the drier pass through the ring. In the drawing, the current going to the drier is shown in red, while the returning current is shown in green. Each of the currents creates a magnetic field that encircles the corresponding wire, according to RHR-2. However, the field lines have opposite directions since the currents have opposite directions. As the drawing shows, the iron ring guides the field lines through the sensing coil. Since the current is ac, the fields from the red and green current are changing. However, no induced voltage appears across the sensing coil, because the red and green field lines have opposite directions and the opposing fields cancel at all times. The net flux through the coil remains zero, and no emf is induced in the coil. Thus, when the drier operates normally, the circuit breaker is not triggered and does not shut down the current to the drier. The picture changes when the drier malfunctions, as when a wire inside the unit breaks and accidentally touches the metal case. When someone touches the case, some of the current begins to pass through the person's body and into the ground, returning to the ac generator *without going through the return wire that passes through the ground fault interrupter.* As a result, the net magnetic field through the sensing coil is no longer zero and changes with time since the current is ac. The changing flux causes an induced voltage to appear across the sensing coil. This voltage triggers the circuit breaker to stop the current. Ground fault interrupters work very fast (in less than a millisecond) and can turn off the current before it reaches a dangerous level.

Conceptual Example 8 discusses another application of electromagnetic induction, namely, how a stove can cook food without getting hot.

THE PHYSICS OF . . .

an induction stove.

Conceptual Example 8 An Induction Stove

Figure 22.13 shows two pots of water that were placed on an induction stove at the same time. There are two interesting features in this drawing. First, the stove itself is cool to the touch. Second, the water in the metal pot is boiling while that in the glass pot is not. How can such a "cool" stove boil water, and why isn't the water in the glass pot boiling?

REASONING The key to this puzzle must be related to the fact that one pot is metal and one is glass. We know that metals are good conductors, while glass is an insulator.

Perhaps the stove causes electricity to flow directly in the metal pot. This is exactly what happens. The stove is called an *induction stove,* because it operates by using electromagnetic induction. Just beneath the cooking surface is a metal coil that carries an ac current. This alternating current produces an alternating magnetic field that extends outward to the location of the metal pan. As the changing magnetic field crosses the bottom surface of the pan, an emf is induced in it. Because the pan is metallic, an induced current is generated by the induced emf. The pan has a finite resistance to the induced current and heats up as energy is dissipated in this resistance. This heat is conducted to the water and eventually raises its temperature to the boiling point. An emf is also induced in the glass pot and the cooking surface of the stove. However, these materials are insulators, so that very little induced current exists within them. Thus, they do not heat up very much and remain cool to the touch.

Related Homework Material: *Question 8*

Figure 22.13 The water in one of the pots is boiling. Yet, the water in the other pot is not boiling, and the stove top is cool to the touch. The stove operates in this way by using electromagnetic induction.

22.5 LENZ'S LAW

THE POLARITY OF THE INDUCED EMF

An induced emf drives current around a circuit just as the emf of a battery does. With a battery, conventional current is directed out of the positive terminal, through the attached device, and into the negative terminal. The same is true for an induced emf, although the location of the positive and negative terminals is generally not as obvious. Therefore, a method is needed for determining the polarity of the induced emf, so the terminals can be identified. As we discuss this method, it will be helpful to keep in mind that the net magnetic field penetrating a coil of wire results from two contributions. One is the original magnetic field that produces the changing flux that leads to the induced emf. The other arises because of the induced current, which, like any current, creates its own magnetic field. The field created by the induced current is called the ***induced magnetic field.***

To determine the polarity of the induced emf, we will use a method based on a discovery made by the Russian physicist Heinrich Lenz (1804–1865). He found that the polarity of an induced emf always leads to an induced current with the following characteristic: The direction of the induced current is such that the induced field either adds to or subtracts from the original field, whichever is necessary to help keep the flux from changing. This fact is known as ***Lenz's law.***

Lenz's Law

The polarity of an induced emf is such that the induced current produces an induced magnetic field that opposes the change in flux causing the emf.

Lenz's law is best illustrated with examples. Each example is worked out according to the following reasoning strategy.

Reasoning Strategy Determining the Polarity of the Induced Emf

1. Determine whether the magnetic flux that penetrates a coil is increasing or decreasing.
2. Find what the direction of the induced magnetic field must be so that it can oppose the ***change*** in flux by adding to or subtracting from the original field.
3. Having found the direction of the induced magnetic field, use RHR-2 (see Section 21.8) to determine the direction of the induced current. Then the polarity of the induced emf can be assigned, because conventional current is directed out of the positive terminal, through the external circuit, and into the negative terminal.

Conceptual Example 9 The Emf Produced by a Moving Magnet

Figure 22.14*a* shows a permanent magnet approaching a loop of wire. The external circuit attached to the loop consists of the resistance R, which could be the resistance of the filament in a light bulb, for instance. Find the direction of the induced current and the polarity of the induced emf.

REASONING In solving this problem, we apply Lenz's law, the essence of which is that the change in magnetic flux must be opposed by the induced magnetic field. The magnetic flux through the loop is increasing, since the magnitude of the magnetic field at the loop is increasing as the magnet approaches. To oppose the increase in the flux, the direction of the induced magnetic field must be opposite to the field of the bar magnet. Since the field of the bar magnet passes through the loop from left to right in part *a* of the drawing, the induced field must pass through the loop from right to left, as in part *b*. To create such an induced field, the induced current must be directed *counterclockwise* around the loop, when viewed from the side nearest the magnet. (See the application of RHR-2 in the drawing.) The loop behaves as a source of emf, just like a battery, with the positive and negative terminals as shown in Figure 22.14*b*.

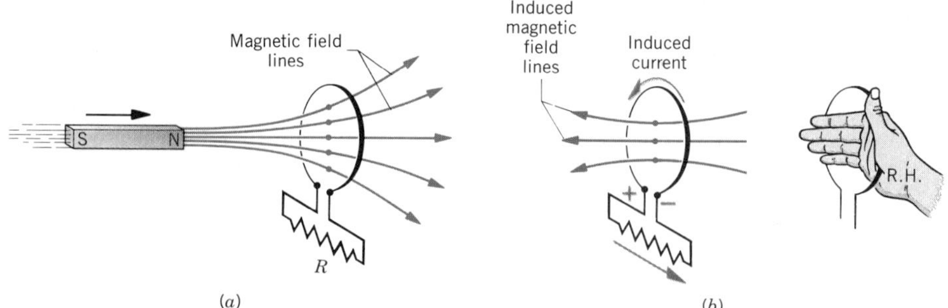

Figure 22.14 (*a*) As the magnet moves to the right, the magnetic flux through the loop increases. The external circuit attached to the loop has a resistance R. (*b*) The polarity of the induced emf is indicated by the + and − symbols.

In Conceptual Example 9 the direction of the induced magnetic field is opposite to the direction of the external field of the bar magnet. However, the induced field is not always opposite to the external field, for Lenz's law requires only that it must oppose the ***change*** in the flux that generates the emf. Conceptual Example 10 illustrates this point.

Conceptual Example 10 The Emf Produced by a Moving Copper Ring

In Figure 22.15 there is a constant magnetic field in a rectangular region of space. This field is directed perpendicularly into the plane of the paper. Outside this region there is no magnetic field. A copper ring slides through the region, from position 1 to position 5. For each of the five positions, determine if an induced current exists in the ring and, if so, find the direction of the current.

REASONING In applying Lenz's law here, we keep in mind that it does not require the induced magnetic field to be opposite to the external field. Rather, the induced field must oppose the change in flux, and sometimes this means that the induced field will reinforce the external field.

Position 1: No flux passes through the ring, because the magnetic field is zero outside the rectangular region. Consequently, there is no change in flux and no induced emf or current.

Position 2: As the ring moves into the region of the magnetic field, the flux increases, and there is an induced emf and an induced current. To determine the direction of the current, we require that the induced magnetic field point out of the plane of the paper, opposite to the external field. With this direction, the induced field will oppose the increase in the flux, in accord with Lenz's law. With the induced field pointing perpendicularly out of the plane of the paper, RHR-2 indicates that the direction of the induced current is counterclockwise, as the drawing indicates.

Position 3: Even though a flux passes through the moving ring, there is no induced emf or current, because the flux remains constant within the rectangular region. To induce an emf, it is not sufficient just to have a flux. The flux must ***change*** to generate an emf.

Position 4: As the ring leaves the magnetic field, the flux decreases. Once again, the induced magnetic field must be in such a direction so as to oppose this change. Since the change is a decrease in flux, the induced field must point in the *same* direction as the external field, namely, into the paper. Only with this orientation will the induced field increase the net magnetic field through the ring and thereby increase the flux. With the induced field pointing into the paper, RHR-2 indicates that the induced current is clockwise around the ring, opposite to what it was in position 2. By comparing the results for positions 2 and 4, it should be clear that the induced magnetic field does not always oppose the external field.

Position 5: As in position 1, there is no induced current, since the magnetic field is everywhere zero.

Related Homework Material: *Problem 33*

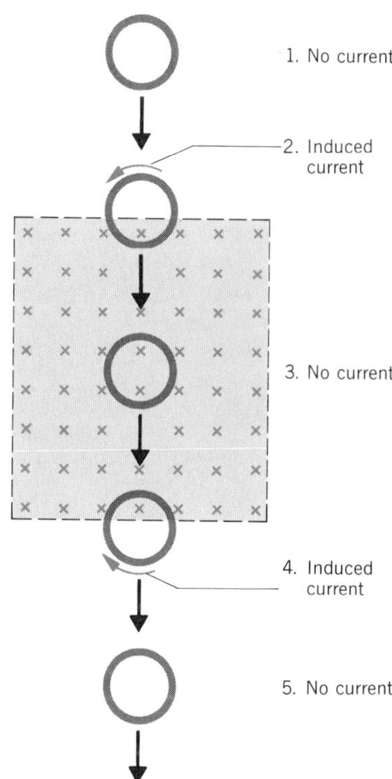

Figure 22.15 A copper ring passes through a rectangular region where a constant magnetic field is directed into the page. The picture shows that a current is induced in the ring at locations 2 and 4.

Lenz's law should not be thought of as an independent law, for it is a consequence of the law of conservation of energy. The connection between energy conservation and induced emf has already been discussed in Conceptual Examples 3 and 4 for the specific case of motional emf. However, the connection is valid for any type of induced emf. In fact, the polarity of the induced emf, as specified by Lenz's law, ensures that energy is conserved.

22.6 APPLICATIONS OF ELECTROMAGNETIC INDUCTION TO THE REPRODUCTION OF SOUND

Electromagnetic induction plays an important role in the technology used for the reproduction of sound. The typical audio system used to reproduce sound looks like that in Figure 22.16. Such systems can include a number of components, and the drawing shows an electric guitar, a cassette deck, and a microphone as examples. As we will now see, each of these generates an induced emf. In general, however, the emf is rather small, so it is strengthened by an amplifier before being sent to the speakers.

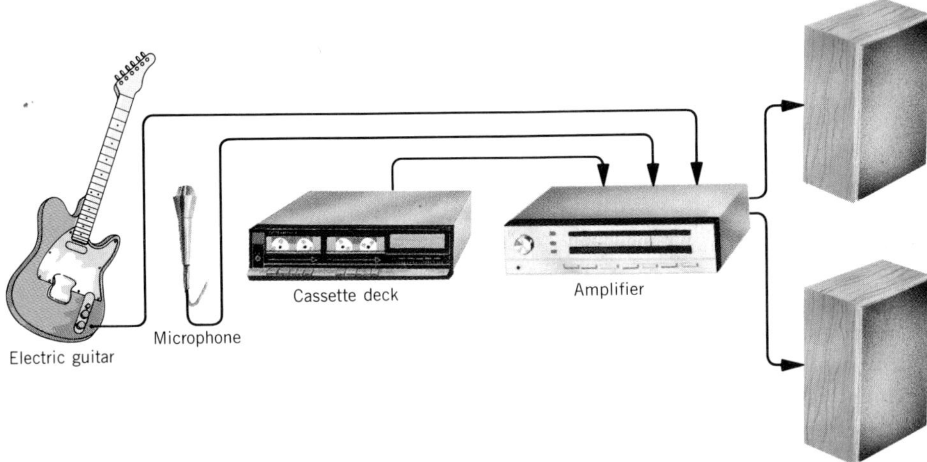

Figure 22.16 The operation of an electric guitar, a microphone, and a cassette deck is based on electromagnetic induction. Each black line represents two conducting wires.

THE PHYSICS OF . . .

the electric guitar pickup.

THE ELECTRIC GUITAR PICKUP

Virtually all electric guitars use electromagnetic pickups in which an induced emf is generated in a coil of wire by a vibrating string. Most guitars have at least two pickups for each string and some, as Figure 22.17 illustrates, have three. These pickups are positioned at different locations under the string, so that each is sensitive to different harmonics produced by the vibrating string.

The guitar string is made from a magnetizable metal. The pickup itself consists of a coil of wire with a permanent magnet located inside the coil. The permanent

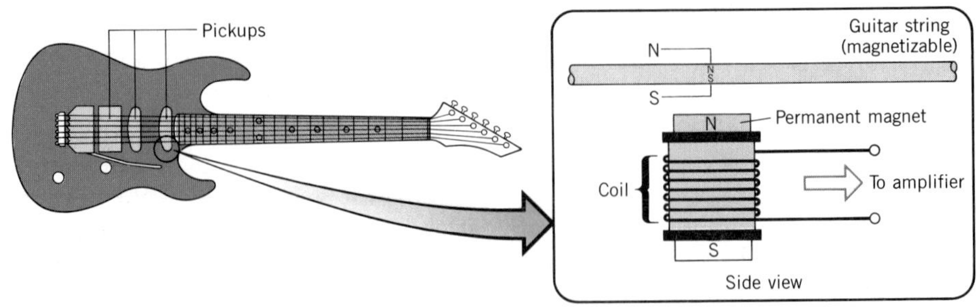

Figure 22.17 When the string of an electric guitar vibrates, an emf is induced in the coil of the pickup. The two ends of the coil are connected to the input of an amplifier.

magnet produces a magnetic field that penetrates the guitar string, causing it to become magnetized with north and south poles. When the magnetized string is plucked, it oscillates above the coil, thereby changing the magnetic flux that passes through the coil. The change in flux induces an emf in the coil. The polarity of the emf reverses with the vibratory motion of the string, so a string vibrating at 440 Hz, for example, induces a 440-Hz ac emf in the coil. This 440-Hz signal, after being amplified, is sent to the speakers, which produce a 440-Hz sound wave (concert A).

THE PLAYBACK HEAD OF A TAPE DECK

THE PHYSICS OF . . .

the playback head of a tape deck.

The playback head of a cassette deck uses a moving tape to generate an emf in a coil of wire. Figure 22.18 shows a section of magnetized tape in which a series of "tape magnets" have been created in the magnetic layer of the tape during the recording process. The tape moves beneath the playback head, which consists of a coil of wire wrapped around an iron core. The iron core has the approximate shape of a horseshoe with a small gap between the two ends. The drawing shows an instant when a tape magnet is under the gap. Some of the magnetic field lines of the tape magnet are routed through the highly magnetizable iron core, and hence through the coil, as they proceed from the north pole to the south pole. Consequently, the flux through the coil changes as the tape moves past the gap. The change in flux leads to an ac emf, which is amplified and sent to the speakers, where the original sound is reproduced.

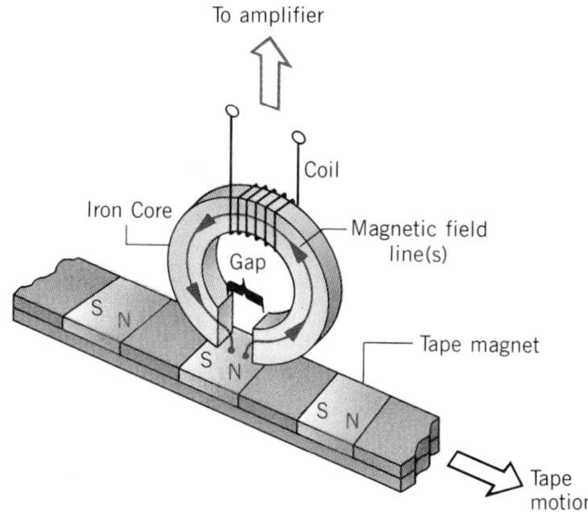

Figure 22.18 The magnetic playback head of a tape deck. As each tape magnet passes by the gap in the iron core, some of the magnetic field lines are routed through the core and the coil. The change in flux through the coil creates an induced emf. The width of the gap has been exaggerated.

THE MICROPHONE

THE PHYSICS OF . . .

a moving coil and a moving magnet microphone.

There are a number of types of microphones, and Figure 22.19 illustrates the one known as a moving coil microphone. When a sound wave strikes the diaphragm of the microphone, the diaphragm vibrates back and forth, and the attached coil moves along with it. Nearby is a stationary magnet. As the coil alternately approaches and recedes from the magnet, the flux through the coil changes. Consequently, an ac emf is induced across the coil. This electrical signal is sent to an amplifier and then to the speakers. In the type of microphone called a moving magnet microphone, the magnet is attached to and moves along with the diaphragm. A nearby coil is fixed in place, and the flux through this coil changes as the magnet moves back and forth. Once again, an ac emf is induced, and sound is converted into electricity.

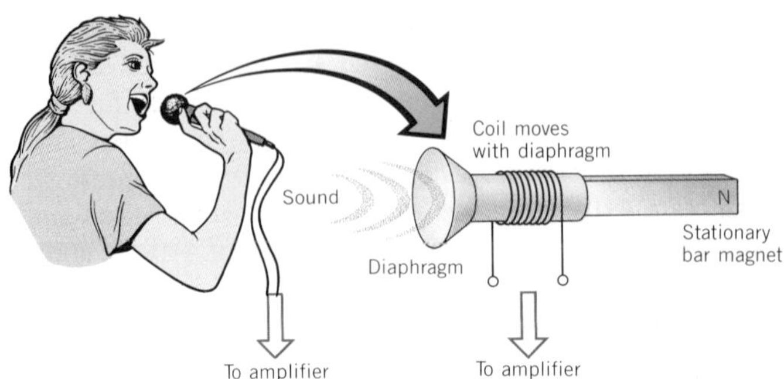

Figure 22.19 A moving coil microphone.

THE PHYSICS OF . . .

the electric generator.

22.7 THE ELECTRIC GENERATOR

HOW A GENERATOR PRODUCES AN EMF

Electric generators are important because they produce virtually all the electrical energy consumed in the world. A generator produces electrical energy from mechanical work, just the opposite of what a motor does. In a motor, an *input* electric current causes a coil to rotate, thereby doing mechanical work on any object attached to the shaft of the motor. In a generator, the shaft is rotated by some mechanical means, such as by an engine or a turbine, and an emf is induced in a coil. If the generator is connected to an external circuit, an electric current is the *output* of the generator.

In its simplest form, an ac generator consists of a coil of wire that is rotated in a uniform magnetic field, as Figure 22.20*a* indicates. Although not shown in the picture, the wire is usually wound around an iron core, and, as in an electric motor, the coil/core combination is called the armature. Each end of the wire forming the coil is connected to the external circuit by means of a metal ring that rotates with the coil. Each ring slides against a stationary carbon brush, to which the external circuit (the lamp in the drawing) is connected.

To see how current is produced by the generator, consider the two vertical sides of the coil in Figure 22.20*b*. Since each is moving in a magnetic field **B**, the magnetic force exerted on the charges in the wire causes them to flow, thus creating a current. With the aid of RHR-1 (fingers of extended right hand point along **B**, thumb along the velocity **v**, palm pushes in the direction of the force on a positive charge), it can be seen that the direction of the current is from bottom to top in the left side and from top to bottom in the right side. Thus, charge flows around the loop. The upper and lower segments of the loop are also moving. However, the magnetic force on the charges in these segments can be ignored, because the force is toward the sides of the wire and not along the length. The emf generated in the coil results only from the magnetic force exerted on the charges in the vertical sides.

Electric generators such as these supply electrical power to homes and industries. Faraday's law of electromagnetic induction describes how generators produce an induced emf.

The magnitude of the motional emf developed in a conductor moving through a magnetic field is given by Equation 22.1. To apply this expression to the left side of the coil, whose length is L (see Figure 22.20*c*), we need to use the velocity component $v_\perp$ that is perpendicular to **B.** Letting θ be the angle between **v** and **B**, it follows that $v_\perp = v \sin\theta$, and the emf can be written as

$$\mathscr{E} = BLv_\perp = BLv \sin\theta$$

The emf induced in the right side has the same magnitude as that for the left side. Since the emfs from both sides drive current in the same direction around the

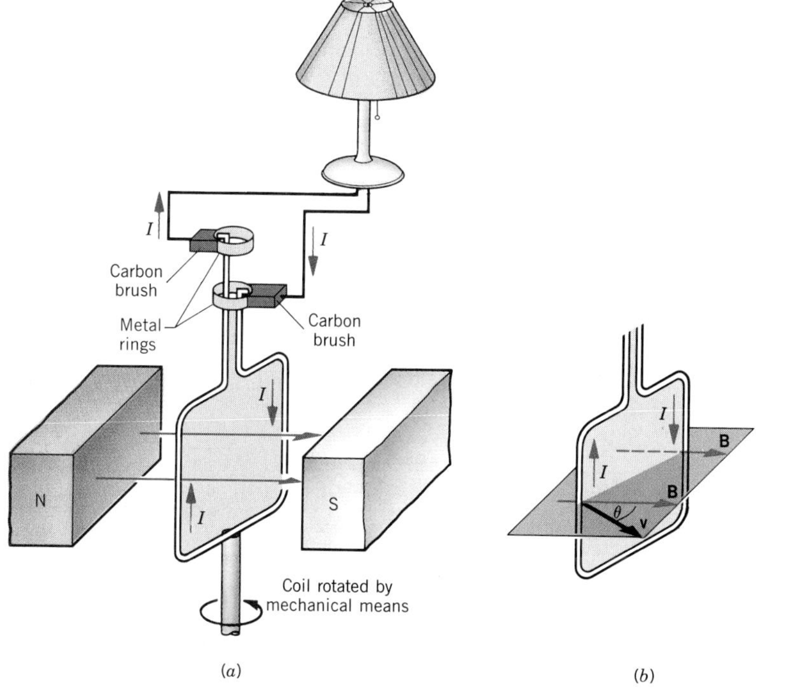

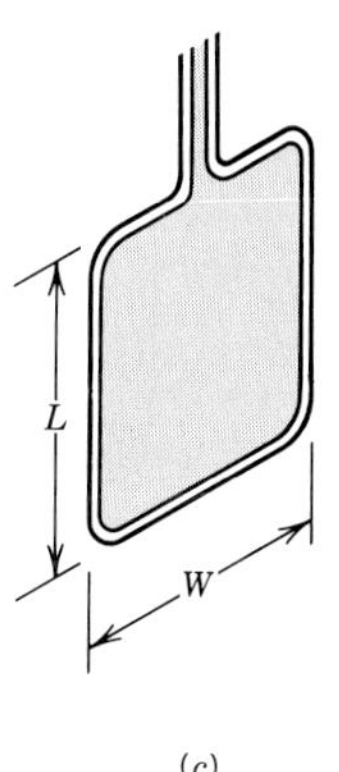

Figure 22.20 (*a*) This electric generator consists of a coil (only one loop is shown) of wire that is rotated in a magnetic field **B** by some mechanical means. (*b*) The current *I* arises because of the magnetic force exerted on the charges in the moving wire. (*c*) The dimensions of the coil.

loop, the emf for the complete loop is $\mathscr{E} = 2BLv \sin\theta$. If the coil consists of N loops, the net emf is N times greater than that of one loop, so

$$\mathscr{E} = N(2BLv \sin\theta)$$

It is convenient to express the variables v and θ in terms of the angular speed ω of the coil. Equation 8.2 shows that the angle θ is the product of the angular speed and the time, $\theta = \omega t$, if it is assumed that $\theta = 0$ when $t = 0$. Furthermore, any point on each vertical side moves on a circular path of radius $r = W/2$, where W is the width of the coil (see Figure 22.20*c*). Thus, the tangential speed v of each side is related to the angular speed ω via Equation 8.9 as $v = r\omega = (W/2)\omega$. Substituting these expressions for θ and v in the equation above for $\mathscr{E}$, and recognizing that the product LW is the area A of the coil, we can write the induced emf as

[Emf induced in a rotating planar coil]

$$\mathscr{E} = NAB\omega \sin\omega t = \mathscr{E}_0 \sin\omega t \tag{22.4}$$

In this equation, the angular speed ω is in radians per second. The expression shows that the emf varies sinusoidally with time. The peak, or maximum, emf $\mathscr{E}_0$ occurs when $\sin\omega t = 1$ and has the value $\mathscr{E}_0 = NAB\omega$. Although Equation 22.4 was derived for a rectangular coil, the result is valid for any planar shape of area A, such as a circle.

The emf of Equation 22.4 is plotted in Figure 22.21, which shows that the emf changes polarity as the coil rotates. This changing polarity is exactly the same as that discussed for an ac voltage in Section 20.5 and illustrated in Figure 20.10. If the external circuit connected to the generator is a closed circuit, an alternating current results that changes direction at the same frequency as the emf changes polarity. Therefore, this electric generator is also called an *alternating current (ac) generator*. The next two examples show how Equation 22.4 is applied.

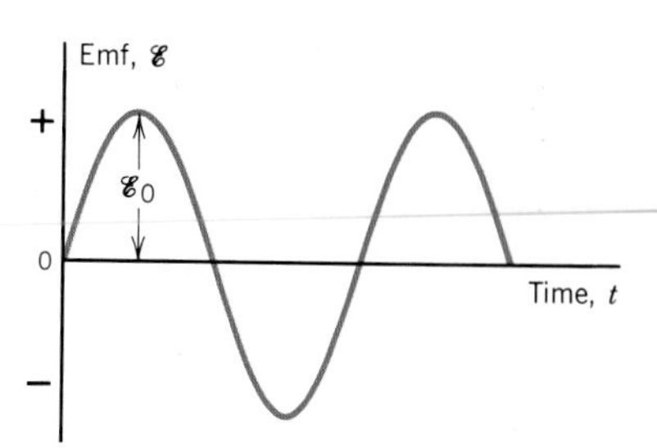

Figure 22.21 An ac generator produces an alternating emf $\mathscr{E}$ that varies as $\mathscr{E} = \mathscr{E}_0 \sin\omega t$.

Example 11 An Ac Generator

In Figure 22.20 the coil of the ac generator rotates at a frequency of 60.0 Hz and develops an emf of 120 V(rms). The coil has an area of $A = 3.0 \times 10^{-3}\ \text{m}^2$ and consists of $N = 500$ turns. Find the magnitude of the magnetic field in which the coil rotates.

REASONING The magnetic field can be found from the relation $\mathscr{E}_0 = NAB\omega$. However, in using this equation we must remember that $\mathscr{E}_0$ is the peak emf, while the given quantity is not a peak value but an rms value of 120 V. The peak emf is related to the rms emf by $\mathscr{E}_0 = \sqrt{2}\,\mathscr{E}_{\text{rms}}$, according to Equation 20.13.

PROBLEM SOLVING INSIGHT

When using the equation $\mathscr{E}_0 = NAB\omega$, remember that the angular frequency ω must be expressed in rad/s, not in Hz. See Equation 10.9 in Section 10.4 to review the relation between rad/s and Hz.

SOLUTION The peak emf is $\mathscr{E}_0 = \sqrt{2}\,\mathscr{E}_{\text{rms}} = \sqrt{2}\,(120\ \text{V}) = 170\ \text{V}$. Since one revolution corresponds to 2π radians, the angular speed of the coil is $\omega = 2\pi(60.0\ \text{Hz}) = 377$ rad/s. The magnitude of the magnetic field is

$$B = \frac{\mathscr{E}_0}{NA\omega} = \frac{170\ \text{V}}{(500)(3.0 \times 10^{-3}\ \text{m}^2)(377\ \text{rad/s})} = \boxed{0.30\ \text{T}}$$

THE PHYSICS OF . . .

a bike generator.

Example 12 A Bike Generator

A generator is mounted on a bicycle to power a headlight. A small wheel on the shaft of the generator is pressed against the bike tire and turns the armature 44 times for each revolution of the tire. The tire has a radius of 0.33 m. The armature has 75 turns, each with an area of $2.6 \times 10^{-3}\ \text{m}^2$, and rotates in a 0.10-T magnetic field. When the peak emf being generated is 6.0 V, what is the linear speed of the bike?

REASONING The relation $\mathscr{E}_0 = NAB\omega$ provides a solution to this problem, if we take advantage of two additional facts. The first is that the angular speed ω of the armature is 44 times larger than the angular speed ω_{tire} of the bike tire. The second is that the tire is rolling, so that ω_{tire} is related to the linear speed v of the bike according to $\omega_{\text{tire}} = v/r$ (Equation 8.12), where r is the radius of the tire.

SOLUTION Using $\omega = 44\omega_{\text{tire}} = 44(v/r)$, we find that the peak emf is $\mathscr{E}_0 = NAB \times 44(v/r)$. Therefore, the linear speed is

$$v = \frac{\mathscr{E}_0 r}{44NAB} = \frac{(6.0\ \text{V})(0.33\ \text{m})}{44(75)(2.6 \times 10^{-3}\ \text{m}^2)(0.10\ \text{T})} = \boxed{2.3\ \text{m/s}}$$

THE ELECTRICAL ENERGY DELIVERED BY A GENERATOR AND THE COUNTERTORQUE

Some power-generating stations burn fossil fuel (coal, gas, or oil) to heat water and produce pressurized steam for turning the blades of a turbine. Others use nuclear fuel or falling water as a source of energy. The shaft of the turbine is linked to that of the generator, so as the generator coil rotates, mechanical work is transformed into electrical energy.

The devices to which the generator supplies electricity are known collectively as the "load," because they place a burden or load on the generator by taking electrical energy from it. If all the devices are switched off, the generator runs under a no-load condition, because there is no current in the external circuit and the generator does not supply electrical energy. Then, work needs to be done on the turbine only to overcome friction and other mechanical losses within the generator itself, and fuel consumption is at a minimum.

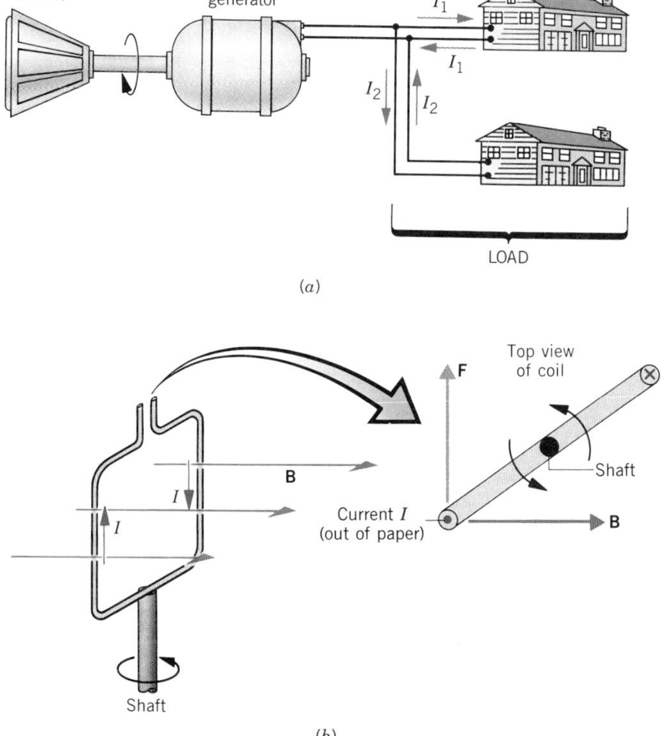

Figure 22.22 (*a*) Current is drawn when a load is connected to the generator. (*b*) The current $I = I_1 + I_2$ in the coil experiences a magnetic force **F** due to the magnetic field **B**. This magnetic force leads to a torque that retards the motion of the coil.

Figure 22.22*a* illustrates a situation in which a load is connected to a generator. Because there is now a current $I = I_1 + I_2$ in the coil, and the coil is situated in a magnetic field, the current experiences a magnetic force **F**. Part *b* of the drawing shows the magnetic force acting on the left side of the coil, the direction of **F** being given by RHR-1. A force of equal magnitude but opposite direction acts on the right side of the coil, although this force is not shown in the drawing. The magnetic force **F** retards the motion of the coil. We encountered such a retarding force in Section 22.2 when discussing the motional emf of a rod sliding along two conducting rails (see Figure 22.5). In the ac generator, the magnetic force **F** gives rise to a *countertorque* that opposes the rotational motion. The greater the current drawn from the generator, the greater the countertorque, and the harder it is for the turbine to turn the coil. To compensate for this countertorque and to keep the coil rotating at a constant angular speed, work must be done on the coil by the turbine, which means more fuel must be burned. This is another example of the law of conservation of energy, since the electrical energy consumed by the load must ultimately come from the energy source used to drive the turbine.

THE BACK EMF GENERATED BY AN ELECTRIC MOTOR

A generator converts mechanical work into electrical energy; in contrast, an electric motor converts electrical energy into mechanical work. Both devices are similar and consist of a coil of wire that rotates in a magnetic field. In fact, as the armature of a motor rotates, the magnetic flux passing through the coil changes and an emf is induced in the coil. Thus, when a motor is operating, two sources of emf are present: (1) the applied emf V that provides current to drive the motor (e.g., from a 120-V outlet), and (2) the emf $\mathscr{E}$ induced by the generator-like action of the rotating coil. The circuit diagram in Figure 22.23 shows these two emfs.

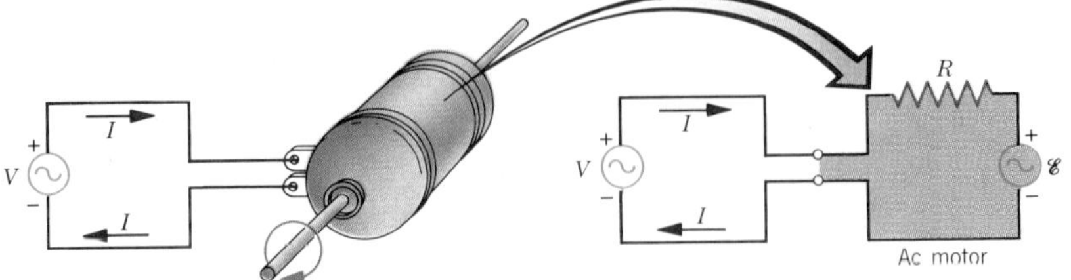

Figure 22.23 The applied emf V supplies the current I to drive the motor. The circuit shows V, along with the electrical equivalent of the motor, including the resistance R of its coil and the back emf $\mathscr{E}$.

Consistent with Lenz's law, the induced emf $\mathscr{E}$ acts to oppose the applied emf V and is called the ***back emf*** or the ***counter emf*** of the motor. The greater the speed of the motor, the greater the flux change through the coil, and the greater is the back emf. Because V and $\mathscr{E}$ have opposite polarities, the net emf in the circuit is $V - \mathscr{E}$. In Figure 22.23, R is the resistance of the wire in the coil, and the current I drawn by the motor is determined from Ohm's law as the net emf divided by the resistance:

$$I = \frac{V - \mathscr{E}}{R} \tag{22.5}$$

The next example uses this result to illustrate that the current in a motor depends on both the applied emf V and the back emf $\mathscr{E}$.

Example 13 Operating a Motor

The coil of an ac motor has a resistance of $R = 4.1\ \Omega$. The motor is plugged into an outlet where $V = 120.0$ volts (rms), and the coil develops a back emf of $\mathscr{E} = 118.0$ volts (rms) when rotating at normal speed. The motor is turning a wheel. Find (a) the current when the motor first starts up and (b) the current when the motor is operating at normal speed.

REASONING Once normal operating speed is attained, the motor need only work to compensate for frictional losses. But in bringing the wheel up to speed from rest, the motor must also do work to increase the wheel's rotational kinetic energy. Thus, bringing the wheel up to speed requires more work, and hence more current, than maintaining the normal operating speed. We expect our answers to parts (a) and (b) to reflect this fact.

SOLUTION

(a) When the motor just starts up, the coil is not rotating, so there is no back emf induced in the coil and $\mathscr{E} = 0$. The start-up current drawn by the motor is

$$I = \frac{V - \mathscr{E}}{R} = \frac{120.0\ \text{V}}{4.1\ \Omega} = \boxed{29\ \text{A}} \tag{22.5}$$

(b) At normal speed, the motor develops a back emf of $\mathscr{E} = 118.0$ volts, so the current is

$$I = \frac{V - \mathscr{E}}{R} = \frac{120.0\ \text{V} - 118.0\ \text{V}}{4.1\ \Omega} = \boxed{0.49\ \text{A}}$$

PROBLEM SOLVING INSIGHT

The current in an electric motor depends on both the applied emf V and any back emf $\mathscr{E}$ developed because the coil of the motor is rotating.

Example 13 illustrates that when a motor is just starting, there is little back emf, and, consequently, a relatively large current exists in the coil. As the motor speeds up, the back emf increases until it reaches a maximum value when the motor is rotating at normal speed. The back emf becomes almost equal to the applied emf, and the current is reduced to a relatively small value. This limiting value of the current is sufficient to provide the torque on the coil to drive the load (such as a fan) and to overcome frictional losses.

22.8 *MUTUAL INDUCTANCE AND SELF-INDUCTANCE*

MUTUAL INDUCTANCE

We have seen that an emf can be induced in a coil by keeping the coil stationary and moving a magnet nearby, or by moving the coil near a stationary magnet. Figure 22.24 illustrates another important method of inducing an emf. Here, two coils of wire are placed close to each other. Coil 1 is connected to an ac generator that sends an alternating current I_1 through the coil. Coil 2 is not attached to a generator, although a voltmeter is connected between the ends of coil 2 to register any emf. It is customary to call coil 1 (the coil connected to the generator) the *primary coil* and coil 2 the *secondary coil.*

The current-carrying primary coil is an electromagnet and creates a magnetic field in the surrounding region. If the two coils are close to each other, a significant fraction of this magnetic field penetrates the secondary coil and produces a magnetic flux. The flux is changing in time, since the current in the primary coil and its associated magnetic field are changing in time. Because of the change in the flux, an emf is induced in the secondary coil.

The effect in which a changing current in one circuit induces an emf in another circuit is called ***mutual induction.*** According to Faraday's law of electromagnetic induction, the emf $\mathcal{E}_2$ induced in the secondary coil is proportional to the change in flux $\Delta\Phi_2$ passing through it. However, $\Delta\Phi_2$ is produced by the change in current ΔI_1 in the primary coil. Therefore, it is convenient to recast Faraday's law into a form that relates $\mathcal{E}_2$ to ΔI_1. To see how this recasting is accomplished, note that the net magnetic flux passing through the secondary coil is $N_2\Phi_2$, where N_2 is the number of loops in the secondary coil and Φ_2 is the flux through one loop (assumed to be the same for all loops). The net flux is proportional to the magnetic field, which, in turn, is proportional to the current I_1 in the primary. Thus, we can

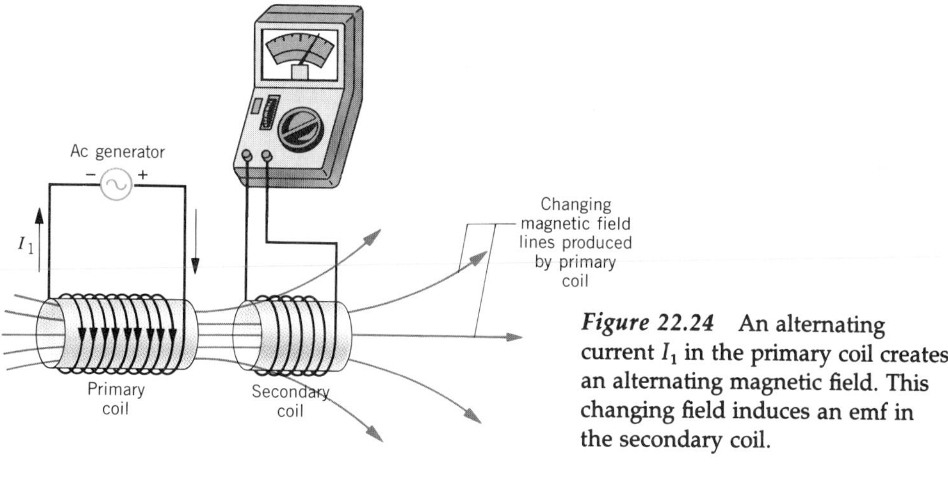

Figure 22.24 An alternating current I_1 in the primary coil creates an alternating magnetic field. This changing field induces an emf in the secondary coil.

write $N_2\Phi_2 \propto I_1$. This proportionality can be converted into an equation in the usual manner by introducing a proportionality constant M, known as the ***mutual inductance:***

$$N_2\Phi_2 = MI_1 \quad \text{or} \quad M = \frac{N_2\Phi_2}{I_1} \tag{22.6}$$

Substituting this equation into Faraday's law, we find that

$$\mathscr{E}_2 = -N_2\frac{\Delta\Phi_2}{\Delta t} = -\frac{\Delta(N_2\Phi_2)}{\Delta t} = -\frac{\Delta(MI_1)}{\Delta t} = -M\frac{\Delta I_1}{\Delta t}$$

Emf due to mutual induction

$$\mathscr{E}_2 = -M\frac{\Delta I_1}{\Delta t} \tag{22.7}$$

Writing Faraday's law in this manner makes it clear that the emf $\mathscr{E}_2$ induced in the secondary coil is due to the change in the current ΔI_1 in the primary coil.

Equation 22.7 shows that the measurement unit for the mutual inductance M is V·s/A, which is called a henry (H) after Joseph Henry: 1 V·s/A = 1 H. The mutual inductance depends on the geometry of the coils and the nature of any ferromagnetic core material that is present. Although M can be calculated for some highly symmetrical arrangements, it is usually measured experimentally. In most situations, values of M are less than 1 H and are often on the order of millihenries (1 mH = 1×10^{-3} H) or microhenries (1 μH = 1×10^{-6} H).

THE PHYSICS OF . . . an induction ammeter.

Because mutual induction permits the transfer of electrical energy from one circuit to another without any physical contact between them, it has many applications. As an example, Figure 22.25*a* shows an induction ammeter, which is a device for measuring alternating current in situations where it would be too time-consuming or risky to disconnect a wire and insert a standard ammeter in the circuit. Suppose that you need to know whether a wire in a broken appliance carries a 60-Hz current, and, if so, how much. Part *b* of the figure indicates that the iron core "jaw" of the induction ammeter is slipped around the wire in question. The alternating current produces a changing magnetic field in the space around the wire. The iron core routes some of the field lines through a coil wrapped around the jaw of the ammeter, as part *b* of the drawing shows. The changing magnetic field induces an emf that is registered by the meter connected to the coil. Since the induced emf is proportional to the current in the appliance wire, the meter can be calibrated to read this current.

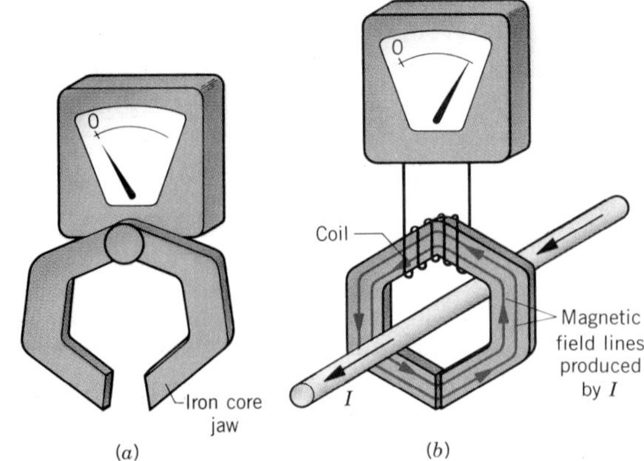

Figure 22.25 An induction ammeter with its iron-core jaw (*a*) open and (*b*) closed around a wire carrying an alternating current I. Some of the magnetic field lines that encircle the wire are routed through the coil by the iron core and lead to an induced emf. The meter detects the emf and is calibrated to display the amount of current in the wire.

SELF-INDUCTANCE

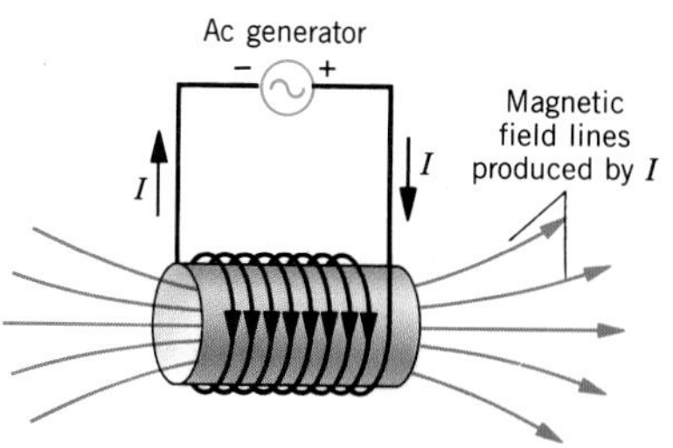

Figure 22.26 The alternating current in the coil generates an alternating magnetic field that induces an emf in the coil.

In all the examples of induced emfs presented so far, the magnetic field has been produced by an external source, such as a permanent magnet or an electromagnet. However, the magnetic field need not arise from an external source. An emf can be induced in a current-carrying coil by a change in the magnetic field that the current itself produces. For instance, Figure 22.26 shows a coil connected to an ac generator. The alternating current creates an alternating magnetic field that, in turn, creates a changing flux through the coil. The change in flux induces an emf in the coil, in accord with Faraday's law. The effect in which a changing current in a circuit induces an emf in the same circuit is referred to as ***self-induction.***

When dealing with self-induction, as with mutual induction, it is customary to recast Faraday's law into a form in which the induced emf is proportional to the change in current in the coil, rather than to the change in flux. If Φ is the magnetic flux that passes through one turn of the coil, then $N\Phi$ is the net flux through a coil of N turns. Since Φ is proportional to the magnetic field, and the magnetic field is proportional to the current I, it follows that $N\Phi \propto I$. By inserting a constant L, called the ***self-inductance*** or simply the ***inductance*** of the coil, we can convert this proportionality into Equation 22.8:

$$N\Phi = LI \quad \text{or} \quad L = \frac{N\Phi}{I} \tag{22.8}$$

Faraday's law of induction now gives the induced emf as

$$\mathscr{E} = -N\frac{\Delta\Phi}{\Delta t} = -\frac{\Delta(N\Phi)}{\Delta t} = -\frac{\Delta(LI)}{\Delta t} = -L\frac{\Delta I}{\Delta t}$$

[Emf due to self-induction]

$$\mathscr{E} = -L\frac{\Delta I}{\Delta t} \tag{22.9}$$

Like mutual inductance, L is measured in henries. The magnitude of L depends on the geometry of the coil and on the core material. By wrapping the coil around a ferromagnetic (iron) core, the magnetic flux—and therefore the inductance—can be increased substantially relative to that for an air core. Because of their self-inductance, coils are known as ***inductors*** and are widely used in electronics. Inductors come in all sizes, typically in the range between millihenries and microhenries. The next example shows how to determine the inductance of a solenoid.

Example 14 The Self-Inductance of a Solenoid

A solenoid of length $\ell = 8.0 \times 10^{-2}$ m and cross-sectional area $A = 5.0 \times 10^{-5}$ m^2 contains $n = 6500$ turns per meter. (a) Find the self-inductance of the solenoid, assuming the core is air. (b) Determine the emf induced in the solenoid when the current increases from 0 to 1.5 A in a time of 0.20 s.

REASONING

(a) The self-inductance can be found by using Equation 22.8, $L = N\Phi/I$, provided the flux Φ can be determined. The flux is given by Equation 22.2 as $\Phi = BA\cos\phi$. In the case of a solenoid, the interior magnetic field is directed perpendicular to the plane of the loops, so $\phi = 0°$ and $\Phi = BA$. The magnetic field inside the solenoid has the value $B = \mu_0 nI$, according to Equation 21.7, where n is the number of turns per unit length.

(b) The emf induced in the solenoid can be obtained from the relation $\mathscr{E} = -L(\Delta I/\Delta t)$, since L, ΔI, and Δt are known.

SOLUTION

(a) The self-inductance of the solenoid is

$$L = \frac{N\Phi}{I} = \frac{N(BA)}{I} = \frac{N(\mu_0 nI)A}{I} = \mu_0 nNA = \mu_0 n^2 A\ell$$

where we have replaced N by $n\ell$. Substituting the given values into this result yields

$$L = \mu_0 n^2 A\ell = (4\pi \times 10^{-7}\ \text{T}\cdot\text{m/A})(6500\ \text{turns/m})^2 \times (5.0 \times 10^{-5}\ \text{m}^2)(8.0 \times 10^{-2}\ \text{m}) = \boxed{2.1 \times 10^{-4}\ \text{H}}$$

(b) The induced emf that results from the increasing current is

$$\mathscr{E} = -L\frac{\Delta I}{\Delta t} = -(2.1 \times 10^{-4}\ \text{H})\left(\frac{1.5\ \text{A}}{0.20\ \text{s}}\right) = \boxed{-1.6 \times 10^{-3}\ \text{V}} \tag{22.9}$$

The negative sign reminds us that the induced emf opposes the increasing current that induces the emf.

THE ENERGY STORED IN AN INDUCTOR

An inductor, like a capacitor, can store energy. This stored energy arises because a generator does work to establish a current in an inductor. Suppose an inductor is connected to a generator whose terminal voltage can be varied continuously from zero to some final value. As the voltage is increased, the current I in the circuit rises continuously from zero to its final value. While the current is rising, an induced emf $\mathscr{E} = -L(\Delta I/\Delta t)$ appears across the inductor. Conforming with Lenz's law, the polarity of the induced emf $\mathscr{E}$ is opposite to that of the generator voltage, so as to oppose the increase in the current. Thus, the generator must do work to push the charges through the inductor against this induced emf. The increment of work ΔW done by the generator in moving a small amount of charge ΔQ through the inductor is $\Delta W = (\Delta Q)\mathscr{E} = (\Delta Q)\, L\,(\Delta I/\Delta t)$, according to Equation 19.4. To ensure that the work done by the generator is positive, as it must be since the generator is driving charge against an opposing emf, the minus sign in front of the $L(\Delta I/\Delta t)$ term has been removed. Since $\Delta Q/\Delta t$ is the current I, the work done by the generator is

$$\Delta W = LI(\Delta I)$$

In this expression ΔW represents the work done by the generator to increase the current in the inductor by an amount ΔI. To determine the total work W done during the time interval when the current is changed from zero to its final value, all the small increments of work ΔW must be added together. This summation is left as an exercise at the end of this chapter (see problem 56). The result is $W = \frac{1}{2}LI^2$, where I represents the final current in the inductor. This work is stored as energy in the inductor, so that

[Energy stored in an inductor]

$$\text{Energy} = \tfrac{1}{2}\, LI^2 \tag{22.10}$$

It is possible to regard the energy in an inductor as being stored in its magnetic field. For the special case of a long solenoid, Example 14 shows that the self-inductance is $L = \mu_0 n^2 A\ell$, where n is the number of turns per unit length, A is the cross-sectional area, and ℓ is the length of the solenoid. As a result, the energy stored in a solenoid is

$$\text{Energy} = \tfrac{1}{2}LI^2 = \tfrac{1}{2}\mu_0 n^2 A\ell I^2$$

Since $B = \mu_0 nI$ at the interior of a long solenoid (Equation 21.7), this energy can be expressed as

$$\text{Energy} = \frac{1}{2\mu_0} B^2 A\ell$$

The term $A\ell$ is the volume inside the solenoid in which the magnetic field exists, so the energy per unit volume or ***energy density*** is

$$\text{Energy density} = \frac{\text{Energy}}{\text{Volume}} = \frac{1}{2\mu_0} B^2 \tag{22.11}$$

This result applies only to magnetic fields in air (or vacuum) or in nonmagnetic materials. Although it was obtained for the special case of a long solenoid, the relation is valid in general for any region of space where a magnetic field exists in air or vacuum. Thus, energy is stored in a magnetic field, just as it is in an electric field.

22.9 TRANSFORMERS

THE PHYSICS OF . . .

transformers.

One of the most important applications of mutual induction and self-induction takes place in a transformer. A ***transformer*** is a device for increasing or decreasing an ac voltage. For example, whenever cordless appliances (e.g., a hand-held vacuum cleaner) are plugged into a wall receptacle to recharge the batteries, a transformer plays a role in reducing the 120-V ac voltage to a much smaller value. Typically, between 3 and 9 V are needed to energize batteries. In another instance, a picture tube in a television set needs about 15 000 V to accelerate the electron beam, and a transformer is used to obtain this high voltage from the 120 V provided at a wall socket.

Figure 22.27 shows a drawing of a transformer. The transformer consists of an iron core on which two coils are wound: a primary coil with N_p turns, and a secondary coil with N_s turns. The primary coil is connected to an ac generator. For the moment, suppose the switch in the secondary circuit is open, so there is no current in this circuit.

The alternating current in the primary coil establishes a changing magnetic field in the iron core. Because iron is easily magnetized, it greatly enhances the magnetic field relative to that in an air core and guides the field lines to the secondary coil. In a well-designed core, nearly all the magnetic flux Φ that passes through each turn of the primary also goes through each turn of the secondary. Since the magnetic field is changing, the flux through the primary and secondary coils is also changing, and consequently an emf is induced in both coils. In the secondary coil the induced emf $\mathcal{E}_s$ arises from mutual induction and is given by Faraday's law as

$$\mathcal{E}_s = -N_s \frac{\Delta\Phi}{\Delta t}$$

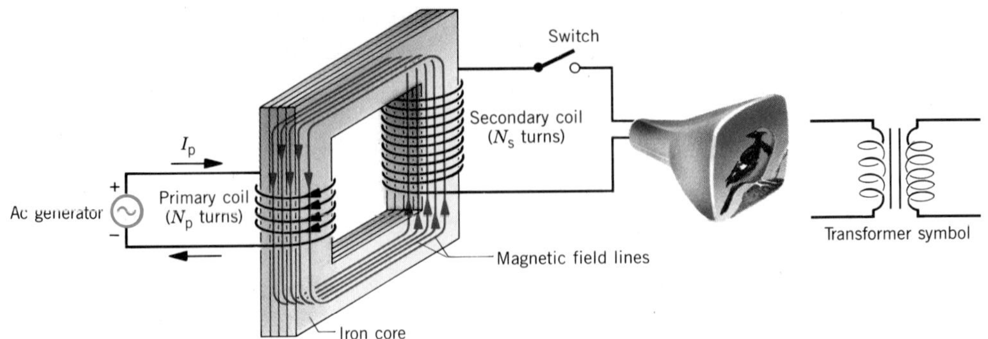

Figure 22.27 A transformer consists of a primary coil and a secondary coil, both wound on an iron core. The changing magnetic flux produced by the current in the primary coil induces an emf in the secondary coil. At the far right is the symbol for a transformer.

In the primary coil the induced emf $\mathscr{E}_p$ is due to self-induction and is specified by Faraday's law as

$$\mathscr{E}_p = -N_p \frac{\Delta\Phi}{\Delta t}$$

The term $\Delta\Phi/\Delta t$ is the same in both of these equations, since the same flux penetrates each turn of both coils. Dividing the two equations shows that

$$\frac{\mathscr{E}_s}{\mathscr{E}_p} = \frac{N_s}{N_p}$$

In a high-quality transformer the resistances of the coils are negligible, so the magnitudes of the emfs, $\mathscr{E}_s$ and $\mathscr{E}_p$, are nearly equal to the terminal voltages, V_s and V_p, across the coils (see Section 20.9 for a discussion of terminal voltage). The relation $\mathscr{E}_s/\mathscr{E}_p = N_s/N_p$ is called the ***transformer equation*** and is usually written in terms of the terminal voltages:

[Transformer equation]

$$\frac{V_s}{V_p} = \frac{N_s}{N_p} \tag{22.12}$$

According to the transformer equation, if N_s is greater than N_p, the secondary (output) voltage is greater than the primary (input) voltage. In this case we have a *step-up* transformer. On the other hand, if N_s is less than N_p, the secondary voltage is less than the primary voltage, and we have a *step-down* transformer. The ratio N_s/N_p is referred to as the *turns ratio* of the transformer. A turns ratio of 8/1 (often written as 8:1) means, for example, that the secondary coil has eight times more turns than does the primary coil. Conversely, a turns ratio of 1:8 implies that the secondary coil has one-eighth as many turns as the primary coil.

A transformer operates with ac electricity and not with steady direct current. A steady direct current in the primary coil produces a flux that does not change, and thus no emf is induced in the secondary coil. The ease with which transformers can change voltages from one value to another is a principal reason why ac is preferred over dc.

Power distribution stations, such as this one in Knoxville, Tennessee, use transformers (in red) to step-up or step-down voltages.

If the switch in the secondary circuit of Figure 22.27 is closed, a current I_s exists in the circuit and electrical energy is fed to the TV tube. This energy comes from the ac generator connected to the primary coil. Although the secondary voltage V_s

may be larger or smaller than the primary voltage V_p, energy is not being created or destroyed by the transformer. Energy conservation requires that the energy delivered to the secondary coil must be the same as the energy delivered to the primary coil, provided no energy is dissipated in heating these coils or is otherwise lost. In a well-designed transformer, less than 1% of the input energy is lost in the form of heat. Noting that power is energy per unit time, and assuming 100% energy transfer, the average power $\overline{P}_p$ delivered to the primary coil is equal to the average power $\overline{P}_s$ delivered to the secondary coil: $\overline{P}_p = \overline{P}_s$. But $P = IV$ (Equation 20.15a), so $I_p V_p = I_s V_s$, or

$$\frac{I_s}{I_p} = \frac{V_p}{V_s} = \frac{N_p}{N_s} \qquad (22.13)$$

Observe that V_s/V_p is equal to the turns ratio N_s/N_p, while I_s/I_p is equal to the inverse turns ratio N_p/N_s. Consequently, ***a transformer that steps up the voltage simultaneously steps down the current, and a transformer that steps down the voltage steps up the current.*** However, the power is neither stepped up nor stepped down, since $\overline{P}_p = \overline{P}_s$. Example 15 emphasizes this fact.

PROBLEM SOLVING INSIGHT

Example 15 A Step-Down Transformer

A step-down transformer inside a stereo receiver has 330 turns in the primary coil and 25 turns in the secondary coil. The plug connects the primary coil to a 120-V wall socket, and there is a current of 0.83 A in the primary coil while the receiver is turned on. Connected to the secondary coil are the transistor circuits of the receiver. Find (a) the voltage across the secondary coil, (b) the current in the secondary coil, and (c) the average electrical power delivered to the transistor circuits.

REASONING AND SOLUTION

(a) The voltage across the secondary coil can be found from the transformer equation:

$$V_s = V_p \frac{N_s}{N_p} = (120\ \text{V})\left(\frac{25}{330}\right) = \boxed{9.1\ \text{V}} \qquad (22.12)$$

(b) The current in the secondary coil follows from Equation 22.13 as

$$I_s = I_p \frac{N_p}{N_s} = (0.83\ \text{A})\left(\frac{330}{25}\right) = \boxed{11\ \text{A}}$$

(c) The average power $\overline{P}_s$ delivered to the secondary is the product of I_s and V_s:

$$\overline{P}_s = I_s V_s = (11\ \text{A})(9.1\ \text{V}) = \boxed{1.0 \times 10^2\ \text{W}} \qquad (20.15\text{a})$$

As a check on our calculation, we verify that the power delivered to the secondary coil is the same as that sent to the primary coil from the wall receptacle: $\overline{P}_p = I_p V_p = (0.83\ \text{A})(120\ \text{V}) = 1.0 \times 10^2\ \text{W}$.

Transformers play an important role in the transmission of power between electrical generating plants and the communities they serve. Whenever electricity is transmitted, there is always some loss of power in the transmission lines themselves due to resistive heating. Since the resistance of the wires is proportional to their length, the longer the wires the greater is the power loss. Power companies reduce this loss by using transformers that step up the voltage to high levels, while reducing the current. A smaller current means less power loss, since $P = I^2R$,

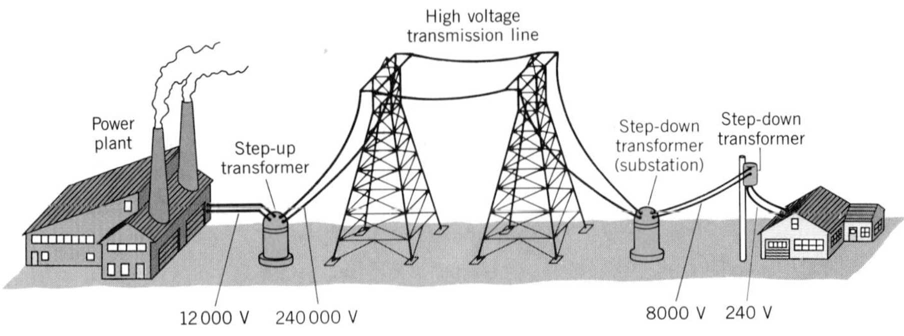

Figure 22.28 Transformers play a key role in the transmission of electric power.

where R is the resistance of the transmission wires. Figure 22.28 shows one possible way of transmitting power. The power plant produces a voltage of 12 000 V. This voltage is then raised to 240 000 V by a 20 : 1 step-up transformer. The high-voltage power is sent over the long-distance transmission line. Upon arrival at the city, the voltage is reduced to about 8000 V at a substation using a 1 : 30 step-down transformer. However, before any domestic use, the voltage is further reduced to 240 V (or possibly 120 V) by another step-down transformer that is often mounted on a utility pole. The power is then distributed to consumers.

INTEGRATION OF CONCEPTS

ELECTROMAGNETIC INDUCTION AND ENERGY CONSERVATION

Electromagnetic induction is energy conservation in action. The conservation of energy principle stipulates that energy can only be transformed from one type to another; it cannot be created or destroyed. And the many applications of electromagnetic induction take advantage of energy transformations. In an ac electrical generator, burning coal or oil, falling water, or nuclear fuel provides the energy to do the work of turning a coil in a magnetic field. As a result, electromagnetic induction converts the chemical energy of the coal or oil, the nuclear energy of uranium, or the gravitational potential energy of the water into electrical energy. In a small bicycle generator, the work of turning a coil in a magnetic field is done by the rider, and electromagnetic induction converts this work into electrical energy. In an electric guitar pickup, electromagnetic induction converts the energy of the vibrating guitar string into electrical energy, which is used to produce sound waves that carry acoustic energy. In a moving coil or moving magnet microphone, the energy in a sound wave causes a diaphragm to vibrate, and electromagnetic induction changes the acoustic energy into electrical energy. In a transformer, the electrical energy delivered to the primary coil is stored in the magnetic field of the coil. From there, it is transferred by electromagnetic induction to the secondary coil, where it appears again as electrical energy in the circuit connected to the secondary coil. Electromagnetic induction is indeed one of the most useful methods available for energy transformation.

SUMMARY

The **magnetic flux Φ** that passes through a surface is $\Phi = BA \cos \phi$, where A is the area of the surface, B is the magnitude of the magnetic field at the surface, and ϕ is the angle between $\mathbf{B}$ and the normal to the surface.

Electromagnetic induction is the phenomenon in which an emf is induced in a coil of wire by a change in the magnetic flux that passes through the coil.

Faraday's law of electromagnetic induction states that the average emf $\mathscr{E}$ induced in a coil of N loops is

$$\mathscr{E} = -N\left(\frac{\Phi - \Phi_0}{t - t_0}\right) = -N\frac{\Delta\Phi}{\Delta t}$$

where $\Delta\Phi$ is the change in magnetic flux through one loop and Δt is the time interval during which the change occurs. For the special case of a conductor of length L moving with speed v perpendicular to a magnetic field $\mathbf{B}$, the induced emf is called motional emf and its value is given by $\mathscr{E} = vBL$.

Lenz's law provides a way to determine the polarity of an induced emf. Lenz's law states that the polarity of an induced emf is such that the induced current produces an induced magnetic field that opposes the change in flux causing the emf. Lenz's law is a consequence of the law of conservation of energy.

In its simplest form, an **electric generator** consists of a coil of N loops that rotates in a uniform magnetic field $\mathbf{B}$. The emf produced by this generator is $\mathscr{E} = NAB\omega \times \sin \omega t = \mathscr{E}_0 \sin \omega t$, where A is the area of the coil, ω is the angular speed (in rad/s) of the coil, and $\mathscr{E}_0$ is the peak emf.

When an electric motor is running, it exhibits a generator-like behavior by producing an induced emf, called a **back emf.** The current I needed to keep the motor running at a constant speed is $I = (V - \mathscr{E})/R$, where V is the emf applied to the motor by an external source, $\mathscr{E}$ is the back emf, and R is the resistance of the coil.

Mutual induction is the effect in which a changing current in the primary coil induces an emf in the secondary coil. The emf $\mathscr{E}_2$ induced in the secondary coil by a change in current ΔI_1 in the primary coil is $\mathscr{E}_2 = -M(\Delta I_1/\Delta t)$, where Δt is the time interval during which the change occurs. The constant M is the **mutual inductance** between the two coils and is measured in henries (H).

Self-induction is the effect in which a change in current ΔI in a coil induces an emf $\mathscr{E} = -L(\Delta I/\Delta t)$ in the same coil. The constant L is the **self-inductance** or **inductance** of the coil and is measured in henries. To establish a current I in an inductor, work must be done by an external agent. This work is stored as energy in the inductor, the amount being Energy $= \frac{1}{2}LI^2$. The energy stored in an inductor can be regarded as being stored in its magnetic field. At any point in air or vacuum or in a nonmagnetic material where a magnetic field $\mathbf{B}$ exists, the **energy density,** or the energy stored per unit volume, is Energy density $= B^2/(2\mu_0)$.

A **transformer** consists of a primary coil of N_p turns and a secondary coil of N_s turns. When an emf $\mathscr{E}_p$ is applied to the primary, an emf $\mathscr{E}_s$ is induced in the secondary according to the relation $\mathscr{E}_s/\mathscr{E}_p = N_s/N_p$. A transformer functions with ac electricity, not with steady dc electricity. If the transformer is 100% efficient in transferring power from the primary coil to the secondary coil, the ratio of the secondary current I_s to the primary current I_p is $I_s/I_p = N_p/N_s$.

QUESTIONS

1. A uniform magnetic field points due east. A horizontal copper rod is perpendicular to this field and is oriented in the north–south direction. The rod falls freely to the earth. (a) Which end of the rod, north or south, becomes positively charged? (b) Which end of the rod becomes positively charged if the rod is initially oriented parallel to the magnetic field? Account for your answers.

2. In the discussion concerning Figure 22.5, we saw that a force of 0.086 N from an external agent was required to keep the rod moving at a constant speed. Suppose the light bulb in the figure is unscrewed from its socket. How much force would now be needed to keep the rod moving at a constant speed? Justify your answer.

3. Eddy currents are electric currents that can arise in a piece of metal when it moves through a region where the magnetic field is not the same everywhere. The picture shows, for example, a metal sheet moving to the right at a velocity $\mathbf{v}$ and a magnetic field $\mathbf{B}$ that is directed perpendicular to the sheet. At the instant represented, the magnetic field only extends over the left half of the sheet. An emf is induced that leads to the eddy current shown. Explain why this current causes the metal sheet to slow down. This action of eddy currents is used in various devices as a brake to damp out unwanted motion.

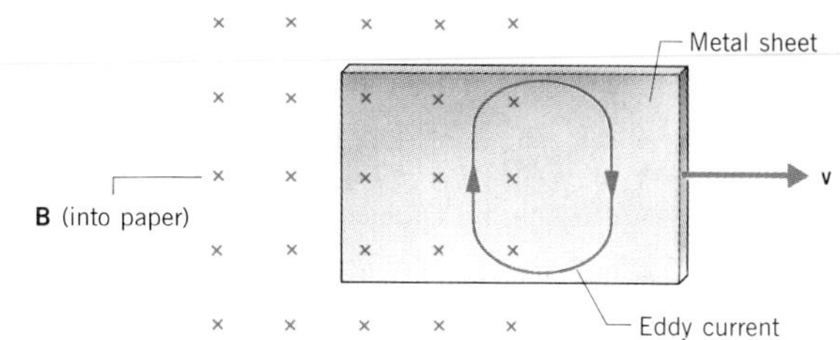

4. A magnetic field is necessary if there is to be a magnetic flux passing through a coil of wire. Yet, just because there is a magnetic field does not mean that a magnetic flux will pass through a coil. Account for this observation.

5. Suppose the magnetic flux through a 1-m^2 flat surface is known to be 2 Wb. From this data alone, is it possible to determine the average magnetic field at the surface? If it is not possible to determine the magnitude of the field, what can be ascertained about the field?

6. A square loop of wire is moving (but not rotating) through a uniform magnetic field. The normal to the loop is oriented parallel to the magnetic field. Is an emf induced in the loop? Give a reason for your answer.

7. Explain how a bolt of lightning can produce a current in the circuit of an electrical appliance, even when the lightning does not directly strike the appliance.

8. Review Conceptual Example 8 before answering this question. A solenoid is connected to an ac source. A copper ring is placed inside the solenoid, with the normal to the ring parallel to the axis of the solenoid. The copper ring gets hot, yet nothing touches it. Why?

9. A robot is designed to move parallel to a cable hidden under the floor. The cable carries a steady direct current I. A sensor mounted on the robot consists of a coil of wire. The coil is near the floor and parallel to it. As long as the robot moves parallel to the cable, with the coil directly over it, no emf is induced in the coil, since the magnetic flux through the coil does not change. But when the robot deviates from the parallel path, an induced emf appears in the coil. The emf is sent to electronic circuits that bring the robot back to the path. Explain why an emf would be induced in the sensor coil.

10. In Figure 22.12, suppose that the ac generator is replaced by a battery. Would the ground fault interrupter still work as a safety device? Justify your answer.

11. In a car, the generator-like action of the alternator occurs while the engine is running and keeps the battery fully charged. The headlights would discharge an old and failing battery quickly if it were not for the alternator. Explain why the engine of a parked car runs more "quietly" with the headlights off than with them on when the battery is in bad shape.

12. In Figure 22.3*b* a coil of wire is being stretched. (a) Using Lenz's law, verify that the induced current in the coil has the direction shown in the drawing. (b) Deduce the direction of the induced current if the direction of the external magnetic field in the figure were reversed.

13. (a) When the switch in the circuit in the drawing is closed, a current is established in the coil and the metal ring "jumps" upward. Explain this behavior. (b) Describe what would happen to the ring if the battery polarity were reversed.

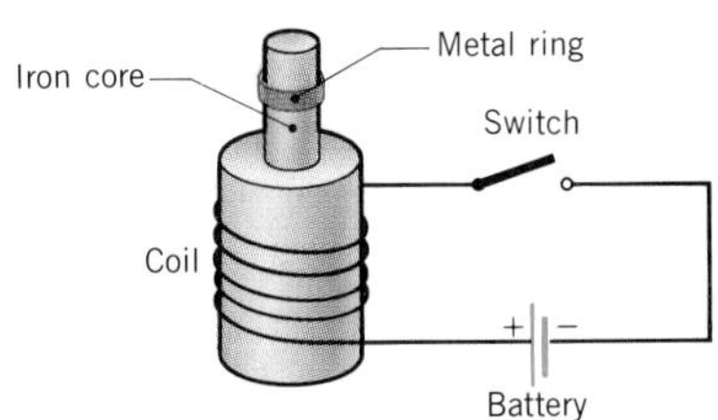

14. The string of an electric guitar vibrates in a standing wave pattern that consists of nodes and antinodes. (Section 17.5 discusses standing waves.) Where should an electromagnetic pickup be located in the standing wave pattern to produce a maximum emf, at a node or an antinode? Why?

15. An electric motor in a hair drier is running at normal speed and, thus, is drawing a relatively small current, as in part (b) of Example 13. What happens to the current drawn by the motor if the shaft is prevented from turning, so the back emf is suddenly reduced to zero? Remembering that the wire in the coil of the motor has some resistance, what happens to the temperature of the coil? Justify your answers.

16. Would a steady direct current in a wire register on the induction ammeter shown in Figure 22.25? Explain.

PROBLEMS

Section 22.2 Motional Emf

1. A spark can jump between two nontouching conductors if the potential difference between them is sufficiently large. A potential difference of approximately 940 V is required to produce a spark in an air gap of 1.0×10^{-4} m. Suppose the light bulb in Figure 22.4*b* is replaced by such a gap. How fast would a 1.6-m rod have to be moving in a magnetic field of 0.85 T to cause a spark to jump across the gap?

2. The wingspan (tip-to-tip) of a Boeing 747 jetliner is 59 m. The plane is flying horizontally at a speed of 220 m/s. The vertical component of the earth's magnetic field is 5.0×10^{-6} T. Find the emf induced between the wing tips.

3. Near San Francisco, where the vertically downward component of the earth's magnetic field is 4.8×10^{-5} T, a car is traveling at 25 m/s. An emf of 2.4×10^{-3} V is induced between the sides of the car. (a) Which side of the car is positive, the driver's side or the passenger's side? (b) What is the width of the car?

4. An emf of 0.35 V is generated between the ends of a metal bar moving through a magnetic field of 0.11 T, as in

Figure 22.4*a*. What field strength would be needed to produce an emf of 1.5 V between the ends of the bar, assuming that all other factors remain the same?

5. The drawing shows three identical rods (A, B, and C) moving in different planes. A constant magnetic field of magnitude 0.45 T is directed along the $+y$ axis. The length of each rod is $L = 1.3$ m, and the speeds are the same, $v_A = v_B = v_C = 2.7$ m/s. For each rod, find the magnitude of the motional emf, and indicate which end (1 or 2) of the rod is positive.

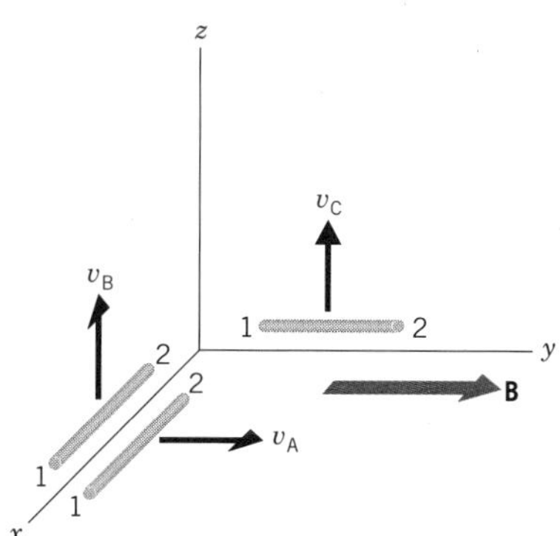

6. A metal rod (length = 0.75 m) moves perpendicular to a magnetic field of 0.15 T. An emf of 0.24 V exists between the ends of the rod. How far does the rod move in 7.0 s?

***7.** The drawing shows a C-shaped conducting rail and a rod sliding along it. The rail and rod have resistivities of $1.7 \times 10^{-8}\ \Omega \cdot$ m and cross-sectional areas of 5.6×10^{-5} m^2. A magnetic field of magnitude 0.22 T is oriented perpendicular to the plane of the rail. The speed of the rod is 3.1 m/s, and at $t = 0$ the rod is located at the left end of the rail. What is the induced current in this circuit when $t = 5.0$ s?

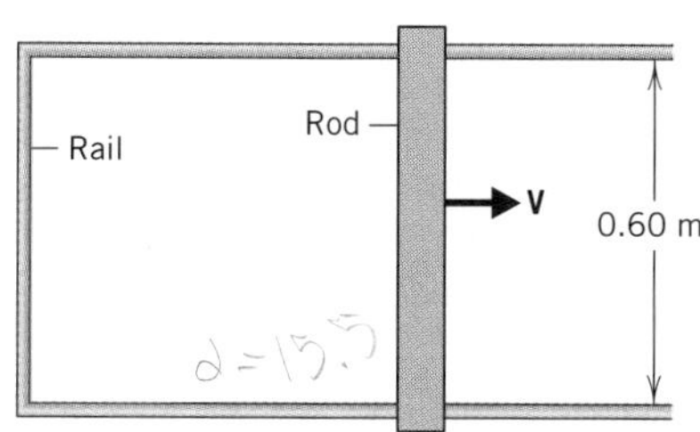

***8.** Suppose the light bulb in Figure 22.4*b* is replaced with a short wire of zero resistance, and the resistance of the rails is negligible. The only resistance is from the moving rod, which is iron (resistivity $= 9.7 \times 10^{-8}\ \Omega \cdot$m). The rod has a cross-sectional area of 3.1×10^{-6} m^2 and moves with a speed of 2.0 m/s. The magnetic field has a magnitude of 0.050 T. What is the current in the rod?

***9.** Suppose the light bulb in Figure 22.4*b* is replaced by a 6.0-Ω electric heater that consumes 15 W of power. The conducting bar moves to the right at a constant speed, the field strength is 2.4 T, and the length of the bar between the rails is 1.2 m. (a) How fast is the bar moving? (b) What force must be applied to the bar to keep it moving to the right at the constant speed found in part (a)?

****10.** Review Conceptual Example 4 as an aid in solving this problem. A conducting rod slides down between two frictionless vertical copper tracks at a constant speed of 5.4 m/s perpendicular to a 0.30-T magnetic field (see the drawing). The resistance of the rod and tracks is negligible. The rod maintains electrical contact with the tracks at all times and has a length of 1.2 m. A 0.50-Ω resistor is attached between the tops of the tracks. (a) What is the mass of the rod? (b) Find the change in the gravitational potential energy that occurs in a time of 0.20 s. (c) Find the electrical energy dissipated in the resistor in 0.20 s.

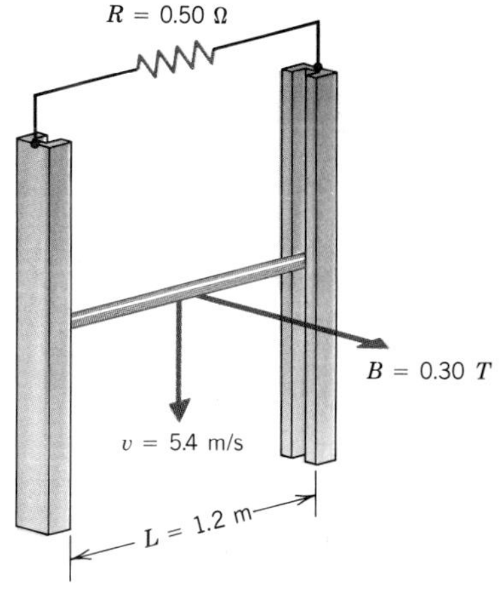

Section 22.3 Magnetic Flux

For problems in this set, assume that the magnetic flux is a positive quantity.

11. A hand is held flat and placed in a uniform magnetic field of magnitude 0.35 T. The hand has an area of 0.0160 m^2 and negligible thickness. Determine the magnetic flux that passes through the hand when the normal to the hand is (a) parallel and (b) perpendicular to the magnetic field.

12. A magnetic field has a magnitude of 0.078 T and is uniform over a circular surface that has a radius of 0.10 m. The field is oriented at an angle of $\phi = 25°$ with respect to the normal to the surface. What is the flux through the surface?

13. A rectangle (0.60 m $\times$ 0.30 m) lies in the *xy* plane. An identical rectangle lies in the *xz* plane. A uniform 0.17-T magnetic field points in the positive *z* direction. Find the flux through each rectangle.

14. The drawing shows three square surfaces, one lying in the xy plane, one in the xz plane, and one in the yz plane. The sides of each square have lengths of 3.0×10^{-2} m. A uniform magnetic field exists in this region, and its components are: $B_x = 0.40$ T, $B_y = 0.50$ T, and $B_z = 0.70$ T. Determine the magnetic flux that passes through the surface that lies in (a) the xy plane, (b) the xz plane, and (c) the yz plane.

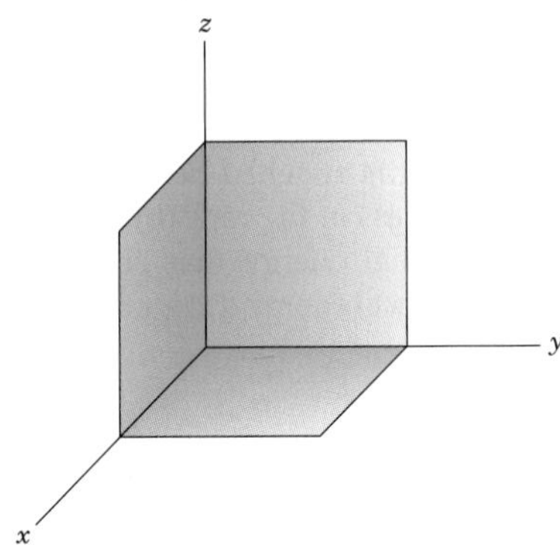

15. A house has a floor area of 112 m² and an outside wall that has an area of 28 m². The earth's magnetic field here has a horizontal component of 2.6×10^{-5} T that points due north and a vertical component of 4.2×10^{-5} T that points straight down, toward the earth. Determine the magnetic flux through the wall if the wall faces (a) north and (b) east. (c) Calculate the magnetic flux that passes through the floor.

***16.** A five-sided object, whose dimensions are shown in the drawing, is placed in a uniform magnetic field. The magnetic field has a magnitude of 0.25 T and points along the positive y direction. Determine the magnetic flux through each of the five sides.

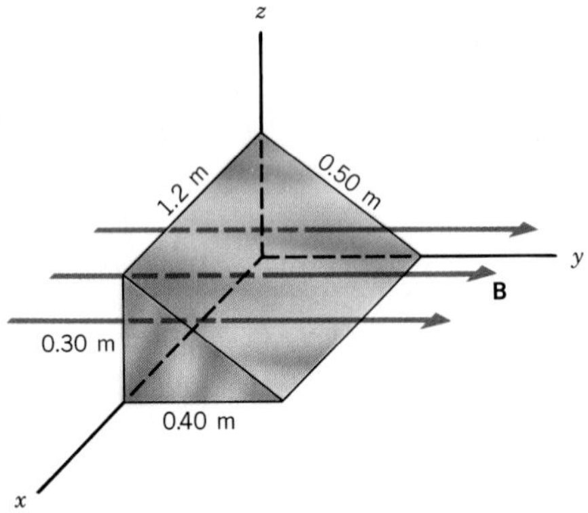

***17.** A rectangular loop of wire is moving toward the bottom of the page with a speed of 0.020 m/s (see the drawing). The loop is leaving a region in which a 2.4-T magnetic field exists; the magnetic field outside this region is zero. During a time of 2.0 s, what is the magnitude of the *change* in the magnetic flux?

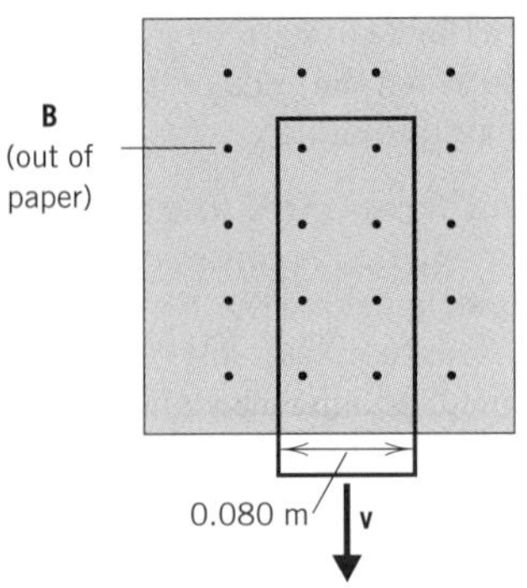

Section 22.4 Faraday's Law of Electromagnetic Induction

18. A magnetic field is perpendicular to a 0.040-m × 0.060-m rectangular coil of wire that has one hundred turns. In a time of 0.050 s, an average emf of magnitude 1.5 V is induced in the coil. What is the magnitude of the change in the magnetic field?

19. The drawing shows a straight wire, a part of which is bent into the shape of a circle. The radius of the circle is 2.0 cm. A constant magnetic field of magnitude 0.55 T is directed perpendicular to the plane of the paper. Someone grabs the ends of the wire and pulls it taut, so the radius of the circle shrinks to zero in a time of 0.25 s. Find the magnitude of the average induced emf between the ends of the wire.

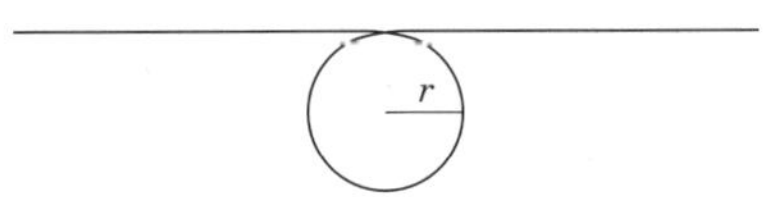

20. A loop of wire is placed in a uniform magnetic field that is parallel to the normal to the loop. The strength of the magnetic field is 3.0 T. The area of the loop begins shrinking at a constant rate of $\Delta A/\Delta t = 0.40$ m²/s. What is the magnitude of the emf induced in the loop while it is shrinking?

21. A circular coil (950 turns, radius = 0.060 m) is rotating in a uniform magnetic field. At $t = 0$, the normal to the coil is perpendicular to the magnetic field. At $t = 0.010$ s, the normal makes an angle of $\phi = 45°$ with the field, because the coil has made one-eighth of a revolution. An average emf of magnitude 0.065 V is induced in the coil. Find the magnitude of the magnetic field at the location of the coil.

22. A 300-turn rectangular loop of wire has an area of 5.0×10^{-3} m². At $t_0 = 0$ a magnetic field is turned on, and its magnitude increases to 0.40 T when $t = 0.80$ s. The field is directed at an angle of $\phi = 30.0°$ with respect to the normal of the loop. (a) Find the magnitude of the average emf induced in the loop. (b) If the loop is a closed circuit whose resistance is 6.0 Ω, determine the average induced current.

23. Magnetic resonance imaging (MRI) is a medical technique for producing "pictures" of the interior of the body. The patient is placed within a strong magnetic field. One safety concern is what would happen to the positively and negatively charged particles in the body fluids if an equipment failure caused the magnetic field to be shut off suddenly. An induced emf could cause these particles to flow, producing an electric current within the body. Suppose the largest surface of the body through which flux passes has an area of 0.032 m^2 and a normal that is parallel to a magnetic field of 1.5 T. Determine the smallest time period during which the field can be allowed to vanish if the magnitude of the average induced emf is to be kept less than 0.010 V.

24. A 1.8-m-long aluminum rod is rotating about an axis that is perpendicular to one end. A 0.27-T magnetic field is directed parallel to the axis. The rod rotates through one-fourth of a circle in 2.0 s. What is the magnitude of the average emf generated between the ends of the rod during this time?

***25.** A conducting coil of 1850 turns is connected to a galvanometer, and the total resistance of the circuit is 45.0 Ω. The area of each turn is 4.70×10^{-4} m^2. This coil is moved from a region where the magnetic field is zero into a region where it is nonzero, the normal to the coil being kept parallel to the magnetic field. The amount of charge that is induced to flow around the circuit is measured to be 8.87×10^{-3} C. Find the magnitude of the magnetic field. (Such a device can be used to measure the magnetic field strength and is called a *flux meter.*)

***26.** A circular loop of wire (radius = 3.4 cm) consists of 1000 turns. A resistor (15 Ω) is connected between the ends of the wire, as the drawing illustrates. A magnetic field is directed perpendicular to the plane of the loop. The field points into the paper and has a magnitude that varies in time as $B = (0.45$ T/s) t, where t is the time. (a) Determine the polarity of point A with respect to point B. (b) What is the electrical energy dissipated in the resistor in 26 s?

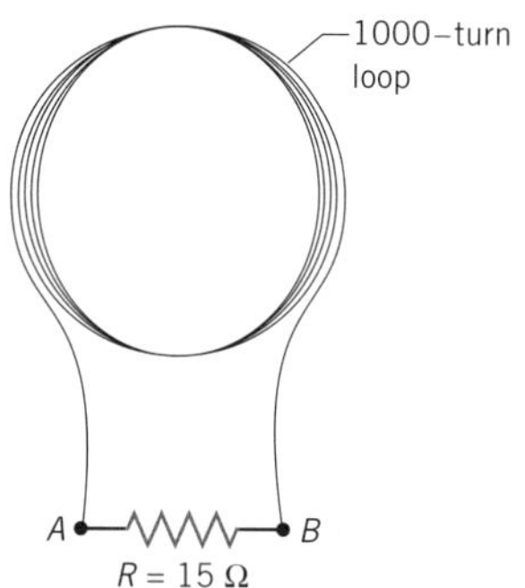

***27.** A copper rod is sliding on two conducting rails that make an angle of 15° with respect to each other, as in the drawing. The rod is moving to the right with a constant speed of 0.40 m/s. A 0.42-T uniform magnetic field is perpendicular to the plane of the paper. Determine the magnitude of the average emf induced in the triangle ABC during the 5.0-s period after the rod has passed point A.

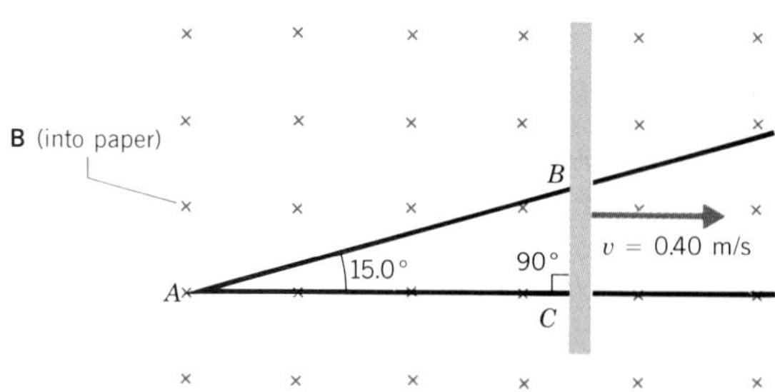

****28.** Two 0.50-m-long conducting rods are rotating at the same speed in opposite directions, and both are perpendicular to a 4.0-T magnetic field. As the drawing shows, the ends of these rods come to within 1.0 mm of each other as they rotate. Moreover, the fixed ends about which the rods are rotating are connected by a wire, so these ends are at the same electric potential. If a potential difference of 4.5×10^3 V is required to cause a 1.0-mm spark in air, what is the angular speed (in rad/s) of the rods when a spark jumps across the gap?

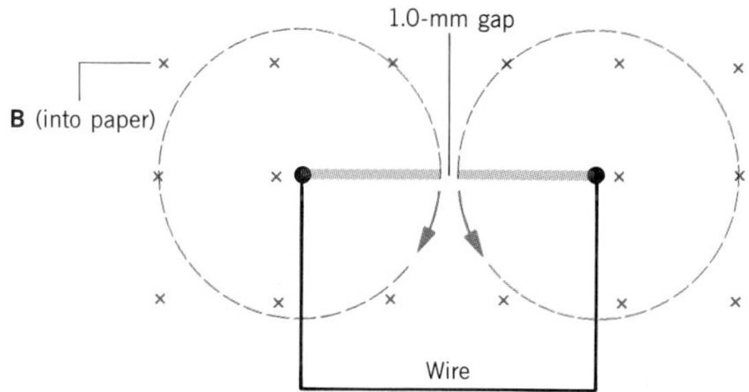

Section 22.5 Lenz's Law

29. Suppose in Figure 22.1 that the bar magnet is held stationary, but the coil of wire is free to move. Which way will current be directed through the ammeter, left-to-right or right-to-left, when the coil is moved (a) to the left and (b) to the right? Explain.

30. What is the direction of the induced current through R in the drawing as the current I decreases to zero? Provide a reason for your answer.

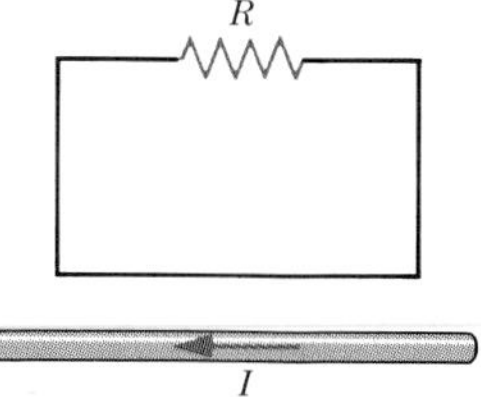

31. As the picture shows, a loop of copper wire is lying flat on a table and is attached to a battery via a switch. The current I in the loop establishes the magnetic field lines shown in color. There are also two smaller conducting loops A and B

lying flat on the table, but not connected to batteries. Determine the direction of the induced current in loops A and B when the switch is (a) opened and (b) closed again. Specify the direction of the currents to be clockwise or counterclockwise when viewed from above the table.

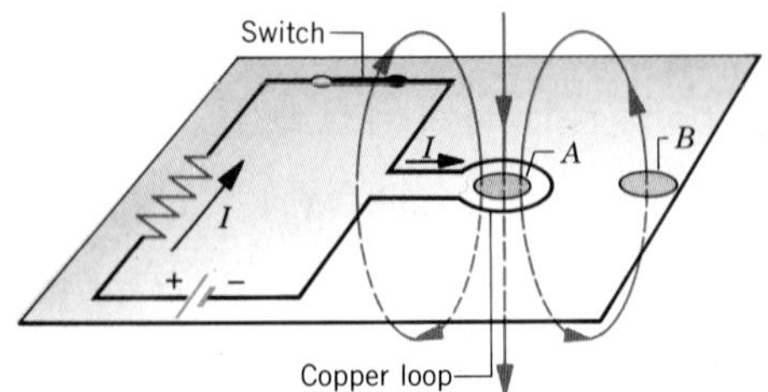

32. The drawing shows that a uniform magnetic field is directed perpendicularly out of the plane of the paper and fills the entire region to the left of the y axis. There is no magnetic field to the right of the y axis. A rigid right triangle ABC is made of copper wire. The triangle rotates counterclockwise about the origin at point C. What is the direction (clockwise or counterclockwise) of the induced current when the triangle is crossing (a) the $+y$ axis, (b) the $-x$ axis, (c) the $-y$ axis, and (d) the $+x$ axis? For each case, justify your answer.

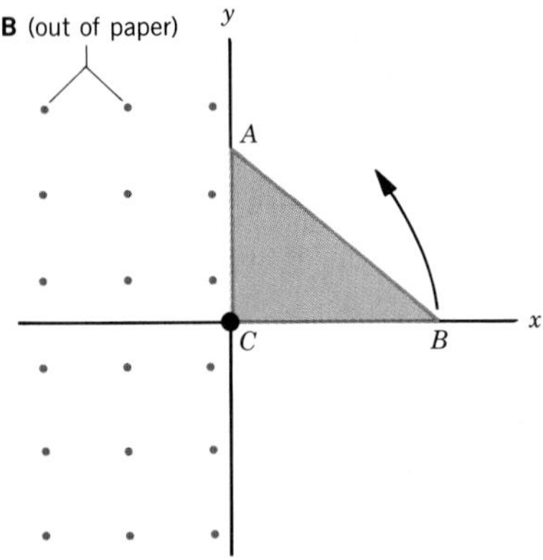

33. Review Conceptual Example 10 as an aid in understanding this problem. A long, straight wire lies on a table and carries a current I. As the drawing shows, a small circular loop of wire is pushed across the top of the table from position 1 to position 2. Determine the direction of the induced current, clockwise or counterclockwise, as the loop moves past each of the positions. Justify your answers.

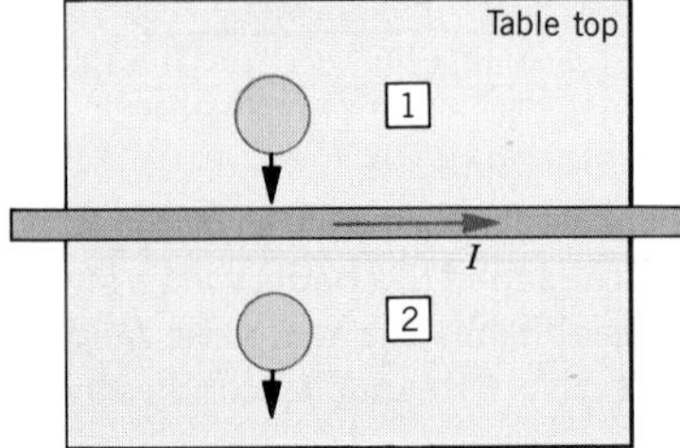

***34.** Indicate the direction of the electric field between the plates of the parallel plate capacitor shown in the drawing if the magnetic field is decreasing in time. Give your reasoning.

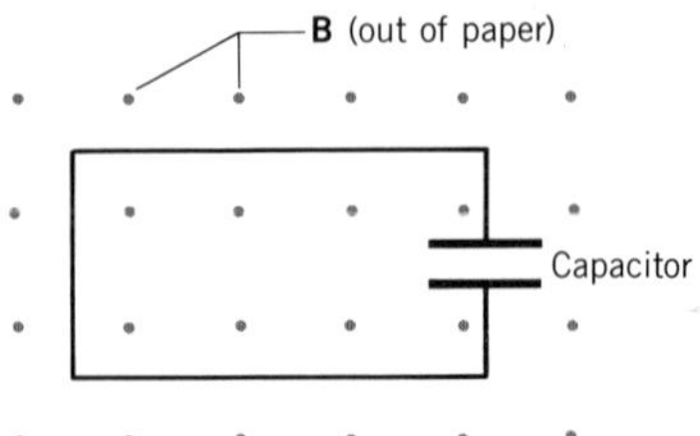

***35.** The drawing shows a bar magnet falling through a metal ring. In part a the ring is solid all the way around, but in part b it has been cut through. (a) Explain why the motion of the magnet in part a is retarded when the magnet is above the ring and below the ring as well. Draw any induced currents that appear in the ring. (b) Explain why the motion of the magnet is unaffected by the ring in part b.

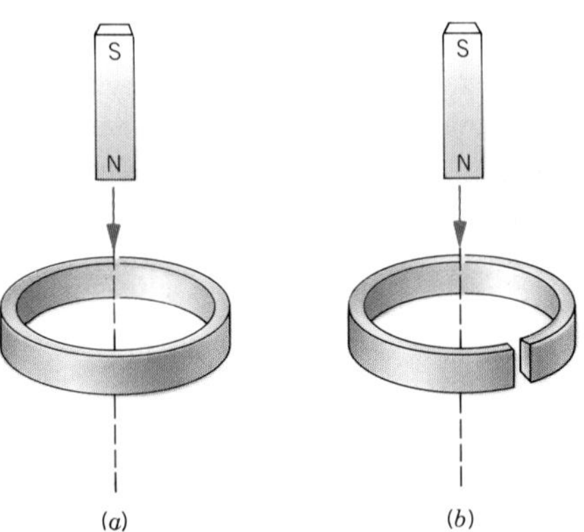

****36.** A wire loop is suspended from a string that is attached to point P in the drawing. When released, the loop swings downward, from left to right, through a uniform magnetic field, with the plane of the loop remaining perpendicular to the plane of the paper at all times. (a) Determine the direction of the current induced in the loop as it swings past the locations labeled I and II. Specify the direction of the current in terms of the points x, y, and z on the loop (e.g., $x \rightarrow y \rightarrow z$ or $z \rightarrow y \rightarrow x$). The points x, y, and z lie behind the plane of the paper. (b) What is the direction of the induced current at the locations II and I when the loop swings back, from right to left? Provide reasons for your answers.

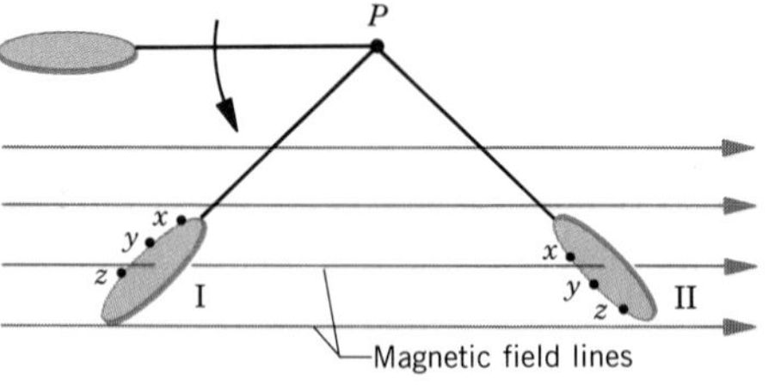

Section 22.7 The Electric Generator

37. A 200-turn rectangular coil has a cross-sectional area of 0.040 m^2. The coil is rotating at an angular speed of 15 rad/s about an axis that is perpendicular to a magnetic field of 1.5 T. Plot one cycle of the induced emf as a function of time, including on the graph numerical values for the maximum emf and the period.

38. You are requested to design a 60.0-Hz ac generator whose maximum emf is to be 5500 V. The generator is to contain a 150-turn coil whose area is 0.85 m^2. What should be the magnitude of the magnetic field in which the coil rotates?

39. The maximum strength of the earth's magnetic field is about 7.0×10^{-5} T near the south magnetic pole. In principle, this field could be used with a rotating coil to generate 60.0-Hz ac electricity. What is the minimum number of turns (area per turn = 0.016 m^2) that the coil must have to produce an rms voltage of 120 V?

40. The coil of an ac generator has an area per turn of 1.2×10^{-2} m^2 and consists of 500 turns. The coil is situated in a 0.13-T magnetic field and is rotating at an angular speed of 34 rad/s. What is the emf induced in the coil at the instant when the normal to the loop makes an angle of 27° with respect to the direction of the magnetic field?

41. A generator has a square coil consisting of 248 turns. The coil rotates at 79.1 rad/s in a 0.170-T magnetic field. The peak output of the generator is 75.0 V. What is the length of one side of the coil?

42. The current in the electric motor of a vacuum cleaner is 2.0 A when the cleaner is plugged into a 120.0-V receptacle and is running at normal speed. The coil resistance of the motor is 24 Ω. Find the back emf generated by the motor.

43. The back emf in a motor is 115 V when the motor is turning at 1800 rev/min. What is the back emf when the motor turns at 3600 rev/min, assuming all other factors remain the same?

*44. The coil of a generator has a radius of 0.14 m. When this coil is unwound, the wire from which it is made has a length of 5.7 m. The magnetic field of the generator is 0.20 T, and the coil rotates at an angular speed of 25 rad/s. What is the peak emf of this generator?

*45. At its normal operating speed, an electric fan motor draws a current of 1.2 A when plugged into a wall socket that provides 120.0 V. However, when the motor just begins to turn the fan blade, it draws a current of 6.0 A. What back emf does the motor generate at its normal operating speed?

**46. A motor is designed to operate on 117 V and draws a current of 12.2 A when it first starts up. At its normal operating speed, the motor draws a current of 2.30 A. Obtain (a) the resistance of the armature coil, (b) the back emf developed at normal speed, and (c) the current drawn by the motor at one-third normal speed.

Section 22.8 Mutual Inductance and Self-Inductance

47. The mutual inductance between two coils is $M =$ 8.0 mH. The current in the primary coil changes at a constant rate from 2.0 to 5.5 A in 0.020 s. Determine the magnitude of the average emf induced in the secondary coil.

48. Mutual induction can be used as the basis for a metal detector. A typical setup uses two large coils that are parallel to each other and have a common axis. Because of mutual induction, the ac generator connected to the primary coil causes an emf of 0.46 V to be induced in the secondary coil. When someone without metal objects walks through the coils, the mutual inductance and, thus, the induced emf do not change much. But when a person carrying a handgun walks through, the mutual inductance increases. If the mutual inductance increases by a factor of three, find the new value of the induced emf. The change in emf can be used to trigger an alarm.

49. Suppose you wish to make a solenoid whose self-inductance is 1.4 mH. The inductor is to have a cross-sectional area of 1.2×10^{-3} m^2 and a length of 0.052 m. How many turns of wire are needed?

50. A coil consists of 275 turns and has a self-inductance of 0.0150 H. The coil carries a current of 0.0170 A. Obtain the magnetic flux through one turn of the coil.

51. Two coils have a mutual inductance of 4.0 mH. In the primary coil the current changes by 3.6 A in 0.030 s. The circuit with the secondary coil has a resistance of 1.5 Ω. Find the magnitude of the average current induced in the secondary coil.

52. How much energy is stored in a 0.085-H inductor that carries a current of 2.5 A?

53. The earth's magnetic field, like any magnetic field, stores energy. The maximum strength of the earth's field is about 7.0×10^{-5} T. Find the maximum magnetic energy stored in the space above a city if the space occupies an area of 5.0×10^{8} m^2 and has a height of 1500 m.

*54. A long, current-carrying solenoid with an air core has 1750 turns per meter of length and a radius of 0.0180 m. A coil of 125 turns is wrapped tightly around the outside of the solenoid. What is the mutual inductance of this system?

*55. A long solenoid is bent into a circular shape so it looks like a doughnut. This wire-wound doughnut is called a toroid. Assume that the diameter of the solenoid is small compared to the radius of the toroid. With this assumption, use the results of Example 14 to determine the self-inductance of the toroid. Express your answer in terms of μ_0, n (the number of turns per unit length), A (the cross-sectional area of the solenoid), and R (the radius of the toroid).

*56. The purpose of this problem is to show that the work W needed to establish a final current I_f in an inductor is $W = \frac{1}{2}LI_f^2$ (Equation 22.10). In Section 22.8 we saw that the amount of work ΔW needed to change the current through an inductor by an amount ΔI is $\Delta W = LI(\Delta I)$, where L is the inductance. The drawing shows a graph of LI versus I. Notice that $LI(\Delta I)$ is the area of the shaded vertical rectangle whose height is LI and whose width is ΔI. Use this fact to show that the total work W needed to establish a current I_f is $W = \frac{1}{2}LI_f^2$.

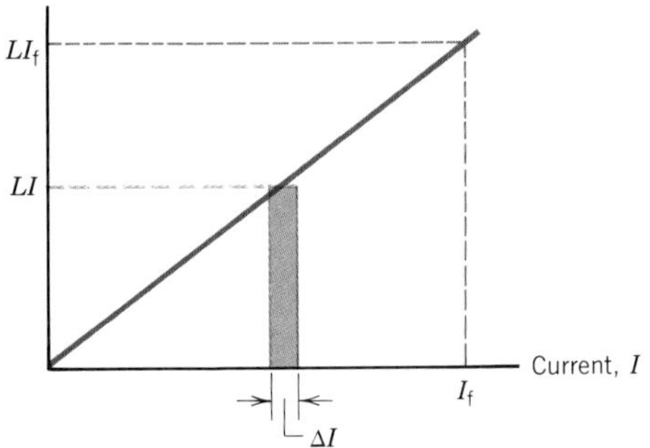

** 57. A solenoid has a cross-sectional area of 6.0×10^{-4} m^2, consists of 400 turns per meter, and carries a current of 0.40 A. A 10-turn coil is wrapped tightly around the circumference of the solenoid. The ends of the coil are connected to a 1.5-Ω resistor. Suddenly, a switch is opened, and the current in the solenoid dies to zero in a time of 0.050 s. Find the average current induced in the coil.

Section 22.9 Transformers

58. The batteries in a portable CD player are recharged by a unit that plugs into a wall socket. Inside the unit is a step-down transformer with a turns ratio of 1:13. The wall socket provides 120 V. What voltage does the secondary coil of the transformer provide?

59. A neon sign requires 12 000 V for its operation. It operates from a 220-V receptacle. (a) What type of transformer, step-up or step-down, is needed? (b) What must be the turns ratio N_s/N_p of the transformer?

60. In some parts of the country, insect "zappers," with their blue lights, are a familiar sight on a summer's night. These devices use a high voltage to electrocute insects. One such device uses an ac voltage of 4150 V, which is obtained from a standard 120.0-V outlet by means of a transformer. If the primary coil has 17 turns, how many turns are in the secondary coil?

61. Electric doorbells found in many homes require 10.0 V to operate. To obtain this voltage from the standard 120-V supply, a transformer is used. Is a step-up or a step-down transformer needed, and what is its turns ratio N_s/N_p?

62. A step-down transformer (turns ratio = 1:8) is used with an electric train to reduce the voltage from the wall receptacle (120.0 V) to a value needed to operate the train. When the train is running, the current in the secondary coil is 3.4 A. What is the current in the primary coil?

63. The input to the primary coil of a transformer is 120 V, while the current in the secondary coil is 0.10 A. (a) When 60.0 W of power are being delivered to the circuit attached to the secondary coil, what is the voltage across the secondary coil? (b) Is the transformer a step-up or a step-down unit, and what is its turns ratio N_s/N_p?

64. The secondary coil of a step-up transformer provides the voltage that operates an electrostatic air filter. The turns ratio of the transformer is 43:1. The primary coil is plugged into a standard 120-V outlet. The current in the secondary coil is 1.5×10^{-3} A. Find the power consumed by the air filter.

* 65. A generating station is producing 1.2×10^6 W of power that is to be sent to a small town located 7.0 km away. Each of the two wires that comprise the transmission line has a resistance per kilometer of length of 5.0×10^{-2} Ω/km. (a) Find the power lost in heating the wires if the power is transmitted at 1200 V. (b) A 100:1 step-up transformer is used to raise the voltage before the power is transmitted. How much power is now lost in heating the wires?

** 66. A generator is connected across the primary coil (N_p turns) of a transformer, while a resistance R_2 is connected across the secondary coil (N_s turns). This circuit is equivalent to a circuit in which a single resistance R_1 is connected directly across the generator, without the transformer. Show that $R_1 = (N_p/N_s)^2 R_2$, by starting with Ohm's law as applied to the secondary coil.

ADDITIONAL PROBLEMS

67. A 75-turn conducting coil has an area of 8.5×10^{-3} m^2 and the normal to the coil is parallel to a magnetic field **B**. The coil has a resistance of 14 Ω. At what rate (in T/s) must the magnitude of **B** change for an induced current of 7.0 mA to exist in the coil?

68. One generator uses a magnetic field of 0.10 T and has a coil area of 0.045 m^2. A second generator has a coil area of 0.015 m^2. The generator coils have the same number of turns and rotate at the same angular speed. What magnetic field should be used in the second generator, so that its peak emf is the same as that of the first generator?

69. In Figure 22.1, suppose the north and south poles of the magnet were interchanged. Determine the direction of the current through the ammeter in parts *b* and *c* of the picture (left-to-right or right-to-left). Give your rationale.

70. A square coil of wire is moving through a region of space where the magnetic field is everywhere the same at all times. The field is directed perpendicular to the plane of the coil. The area and speed of the coil are 1.3×10^{-4} m^2 and 2.2 m/s. The magnitude of the magnetic field is .033 T. What is the emf induced in the coil?

71. The coil of an electromagnet carries a steady direct current of 8.0 A and has a self-inductance of 0.150 H. Suddenly a switch is opened, and the current decreases to zero in 7.0×10^{-3} s. Obtain the magnitude of the average emf induced in the coil during this time.

72. The resistances of the primary and secondary coils of a transformer are 56 and 14 Ω, respectively. Both coils are made from lengths of the same copper wire. The circular turns of each coil have the same diameter. Find the turns ratio N_s/N_p.

73. A generator produces a peak emf of 12.0 V when the armature rotates at 750 rev/min. What is the peak emf when the armature rotates at 2250 rev/min, assuming everything else remains the same?

***74.** A 3.0-μF capacitor has a voltage of 35 V between its plates. What must be the current in a 5.0-mH inductor, such that the energy stored in the inductor equals the energy stored in the capacitor?

***75.** A magnetic field has a magnitude of 12 T. What is the magnitude of an electric field that stores the same energy per unit volume as this magnetic field?

***76.** A large circular loop carries a current I. A much smaller circular loop is held above the center of the large loop, with the planes of the loops parallel. The small loop is released and falls downward through the large loop, all the while maintaining its parallel orientation. The center of the small loop remains in line with the center of the large loop at all times. Is the direction of the current induced in the small loop the same as I or opposite to I when (a) the small loop is above the large loop and (b) the small loop has fallen below the large loop? *(Hint: With the aid of Figure 21.32, first identify the direction of the magnetic field along the axis of the large loop.)* Justify your answers.

***77.** A magnetic field is passing through a loop of wire whose area is 0.018 m². The direction of the magnetic field is parallel to the normal to the loop, and the magnitude of the field is increasing at the rate of 0.20 T/s. (a) Determine the magnitude of the emf induced in the loop. (b) Suppose the area of the loop can be enlarged or shrunk. If the magnetic field is increasing as in part (a), at what rate (in m²/s) should the area be changed at the instant when $B = 1.8$ T if the induced emf is to be zero? Explain whether the area is to be enlarged or shrunk.

****78.** The armature of an electric drill motor has a resistance of 15.0 Ω. When connected to a 120.0-V outlet, the motor rotates at its normal speed and develops a back emf of 108 V. (a) What is the current through the motor? (b) If the armature "freezes up" due to a lack of lubrication in the bearings and can no longer rotate, what is the current in the stationary armature? (c) What is the current when the motor runs at only half speed?

****79.** Coil 1 is a flat circular coil that has N_1 turns and a radius R_1. At its center is a much smaller flat, circular coil that has N_2 turns and radius R_2. The planes of the coils are parallel. Assume that coil 2 is so small that the magnetic field due to coil 1 has nearly the same value at all points covered by the area of coil 2. Determine an expression for the mutual inductance between these two coils in terms of μ_0, N_1, R_1, N_2, and R_2.

****80.** The drawing shows a copper wire (negligible resistance) bent into a circular shape with a radius of 0.50 m. The radial section BC is fixed in place, while the copper bar AC sweeps around at an angular speed of 15 rad/s. The bar makes electrical contact with the wire at all times. A uniform magnetic field exists everywhere, is perpendicular to the plane of the circle, and has a magnitude of 3.8×10^{-3} T. Find the magnitude of the current induced in the loop ABC.

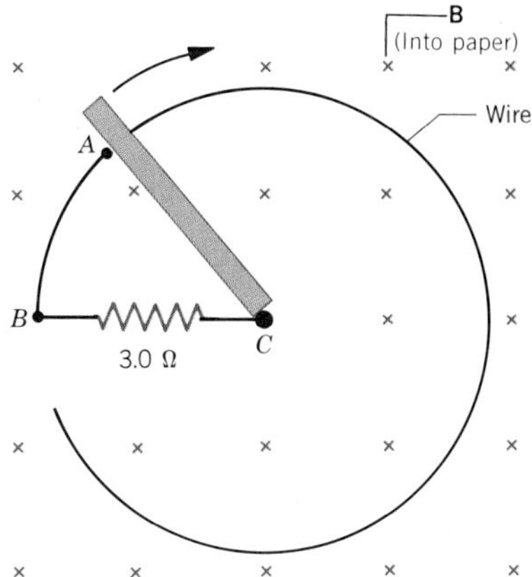

CHAPTER 23 ALTERNATING CURRENT CIRCUITS

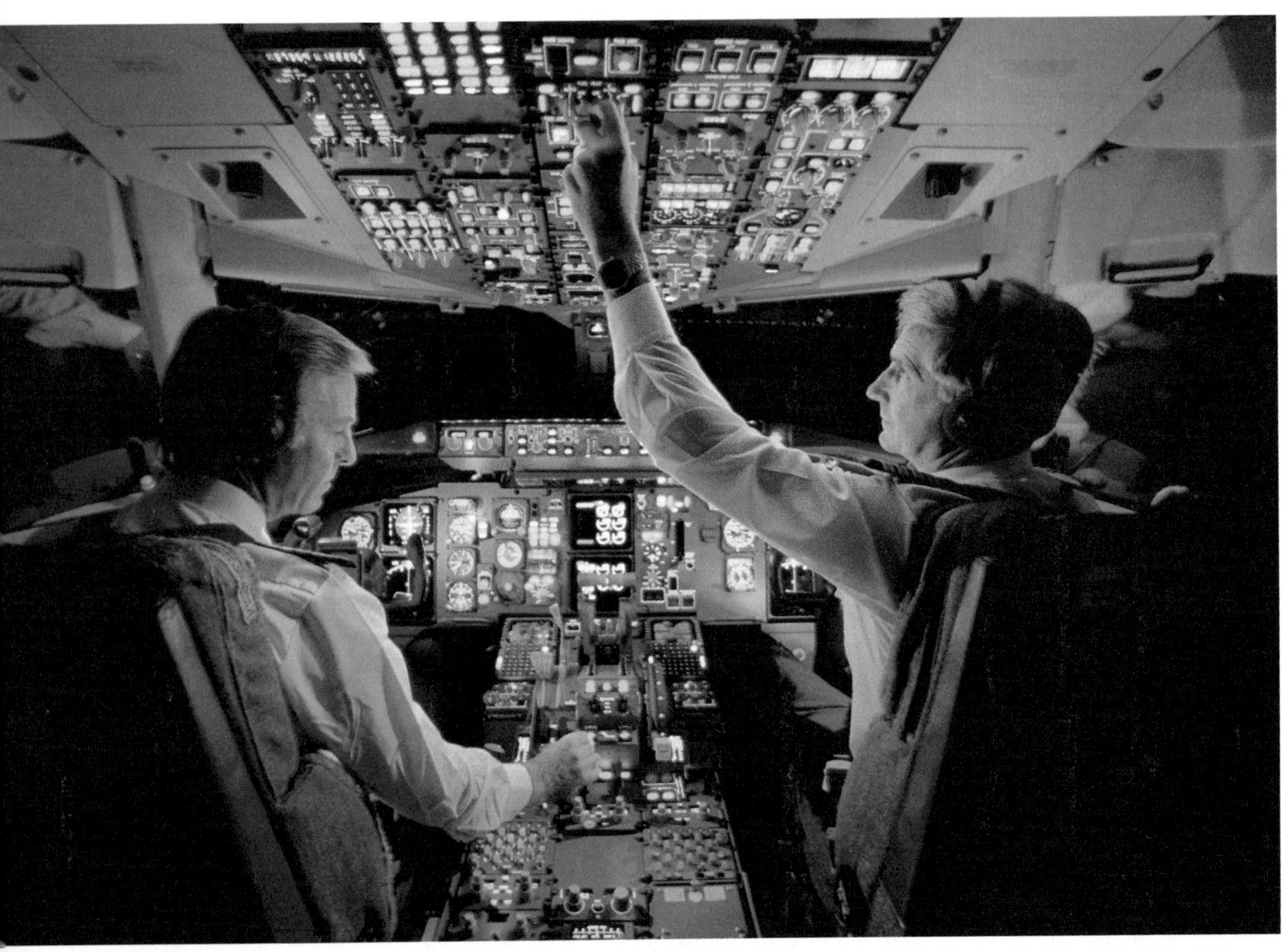

The applications of alternating current (ac) circuits are so widespread that it is difficult to imagine living without them. This airplane cockpit, for instance, is filled with circuits that depend on alternating current, and commercial aviation in its present form could not exist without them. We use ac circuits in every room of our houses: lights, television, microwave oven, refrigerator, heating system, toaster, hair dryer—the list goes on and on. In such applications, the basic elements of ac electricity are frequency, rms voltage, rms current, and power, as Section 20.5 has already discussed. However, our discussion there focused on circuits that contain only resistors. In the present chapter, we deal with a number of important additional circuit components, including capacitors, inductors, diodes, and transistors, that make ac electricity vastly more useful.

23.1 CAPACITORS AND CAPACITIVE REACTANCE

Our experience with capacitors so far has been in dc circuits. As we have seen in Section 20.13, charge flows in a dc circuit only for the brief period after the battery voltage is applied across the capacitor. In other words, charge flows only while the capacitor is charging up. After the capacitor becomes fully charged, no more charge leaves the battery. However, suppose the battery connections to the fully charged capacitor were suddenly reversed, with the positive terminal being connected to the negative plate and the negative terminal being connected to the positive plate. Then charge would flow again, but in the reverse direction, until the battery recharged the capacitor according to the new connections. What happens in an ac circuit is similar. The polarity of the voltage applied to the capacitor continually switches back and forth, and, in response, charges flow first one way around the circuit and then the other way. This flow of charge, surging back and forth, constitutes an alternating current. Thus, charge flows continuously in an ac circuit containing a capacitor.

To help set the stage for the present discussion, recall that $V_{rms} = I_{rms}R$ (Equation 20.14) for a purely resistive ac circuit. The resistance R has the same value for any frequency. Figure 23.1 emphasizes this fact by showing that a graph of resistance versus frequency is a horizontal straight line.

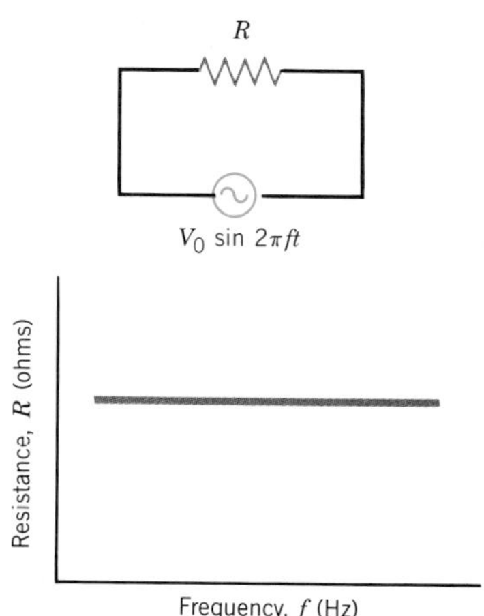

Figure 23.1 The resistance in a purely resistive circuit has the same value at all frequencies. The maximum emf of the generator is V_0.

For the rms voltage across a capacitor the following expression applies, which is analogous to $V_{rms} = I_{rms}R$:

$$V_{rms} = I_{rms}X_C \tag{23.1}$$

The term X_C appears in place of the resistance R and is called the ***capacitive reactance.*** The capacitive reactance, like resistance, is measured in *ohms* and determines how much rms current exists in a capacitor in response to a given rms voltage across the capacitor. It is found experimentally that the capacitive reactance X_C is inversely proportional to both the frequency f and the capacitance C, according to the following equation:

$$X_C = \frac{1}{2\pi fC} \tag{23.2}$$

For a fixed value of the capacitance C, Figure 23.2 gives a plot of X_C versus frequency, according to Equation 23.2. A comparison of this drawing with Figure 23.1 reveals that a capacitor behaves differently than a resistor. As the frequency becomes very large, Figure 23.2 shows that X_C approaches zero, signifying that a capacitor offers only a negligibly small opposition to the alternating current. In contrast, in the limit of zero frequency (i.e., dc current), X_C becomes infinitely large, and a capacitor provides so much opposition to the motion of charges that there is no current. Example 1 illustrates how frequency and capacitance determine the amount of current in an ac circuit.

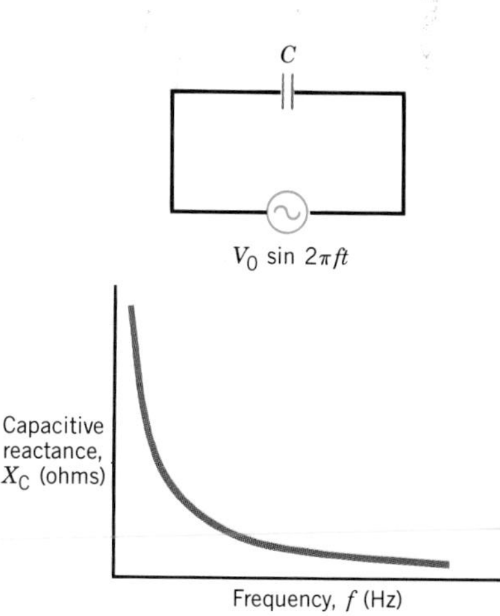

Figure 23.2 The capacitive reactance X_C is inversely proportional to the frequency f according to $X_C = 1/(2\pi fC)$.

Example 1 A Capacitor in an Ac Circuit

For the circuit in Figure 23.2, the capacitance of the capacitor is 1.50 μF, and the rms voltage of the generator is 25.0 V. What is the rms current in the circuit when the frequency of the generator is (a) 1.00×10^2 Hz and (b) 5.00×10^3 Hz?

REASONING The current can be found from $I_{rms} = V_{rms}/X_C$, once the capacitive reactance is determined. The values for the capacitive reactance will reflect the fact that the capacitor provides more opposition to the current when the frequency is smaller.

SOLUTION
(a) At a frequency of 1.00×10^2 Hz, we find

$$X_C = \frac{1}{2\pi fC} = \frac{1}{2\pi(1.00 \times 10^2 \text{ Hz})(1.50 \times 10^{-6} \text{ F})} = 1060\ \Omega \qquad (23.2)$$

$$I_{rms} = \frac{V_{rms}}{X_C} = \frac{25.0 \text{ V}}{1060\ \Omega} = \boxed{0.0236 \text{ A}} \qquad (23.1)$$

(b) When the frequency is 5.00×10^3 Hz, the calculations are similar:

$$X_C = \frac{1}{2\pi fC} = \frac{1}{2\pi(5.00 \times 10^3 \text{ Hz})(1.50 \times 10^{-6} \text{ F})} = 21.2\ \Omega$$

$$I_{rms} = \frac{V_{rms}}{X_C} = \frac{25.0 \text{ V}}{21.2\ \Omega} = \boxed{1.18 \text{ A}}$$

PROBLEM SOLVING INSIGHT

The capacitive reactance X_C is inversely proportional to the frequency f of the current. If the frequency increases by a factor of 50, for example, the capacitive reactance decreases by a factor of 50.

We now consider the behavior of the instantaneous (not rms) voltage and current. For comparison, Figure 23.3 shows graphs of voltage and current versus time in a resistive circuit. These graphs indicate that, when only resistance is present, the voltage and current are proportional to each other at every moment. For example, when the voltage increases from A to B on the graph, the current follows along in step, increasing from A' to B' during the same time. Likewise, when the voltage decreases from B to C, the current decreases from B' to C'. For this reason, the current in a resistance R is said to be ***in phase*** with the voltage across the resistance.

For a capacitor, this in-phase relation between instantaneous voltage and current does *not* exist. Figure 23.4 shows graphs of the ac voltage and current versus time for a circuit that contains only a capacitor. As the voltage increases from A to B, the charge on the capacitor increases and reaches its full value at B. The current, however, is not the same thing as the charge. The current is the rate of flow of charge and has a maximum positive value at the start of the charging process at A'. It is a maximum, because there is no charge on the capacitor at the start and hence no capacitor voltage to oppose the generator voltage. When the capacitor is fully charged at B, the capacitor voltage has a magnitude equal to that of the generator and completely opposes the generator voltage. The result is that the current decreases to zero at B'. While the capacitor voltage decreases from B to C, the charges flow out of the capacitor in a direction opposite to that of the charging current, as indicated by the negative current from B' to C'. Thus, voltage and current are not in phase but are, in fact, one-quarter wave cycle out of step, or out of phase. More specifically, assuming that the voltage fluctuates as $V_0 \sin(2\pi ft)$, the current varies as $I_0 \sin(2\pi ft + \pi/2) = I_0 \cos(2\pi ft)$. Since $\pi/2$ radians correspond to 90° and since the current reaches its maximum value *before* the voltage does, it is said that the current through a capacitor *leads* the voltage across the capacitor by a phase angle of 90°.

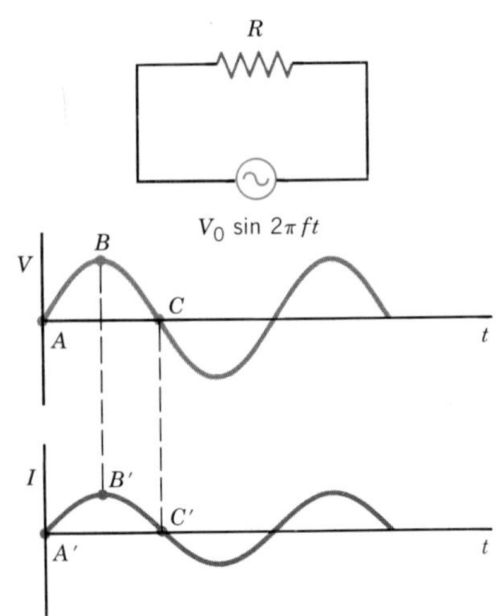

Figure 23.3 The instantaneous voltage V and current I in a purely resistive circuit are *in phase,* which means that they increase and decrease in step with one another.

The fact that the current and voltage for a capacitor are 90° out of phase has an important consequence from the point of view of electric power, since power is the product of current and voltage. For the time interval between points A and B (or A' and B') in Figure 23.4, both current and voltage are positive. Therefore, the instantaneous power is also positive, meaning that the generator is delivering energy to the capacitor. However, during the period between B and C (or B' and C'), the current is negative while the voltage remains positive, and the power, being the product of the two, is negative. During this period, the capacitor is returning energy to the generator. Thus, the power alternates between positive and negative values for equal periods of time. In other words, the capacitor alternately absorbs and releases energy. Consequently, ***on the average, the power is zero and a capacitor uses no energy in an ac circuit.***

It will prove useful later on to use a model for the voltage and current in ac circuits. In this model, voltage and current are represented by rotating arrows, often called *phasors,* whose lengths correspond to the maximum voltage V_0 and maximum current I_0, as Figure 23.5 indicates. These phasors rotate counterclockwise at a frequency f. For a resistor, the phasors are colinear as they rotate (see part a of the drawing), because voltage and current are in phase. For a capacitor (see part b), the phasors remain perpendicular while rotating, because the phase angle between the current and the voltage is 90°. Since current leads voltage for a capacitor, the current phasor is ahead of the voltage phasor in the direction of rotation. In both cases in Figure 23.5, the instantaneous voltage and current are given by the vertical components of the phasors.

Figure 23.4 In a circuit containing only a capacitor, the instantaneous voltage and current are not in phase, as they are in a purely resistive circuit. Instead, the current *leads* the voltage by a phase angle of 90° (one-quarter of a cycle).

23.2 INDUCTORS AND INDUCTIVE REACTANCE

As Section 22.8 discusses, an inductor is usually a coil of wire, and the basis of its operation is Faraday's law of electromagnetic induction. According to Faraday's law, an inductor develops a voltage that opposes a change in the current. This voltage V is given by $V = -L(\Delta I/\Delta t)$ (see Equation 22.9*), where $\Delta I/\Delta t$ is the rate at which the current changes and L is the inductance of the inductor. In an ac circuit the current is always changing, and Faraday's law can be used to show that the rms voltage across an inductor is

$$V_{\text{rms}} = I_{\text{rms}} X_{\text{L}} \tag{23.3}$$

Equation 23.3 is analogous to $V_{\text{rms}} = I_{\text{rms}} R$, with the term X_{L} appearing in place of the resistance R and being called the ***inductive reactance.*** The inductive reactance is measured in ohms and determines how much rms current exists in an inductor for a given rms voltage across the inductor. It is found experimentally that the inductive reactance X_{L} is directly proportional to the frequency f and the inductance L, as indicated in the following equation:

$$X_{\text{L}} = 2\pi f L \tag{23.4}$$

This relation indicates that the larger the inductance, the larger the inductive reactance. Note that the inductive reactance is directly proportional to the frequency ($X_{\text{L}} \propto f$), in contrast to the capacitive reactance, which is inversely proportional to the frequency ($X_{\text{C}} \propto 1/f$).

Figure 23.6 shows a graph of the inductive reactance versus frequency for a fixed value of the inductance, according to Equation 23.4. As frequency becomes very large, X_{L} also becomes very large. In such a situation, an inductor provides a large opposition to the alternating current. In the limit of zero frequency (i.e.,

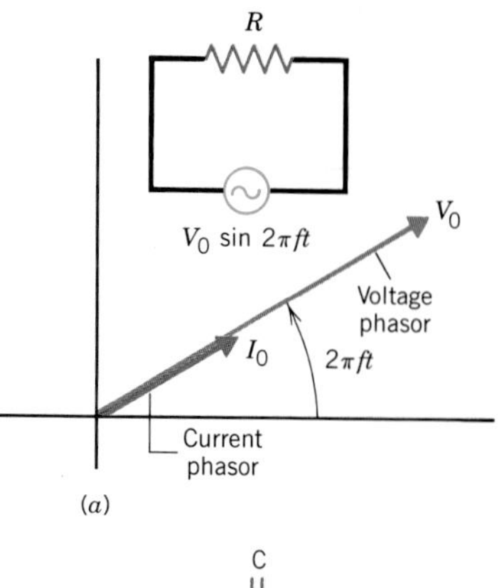

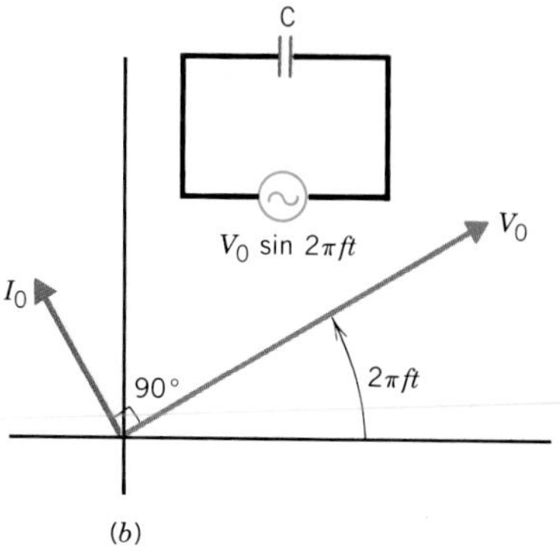

Figure 23.5 These rotating arrow models represent the voltage and the current in ac circuits that contain (a) only a resistor and (b) only a capacitor.

* When an inductor is used in a circuit, the notation is simplified if we designate the potential difference across the inductor as the voltage V, rather than the emf $\mathscr{E}$.

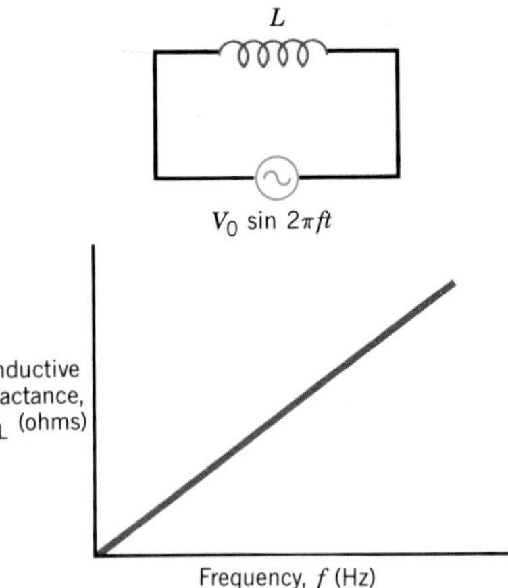

Figure 23.6 In an ac circuit the inductive reactance X_L is directly proportional to the frequency f, according to $X_L = 2\pi fL$.

PROBLEM SOLVING INSIGHT

The inductive reactance X_L is directly proportional to the frequency f of the current. If the frequency increases by a factor of 50, for example, the inductive reactance increases by a factor of 50.

direct current), X_L becomes zero, indicating that an inductor does not oppose direct current at all. The next example demonstrates the effect of inductive reactance on the current in an ac circuit.

Example 2 An Inductor in an Ac Circuit

The circuit in Figure 23.6 contains a 3.60-mH inductor. The rms voltage of the generator is 25.0 V. Find the rms current in the circuit when the generator frequency is (a) 1.00×10^2 Hz and (b) 5.00×10^3 Hz.

REASONING The current can be calculated from $I_{rms} = V_{rms}/X_L$, provided the inductive reactance is obtained first. The inductor offers more opposition to the changing current when the frequency is larger, and the values for the inductive reactance will reflect this fact.

SOLUTION

(a) At a frequency of 1.00×10^2 Hz, we find

$$X_L = 2\pi fL = 2\pi(1.00 \times 10^2 \text{ Hz})(3.60 \times 10^{-3} \text{ H}) = 2.26\ \Omega \qquad (23.4)$$

$$I_{rms} = \frac{V_{rms}}{X_L} = \frac{25.0 \text{ V}}{2.26\ \Omega} = \boxed{11.1 \text{ A}} \qquad (23.3)$$

(b) The calculation is similar when the frequency is 5.00×10^3 Hz:

$$X_L = 2\pi fL = 2\pi(5.00 \times 10^3 \text{ Hz})(3.60 \times 10^{-3} \text{ H}) = 113\ \Omega$$

$$I_{rms} = \frac{V_{rms}}{X_L} = \frac{25.0 \text{ V}}{113\ \Omega} = \boxed{0.221 \text{ A}}$$

By virtue of its inductive reactance, an inductor affects the amount of current in an ac circuit. The inductor also influences the current in another way, as Figure 23.7 shows. This figure displays graphs of voltage and current versus time for a circuit containing only an inductor. At a maximum or minimum on the current graph, the current does not change much with time, so the voltage generated by the inductor to oppose a change in the current is zero. At the points on the current graph where the current is zero, the graph is at its steepest, and the current has the largest rate of increase or decrease. Correspondingly, the voltage generated by the inductor to oppose a change in the current has the largest positive or negative value. Thus, current and voltage are not in phase but are one-quarter of a wave cycle out of phase. If the voltage varies as $V_0 \sin (2\pi ft)$, the current fluctuates as $I_0 \sin (2\pi ft - \pi/2) = -I_0 \cos (2\pi ft)$. The current reaches its maximum *after* the voltage does, and it is said that the current *lags behind* the voltage by a phase angle of 90° ($\pi/2$ radians). In a purely capacitive circuit, in contrast, the current leads the voltage by 90°.

In an inductor the 90° phase difference between current and voltage leads to the same result for average power that it does in a capacitor. An inductor alternately absorbs and releases energy for equal periods of time, so ***on the average, the power is zero and an inductor uses no energy in an ac circuit.***

As an alternative to the graphs in Figure 23.7, Figure 23.8 uses phasors to describe the instantaneous voltage and current in a circuit containing only an inductor. The voltage and current phasors remain perpendicular as they rotate, for there is a 90° phase angle between them. The current phasor lags behind the voltage phasor, relative to the direction of rotation, in contrast to the equivalent

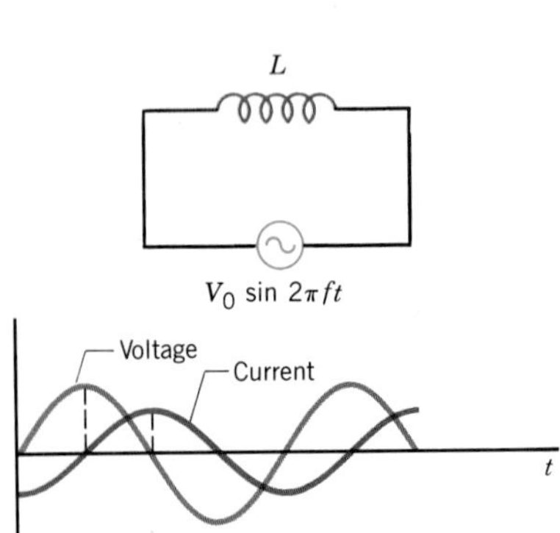

Figure 23.7 The instantaneous voltage and current in a circuit containing only an inductor are not in phase. The current *lags behind* the voltage by a phase angle of 90° (one-quarter of a cycle).

picture for a capacitor. Once again, the instantaneous values for voltage and current are given by the vertical components of the phasors.

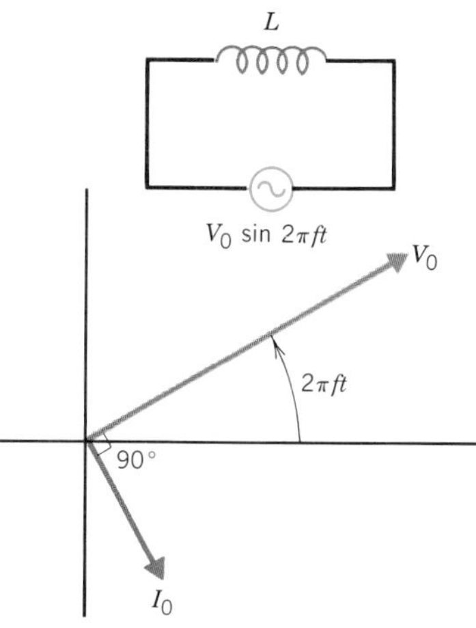

Figure 23.8 This phasor model represents the voltage and current in a circuit that contains only an inductor.

23.3 CIRCUITS CONTAINING RESISTANCE, CAPACITANCE, AND INDUCTANCE

Capacitors and inductors can be combined along with resistors in a single circuit. The simplest combination contains a resistor, a capacitor, and an inductor in series, as Figure 23.9 shows. In a series RCL circuit the total opposition to the flow of charge is called the ***impedance*** of the circuit and comes partially from (1) the resistance R, (2) the capacitive reactance X_C, and (3) the inductive reactance X_L. Figure 23.10 shows a graph of impedance versus frequency and emphasizes the frequency regions where each circuit component dominates. At low frequencies X_C becomes very large, and so does the impedance, with X_C making a much greater contribution than either X_L or R. At high frequencies X_L becomes very large, leading once again to a large impedance. However, in this case X_L dominates over X_C and R. At intermediate frequencies the impedance is smaller than it is at either extreme. In fact, we will see that there is a single frequency where the capacitive and inductive reactances cancel, leaving the frequency-independent resistance to dominate.

When the resistor, the capacitor, and the inductor are wired in series, it is tempting to follow the analogy of a series combination of resistors and calculate the impedance by simply adding together R, X_C, and X_L. However, such a procedure is not correct. Instead, the phasors shown in Figure 23.11 must be used. The lengths of the voltage phasors in this drawing represent the maximum voltages V_R, V_C, and V_L across the resistor, the capacitor, and the inductor, respectively. The current is the same for each device, since the circuit is wired in series. The length of the current phasor represents the maximum current I_0. Notice that the drawing shows the current phasor to be (1) in phase with the voltage phasor for the resistor, (2) ahead of the voltage phasor for the capacitor by 90°, and (3) behind the voltage phasor for the inductor by 90°. These three facts are consistent with our earlier discussion in Sections 23.1 and 23.2.

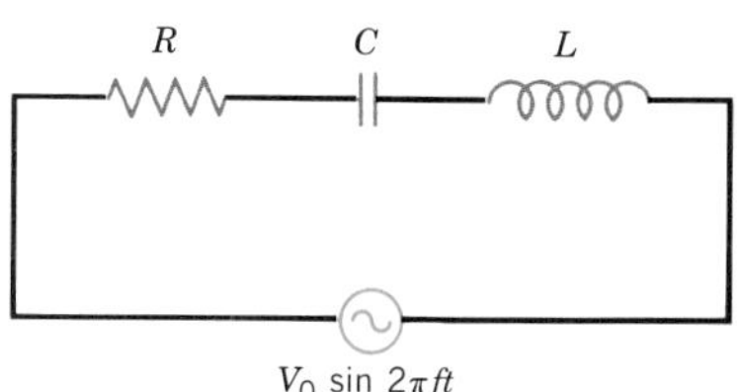

Figure 23.9 A series RCL circuit contains a resistor, a capacitor, and an inductor.

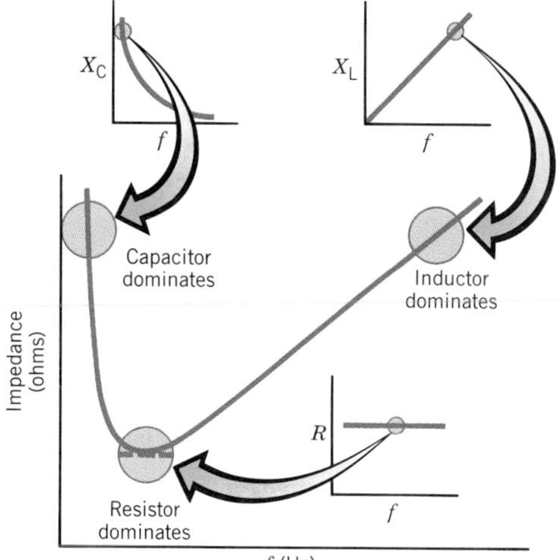

Figure 23.10 In a series RCL circuit the impedance varies with frequency, as this graph shows.

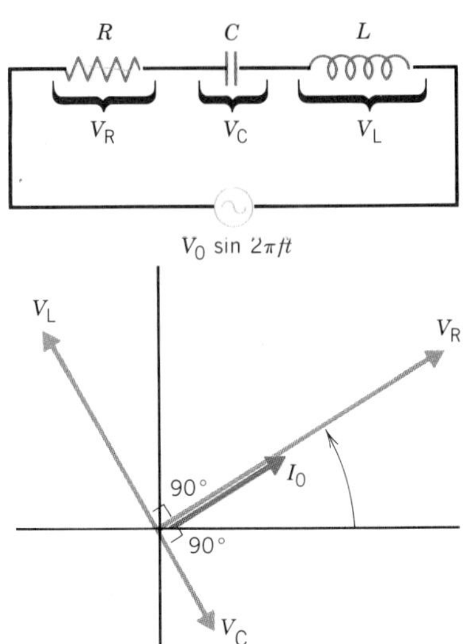

Figure 23.11 The relation between the three voltage phasors (V_R, V_C, and V_L) and the current phasor (I_0) for a series RCL circuit.

The basis for dealing with the voltage phasors in Figure 23.11 is Kirchhoff's loop rule. In an ac circuit this rule applies to the *instantaneous* voltages across each circuit component and the generator. Therefore, it is necessary to take into account the fact that these voltages do not have the same phase; that is, the phasors V_R, V_C, and V_L point in different directions in the drawing. Kirchhoff's loop rule indicates that the phasors add together to give the total voltage V_0 that is supplied to the circuit by the generator. The addition, however, must be like a vector addition, to take into account the different directions. Since V_L and V_C point in opposite directions, they combine to give a resultant phasor of $V_L - V_C$, as Figure 23.12 shows. In this drawing the resultant $V_L - V_C$ is perpendicular to V_R and may be combined with it to give the total voltage V_0. Using the Pythagorean theorem, we find

$$V_0^2 = V_R^2 + (V_L - V_C)^2$$

In this equation each of the symbols stands for a maximum voltage and when divided by $\sqrt{2}$ gives the corresponding rms voltage. Therefore, it is possible to divide both sides of the equation by $(\sqrt{2})^2$ and obtain a result for $V_{rms} = V_0/\sqrt{2}$. This result has exactly the same form as that above, but involves the rms voltages $V_{R\text{-}rms}$, $V_{C\text{-}rms}$, and $V_{L\text{-}rms}$. However, to avoid such awkward symbols, we simply interpret V_R, V_C, and V_L as rms quantities in the following expression:

$$V_{rms}^2 = V_R^2 + (V_L - V_C)^2 \tag{23.5}$$

The last step in determining the impedance of the circuit is to remember that $V_R = I_{rms}R$, $V_C = I_{rms}X_C$, and $V_L = I_{rms}X_L$. With these substitutions Equation 23.5 can be written as

$$V_{rms} = I_{rms}\sqrt{R^2 + (X_L - X_C)^2}$$

Therefore, for the entire RCL circuit, it follows that

$$V_{rms} = I_{rms}Z \tag{23.6}$$

where the impedance Z of the circuit is

[**Series RCL combination**]
$$Z = \sqrt{R^2 + (X_L - X_C)^2} \tag{23.7}$$

The impedance of the circuit, like R, X_C, and X_L, is measured in ohms. In Equation 23.7, $X_L = 2\pi fL$ and $X_C = 1/(2\pi fC)$, and a plot of Z versus frequency f gives the graph shown earlier in Figure 23.10. The minimum in the graph can now be seen to occur when $X_L = X_C$, so that at this point $Z = R$, the resistance in the circuit.

The phase angle between the current in and the voltage across a series RCL combination is the angle ϕ between the current phasor I_0 and the voltage phasor V_0 in Figure 23.12. According to the drawing, the tangent of this angle is

$$\tan\phi = \frac{V_L - V_C}{V_R} = \frac{I_{rms}X_L - I_{rms}X_C}{I_{rms}R}$$

[**Series RCL combination**]
$$\tan\phi = \frac{X_L - X_C}{R} \tag{23.8}$$

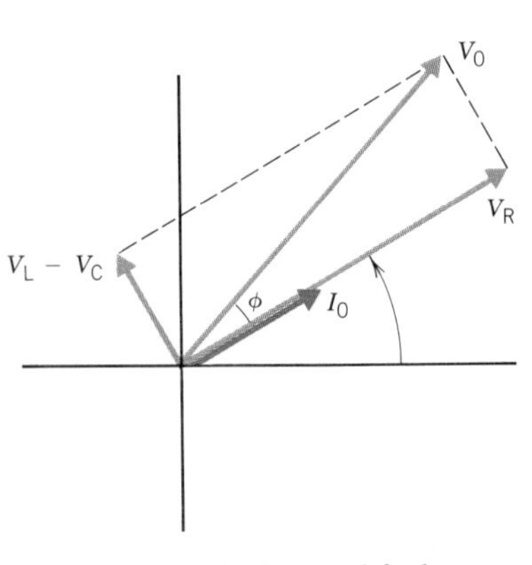

Figure 23.12 This simplified version of Figure 23.11 results when the phasors V_L and V_C, which point in opposite directions, are combined to give a resultant of $V_L - V_C$.

The phase angle ϕ is important, because it has a major effect on the power dissipated by the circuit. Remember, on the average, only the resistance consumes power; that is, $\overline{P} = I_{rms}^2R$ (Equation 20.15b). According to Figure 23.12, $\cos\phi = V_R/V_0 = (I_{rms}R)/(I_{rms}Z) = R/Z$, so that $R = Z\cos\phi$. Therefore,

$$\overline{P} = I_{rms}^2 Z \cos\phi = I_{rms}(I_{rms}Z)\cos\phi$$

$$\overline{P} = I_{rms}V_{rms}\cos\phi \qquad (23.9)$$

where V_{rms} is the rms voltage of the generator. The term $\cos\phi$ is called the *power factor* of the circuit. As a check on the validity of Equation 23.9, note that if no resistance is present, $R = 0$, and $\cos\phi = R/Z = 0$. Consequently, $\overline{P} = I_{rms}V_{rms} \times \cos\phi = 0$, a result that is expected since neither a capacitor nor an inductor uses energy on the average. Conversely, if only resistance is present, $Z = \sqrt{R^2 + (X_L - X_C)^2} = R$, and $\cos\phi = R/Z = 1$. In this case, $\overline{P} = I_{rms}V_{rms} \times \cos\phi = I_{rms}V_{rms}$, which is the expression for the average power dissipated in a resistor. Example 3 deals with the current, voltages, and power for a series RCL circuit.

Example 3 Current, Voltages, and Power in a Series RCL Circuit

A series RCL circuit contains a 148-Ω resistor, a 1.50-μF capacitor, and a 35.7-mH inductor. The generator has a frequency of 512 Hz and an rms voltage of 35.0 V. Obtain (a) the rms voltage across each circuit element and (b) the electrical power consumed by the circuit.

REASONING The rms voltages across each circuit element can be determined from $V_R = I_{rms}R$, $V_C = I_{rms}X_C$, and $V_L = I_{rms}X_L$, as soon as the rms current and the reactances X_C and X_L are known. Since the rms current can be found from $I_{rms} = V_{rms}/Z$, the first step in the solution is to find the impedance Z from the individual reactances. The power consumed is given by $\overline{P} = I_{rms}V_{rms}\cos\phi$, where the phase angle ϕ can be obtained from $\tan\phi = (X_L - X_C)/R$.

SOLUTION

(a) The individual reactances are

$$X_C = \frac{1}{2\pi fC} = \frac{1}{2\pi(512\ \text{Hz})(1.50 \times 10^{-6}\ \text{F})} = 207\ \Omega \qquad (23.2)$$

$$X_L = 2\pi fL = 2\pi(512\ \text{Hz})(35.7 \times 10^{-3}\ \text{H}) = 115\ \Omega \qquad (23.4)$$

The impedance of the circuit is

$$Z = \sqrt{R^2 + (X_L - X_C)^2} = \sqrt{(148\ \Omega)^2 + (115\ \Omega - 207\ \Omega)^2} = 174\ \Omega \qquad (23.7)$$

The current through each circuit element is

$$I_{rms} = \frac{V_{rms}}{Z} = \frac{35.0\ \text{V}}{174\ \Omega} = 0.201\ \text{A} \qquad (23.6)$$

The rms voltages across each circuit element now follow immediately:

$$V_R = I_{rms}R = (0.201\ \text{A})(148\ \Omega) = \boxed{29.7\ \text{V}} \qquad (20.14)$$

$$V_C = I_{rms}X_C = (0.201\ \text{A})(207\ \Omega) = \boxed{41.6\ \text{V}} \qquad (23.1)$$

$$V_L = I_{rms}X_L = (0.201\ \text{A})(115\ \Omega) = \boxed{23.1\ \text{V}} \qquad (23.3)$$

Observe that these three rms voltages do not add up to give the generator's rms voltage, which is 35.0 V. Instead, the rms voltages satisfy Equation 23.5. It is the sum of the *instantaneous* voltages across R, C, and L, rather than the sum of the rms voltages, that add up to give the generator's *instantaneous* voltage, according to Kirchhoff's loop rule.

PROBLEM SOLVING INSIGHT

In an RCL series circuit, the rms voltages across the resistor, capacitor, and inductor do not add up to equal the rms voltage across the generator.

(b) The power consumed by the circuit is $\overline{P} = I_{rms}V_{rms} \cos \phi$. Therefore, a value for the phase angle ϕ is needed and can be obtained as follows:

$$\tan \phi = \frac{X_L - X_C}{R} = \frac{115\ \Omega - 207\ \Omega}{148\ \Omega} = -0.62 \quad (23.8)$$

$$\phi = \tan^{-1}(-0.62) = -32°$$

The phase angle is negative since the circuit is more capacitive than inductive (X_C is greater than X_L), and the current leads the voltage. The average power consumed is

$$\overline{P} = I_{rms}V_{rms} \cos \phi = (0.201\ \text{A})(35.0\ \text{V}) \cos (-32°) = \boxed{6.0\ \text{W}} \quad (23.9)$$

In addition to the series RCL circuit, there are many different ways to connect resistors, capacitors, and inductors. In analyzing these additional possibilities, it helps to keep in mind the behavior of capacitors and inductors at the extreme limits of the frequency. When the frequency approaches zero (i.e., dc conditions), the reactance of a capacitor becomes so large that no charge can flow through the capacitor. It is as if the capacitor were cut out of the circuit, leaving an open gap in the connecting wire. In the limit of zero frequency the reactance of an inductor is vanishingly small. The inductor offers no opposition to a dc current. It is as if the inductor were replaced with a wire of zero resistance. In the limit of very large frequency, the behaviors of a capacitor and an inductor are reversed. The capacitor has a very small reactance and offers little opposition to the current, as if it were replaced by a wire with zero resistance. In contrast, the inductor has a very large reactance when the frequency is very large. The inductor offers so much opposition to the current that it might as well be cut out of the circuit, leaving an open gap in the connecting wire. Conceptual Example 4 illustrates how to gain insight into more complicated circuits using the limiting behaviors of capacitors and inductors.

Conceptual Example 4 The Limiting Behavior of Capacitors and Inductors

Figure 23.13*a* shows two circuits. The rms voltage of the generator is the same in each case. The values of the resistance *R*, the capacitance *C*, and the inductance *L* in these circuits are the same. The frequency of the ac generator is very near zero. In which circuit does the generator supply more rms current?

REASONING According to Equation 23.6, the rms current is given by $I_{rms} = V_{rms}/Z$. Since V_{rms} is the same in each case, the greater current is delivered to the circuit with the smaller impedance Z. In the limit of very small frequency, the capacitors behave as if they were cut out of the circuit, leaving gaps in the connecting wires. In this limit, however, the inductors behave as if they were replaced by wires with zero resistance. Figure 23.13*b* shows the circuits as they would appear according to these changes. It is clear that circuit I behaves as if it contained only two identical resistors in series, with a total impedance of $Z = 2R$. In contrast, circuit II behaves as if it contained two identical resistors in parallel, in which case the total impedance is given by $1/Z = 1/R + 1/R$, or $Z = R/2$. At a frequency very near zero, then, circuit II has the smaller impedance and its generator supplies the greater rms current.

Related Homework Material: *Question 9, Problems 24, 25, and 42*

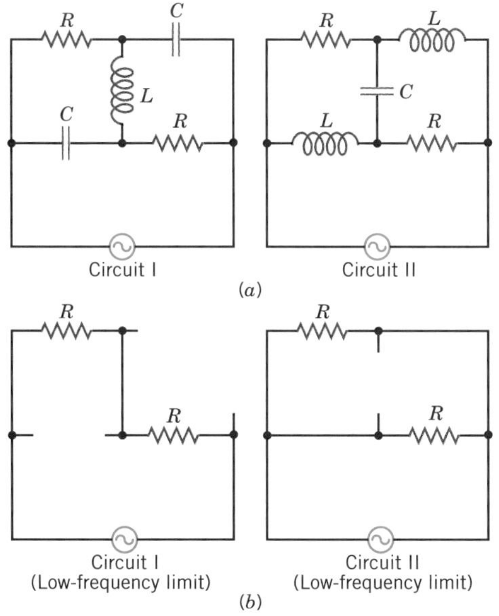

Figure 23.13 (*a*) These circuits are discussed in the limit of very small frequency in Conceptual Example 4. (*b*) For a frequency very near zero, the circuits in part *a* behave as if they were as shown here.

23.4 RESONANCE IN ELECTRIC CIRCUITS

The behavior of current and voltage in a series RCL circuit can give rise to a condition of ***resonance***. Resonance occurs when the frequency of a vibrating force exactly matches a natural frequency of the object to which the force is applied, as discussed in Section 10.8. In the electric case the vibrating force is provided by the oscillating electric field that is related to the voltage supplied by the generator.

Figure 23.14 helps us to understand why there is a resonance frequency for an ac circuit. This drawing presents an analogy between the electrical case (ignoring resistance) and the mechanical case of an object on a horizontal spring (ignoring friction). Part *a* shows a fully stretched spring that has just been released, with the

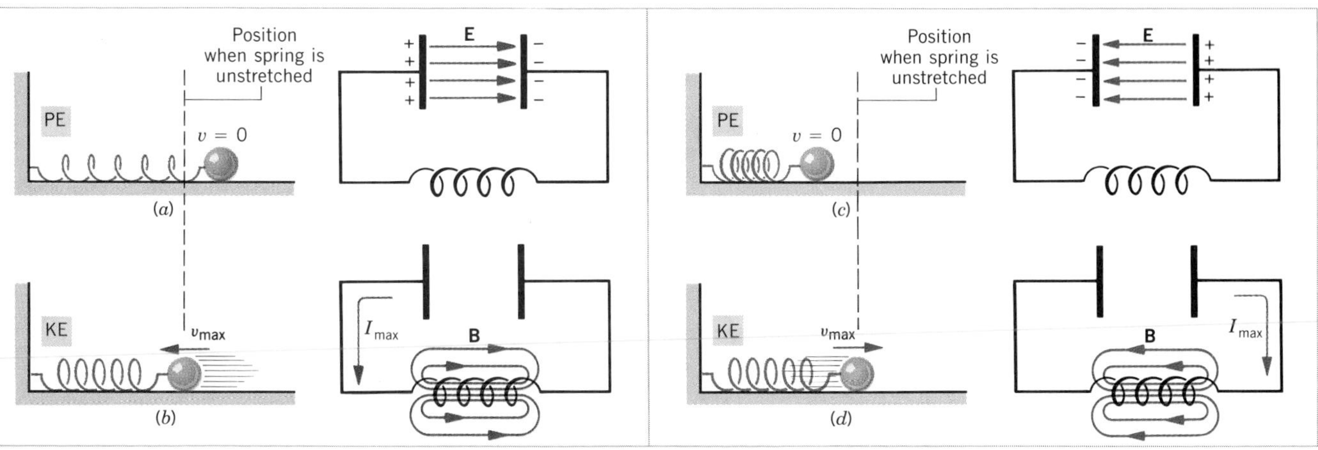

Figure 23.14 The oscillation of an object on a spring is analogous to the oscillation of the electric and magnetic fields that occur, respectively, in a capacitor and in an inductor.

initial speed v of the object being zero. All the energy is stored in the form of elastic potential energy. When the object begins to move, it gradually loses potential energy and picks up kinetic energy. In part b, the object moves with maximum kinetic energy through the position where the spring is unstretched (zero potential energy). Because of its inertia, the moving object coasts through this position and eventually comes to a halt in part c when the spring is fully compressed and all kinetic energy has been converted back into elastic potential energy. Part d of the picture is like part b, except the direction of motion is reversed. The resonance frequency f_0 of the object on the spring is the natural frequency at which the object vibrates and is given as $f_0 = [1/(2\pi)]\sqrt{k/m}$ according to Equation 10.14. In this expression, m is the mass of the object, and k is the spring constant.

In the electrical case, Figure 23.14a begins with a fully charged capacitor that has just been connected to an inductor. At this instant the energy is stored in the electric field between the capacitor plates. As the capacitor discharges, the electric field between the plates decreases, while a magnetic field builds up around the inductor because of the increasing current in the circuit. The maximum current and the maximum magnetic field exist at the instant the capacitor is completely discharged, as in part b of the figure. Energy is now stored entirely in the magnetic field of the inductor. The voltage induced in the inductor keeps the charges flowing until the capacitor again becomes fully charged, but now with reverse polarity, as in part c. Once again, the energy is stored in the electric field between the plates and no energy resides in the magnetic field of the inductor. Part d of the cycle repeats part b, but with reversed directions of current and magnetic field. Thus, an ac circuit can have a resonance frequency, because there is a natural tendency for energy to shuttle back and forth between the electric field of the capacitor and the magnetic field of the inductor.

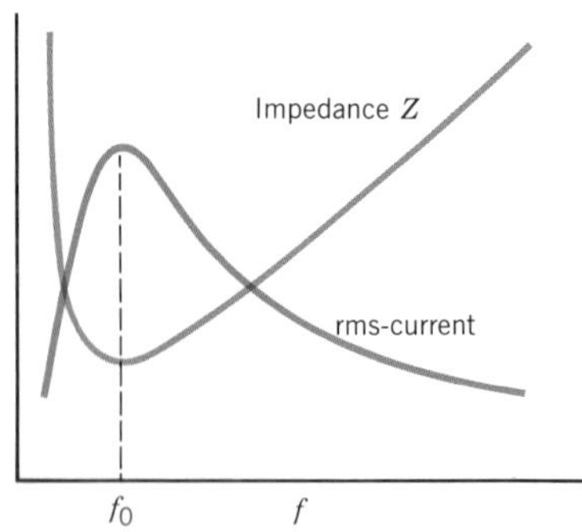

Figure 23.15 In a series RCL circuit the impedance is a minimum, and the current is a maximum, when the frequency f equals the resonance frequency f_0 of the circuit.

To determine the resonance frequency at which energy shuttles back and forth between the capacitor and the inductor, we note that the current in an RCL circuit is $I_{rms} = V_{rms}/Z$. In this expression Z is the impedance of the circuit and is given by $Z = \sqrt{R^2 + (X_L - X_C)^2}$. As Figure 23.15 illustrates, the rms current is a maximum when the impedance is a minimum, assuming a given generator voltage. The minimum impedance occurs when the frequency is f_0, such that $X_L = X_C$ or $2\pi f_0 L = 1/(2\pi f_0 C)$. This result can be solved for f_0, which is the resonance frequency:

$$f_0 = \frac{1}{2\pi\sqrt{LC}} \tag{23.10}$$

The resonance frequency is determined by the inductance and the capacitance, but not the resistance.

The effect of resistance on electrical resonance is to make the "sharpness" of the circuit response less pronounced, as Figure 23.16 indicates. When the resistance is small, the current-versus-frequency graph falls off suddenly on either side of the maximum current. When the resistance is large, the falloff is more gradual, and there is less current at the maximum.

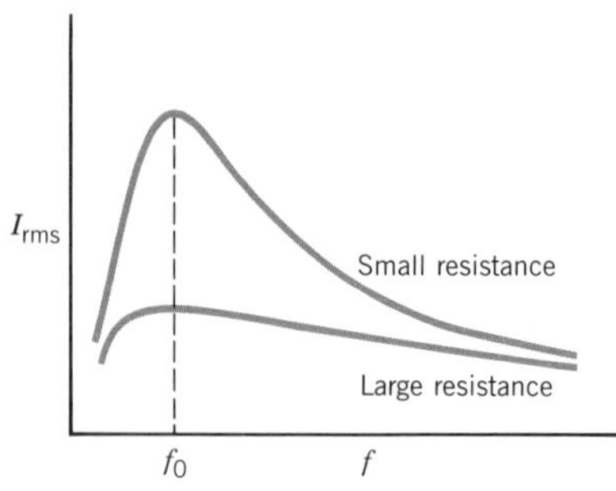

Figure 23.16 The effect of resistance on the current in a series RCL circuit.

Conceptual Example 5 focuses on what resonance means in an RCL circuit.

Conceptual Example 5 Resonance

A light bulb is essentially a resistor that glows more brightly when there is more current in it. Figure 23.17a shows a light bulb connected to an ac generator. When a capacitor is added in series with the bulb, as in part *b* of the drawing, the brightness decreases, no matter what the value of the capacitance *C* is. However, if an inductor is added in series with the bulb and the capacitor, as in part *c*, the brightness may either increase or decrease, depending on the value of the inductance *L*. Why?

REASONING The current in the circuit controls the brightness of the bulb. Therefore, it is necessary to understand what happens to the current as the capacitor and then the inductor are put into the circuit. In Figure 23.17*b*, the capacitor adds capacitive reactance and can only increase the circuit impedance. A greater impedance causes the current to decrease, and along with it, the brightness of the bulb. But how do we know that the capacitive reactance can cause only an increase in the impedance? We know, because the only way a decrease could occur is for there to be resonance, in which capacitive and inductive reactance offset one another and decrease the circuit impedance to a minimum. But resonance is possible only when capacitance and inductance are both present in a circuit, and there is no inductance in Figure 23.17*b*. In contrast, when the inductor is added in part *c*, resonance becomes possible. Then, depending on the value of the inductance, the inductive reactance and the capacitive reactance may cancel, the circuit impedance being reduced to a minimum. As a result, the current and the brightness of the bulb would increase. If the inductance were very large, however, the inductive reactance would dominant over the resistance and the capacitive reactance to such an extent that the circuit impedance would increase. In this event, a reduction in the current and the brightness of the bulb would occur.

Related Homework Material: *Problem 31*

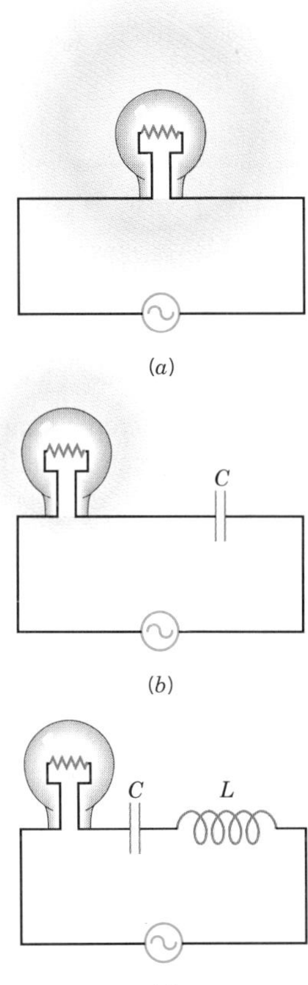

Figure 23.17 (*a*) The filament in a light bulb is a resistor that glows when heated by the current in it. (*b*) The brightness of the bulb decreases when a capacitor is added in series with the bulb. (*c*) The brightness may increase or decrease when an inductor is added in series with the bulb and the capacitor.

The following example deals with one application of resonance in electrical circuits. In this example the focus is on the oscillation of energy between a capacitor and an inductor. Once a capacitor/inductor combination is energized, the energy will oscillate indefinitely as in Figure 23.14, provided there is some provision to replace any dissipative losses that occur because of resistance. Circuits that include this type of provision are called oscillator circuits.

THE PHYSICS OF . . .

a heterodyne metal detector.

Example 6 A Heterodyne Metal Detector

Figure 23.18 shows a heterodyne metal detector being used. As Figure 23.19 illustrates, this device utilizes two capacitor/inductor oscillator circuits, *A* and *B*. Each circuit produces its own resonance frequency, $f_{0A} = 1/(2\pi\sqrt{L_A C})$ and $f_{0B} = 1/(2\pi\sqrt{L_B C})$. Any difference between these two frequencies is detected through earphones as a beat frequency $f_{0B} - f_{0A}$, similar to the beat frequency that two musical tones produce. In the absence of any nearby metal object, the inductances L_A and L_B are the same, and f_{0A} and f_{0B} are identical. There is no beat frequency. When inductor *B* (the search coil) comes near a piece of metal, the inductance L_B decreases, the corresponding oscillator frequency f_{0B} increases, and a beat frequency is heard. Suppose that initially each inductor is adjusted so $L_B = L_A$, and each oscillator has a resonance frequency of 855.5 kHz. Assuming that the inductance of search coil *B* decreases by 1.00% due to a nearby piece of metal, determine the beat frequency heard through the earphones.

REASONING To find the beat frequency $f_{0B} - f_{0A}$, we need to determine the amount by which the resonance frequency f_{0B} changes because of a 1.00% decrease in the inductance L_B.

Figure 23.18 A heterodyne metal detector is used to locate buried metal objects.

SOLUTION We begin by obtaining the ratio of f_{0B} to f_{0A}:

$$\frac{f_{0B}}{f_{0A}} = \frac{\dfrac{1}{2\pi\sqrt{L_B C}}}{\dfrac{1}{2\pi\sqrt{L_A C}}} = \sqrt{\frac{L_A}{L_B}}$$

But due to the nearby piece of metal $L_B = 0.9900L_A$, so that

$$\frac{f_{0B}}{f_{0A}} = \sqrt{\frac{L_A}{0.9900L_A}} = 1.005$$

Therefore, the new value for f_{0B} is $f_{0B} = 1.005f_{0A} = 1.005 \times (855.5 \text{ kHz}) = 859.8 \text{ kHz}$. As a result, the detected beat frequency is

$$f_{0B} - f_{0A} = 859.8 \text{ kHz} - 855.5 \text{ kHz} = \boxed{4.3 \text{ kHz}}$$

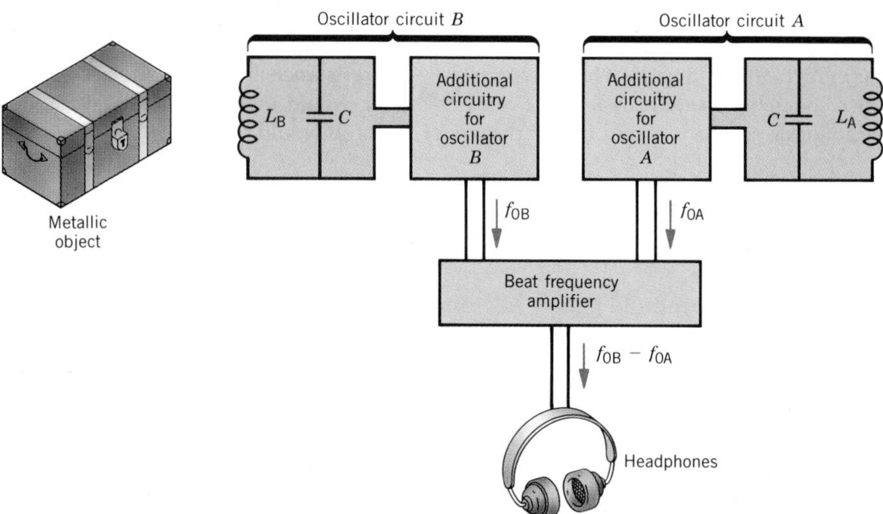

Figure 23.19 A heterodyne metal detector uses two electrical oscillators, *A* and *B*, in its operation. When the resonance frequency of oscillator *B* is changed due to the proximity of a metallic object, a beat frequency, whose value is $f_{0B} - f_{0A}$, is heard in the headphones.

23.5 SEMICONDUCTOR DEVICES

Semiconductor devices such as diodes and transistors are widely used in modern electronics, and Figure 23.20 illustrates one application. The drawing shows an audio system in which small ac voltages (originating in a compact disc player, etc.) are amplified so they can drive the speaker(s). The electric circuits that accomplish the amplification do so with the aid of a dc voltage provided by the power supply. In portable units the power supply is simply a battery. In nonportable units, however, the power supply is a separate electric circuit containing diodes, along with other elements. As we will see, the diodes convert the 60-Hz ac voltage present at a wall outlet into the dc voltage needed by the amplifier, which, in turn, performs its job of amplification with the aid of transistors.

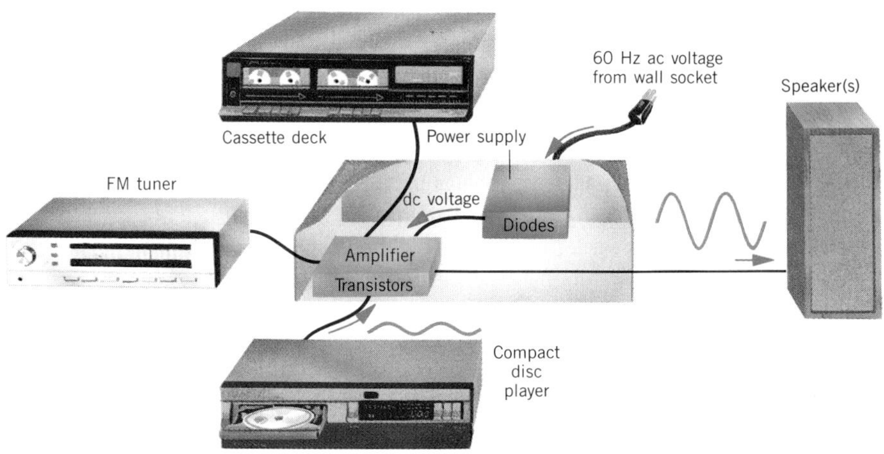

Figure 23.20 In a typical audio system, diodes are used in the power supply to create a dc voltage from the ac voltage present at the wall socket. This dc voltage is necessary so the transistors in the amplifier can perform their task of enlarging the small ac voltages originating in the compact disc player, etc.

n-TYPE AND p-TYPE SEMICONDUCTORS

The materials used in diodes and transistors are semiconductors, such as silicon and germanium. However, they are not pure materials, because small amounts of "impurity" atoms (about one part in a million) have been added to them to change their conductive properties. For instance, Figure 23.21*a* shows an array of atoms that symbolizes the crystal structure in pure silicon. Each silicon atom has four outer-shell* electrons, and each electron participates with electrons from neighboring atoms in forming the bonds that hold the crystal together. Since they participate in forming bonds, these electrons generally do not move throughout the crystal. Consequently, pure silicon and germanium are not good conductors of electricity. It is possible, however, to increase their conductivities by adding tiny amounts of impurity atoms, such as phosphorus or arsenic, whose atoms have five outer-shell electrons. For example, when a phosphorus atom replaces a silicon atom in the crystal, only four of the five outer-shell electrons of phosphorus fit into the crystal structure. The extra fifth electron does not fit in and is relatively free to diffuse throughout the crystal, as part *b* of the drawing suggests. A semiconductor containing small amounts of phosphorus can, therefore, be envisioned as containing immobile, positively charged phosphorus atoms and a pool of electrons that are free to wander throughout the material. These mobile electrons allow the semiconductor to conduct electricity.

The process of adding impurity atoms is called *doping*. A semiconductor doped with an impurity that contributes mobile electrons is called an ***n-type semiconductor***, since the mobile charge carriers have a **n**egative charge. Note that an n-type semiconductor is overall electrically neutral, since it contains equal numbers of positive and negative charges.

It is also possible to dope a silicon crystal with an impurity whose atoms have only three outer-shell electrons (e.g., boron or gallium). Because of the missing fourth electron, there is a "hole" in the lattice structure at the boron atom, as Figure 23.21*c* illustrates. An electron from a neighboring silicon atom can move into this hole, in which event the region around the boron atom, having acquired the electron, becomes negatively charged. Of course, when a nearby electron does move, it leaves behind a hole. This hole is positively charged, since it results from the removal of an electron from the vicinity of a neutral silicon atom. The vast

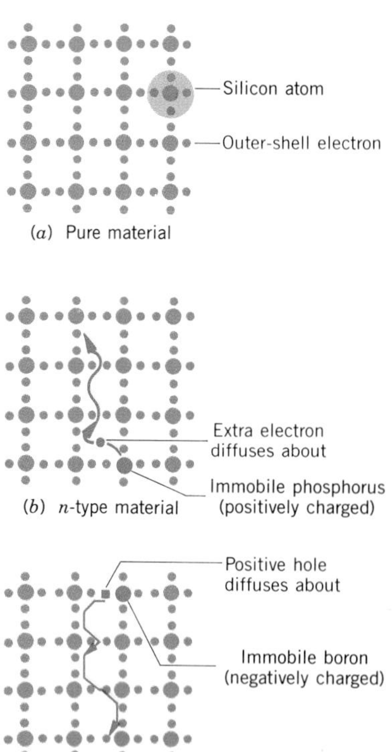

Figure 23.21 A silicon crystal that is (*a*) undoped or pure, (*b*) doped with phosphorus to produce an n-type material, and (*c*) doped with boron to produce a p-type material.

* Section 30.6 discusses the electronic structure of the atom in terms of "shells."

Much of the energy needed to operate this house comes from solar cells mounted to form large panels. The panels can be seen on the roof, just above the slanted glass surface of the greenhouse. In a solar cell a *p-n* junction is used to convert sunlight directly into electricity.

majority of atoms in the lattice are silicon, so the hole is almost always next to another silicon atom. Consequently, an electron from one of these adjacent atoms can move into the hole, with the result that the hole moves to yet another location. In this fashion, a positively charged hole can wander through the crystal. This type of semiconductor can, therefore, be viewed as containing immobile, negatively charged boron atoms and an equal number of positively charged, mobile holes. Because of the mobile holes, the semiconductor can conduct electricity. In this case the charge carriers are positive, as can be verified by measuring the Hall emf (see Section 21.5). A semiconductor doped with an impurity that introduces mobile positive holes is called a ***p-type semiconductor.***

THE SEMICONDUCTOR DIODE

A ***p-n junction diode*** is a device formed from a *p*-type semiconductor and an *n*-type semiconductor. The *p-n* junction between the two materials is of fundamental importance to the operation of diodes and transistors. Figure 23.22 shows separate *p*-type and *n*-type semiconductors, each electrically neutral. Figure 23.23*a* shows them joined together to form a diode. Mobile electrons from the *n*-type semiconductor and mobile holes from the *p*-type semiconductor flow across the junction and combine. This process leaves the *n*-type material with a positive charge layer and the *p*-type material with a negative charge layer, as part *b* of the drawing indicates. The positive and negative charge layers on the two sides of the junction set up an electric field **E**, much like that in a parallel plate capacitor. This electric field tends to prevent any further movement of charge across the junction, and all charge flow quickly stops.

THE PHYSICS OF . . .

a semiconductor diode.

Suppose now that a battery is connected across the *p-n* junction, as in Figure 23.24*a*, where the negative terminal of the battery is attached to the *n*-material, and the positive terminal is attached to the *p*-material. In this situation the junction is said to be in a condition of ***forward bias,*** and as a result, there is a current in the circuit. The mobile electrons in the *n*-material are repelled by the negative terminal of the battery and move toward the junction. Likewise, the positive holes in the *p*-material are repelled by the positive terminal of the battery and also move toward the junction. At the junction the electrons fill the holes. In the meantime, the negative terminal of the battery provides a fresh supply of electrons to the *n*-material, and the positive terminal pulls off electrons from the *p*-material, forming new holes in the process. Consequently, a continual flow of charge, and hence a current, is maintained.

In Figure 23.24*b* the battery polarity has been reversed, and the *p-n* junction is in a condition known as ***reverse bias.*** The battery forces electrons in the *n*-material and holes in the *p*-material away from the junction. As a result, the potential across the junction builds up until it opposes the battery potential, and very little

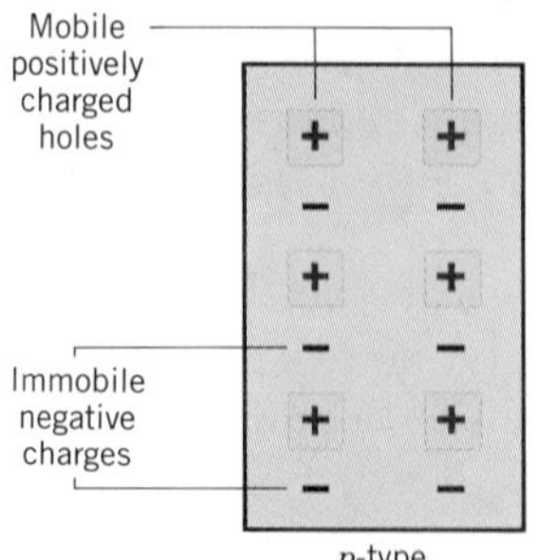

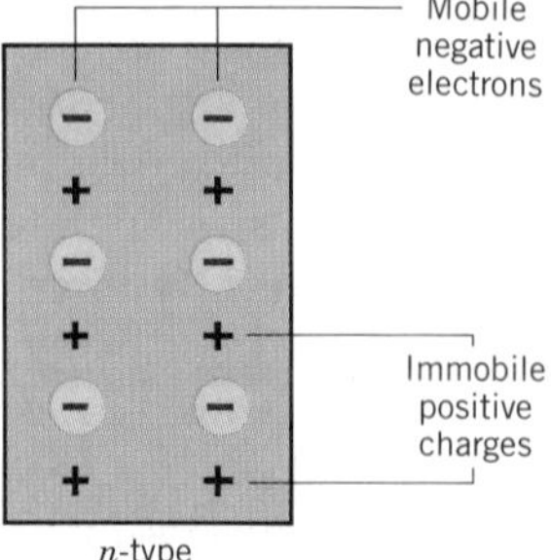

Figure 23.22 A *p*-type semiconductor and an *n*-type semiconductor.

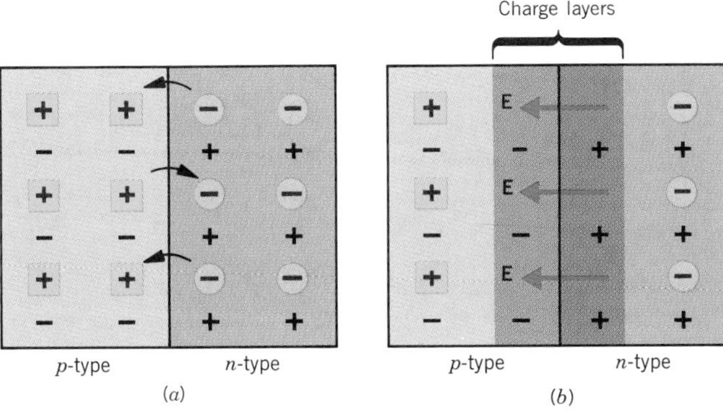

Figure 23.23 (*a*) At the junction between *n* and *p* materials, mobile electrons and holes combine and (*b*) create positive and negative charge layers. The electric field produced by the charge layers is **E**.

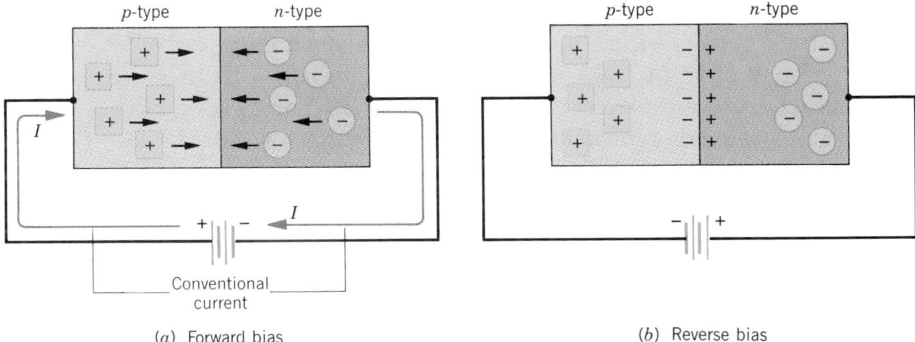

Figure 23.24 (*a*) There is an appreciable current through the diode when the diode is forward biased. (*b*) Under a reverse bias condition, there is almost no current through the diode.

current can be sustained through the diode. The diode, then, is a unidirectional device, for it allows current to pass only in one direction.

The graph in Figure 23.25 shows the dependence of the current on the magnitude and polarity of the voltage applied across a *p-n* junction diode. The exact values of the current depend on the nature of the semiconductor and the extent of the doping. Also shown in the drawing is the symbol used for a diode (—▶|—). The direction of the arrowhead in the symbol indicates the direction of the conventional current in the diode under a forward bias condition. In a forward bias condition, the side of the symbol that contains the arrowhead has a positive potential relative to the other side.

Because diodes are unidirectional devices, they are commonly used in ***rectifier circuits,*** which convert an ac voltage into a dc voltage. For instance, Figure 23.26 shows a circuit in which charges flow through the resistance *R* only while the ac generator biases the diode in the forward direction. Since current occurs only

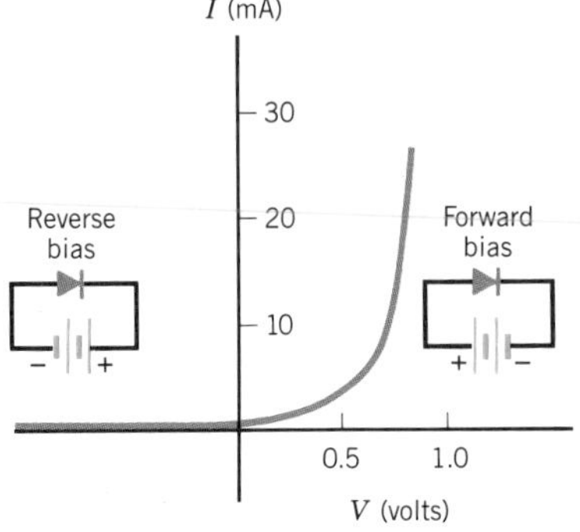

Figure 23.25 The current-versus-voltage characteristics of a typical *p-n* junction diode.

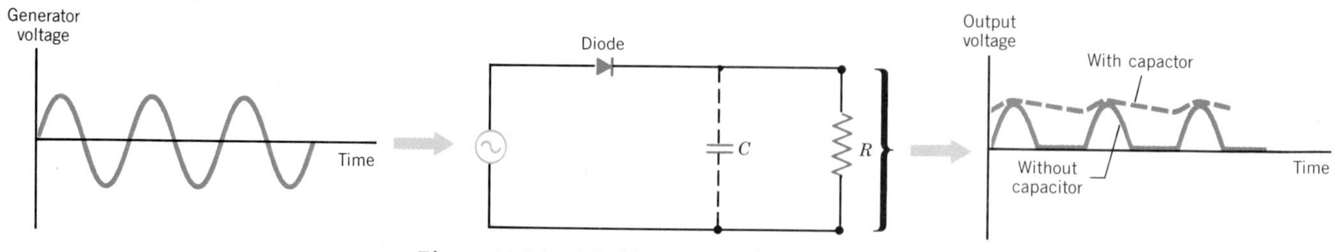

Figure 23.26 A half-wave rectifier circuit, together with a capacitor, constitutes a dc power supply, because the rectifier converts an ac voltage into a dc voltage.

during one-half of every generator voltage cycle, the circuit is called a half-wave rectifier. A plot of the output voltage across the resistor reveals that only the positive halves of each cycle are present. If a capacitor is added in parallel with the resistor, as indicated in the drawing, the capacitor charges up and keeps the voltage from dropping to zero between each positive half-cycle. It is also possible to construct full-wave rectifier circuits, in which both halves of every cycle of the generator voltage drive current through the load resistor in the same direction, as Conceptual Example 7 discusses.

THE PHYSICS OF . . .

a full-wave rectifier.

Conceptual Example 7 A Full-Wave Rectifier

Figure 23.27*a* shows the kind of full-wave rectifier that is known as a bridge rectifier. It uses four diodes. The direction of the current through the load resistance *R* is the same for both positive and negative halves of the generator's voltage cycle. Through which diodes does charge flow on the positive half and on the negative half of the cycle, and what is the direction of the current through the load resistance?

REASONING Our reasoning is based on the fact that charge flows through a diode only when the diode is in a forward bias condition. In a forward bias condition, the side of the diode symbol (—▶|—) that contains the arrowhead has a positive potential relative to the other side. Figure 23.27*b* shows the circuit for the positive half of the generator's voltage cycle, during which point *A* has the positive potential and point *B* the negative potential. Since point *A* is positive and is connected directly to the arrowhead for diode 1, that diode is in a forward bias condition. In contrast, it is not the arrowhead for diode 2 that is positive but the opposite side, so diode 2 is in a reverse bias condition. Since point *B* is negative and is connected directly to the side of diode 3 opposite the arrowhead, the arrowhead itself must be positive, and diode 3 is in a forward bias condition. In contrast, it is the arrowhead of diode 4 that is negative, so diode 4 is in a reverse bias condition. We see, then, that during the positive half of the voltage cycle, diodes 1 and 3 are forward biased. In Figure 23.27*b*, these diodes are highlighted by colored rectangles and allow charge to flow from left to right through the resistance *R*.

Figure 23.27*c* shows the circuit for the negative half of the voltage cycle, during which point *B* has the positive potential and point *A* the negative potential. Reasoning similar to that just discussed reveals that diodes 2 and 4 are now forward biased, while diodes 1 and 3 are reverse biased. The forward biased diodes are highlighted in Figure 23.27*c* by colored rectangles and again allow charge to flow from left to right through the resistance *R*.

When a circuit such as that in Figure 23.26 or 23.27 includes a capacitor and also a transformer to establish the desired voltage level, the circuit is called a power supply. In the audio system in Figure 23.20, the power supply receives the

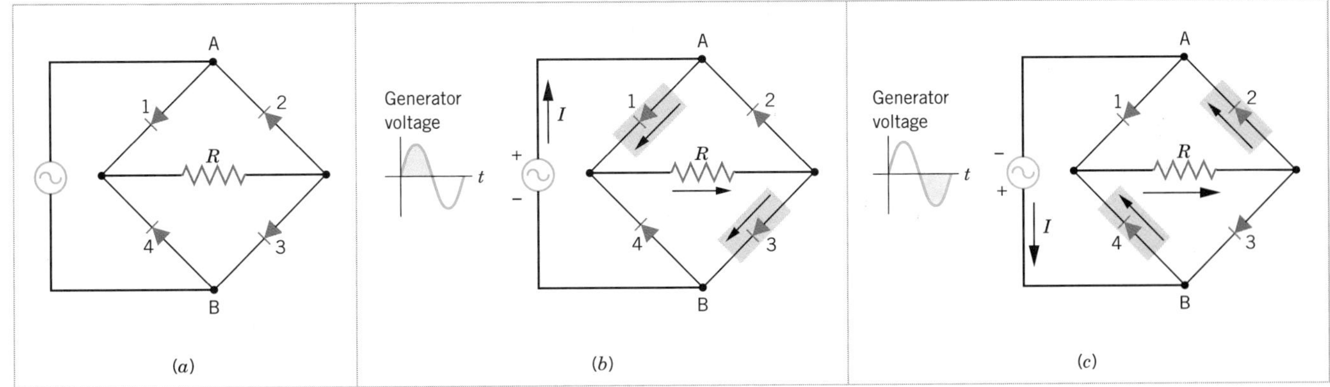

Figure 23.27 (*a*) A full-wave rectifier circuit. The condition of the circuit during (*b*) the positive half and (*c*) the negative half of the generator's voltage cycle.

60-Hz ac voltage from a wall socket and produces a dc output voltage that is used for the transistors within the amplifier. Power supplies using diodes are also found in virtually all electronic appliances, such as televisions and microwave ovens.

SOLAR CELLS

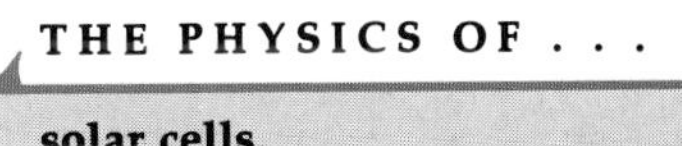

Solar cells use *p-n* junctions to convert sunlight directly into electricity, as Figure 23.28 illustrates. The solar cell in this drawing consists of a *p*-type semiconductor surrounding an *n*-type semiconductor. As discussed earlier, charge layers form at the junction between the two types of semiconductors, leading to an electric field **E** pointing from the *n*-type toward the *p*-type layer. The outer covering of *p*-type material is so thin that sunlight penetrates into the charge layers and ionizes some of the atoms there. In the process of ionization, the energy of the sunlight causes a negative electron to be ejected from the atom, leaving behind a positive hole. As the drawing indicates, the electric field in the charge layers causes the electron and the hole to move away from the junction. The electron moves into the *n*-type material, and the hole moves into the *p*-type material. As a result, the sunlight causes the solar cell to develop negative and positive terminals, much like the terminals of a battery. The current that a single solar cell can provide is small, so applications of solar cells often use many of them mounted to form large panels.

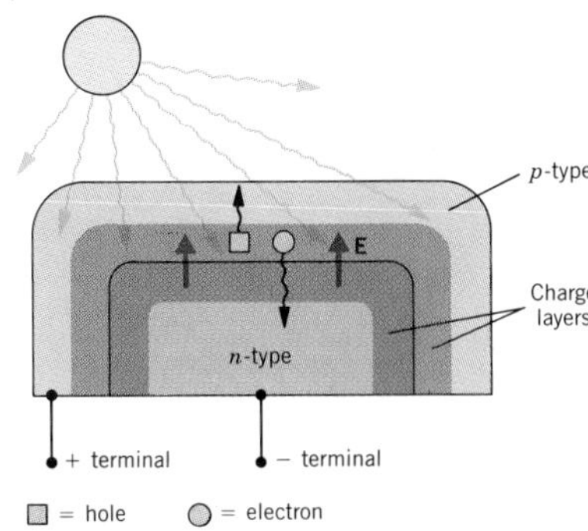

Figure 23.28 A solar cell formed from a *p-n* junction.

TRANSISTORS

A number of different kinds of transistors are in use today. One type is the ***bipolar junction transistor,*** which consists of two *p-n* junctions formed by three layers of doped semiconductors. As Figure 23.29 indicates, there are *pnp* and *npn* transistors. In either case, the middle region is made very thin compared to the outer regions.

THE PHYSICS OF . . .
transistors.

A transistor is useful because it can be used in circuits that amplify a smaller voltage into a larger one. A transistor plays the same kind of role in an amplifier circuit that a valve does when it controls the flow of water through a pipe. A small change in the valve setting produces a large change in the amount of water per second that flows through the pipe. In other words, a small change in the voltage input to a transistor produces a large change in the output from the transistor.

Figure 23.30 shows a *pnp* transistor connected to two batteries, labeled V_E and

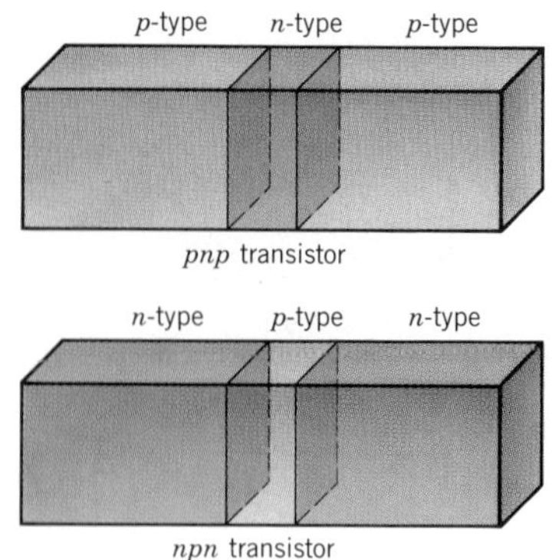

Figure 23.29 There are two kinds of bipolar junction transistors, *pnp* and *npn*.

V_C. The voltages V_E and V_C are applied in such a way that the *p-n* junction on the left has a forward bias, while the *p-n* junction on the right has a reverse bias. Moreover, the voltage V_C is usually much larger than V_E for a reason to be discussed shortly. The drawing also shows the standard symbol and nomenclature for the three sections of the transistor, namely, the *emitter,* the *base,* and the *collector.* The arrowhead in the symbol points in the direction of the conventional current through the emitter.

The positive terminal of V_E pushes the mobile positive holes in the *p*-type material of the emitter toward the emitter/base junction. And since this junction has a forward bias, the holes enter the base region readily. Once in the base region, the holes come under the strong influence of V_C and are attracted to its negative terminal. Since the base is so thin (about 10^{-6} m or so), approximately 98% of the holes are drawn through the base and into the collector. The remaining 2% of the holes combine with free electrons in the base region, thereby giving rise to a small base current I_B. As the drawing shows, the moving holes in the emitter and collector constitute currents that are labeled I_E and I_C, respectively. From Kirchhoff's junction rule it follows that $I_C = I_E - I_B$.

Because the base current I_B is small, the collector current is determined primarily by current from the emitter ($I_C = I_E - I_B \approx I_E$). This means that a change in I_E will cause a change in I_C of nearly the same amount. Furthermore, a substantial change in I_E can be caused by only a small change in the forward bias voltage V_E. To see that this is the case, look back at Figure 23.25 and notice how steep the current-versus-voltage curve is for a *p-n* junction; small changes in the forward bias voltage give rise to large changes in the current.

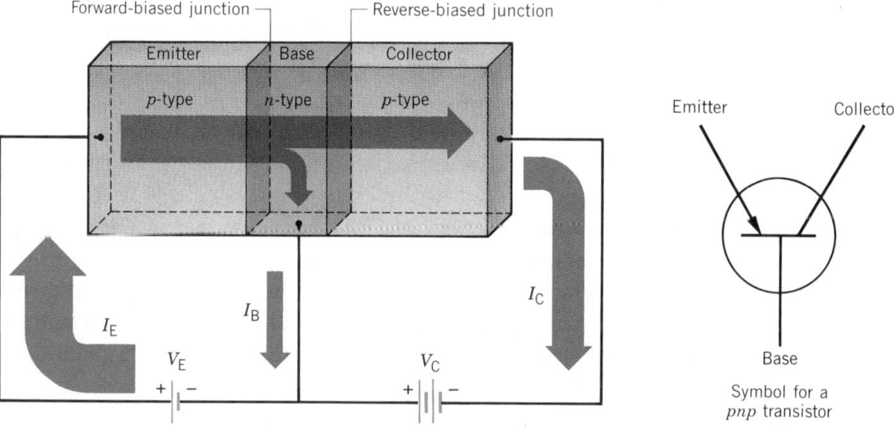

Figure 23.30 A *pnp* transistor, along with its bias voltages V_E and V_C. On the symbol for the *pnp* transistor, the emitter is marked with an arrowhead that denotes the direction of conventional current through the emitter.

With the help of Figure 23.31 we can now appreciate what was meant by the earlier statement that a small change in the voltage input to a transistor leads to a large change in the output. This picture shows an ac generator connected in series with the battery V_E and a resistance R connected in series with the collector. The generator voltage could originate from many sources, such as an electric guitar pickup or a compact disc player, while the resistance R could represent a loudspeaker. The generator introduces small voltage changes in the forward bias across the emitter/base junction and, thus, causes large corresponding changes in the current I_C leaving the collector and passing through the resistance R. As a result, the output voltage across R is an enlarged or amplified version of the input voltage of the generator. The operation of an *npn* transistor is similar to that of a

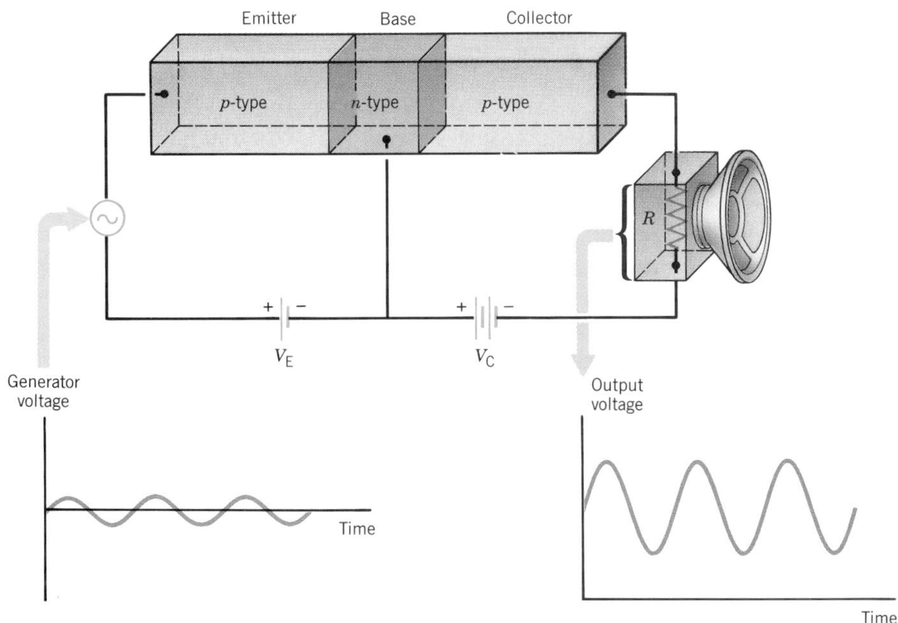

Figure 23.31 The basic *pnp* transistor amplifier in this drawing amplifies a small generator voltage to produce an enlarged voltage across the resistance *R*.

pnp transistor. The main difference is that the bias voltages and current directions are reversed.

It is important to realize that the increased power available at the output of a transistor amplifier does *not* come from the transistor itself. Rather, it comes from the power provided by the voltage source V_C. The transistor, acting like an automatic valve, merely allows the small, weak signals from the input generator to control the power taken from the source V_C and delivered to the resistance *R*.

Today it is possible to combine arrays of thousands of transistors, diodes, resistors, and capacitors on a tiny "chip" of silicon that usually measures less than a centimeter on a side. These arrays are called integrated circuits (ICs) and can be designed to perform almost any desired electronic function. Integrated circuits, such as the type in Figure 23.32, have revolutionized the electronics industry and lie at the heart of computers, hand-held calculators, digital watches, and programmable appliances.

Figure 23.32 Integrated circuit (IC) chips are manufactured on wafers of semiconductor material. Shown here is a close-up of a single wafer. It contains a number of the new PowerPC 601 chips being used by the Apple Corporation in its newest generation of computers. In the photograph one of these chips is entirely visible, surrounded by portions of nine others.

SUMMARY

In an ac circuit the rms voltage across a capacitor is related to the rms current according to $V_{rms} = I_{rms}X_C$, where X_C is the **capacitive reactance.** The capacitive reactance is measured in ohms and is given by $X_C = 1/(2\pi fC)$ for a capacitance C and a frequency f. The current in a capacitor leads the voltage across the capacitor by a phase angle of 90°, and as a result, a capacitor consumes no power, on the average.

For an inductor the rms voltage and the rms current are related by $V_{rms} = I_{rms}X_L$, where X_L is the **inductive reactance.** For an inductance L and frequency f the inductive reactance is given in ohms as $X_L = 2\pi fL$. The ac current in an inductor lags behind the voltage by a phase angle of 90°. Consequently, an inductor, like a capacitor, consumes no power, on the average.

When a resistor, a capacitor, and an inductor are connected in series, the rms voltage across the combination is related to the rms current according to $V_{rms} = I_{rms}Z$, where Z is the **impedance** of the combination. The impedance (in ohms) for the series combination is $Z = \sqrt{R^2 + (X_L - X_C)^2}$. The phase angle ϕ between current and voltage for a series RCL combination is given by $\tan \phi = (X_L - X_C)/R$. Only the resistor in the combination dissipates power on the average, according to the relation $\overline{P} = I_{rms}V_{rms} \cos \phi$. The term $\cos \phi$ is the **power factor** of the circuit.

A series RCL circuit has a **resonance frequency** f_0 that is $f_0 = 1/(2\pi\sqrt{LC})$. At resonance the impedance of the circuit has a minimum value equal to the resistance R, and the rms current has a maximum value.

In an ***n*-type semiconductor,** mobile negative electrons carry the current. An n-type material is produced by doping a semiconductor such as silicon with a small amount of impurity such as phosphorus. In a ***p*-type semiconductor,** mobile positive holes in the crystal structure carry the current. A p-type material is made by doping a semiconductor with an impurity such as boron. These two types of semiconductors are used in the ***p-n* junction diode** and in ***pnp* and *npn* bipolar junction transistors.**

QUESTIONS

1. A flashlight uses a battery to produce the light. (a) Describe what would happen to the brightness of the bulb if a capacitor were inserted in the circuit in series with the bulb. (b) Repeat (a), assuming that an inductor is inserted instead of a capacitor. In each case, give your reasoning.

2. A light bulb is connected directly to the 60-Hz ac voltage present at a wall outlet. (a) Describe what would happen to the brightness of the bulb, if a parallel plate capacitor (without a dielectric between the plates) were inserted in series between the light bulb and the wall outlet. (b) Describe what happens to the brightness when a dielectric material is inserted into the space between the plates. In both (a) and (b) explain your reasoning.

3. The ends of a long, straight wire are connected to the terminals of an ac generator, and the current is measured. The wire is then disconnected, wound into the shape of a multiple-turn coil, and reconnected to the generator. In which case does the generator deliver a larger current? Explain.

4. An air-core inductor is connected in series with a light bulb, and this circuit is plugged into an electrical outlet. What happens to the brightness of the bulb when a piece of iron is inserted inside the inductor? Give a reason for your answer.

5. Consider two ac circuits. In each case, the generators are identical (same frequency, same rms voltage). In circuit I only a resistor is connected across the generator. In circuit II the same resistor is in series with an inductor. Which circuit uses more power? Give your reasoning.

6. In a series circuit a resistor and an inductor are connected to a generator whose rms voltage is 160 V. It is determined that the rms voltage across the resistor is 113 V, while the rms voltage across the inductor is also 113 V. Notice that 113 V + 113 V is greater than the 160 V provided by the generator. Does this situation involving rms voltages violate Kirchhoff's loop rule? Explain.

7. Can an ac circuit, such as the one in Figure 23.9, have the same impedance at two different generator frequencies? Justify your answer.

8. An inductor and a capacitor are connected in parallel across the terminals of a generator. What happens to the current from the generator as the frequency becomes (a) very large and (b) very small? Give your reasoning.

9. Review Conceptual Example 4. For which of the two circuits discussed there does the generator deliver more current when the frequency is very large? Account for your answer.

10. Is it possible for two series RCL circuits to have the same resonance frequencies and yet have (a) different R values and (b) different C and L values? Justify your answers.

11. Suppose the generator connected to a series RCL circuit has a frequency that is greater than the resonance frequency of the circuit. Suppose, in addition, that it is necessary to match the resonance frequency of the circuit to the frequency of the generator. To accomplish this, should you add a second capacitor in series or in parallel with the one already present? Explain your choice.

PROBLEMS

Note: For problems in this set, the ac current and voltage are rms values, and the power is an average value, unless indicated otherwise.

Section 23.1 Capacitors and Capacitive Reactance

1. At what frequency does a 7.50-μF capacitor have a reactance of 168 Ω?

2. What voltage is needed to create a current of 29.0 mA in a circuit containing only a 0.565-μF capacitor, when the frequency is 2.60 kHz?

3. Three capacitors are connected in parallel across the terminals of a 440-Hz generator that supplies a voltage of 17 V. The capacitances are 2.0, 4.0, and 7.0 μF. (a) Find the equivalent capacitance of these capacitors. (b) What is the total current supplied by the generator?

4. A circuit consists of a 3.00-μF and a 6.00-μF capacitor connected in series across the terminals of a 510-Hz generator. The voltage of the generator is 120 V. (a) Determine the equivalent capacitance of the two capacitors. (b) Find the current in the circuit.

5. A capacitor is attached to a 5.00-Hz generator. The current is observed to reach a maximum value at a certain time. What is the least amount of time that passes before the voltage across the capacitor reaches its maximum value?

***6.** A capacitor consists of two square metal plates that are parallel, each having an area of 1.0×10^{-4} m^2. This capacitor is connected to a generator that has a frequency of 11 kHz and a voltage of 150 V. The current in the circuit is measured to be 9.4 μA. Assuming that there is air between the plates, determine the distance between them.

***7.** Two identical and empty parallel plate capacitors are connected in parallel across the terminals of an ac generator. The generator delivers a total current of 0.24 A. By how much does the current *change* when a material with a dielectric constant of 3.3 is inserted between the plates of *one* of the capacitors? Does the current increase or decrease?

****8.** A capacitor (capacitance C_1) is connected across the terminals of an ac generator. Without changing the voltage or frequency of the generator, a second capacitor (capacitance C_2) is added in series with the first one. As a result, the current delivered by the generator decreases by a factor of three. Suppose the second capacitor had been added in parallel with the first one, instead of in series. By what factor would the current delivered by the generator have increased?

Section 23.2 Inductors and Inductive Reactance

9. What is the inductance of an inductor that has a reactance of 1.8 kΩ at a frequency of 4.2 kHz?

10. The current in an inductor is 0.20 A, and the frequency is 750 Hz. If the inductance is 0.080 H, what is the voltage across the inductor?

11. The reactance of an inductor is 480 Ω when the frequency is 1350 Hz. What is the reactance when the frequency is 450 Hz?

12. The transformer for an electric toy train has a primary winding whose inductance is 2.4 H. A voltage of 120 V (frequency = 60.0 Hz) is applied to the primary coil when the transformer is plugged into an electrical outlet. Assuming the train is not connected to the secondary of the transformer, find the current in the primary.

13. A 0.047-H inductor is wired across the terminals of a generator that has a voltage of 2.1 V and supplies a current of 0.023 A. Find the frequency of the generator.

***14.** A 0.313-H and a 0.127-H inductor are connected in parallel across the terminals of a generator. The generator has a voltage of 9.00 V and a frequency of 266 Hz. What is the total current that the generator delivers?

***15.** The rms current in a solenoid is 0.036 A when the solenoid is connected to an 18-kHz generator. The solenoid has a cross-sectional area of 3.1×10^{-5} m^2 and a length of 2.5 cm. The solenoid has 135 turns. Determine the *peak voltage* of the generator.

****16.** Two inductors are connected in parallel across the terminals of a generator. One inductor has an inductance of $L_1 = 0.030$ H, while the other inductor has an inductance of $L_2 = 0.060$ H. A single inductor, with an inductance L, is connected across the terminals of a second generator that has the same frequency and voltage as the first one. The current delivered by the second generator is equal to the *total* current delivered by the first generator. Find L.

Section 23.3 Circuits Containing Resistance, Capacitance, and Inductance

17. A series RCL circuit includes a resistance of 275 Ω, an inductive reactance of 648 Ω, and a capacitive reactance of 415 Ω. The current in the circuit is 0.233 A. What is the voltage of the generator?

18. The purpose of this problem is to verify the shapes of the graphs of capacitive and inductive reactance versus frequency, which are shown in Figures 23.2 and 23.6. Plot these graphs for a 20.0-μF capacitor and a 5.00-mH inductor. Use a frequency of 10 Hz and five equally spaced frequencies between 100 and 1000 Hz.

19. A series RCL circuit contains a 47.0-Ω resistor, a 2.00-μF capacitor, and a 4.00-mH inductor. When the frequency is 2550 Hz, what is the power factor of the circuit?

20. An ac generator has a frequency of 5.60 kHz and produces a current of 0.0530 A in a series circuit that contains only a 218-Ω resistor and a 0.100-μF capacitor. Obtain (a) the voltage of the generator and (b) the phase angle between the current and the voltage.

21. A circuit consists of a 215-Ω resistor and a 0.200-H inductor. These two elements are connected in series across a generator that has a frequency of 106 Hz and a voltage of 234 V. (a) What is the current in the circuit? (b) Determine the phase angle between the current and the voltage.

22. A 2700-Ω resistor and a 1.1-μF capacitor are connected in series across a generator (60.0 Hz, 120 V). Determine the power dissipated in the circuit.

23. A series circuit has an impedance of 192 Ω, and the current leads the voltage by 75.0°. The circuit contains two different elements. (a) From the phase angle between current and voltage, decide which elements are present, R and C, R and L, or C and L. (b) Find values for the appropriate quantities, R and X_C, or R and X_L, or X_C and X_L.

***24.** Refer to Conceptual Example 4 as background for this problem. For the circuit shown in the drawing, find the current provided by the generator when the frequency is (a) very large and (b) very small.

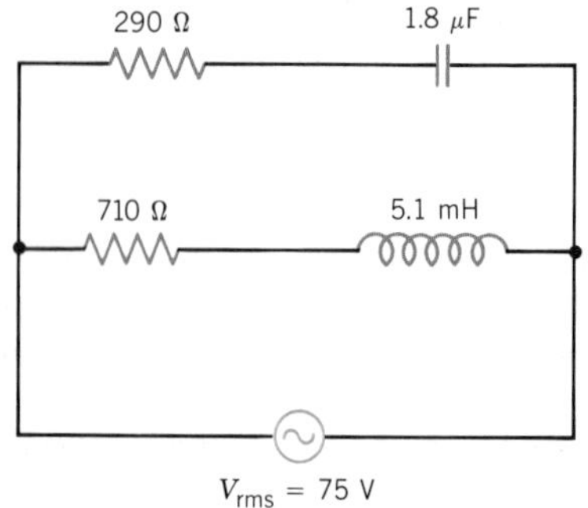

***25.** Review Conceptual Example 4 before attempting this problem. Then refer to the drawing below, in which the generator delivers four times more current at very low frequencies than it does at very high frequencies. Find the ratio R_2/R_1 of the resistances.

***26.** In reality, there is some resistance R in the wire from which an inductor is made. Therefore, an actual inductor should be represented as a resistor in series with an ideal (resistanceless) inductor. With this in mind, suppose the current in a 2.8-mH inductor is I_0 when the inductor is connected to a 12-V battery. However, when the battery is re-

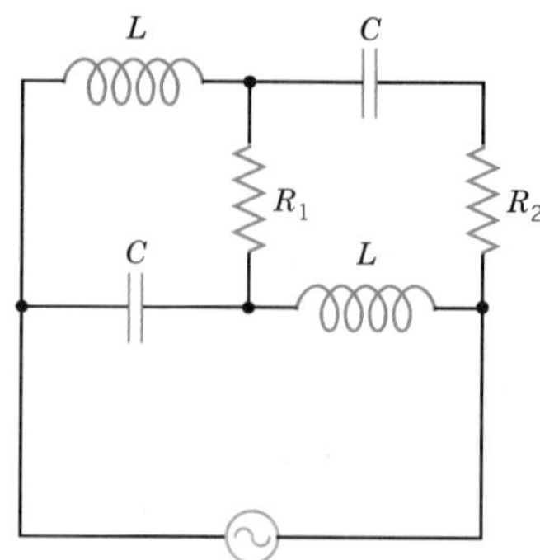

Problem 25

placed with a 1500-Hz generator whose voltage is 12 V, the current is $I_0/3$. What is the resistance R of the wire?

***27.** A series circuit contains only a resistor and an inductor. The voltage V of the generator is fixed. If $R = 16\ \Omega$ and $L = 4.0$ mH, find the frequency at which the current is one-half of its value at zero frequency.

****28.** When a resistor is connected by itself to an ac generator, the average power dissipated in the resistor is 1.000 W. When a capacitor is added in series with the resistor, the power dissipated is 0.500 W. When an inductor is added in series with the resistor (without the capacitor), the power dissipated is 0.250 W. Determine the power dissipated when both the capacitor and the inductor are added in series with the resistor.

Section 23.4 Resonance in Electric Circuits

29. A series RCL circuit has a capacitance of 1.20 μF and an inductance of 2.00 mH. What is the resonance frequency of the circuit?

30. A series RCL circuit has a resonance frequency of 690 kHz. If the value of the inductance is 26 μH, what is the value of the capacitance?

31. Review Conceptual Example 5. A light bulb has a resistance of 240 Ω. It is connected to a standard wall socket (120 V, 60.0 Hz). (a) Determine the current in the bulb. (b) Determine the current in the bulb after a 10.0-μF capacitor is added in series in the circuit. (c) It is possible to return the current in the bulb to the value calculated in part (a) by adding an inductor in series with the bulb and the capacitor. What is the value of the inductance of this inductor?

32. The resistor in a series RCL circuit has a resistance of 92 Ω, while the voltage of the generator is 3.0 V. At resonance, what is the average power dissipated in the circuit?

33. A series RCL circuit is at resonance and contains a variable resistor that is set to 175 Ω. The power dissipated in the circuit is 2.6 W. Assuming that the voltage remains constant, how much power is dissipated when the variable resistor is set to 562 Ω?

34. The resonance frequency of a series RCL circuit is 7.8 kHz. The inductance and capacitance of the circuit are each doubled. What is the new resonance frequency?

35. The power dissipated in a series RCL circuit is 65.0 W, and the current is 0.530 A. The circuit is at resonance. Determine the voltage of the generator.

*36. The ratio of the inductive reactance to the capacitive reactance is observed to be 5.36 in a series RCL circuit. The resonance frequency of the circuit is 225 Hz. What is the frequency of the generator that is connected to the circuit?

*37. Suppose you have a number of capacitors. Each of these capacitors is identical to the capacitor that is already in a series RCL circuit. How many of these additional capacitors must be inserted in series in the circuit, so the resonance frequency triples?

**38. In a series RCL circuit the dissipated power drops by a factor of two when the frequency of the generator is changed from the resonance frequency to a nonresonance frequency. The peak voltage is held constant while this change is made. Determine the power factor of the circuit at the nonresonance frequency.

**39. When the frequency is twice the resonance frequency, the impedance of a series RCL circuit is twice the value of the impedance at resonance. Obtain the ratios of the inductive and capacitive reactances to the resistance; that is, obtain X_L/R and X_C/R.

ADDITIONAL PROBLEMS

40. In a series circuit, a generator (1350 Hz, 15.0 V) is connected to a 16.0-Ω resistor, a 4.10-μF capacitor, and a 5.30-mH inductor. Find the voltage across each circuit element.

41. Two identical capacitors are connected in series to an ac generator that has a frequency of 620 Hz and produces a voltage of 24 V. The current in the circuit is 0.16 A. What is the capacitance of each capacitor?

42. Review Conceptual Example 4 and Figure 23.13. Find the ratio of the current in circuit I to that in circuit II in the high frequency limit for the same generator voltage.

43. A simple metal detector consists of a series circuit formed by a 1.70-mH inductor, a 3.00-μF capacitor, and a generator with a voltage of 9.00 V. The inductor has the shape of a large coil (see the drawing). The resistance of the wire in the coil is 3.50 Ω. The frequency of the generator is held constant at the resonance frequency that applies when there is no metal passing through the coil. When a person with a metal object walks through the coil, the inductance increases and, consequently, the current in the circuit changes. The change in current can be used to sound a warning. Determine the current in the circuit (a) when no metal is present and (b) when a metal object causes a 4.0% increase in the inductance. (c) Find the change in current. Is it an increase or a decrease?

44. An 8.2-mH inductor is connected to an ac generator (10.0 V rms, 620 Hz). Determine the *peak value* of the current supplied by the generator.

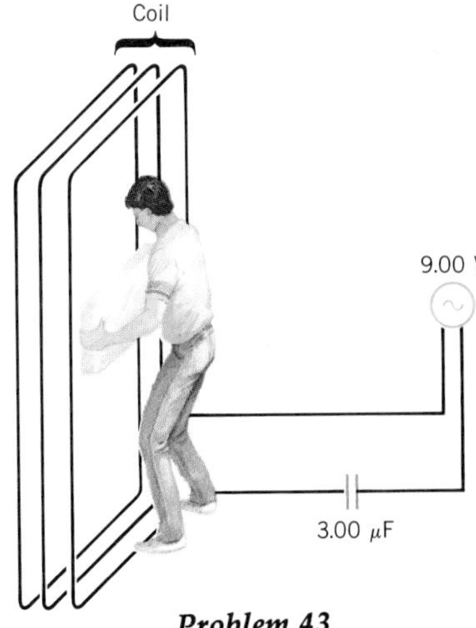

Problem 43

45. Two series RCL circuits have the same resonance frequency, yet circuit *A* has a capacitance of 2.3 μF while circuit *B* has a capacitance of 3.9 μF. Find the ratio L_A/L_B of the inductances in these circuits.

46. A circuit contains a resistor in series with either an inductor or a capacitor. The impedance of the circuit is 368 Ω. The current lags behind the voltage by 62.0°. (a) From the phase angle between the current and the voltage decide whether the resistor is in series with an inductor or a capacitor. (b) Find the values of *R* and X_L or *R* and X_C, whichever are appropriate.

47. The elements in a series RCL circuit are a 106-Ω resistor, a 3.30-μF capacitor, and a 0.0310-H inductor. What is the impedance of the circuit and the phase angle between the current and the voltage when the frequency is 609 Hz?

*48. The resonance frequency of a series RCL circuit is 13 MHz. The capacitor is an empty parallel plate capacitor. What is the resonance frequency when the capacitor is filled with a material whose dielectric constant is 5.2?

*49. A series RCL circuit contains a 5.10-μF capacitor and a generator whose voltage is 11.0 V. At a resonance frequency of 1.30 kHz the power dissipated in the circuit is 25.0 W. Find the values of (a) the inductance and (b) the resistance. (c) Calculate the power factor when the generator frequency is 2.31 kHz.

*50. A circuit consists of an 85-Ω resistor in series with a 4.0-μF capacitor, the two being connected between the terminals of an ac generator. The voltage of the generator is fixed. At what frequency is the current in the circuit one-half the value that exists when the frequency is very large?

**51. A 108-Ω resistor, a 0.200-μF capacitor, and a 5.42-mH inductor are connected in series to a generator whose voltage is 26.0 V. The current in the circuit is 0.141 A. Because of the shape of the current-versus-frequency graph (see Figure 23.15), there are two possible values for the frequency that correspond to this current. Obtain these two values.

INTERSECTION

Physics & the Environment

Since the beginning of the Industrial Revolution in the eighteenth century, factories and the consumers of the products they make have been throwing out waste products. Rivers were quickly polluted by the effluent from early mills, but the general assumption until recently was that the earth was so vast that the refuse would not affect it globally or permanently: "The solution to pollution is dilution," went the old adage. By the middle of this century the output of waste had increased to the point where its impact could no longer be ignored.

Imagine that you are an investigator for your state's Department of Environmental Protection, with the responsibility of monitoring water quality in drinking water reservoirs. Some days you spend scrambling around shorelines, or rowing out into deeper water, taking water samples, but more often than not you are in your laboratory analyzing and evaluating the samples that have been collected. How does basic physics affect your work?

Your lab is filled with instruments that analyze the chemicals contained in the water samples. All of these use the fundamental physics of matter to perform their functions. Spectrophotometers measure the unique frequencies of light absorbed by various elements, and fluorometers measure the fluorescent light emitted by the atoms and molecules. Today you are using a gas chromatograph. It indicates the chemical composition of the samples, using the fact that different elements and compounds have different masses, sizes, and boiling points. The time it takes different substances to get through the chromatograph varies with these factors, and the intensity of emission from the chromatograph tells how much there is of each substance in the water.

You compare the output of the chromatograph to a reference table, identifying decay products of herbicides and pesticides from agricultural activities upstream of the reservoir. Nitrates from fertilizer and phosphates from detergents and manufacturing are common pollutants. (In the former Soviet Union, the nitrate pollution is so bad in the rivers feeding the Aral Sea that the water is undrinkable and lethal to fish. This area is considered to be one of the most polluted and dangerous bodies of water in the world.)

All the nitrate and phosphate readings are within Environmental Protection Agency (EPA) limits. All routine . . . no, what's that peak on the graph? It wasn't there last week. Checking the reference table makes it clear: dioxin! You look further, identifying three other industrial by-products banned by the EPA. Lethal wastes are leaching into the reservoir, and a million and one-half people drink this water.

You study topological charts showing the water flowing into the reservoir. The water comes in from a river, let's call it the Darian, that in turn is fed by six smaller rivers and countless streams. To locate the dump site more water samples are needed, and fast. Three hours and several urgent phone calls later the first samples begin arriving. Twenty-five miles up river, beyond where the Acticala River joins the Darian, the samples show no dioxin. Below the Acticala, the dioxin is in the river. You direct your field teams to sample the water along the Acticala.

Here's where it gets tough. Samples show dioxin 8 miles up the Acticala, but the first industries along the river are 19 miles up river. Where the dioxin appears to be entering it, the river is lined with dairy farms and woodlands. Reconnaissance photos along the entire river reveal neither signs of illegal dumping nor the scruffy vegetation characteristic of areas containing pollutants. You conclude that the effluent is coming into the river through underground water, which seeps through porous rock *(the aquifer)*.

If the pollutants are seeping into the aquifer and are being carried downhill by gravity into the Acticala, then the farther from their source they get, the more diluted they become. Maps show that there are plenty of wells for drinking and irrigation up from the river. You call for water samples from the wells. By noon the next day your map is covered with dioxin concentration numbers. As you hoped, the concentrations are pyramid-shaped, leading to an apex 3 miles up from the river.

During the Gulf war, many oil wells in Kuwait were intentionally set on fire, and pipelines were opened to the sea. A number of the techniques used for dealing with the resulting pollution have their basis in physics principles.

You order a helicopter, load it with equipment, and head for that location. Hovering over the lightly wooded hillside, you see nothing suspicious. But somewhere there must be a buried, illegal dump. You point to a clearing one-half mile away, and the 'copter sets down. You are wearing hot, confining, protective clothing, which makes you irritable and all the more eager to find the dump. A magnetometer is unloaded from the helicopter, which will measure the magnetic field around the suspected dump site. For an hour you systematically crisscross an acre of ground. Finally, you are rewarded by a signal from the magnetometer that suggests buried metal. Carefully, the crew begins digging and not 3 feet down they encounter the top of a 55-gallon drum. It is not the only one, and several of them are leaking. While the cleanup is being done, you spend time back at your lab analyzing the contents of the drums for clues as to who manufactured it.

FOR INFORMATION ABOUT CAREERS RELATED TO THE ENVIRONMENT

Entrance into the environmental field in colleges and universities without specialized environmental studies programs is often through civil engineering or similar fields.

Encyclopedia of Careers and Vocational Guidance, Vol. 3, 8th edition, 1990, William E. Hopke, ed., J. G. Ferguson, publisher, has an in-depth description of the working conditions groundwater professionals encounter.

Career Information Center, Vol. II, 4th edition, 1990, Glencoe/MacMillan, publishers, under the title "Environmentalists."

Professional Careers Sourcebook, K. M. Savage and C. A. Dorgan, eds., 1990, Gale Research, Inc., publisher, provides addresses and phone numbers for dozens of guides and other resources about careers in environmental-related fields under the title "Civil Engineering."

You can also write to the following, among many others listed in the books above:

- Association of Environmental Scientists and Engineers, 2718 Southwest Kelly, Portland, OR 97204
- National Association of Environmental Professionals, Box 9400, Washington, DC 20016
- American Society of Civil Engineers, 345 East 47th St., New York, NY 10017

PART FIVE

LIGHT AND OPTICS

Light! We see by it, feel its warmth on a sunny day, and depend on it for our food supply. Plants convert sunlight into chemical energy by means of photosynthesis, and this energy sustains life for the plants themselves and the animals that feed on them.

We have also learned how to use light in a number of ways, some of which are so commonplace that they are taken for granted. For example, mirrors are designed to reflect light, and they are familiar items in homes and automobiles. Anyone who is nearsighted or farsighted uses eyeglasses or contact lenses to see better. These marvelous devices bend the light coming into the eye, so as to compensate for vision problems. Microscopes, telescopes, and binoculars use a combination of mirrors and/or lenses to enlarge objects that would otherwise be too small to see. These instruments have immensely broadened our knowledge of the microscopic world and the ultramacroscopic world of planets and stars. Cameras and video camcorders use lenses to record the good times with friends and family. Light is also used in the fields of communications and medicine. Many telephone cables are now being replaced by fiber-optic bundles, which can carry an enormous number of high-quality telephone conversations in a very small space. And the light from lasers is used routinely in a number of surgical techniques.

In many circumstances, light behaves as a wave. And because it has an electromagnetic nature, we begin in Chapter 24 by considering some of the fundamental characteristics of electromagnetic waves. In Chapters 25 and 26 we will explore the physics of how light is reflected and bent and will discuss a number of optical devices.

Since light can behave as a wave, interesting interference effects can arise when two or more light waves occupy the same space at the same time. In Chapter 27 we will investigate the interference of light, and show how it leads to colorful thin films (such as soap bubbles), to applications in compact disc technology, and to a limitation on the ability of the human eye to detect separately two objects that are close together. The basic ideas about interference will be familiar from our earlier discussion in Chapter 17 concerning the interference between sound waves.

CHAPTER 24 ELECTROMAGNETIC WAVES

It was the great Scottish physicist James Clerk Maxwell (1831–1879) who showed that electric and magnetic fields fluctuating together can form a propagating wave, appropriately called an ***electromagnetic wave.*** *Sunlight is a mixture of electromagnetic waves having different wavelengths. Each wavelength in the light produces a different color sensation in our eyes, from red to green to violet. The two rainbow lorikeets in this photograph have beautiful colors, and yet the colors do not belong to the birds themselves. The colors are present in the sunlight, in the sense that sunlight is a mixture of different wavelengths. When this mixture strikes the birds, different parts of the feathers reflect certain wavelengths more strongly than others, giving each species its colorful pattern. Thus, the colors of a bird "belong" to the electromagnetic waves that reflect from it. This chapter examines the general properties of electromagnetic waves and explores their many uses in our lives.*

24.1 THE NATURE OF ELECTROMAGNETIC WAVES

Figure 24.1 illustrates one way to create an electromagnetic wave. The setup consists of two straight metal wires that are connected to the terminals of an ac generator and serve as an antenna. The potential difference between the terminals changes sinusoidally with time and has a period T. Part *a* shows the instant when there is no charge at the ends of either wire. Since there is no charge, there is no electric field at the point P just to the right of the antenna. As time passes, the top wire becomes positively charged and the bottom wire negatively charged. One-quarter of a cycle later ($t = \frac{1}{4}T$), the charges have attained their maximum values, as indicated in part *b* of the drawing. Correspondingly, the electric field at point P has increased to its maximum strength in the downward direction.* This maximum field is represented by the red arrow in part *b*. Part *b* also indicates that the electric field created at earlier times (i.e., the black arrow in the picture) has not disappeared, but has moved to the right. Here lies the crux of the matter. At points far removed from the wire, the effect of the charges in creating the electric field is not felt immediately. Instead, the field is created first near the wires and then, like the effect of a pebble dropped into a pond, moves outward as a wave in all directions. Only the field moving to the right is shown in the picture for the sake of clarity. Eventually, after a time determined by the speed at which the wave travels, the field reaches distant points.

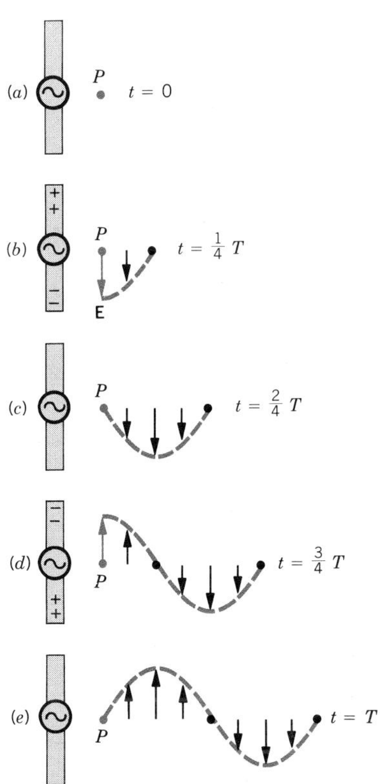

Figure 24.1 In each part of the drawing, the red arrow represents the electric field **E** produced at point P by the oscillating charges on the antenna at the indicated time. The black arrows represent the electric fields created at earlier times. All the fields propagate to the right. Fields also are produced and propagate in other directions, but these fields are omitted for simplicity.

Parts *c–e* of Figure 24.1 show the creation of the electric field at point P (red arrow) at later times during the generator cycle. In each part, the fields produced earlier in the sequence (black arrows) continue propagating toward the right. Part *d* shows the charges on the wires when the polarity of the generator has reversed, so the top wire is negative and the bottom wire is positive. As a result, the electric field at P has reversed its direction and points upward. In part *e* of the sequence, a complete sine wave has been drawn through the tips of the electric field vectors to emphasize that the field changes sinusoidally.

So far, our focus has been on the electric field created by the charges on the antenna wires. However, a magnetic field is also created, because charges flowing in the antenna constitute an electric current, and an electric current creates a magnetic field. Figure 24.2 illustrates the magnetic field direction at point P at the instant when the current in the antenna wire is upward. With the aid of Right-Hand Rule No. 2 (thumb of right hand points along I, fingers curl in the direction of **B**), the magnetic field at P can be seen to point into the page. As the oscillating current changes in magnitude and direction, the magnetic field at P changes accordingly. As with electric fields, the magnetic fields created at earlier times propagate outward as a wave. Moreover, the magnetic field is zero when the electric field has its maximum positive or negative value, and the magnetic field has its maximum positive or negative value when the electric field is zero. In other words, the two fields are 90° out of phase with one another.

It is important to notice that the magnetic field in Figure 24.2 is perpendicular to the page, whereas the electric field in Figure 24.1 lies in the plane of the page. Thus, the electric and magnetic fields created by the antenna wires are mutually perpendicular, and they remain so as they move away from the antenna. Moreover, both fields are perpendicular to the direction of travel. These perpendicular electric and magnetic fields, moving together, constitute an electromagnetic wave.

* The direction of the electric field can be obtained by imagining a positive test charge at P and determining the direction in which it would be pushed because of the charges on the wires.

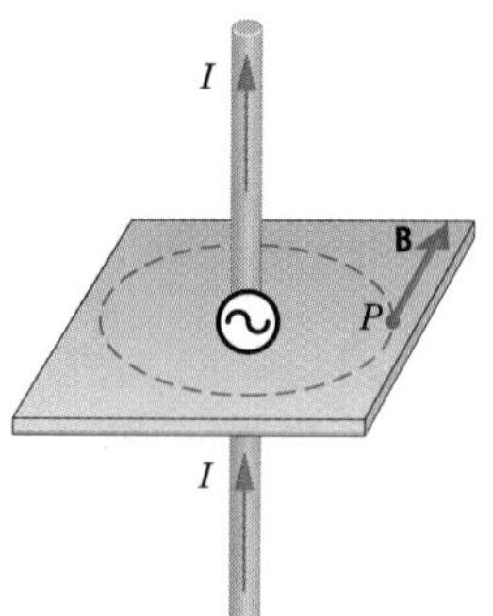

Figure 24.2 The oscillating current I in the antenna wires creates a magnetic field **B** at point P. The direction of **B** is tangent to a circle centered on the wires. When the current is upward, the magnetic field at point P is directed into the page. At a later time, when the current is downward, the field is directed out of the page.

The electric and magnetic fields illustrated in the previous drawings decrease to zero rapidly with increasing distance from the antenna. Therefore, they exist mainly near the antenna and together are called the ***near field.*** Electric and magnetic fields do form an electromagnetic wave at large distances from the antenna, however. These fields arise from an effect that is different from that which produces the near field and are referred to as the ***radiation field.*** Faraday's law of electromagnetic induction provides part of the basis for the radiation field. As Section 22.4 discusses, this law describes the emf or potential difference produced by a changing magnetic field. And, as Section 19.4 explains, a potential difference can be related to an electric field. Thus, a changing magnetic field produces an electric field. Maxwell predicted that the reverse effect also occurs, namely, that a changing electric field produces a magnetic field. The radiation field arises, then, because the changing magnetic field creates an electric field that fluctuates in time and the changing electric field creates the magnetic field.

Figure 24.3 shows the electromagnetic wave of the radiation field far from the antenna. The picture shows only the part of the wave traveling along the $+x$ axis. The part traveling in the other directions has been omitted for clarity. Note that for the radiation field, the electric and magnetic parts of the wave reach a maximum together. In other words, the fluctuating electric and magnetic fields are in phase, in contrast to the 90° phase relation for the near field.

It should be clear from Figure 24.3 that ***an electromagnetic wave is a transverse wave,*** because the electric and magnetic fields are both perpendicular to the direction in which the wave travels. Moreover, this kind of transverse wave, unlike a wave on a string, does not require a medium in which to propagate. ***Electromagnetic waves can travel through a vacuum or a material substance,*** since electric and magnetic fields can exist in either one.

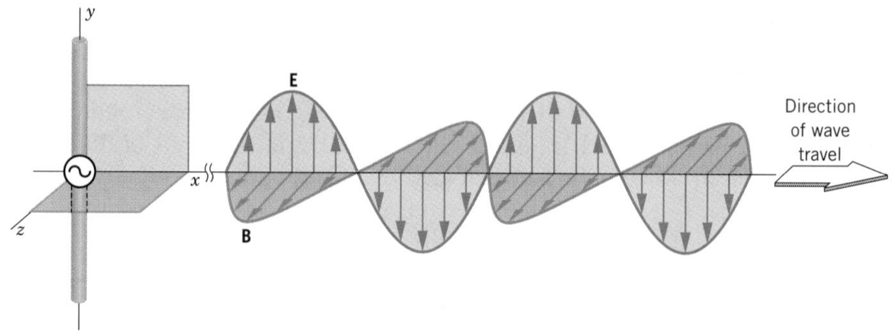

Figure 24.3 This picture shows the wave of the radiation field far away from the antenna. Observe that **E** and **B** are perpendicular to each other, and both are perpendicular to the direction of travel. The wave also travels outward in other directions, but for clarity only the part traveling along the positive x axis is shown.

Electromagnetic waves can be produced in situations that do not involve a wire antenna. In general, any electric charge that is accelerating emits an electromagnetic wave, whether the charge is inside a wire or not. In an alternating current, an electron oscillates in simple harmonic motion along the length of the wire and is one example of an accelerating charge.

All electromagnetic waves move through a vacuum at the same speed, and the symbol c is used to denote its value. This speed is called the ***speed of light in a vacuum*** and is $c = 3.00 \times 10^8$ m/s. In air, electromagnetic waves travel at nearly the same speed as they do in a vacuum, but, in general, they move through a substance such as glass at a speed that is less than c.

The frequency of an electromagnetic wave is determined by the oscillation frequency of the electric charges at the source of the wave. In Figures 24.1–24.3 the wave frequency would equal the frequency of the ac generator. Suppose, for example, that the antenna is broadcasting electromagnetic waves known as radio waves. The frequencies of AM radio waves lie between 545 and 1605 kHz, these numbers corresponding to the limits of the AM broadcast band on the dial. The frequencies of FM radio waves lie between 88 and 108 MHz on the dial. Television channels 2–6, on the other hand, utilize electromagnetic waves with frequencies between 54 and 88 MHz, while channels 7–13 use frequencies between 174 and 216 MHz.

Both straight and loop antennas are being used on this cruise ship.

THE PHYSICS OF . . .

radio and television reception.

Radio and television reception involves a process that is the reverse of that outlined earlier for the creation of electromagnetic waves. When broadcasted waves reach a receiving antenna, they interact with the electric charges in the antenna wires. Either the electric field or the magnetic field of the waves can be used. To take full advantage of the electric field, the wires of the receiving antenna must be parallel to the electric field, as Figure 24.4 indicates. The electric field acts on the electrons in the wire, forcing them to oscillate back and forth along the length of the wire. Consequently, an ac current exists in the antenna and the circuit connected to it. The variable-capacitor C (—|⁄|—) and the inductor L in the circuit provide one way to select the frequency of the desired electromagnetic wave. By adjusting the value of the capacitance, it is possible to adjust the corresponding resonance frequency f_0 of the circuit [$f_0 = 1/(2\pi\sqrt{LC})$, Equation 23.10] to match the frequency of the wave. Under the condition of resonance there will be a maximum oscillating current in the inductor. Because of mutual inductance, this current creates a maximum voltage in the second coil in the drawing, and this voltage can then be amplified and processed by the remaining radio or television circuitry.

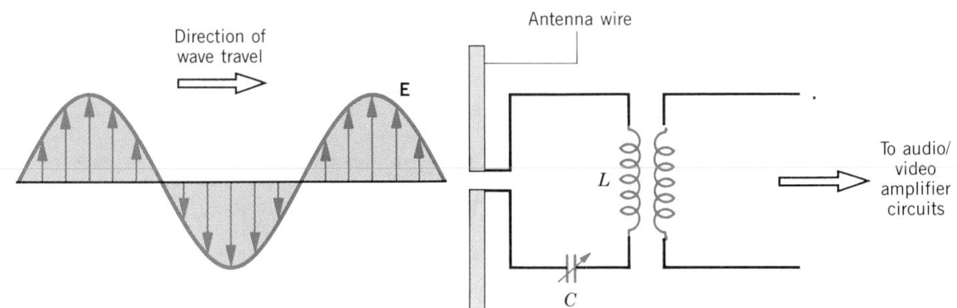

Figure 24.4 A radio wave can be detected with a receiving antenna wire that is parallel to the electric field of the wave. The magnetic field of the radio wave has been omitted for simplicity.

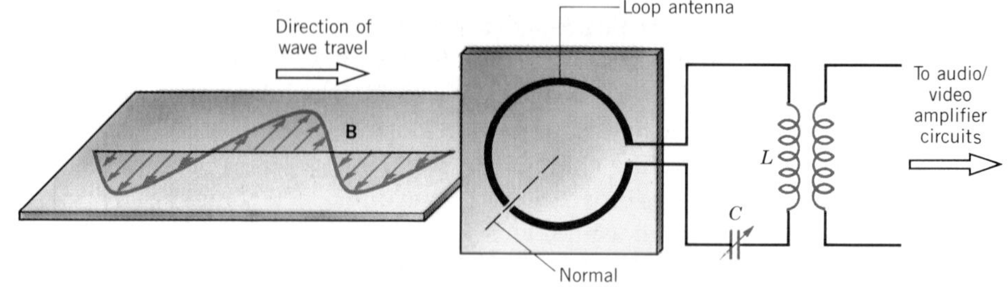

Figure 24.5 With a receiving antenna in the form of a loop, the magnetic field of a broadcasted radio wave can be detected. The normal to the plane of the loop should be parallel to the magnetic field for best reception. For clarity, the electric field of the radio wave has been omitted.

To detect the magnetic field of a broadcasted radio wave, a receiving antenna in the form of a loop can be used, as Figure 24.5 shows. For best reception, the normal to the plane of the wire loop is oriented parallel to the magnetic field. Then, as the wave sweeps by, the magnetic field penetrates the loop, and the changing magnetic flux induces a voltage and a current in the loop, in accord with Faraday's law. Once again, the resonance frequency of a capacitor/inductor combination can be adjusted to match the frequency of the desired electromagnetic wave.

Radio waves are only one part of the broad spectrum of electromagnetic waves that has been discovered. The next section discusses the entire spectrum.

24.2 THE ELECTROMAGNETIC SPECTRUM

An electromagnetic wave, like any periodic wave, has a frequency f and a wavelength λ that are related to the speed v of the wave by $v = f\lambda$ (Equation 16.1). For electromagnetic waves traveling through a vacuum or, to a good approximation, through air, the speed is $v = c$, so $c = f\lambda$.

As Figure 24.6 shows, electromagnetic waves exist with an enormous range of frequencies, from values less than 10^4 Hz to greater than 10^{22} Hz. The series of electromagnetic waves depicted in the drawing, arranged in order of their frequencies, is called the ***electromagnetic spectrum.*** Since all these waves travel

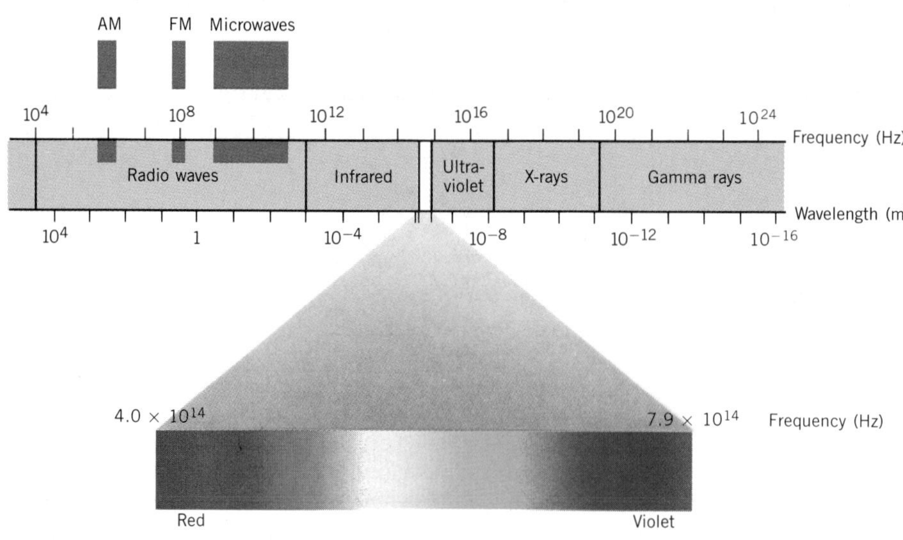

Figure 24.6 The electromagnetic spectrum.

through a vacuum at the same speed of $c = 3.00 \times 10^8$ m/s, Equation 16.1 can be used to find the correspondingly wide range of wavelengths that the picture also displays. Historically, regions of the electromagnetic spectrum have been given names such as radio waves and infrared waves. Although the boundary between two regions is drawn as a sharp line in the drawing, the boundary is not so well defined in practice, and the regions often overlap.

Beginning on the left in Figure 24.6, we find radio waves. Lower-frequency radio waves are generally produced by electric oscillator circuits, while higher-frequency radio waves (called microwaves) are usually generated using electron tubes called klystrons. Infrared radiation, sometimes loosely called heat waves, originates with the vibration and rotation of molecules within a material. Visible light is emitted by hot objects, such as the sun, a burning log, or the filament of an incandescent light bulb, when the temperature is high enough to excite the electrons within an atom. Ultraviolet frequencies can be produced from the discharge of an electric arc. X-rays are produced by the sudden deceleration of high-speed electrons. And, finally, gamma rays are radiation from nuclear decay.

Of all the frequency ranges in the electromagnetic spectrum, the most familiar is that of visible light, although it is the smallest range indicated in Figure 24.6. Only waves with frequencies between about 4.0×10^{14} Hz and 7.9×10^{14} Hz are perceived by the human eye as visible light. Usually visible light is discussed in terms of wavelengths (in vacuum) rather than frequencies. As Example 1 indicates, the wavelengths of visible light are extremely small and, therefore, are normally expressed in *nanometers* (nm); 1 nm $= 10^{-9}$ m. An obsolete (non-SI) unit occasionally used for wavelengths is the *angstrom* (Å); 1 Å $= 10^{-10}$ m.

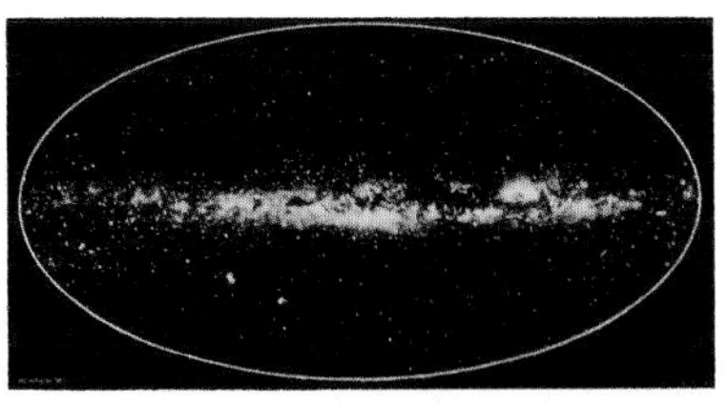

(*a*)

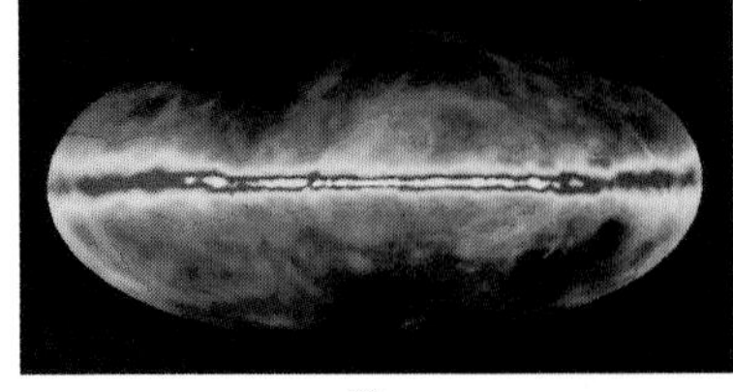

(*b*)

Two views of our galaxy (the Milky Way) as observed in two different regions of the electromagnetic spectrum: (*a*) the visible region, and (*b*) the infrared region. (The infrared region is invisible to the eye, so the colors in this picture have been computer generated.)

Example 1 The Wavelengths of Visible Light

Find the range in wavelengths (in vacuum) for visible light in the frequency range between 4.0×10^{14} Hz (red light) and 7.9×10^{14} Hz (violet light). Express the answers in nanometers.

REASONING According to Equation 16.1, the wavelength (in vacuum) λ of a light wave is equal to the speed of light c in a vacuum divided by the frequency f of the wave, $\lambda = c/f$.

SOLUTION The wavelength corresponding to a frequency of 4.0×10^{14} Hz is

$$\lambda = \frac{c}{f} = \frac{3.00 \times 10^8 \text{ m/s}}{4.0 \times 10^{14} \text{ Hz}} = 7.5 \times 10^{-7} \text{ m}$$

Since 1 nm $= 10^{-9}$ m, it follows that $\boxed{\lambda = 750 \text{ nm}}$.

The calculation for a frequency of 7.9×10^{14} Hz is similar:

$$\lambda = \frac{c}{f} = \frac{3.00 \times 10^8 \text{ m/s}}{7.9 \times 10^{14} \text{ Hz}} = 3.8 \times 10^{-7} \text{ m} \quad \text{or} \quad \boxed{\lambda = 380 \text{ nm}}$$

The eye/brain recognizes light of different wavelengths as different colors. A wavelength of 750 nm (in vacuum) is approximately the longest wavelength of red light, whereas 380 nm (in vacuum) is approximately the shortest wavelength of violet light. Between these limits are found the other familiar colors, as Figure 24.6 indicates.

The association between color and wavelength in the visible part of the electromagnetic spectrum is well known. But, just as important, the wavelength plays a central role in governing the behavior and use of electromagnetic waves in all regions of the spectrum. For instance, Conceptual Example 2 considers the influence of the wavelength on diffraction.

Conceptual Example 2 The Diffraction of AM and FM Radio Waves

As we have seen in Chapter 17, diffraction is the ability of a wave to bend around an obstacle or the edges of an opening. Based on the discussion in that chapter, would you expect AM or FM radio waves to bend more readily around an obstacle such as a building?

REASONING Section 17.3 points out that, other things being equal, sound waves exhibit diffraction to a greater extent when the wavelength is longer than when it is shorter. Based on this information, we expect that longer-wavelength electromagnetic waves will bend more readily around obstacles than will shorter-wavelength waves. Figure 24.6 shows that AM radio waves have considerably longer wavelengths than FM waves. Therefore, AM waves have a greater ability to bend around buildings than do FM waves. The inability of FM waves to diffract around obstacles is a major reason why FM stations broadcast their signals essentially in a "line-of-sight" fashion.

The picture of light as a wave is supported by experiments that will be discussed in Chapter 27. However, there are also experiments indicating that light can behave as if it were composed of discrete particles, rather than waves. These experiments will be discussed in Chapter 29. Wave theories and particle theories of light have been around for hundreds of years, and it is now widely accepted that light, as well as other electromagnetic radiation, exhibits a dual nature. Either wave-like or particle-like behavior can be observed, depending on the kind of experiment being performed.

24.3 THE SPEED OF LIGHT

At a speed of 3.00×10^8 m/s, light travels from the earth to the moon in a little over a second, so the time required for light to travel between two places on earth is very short. In fact, early attempts at measuring the speed of light ran into just this problem. Galileo (1564–1642), for example, attempted to measure the speed of light at night by stationing a helper at a distant point. The helper was to uncover a lamp as soon as he saw the light from a lamp that Galileo had uncovered. Galileo hoped to measure the time it took for light to travel from his lamp to his assistant and for the return trip. Of course, the time was so short that the light appeared to move instantaneously from one point to the other, leaving Galileo only the conclusion that light travels very rapidly indeed.

In the first measurement of the speed of light that did not depend on astronomical observations, the French scientist Armand Fizeau (1819–1896) used a rotating notched wheel, as Figure 24.7 illustrates. The light from the source passes through one notch on its way to a mirror located some distance away (8633 m in Fizeau's experiment). After reflection, the light travels back and passes through another notch in the wheel only if the rotational speed of the wheel is appropriate.

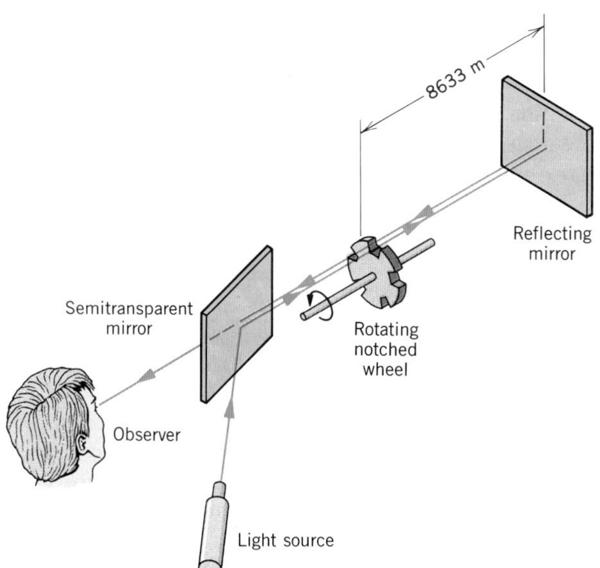

Figure 24.7 A version of the experiment that Fizeau carried out to measure the speed of light.

The rotational speed must be such that the time it takes for the light to travel the round-trip must match the time it takes for the second notch (or third or fourth, etc.) to rotate into the position previously occupied by the first notch. From a knowledge of the rotational speed of the wheel that permits the light to pass back through the second notch, Fizeau was able to determine the speed of light as 3.13×10^8 m/s.

More accurate measurements of the speed of light were performed later using a rotating mirror instead of a notched wheel. Figure 24.8 shows a simplified version of this setup. It was used first by the French scientist Jean Foucault (1819–1868) and later in a more refined version by the American physicist Albert Michelson (1852–1931), who obtained the value of $c = (2.997\,96 \pm 0.000\,04) \times 10^8$ m/s in 1926. The next example illustrates Michelson's method.

Example 3 The Angular Speed of Michelson's Rotating Mirror

If the angular speed of the rotating eight-sided mirror in Figure 24.8 is adjusted correctly, light reflected from one side travels to the fixed mirror, reflects, and can be detected after reflecting from another side that has rotated into place at just the right time. For one of his experiments, Michelson placed mirrors on Mt. San Antonio and Mt. Wilson in California, a distance of 35 km apart. Knowing that $c = 3.00 \times 10^8$ m/s, obtain the minimum angular speed for the eight-sided mirror in Michelson's experiment.

REASONING The minimum angular speed is that at which one side of the mirror rotates one-eighth of a revolution during the time it takes for the light to make the round-trip between Mt. San Antonio and Mt. Wilson. The angular speed of the mirror can be determined by dividing its angular displacement ($\frac{1}{8}$ revolution) by the round-trip time. The round-trip time is the round-trip distance divided by the speed of light.

SOLUTION The round-trip travel time is $t = 2(35 \times 10^3 \text{ m})/(3.00 \times 10^8 \text{ m/s}) = 2.3 \times 10^{-4}$ s. The minimum angular speed ω is

$$\omega = \frac{\frac{1}{8} \text{ revolution}}{2.3 \times 10^{-4} \text{ s}} = \boxed{540 \text{ rev/s}}$$

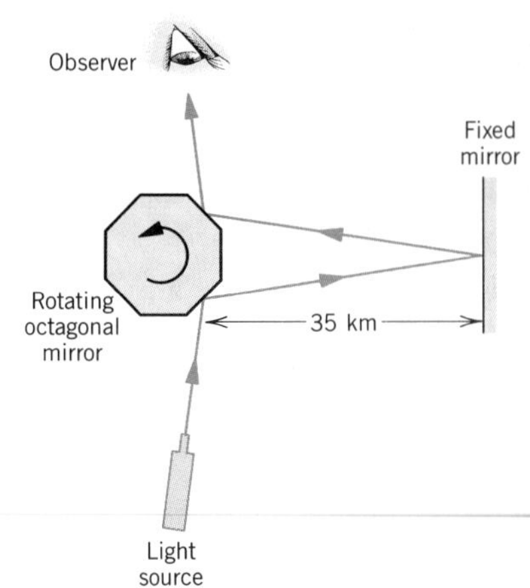

Figure 24.8 Between 1878 and 1931 Michelson used a rotating eight-sided mirror to measure the speed of light. This is a simplified version of the setup.

(a)

(b)

Figure 24.9 A view of the sky (*a*) before and (*b*) a few hours after the 1987 supernova.

Today, the speed of light has been determined with such high accuracy that it is used to define the meter. As discussed in Section 1.2, the speed of light is now *defined* to be $c = 299\ 792\ 458$ m/s (although a value of 3.00×10^8 m/s is adequate for most calculations). The second is defined in terms of a cesium clock, and the meter is then defined as the distance light travels in a vacuum during a time of 1/(299 792 458) s. Although the speed of light in a vacuum is large, it is finite, so it takes a finite amount of time for light to travel from one place to another. This time is especially long for light traveling between astronomical objects, as Conceptual Example 4 discusses.

Conceptual Example 4 Looking Back in Time

A supernova is a violent explosion that occurs at the death of certain stars. For a few days after the explosion, the intensity of the emitted light can become a billion times greater than that of our own sun. But after several years, the intensity usually returns to zero. Supernovae are relatively rare events in the universe, for only six have been observed in our galaxy within the past 400 years. One of them was recorded in 1987. It occurred in a neighboring galaxy, approximately 1.66×10^{21} m away. Figure 24.9 shows a photograph of the sky (*a*) before the explosion, and (*b*) a few hours after the explosion. Even though this supernova appeared recently, astronomers say that viewing it is like "looking back in time." Why do they say that?

REASONING The light from the supernova traveled to earth at a speed of $c = 3.00 \times 10^8$ m/s. Even at such a speed, the travel time is large, because the distance $d = 1.66 \times 10^{21}$ m is enormous. The time t is $t = d/c = (1.66 \times 10^{21}\ \text{m})/(3.00 \times 10^8\ \text{m/s}) = 5.53 \times 10^{12}$ s. This corresponds to 175 000 years. So when astronomers saw the explosion in 1987, they were actually seeing the light that left the supernova 175 000 years earlier; in other words, they were "looking back in time." In fact, whenever we view any celestial object, such as a star, we are seeing it as it was a long time ago. The farther the object is from earth, the longer it takes for the light to reach us, and the further back in time we are looking.

Related Homework Material: *Problem 19*

In 1865 Maxwell determined theoretically that electromagnetic waves propagate through a vacuum at a speed given by

$$c = \frac{1}{\sqrt{\epsilon_0 \mu_0}} \tag{24.1}$$

where $\epsilon_0 = 8.85 \times 10^{-12}\ \text{C}^2/(\text{N} \cdot \text{m}^2)$ is the (electric) permittivity of free space and $\mu_0 = 4\pi \times 10^{-7}\ \text{T} \cdot \text{m/A}$ is the (magnetic) permeability of free space. Originally ϵ_0 was introduced in Section 18.5 as an alternative way of writing the proportionality constant k in Coulomb's law $[k = 1/(4\pi\epsilon_0)]$ and, hence, plays a basic role in determining the strengths of the electric fields created by point charges. The role of μ_0 is similar for magnetic fields; it was introduced in Section 21.8 as part of a proportionality constant in the expression for the magnetic field created by the current in a long straight wire. Substituting the values for ϵ_0 and μ_0 into Equation 24.1 shows that

$$c = \frac{1}{\sqrt{[8.85 \times 10^{-12}\ \text{C}^2/(\text{N} \cdot \text{m}^2)][4\pi \times 10^{-7}\ \text{T} \cdot \text{m/A}]}} = 3.00 \times 10^8\ \text{m/s}$$

The experimental and theoretical values for c agree. Maxwell's success in predicting c provided a basis for inferring that light behaves as a wave consisting of oscillating electric and magnetic fields.

24.4 THE ENERGY CARRIED BY ELECTROMAGNETIC WAVES

Electromagnetic waves, like water waves or sound waves, carry energy. The energy is carried by the electric and magnetic fields that comprise the wave. It is because of this energy, for example, that microwaves can cook a dinner. In a microwave oven, microwaves pass directly through the food and deliver their energy to it in the process, as illustrated in Figure 24.10. The electric field of the microwaves is largely responsible for delivering the energy, and water molecules, one of the most abundant ingredients in food, readily absorb it. The reason water absorbs microwaves so readily is that water molecules have a permanent dipole moment; that is, one end of a water molecule has a slight positive charge and the other end a negative charge of equal magnitude. The electric field of the microwaves exerts forces on the positive and negative ends of the molecule, causing it to spin. Because the field is oscillating rapidly—about 2.4×10^9 times a second—the water molecules are kept spinning at a high rate. In the process, the energy from the microwaves is converted into heat, which is passed on to neighboring food molecules. Because microwaves penetrate all parts of the food, the heating is uniform as well as fast.

THE PHYSICS OF . . .

a microwave oven.

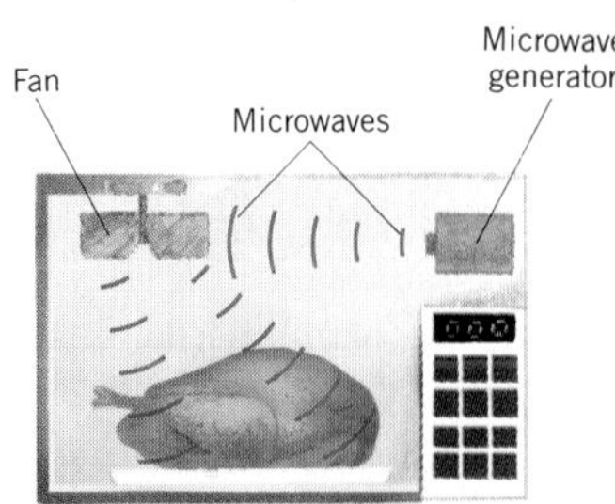

Figure 24.10 A microwave oven. The rotating fan blades reflect the microwaves to all parts of the oven.

A measure of the energy stored in the electric field **E** of an electromagnetic wave, such as a microwave, is provided by the electric energy density. As we saw in Section 19.5, this density is the electric energy per unit volume of space in which the electric field exists:

$$\text{Electric energy density} = \frac{\text{Electric energy}}{\text{Volume}} = \tfrac{1}{2}\epsilon_0 E^2 \qquad (19.12)$$

where the dielectric constant κ has been set equal to unity, since we are dealing with an electric field in a vacuum (or air). From Section 22.8, the analogous expression for the magnetic energy density is

$$\text{Magnetic energy density} = \frac{\text{Magnetic energy}}{\text{Volume}} = \frac{1}{2\mu_0} B^2 \qquad (22.11)$$

The ***total energy density*** u of an electromagnetic wave in a vacuum is the sum of these two energy densities:

$$u = \frac{\text{Total energy}}{\text{Volume}} = \frac{1}{2}\epsilon_0 E^2 + \frac{1}{2\mu_0} B^2 \qquad (24.2)$$

In an electromagnetic wave propagating through a vacuum or air, the electric field and the magnetic field carry equal amounts of energy per unit volume of space. The fact that the two energy densities are equal means that the electric field is related to the magnetic field through the relation $E = cB$. To demonstrate this result, we set the electric energy density equal to the magnetic energy density:

$$\frac{1}{2}\epsilon_0 E^2 = \frac{1}{2\mu_0} B^2$$

Taking the square root of both sides of this equation and solving for E, we get $E = (1/\sqrt{\mu_0\epsilon_0})B$. But $c = 1/\sqrt{\mu_0\epsilon_0}$ according to Equation 24.1, so we have the result that

$$E = cB \tag{24.3}$$

Since $\frac{1}{2}\epsilon_0 E^2 = \frac{1}{2}(B^2/\mu_0)$, it is possible to rewrite Equation 24.2 for the total energy density in two additional, but equivalent, forms:

$$u = \frac{1}{2}\epsilon_0 E^2 + \frac{1}{2\mu_0} B^2 \tag{24.2a}$$

$$u = \epsilon_0 E^2 \tag{24.2b}$$

$$u = \frac{1}{\mu_0} B^2 \tag{24.2c}$$

In an electromagnetic wave, the electric and magnetic fields fluctuate sinusoidally in time, so Equations 24.2 give the energy density of the wave at any instant in time. If an average value $\bar{u}$ for the total energy density is desired, average values are needed for E^2 and B^2. In Section 20.5 we faced a similar situation for alternating currents and voltages and introduced rms quantities. Using an analogous procedure here, it follows that the rms values for the electric and magnetic fields, E_{rms} and B_{rms}, are related to the maximum values of these fields, E_0 and B_0, by

$$E_{rms} = \frac{1}{\sqrt{2}} E_0 \quad \text{and} \quad B_{rms} = \frac{1}{\sqrt{2}} B_0$$

Equations 24.2a–c can now be interpreted as giving the average energy density $\bar{u}$, provided the symbols E and B are interpreted to mean the rms values given above. The average energy density of the sunlight reaching the earth is determined in the next example.

Example 5 The Average Energy Density of Sunlight

Sunlight enters the top of the earth's atmosphere with an electric field whose rms value is $E_{rms} = 720$ N/C. Find (a) the rms value of the sunlight's magnetic field and (b) the average total energy density of this electromagnetic wave.

REASONING Since the magnitudes of the magnetic and electric fields are related according to Equation 24.3, the rms value of the magnetic field is $B_{rms} = E_{rms}/c$. The average energy density $\bar{u}$ in the sunlight can be obtained from Equation 24.2b, provided the rms value is used for the electric field.

SOLUTION

(a) The rms magnetic field is

$$B_{rms} = \frac{E_{rms}}{c} = \frac{720\ \text{N/C}}{3.0 \times 10^8\ \text{m/s}} = \boxed{2.4 \times 10^{-6}\ \text{T}} \tag{24.3}$$

(b) The average energy density is

$$\bar{u} = \epsilon_0 E_{rms}^2 \tag{24.2b}$$

$$\bar{u} = (8.85 \times 10^{-12}\ \text{C}^2/\text{N}\cdot\text{m}^2)(720\ \text{N/C})^2 = \boxed{4.6 \times 10^{-6}\ \text{J/m}^3}$$

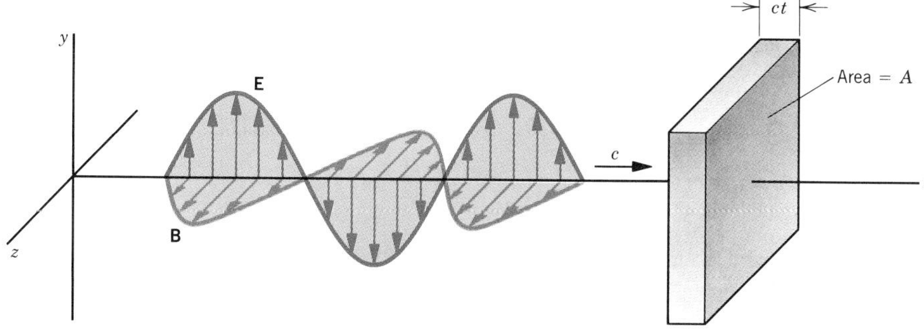

Figure 24.11 In a time t, an electromagnetic wave moves a distance ct along the x axis and passes through a surface of area A.

As an electromagnetic wave moves through space, it carries its energy from one region to another. This energy transport is characterized by the ***intensity*** of the wave. Recall that the intensity of a sound wave is the sound power that passes perpendicularly through a surface divided by the area of the surface (see Equation 16.8). In other words, intensity is power per unit area or energy per unit time per unit area. To help us apply this definition of intensity in the present situation, Figure 24.11 shows an electromagnetic wave traveling in a vacuum along the x axis. In a time t the wave travels the distance ct, passing through the surface of area A. Consequently, the volume of space through which the wave passes is ctA. The total (electric and magnetic) energy in this volume is

$$\text{Total energy} = (\text{Total energy density}) \times \text{Volume} = u(ctA)$$

Using this expression, it can be seen that the intensity S and the total energy density u are related as follows:

$$S = \frac{\text{Total energy}}{\text{Time} \cdot \text{Area}} = \frac{uctA}{tA} = cu \tag{24.4}$$

Substituting Equations 24.2a–c, one at a time, into Equation 24.4 shows that the intensity of an electromagnetic wave depends on the electric and magnetic fields according to the following equivalent relations:

$$S = cu = \frac{1}{2}c\epsilon_0 E^2 + \frac{c}{2\mu_0}B^2 \tag{24.5a}$$

$$S = c\epsilon_0 E^2 \tag{24.5b}$$

$$S = \frac{c}{\mu_0}B^2 \tag{24.5c}$$

If the rms values for the electric and magnetic fields are used in Equations 24.5a–c, the intensity becomes an average intensity $\overline{S}$, as Example 6 illustrates.

Example 6 A Neodymium-Glass Laser

A neodymium-glass laser emits short pulses of high-intensity electromagnetic waves. The electric field has an rms value of $E_{rms} = 2.0 \times 10^9$ N/C. Find the average power of each pulse that passes through a 1.6×10^{-5}-m² surface that is perpendicular to the laser beam.

PROBLEM SOLVING INSIGHT

The concepts of power and intensity are similar, but they are not the same. Intensity is the power that passes perpendicularly through a surface divided by the area of the surface.

REASONING Since the intensity of a wave is the power per unit area that passes perpendicularly through a surface, the average power $\overline{P}$ of the wave is the product of the average intensity $\overline{S}$ and the area A.

SOLUTION The average power is $\overline{P} = \overline{S}A$. But from Equation 24.5b, $\overline{S} = c\epsilon_0 E_{\text{rms}}^2$ so that

$$\overline{P} = c\epsilon_0 E_{\text{rms}}^2 A$$

$$\overline{P} = (3.0 \times 10^8 \text{ m/s})(8.85 \times 10^{-12} \text{ C}^2/\text{N}\cdot\text{m}^2)(2.0 \times 10^9 \text{ N/C})^2(1.6 \times 10^{-5} \text{ m}^2)$$

$$= \boxed{1.7 \times 10^{11} \text{ W}}$$

24.5 *POLARIZATION*

POLARIZED ELECTROMAGNETIC WAVES

One of the essential features of electromagnetic waves is that they are transverse waves, and because of this feature they can be polarized. Figure 24.12 illustrates the idea of polarization by showing a transverse wave as it travels along a rope toward a slit. The wave is said to be ***linearly polarized,*** which means that its vibrations always occur along one direction. This direction is called the direction of polarization. In part *a* of the picture, the direction of polarization is vertical, and the slit is parallel to it. Consequently, the wave passes through easily. However, when the slit is turned perpendicular to the direction of polarization, as in part *b*, the wave cannot pass, because the slit prevents the rope from oscillating. Note that for longitudinal waves, such as sound waves, the notion of polarization has no meaning. In a longitudinal wave the direction of vibration is along the direction of travel, and, thus, the orientation of the slit would have no effect on the wave.

In an electromagnetic wave such as that in Figure 24.3, the electric field oscillates along the y axis. Similarly, the magnetic field oscillates along the z axis. Therefore, the wave is linearly polarized, with the direction of polarization taken arbitrarily to be that along which the electric field oscillates. If the wave is a radio wave generated by a straight-wire antenna, the direction of polarization is deter-

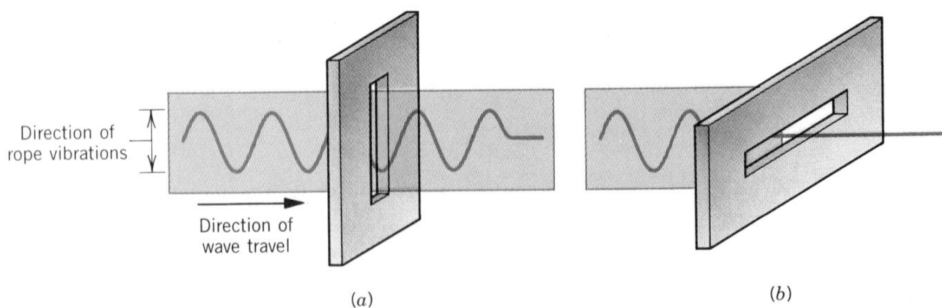

Figure 24.12 A transverse wave is linearly polarized when its vibrations always occur along one direction. (*a*) A linearly polarized wave on a rope can pass through a slit that is parallel to the direction of the rope vibrations, but (*b*) cannot pass through a slit that is perpendicular to the vibrations.

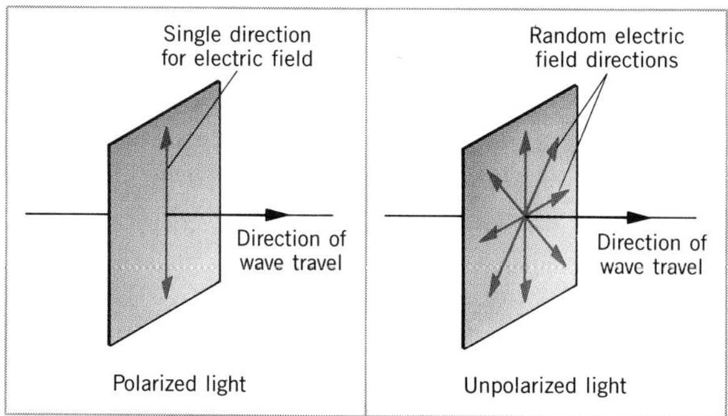

Figure 24.13 Polarized light consists of an electromagnetic wave in which the electric field fluctuates along a single direction. Unpolarized light consists of short bursts of electromagnetic waves emitted by many different atoms. The electric field directions of these bursts are perpendicular to the direction of wave travel but are distributed randomly about it.

mined by the orientation of the antenna. In comparison, the electromagnetic waves given off as light by an incandescent light bulb are completely unpolarized. In this case the light waves are emitted by a large number of atoms in the hot filament of the bulb. When an electron in an atom oscillates, the atom behaves as a miniature antenna that broadcasts light for brief periods of time, about 10^{-8} seconds. However, the directions of these atomic antennas change randomly as a result of collisions. Unpolarized light, then, consists of many individual waves, emitted in short bursts by many "atomic antennas," each with its own direction of polarization. Figure 24.13 compares polarized and unpolarized light. In the unpolarized case, the arrows shown around the direction of wave travel symbolize the random directions of polarization of the individual waves that comprise the light.

Linearly polarized light can be produced from unpolarized light with the aid of certain materials. One commercially available material goes under the name of Polaroid. Such materials allow only the component of the electric field along one direction to pass through, while absorbing the field component perpendicular to this direction. As Figure 24.14 indicates, the direction of polarization that a polarizing material allows through is called the ***transmission axis.*** No matter how the transmission axis is oriented, the intensity of the transmitted polarized light is one-half that of the incident unpolarized light. The reason for this is that the unpolarized light contains all polarization directions to an equal extent. Moreover, the electric field for each direction can be resolved into components perpendicular and parallel to the transmission axis, with the result that the average components perpendicular and parallel to the transmission axis are equal. As a result, the polarizing material absorbs as much of the electric (and magnetic) field strength as it transmits, and the intensity of the transmitted polarized light is one-half the intensity of the incident unpolarized light.

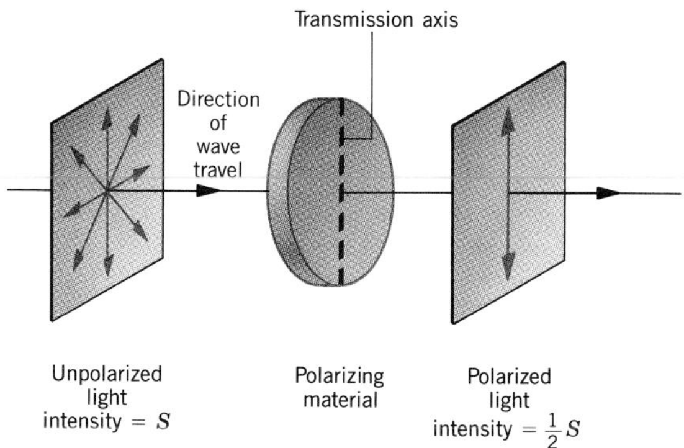

Figure 24.14 With the aid of a piece of polarizing material, polarized light may be produced from unpolarized light. The transmission axis of the material is the direction of polarization of the light that passes through the material.

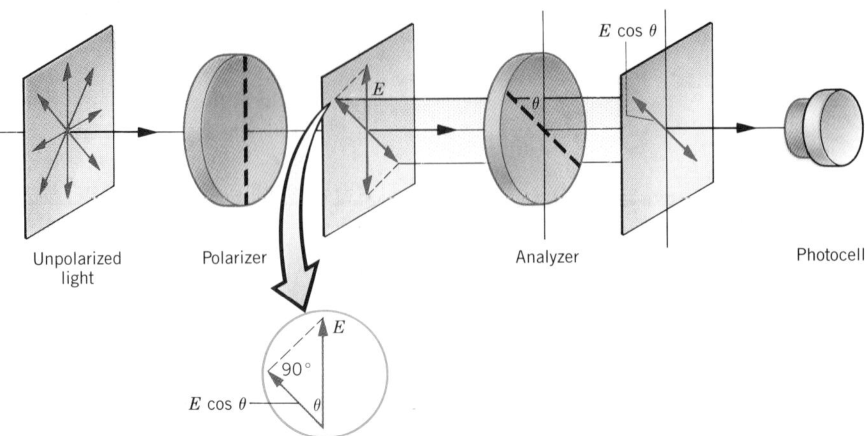

Figure 24.15 Two sheets of polarizing material, called the polarizer and the analyzer, may be used to adjust the polarization direction and intensity of the light reaching the photocell. This can be done by changing the angle θ between the transmission axes of the polarizer and analyzer.

MALUS' LAW

Once polarized light has been produced with a piece of polarizing material, it is possible to use a second piece to change the polarization direction and to adjust the intensity of the light. Figure 24.15 shows how. In this picture the first piece of polarizing material is called the ***polarizer,*** while the second piece is referred to as the ***analyzer.*** The transmission axis of the analyzer is oriented at an angle θ relative to the transmission axis of the polarizer. If the electric field strength of the polarized light incident on the analyzer is E, the field strength passing through is the component parallel to the transmission axis, or $E \cos \theta$. According to Equation 24.5b, the intensity is proportional to the square of the electric field strength. Consequently, the average intensity of polarized light passing through the analyzer is proportional to $\cos^2 \theta$. Thus, both the polarization direction and the intensity of the light can be adjusted by rotating the transmission axis of the analyzer relative to that of the polarizer. The average intensity $\bar{S}$ of the light leaving the analyzer, then, is

[Malus' law]
$$\bar{S} = \bar{S}_0 \cos^2 \theta \tag{24.6}$$

where $\bar{S}_0$ is the average intensity of the light entering the analyzer. Equation 24.6 is sometimes called ***Malus' law,*** for it was discovered by the French engineer Etienne-Louis Malus (1775–1812). Example 7 illustrates the use of Malus' law.

Example 7 Using Polarizers and Analyzers

What value of θ should be used in Figure 24.15, so the average intensity of the polarized light reaching the photocell is one-tenth the average intensity of the unpolarized light?

REASONING Both the polarizer and the analyzer reduce the intensity of the light. The polarizer reduces the intensity by a factor of one-half, as discussed earlier. Therefore, if the average intensity of the unpolarized light is $\bar{I}$, the average intensity of the polarized light leaving the polarizer and striking the analyzer is $\bar{S}_0 = \bar{I}/2$. The angle θ must now be selected so the average intensity of the light leaving the analyzer is $\bar{S} = \bar{I}/10$. The solution follows directly from Malus' law.

SOLUTION Using $\bar{S}_0 = \bar{I}/2$ and $\bar{S} = \bar{I}/10$ in Malus' law, we find that $\bar{I}/10 = (\bar{I}/2) \cos^2 \theta$. Solving this relation for $\cos \theta$ yields

$$\cos \theta = \sqrt{\tfrac{1}{5}} = 0.447 \quad \text{and} \quad \theta = \cos^{-1}(0.447) = \boxed{63.4^\circ}$$

PROBLEM SOLVING INSIGHT

Remember that when unpolarized light strikes a polarizer, only one-half of the incident light is transmitted, the other half being absorbed by the polarizer.

Figure 24.16 When Polaroid sunglasses are uncrossed (left photograph), the transmitted light is dimmed due to the extra thickness of tinted plastic. However, when they are crossed (right photograph), the transmitted light is reduced to zero because of the effects of polarization.

When $\theta = 90°$ in Figure 24.15, the polarizer and analyzer are said to be ***crossed,*** and no light is transmitted by the polarizer/analyzer combination. As an illustration of this effect, Figure 24.16 shows two pairs of Polaroid sunglasses in uncrossed and crossed configurations. Conceptual Example 8 illustrates an interesting result that occurs when a piece of polarizing material is inserted between a crossed polarizer and analyzer.

Conceptual Example 8 How Can a Crossed Analyzer and Polarizer Transmit Light?

Figure 24.17*a* shows unpolarized light falling on a crossed ($\theta = 90°$) analyzer and polarizer. As explained earlier in the text, no light reaches the photocell when the polarizer and analyzer are crossed. Suppose now, that a third piece of polarizing material is inserted between the polarizer and analyzer, as in part *b* of the drawing. Does light now reach the photocell?

REASONING Surprisingly, the answer is yes, even though the polarizer and analyzer are crossed. To show why, we note that if any light is to pass through the analyzer, it must have an electric field component parallel to the transmission axis of the analyzer. Without the insert, there is no such component. But with the insert, there is, as part *b* of the drawing illustrates. The light passing through the insert has its electric field oriented parallel to the insert's transmission axis, so that there is a component of that field parallel to the analyzer's axis. Part *c* of the drawing shows that the electric field E of the light leaving the polarizer makes an angle θ with respect to the transmission axis of the insert. The component of the electric field that is parallel to the insert's transmission axis is $E \cos \theta$, and this component passes through the insert and on to the analyzer. Part *d* of the drawing illustrates that the electric field ($E \cos \theta$) incident on the analyzer has a component parallel to the transmission axis of the analyzer, namely $(E \cos \theta) \sin \theta$. This component passes through the analyzer, so that light now reaches the photocell.

Light will reach the photocell as long as the angle θ is between 0 and 90°. If the angle is 0 or 90°, however, no light reaches the photocell. Can you explain, without using any equations, why no light reaches the photocell under either of these conditions?

Related Homework Material: *Problems 39 and 40*

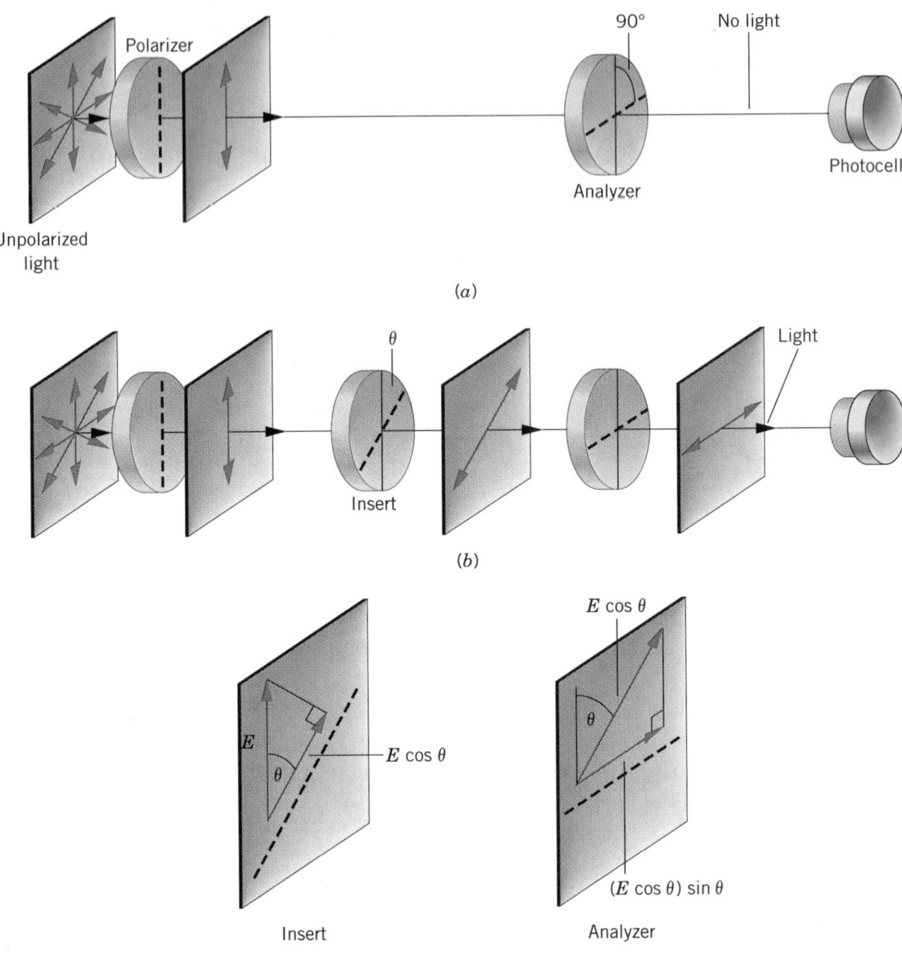

Figure 24.17 (*a*) A crossed polarizer and analyzer. (*b*) Light reaches the photocell when a piece of polarizing material is inserted between the polarizer and analyzer. (*c*) The component of the electric field parallel to the transmission axis of the insert is $E \cos \theta$. (*d*) Light incident on the analyzer has a component $(E \cos \theta) \sin \theta$ parallel to its transmission axis.

THE PHYSICS OF . . .

a liquid crystal display.

An application of a crossed polarizer/analyzer combination occurs in one kind of liquid crystal display (LCD). LCDs are widely used in pocket calculators and digital watches. The display usually consists of blackened numbers and letters set against a light gray background. As Figure 24.18 indicates, each number or letter is formed from a combination of liquid crystal segments that have been "turned on" and appear black. Let us now see what it means for a liquid crystal to be turned on and how polarized light is used.

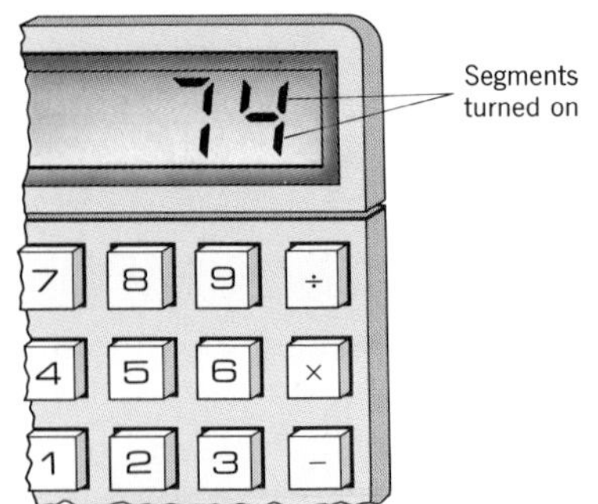

Figure 24.18 Liquid crystal displays use liquid crystal "segments" to form the numbers.

The liquid crystal part of an LCD segment consists of the liquid crystal material sandwiched between two transparent electrodes, as in Figure 24.19. When a voltage is applied between the electrodes, the liquid crystal is said to be "on." Part *a* of the picture shows that linearly polarized incident light passes through the "on" material without having its direction of polarization affected. When the voltage is removed, as in part *b*, the liquid crystal is said to be "off" and now rotates the direction of polarization by 90°.

A complete LCD segment includes a crossed polarizer/analyzer combination, as Figure 24.20 illustrates. The polarizer, analyzer, electrodes, and liquid crystal material are packaged as a single unit. The polarizer produces polarized light from incident unpolarized light. With the display segment turned on, as in part *a*, the polarized light emerges from the liquid crystal only to be absorbed by the ana-

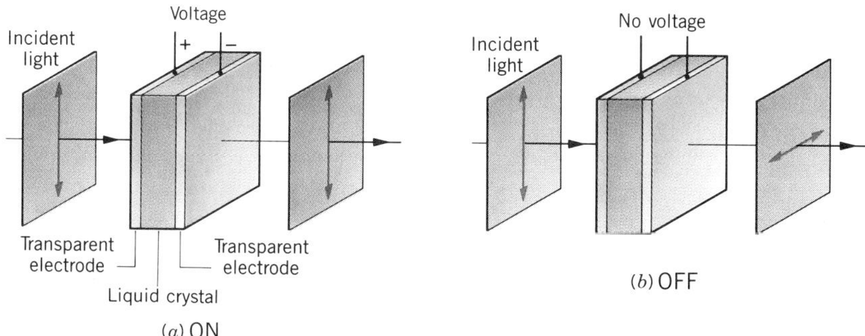

Figure 24.19 A liquid crystal in its (*a*) "on" state and (*b*) "off" state.

lyzer, since the light is polarized perpendicular to the transmission axis of the analyzer. Since no light emerges from the analyzer, an observer sees a black segment against a light gray background, as in Figure 24.18. On the other hand, the segment is turned off when the voltage is removed, in which case the liquid crystal rotates the direction of polarization by 90° to coincide with the axis of the analyzer, as in Figure 24.20*b*. The light now passes through the analyzer and enters the eye of the observer. However, the light coming from the segment has been designed to have the same color and shade (light gray) as the background of the display, so the segment becomes indistinguishable from the background.

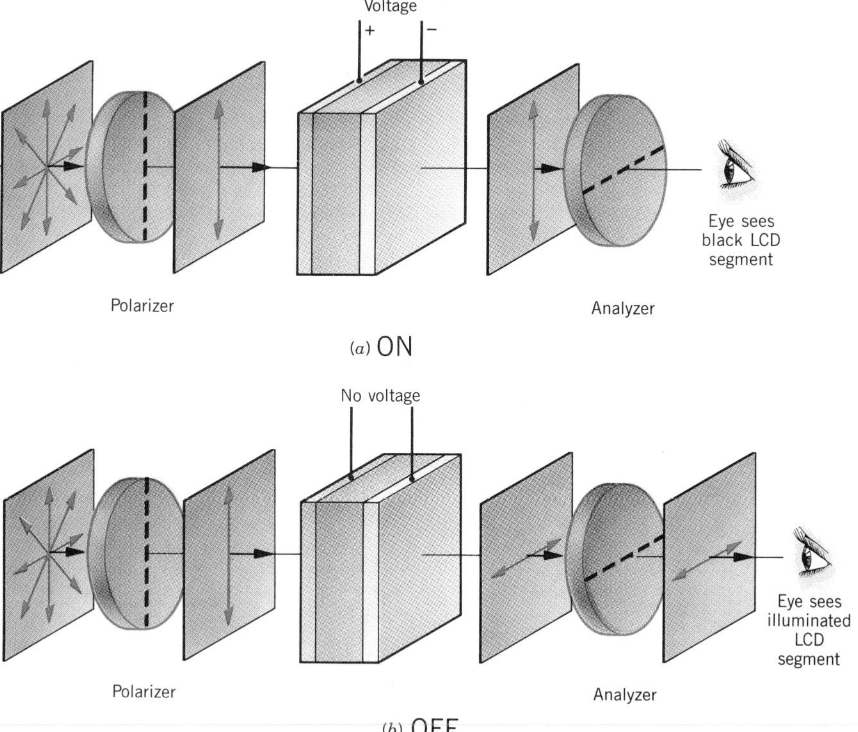

Figure 24.20 A liquid crystal display (LCD) incorporates a crossed polarizer/analyzer combination. (*a*) When the LCD segment is turned on, a voltage is applied to the electrodes, no light is transmitted through the analyzer, and the observer sees a black segment. (*b*) The LCD segment is turned off (no voltage), and light from the segment, which is the same color as the background light, reaches the observer. The segment is invisible, however, due to a lack of contrast between it and the background.

THE OCCURRENCE OF POLARIZED LIGHT IN NATURE

THE PHYSICS OF . . .

Polaroid sunglasses.

Polaroid is a familiar material because of its widespread use in sunglasses. Such sunglasses are designed so that the transmission axis of the Polaroid is oriented vertically when the glasses are worn in the usual fashion. Thus, the glasses prevent any light that is polarized horizontally from reaching the eye. Light from the sun is unpolarized, but a considerable amount of horizontally polarized sunlight originates by reflection from horizontal surfaces such as that of a lake. Section 26.4 discusses this effect. Polaroid sunglasses reduce glare by preventing the horizontally polarized reflected light from reaching the eyes.

"Fido" learns about polarizing sunglasses.

Polarized sunlight also originates from the scattering of light by molecules in the atmosphere. Figure 24.21 shows light being scattered by a single atmospheric molecule. The electric fields in the unpolarized sunlight cause the electrons in the molecule to vibrate perpendicular to the direction in which the light is traveling. The electrons, in turn, reradiate the electromagnetic waves in different directions, as the drawing illustrates. The light radiated straight ahead in direction *A* is unpolarized, just like the incident light. But light radiated perpendicular to the incident light in direction *C* is polarized. Light radiated in the intermediate direction *B* is partially polarized.

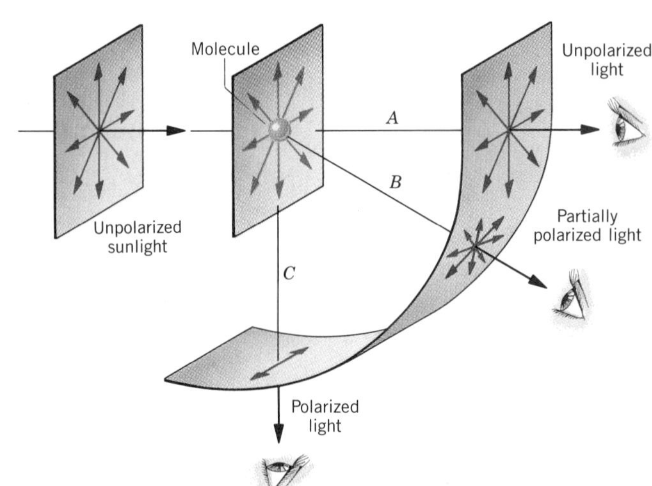

Figure 24.21 In the process of being scattered from atmospheric molecules, unpolarized light from the sun becomes partially polarized.

INTEGRATION OF CONCEPTS

ELECTROMAGNETIC WAVES AND OTHER KINDS OF WAVES

A wave is a disturbance that travels from place to place, carrying energy as it goes. The waves that we have discussed so far include water waves, waves on a string, and sound waves. To this list we now add electromagnetic waves. As an electromagnetic wave passes by, the electric and magnetic fields of the wave can disturb electric charges and currents by exerting forces on them. Electromagnetic waves share with all waves the ability to transport energy. And like other waves, they can exhibit diffraction. To describe electromagnetic waves, we use the same concepts that apply to other waves, namely, wavelength, frequency, amplitude, speed, and intensity. As is the case for other waves, the product of the wavelength and frequency of an electromagnetic wave equals the wave speed. Like a wave on a string, an electromagnetic wave is a transverse wave, for its electric and magnetic fields are perpendicular to the direction in which the wave travels. Because of its transverse nature, an electromagnetic wave can be polarized. In contrast, a sound wave is a longitudinal wave and, for such waves, the idea of polarization has no meaning. Electromagnetic waves also have some characteristics that distinguish them from other kinds of waves. Most significantly, they are the only waves that can travel through a vacuum. Compared to other kinds of waves, electromagnetic waves travel very fast (3.00×10^8 m/s, in a vacuum). And they encompass a remarkably wide range of frequencies, from less than 10^4 Hz to greater than 10^{22} Hz. These distinguishing characteristics have led to the widespread use of electromagnetic waves to carry information in the field of telecommunications.

SUMMARY

An **electromagnetic wave** in a vacuum consists of mutually perpendicular and oscillating electric and magnetic fields. The wave is a transverse wave, since the fields are perpendicular to the direction in which the wave travels. All electromagnetic waves, regardless of their frequency, travel through a vacuum at the same speed, the speed of light c ($c = 3.00 \times 10^8$ m/s).

The frequency f and wavelength λ of an electromagnetic wave in a vacuum are related to its speed through the relation $c = f\lambda$. The series of electromagnetic waves, arranged in order of their frequencies, is called the **electromagnetic spectrum.** The electromagnetic spectrum is composed of groups of waves that are known as radio waves, infrared radiation, visible light, ultraviolet radiation, X-rays, and gamma rays. **Visible light** has frequencies between about 4.0×10^{14} Hz and 7.9×10^{14} Hz. The human eye and brain perceive different frequencies or wavelengths as different colors.

Maxwell calculated that the speed of light in a vacuum is $c = 1/\sqrt{\epsilon_0 \mu_0}$, where ϵ_0 is the (electric) permittivity of free space and μ_0 is the (magnetic) permeability of free space.

The **total energy density** u of an electromagnetic wave is the total energy per unit volume of the wave and, in a vacuum, is given by $u = \frac{1}{2}\epsilon_0 E^2 + B^2/(2\mu_0)$, where E and B are the magnitudes of the electric and magnetic fields. In a vacuum, E and B are related by $E = cB$, and the electric and magnetic parts of the total energy density are equal.

The **intensity** of an electromagnetic wave is the power that the wave carries perpendicularly through a surface divided by the area of the surface. In a vacuum, the intensity S is related to the total energy density u according to $S = cu$.

A **linearly polarized** electromagnetic wave is one in which all oscillations of the electric field occur along one direction, which is taken to be the direction of polarization. In **unpolarized light** the direction of polarization does not remain fixed, but fluctuates randomly in time.

Polarizing materials allow only the component of the wave's electric field along one direction to pass through them. The preferred transmission direction for the electric field is called the **transmission axis** of the material. When unpolarized light is incident on a piece of polarizing material, the transmitted polarized light has an intensity that is one-half that of the incident light. When two pieces of polarizing material are used one after the other, the first one is called the polarizer, while the second one is referred to as the analyzer. If the average intensity of polarized light falling on an analyzer is $\bar{S}_0$, the average intensity $\bar{S}$ of the light leaving the analyzer is given by **Malus' law** as $\bar{S} = \bar{S}_0 \cos^2 \theta$, where θ is the angle between the transmission axes of the polarizer and analyzer. When $\theta = 90°$, the polarizer and the analyzer are said to be "crossed," and no light passes through the analyzer.

QUESTIONS

1. Compare the properties of electromagnetic waves and sound waves by answering "yes" or "no" to the questions in the table below.

Property	Electromagnetic Wave	Sound Wave
Transverse?		
Longitudinal?		
Can be polarized?		
Can travel through a vacuum?		
Can travel through a material such as glass?		
Involves electric and magnetic fields?		
Involves pressure oscillations?		

2. Refer to Figure 24.1. Between the times indicated in parts *c* and *d* in the drawing, what is the direction of the magnetic field at the point *P* for the electromagnetic wave being generated? Is it directed into or out of the plane of the paper? Justify your answer.

3. A transmitting antenna is located at the origin of an x, y, z axis system and broadcasts an electromagnetic wave whose electric field oscillates along the y axis. The wave travels along the $+x$ axis. Three possible wire loops can be used with an LC-tuned circuit to detect this wave: One loop lies in the xy plane, another in the xz plane, and the third in the yz plane. Which of the loops will detect the wave? Why?

4. Why does the peak value of the emf induced in a loop antenna (see Figure 24.5) depend on the frequency of the electromagnetic wave, while the peak value of the emf induced in a straight-wire antenna (see Figure 24.4) does not?

5. Suppose that the electric field of an electromagnetic wave decreases in magnitude. Does the magnetic field increase, decrease, or remain the same? Account for your answer.

6. Refer to Figure 24.8 and Example 3 in the text. Would light be detected by the observer if the eight-sided mirror were to rotate at an angular speed of 1080 rev/s, instead of 540 rev/s? Explain.

7. Is there any real difference between a polarizer and an analyzer? In other words, can a polarizer be used as an analyzer, and vice versa?

8. Malus' law applies to the setup in Figure 24.15, which shows the analyzer rotated through an angle θ and the polarizer held fixed. Does Malus' law apply when the analyzer is held fixed and the polarizer is rotated? Give your reasoning.

9. In Example 7, we saw that, when the angle between the polarizer and analyzer is 63.4°, the intensity of the transmitted light drops to one-tenth of that of the incident unpolarized light. What happens to the light intensity that is not transmitted?

10. Light is incident from the left on two pieces of polarizing material, 1 and 2. As part *a* of the drawing illustrates, the transmission axis of material 1 is along the vertical direction, while that of material 2 makes an angle of θ with respect to the vertical. In part *b* of the drawing the two polarizing materials are interchanged. (a) Assume that the incident light is unpolarized and determine whether the intensity of the transmitted light in part *a* is greater than, equal to, or less than that in part *b*. (b) Repeat part (a), assuming that the incident light is linearly polarized along the vertical direction. Justify your answers to both parts (a) and (b).

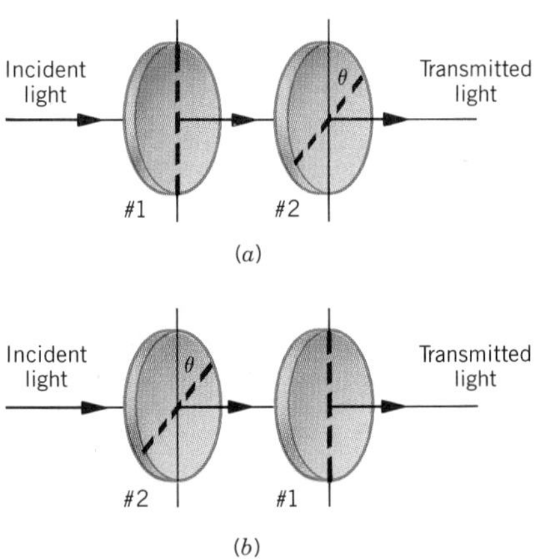

11. You are sitting upright on the beach near a lake on a sunny day, wearing Polaroid sunglasses. When you lay down on your side, facing the lake, the sunglasses don't work as well as they did while you were sitting upright. Why not?

PROBLEMS

Section 24.1 The Nature of Electromagnetic Waves

1. In 1980 and 1981, two Voyager spacecraft sent back beautiful photographs of Saturn via radio transmission. If the distance between earth and Saturn was 1.277×10^{12} m, how much time (in minutes) was required for the transmission?

2. In astronomy, distances are often expressed in light-years. One light-year is the distance traveled by light in one year. The distance to Alpha Centauri, the closest star other than our own sun that can be seen by the naked eye, is 4.3 light-years. Express this distance in meters.

3. In Figure 24.4 the value of the inductance is 2.6×10^{-4} H. For an AM radio station broadcasting at a frequency of 1200 kHz, find the value to which the capacitance must be adjusted.

4. For an FM radio station broadcasting at a frequency of 88.0 MHz, the capacitance in Figure 24.4 must be adjusted to a value of 23.0×10^{-12} F. Assuming the inductance does not change, determine the value of the capacitance for an FM station broadcasting at 108.0 MHz.

***5.** Equation 16.3, $y = A \sin(2\pi ft - 2\pi x/\lambda)$, gives the mathematical representation of a wave oscillating in the y direction and traveling in the positive x direction. Let y in this equation equal the electric field of an electromagnetic wave traveling in a vacuum. Assuming that the maximum electric field is $A = 156$ N/C and that the frequency is $f = 1.50 \times 10^8$ Hz, plot a graph of the electric field strength versus position, using for x the following values: 0, 0.50, 1.00, 1.50, and 2.00 m. Plot this graph for (a) a time $t = 0$ and (b) a time t that is one-fourth of the wave's period.

****6.** An electromagnetic wave is traveling in a vacuum along the $+x$ axis. The electric field of the wave is represented mathematically as

$$E = (31 \text{ N/C}) \sin[(5.4 \times 10^8 \text{ s}^{-1})t - (1.8 \text{ m}^{-1})x]$$

where t is the time and x is the distance. (a) What is the amplitude of the magnetic field of the electromagnetic wave? (b) What is the frequency of the wave? (c) What is the wavelength of the wave? (d) Write an equation, similar to the one above, for the magnetic field of the wave.

Section 24.2 The Electromagnetic Spectrum

7. Some of the X-rays produced in an X-ray machine have a wavelength of 2.1 nm. What is the frequency of these electromagnetic waves?

8. A truck driver is broadcasting at a frequency of 26.965 MHz with a CB (citizen's band) radio. Determine the wavelength of the electromagnetic wave being used.

9. Television sets often use a "rabbit-ears" antenna. A rabbit-ears antenna consists of a pair of metal rods. The length of each rod can be adjusted to be one-quarter of a wavelength of an electromagnetic wave whose frequency is 60.0 MHz. How long is each rod?

10. What is the ratio of the wavelength of microwaves ($f = 2.4 \times 10^9$ Hz) used in a microwave oven to radio waves ($f = 6.0 \times 10^5$ Hz) of an AM radio station?

11. A radio station broadcasts a radio wave whose wavelength is 274 m. (a) What is the frequency of the wave? (b) Is this radio wave AM or FM? (See Figure 24.6.)

12. Determine the range of wavelengths for FM radio waves with frequencies between 88.0 and 108.0 MHz.

13. The human eye is most sensitive to light having a frequency of about 5.5×10^{14} Hz, which is in the yellow-green region of the electromagnetic spectrum. How many wavelengths of this light can fit across the width of your thumb, a distance of about 2.0 cm?

***14.** Section 17.5 deals with transverse standing waves on a string. Electromagnetic waves also can form standing waves. In a standing wave pattern formed from microwaves, the distance between a node and an adjacent antinode is 0.50 cm. What is the microwave frequency?

Section 24.3 The Speed of Light

15. Ghost images are formed in a TV picture when the electromagnetic wave from the broadcasting antenna reflects from a building or other large object and arrives at the TV set shortly after the wave coming directly from the broadcasting antenna. If the reflected wave arrives 4.0×10^{-6} s after the direct wave, what is the difference in the distances traveled by the two waves?

16. A communications satellite is in a synchronous orbit that is 3.6×10^7 m above the surface of the earth. The satellite is located midway between Quito, Equador, and Belém, Brazil, two cities almost on the equator that are separated by a distance of 3.5×10^6 m. Find the time it takes for a telephone call to go by way of satellite between these cities. Ignore the curvature of the earth.

17. (a) Neil A. Armstrong was the first person to walk on the moon. The distance between the earth and the moon is 3.85×10^8 m. Find the time it took for his voice to reach earth via radio waves. (b) Determine the communication time for the first person who will some day walk on Mars, which is 5.6×10^{10} m from earth at the point of closest approach.

18. In Fizeau's experiment (see Figure 24.7) the rotating wheel contained 720 notches. Knowing that the speed of light in air is 3.00×10^8 m/s, obtain the angular speed (in rev/s) of the wheel if the incident light is to pass through one notch and the reflected light is to pass through an adjacent notch.

19. Review Conceptual Example 4 before attempting this problem. The brightest star in the night sky is Sirius, which is at a distance of 8.3×10^{16} m. When we look at this star, how far back in time are we seeing it? Express your answer in years. (There are $365\frac{1}{4}$ days in one year.)

20. The distance between earth and the moon can be determined from the time it takes for a laser beam to travel from earth to a reflector on the moon and back. If the round-trip time can be measured to an accuracy of one-tenth of a nanosecond (1 ns $= 10^{-9}$ s), what is the corresponding error in the earth–moon distance?

***21.** A mirror faces a cliff, located some distance away. Mounted on the cliff is a second mirror, directly opposite the first mirror and facing toward it. A gun is fired very close to the first mirror. The speed of sound is 343 m/s. How many times does the flash of the gunshot travel the round-trip distance between the mirrors before the echo of the gunshot is heard?

*22. A celebrity holds a press conference, which is televised live. A television viewer hears the sound picked up by a microphone directly in front of the celebrity. This viewer is seated 2.0 m from the television set. A reporter at the press conference is located 5.0 m from the microphone and hears the words directly *at the very same instant* that the television viewer hears them. Using a value of 343 m/s for the speed of sound, determine the maximum distance between the television viewer and the celebrity.

Section 24.4 The Energy Carried by Electromagnetic Waves

23. A laser emits a narrow beam of light. The radius of the beam is 1.0×10^{-3} m, and the power is 1.2×10^{-3} W. What is the intensity of the laser beam?

24. Suppose the electric field in an electromagnetic wave has a maximum strength of 2140 N/C. What is the maximum strength of the magnetic field of the wave?

25. The microwave radiation left over from the Big Bang explosion of the universe has an average energy density of 4×10^{-14} J/m^3. What is the rms value of the electric field of this radiation?

26. The rms value of the electric field in an electromagnetic wave is 123 N/C. The wave passes perpendicularly through a surface of area 0.350 m^2. How much energy does this wave carry across the surface in one minute?

27. An industrial laser is used to burn a hole through a piece of metal. The average intensity of the light is $\bar{S} = 1.23 \times 10^9$ W/m^2. What is the rms value of (a) the electric field and (b) the magnetic field in the electromagnetic wave emitted by the laser?

28. On a cloudless day, the sunlight that reaches the surface of the earth has an average intensity of about 1.0×10^3 W/m^2. What is the average electromagnetic energy contained in 2.0 m^3 of space just above the earth's surface?

29. Show that, in addition to Equations 24.2a–c, the total energy density for an electromagnetic wave can be expressed as $u = (\sqrt{\epsilon_0/\mu_0})EB$.

*30. An argon-ion laser produces a cylindrical beam of light whose average power is 0.750 W. How much energy is contained in a 2.50-m length of the beam?

*31. In an electromagnetic wave the magnetic field strength has an rms value of 9.11×10^{-8} T. Find (a) the rms value of the electric field strength, (b) the average total energy density of the wave, and (c) the average intensity of the wave.

*32. The mean distance between earth and the sun is 1.50×10^{11} m. The average intensity of solar radiation incident on the upper atmosphere of the earth is 1390 W/m^2. Assuming the sun emits radiation uniformly in all directions, determine the total power radiated by the sun.

**33. The power radiated by the sun is 3.9×10^{26} W. The earth orbits the sun in a nearly circular orbit of radius 1.5×10^{11} m. The earth's axis of rotation is tilted by 27° relative to the plane of the orbit (see the drawing), so sunlight does not strike the equator perpendicularly. What is the intensity of the light striking the equator?

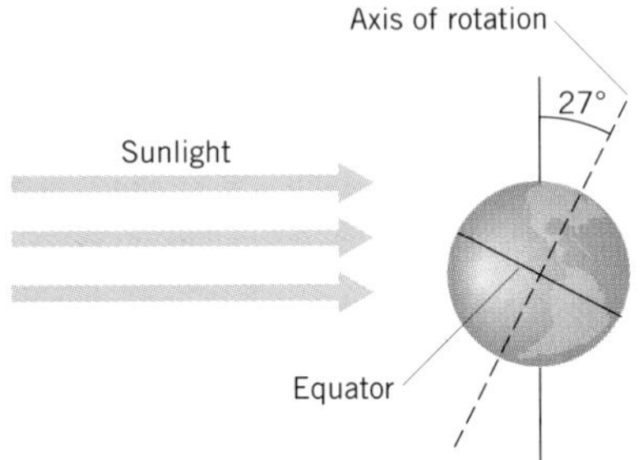

Section 24.5 Polarization

34. Polarized light strikes a piece of polarizing material. The incident light is polarized at an angle of 30.0° relative to the transmission axis of the material. What percentage of the light intensity is transmitted?

35. Unpolarized light whose intensity is 1.10 W/m^2 is incident on the polarizer in Figure 24.15. (a) What is the intensity of the light leaving the polarizer? (b) If the analyzer is set at an angle of $\theta = 75°$ with respect to the polarizer, what is the intensity of the light that reaches the photocell?

36. What should be the angle between the transmission axes of the polarizer and the analyzer in Figure 24.15, so the polarized light reaching the photocell has an intensity that is (a) one-half and (b) one-fourth of the intensity of the incident unpolarized light?

37. In the polarizer/analyzer combination in Figure 24.15, 90.0% of the light intensity falling on the analyzer is absorbed. Determine the angle between the transmission axes of the polarizer and the analyzer.

38. The orientation of the transmission axis for each of the three sheets of polarizing material in the drawing is labeled relative to the vertical. A beam of light, polarized in the vertical direction, is incident on the first sheet. The intensity of the incident beam is 1550 W/m^2. Obtain the intensity of the beam transmitted through the three sheets when: (a) $\theta_1 = 0°$, $\theta_2 = 40.0°$, $\theta_3 = 75.0°$; (b) $\theta_1 = 30.0°$, $\theta_2 = 30.0°$, $\theta_3 = 70.0°$.

*39. Review Conceptual Example 8 as an aid in solving this problem. Suppose unpolarized light of intensity 150 W/m^2 falls on the polarizer in Figure 24.17*b*, and the angle θ in the drawing is 30.0°. What is the intensity of the light that reaches the photocell?

*40. Before attempting this problem, review Conceptual Example 8. The polarizer and the analyzer in Figure 24.17*a* are

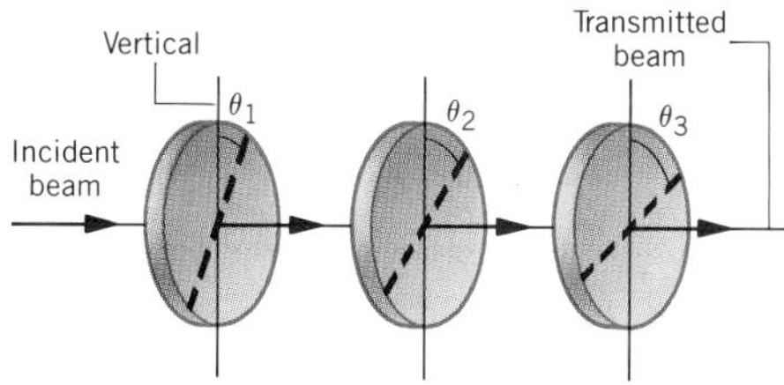

Problem 38

crossed and no light falls on the photocell. Then, a third piece of polarizing material is inserted between the polarizer and the analyzer, with its transmission axis oriented at $\theta = 45.0°$ relative to the transmission axes of the polarizer and the analyzer. If the unpolarized light intensity incident on the polarizer is I, what fraction of I now falls on the photocell?

****41.** More than one analyzer can be used in a setup like that in Figure 24.15, each analyzer following the previous one. Suppose that the transmission axis of the first analyzer is rotated 27° relative to the transmission axis of the polarizer, and that the transmission axis of each additional analyzer is rotated 27° relative to the transmission axis of the previous one. What is the minimum number of analyzers needed, so the light reaching the photocell has an intensity that is reduced by at least a factor of one hundred relative to that striking the first analyzer?

ADDITIONAL PROBLEMS

42. Obtain the wavelengths in vacuum for blue light with a frequency of 6.34×10^{14} Hz and for orange light with a frequency of 4.95×10^{14} Hz. Express your answers in nanometers (nm).

43. Suppose that the amplitude E of the electric field in an electromagnetic wave is tripled. (a) By what factor does the amplitude B of the magnetic field increase? (b) By what factor does the average intensity $\bar{S}$ of the wave increase?

44. A future space station in orbit about the earth is being powered by an electromagnetic beam from the earth. The beam has a cross-sectional area of 135 m^2 and transmits an average power of 1.20×10^4 W. What are the rms values of the (a) electric and (b) magnetic fields?

45. The mean distance from the sun to the earth is 1.50×10^{11} m. How long does it take for sunlight to reach the earth?

46. The average intensity of sunlight at the top of the earth's atmosphere is 1390 W/m^2. What is the maximum energy that a 25-m × 45-m solar panel could collect in one hour in this sunlight?

47. The intensity of sunlight at the top of the earth's atmosphere is about 1390 W/m^2. The distance between the sun and earth is 1.50×10^{11} m, while that between the sun and Mars is 2.28×10^{11} m. What is the intensity of sunlight at the surface of Mars?

48. TV channel 3 (VHF) broadcasts at a frequency of 63.0 MHz. TV channel 23 (UHF) broadcasts at a frequency of 527 MHz. Find the ratio (VHF/UHF) of the wavelengths for these channels.

49. Linearly polarized light is incident on a piece of polarizing material. What is the ratio of the transmitted light intensity to the incident light intensity when the angle between the transmission axis and the incident electric field is (a) 25° and (b) 65°?

50. As Section 24.3 discusses, Maxwell calculated that the speed of light in a vacuum is given by $c = 1/\sqrt{\epsilon_0\mu_0}$. The unit for ϵ_0 is $C^2/(N \cdot m^2)$ and the unit for μ_0 is T·m/A. Show that the unit for $1/\sqrt{\epsilon_0\mu_0}$ is meters per second.

***51.** In experiment 1, unpolarized light falls on the polarizer in Figure 24.15, with $\theta = 60.0°$. In experiment 2, the unpolarized light is replaced with light of the same intensity, but light that is polarized along the direction of the polarizer transmission axis. By how many *additional* degrees and in what direction must the analyzer be rotated, so the light falling on the photocell has the same intensity as it did in experiment 1?

***52.** A flat coil of wire is used with an LC-tuned circuit as a receiving antenna. The coil has a radius of 0.25 m and consists of 450 turns. The transmitted radio wave has a frequency of 1.2 MHz. The magnetic field of the wave is parallel to the normal to the coil and has a maximum value of 2.0×10^{-13} T. Using Faraday's law of electromagnetic induction and the fact that the magnetic field changes from zero to its maximum value in one-quarter of a wave period, find the magnitude of the average emf induced in the antenna during this time.

***53.** The average intensity of sunlight reaching the earth is 1390 W/m^2. A charge of 2.6×10^{-8} C is placed in the path of this electromagnetic wave. (a) What is the magnitude of the maximum electric force that the charge experiences? (b) If the charge is moving at a speed of 3.7×10^4 m/s, what is the magnitude of the maximum magnetic force that the charge could experience?

***54.** A source is radiating electromagnetic waves uniformly in all directions. At a distance of 2.0×10^4 m from the source, the rms value of the electric field is 0.25 N/C. What is the average total power emitted by the source?

***55.** A tiny source of light emits light uniformly in all directions. The average power emitted is 60.0 W. For a point located 8.00 m away from this source, determine the rms electric and magnetic field strengths in the light waves.

****56.** Suppose that the light falling on the polarizer in Figure 24.15 is partially polarized (average intensity $= \bar{S}_P$) and partially unpolarized (average intensity $= \bar{S}_U$). The total incident intensity is $\bar{S}_P + \bar{S}_U$, and the percentage polarization is $100\bar{S}_P/(\bar{S}_P + \bar{S}_U)$. When the polarizer is rotated in such a situation, the intensity reaching the photocell varies between a minimum value of $\bar{S}_{min}$ and a maximum value of $\bar{S}_{max}$. Show that the percentage polarization can be expressed as $100(\bar{S}_{max} - \bar{S}_{min})/(\bar{S}_{max} + \bar{S}_{min})$.

MYSTERIES SOLVED BY PHYSICS

Discovering the Lost City of Ubar

More than 4700 years ago, the city of Ubar was built somewhere on the Arabian peninsula. For nearly three thousand years Ubar was the center of the frankincense trade. Frankincense, used as incense and in perfumery, embalming, and fumigation, is an aromatic gum resin from trees found in the Middle East and Africa. Thousands of years ago it was also a symbol of great holiness and wealth. Ubar was renowned for its architecture, fruit trees, trees yielding frankincense, and opulence. Frankincense was so valuable that it was called "gold" in Sumerian literature. Then suddenly, around two thousand years ago, Ubar disappeared. Its remains were covered by shifting, blowing sands. Its location was forgotten, but its allure was not.

Islamic graves at the site in Oman now believed to be the legendary city of Ubar.

Where is Ubar? Only death kept T.E. Lawrence (Lawrence of Arabia) from seeking it. Another British explorer, Bertrand Thomas, searched the Rub'al-Khali desert for Ubar in 1929–1930. Guided by historical documents and local lore, he actually came near it, possibly even walking over what is now believed to be the site. But, missing the big picture of the area he was searching, he failed to identify the city. In 1981, Los Angeles filmmaker Nicholas Clapp became intrigued with finding Ubar. He learned about NASA's ability to make images of the earth from space using radio waves. Because of its very low dielectric constant, sand allows radio signals to penetrate several meters through it. The radio waves then reflect off of harder surfaces below the ever-changing desert floor. Clapp recruited a team of experts in several fields to join the search, including scientists at NASA's Jet Propulsion Laboratory in Pasadena, California, who had developed the imaging radar system for the space shuttle. In a 1984 shuttle flight, the imaging radar took aim on the southern Arabian peninsula, the same area explored by Thomas (and also described in vague terms as the location of Ubar by geographer Claudius Ptolemy in the second century, A.D.).

The radar images were digitized and combined by computer with infrared and optical images taken by the French SPOT and American Landsat satellites. The data from the infrared and visible radiation provided information about the soil and sand, such as the fact that camels have ground the sand into fine-grained pieces different from the rest of the desert. The trampled sand reflects different amounts of infrared and visible radiation than does loosely packed sand. As a result, the combined images of the desert showed many previously unknown caravan trails. While the satellite images did not reveal the ruins of a city, the group of physicists and geologists working on the project at the Jet Propulsion Laboratory observed that many of the old caravan trails converged on the Ash Shisar water hole in Oman, inland from the south-central coast of the Arabian peninsula.

In 1991, using satellite positioning equipment, archaeologists, geologists, and their support team traveled to the area and located some of the caravan trails identified from space. They found themselves near the area Thomas had explored, but unlike him, they knew from the satellite images that they were at the hub of a vast trading center. Persisting in their exploration and digging, the team brought together by Nicholas Clapp began uncovering a vast, elaborate city. While it is still uncertain whether this is Ubar, the evidence in the form of architectural wonders, pottery, and metal fragments mounts. If this is Ubar, why did it vanish? The answer comes from below the city; the explorers found that it was built over a vast limestone cave that eventually collapsed, plunging the city down into the earth. The ubiquitous Arabian sand quickly covered it over, where it has remained for nearly two millennia.

For Further Reading See: *Archaeology,* Vol. 45, No. 4, p. 6, July–August 1992.

The New York Times, Vol. 141, p. 1, Col. 1, 5 February 1992.

THE REFLECTION OF LIGHT: MIRRORS

CHAPTER 25

Ever watchful, this lion drinks from a quiet pool, whose smooth mirror-like surface acts to double his presence. Mirrors produce images because the light that strikes them is reflected, rather than absorbed. Reflected light does much more than produce dramatic and beautiful pictures, however. For example, it allows our sense of vision to be remarkably useful. When we "see" an object, light from it enters our eyes and evokes the sensation of vision. Some objects themselves produce the light that we see, like the sun, a flame, or a light bulb. Most objects, though, reflect into our eyes light that originates elsewhere, and if it were not for reflection, we would not be able to see them. Reflection also plays a major part, as we will learn, in some of the techniques being used to harness solar energy as an alternative energy source. And we have already seen in Section 18.10 that the reflection of light plays a role in the operation of laser printers. In fact, the reflection of laser light is used widely, in applications ranging from measuring the speeds of automobiles to reading price information from bar codes at the supermarket. We begin our discussion of reflection by introducing the concepts of a wave front and a ray of light. A discussion of the law of reflection follows, which is then used to explain how a flat or plane mirror works. A plane mirror is the kind that the thirsty lion is crouching over and is also the kind that you look into in the morning. Plane mirrors are only one of several types of mirrors, however. Another kind has the shape of a section cut from a sphere. We will discuss two kinds of spherical mirrors, concave and convex, and learn how to use the law of reflection to locate the images that such mirrors produce. A number of applications of spherical mirrors will also be presented.

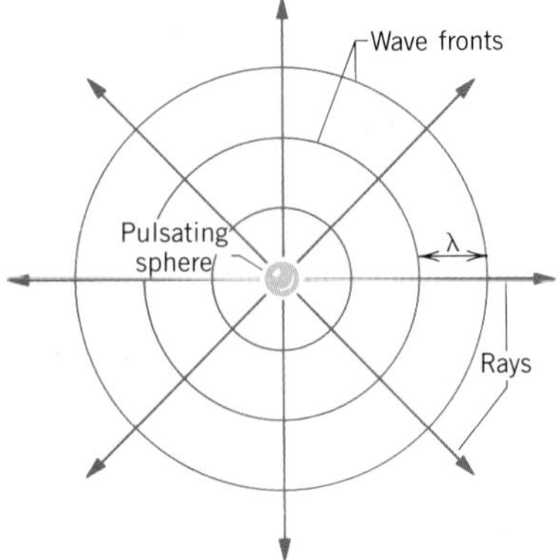

Figure 25.1 A cross-sectional view of a sound wave emitted by a pulsating sphere. The wave fronts are drawn through the condensations of the wave, so the distance between two successive wave fronts is the wavelength λ. The rays are perpendicular to the wave fronts and point in the direction of the velocity of the wave.

25.1 WAVE FRONTS AND RAYS

To introduce the concepts of a wave front and a ray, we take advantage of a topic that we have studied before, sound waves. Both sound and light are kinds of waves. Sound is a mechanical wave, while light is electromagnetic in nature. However, the wave properties of each are similar. In particular, the ideas of a wave front and a ray apply to both.

Consider a small spherical object whose surface is pulsating in simple harmonic motion. A sound wave is emitted that moves spherically outward from the object at a constant speed. To represent this wave, we draw surfaces through all points of the wave that are in the same phase of motion. These surfaces of constant phase are called ***wave fronts.*** Figure 25.1 shows a two-dimensional view of the wave fronts. In this view the wave fronts appear as concentric circles about the vibrating object. If the wave fronts are drawn through the condensations, or crests, of the sound wave, as they are in the picture, the distance between adjacent wave fronts equals the wavelength λ. The radial lines pointing outward from the source and perpendicular to the wave fronts are called ***rays.*** The rays point in the direction of the velocity of the wave.

Figure 25.2*a* shows small sections of two adjacent spherical wave fronts. At large distances from the source, the wave fronts become less and less curved and approach the shape of flat surfaces, as part *b* of the drawing shows. Waves whose wave fronts are flat surfaces (i.e., planes) are known as ***plane waves*** and are important in understanding the properties of mirrors and lenses. Since rays are perpendicular to the wave fronts, the rays for a plane wave are parallel to each other.

The concepts of wave fronts and rays can also be applied to light waves. For light waves, the ray concept is particularly convenient for showing the path taken by the light. We will make frequent use of light rays, and they can be regarded essentially as narrow beams of light.

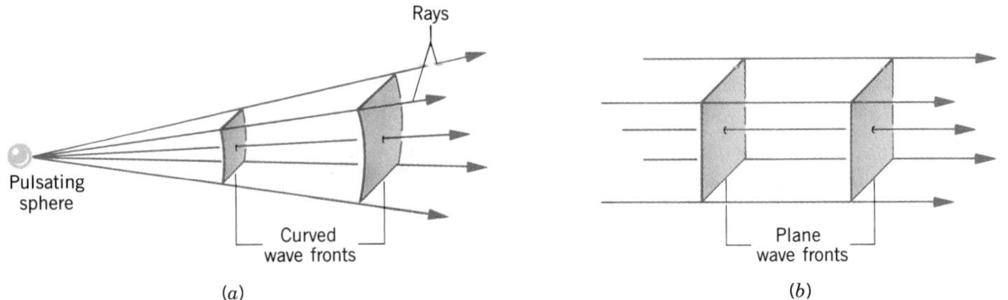

Figure 25.2 (*a*) Portions of two spherical wave fronts are shown. The rays are perpendicular to the wave fronts and diverge. (*b*) For a plane wave, the wave fronts are flat surfaces, and the rays are parallel to each other.

25.2 THE REFLECTION OF LIGHT

Most objects reflect a certain portion of the light falling on them. Suppose a ray of light is incident on a flat, shiny surface, such as the mirror in Figure 25.3. As the drawing shows, the *angle of incidence* θ_i is the angle that the incident ray makes with respect to the normal, which is a line drawn perpendicular to the surface at the point of incidence. The *angle of reflection* θ_r is the angle that the reflected ray makes with the normal. The ***law of reflection*** describes the behavior of the incident and reflected rays.

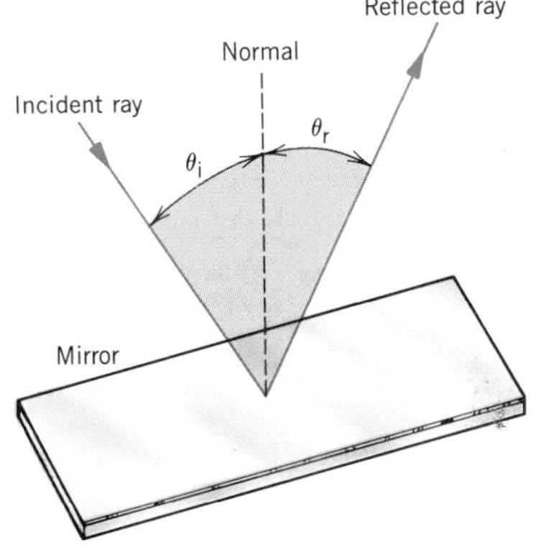

Figure 25.3 The angle of reflection θ_r equals the angle of incidence θ_i. These angles are measured with respect to the normal, which is a line drawn perpendicular to the surface of the mirror at the point of incidence.

Law of Reflection

The incident ray, the reflected ray, and the normal to the surface all lie in the same plane, and the angle of reflection θ_r equals the angle of incidence θ_i:

$$\theta_r = \theta_i$$

When parallel light rays strike a smooth, plane surface, such as that in Figure 25.4*a*, the reflected rays are parallel to each other. This type of reflection is known as *specular reflection* and is important in determining the properties of mirrors. Most surfaces, however, are not perfectly smooth, for they contain irregularities the sizes of which are equal to or greater than the wavelength of light. The irregular surface reflects the light rays in various directions, as part *b* of the drawing suggests. This type of reflection is known as *diffuse reflection.* Common surfaces that give rise to diffuse reflection are most papers, wood, nonpolished metals, and walls covered with a "flat" (nongloss) paint.

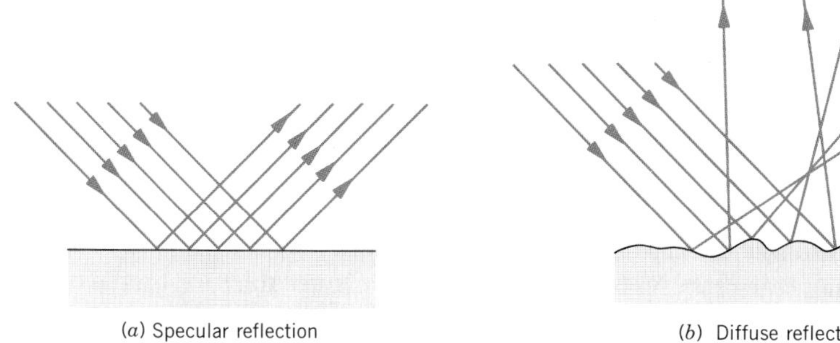

Figure 25.4 (*a*) The drawing shows specular reflection from a polished plane surface, such as a mirror. The reflected rays are parallel to each other. (*b*) The rough surface reflects the light rays in all directions; this type of reflection is known as diffuse reflection.

25.3 THE FORMATION OF IMAGES BY A PLANE MIRROR

When you look into a plane (flat) mirror, you see an image of yourself that has three fundamental properties:

1. The image is upright.
2. The image is the same size as you are.
3. The image is located as far behind the mirror as you are in front of it.

As Figure 25.5*a* illustrates, the image of yourself in the mirror is also reversed left to right. If you wave your *right* hand, it is the *left* hand of the image that waves back. Similarly, letters and words held up to a mirror are reversed. Ambulances and other emergency vehicles are often lettered in reverse, as in Figure 25.5*b*, so that the letters will appear normal when seen in the rearview mirror of a car.

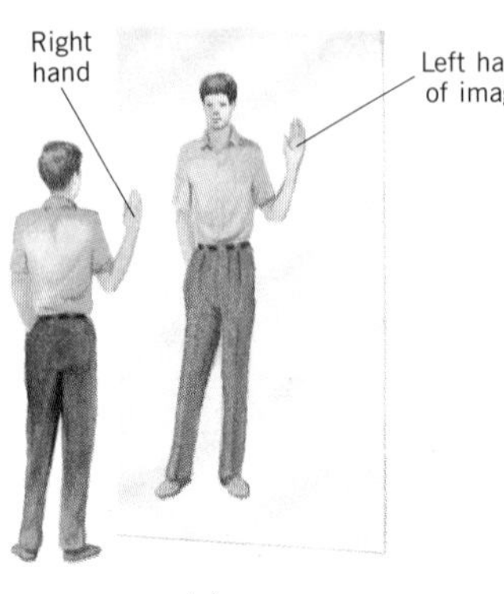

Figure 25.5 (*a*) The person's right hand becomes the image's left hand when viewed in a plane mirror. (*b*) Many emergency vehicles are reverse-lettered so the lettering appears normal when viewed through the rearview mirror of a car.

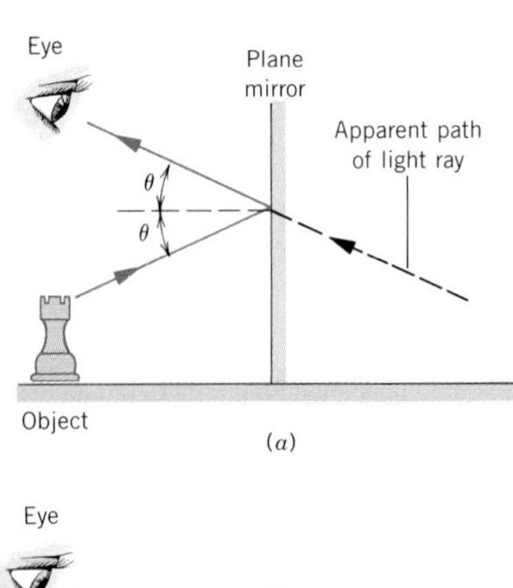

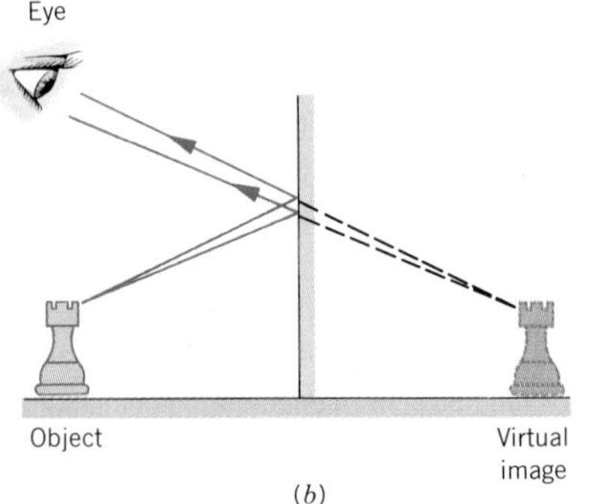

Figure 25.6 (*a*) A ray of light from the top of the chess piece reflects from the mirror. To the eye, the ray seems to come from behind the mirror. (*b*) The bundle of rays from the top of the object appears to originate from the image behind the mirror.

To illustrate why an image appears to originate from behind a plane mirror, Figure 25.6*a* shows a light ray leaving the top of an object. This ray reflects from the mirror (angle of reflection equals angle of incidence) and enters the eye. To the eye, it appears that the ray originates from behind the mirror, somewhere back along the dashed line. Actually, rays going in all directions leave each point on the object. But only a small bundle of such rays is intercepted by the eye. Part *b* of the figure shows a bundle of two rays leaving the top of the object. All the rays that leave a given point on the object, no matter what angle θ they have when they strike the mirror, appear to originate from a corresponding point on the image behind the mirror (see the dashed lines in part *b*). For each point on the object, there is a single corresponding point on the image, and it is this fact that makes the image in a plane mirror a sharp and undistorted one.

Although rays of light *seem* to come from the image, it is evident from Figure 25.6*b* that no light emanates from behind the plane mirror where the image appears to be. Because the rays of light do not actually emanate from the image, it is called a ***virtual image.*** In this text the parts of the light rays that appear to come from a virtual image are represented by dashed lines. *Curved* mirrors, on the other hand, can produce images from which light rays actually do emanate. Such images are known as ***real images*** and are discussed in later sections.

With the aid of the law of reflection, it is possible to show that the image is located as far behind a plane mirror as the object is in front of it. In Figure 25.7 the object distance is d_o and the image distance is d_i. A ray of light leaves the base of the object, strikes the mirror at an angle of incidence θ, and is reflected at the same angle. To the eye, this ray appears to come from the base of the image. For the angles β_1 and β_2 in the drawing it follows that $\theta + \beta_1 = 90°$ and $\alpha + \beta_2 = 90°$. But the angle α is equal to the angle of reflection θ, since the two are opposite angles formed by intersecting lines. Therefore, $\beta_1 = \beta_2$. As a result, triangles *ABC* and *DBC* are identical (congruent), because they share a common side *BC* and have equal angles ($\beta_1 = \beta_2$) at the top and equal angles (90°) at the base. Thus, the object distance d_o equals the image distance d_i.

By starting with a light ray from the top of the object, rather than from the bottom, we can extend the line of reasoning given above to show that the height of the image also equals the height of the object.

Conceptual Examples 1 and 2 discuss some interesting features of plane mirrors.

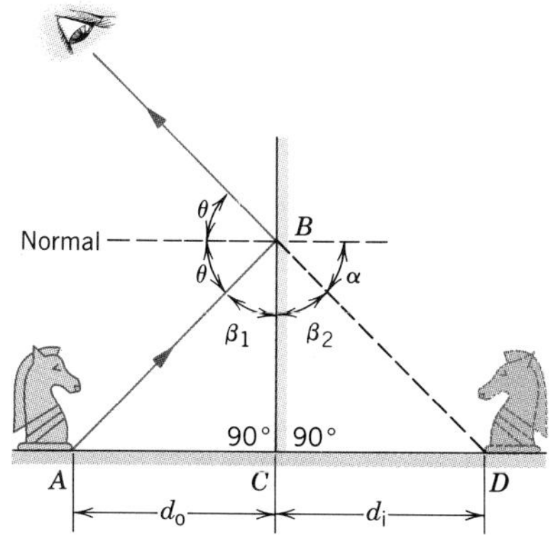

Figure 25.7 This drawing illustrates the geometry used to show that for a plane mirror the image distance d_i equals the object distance d_o.

Conceptual Example 1 Full-Length Versus Half-Length Mirrors

In Figure 25.8 a woman is standing in front of a plane mirror. What is the minimum mirror height necessary for her to see her full image?

REASONING The mirror is labeled *ABCD* in the drawing and is the same height as the woman. Light emanating from her body is reflected by the mirror, and some of this light enters her eyes. Let us consider a ray of light from her foot *F*. This ray strikes the mirror at *B* and enters her eyes at *E*. According to the law of reflection, the angles of incidence and reflection are both θ. Any light from her foot that strikes the mirror below *B* is reflected toward a point on her body that is below her eyes. Since light striking the mirror below *B* does not enter her eyes, the part of the mirror between *B* and *A* may be removed. The section *BC* of the mirror that produces the image is one-half the woman's height between *F* and *E*. This follows, because the right triangles *FBM* and *EBM* are identical. They are identical, because they share a common side *BM* and have two angles, θ and 90°, that are the same.

The blowup in Figure 25.8 illustrates a similar line of reasoning, starting with a ray from the woman's head at *H*. This ray is reflected from the mirror at *P* and enters her eyes. The top mirror section *PD* can be removed without disturbing this reflection. The necessary section *CP* is one-half the woman's height between her head at *H* and her eyes at *E*. We find, then, that only the sections *BC* and *CP* are needed for the woman to see her full height. The height of section *BC* plus section *CP* is exactly one-half the woman's height. Thus, to view one's full length in a mirror, only a half-length mirror is needed. Note that the conclusions here are valid regardless of how far the person stands from the mirror.

Related Homework Material: *Problem 3*

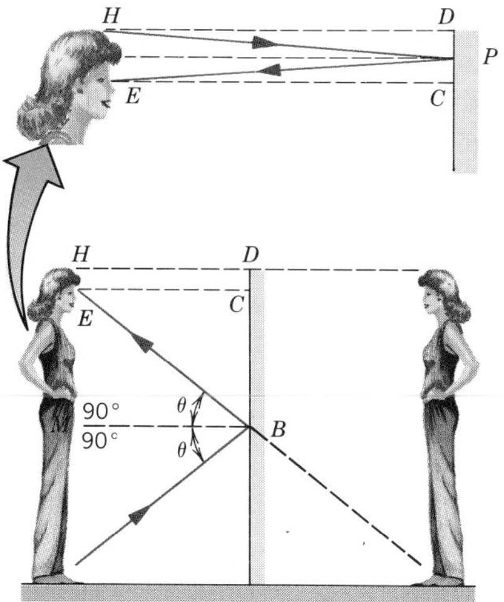

Figure 25.8 For the woman to see her full-sized image, only a half-sized mirror is needed.

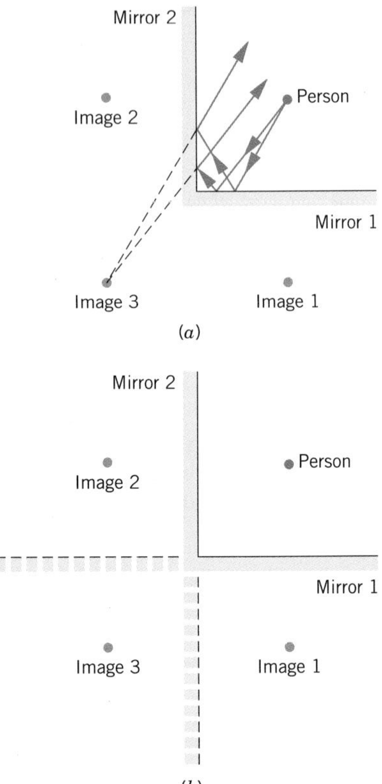

Figure 25.9 Top views of a person standing in front of two mirrors. (*a*) For two mirrors that are perpendicular, a "double" reflection gives rise to a third image. (*b*) It is possible to understand the third image by considering the fact that each mirror is itself reflected in the other. The reflected mirrors are shown here as dashes.

These two perpendicular plane mirrors produce three images of the person in front of them, the third image being due to light that has undergone two reflections, as Conceptual Example 2 discusses.

Conceptual Example 2 Multiple Reflections

A person tries on a new outfit in a clothing store. To examine the fit, he stands in front of two mirrors that intersect at a 90° angle. Why does he see three images of himself?

REASONING Figure 25.9*a* shows a top view of the person in front of the mirrors. It is a straightforward matter to understand two of the images that he sees. These are the images that are normally seen when one stands in front of a mirror. Standing in front of mirror 1, he sees image 1, which is located as far behind that mirror as he is in front of it. He also sees image 2 behind mirror 2, at a distance that matches his distance in front of that mirror. Each of these images arises from light emanating from his body and reflecting from a single mirror. However, it is also possible for light to undergo two reflections in sequence, first from one mirror and then from the other. When a double reflection occurs, an additional image becomes possible. Figure 25.9*a* shows two rays of light that strike mirror 1. Each one, according to the law of reflection, has an angle of reflection that equals the angle of incidence. The rays then strike mirror 2, where they again are reflected according to the law of reflection. When the outgoing rays are extended backward (see the dashed lines in the drawing), they intersect and appear to originate from image 3. Because the mirrors are perpendicular, image 3 can also be located at the same place by starting with two rays that strike mirror 2 first and then mirror 1.

It is also possible to understand why image 3 exists by a different, but equivalent, line of reasoning. Each of the mirrors is reflected in the other. Figure 25.9*b* shows the reflected mirrors as dashes. In the picture, image 1 is "in front of" the image of mirror 2. Thus, we can regard the third image as the reflection of image 1 in the image of mirror 2. Similarly, we can regard the third image as the reflection of image 2 in the image of mirror 1.

Related Homework Material: *Problem 4*

25.4 SPHERICAL MIRRORS

The most common type of curved mirror is a spherical mirror. As Figure 25.10 shows, a spherical mirror has the shape of a section from the surface of a sphere. If the inside or concave surface of the mirror is polished, it is a ***concave mirror.*** If the outside or convex surface is polished, it is a ***convex mirror.*** The drawing shows both types of mirrors, with a light ray reflecting from the polished surface. The law of reflection applies, just as it does for a plane mirror. For either type of spherical mirror, the normal is drawn perpendicular to the mirror at the point of incidence. For each type, the center of curvature is located at point *C*, and the radius of curvature is *R*. The ***principal axis*** of the mirror is a straight line drawn through *C* and the midpoint of the mirror.

Figure 25.11 shows a tree in front of a concave mirror. A point on this tree lies on the principal axis of the mirror and is beyond the center of curvature *C*. Light rays emanate from this point and reflect from the mirror, consistent with the law of reflection. If the rays are near the principal axis, they cross it at a common point after reflection. This point is called the image point. The rays continue to diverge from the image point as if there were an object there. Since light rays actually come from the image point, the image is a real image.

If the tree in Figure 25.11 is infinitely far from the mirror, the rays are parallel to each other and to the principal axis as they approach the mirror. Figure 25.12

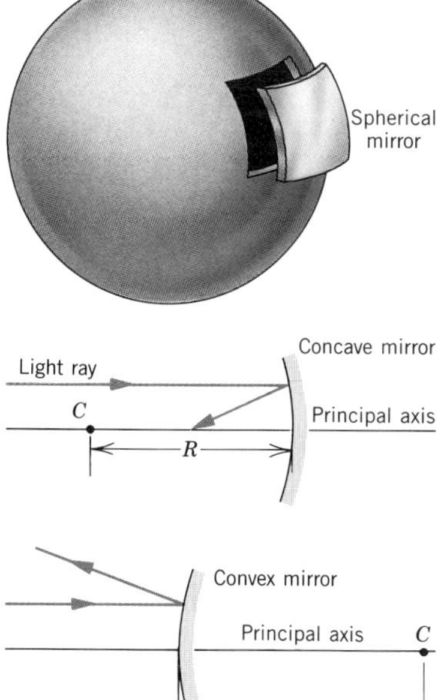

Figure 25.10 A spherical mirror has the shape of a segment of a spherical surface. The center of curvature is point C and the radius is R. For a concave mirror, the reflecting surface is the inner one, while for a convex mirror it is the outer one.

shows rays near and parallel to the principal axis, as they reflect from the mirror and pass through an image point. In this special case the image point is referred to as the ***focal point*** F of the mirror. Therefore, an object infinitely far away on the principal axis gives rise to an image at the focal point of the mirror. The distance between the focal point and the middle of the mirror is the ***focal length*** f of the mirror.

We can show that the focal point F lies halfway between the center of curvature C and the middle of a concave mirror. In Figure 25.13, a light ray parallel to the principal axis strikes the mirror at point A. The line CA is the radius of the mirror and, therefore, is the normal to the spherical surface at the point of incidence. The ray reflects from the mirror such that the angle of reflection θ equals the angle of

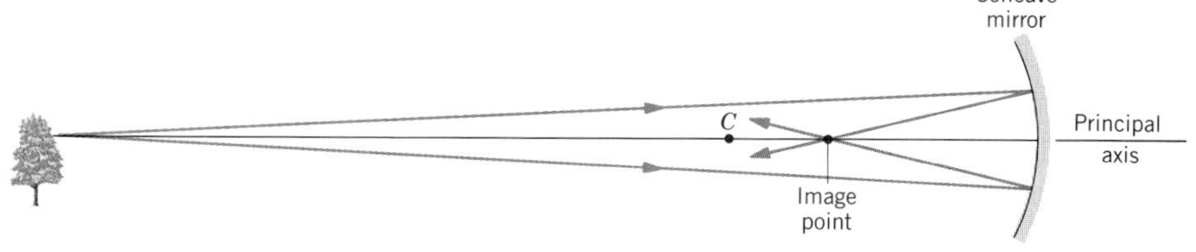

Figure 25.11 A point on the tree lies on the principal axis of the concave mirror. Rays from this point that are near the principal axis are reflected from the mirror and cross the axis at the image point.

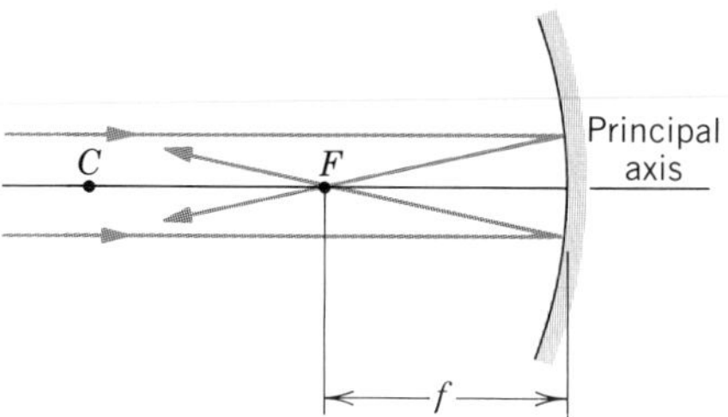

Figure 25.12 Light rays that are near and parallel to the principal axis are reflected from a concave mirror and converge at the focal point F. The focal length f of the mirror is the distance between F and the middle of the mirror.

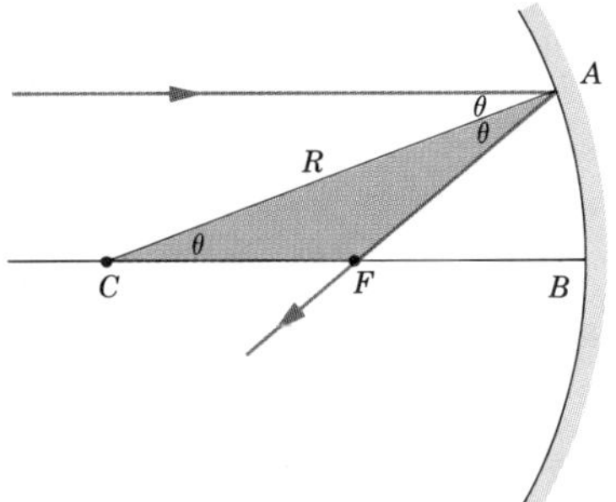

Figure 25.13 This drawing is used to show that the focal point F of a concave mirror is halfway between the center of curvature C and the mirror at point B.

incidence. Furthermore, the angle ACF is also θ, because the radial line CA is a transversal of two parallel lines. Since two of its angles are equal, the colored triangle CAF is an isosceles triangle; thus, sides CF and FA are equal. But when the incoming ray lies close to the principal axis, the angle of incidence θ is small, and the distance FA does not differ appreciably from the distance FB. Therefore, in the limit that θ is small, $CF = FA = FB$, and so the focal point F lies halfway between the center of curvature and the mirror. In other words, the focal length f is one-half of the radius R:

[Focal length of a concave mirror]
$$f = \tfrac{1}{2}R \tag{25.1}$$

Rays that lie close to the principal axis are known as ***paraxial rays,**** and Equation 25.1 is valid only for such rays. Rays that are far from the principal axis do not converge to a single point after reflection from the mirror, as Figure 25.14 shows. The result is a blurred image. The fact that a spherical mirror does not bring all rays parallel to the axis to a single image point is known as ***spherical aberration.*** Spherical aberration can be minimized by using a mirror whose height is small compared to the radius of curvature.

THE PHYSICS OF . . .

capturing solar energy with mirrors.

A sharp image point can be obtained with a large mirror, if the mirror is parabolic in shape instead of spherical. The shape of a parabolic mirror is such that all light rays parallel to the principal axis, regardless of their distance from the axis, are reflected through a single image point. However, parabolic mirrors are costly to manufacture and are used where the sharpest images are required, as in research-quality telescopes. Parabolic mirrors are also used in one method of capturing solar energy for commercial purposes. Figure 25.15 shows a solar energy "farm" consisting of long rows of concave parabolic mirrors that reflect the sun's rays to the focal point. Located at the focal point and running the length of each row is an oil-filled pipe. The focused rays of the sun heat the oil, the heat from which is used to generate steam. The steam, in turn, drives a turbine connected to an electric generator.

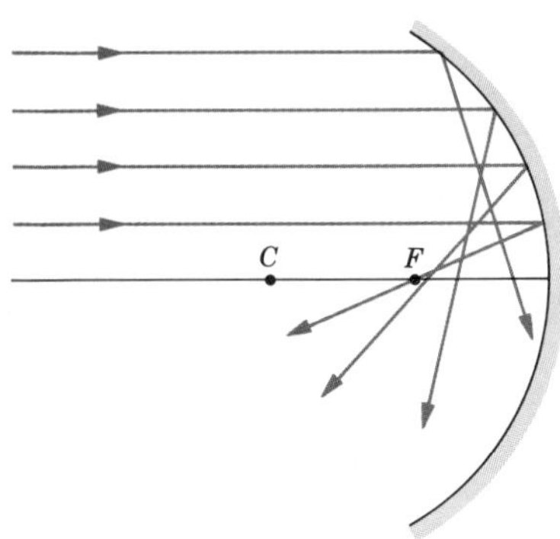

Figure 25.14 Rays that are farthest from the principal axis have the greatest angle of incidence and miss the focal point F after reflection from the mirror.

Figure 25.15 This solar-thermal electric plant in the Mojave Desert uses long rows of parabolic mirrors to focus the sun's rays on oil-filled pipes, which are located at the focal point of each mirror.

* Paraxial rays are close to the principal axis, but not necessarily parallel to it.

A convex mirror also has a focal point, and Figure 25.16 illustrates its meaning. In this picture, parallel rays are incident on a convex mirror. Clearly, the rays diverge after being reflected. If the incident parallel rays are paraxial, the reflected rays seem to come from a single point *F* behind the mirror. This point is the focal point of the convex mirror, and its distance from the midpoint of the mirror is the focal length *f*. The focal length of a convex mirror is also one-half of the radius of curvature, just as it is for a concave mirror. However, we assign the focal length of a convex mirror a negative value, because it will be convenient later on to do so:

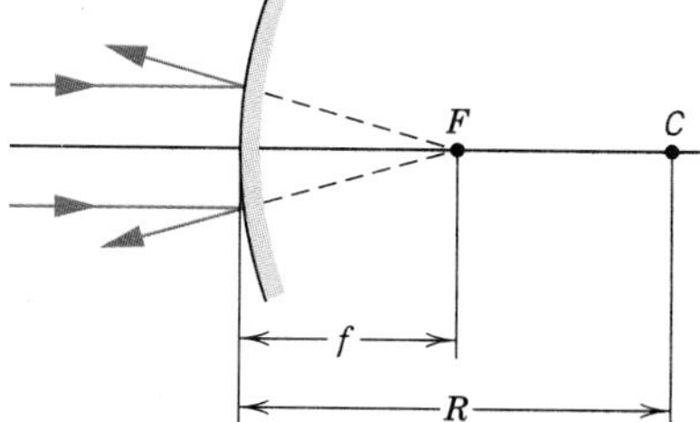

Figure 25.16 **When paraxial light rays that are parallel to the principal axis strike a convex mirror, the reflected rays appear to originate from the focal point *F*. The radius of curvature is *R* and the focal length is *f*.**

$$\left[\begin{array}{l}\textbf{Focal length of}\\ \textbf{a convex mirror}\end{array}\right] \qquad f = -\tfrac{1}{2}R \qquad (25.2)$$

25.5 THE FORMATION OF IMAGES BY SPHERICAL MIRRORS

IMAGE FORMATION BY A CONCAVE MIRROR

As we have seen, some of the light rays emitted from an object in front of a mirror strike the mirror, reflect from it, and form an image. To determine the location and size of the image for a concave mirror, we take advantage of the following fact: paraxial rays leave a point on the object and intersect at a corresponding point on the image after reflection. Three specific paraxial rays are especially convenient to use. In Figure 25.17 they leave a point on the top of the object and are labeled 1, 2, and 3. When tracing their paths, we follow the reasoning strategy outlined below.

Reasoning Strategy Ray Tracing for a Concave Mirror

Ray 1. This ray is initially parallel to the principal axis and therefore passes through the focal point *F* after reflection from the mirror.

Ray 2. This ray passes through the focal point *F* and is reflected parallel to the principal axis. Ray 2 is analogous to ray 1, except the order of the incident and reflected rays is interchanged.

Ray 3. This ray travels along a line that passes through the center of curvature *C* and follows a radius of the spherical mirror; as a result, the ray strikes the mirror perpendicularly and reflects back on itself.

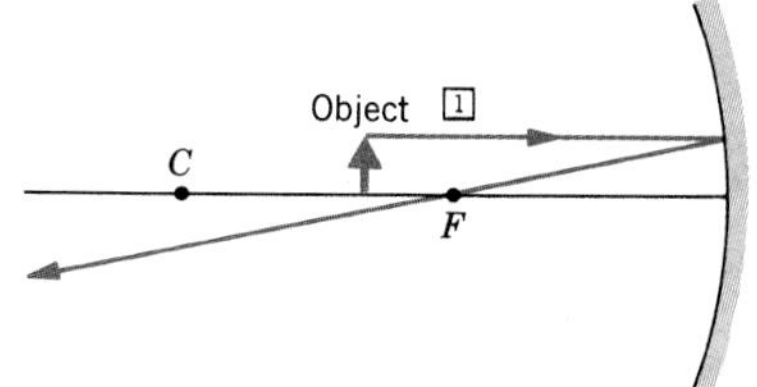

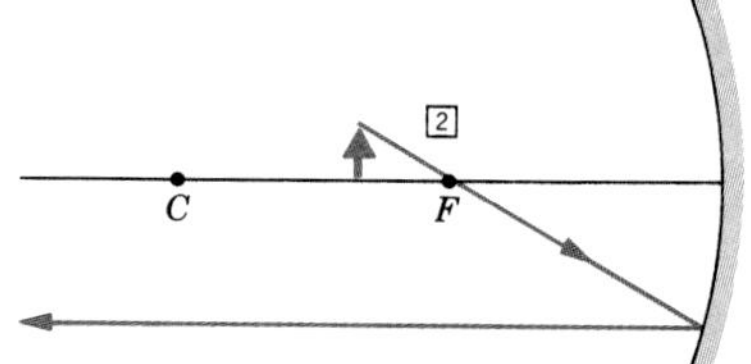

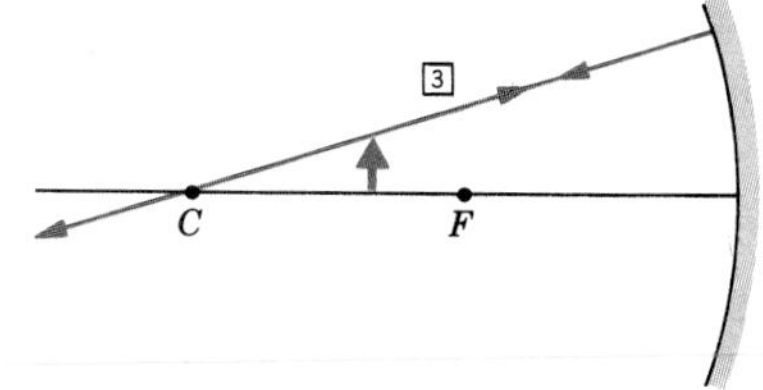

Figure 25.17 **The rays labeled 1, 2, and 3 are useful in locating the image of an object that stands in front of a concave spherical mirror. The object is represented as a vertical arrow.**

If rays 1, 2, and 3 are superimposed on a scale drawing, they converge at a point on the top of the image, as can be seen in Figure 25.18*a*.* Although three rays have been used here to locate the image, only two are really needed; the third ray is usually drawn to serve as a check. In a similar fashion, rays from all other points on the object locate corresponding points on the image, and the mirror forms a complete image of the object. If you place your eye as shown in the drawing, you will see an image that is *larger* and *inverted* relative to the object. The image is real, because the light rays actually pass through the image.

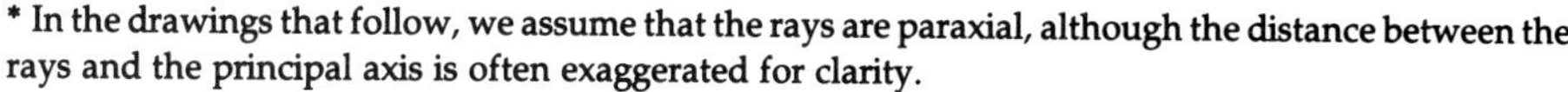
* In the drawings that follow, we assume that the rays are paraxial, although the distance between the rays and the principal axis is often exaggerated for clarity.

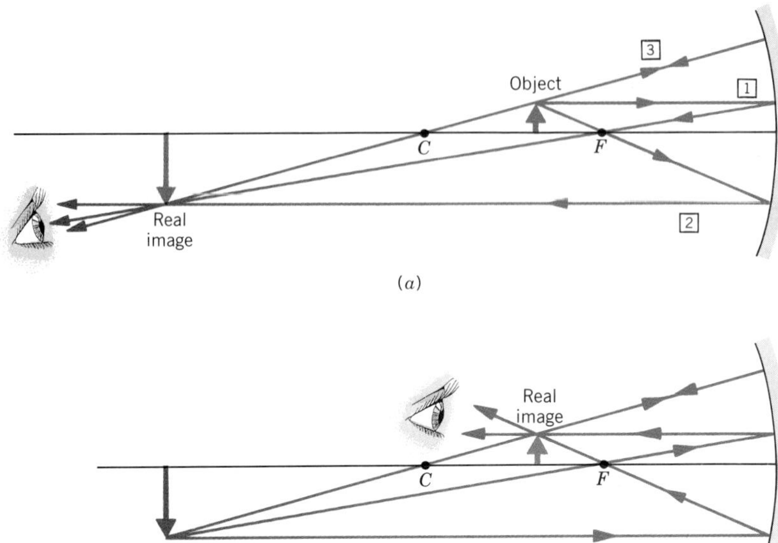

Figure 25.18 (*a*) When an object is placed between the focal point *F* and the center of curvature *C* of a concave mirror, a real image is formed. The image is enlarged and inverted with respect to the object. (*b*) When the object is located beyond the center of curvature *C*, a real image is created that is reduced in size and inverted with respect to the object.

If the object and image in Figure 25.18*a* are interchanged, the situation in part *b* of the drawing results. The three rays in part *b* are the same as those in part *a*, except the directions are reversed. These drawings illustrate the ***principle of reversibility,*** which states that ***if the direction of a light ray is reversed, the light retraces its original path.*** This principle is quite general and is not restricted just to reflection from mirrors.

When the object is placed between the focal point *F* and the mirror, as in Figure 25.19*a*, three rays can again be drawn to find the image. But now ray 2 does not go through the focal point on its way to the mirror, since the object is inside the focal point. However, when projected backward, ray 2 appears to come from the focal point. Therefore, after reflection, ray 2 is directed parallel to the principal axis. In this case the three reflected rays diverge from each other and do not converge to a common point. However, when projected behind the mirror, the three rays appear to come from a point behind the mirror; thus, a virtual image is formed. This virtual image is larger than the object and upright. Makeup and shaving mirrors are concave mirrors. When you place your face between the mirror and its focal point, you see an enlarged virtual image of yourself, as part *b* of the drawing shows.

THE PHYSICS OF . . .

makeup and shaving mirrors.

Concave mirrors are also used in a new method for displaying the speed of a car. The method presents a digital readout (e.g., 55 mph) that the driver sees when looking directly through the windshield, as in Figure 25.20*a*. The advantage of the method, which is called a head-up display (HUD), is that the driver does not need to take his or her eyes off the road to monitor the speed. Figure 25.20*b* shows how a HUD works. Located below the windshield is a readout device that displays the speed in digital form. This device is located in front of a concave mirror, but within its focal point. The arrangement is similar to that in Figure 25.19*a* and produces a virtual, upright, and enlarged image of the speed readout, just as a makeup mirror would (see virtual image 1 in Figure 25.20*b*). Light rays that appear to come from this image strike the windshield at a place where a so-called "combiner" is located. The purpose of the combiner is to combine the digital readout informa-

THE PHYSICS OF . . .

a head-up display for automobiles.

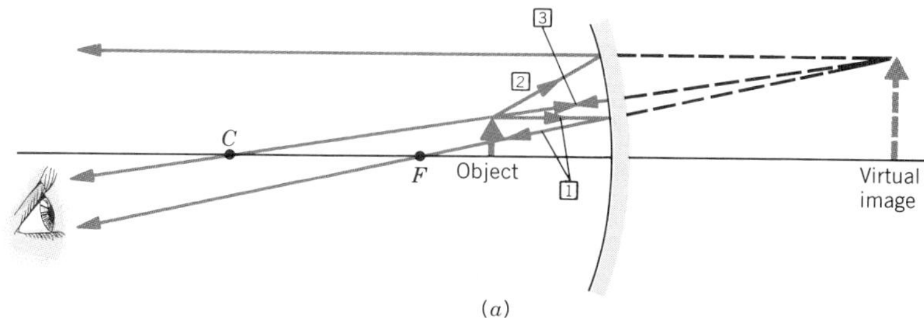

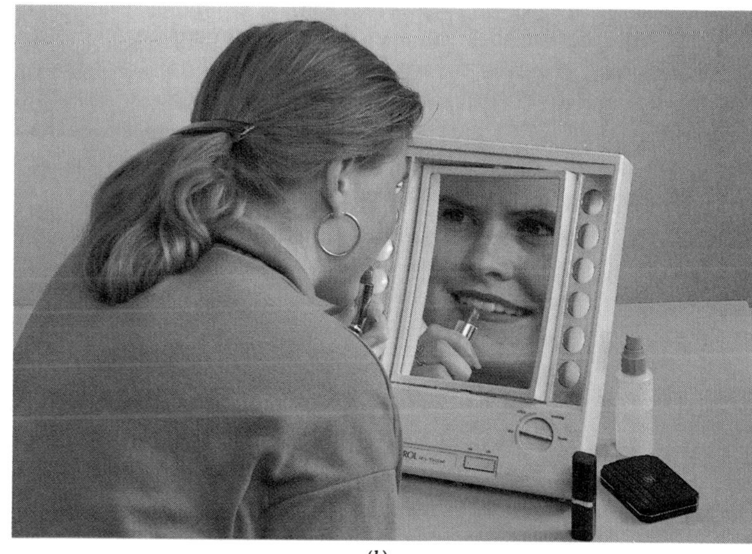
(b)

Figure 25.19 (*a*) When an object is located between a concave mirror and its focal point *F*, an enlarged, upright, virtual image is produced. (*b*) A makeup mirror is concave, and a virtual image is formed when the object is within the focal point of the mirror.

tion with the field of view that the driver sees through the windshield. The combiner is virtually undetectable by the driver, because it allows all colors except one to pass through it unaffected. The one exception is the color produced by the digital readout device. For this color, the combiner behaves as a plane mirror and reflects the light that appears to originate from image 1. Thus, the combiner

(*a*)

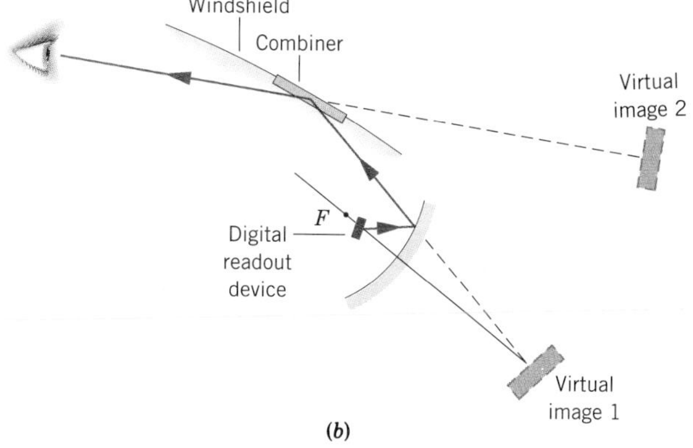

(*b*)

Figure 25.20 (*a*) A head-up display (HUD) presents the driver with a digital readout of his or her speed in the field of view seen through the windshield. (*b*) One version of a HUD uses a concave mirror.

produces image 2, which is what the driver sees. The location of image 2 is determined by how far behind the concave mirror image 1 is. This, in turn, is determined by the position of the digital readout device in front of the mirror and the mirror's focal length. The system is designed to place image 2 near the front bumper. The driver can then read the speed with his eyes focused just as they are to see the road. HUDs have been used for a number of years in fighter aircraft and are now beginning to appear in automobiles.

IMAGE FORMATION BY A CONVEX MIRROR

The procedure for determining the location and size of an image in a convex mirror is similar to that for a concave mirror. The same three rays are used. However, the focal point and center of curvature of a convex mirror lie behind the mirror, not in front of it. Figure 25.21*a* shows the rays. When tracing their paths, we follow the reasoning strategy outlined below, which takes into account the different locations of the focal point and center of curvature.

Reasoning Strategy Ray Tracing for a Convex Mirror

Ray 1. This ray is initially parallel to the principal axis and therefore appears to originate from the focal point *F* after reflection from the mirror.

Ray 2. This ray heads toward *F*, emerging parallel to the principal axis after reflection. Ray 2 is analogous to ray 1, except the order of the incident and reflected rays is interchanged.

Ray 3. This ray travels toward the center of curvature *C*; as a result, the ray strikes the mirror perpendicularly and reflects back on itself.

The three rays in Figure 25.21*a* appear to come from a point on a virtual image that is behind the mirror. The virtual image is diminished in size and upright, relative to the object. A convex mirror *always* forms a virtual image of the object,

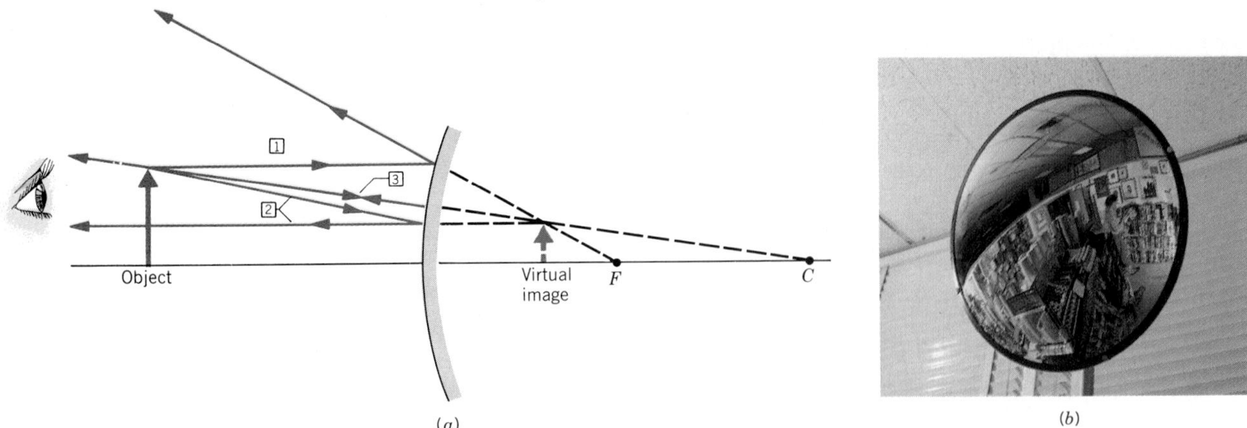

Figure 25.21 (*a*) An object placed in front of a convex mirror produces a virtual image behind the mirror. The virtual image is reduced in size and upright. (*b*) Convex mirrors are often used for security purposes.

no matter where in front of the mirror the object is placed. Because of the shape of convex mirrors, they give a wider field of view than do other types of mirrors. Therefore, they are often used for security purposes, as in Figure 25.21*b*. A mirror with a wide field of view is also needed to give a driver a good rear view. Thus, the outside mirror on the passenger side is often a convex mirror. Printed on such a mirror is usually the warning "VEHICLES IN MIRROR ARE CLOSER THAN THEY APPEAR." The reason for the warning is that the virtual image in Figure 25.21*a* is reduced in size and therefore looks smaller, just as a distant object would appear in a plane mirror. An unwary driver, thinking that the side-view mirror is a plane mirror, might incorrectly deduce from the small size of the image that the car behind is far enough away to ignore.

THE PHYSICS OF . . .
passenger-side automobile mirrors.

25.6 THE MIRROR EQUATION AND THE MAGNIFICATION EQUATION

CONCAVE MIRRORS

Ray diagrams drawn to scale are useful for determining the location and size of the image formed by a mirror. However, for an accurate description of the image, a more analytical technique is needed. It is possible to derive an equation, called the mirror equation, that gives the image distance if the object distance and the focal length of the mirror are known. The image distance is the distance between the image and the mirror, while the object distance is the distance between the object and the mirror. In Figure 25.22 the image and object distances are labeled d_i and d_o, respectively. The height of the image is h_i, and the height of the object is h_o. Part *a* of the drawing shows a ray leaving the top of the object and striking the mirror at the point where the principal axis intersects the mirror. Since the principal axis is perpendicular to the mirror, it is also the normal at the point of incidence. Therefore, the ray reflects at an equal angle and passes through the image. The pink and the green triangles are similar because they have equal angles, so

$$\frac{h_o}{-h_i} = \frac{d_o}{d_i}$$

The minus sign appears on the left in this equation, because the image is inverted in Figure 25.22*a*. In part *b* another ray leaves the top of the object, this time passing through the focal point *F*, reflecting parallel to the principal axis, and then

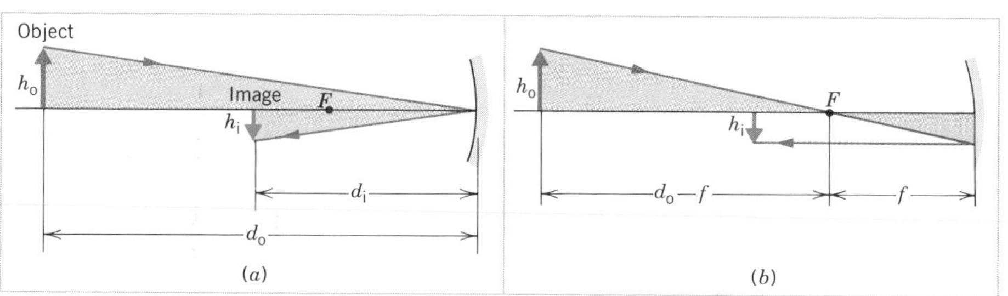

Figure 25.22 These diagrams are used to derive the mirror equation and the magnification equation. (*a*) The triangle shaded in pink is similar to the triangle shaded in green. (*b*) If the ray is close to the principal axis, the two shaded regions are almost similar triangles.

passing through the image. Provided the ray remains close to the axis, the pink triangle and the green area can be considered to be similar triangles, with the result that

$$\frac{h_o}{-h_i} = \frac{d_o - f}{f}$$

Setting the two equations above equal to each other yields $d_o/d_i = (d_o - f)/f$. Rearranging this result gives the ***mirror equation:***

[Mirror equation]

$$\frac{1}{d_o} + \frac{1}{d_i} = \frac{1}{f} \qquad (25.3)$$

We have derived this equation for a real image formed in front of a concave mirror. In this case, the image distance is a positive quantity, as are the object distance and the focal length. However, we have seen in the last section that a concave mirror can also form a virtual image, if the object is located between the focal point and the mirror. Equation 25.3 can also be applied to such a situation, provided that we adopt the convention that d_i is negative for an image behind the mirror, as it is for a virtual image. ***When using the mirror equation under any circumstances, it is useful to construct a ray diagram to guide your thinking and check on your calculation.***

PROBLEM SOLVING INSIGHT

The ***magnification*** m of a mirror is defined as the ratio of the image height to the object height: $m = h_i/h_o$. If the image height is less than the object height, the magnitude of m is less than one. Conversely, if the image is larger than the object, the magnitude of m is greater than one. Since $h_i/h_o = -d_i/d_o$, it follows that

[Magnification equation]

$$m = \frac{\text{Image height}}{\text{Object height}} = -\frac{d_i}{d_o} \qquad (25.4)$$

As Examples 3 and 4 show, the value of m is positive if the image is upright and negative if the image is inverted.

Example 3 A Real Image Formed by a Concave Mirror

A 2.0-cm-high object is placed 7.10 cm from a concave mirror whose radius of curvature is 10.20 cm. Find (a) the location of the image and (b) its size.

REASONING Since $f = \frac{1}{2}R = (10.20\text{ cm})/2 = 5.10\text{ cm}$, the object is located between the focal point F and the center of curvature C of the mirror, as in Figure 25.18*a*. Based on this figure, we expect that the image is real and that, relative to the object, it is farther away from the mirror, inverted, and larger.

SOLUTION

(a) With $d_o = 7.10$ cm and $f = 5.10$ cm, the mirror equation can be used to find the image distance:

$$\frac{1}{d_i} = \frac{1}{f} - \frac{1}{d_o} = \frac{1}{5.10\text{ cm}} - \frac{1}{7.10\text{ cm}} = 0.055\text{ cm}^{-1} \quad \text{or} \quad \boxed{d_i = 18\text{ cm}}$$

In this calculation, f and d_o are positive numbers, indicating that the focal point and the object are in front of the mirror. The positive answer for d_i means that the image is also in front of the mirror, and the reflected rays actually pass through the image, as Figure 25.18*a* shows. In other words, the positive value for d_i indicates that the image is a real image.

PROBLEM SOLVING INSIGHT

According to the mirror equation, the image distance d_i has a reciprocal given by $d_i^{-1} = f^{-1} - d_o^{-1}$, where f is the focal length and d_o is the object distance. After combining the reciprocals f^{-1} and d_o^{-1}, do not forget to take the reciprocal of the result to find d_i.

(b) The height of the image can be determined once the magnification m of the mirror is known. The magnification equation can be used to find m:

$$m = -\frac{d_i}{d_o} = -\frac{18\ \text{cm}}{7.10\ \text{cm}} = -2.5$$

The image height is $h_i = mh_o = (-2.5)(2.0\ \text{cm}) = \boxed{-5.0\ \text{cm}}$. The image is 2.5 times larger than the object, the negative values for m and h_i indicating that the image is inverted with respect to the object, as in Figure 25.18*a*.

Mirrors work equally well in and out of water.

Example 4 A Virtual Image Formed by a Concave Mirror

An object is placed 6.00 cm in front of a concave mirror that has a 10.0-cm focal length. (a) Determine the location of the image. (b) If the object is 1.2 cm high, find the image height.

REASONING The object is located between the focal point and the mirror, as in Figure 25.19*a*. The setup is analogous to a person using a makeup or shaving mirror. Therefore, we expect that the image is virtual and that, relative to the object, it is upright and larger.

SOLUTION

(a) Using the mirror equation with $d_o = 6.00$ cm and $f = 10.0$ cm, we have

$$\frac{1}{d_i} = \frac{1}{f} - \frac{1}{d_o} = \frac{1}{10.0\ \text{cm}} - \frac{1}{6.00\ \text{cm}} = -0.067\ \text{cm}^{-1} \quad \text{or} \quad \boxed{d_i = -15\ \text{cm}}$$

The answer for d_i is negative, indicating that the image is *behind* the mirror. Thus, as expected, the image is a virtual image.

(b) The image height h_i can be found from the magnification and the object height h_o:

$$m = -\frac{d_i}{d_o} = -\frac{(-15\ \text{cm})}{6.00\ \text{cm}} = 2.5$$

The image height is $h_i = mh_o = (2.5)(1.2\ \text{cm}) = \boxed{3.0\ \text{cm}}$. The image is larger than the object, and the positive values for m and h_i indicate that the image is upright (see Figure 25.19*a*).

Surprising effects can sometimes be created by combining more than one mirror. Conceptual Example 5 discusses one such case.

Conceptual Example 5 A Mirage

A toy is designed from two identical concave mirrors that are joined with their reflecting surfaces facing each other. The toy is designed to be a conversation piece that sits on a coffee table. The top mirror has a hole cut into it, so that an object can be put inside to rest on the bottom mirror. The mirrors are joined such that the distance between them is equal to the focal length f of each mirror. As Figure 25.23*a* illustrates, the toy is intended to create a mirage of the object within it. The mirage is an image of the object that appears to float just above the hole, realistic enough to fool people into reaching for it. How does the toy work?

REASONING For convenience, parts *b* and *c* of the drawing show the toy oriented with the mirrors vertical, which matches the way we have drawn them in this chapter. The bottom mirror is on the left. The key to understanding how the toy works is to realize that the object forms a first image in one of the mirrors and that this image serves as the object for the second mirror. The second mirror then forms the final image, or mirage. Thus, we need to consider the mirrors one at a time. In part *b* we consider only the mirror on the right, which is the top mirror (the hole has been omitted for simplicity). Because of the way the mirrors are joined together, the focal point *F* of this mirror is located just at the surface of the other mirror, so that the object lies just within it. This situation is similar to that of the makeup mirror in Figure 25.19*a*, leading to a virtual first image, which is enlarged and upright. As Figure 25.23*c* shows, this image serves as the object for the mirror on the left, which is the bottom mirror in the toy. Because of the way the mirrors are joined and the value chosen for the focal length, the object in part *c* is located beyond the center of curvature C. Except for being switched left to right, the arrangement is just like that in Figure 25.18*b*. As a result, a real final image is formed between the focal point and the center of curvature of the mirror on the left, which is just above the hole in the top of the toy.

Related Homework Material: *Problem 15*

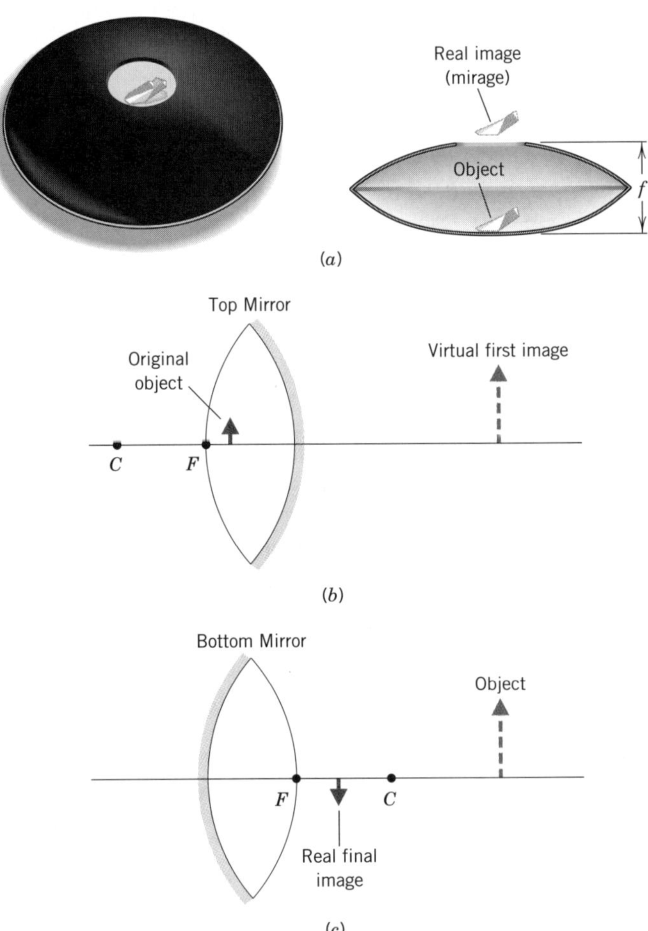

Figure 25.23 (*a*) A mirage toy consists of two concave mirrors mounted face to face. An object that is put into the toy seems to float in midair above the hole in the top mirror. (*b*) One mirror produces a virtual first image that serves as (*c*) the object for the other mirror, which then produces the real final image.

CONVEX MIRRORS

The mirror equation and the magnification equation can also be used with convex mirrors, provided the focal length f is taken to be a *negative number,* as indicated earlier in Equation 25.2. One way to remember this is to recall that the focal point of a convex mirror lies *behind* the mirror. Example 6 deals with a convex mirror.

Example 6 A Virtual Image Formed by a Convex Mirror

A convex mirror is used to reflect light from an object placed 66 cm in front of the mirror. The focal length of the mirror is $f = -46$ cm (note the minus sign). Find (a) the location of the image and (b) the magnification.

REASONING We have seen that a convex mirror always forms a virtual image, as in Figure 25.21*a*, where the image is upright and smaller than the object. These characteristics should also be indicated by the results of our analysis here.

SOLUTION
(a) With $d_o = 66$ cm and $f = -46$ cm, the mirror equation gives

$$\frac{1}{d_i} = \frac{1}{f} - \frac{1}{d_o} = \frac{1}{-46 \text{ cm}} - \frac{1}{66 \text{ cm}} = -0.037 \text{ cm}^{-1} \quad \text{or} \quad \boxed{d_i = -27 \text{ cm}}$$

The negative sign for d_i indicates that the image is behind the mirror and, therefore, is a virtual image.

(b) According to the magnification equation, the magnification is

$$m = -\frac{d_i}{d_o} = -\frac{(-27 \text{ cm})}{66 \text{ cm}} = \boxed{0.41}$$

The image is smaller (m is less than one) and upright (m is positive) with respect to the object.

This enormous polished sphere (called La Geode) is in Paris, France. Its surface acts like a convex mirror.

Convex mirrors, like plane (flat) mirrors, always produce virtual images behind the mirror. However, the virtual image in a convex mirror is closer to the mirror than it would be if the mirror were planar, as Example 7 illustrates.

Example 7 A Convex Versus a Plane Mirror

An object is placed 9.00 cm in front of a mirror. The image is 3.00 cm closer to the mirror when the mirror is convex than when it is planar. Find the focal length of the convex mirror.

REASONING For a plane mirror, the image and the object are the same distance on either side of the mirror. Thus, the image would be 9.00 cm behind a plane mirror. If the image in a convex mirror is 3.00 cm closer than this, the image must be located 6.00 cm behind the convex mirror. In other words, when the object distance is $d_o = 9.00$ cm, the image distance for the convex mirror is $d_i = -6.00$ cm (negative, because the image is virtual). The mirror equation can be used to find the focal length of the mirror.

SOLUTION According to the mirror equation, the reciprocal of the focal length is

$$\frac{1}{f} = \frac{1}{d_o} + \frac{1}{d_i} = \frac{1}{9.00 \text{ cm}} + \frac{1}{(-6.00 \text{ cm})} = -0.056 \text{ cm}^{-1} \quad \text{or} \quad \boxed{f = -18 \text{ cm}}$$

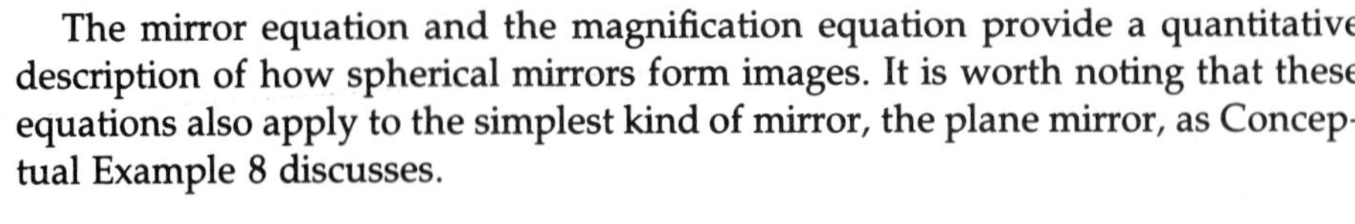

The mirror equation and the magnification equation provide a quantitative description of how spherical mirrors form images. It is worth noting that these equations also apply to the simplest kind of mirror, the plane mirror, as Conceptual Example 8 discusses.

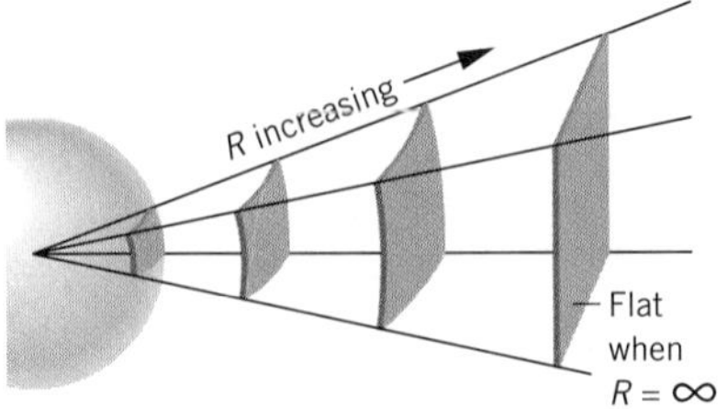

Figure 25.24 As the radius R of a sphere increases, the surface of the sphere in any localized region becomes flatter and flatter. In the limit of an infinitely large radius ($R = \infty$), the surface of a sphere becomes indistinguishable from a plane.

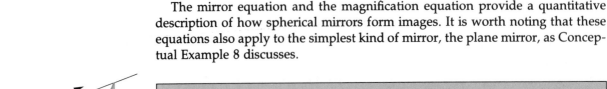

Conceptual Example 8 Plane Mirrors as Limiting Cases of Spherical Mirrors

Why can a plane mirror be regarded as a spherical mirror in the limit that the radius R of the mirror is infinitely large?

REASONING Figure 25.24 illustrates why a plane mirror can be regarded as a spherical mirror with an infinitely large radius ($R = \infty$). As the radius becomes larger and larger, the surface of a sphere in any localized region becomes flatter and flatter, eventually becoming indistinguishable from that of a planar surface. We now explore how this limit affects the mirror equation and the magnification equation, in order to see if they correctly predict the behavior of a plane mirror.

In the limit of $R = \infty$, the focal length of a concave mirror becomes $f = \frac{1}{2}R = \infty$, according to Equation 25.1. However, it is the reciprocal of the focal length, or $1/f$, that appears in the mirror equation, and $1/f = 1/\infty = 0$. Thus, the mirror equation becomes

$$\frac{1}{d_o} + \frac{1}{d_i} = \frac{1}{f} = 0 \quad \text{or} \quad d_i = -d_o$$

The relation $d_i = -d_o$ correctly describes the behavior of a plane mirror in producing a virtual image that is exactly as far behind the mirror as the object is in front of it. With this result, the magnification equation becomes

$$m = -\frac{d_i}{d_o} = -\frac{-d_o}{d_o} = 1$$

The relation $m = 1$ correctly describes the behavior of a plane mirror in producing an upright image that has the same height as the object. We see, then, that in the limit of an infinitely large radius, the equations that describe spherical mirrors also describe plane mirrors.

Related Homework Material: *Problem 19*

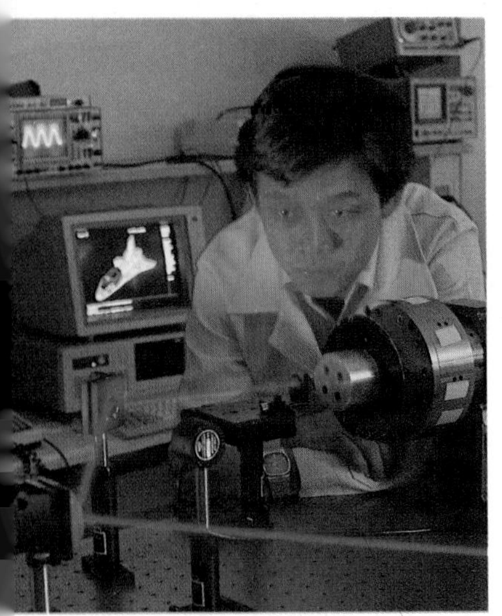

This researcher is using plane mirrors to reflect a red laser beam.

SUMMARY OF SIGN CONVENTIONS

We conclude this section by summarizing the sign conventions that are used with the mirror equation and the magnification equation. These conventions apply to both concave and convex mirrors:

Object distance

d_o is + if the object is in front of the mirror (real object).

d_o is − if the object is behind the mirror (virtual object).*

Image distance

d_i is + if the image is in front of the mirror (real image).

d_i is − if the image is behind the mirror (virtual image).

* Sometimes optical systems use two (or more) mirrors, and the image formed by the first mirror serves as the object for the second mirror. Occasionally, such an object falls *behind* the second mirror. In this case the object distance is negative, and the object is said to be a virtual object.

Focal length
f is + for a concave mirror.
f is − for a convex mirror.
Magnification
m is + for an image that is upright with respect to the object.
m is − for an image that is inverted with respect to the object.

SUMMARY

Wave fronts are surfaces on which all points of a wave are in the same phase of motion. If the wave fronts are flat surfaces, the wave is called a **plane wave. Rays** are lines that are perpendicular to the wave fronts and point in the direction of the velocity of the wave.

When light reflects from a smooth surface, the reflected light obeys the **law of reflection,** which states that (a) the incident ray, the reflected ray, and the normal to the surface all lie in the same plane, and (b) the angle of reflection equals the angle of incidence. The law of reflection explains how mirrors form images. A **virtual image** is one from which rays of light do not actually come, but only appear to do so. A **real image** is one from which rays of light actually emanate. A **plane mirror** forms an upright, virtual image that is located as far behind the mirror as the object is in front of it. In addition, the heights of the image and the object are equal.

A **spherical mirror** has the shape of a section from the surface of a sphere. The **principal axis** of a mirror is a straight line drawn through the center of curvature and the middle of the mirror's surface. Rays that lie close to the principal axis are known as **paraxial rays.** The **radius of curvature** R of the mirror is the distance from the center of curvature to the mirror. The **focal point** of a concave spherical mirror is a point on the principal axis, in front of the mirror. Incident paraxial rays that are parallel to the principal axis converge to the focal point after being reflected from the concave mirror. The focal point of a convex spherical mirror is a point on the principal axis behind the mirror. For a convex mirror, paraxial rays that are parallel to the principal axis seem to diverge from the focal point after reflecting from the mirror. The **focal length** f of a mirror is the distance along the principal axis between the focal point and the mirror. The focal length and the radius of curvature are related by $f = \frac{1}{2}R$ for a concave mirror and $f = -\frac{1}{2}R$ for a convex mirror.

The image produced by a concave mirror can be located by the **ray technique,** using the three rays shown in Figure 25.17. Similarly, the image produced by a convex mirror can be found using the three rays shown in Figure 25.21.

The **mirror equation** can be used with either concave or convex mirrors and specifies the relation between the image distance d_i, the object distance d_o, and the focal length f of the mirror: $1/d_o + 1/d_i = 1/f$. The **magnification** m of a mirror is the ratio of the image height h_i to the object height h_o. The magnification is also related to d_i and d_o by the **magnification equation:** $m = h_i/h_o = -d_i/d_o$. The algebraic sign conventions for the variables appearing in these equations are summarized at the end of Section 25.6.

QUESTIONS

1. A sign painted on a store window is reversed when viewed from inside the store. If a person inside the store views the reversed sign in a plane mirror, does the sign appear as it would when viewed from outside the store? (Try it by writing some letters on a transparent sheet of paper and then holding the back side of the paper up to a mirror.) Explain.

2. Which kind of spherical mirror, concave or convex, can be used to start a fire with sunlight? For the best results, how far from the mirror should the paper to be ignited be placed? Explain.

3. The photograph shows an experimental device at Sandia National Laboratories in New Mexico. This device is a mirror that focuses sunlight to heat sodium to a boil, which then heats helium gas in an engine. The engine does the work of driving a generator to produce electricity. The sodium unit and the engine are labeled in the photo. What kind of mirror is

being used, and where is the sodium unit located relative to the mirror? Express your answer in terms of the focal length of the mirror. Give your reasoning.

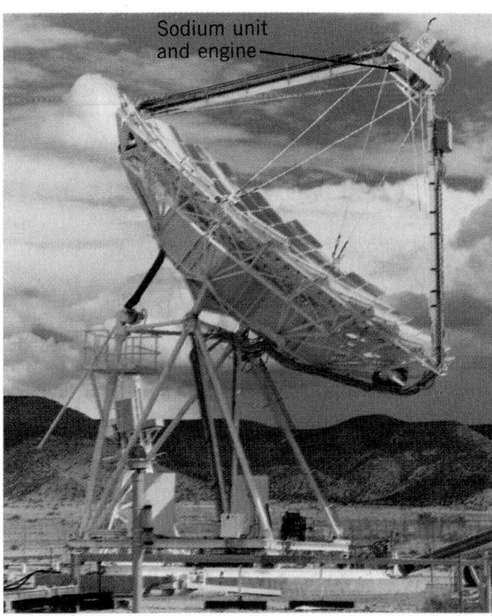

4. Why is your image blurred when you look at yourself in a small shiny sphere, such as a Christmas tree ornament?

5. Refer to Figure 25.14 and the related discussion about spherical aberration. To bring the top ray closer to the focal point F after reflection, describe how you would change the shape of the mirror. Would you open it up to produce a more gently curving shape or close it down to bring the top and bottom edges closer to the principal axis? Account for your answer using the law of reflection.

6. (a) Can the image formed by a concave mirror ever be projected directly onto a screen, without the help of other mirrors or lenses? If so, specify where the object should be placed relative to the mirror. (b) Repeat part (a) assuming that the mirror is convex.

7. When you look at the back side of a shiny teaspoon, held at arm's length, you see yourself upright. When you look at the other side of the spoon, you see yourself upside down. Why?

8. Suppose you wish to design a searchlight that produces a nearly parallel beam of light. The searchlight consists of a light bulb in front of a concave spherical mirror. Where should the bulb be positioned along the principal axis of the mirror? Give your reasoning.

9. If you stand between two parallel plane mirrors, you see an infinite number of images of yourself. This occurs because an image in one mirror is reflected in the other mirror to produce another image, which is then re-reflected, and so forth. The multiple images are equally spaced. Suppose that you are facing a convex mirror, with a plane mirror behind you. Describe what you would see and comment about the spacing between any multiple images. Explain your reasoning.

10. Sometimes news personnel covering an event use a microphone arrangement that is designed to increase the ability of the mike to pick up weak sounds. The drawing shows that the arrangement consists of a "hollowed-out" shell behind the mike. The shell acts like a mirror for sound waves. Explain how this arrangement enables the mike to detect weak sounds.

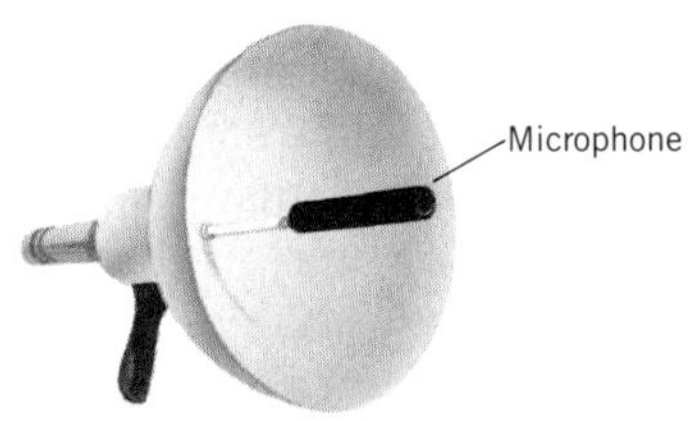

11. When you see the image of yourself formed by a mirror, it is because (1) light rays actually coming from a real image enter your eyes or (2) light rays appearing to come from a virtual image enter your eyes. If light rays from the image do not enter your eyes, you do not see yourself. Are there any places on the principal axis where you cannot see yourself when you are in front of a mirror that is (a) convex and (b) concave? If so, where are these places?

12. Plane mirrors and convex mirrors form virtual images. With a plane mirror, the image may be infinitely far behind the mirror, depending on where the object is located in front of the mirror. For an object in front of a single convex mirror, what is the greatest distance behind the mirror at which the image can be found? Justify your answer.

13. Suppose you stand in front of a spherical mirror (concave or convex). Is it possible for your image to be (a) real and upright or (b) virtual and inverted? Justify your answers.

PROBLEMS

Section 25.2 The Reflection of Light, Section 25.3 The Formation of Images by a Plane Mirror

1. Two plane mirrors are separated by 120°, as the drawing illustrates. If a ray strikes mirror M_1 at a 65° angle of incidence, at what angle θ does it leave mirror M_2?

2. A person stands 3.6 m in front of a wall that is covered floor-to-ceiling with a plane mirror. His eyes are 1.8 m above the floor. He holds a flashlight between his feet and manages to point it at the mirror. At what angle of incidence must the light strike the mirror so the light will reach his eyes?

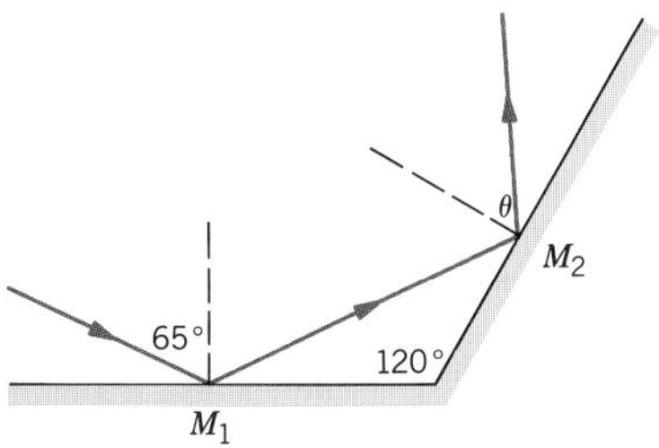

Problem 1

3. Review Conceptual Example 1 before attempting this problem. A person whose eyes are 1.50 m above the floor stands in front of a plane mirror. The top of his head is 0.10 m above his eyes. (a) What is the height of the shortest mirror in which he can see his entire image? (b) How far above the floor should the bottom edge of the mirror be placed?

4. Review Conceptual Example 2. Suppose that in Figure 25.9 the two perpendicular plane mirrors are represented by the $+x$ and $+y$ axes of an x, y coordinate system. An object is in front of these mirrors at a point whose coordinates are $x = +2.0$ m and $y = +1.0$ m. Find the coordinates that locate each of the three images.

5. Two diverging light rays, originating from the same point, have an angle of 10° between them. After the rays reflect from a plane mirror, what is the angle between them? Construct one possible ray diagram that supports your answer.

6. Suppose you walk with a speed of 0.90 m/s toward a plane mirror. What is the speed of your image *relative to you,* when your velocity is (a) perpendicular to the mirror and (b) at an angle of 50.0° with respect to the normal to the mirror?

***7.** Two plane mirrors are facing each other. They are parallel, 3.00 cm apart, and 17.0 cm in length, as the picture indicates. A laser beam is directed at the top mirror, from the left edge of the bottom mirror. What is (a) the smallest and (b) largest angle of incidence θ that allows the beam to hit each mirror once or less?

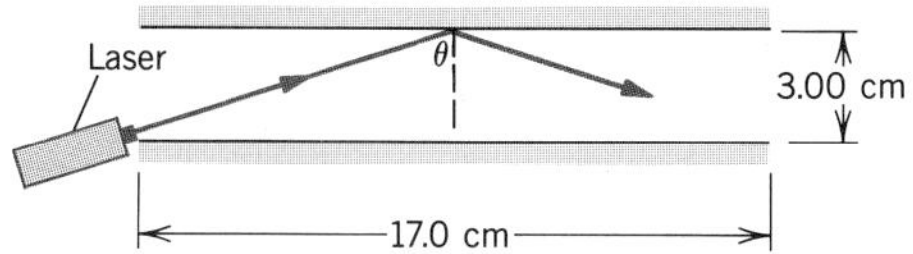

***8.** A ray of light strikes a plane mirror at a 45° angle of incidence. The mirror is then rotated by 15° into the position shown in red in the drawing, while the incident ray is kept fixed. (a) Through what angle ϕ does the reflected ray rotate? (b) What is the answer to part (a) if the angle of incidence is 60° instead of 45°?

****9.** A lamp is twice as far in front of a plane mirror as a person is. Light from the lamp reaches the person via two paths. It strikes the mirror at a 30.0° angle of incidence and reflects from it before reaching the person. It also travels directly to the person without reflecting. Find the ratio of the travel time along the reflected path to the travel time along the direct path.

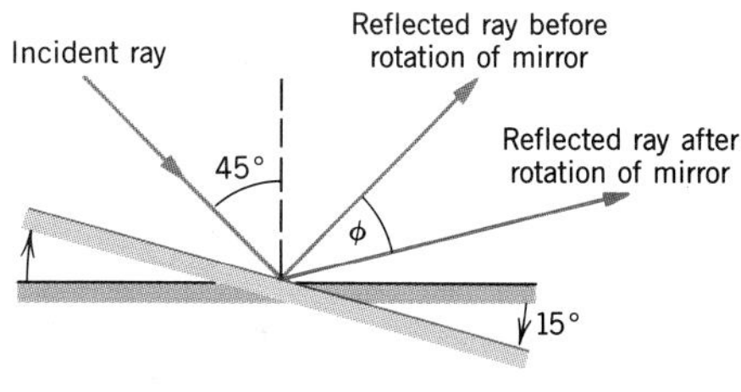

Problem 8

Section 25.4 Spherical Mirrors, Section 25.5 The Formation of Images by Spherical Mirrors

10. A 2.0-cm-high object is situated 15.0 cm in front of a concave mirror that has a radius of curvature of 10.0 cm. Using a ray diagram drawn to scale, measure the location and the height of the image. The mirror must be drawn to scale.

11. Repeat problem 10 for a concave mirror with a focal length of 20.0 cm, an object distance of 12.0 cm, and a 2.0-cm-high object.

12. A convex mirror has a focal length of −50.0 cm. A 15.0-cm-tall object is located 50.0 cm in front of this mirror. Using a ray diagram drawn to scale, determine the location and size of the image. Note that the mirror must be drawn to scale.

13. Repeat problem 10 for a convex mirror with a radius of curvature of 1.00×10^2 cm, an object distance of 25.0 cm, and a 10.0-cm-high object.

14. Repeat problem 10 for a concave mirror with a focal length of 7.50 cm, an object distance of 11.0 cm, and a 1.0-cm-high object.

***15.** Review Conceptual Example 5. Assume that the focal length of each concave mirror is 10.0 cm. Construct the ray diagrams to scale that correspond to Figure 25.23*b* and 25.23*c*. Assume that the original object is 1.0 cm tall and is located 2.0 cm from the bottom mirror. Find the magnitude of the height of the final image. Note that the mirrors must be drawn to scale.

Section 25.6 The Mirror Equation and the Magnification Equation

16. The focal length of a concave mirror is 17 cm. An object is located 38 cm in front of this mirror. Where is the image located?

17. An object is 14 cm in front of a convex mirror that has a focal length of −23 cm. Determine the location of the image.

18. A concave mirror has a focal length of 42 cm. The image formed by this mirror is 97 cm in front of the mirror. What is the object distance?

19. Review Conceptual Example 8 as background for this problem. An object is located 15.0 cm in front of a mirror. If the mirror is a plane mirror, the image distance would be $d_i = -15.0$ cm. However, the mirror is convex. Verify that the larger the radius of the mirror is, the closer the image distance is to the value of -15.0 cm. Do this by calculating the image distance for a radius of (a) 106 cm and (b) 956 cm.

20. A clown is using a concave makeup mirror to get ready for a show and is 27 cm in front of the mirror. The image is 65 cm *behind* the mirror. Find (a) the focal length of the mirror and (b) its magnification.

21. The image behind a convex mirror (radius of curvature = 68 cm) is located 22 cm from the mirror. (a) Where is the object located and (b) what is the magnification of the mirror? Determine whether the image is (c) upright or inverted and (d) larger or smaller than the object.

22. The image formed by a convex mirror is located a distance of 22 cm behind the mirror, when the object is 34 cm in front of the mirror. What is the focal length of the mirror?

23. Convex mirrors are being used to monitor the aisles in a store. The mirrors have a radius of curvature of 4.0 m. (a) What is the image distance if a customer is 15 m in front of the mirror? (b) Is the image real or virtual? (c) If a customer is 1.6 m tall, how tall is the image?

24. A dentist's mirror is placed 1.5 cm from a tooth. The *enlarged* image is located 4.3 cm behind the mirror. (a) What kind of mirror (plane, concave, or convex) is being used? (b) Determine the focal length of the mirror. (c) What is the magnification? (d) How is the image oriented relative to the object?

25. A small postage stamp is placed in front of a concave mirror (radius = R), such that the image distance equals the object distance. (a) In terms of R, what is the object distance? (b) What is the magnification of the mirror? (c) State whether the image is upright or inverted relative to the object. Draw a ray diagram to guide your thinking.

***26.** In Figure 25.20*b* the head-up display is designed so that the distance between the digital readout device and virtual image 1 is 2.00 m. The magnification of virtual image 1 is 4.00. Find the focal length of the concave mirror. (*Hint: Remember that the image distance for virtual image 1 is a negative quantity.*)

***27.** A concave shaving mirror is designed so the virtual image is twice the size of the object, when the distance between the object and the mirror is 15 cm. (a) Determine the radius of curvature of the mirror. (b) Draw a ray diagram to scale showing this situation.

***28.** Show that to produce an image with magnification m, using a mirror whose focal length is f, the object must be placed at a distance d_o from the mirror, where $d_o = (m-1)f/m$.

***29.** An object is located 14.0 cm in front of a convex mirror, the image being 7.00 cm behind the mirror. A second object, twice as tall as the first one, is placed in front of the mirror, but at a different location. The image of this second object has the same height as the other image. How far in front of the mirror is the second object located?

****30.** A spherical mirror is polished on both sides. When used as a convex mirror, the magnification is $+1/4$. What is the magnification when used as a concave mirror, the object remaining the same distance from the mirror?

****31.** Using the mirror equation and the magnification equation, show that for a convex mirror the image is always (a) virtual (i.e., d_i is always negative) and (b) upright and smaller, relative to the object (i.e., m is positive and less than one).

ADDITIONAL PROBLEMS

32. The image of a very distant car is located 12 cm behind a convex mirror. (a) What is the radius of curvature of the mirror? (b) Draw a ray diagram to scale showing this situation.

33. A concave mirror ($R = 64.0$ cm) is used to project a transparent slide onto a wall. The slide is located at a distance of 38.0 cm from the mirror, and a small flashlight shines light through the slide and onto the mirror. The setup is similar to that in Figure 25.18*a*. (a) How far from the wall should the mirror be located? (b) The height of the object on the slide is 1.20 cm. What is the height of the image? (c) How should the slide be oriented, so that the picture on the wall looks normal?

34. A coin is placed 8.0 cm in front of a concave mirror. The mirror produces a real image that has a diameter 4.0 times larger than that of the coin. What is the image distance?

35. The intent of this problem is to illustrate the phenomenon of spherical aberration. Draw a semicircle with a radius of 10 cm to represent a concave spherical mirror. Recall that a radial line drawn between the center of curvature and any point on the arc is perpendicular to the arc and, hence, is a normal to the mirror. (a) Draw a ray parallel to the principal axis at a distance of 5 cm from the axis. Where the ray strikes the mirror, draw the normal and the reflected ray, such that the angle of reflection equals the angle of incidence. Extend the reflected ray until it intersects the principal axis. This ray should cross the axis just inside the focal point. (b) Repeat (a) with a ray drawn at a distance of 7.5 cm from the principal axis and note how much farther from the focal point this ray crosses the axis after reflection.

36. A convex mirror produces an image that is half the size of an object that is placed 13 cm in front of it. What is the focal length of the mirror?

37. The image produced by a concave mirror is located 26 cm in front of the mirror. The focal length of the mirror is 12 cm. How far in front of the mirror is the object located?

***38.** The drawing shows a top view of a square room. One wall is missing, and the other three are each mirrors. From point P in the center of the open side, a laser is fired, with the intent of hitting a small target located at the center of one wall. Identify five directions in which the laser can be fired and score a hit, assuming that the light does not strike any mirror more than once. Draw the rays to confirm your choices.

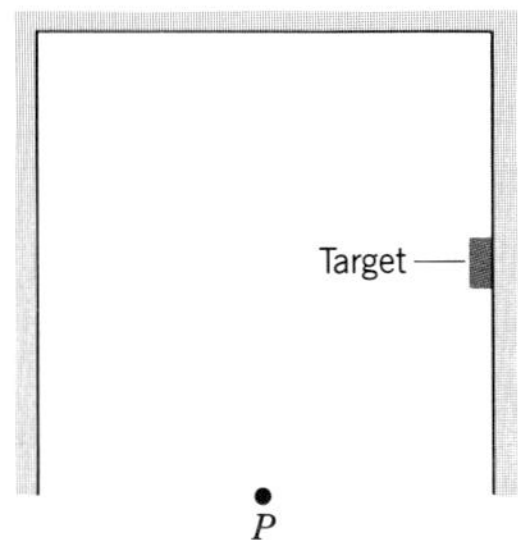

***39.** A gemstone is placed 20.0 cm in front of a concave mirror and is within the focal point. When the concave mirror is replaced with a plane mirror, the image moves 15.0 cm toward the mirror. Find the focal length of the concave mirror.

***40.** (a) Where should a diamond ring be placed in front of a concave mirror, such that the image is twice the size of the ring? There are two answers, depending on whether the image is upright or inverted. Express your answers in terms of the radius of curvature R. (b) Draw ray diagrams to confirm your answers.

***41.** An image formed by a convex mirror ($f = -24.0$ cm) has a magnification of 0.150. Which way and by how much should the object be moved to double the size of the image?

****42.** A concave mirror has a focal length of 30.0 cm. The distance between an object and its image is 45.0 cm. Find the object and image distances assuming that (a) the object lies beyond the center of curvature and (b) the object lies within the focal point.

****43.** In the drawing for problem 38, a laser is fired from point P in the center of the open side of the square room. The laser is pointed at the mirrored wall on the right. At what angle of incidence must the light strike the right-hand wall, so that after being reflected, the light hits the left corner of the back wall?

CHAPTER 26 THE REFRACTION OF LIGHT: LENSES AND OPTICAL INSTRUMENTS

Light can travel through many different media, such as solids, liquids, and gases, although it does so at different speeds. In this chapter, we will see that when light passes from one medium, such as air, into another medium, such as glass, the difference in speeds leads to a change in the direction of travel. This directional change or bending of the light lies at the heart of some remarkable effects, depending on the nature of the materials and their shapes. For example, the beautiful glass sculpture shown in the photograph is alive with color, because different wavelengths of light are bent by different amounts as they enter the sculpture, pass through it, and exit into the air. The change in direction of travel is also responsible for rainbows and the sparkle of diamonds. It is the basis for the important field of fiber optics, which has revolutionized telecommunications and is a key element in the development of the so-called information superhighway. In medicine, fiber optics also plays a significant role in a number of surgical and diagnostic tools. Our eyes contain various fluids, as well as the lens itself, that bend the incoming light to form a sharp image on the retina and give us clear vision. In addition, we often use eyeglasses or contact lenses that bend the incoming light by just the right amount to correct for deficiencies in our vision. The operation of cameras, microscopes, and telescopes also depends on the change in travel direction that occurs when light enters and leaves the lenses of these instruments. In this chapter we will investigate these devices and show how they use light in their operation. To see what happens when light passes from one medium into another, let us begin with a discussion of how the speed of light changes in different materials.

26.1 THE INDEX OF REFRACTION

Many materials, such as air, water, and glass are transparent to light. As it goes from one transparent material into another, a ray of light will deviate from its incident direction (see Figure 26.1*a*), unless it enters perpendicular (normal incidence) to the surface between the materials. This change in direction as light passes from one medium into another is called ***refraction.*** Refraction plays a central role in determining the properties of the lenses used in a wide variety of optical instruments, including eyeglasses, cameras, microscopes, telescopes, and even the human eye itself.

As we will see, refraction depends on the speed of light in a material, and when light travels through a solid, a liquid, or a gas, its speed is different from that in a vacuum. The ***index of refraction*** (or ***refractive index***) n is the ratio of the speed of light c in a vacuum to the speed of light v in the material:

[Index of refraction]

$$n = \frac{\text{Speed of light in a vacuum}}{\text{Speed of light in the material}} = \frac{c}{v} \tag{26.1}$$

Table 26.1 lists the refractive indices for some common substances. As the table indicates, the values of n are greater than unity, so the speed of light in a material medium is less than it is in a vacuum. For example, the index of refraction for diamond is $n = 2.419$, so the speed of light in diamond is $v = c/n = (3.00 \times 10^8 \text{ m/s})/2.419 = 1.24 \times 10^8$ m/s. In contrast, the index of refraction for air and other gases is so close to unity that $n_{\text{air}} = 1$ for most purposes. The index of refraction depends slightly on the wavelength of the light, and the values in Table 26.1 correspond to a wavelength of $\lambda = 589$ nm in a vacuum.

26.2 SNELL'S LAW AND THE REFRACTION OF LIGHT

SNELL'S LAW

When light strikes the interface between two transparent materials, such as air and water, the light generally divides into two parts, as Figure 26.1*a* illustrates. Part of the light is reflected, with the angle of reflection equaling the angle of incidence. The remainder of the light is transmitted across the interface. If the incident ray does not strike the interface at normal incidence, the transmitted ray has a different direction than the incident ray. The ray that enters the second material is said to be refracted.

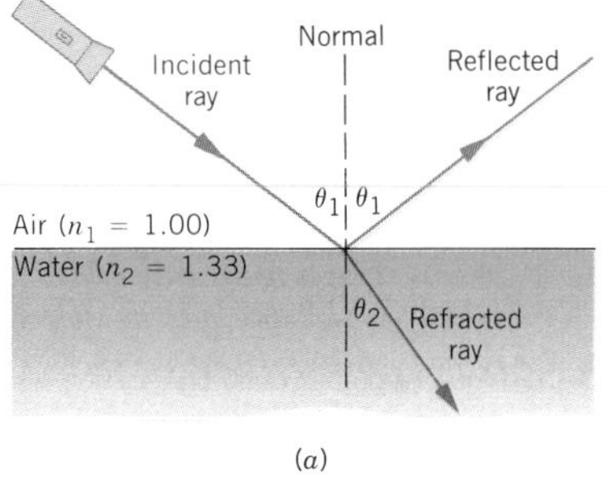

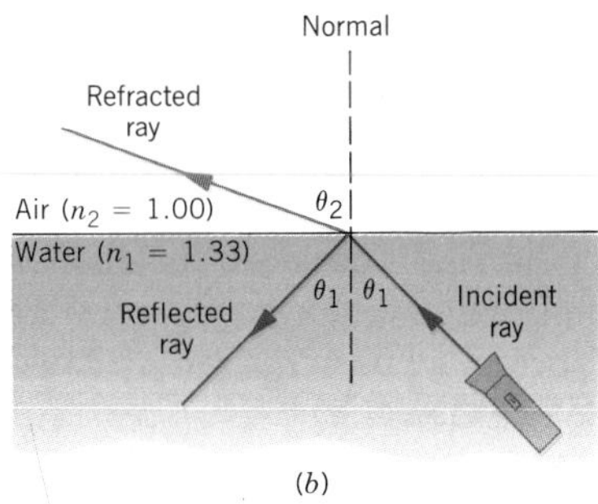

Figure 26.1 (*a*) When a ray of light is directed from air into water, part of the light is reflected at the surface and the remainder is refracted into the water. The refracted ray is bent *toward* the normal ($\theta_2 < \theta_1$). (*b*) When a ray of light is directed from water into air, the refracted ray in air is bent *away* from the normal ($\theta_2 > \theta_1$).

Table 26.1 Index of Refraction[a] for Various Substances

Substance	Index of Refraction, n
Solids at 20 °C	
Diamond	2.419
Glass, crown	1.523
Ice (0 °C)	1.309
Sodium chloride	1.544
Quartz	
Crystalline	1.544
Fused	1.458
Liquids at 20 °C	
Benzene	1.501
Carbon disulfide	1.632
Carbon tetrachloride	1.461
Ethyl alcohol	1.362
Water	1.333
Gases at 0 °C, 1 atm	
Air	1.000 293
Carbon dioxide	1.000 45
Oxygen, O_2	1.000 271
Hydrogen, H_2	1.000 139

[a] Measured with light whose wavelength in a vacuum is 589 nm.

In Figure 26.1*a* the light travels from a medium where the refractive index is smaller (air) into a medium where it is larger (water), and the refracted ray is bent *toward* the normal. Both the incident and refracted rays obey the principle of reversibility, so their directions can be reversed to give a situation like that in part *b* of the drawing. Here light travels from a material with a greater refractive index (water) into one with a smaller refractive index (air), and the refracted ray is bent *away* from the normal. In this case the reflected ray lies in the water, rather than in the air. In both parts of the drawing the angles of incidence, refraction, and reflection are measured relative to the normal. Note that the index of refraction of air is labeled n_1 in part *a*, while it is labeled n_2 in part *b*, because *we label all variables associated with the incident (and reflected) ray with a subscript 1 and all variables associated with the refracted ray with a subscript 2.*

The angle of refraction θ_2 depends on the angle of incidence θ_1 and on the indices of refraction, n_2 and n_1, of the two media. The relation between these quantities is known as ***Snell's law of refraction,*** after the Dutch mathematician Willebrord Snell (1591–1626) who discovered it experimentally. A proof of Snell's law is presented at the end of this section.

Snell's Law of Refraction

When light travels from a material with refractive index n_1 into a material with refractive index n_2, the refracted ray, the incident ray, and the normal to the surface all lie in the same plane. The angle of refraction θ_2 is related to the angle of incidence θ_1 by

$$n_1 \sin \theta_1 = n_2 \sin \theta_2 \qquad (26.2)$$

Example 1 illustrates the use of Snell's law.

Example 1 Determining the Angle of Refraction

A light ray strikes an air/water surface at an angle of 46° with respect to the normal. Find the angle of refraction when the direction of the ray is (a) from air to water and (b) from water to air.

REASONING Snell's law of refraction applies to both part (a) and part (b). However, in part (a) the refracted ray is in water, while in part (b) it is in air. We keep track of this difference by always labeling the variables associated with the incident ray with a subscript 1 and the variables associated with the refracted ray with a subscript 2.

SOLUTION
(a) The incident ray is in air, so $\theta_1 = 46°$ and $n_1 = 1.00$. The refracted ray is in water, so $n_2 = 1.33$. Snell's law can be used to find the angle of refraction:

$$\sin\theta_2 = \frac{n_1 \sin\theta_1}{n_2} = \frac{(1.00)\sin 46°}{1.33} = 0.54 \qquad (26.2)$$

$$\theta_2 = \sin^{-1}(0.54) = \boxed{33°}$$

The refracted ray is bent *toward* the normal, since θ_2 is less than θ_1, as Figure 26.1*a* shows.

(b) With the refracted ray in air, we find that

$$\sin\theta_2 = \frac{n_1 \sin\theta_1}{n_2} = \frac{(1.33)\sin 46°}{1.00} = 0.96$$

$$\theta_2 = \sin^{-1}(0.96) = \boxed{74°}$$

Since θ_2 is greater than θ_1, the refracted ray is bent *away* from the normal, as Figure 26.1*b* indicates.

PROBLEM SOLVING INSIGHT

The angle of incidence θ_1 and the angle of refraction θ_2 that appear in Snell's law are measured with respect to the normal to the surface. They are not measured with respect to the surface itself.

The simultaneous reflection and refraction of light at an interface has applications in a number of devices. For instance, many cars come equipped with an interior rearview mirror that has an adjustment lever. One position of the lever sets the mirror for day driving, while another position sets it for night driving. The night setting is useful for reducing glare from the headlights of the car behind. As Figure 26.2*a* indicates, this kind of mirror is a glass wedge, the back side of which is silvered and highly reflecting. Part *b* of the picture shows the day setting. Light from the car behind follows the path *ABCD* in reaching the driver's eye. At points *A* and *C*, where the light strikes the air–glass surface, there are both reflected and refracted rays. The reflected rays are drawn as thin lines, the thinness denoting that only a small percentage (about 10%) of the light during the day is reflected at *A* and *C*. The weak reflected rays at *A* and *C* do not reach the driver's eye. In contrast, almost all the light reaching the silvered back surface at *B* is reflected toward the driver. Since most of the light follows the path *ABCD*, the driver sees a bright image of the car behind during the day.

THE PHYSICS OF . . .
rearview mirrors that have a day–night adjustment.

During the night, the adjustment lever can be used to rotate the mirror clockwise (see part *c* of the drawing), away from the driver. Now, most of the light from the headlights behind follows the path *ABC* and does not reach the driver. Only the light that is weakly reflected from the front surface along path *AD* is seen. As a result, there is significantly less glare.

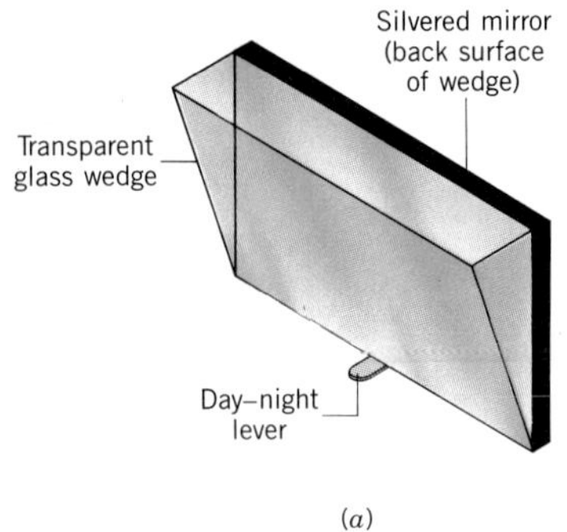

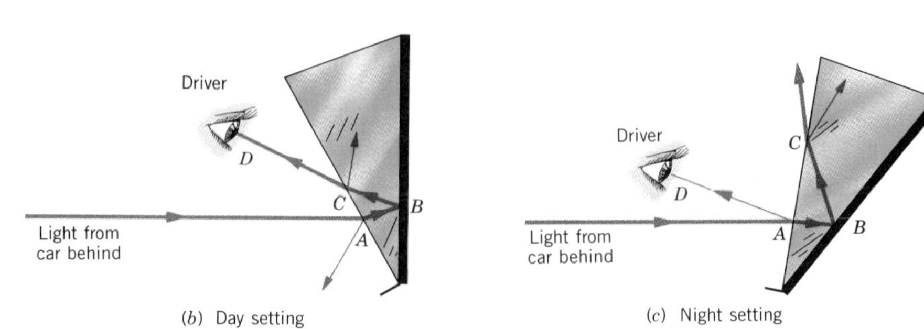

Figure 26.2 An interior rearview mirror with a day–night adjustment lever.

APPARENT DEPTH

One interesting consequence of refraction is that an object lying under water appears to be closer to the surface than it actually is. Example 2 sets the stage for explaining why, by showing what must be done to shine a light on such an object.

Example 2 Finding a Sunken Chest

A searchlight on a yacht is being used at night to illuminate a sunken chest, as in Figure 26.3. At what angle of incidence θ_1 should the light be aimed?

REASONING AND SOLUTION The angle of incidence θ_1 can be determined from Snell's law, provided the angle of refraction θ_2 can be found. From the data in the drawing it follows that $\tan \theta_2 = (2.0\text{ m})/(3.3\text{ m})$, so $\theta_2 = 31°$. With $n_1 = 1.00$ for air and $n_2 = 1.33$ for water, Snell's law gives

$$\sin \theta_1 = \frac{n_2 \sin \theta_2}{n_1} = \frac{(1.33)\sin 31°}{1.00} = 0.69$$

$$\theta_1 = \sin^{-1}(0.69) = \boxed{44°}$$

PROBLEM SOLVING INSIGHT

Remember that the refractive indices are written as n_1 for the medium in which the incident light travels and n_2 for the medium in which the refracted light travels.

When the sunken chest in Example 2 is viewed from the boat (Figure 26.4*a*), light rays from the chest pass upward through the water, refract away from the normal when they enter the air, and then travel to the observer. This picture is similar to Figure 26.3, except the direction of the rays is reversed and the searchlight is replaced by an observer. When the rays entering the air are extended back into the water (see the dashed lines), they indicate that the observer sees a virtual image of the chest at an *apparent depth* that is less than the actual depth. The image is virtual, because light rays do not actually pass through it. For the situation shown in Figure 26.4*a*, it is difficult to determine the apparent depth. A case that is much simpler is shown in part *b* of the drawing, where the observer is *directly above* the submerged object. In this case, the apparent depth d' is related to the actual depth d by

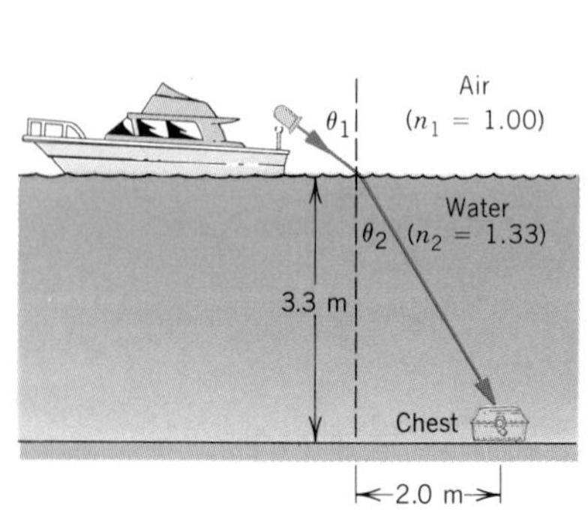

Figure 26.3 The beam from the searchlight is refracted when it enters the water.

Apparent depth with observer directly above object

$$d' = d\left(\frac{n_2}{n_1}\right) \tag{26.3}$$

In this result, n_1 is the refractive index of the medium associated with the incident ray (the medium in which the object is located), while n_2 refers to the medium associated with the refracted ray (the medium in which the observer is situated). The proof of Equation 26.3 is left as problem 20 at the end of the chapter. Example 3 illustrates that the effect of apparent depth is quite noticeable in water.

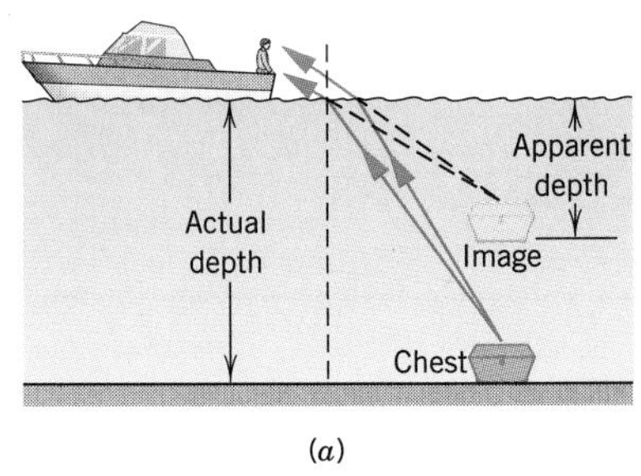

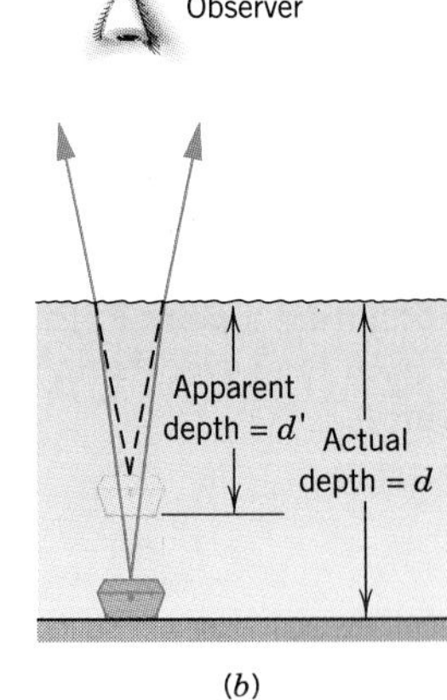

Figure 26.4 **(*a*) Because light from the chest is refracted away from the normal when the light enters the air, the apparent depth of the image is less than the actual depth. (*b*) The observer is viewing the submerged object from directly overhead.**

Example 3 The Apparent Depth of a Swimming Pool

A swimmer is treading water (with her head above the water) at the surface of a 3.00-m-deep pool. She sees a coin on the bottom directly below. How deep does the coin appear to be?

REASONING Equation 26.3 may be used to find the apparent depth, provided we remember that the light rays travel from the coin to the swimmer. Therefore, the incident ray is coming from the coin under the water ($n_1 = 1.33$), while the refracted ray is in the air ($n_2 = 1.00$).

SOLUTION The apparent depth d' of the coin is related to its actual depth d by Equation 26.3:

$$d' = d\left(\frac{n_2}{n_1}\right) = (3.00\text{ m})\left(\frac{1.00}{1.33}\right) = \boxed{2.26\text{ m}}$$

The coin appears to be closer than it actually is.

In Example 3, a person looking at a coin on the bottom of a pool of water sees the coin at an apparent depth that is less than the actual depth. Conceptual Example 4 considers the reverse situation, namely, a person looking from under the water at a coin in the air.

Conceptual Example 4 On the Inside Looking Out

A swimmer is under water and looking up at the surface. Someone holds a coin in the air, directly above the swimmer's eyes. To the swimmer, the coin appears to be at a certain height above the water. Is the apparent height of the coin greater than, less than, or the same as its actual height?

REASONING Figure 26.5 shows the swimmer's eye under water and the coin. The drawing also shows that the apparent height of the coin above the surface of the water is greater than the actual height. This situation is the opposite of that illustrated in Figure 26.4*b*, where an object beneath the water appears to be closer to the surface than it actually is. Consider two rays of light leaving a point P on the coin. When the rays enter the water, they are refracted toward the normal because water has a larger index of refraction than air. By extending the refracted rays backward (see the dashed lines in the drawing), we find that, to the swimmer, the rays appear to originate from a point P'. The point P' is farther from the water than point P, so the swimmer sees the coin at an apparent height d' that is *greater* than the actual height d. Equation 26.3, $d' = d(n_2/n_1)$, reveals the same result, if, as usual, we let n_1 represent the medium (air) associated with the incident ray, and n_2 represent the medium (water) associated with the refracted ray. Since n_2 for water is greater than n_1 for air, the ratio n_2/n_1 is greater than one and d' is larger than d.

Related Homework Material:* *Problem 14

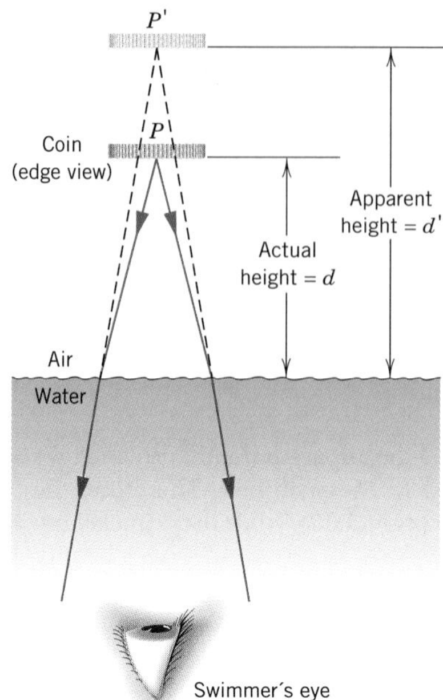

Figure 26.5 An underwater swimmer sees a coin in the air above the water to be farther from the water than it actually is. This is because rays from the coin refract toward the normal as they enter the water. The swimmer perceives the rays as originating from a point P' that is farther from the water than the actual point P.

THE DISPLACEMENT OF LIGHT BY A TRANSPARENT SLAB OF MATERIAL

A common use of a transparent material, such as glass, is for windows. A window pane consists of a plate of glass with parallel surfaces. When a ray of light passes through the glass, the emergent ray is parallel to the incident ray, but displaced from it, as Figure 26.6 shows. This result can be verified by applying Snell's law to each of the two glass surfaces, with the result that $n_1 \sin \theta_1 = n_2 \sin \theta_2 = n_3 \sin \theta_3$. Since air surrounds the glass, $n_1 = n_3$, and it follows that $\sin \theta_1 = \sin \theta_3$. Therefore, $\theta_1 = \theta_3$, and the emergent ray is parallel to the incident ray. However, the drawing shows that the emergent ray is displaced laterally relative to the incident ray. The extent of the lateral displacement depends on the angle of incidence and on the thickness and refractive index of the glass.

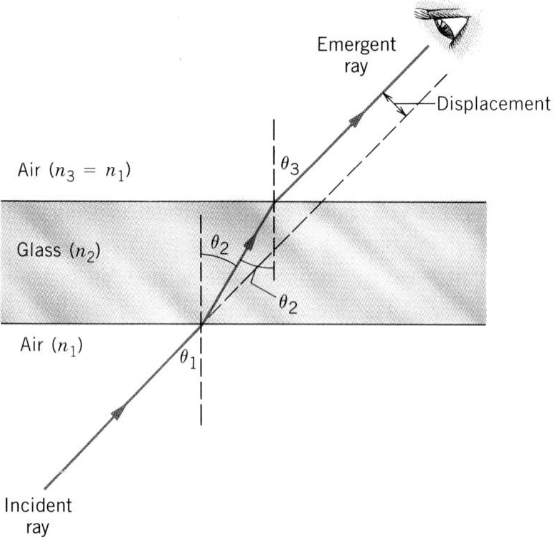

Figure 26.6 When a ray of light passes through a pane of glass that has parallel surfaces and is surrounded by air, the emergent ray is parallel to the incident ray ($\theta_3 = \theta_1$), but is displaced from it.

DERIVATION OF SNELL'S LAW

Snell's law can be derived by considering what happens to the wave fronts when the light passes from one medium into another. Figure 26.7*a* shows light propagating from medium 1, where the speed is relatively large, into medium 2, where the speed is smaller; therefore, n_1 is less than n_2. The plane wave fronts in this picture are drawn perpendicular to the incident and refracted rays. Since the part of each wave front that penetrates medium 2 slows down first, the wave fronts in medium 2 are rotated clockwise relative to those in medium 1. Correspondingly, the refracted ray in medium 2 is bent toward the normal, as the drawing shows.

Although the incident and refracted waves have different speeds, *they have the same frequency f*. The fact that the frequency of the light does not change can be understood in terms of the atomic mechanism underlying the generation of the refracted wave. When the electromagnetic wave strikes the surface, the oscillating electric field forces the electrons in the molecules of medium 2 to oscillate at the same frequency as the wave. The accelerating electrons behave like atomic antennas that radiate "extra" electromagnetic waves, which combine with the original wave. The net electromagnetic wave within medium 2 is then a superposition of the original wave plus the extra radiated waves, and it is this superposition that constitutes the refracted wave. Since the extra waves are radiated at the same frequency as the original wave, the refracted wave also has the same frequency as the original wave.

The distance between successive wave fronts in Figure 26.7*a* has been chosen to be the wavelength λ. Since the frequencies are the same in both media, but the speeds are different, it follows from Equation 16.1 that the wavelengths are different: $\lambda_1 = v_1/f$ and $\lambda_2 = v_2/f$. Since v_1 is assumed to be larger than v_2, λ_1 is larger than λ_2, and the wave fronts are farther apart in medium 1.

Figure 26.7*b* shows an enlarged view of the incident and refracted wave fronts at the surface. The angles θ_1 and θ_2 within the colored right triangles are, respectively, the angles of incidence and refraction. In addition, the triangles share the same hypotenuse h. Therefore,

$$\sin\theta_1 = \frac{\lambda_1}{h} = \frac{(v_1/f)}{h} = \frac{v_1}{hf} \quad \text{and} \quad \sin\theta_2 = \frac{\lambda_2}{h} = \frac{(v_2/f)}{h} = \frac{v_2}{hf}$$

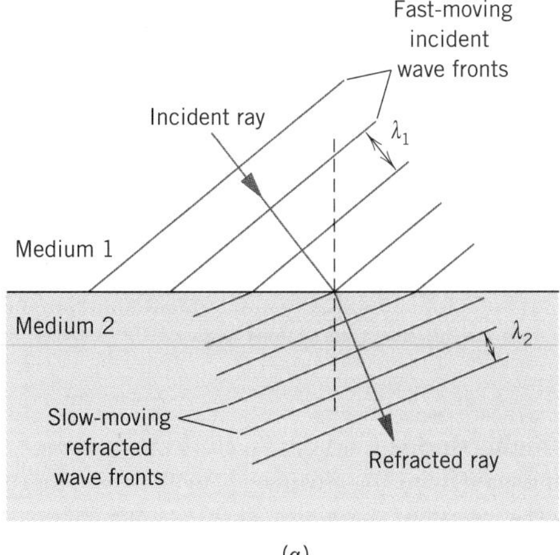

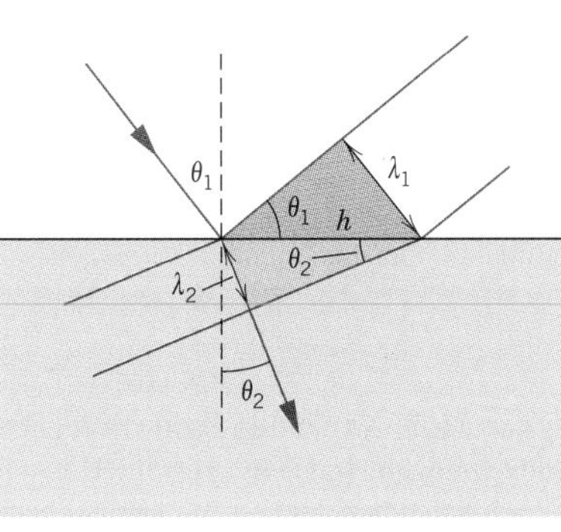

Figure 26.7 (*a*) The wave fronts are refracted as the light ray passes from medium 1 into medium 2. (*b*) An enlarged view of the incident and refracted wave fronts at the surface.

Combining these two equations into a single equation by eliminating the common term hf gives

$$\frac{\sin\theta_1}{v_1} = \frac{\sin\theta_2}{v_2}$$

By multiplying each side of this result by c, the speed of light in a vacuum, and recognizing that the ratio c/v is the index of refraction n, we arrive at Snell's law of refraction: $n_1 \sin\theta_1 = n_2 \sin\theta_2$.

26.3 TOTAL INTERNAL REFLECTION

THE CRITICAL ANGLE AND TOTAL INTERNAL REFLECTION

When light passes from a medium of larger refractive index into one of smaller refractive index—for example, from water to air—the refracted ray bends *away* from the normal, as in Figure 26.8*a*. As the angle of incidence increases, the angle of refraction also increases. When the angle of incidence reaches a certain value, called the ***critical angle*** θ_c, the angle of refraction is 90°. Then the refracted ray points along the surface. Part *b* illustrates what happens at the critical angle. When the angle of incidence exceeds the critical angle, as in part *c* of the drawing, there is no refracted light. All the incident light is reflected back into the medium from which it came, a phenomenon called ***total internal reflection.***

Total internal reflection occurs only when light travels from a higher-index medium toward a lower-index medium. Total internal reflection does not occur when light propagates in the reverse direction—for example, from air to water. In this situation, the refracted ray bends toward the normal, rather than away from it, so there is always a refracted ray, regardless of the angle of incidence.

An expression for the critical angle θ_c can be obtained from Snell's law by setting $\theta_1 = \theta_c$ and $\theta_2 = 90°$:

[Critical angle]

$$\sin\theta_c = \frac{n_2 \sin 90°}{n_1} = \frac{n_2}{n_1} \qquad (n_1 > n_2) \tag{26.4}$$

For example, the critical angle for light traveling from water ($n_1 = 1.33$) to air ($n_2 = 1.00$) is $\theta_c = \sin^{-1}(1.00/1.33) = 48.8°$. For incident angles greater than 48.8°, Snell's law predicts that $\sin\theta_2$ is greater than unity, a value that is not

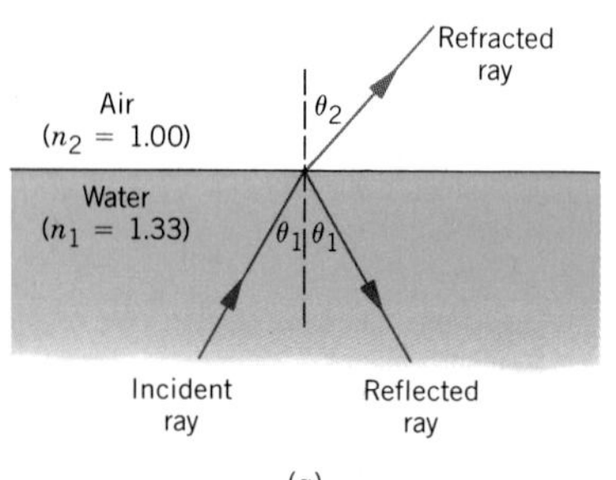

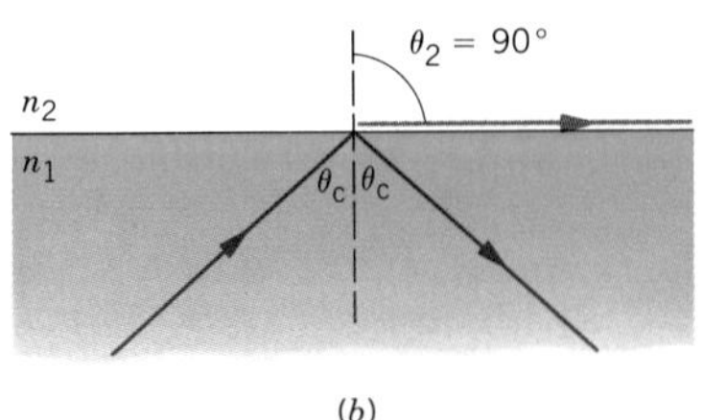

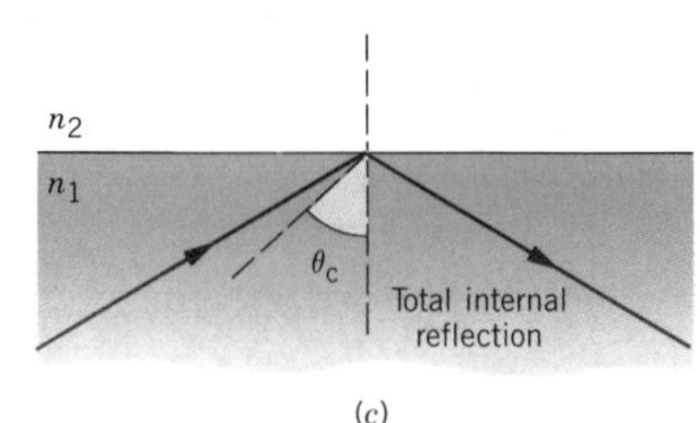

Figure 26.8 (*a*) When light travels from a higher-index medium (water) into a lower-index medium (air), the refracted ray is bent away from the normal. (*b*) When the angle of incidence is equal to the critical angle θ_c, the angle of refraction is 90°. (*c*) If θ_1 is greater than θ_c, there is no refracted ray, and total internal reflection occurs.

possible. Thus, for all light rays with incident angles exceeding 48.8° there is no refracted light, and the light is totally reflected back into the water, as Figure 26.8*c* indicates. The next example illustrates how the critical angle changes when the indices of refraction change.

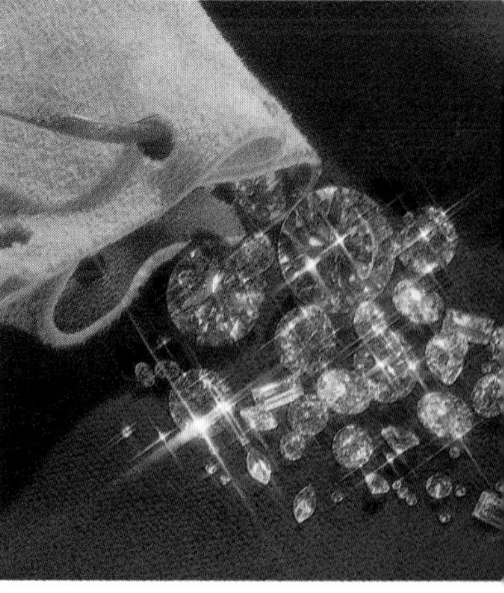

A diamond sparkles because much of the light that enters the gem is reflected back out the top due to the phenomenon of total internal reflection.

Example 5 Total Internal Reflection

A beam of light is propagating through diamond ($n_1 = 2.42$) and strikes a diamond–air interface at an angle of incidence of 28°. (a) Will part of the beam enter the air ($n_2 = 1.00$) or will the beam be totally reflected at the interface? (b) Repeat part (a), assuming that the diamond is surrounded by water ($n_2 = 1.33$).

REASONING Total internal reflection occurs only when the beam of light has an angle of incidence that is equal to or greater than the critical angle θ_c. The critical angle is different in parts (a) and (b), since it depends on the ratio n_2/n_1 of the indices of refraction of the incident (n_1) and refracting (n_2) media.

SOLUTION

(a) The critical angle θ_c for total internal reflection at the diamond–air interface is given by Equation 26.4 as

$$\theta_c = \sin^{-1}\left(\frac{n_2}{n_1}\right) = \sin^{-1}\left(\frac{1.00}{2.42}\right) = 24.4°$$

Because the angle of incidence of 28° is greater than the critical angle, there is no refraction, and the light is totally reflected back into the diamond.

(b) If water, rather than air, surrounds the diamond, the critical angle for total internal reflection becomes larger:

$$\theta_c = \sin^{-1}\left(\frac{n_2}{n_1}\right) = \sin^{-1}\left(\frac{1.33}{2.42}\right) = 33.3°$$

Now a beam of light that has an angle of incidence of 28° (less than the critical angle) at the diamond–water interface is refracted into the water.

The critical angle plays an important role in why a diamond sparkles, as Conceptual Example 6 discusses.

THE PHYSICS OF . . .

why a diamond sparkles.

Conceptual Example 6 The Sparkle of a Diamond

A diamond gemstone is famous for its sparkle, because the light coming from it glitters as it is moved about. Why does a diamond exhibit such brilliance? And why does a diamond lose much of its brilliance when placed under water?

REASONING A diamond sparkles because, when it is held a certain way, the intensity of the light coming from it is greatly enhanced. What causes this enhancement? We know from Table 26.1 that diamond has an index of refraction that is considerably larger than most materials. Is the sparkle of a diamond related to its large index of refraction? The answer is yes, and Figure 26.9 illustrates why. Part *a* of the drawing shows a ray of light striking a bottom facet of the diamond at an angle of incidence that is greater than the critical angle. This ray, and all other rays whose angle

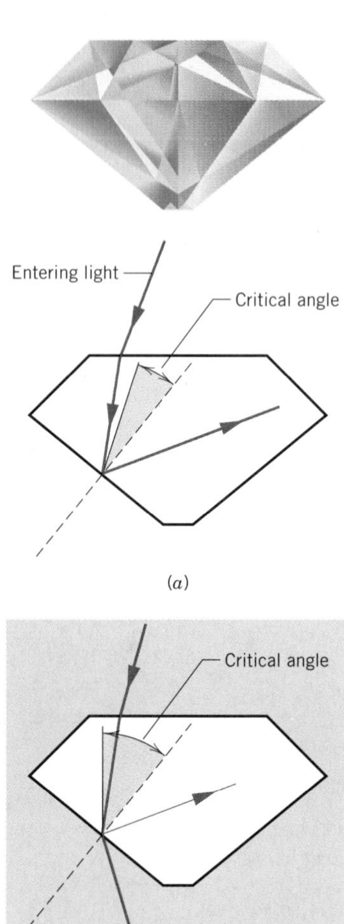

Figure 26.9 (*a*) When this light ray strikes the bottom facet of a diamond–air interface, its angle of incidence is greater than the critical angle, so the light is totally internally reflected. (*b*) When the same ray strikes a diamond–water interface, only part of the ray is reflected back into the diamond, the remainder being refracted into the water.

of incidence exceed the critical angle, are totally reflected back into the diamond, eventually exiting the top surface to give the diamond its sparkle. Many of the rays striking a bottom facet behave in this fashion, because diamond has a relatively small critical angle in air. As part (a) of Example 5 shows, the critical angle is 24.4° and is small because the index of refraction of diamond is large compared to that of air.

Now consider what happens to the same ray of light shown in Figure 26.9*a* when the diamond is placed in water. Because water has a larger index of refraction than air, the critical angle increases to 33.3°, as part (b) of Example 5 shows. Therefore, this particular ray is no longer totally internally reflected. As Figure 26.9*b* illustrates, only some of the light is reflected back into the diamond, and the remainder escapes into the water. Consequently, less light exits from the top of the diamond, causing it to lose much of its brilliance.

Most other materials, such as glass and quartz, do not sparkle nearly as much as diamond. This is because their indices of refraction are much less than that of diamond. Thus, their critical angles are considerably greater than that of diamond, so less light is totally internally reflected. In this respect, these materials behave like a diamond that has been placed in water.

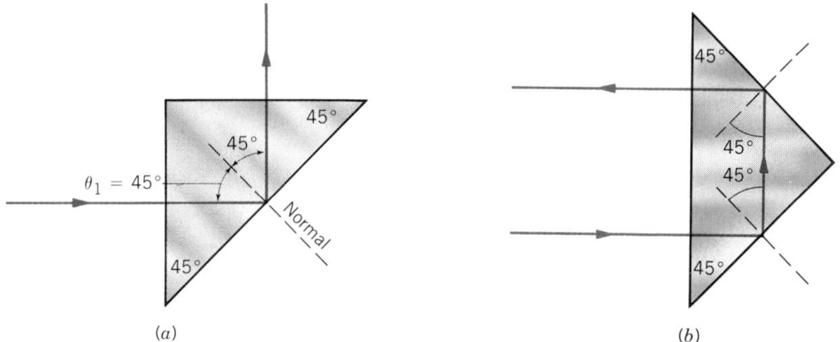

Figure 26.10 Total internal reflection at a glass–air interface can be used to turn a ray of light through an angle of (*a*) 90° or (*b*) 180°.

PRISMS AND TOTAL INTERNAL REFLECTION

Many optical instruments, such as binoculars, periscopes, and telescopes, use glass prisms to turn a beam of light through 90° or 180°. Figure 26.10*a* shows a light ray striking a 45°–45°–90° glass prism ($n_1 = 1.5$). Most of the light enters the prism and is directed toward the hypotenuse of the prism with an angle of incidence of $\theta_1 = 45°$. The critical angle for a glass–air interface is $\theta_c = \sin^{-1}(n_2/n_1) = \sin^{-1}(1.0/1.5) = 42°$. Since the angle of incidence is greater than the critical angle, the light is totally reflected at the hypotenuse and is directed

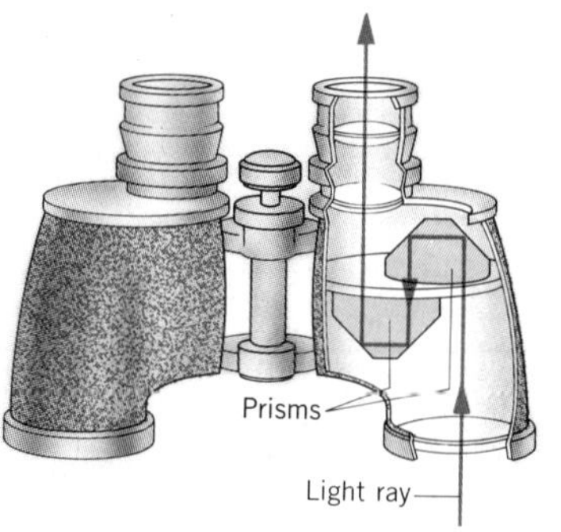

Figure 26.11 Two prisms, each reflecting the light twice by total internal reflection, are sometimes used in binoculars to produce a lateral displacement of a light ray.

vertically upward in the drawing, having been turned through an angle of 90°. Part *b* of the picture shows how the same prism can turn the beam through 180° when total internal reflection occurs twice. Prisms can also be used in tandem to produce a lateral displacement of a light ray, while leaving its initial direction unaltered. Figure 26.11 illustrates such an application in binoculars.

FIBER OPTICS

THE PHYSICS OF . . .

fiber optics.

Another application of total internal reflection is in fiber optics, where hair-thin threads of glass or plastic, called optical fibers, "pipe" light from one place to another. Figure 26.12*a* shows that an optical fiber consists of a cylindrical inner *core* that carries the light and an outer concentric shell, the *cladding*. The core is made from transparent glass or plastic that has a relatively high index of refraction. The cladding, also made of glass, has a relatively low index of refraction. Light enters one end of the core, strikes the core/cladding surface at an angle of incidence greater than the critical angle, and, therefore, is reflected back into the core. Light, then, travels inside the optical fiber along a zigzag path. In a well-designed fiber, little light is lost as a result of absorption by the core, so light can travel many kilometers before its intensity diminishes appreciably. Fibers are often bundled together to produce cables. Because the fibers themselves are so thin, the cables are relatively small and flexible, and they can fit into places inaccessible to larger wire cables.

Optical fibers are revolutionizing video, telephone, and computer-data communications, because a light beam can carry information through an optical fiber just as electricity carries information through copper wires and radio waves carry information through space. The information-carrying capacity of light, however, is thousands of times greater than that of electricity or radio waves. A laser beam traveling through a single optical fiber can carry tens of thousands of telephone conversations and several TV programs simultaneously. Optical fiber cables are the medium of choice for high-quality telecommunications, because the cables are relatively immune to external electrical interference.

Flexible, fiber optic cables are also used in medicine. For instance, such a cable can be passed through the esophagus into the stomach to search for ulcers and

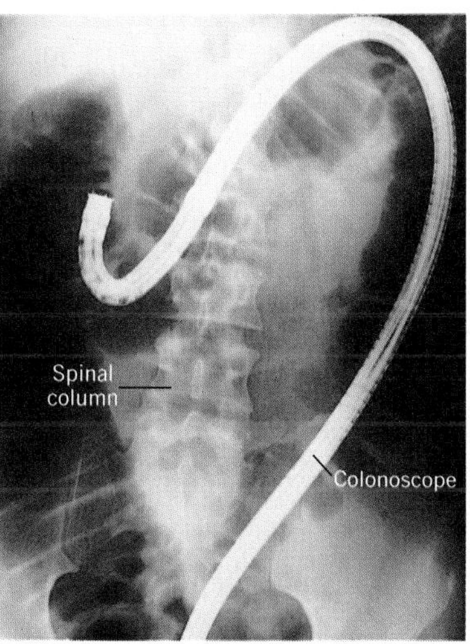

Fiber optics has led to the development of the colonoscope, which is used to examine the interior of the colon. The colonoscope provides one of the best ways to detect colon cancer in its early stages.

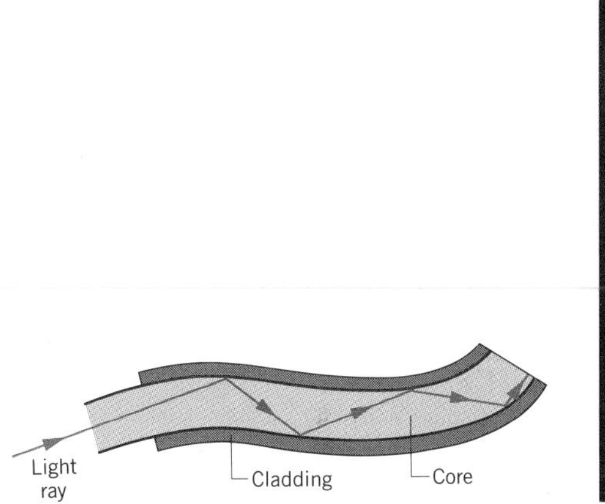

(*a*)

(*b*)

Figure 26.12 (*a*) Light can travel with little loss in a curved optical fiber, because the light is totally reflected whenever it strikes the core–cladding interface and because the absorption of light by the glass core itself is small. (*b*) The thickness of the optical fiber (core plus cladding) is about the thickness of a human hair.

This technician is monitoring the production of fiber-optic material.

other abnormalities. Light is carried into the stomach by the outer fibers of the cable, reflected back by the stomach wall, and transmitted out of the stomach by the inner fibers of the same cable. The image can be displayed on a TV monitor or recorded on film. In arthroscopic surgery, a small surgical instrument, several millimeters in diameter, is mounted at the end of an optical fiber cable. The surgeon can insert the instrument and cable into a joint, such as the knee, with only a tiny incision and minimal damage to the surrounding tissue.

26.4 POLARIZATION AND THE REFLECTION AND REFRACTION OF LIGHT

For incident angles other than 0°, unpolarized light becomes partially polarized in reflecting from a nonmetallic surface, such as water. To demonstrate this fact, rotate a pair of Polaroid sunglasses in the sunlight reflected from a lake. You will see that the light intensity transmitted through the glasses is a minimum when the glasses are oriented as they are normally worn. Since sunglasses are built with the transmission axis aligned in the vertical direction, it follows that the light reflected from the lake is partially polarized in the horizontal direction.

There is one special angle of incidence at which the reflected light is completely polarized parallel to the surface, the refracted ray being only partially polarized. This angle is called the ***Brewster angle*** θ_B. Figure 26.13 summarizes what happens when unpolarized light strikes a nonmetallic surface at the Brewster angle. The value of θ_B is given by ***Brewster's law,*** in which n_1 and n_2 are, respectively, the refractive indices of the materials in which the incident and refracted rays propagate:

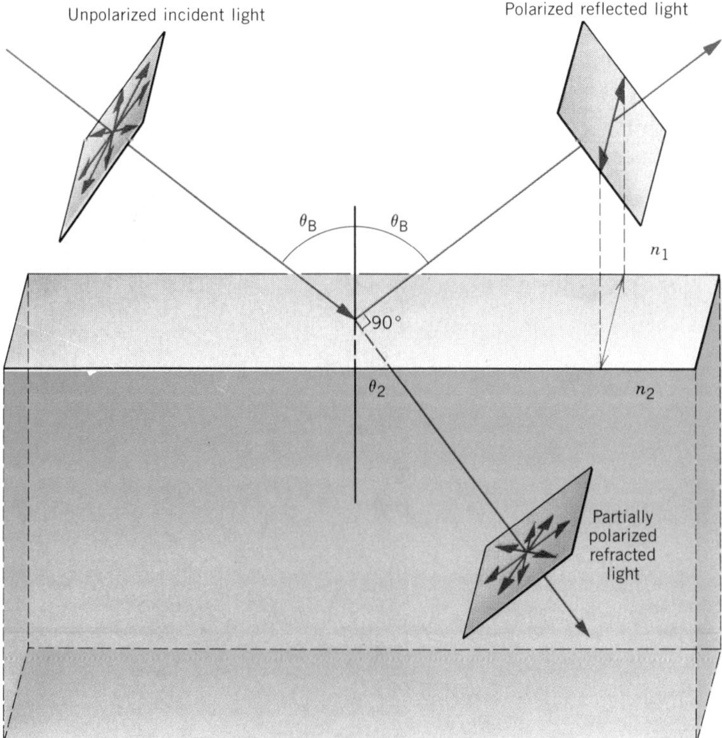

Figure 26.13 When unpolarized light is incident on a nonmetallic surface at the Brewster angle θ_B, the reflected light is 100% polarized in a direction parallel to the surface. The angle between the reflected and refracted rays is 90°.

[Brewster's law]

$$\tan \theta_B = \frac{n_2}{n_1} \tag{26.5}$$

This relation is named after the Scotsman David Brewster (1781–1868), who discovered it. Figure 26.13 also indicates that the reflected and refracted rays are perpendicular to each other when light strikes the surface at the Brewster angle. This result is proved in problem 39 at the end of the chapter.

26.5 THE DISPERSION OF LIGHT: PRISMS AND RAINBOWS

Figure 26.14*a* shows a ray of light passing through a glass prism that is surrounded by air. When the light enters the prism at the left face, the refracted ray is bent toward the normal, for the refractive index of glass is greater than that of air. Conversely, when the light leaves the prism at the right face and enters the air, the light is refracted away from the normal. Thus, the net effect of the prism is to change the direction of the ray, causing it to bend downward upon entering the prism, and downward again upon leaving. Because the refractive index of the glass depends on wavelength (see Table 26.2), the rays corresponding to different colors are bent by different amounts by the prism and depart traveling in different directions. The greater the index of refraction for a given color, the greater the

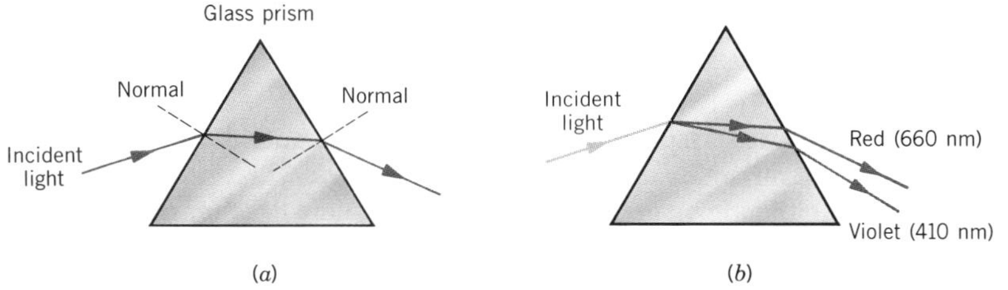

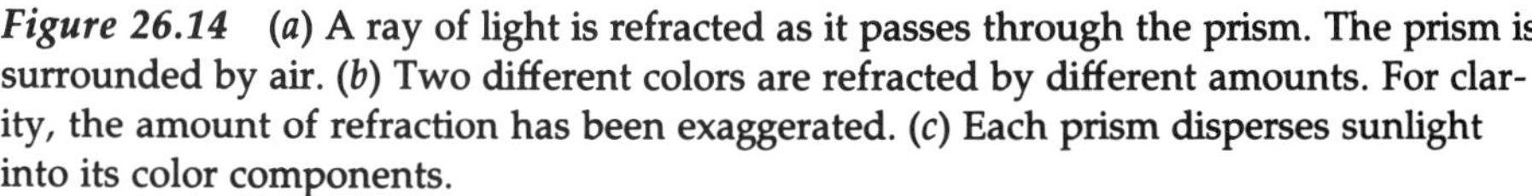

Figure 26.14 (*a*) A ray of light is refracted as it passes through the prism. The prism is surrounded by air. (*b*) Two different colors are refracted by different amounts. For clarity, the amount of refraction has been exaggerated. (*c*) Each prism disperses sunlight into its color components.

Table 26.2 Indices of Refraction *n* of Selected Materials at Various Wavelengths

Approximate Color	Wavelength in Vacuum (nm)	Crown Glass	Flint Glass	Diamond
Red	660	1.520	1.662	2.410
Orange	610	1.522	1.665	2.415
Yellow	580	1.523	1.667	2.417
Green	550	1.526	1.674	2.426
Blue	470	1.531	1.684	2.444
Violet	410	1.538	1.698	2.458

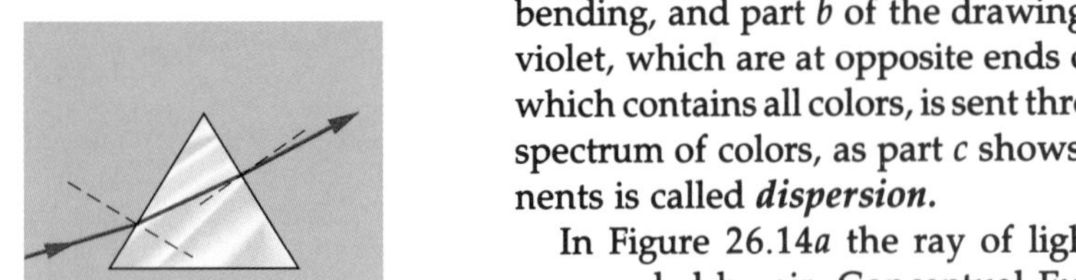

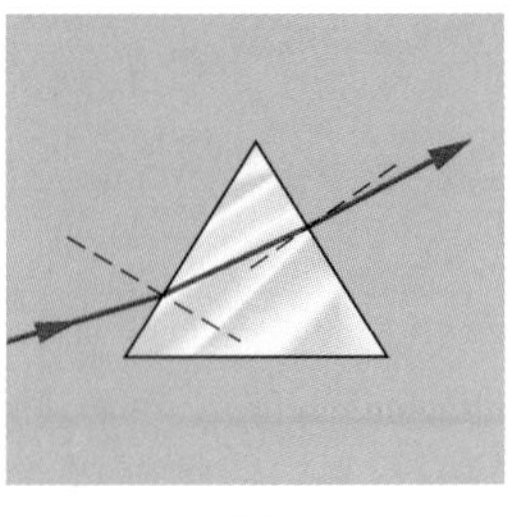

(a)

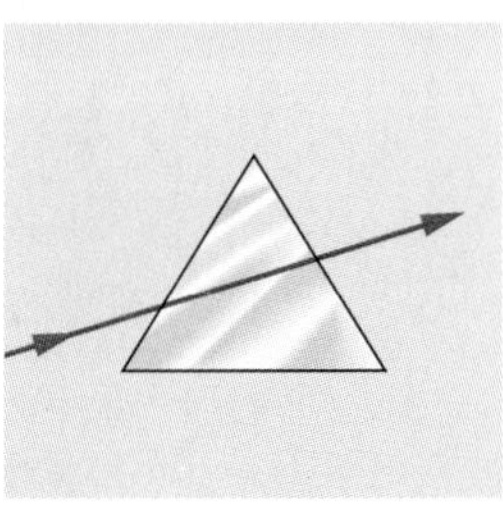

(b)

Figure 26.15 A ray of light passes through identical prisms, each surrounded by a different fluid. The ray of light is (*a*) refracted upward and (*b*) not refracted at all.

bending, and part *b* of the drawing shows the refractions for the colors red and violet, which are at opposite ends of the visible spectrum. If a beam of sunlight, which contains all colors, is sent through the prism, the sunlight is separated into a spectrum of colors, as part *c* shows. The spreading of light into its color components is called ***dispersion.***

In Figure 26.14*a* the ray of light is refracted twice by a glass prism that is surrounded by air. Conceptual Example 7 explores what happens to the light when the prism is surrounded by materials other than air.

Conceptual Example 7 The Refraction of Light Depends on Two Refractive Indices

In Figure 26.14*a* the glass prism is surrounded by air and bends the ray of light downward. It is also possible for the prism to bend the ray upward, as in Figure 26.15*a*, or to not bend the ray at all, as in part *b* of the drawing. How can the situations illustrated in Figure 26.15 arise?

REASONING It is important to realize that the effect of the prism on the light does not depend on the prism alone. After all, Snell's law of refraction includes the refractive indices of *both* materials on either side of an interface. With this in mind, we note that the ray bends upward, or away from the normal, as it enters the prism in Figure 26.15*a*. A ray bends away from the normal when it travels from a medium with a larger refractive index into a medium with a smaller refractive index. When the ray leaves the prism, it again bends upward, which is toward the normal at the point of exit. A ray bends toward the normal when traveling from a smaller toward a larger refractive index. Thus, the situation in Figure 26.15*a* could arise if the prism were immersed in a fluid, such as carbon disulfide, which has a larger refractive index than does glass (see Table 26.1).

We have seen in Figures 26.14*a* and 26.15*a* that a glass prism can bend a ray of light either downward or upward, depending on whether the surrounding fluid has a smaller or larger index of refraction than the glass. It seems logical to conclude that the ray will not bend at all, either down or up, if the surrounding fluid has the *same* index of refraction as the glass—a condition known as *index matching*. This is exactly what is happening in Figure 26.15*b*, for the ray proceeds straight through the prism as if it were not even there. We can use Snell's law, $n_1 \sin \theta_1 = n_2 \sin \theta_2$, to show that the ray is not bent upon entering or leaving the prism under index matching conditions. If the index of refraction of the surrounding fluid equals that of the glass prism, then $n_1 = n_2$, and Snell's law reduces to $\sin \theta_1 = \sin \theta_2$. Therefore, the angle of refraction equals the angle of incidence and no bending of the light occurs.

Related Homework Material: *Questions 18 and 21, Problems 45 and 109*

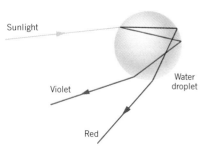

Figure 26.16 When sunlight emerges from a water droplet, the light is dispersed into its constituent colors, of which only two are shown.

Another example of dispersion occurs in rainbows, in which refraction by water droplets gives rise to the colors. Rainbows are often seen just as a storm is leaving, if we look at the departing rain with the sun at our backs. When light from the sun enters a spherical raindrop, as in Figure 26.16, light of each color is refracted or bent by an amount that depends on the refractive index of water for that wavelength. After reflection from the back surface of the droplet, the different colors are again refracted as they reenter the air. Although each droplet disperses the light into its full spectrum of colors, the observer in Figure 26.17 sees only one color of light coming from any given droplet, since only one color travels in the right direction to reach the observer's eyes. However, all colors are visible in a rainbow, because each color originates from different droplets at different angles of elevation.

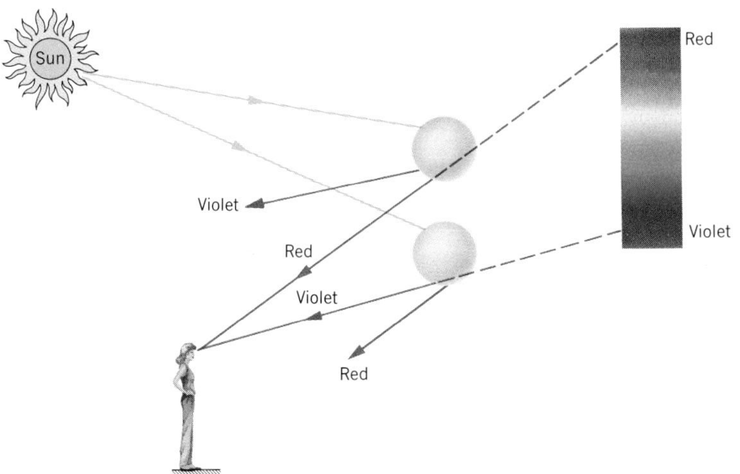

Figure 26.17 The different colors seen in a rainbow originate from water droplets at different angles of elevation.

THE PHYSICS OF . . .

rainbows.

26.6 LENSES

CONVERGING LENSES

The lenses used in optical instruments, such as eyeglasses, cameras, and telescopes, are made from transparent materials that refract light. They refract the light in such a way that an image of the source of the light is formed. Figure 26.18*a* shows a crude lens formed from two glass prisms. Suppose an object, centered on the principal axis, is infinitely far from the lens so the rays from the object are parallel to the principal axis. In passing through the prisms, these rays are bent toward the axis because of refraction. Unfortunately, the rays do not all cross the axis at the same place, and, therefore, such a crude lens gives rise to a "blurred" image of the object.

A better lens can be constructed from a single piece of transparent material with properly curved surfaces, often spherical, as in part *b* of the drawing. With

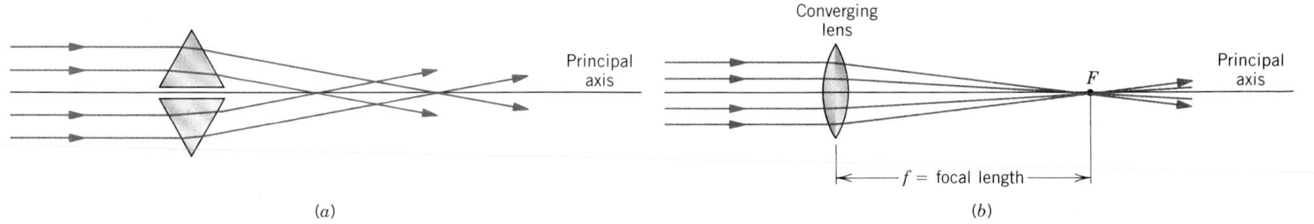

Figure 26.18 (*a*) These two prisms cause rays of light that are parallel to the principal axis to change direction and cross the axis at different points. (*b*) With a converging lens, paraxial rays that are parallel to the principal axis converge to the focal point F after passing through the lens.

this improved lens, rays that are near the principal axis (paraxial rays) and parallel to it converge to a single point on the axis after emerging from the lens. This point is called the ***focal point*** F of the lens. Thus, an object located infinitely far away on the principal axis leads to an image at the focal point of the lens. The distance between the focal point and the lens is the ***focal length*** f. In what follows, we assume the lens is sufficiently thin compared to f that it makes no difference whether f is measured between the focal point and either surface of the lens or the center of the lens. The type of lens in Figure 26.18*b* is known as a ***converging lens,*** because it causes incident parallel rays to converge at the focal point.

DIVERGING LENSES

Another type of lens found in optical instruments is a ***diverging lens,*** which causes incident parallel rays to diverge after exiting the lens. Two prisms can also be used to form a crude diverging lens, as in Figure 26.19*a*. In a properly designed diverging lens, such as that in part *b* of the picture, paraxial rays that are parallel to the principal axis appear to originate from a single point on the axis after passing through the lens. This point is the focal point F of the diverging lens, and its distance f from the lens is the focal length. Again, we assume that the lens is thin compared to the focal length.

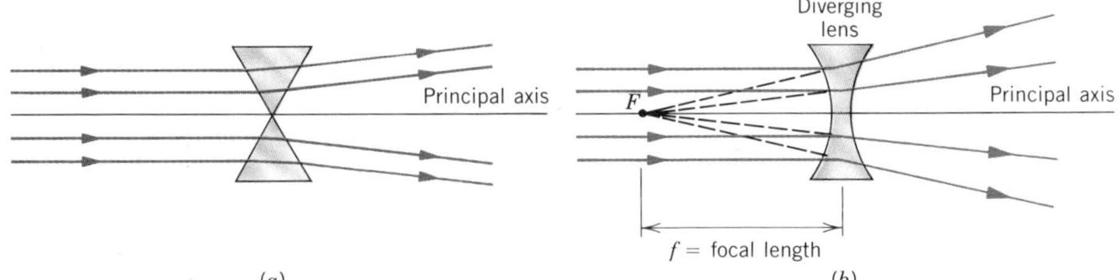

Figure 26.19 (*a*) These two prisms cause parallel rays to diverge. (*b*) With a diverging lens, paraxial rays that are parallel to the principal axis appear to originate from the focal point F after passing through the lens.

Converging and diverging lenses come in a variety of shapes, as Figure 26.20 illustrates. Observe that converging lenses are thicker at the center than at the edges, whereas diverging lenses are thinner at the center.

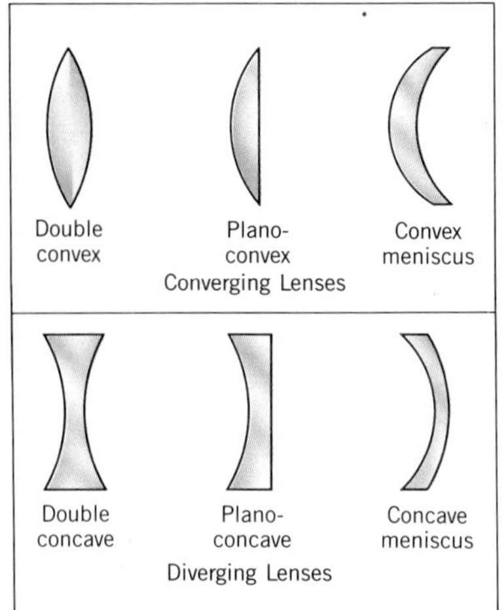

Figure 26.20 Converging and diverging lenses come in a variety of shapes.

26.7 THE FORMATION OF IMAGES BY LENSES

RAY DIAGRAMS

Each point on an object emits light rays in all directions, and when some of these rays pass through a lens, they form an image. As with mirrors, ray diagrams can be drawn to determine the location and size of the image. Lenses differ from mirrors, however, in that light can pass through a lens from left to right or from right to left. Therefore, when constructing ray diagrams, begin by locating a focal point F on *each side of the lens;* each point lies on the principal axis at the same distance f from the lens. The lens is assumed to be a thin lens, in that its thickness is small compared with the focal length and the distances of the object and the image from the lens. For convenience, it is also assumed that the object is located to the left of the lens and is oriented perpendicular to the principal axis. There are three paraxial rays that leave a point on the top of the object and are especially helpful in drawing ray diagrams. They are labeled 1, 2, and 3 in Figure 26.21. When tracing their paths, we follow the reasoning strategy outlined below.

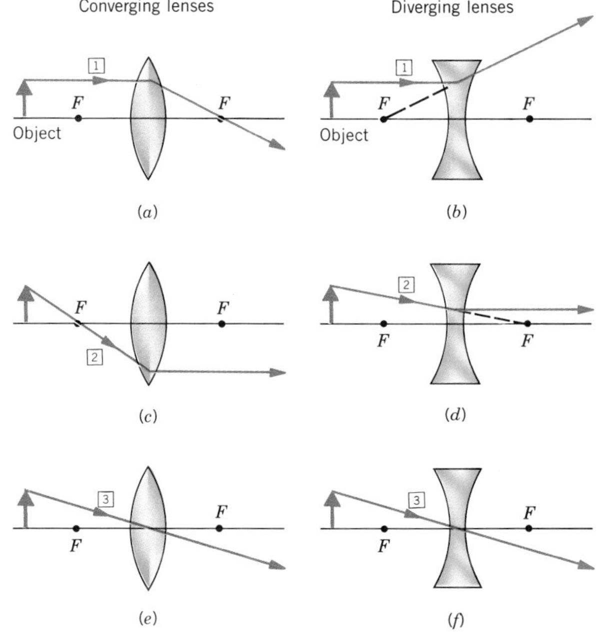

Figure 26.21 The rays shown here are useful in determining the nature of the images formed by converging and diverging lenses.

Reasoning Strategy—Ray Tracing for Converging and Diverging Lenses

Converging Lens	Diverging Lens
Ray 1	
This ray initially travels parallel to the principal axis. In passing through a converging lens, the ray is refracted toward the axis and travels through the focal point on the right side of the lens, as Figure 26.21*a* shows.	This ray initially travels parallel to the principal axis. In passing through a diverging lens, the ray is refracted away from the axis, and *appears* to have originated from the focal point on the left of the lens. The dashed line in Figure 26.21*b* represents the apparent path of the ray.
Ray 2	
This ray first passes through the focal point on the left and then is refracted by the lens in such a way that it leaves traveling parallel to the axis, as in part *c*.	This ray leaves the object and moves toward the focal point on the right of the lens. Before reaching the focal point, however, the ray is refracted by the lens so as to exit parallel to the axis. See part *d*, where the dashed line indicates the ray's path in the absence of the lens.
Ray 3	
This ray travels directly through the center of the thin lens without any appreciable bending, as in part *e*.	This ray travels directly through the center of the thin lens without any appreciable bending, as in part *f*.

The water droplets in a rainbow have much in common with lenses, because they both refract the light that enters them.

Ray 3 does not bend as it proceeds through the lens, because the front and back surfaces of each type of lens are nearly parallel at the center. Thus, in either case, the lens behaves as a transparent slab. As Figure 26.6 shows, the rays incident on and exiting from a slab travel in the same direction with only a lateral displacement. If the lens is sufficiently thin, the displacement is negligibly small.

IMAGE FORMATION BY A CONVERGING LENS

THE PHYSICS OF . . . a camera.

Figure 26.22*a* illustrates the formation of a real image by a converging lens. Here the object is located at a distance from the lens that is greater than twice the focal length (beyond the point labeled 2*F*). To locate the image, any two of the three special rays can be drawn from the tip of the object, although all three are shown in the drawing. The point on the right side of the lens where these rays intersect locates the tip of the image. The ray diagram indicates that the image is real, inverted, and smaller than the object. This optical arrangement is similar to that used in a camera, where a piece of film records the image (see part *b* of the drawing).

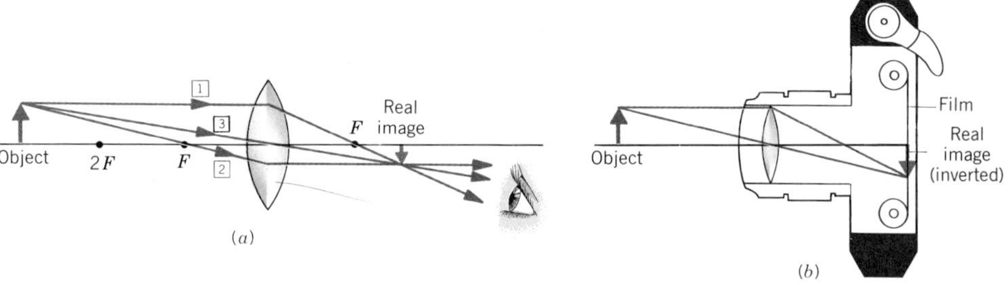

Figure 26.22 (*a*) When the object is placed to the left of the point labeled 2*F*, a real, inverted, and smaller image is formed. (*b*) The arrangement in part *a* is like that used in a camera.

THE PHYSICS OF . . . a slide or film projector.

When the object is placed between 2*F* and *F*, as in Figure 26.23*a*, the image is still real and inverted; however, the image is now larger than the object. This optical system is used in a slide or film projector in which a small piece of film is the object and the enlarged image falls on a screen. However, to obtain an image that is right-side up, the film must be placed in the projector upside down.

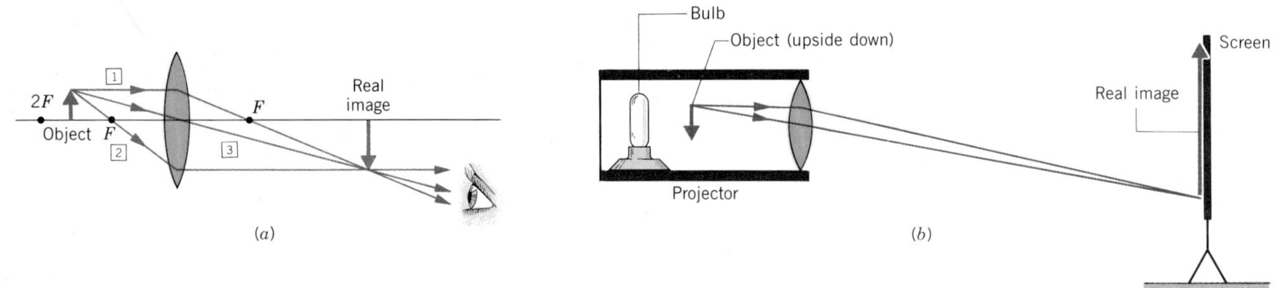

Figure 26.23 (*a*) When the object is placed between 2*F* and *F*, the image is real, inverted, and larger than the object. (*b*) This arrangement is found in projectors.

When the object is located between the focal point and the lens, as in Figure 26.24, the rays diverge after leaving the lens. To a person viewing the diverging rays, they appear to come from an image behind (to the left) of the lens. Because the rays do not actually come from the image, it is a virtual image. The ray diagram shows that the virtual image is upright and enlarged. A magnifying glass uses this arrangement, as can be seen in part *b* of the drawing.

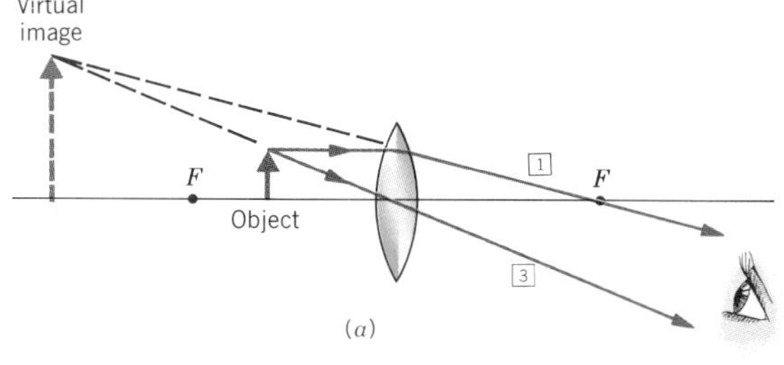

Figure 26.24 (*a*) When an object is placed within the focal point *F* of a converging lens, an upright, enlarged, and virtual image is created. (*b*) Such an image is seen when looking through a magnifying glass.

IMAGE FORMATION BY A DIVERGING LENS

Light rays diverge upon leaving a diverging lens, as Figure 26.25 shows, and the ray diagram indicates that a virtual image is formed on the left side of the lens. In fact, regardless of the position of a real object, a diverging lens always forms a virtual image that is upright and smaller relative to the object. Conceptual Example 8 explores the fundamental difference between a diverging lens and a converging lens.

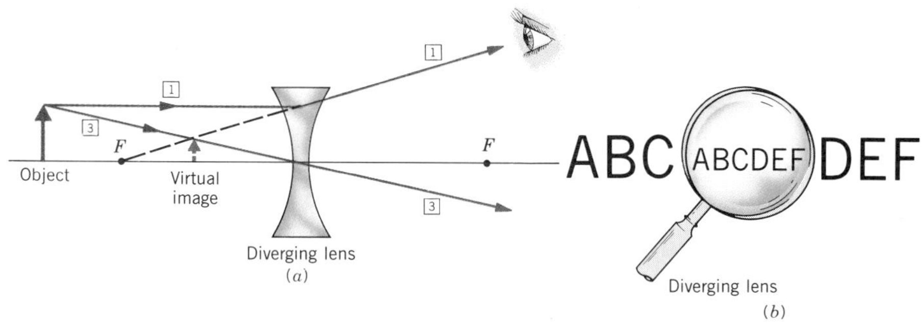

Figure 26.25 (*a*) A diverging lens always forms a virtual image of a real object. The image is upright and smaller relative to the object. (*b*) The image seen through a diverging lens.

Conceptual Example 8 Starting a Fire with Sunlight

Imagine that you are on a camping trip. It's a sunny day, so you decide to start a fire by using a lens to focus the sun's rays onto a piece of paper (Figure 26.26). What type of lens could be used to kindle the fire, a diverging lens or a converging lens? And where should the lens be placed relative to the paper in order to ignite it as quickly as possible?

REASONING To ignite the paper, the sun's rays must be concentrated so as to produce the largest possible light intensity. With such a large intensity, the paper heats up rapidly, ultimately reaching its kindling temperature. A diverging lens cannot be used to concentrate the rays, because, as Figure 26.19 illustrates, they diverge after leaving such a lens. The diverging rays give rise to a smaller light intensity, since they spread out over larger areas after leaving the lens.

On the other hand, the sun's rays can be concentrated after leaving a converging lens, as Figure 26.18 shows. It can be seen in this drawing that, with the sun so far away, the point of greatest concentration occurs at the focal point of the lens, where all the rays cross the principal axis. Thus, to start the fire in the quickest time possible, use a converging lens and position it so that the paper is at the focal point of the lens.

Related Homework Material: *Question 23, Problem 57*

Figure 26.26 A lens is being used to start a fire by focusing the sun's rays onto a piece of paper. Is the lens a diverging or a converging lens?

26.8 THE THIN-LENS EQUATION AND THE MAGNIFICATION EQUATION

For an object in front of a spherical mirror, it is possible to determine the location, size, and nature of the image with the mirror equation and the magnification equation. A similar analysis can be carried out for thin lenses, and the resulting equations are identical to those used with mirrors. These equations are

[Thin-lens equation]

$$\frac{1}{d_o} + \frac{1}{d_i} = \frac{1}{f} \tag{26.6}$$

[Magnification equation]

$$m = \frac{\text{Image height}}{\text{Object height}} = \frac{h_i}{h_o} = -\frac{d_i}{d_o} \tag{26.7}$$

Figure 26.27 defines the symbols in these expressions with the aid of a thin converging lens, but the expressions also apply to a diverging lens, if it is thin. The derivations of these equations are presented at the end of this section.

Certain sign conventions accompany the use of the thin-lens and magnification equations, and the conventions are similar to those used with mirrors in Section 25.6. These conventions allow the equations to convey information about whether the image is real or virtual, upright or inverted, and enlarged or reduced with respect to the object. The issue of real-versus-virtual images, however, is slightly different with lenses than with mirrors. With a mirror, a real image is formed on the *same side* of the mirror as the object, in which case the image distance d_i is a positive number (see Figure 25.18). With a lens, a positive value for d_i also means the image is real. But, starting with an actual object, a real image is formed on the *opposite side* of the lens as the object (see Figure 26.27). The ***sign conventions*** listed below apply to light rays traveling from left to right from a real object.

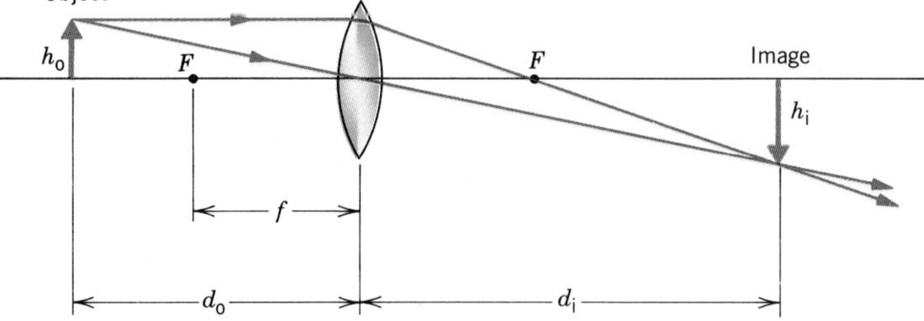

Figure 26.27 **The drawing shows the focal length *f*, the object distance d_o, and the image distance d_i, for a converging lens. The object and image heights are, respectively, h_o and h_i.**

PROBLEM SOLVING INSIGHT

When using the thin-lens and magnification equations, be sure to follow the sign conventions summarized here.

Object distance

d_o is + if the object is to the left of the lens (real object), as is usually the case.
d_o is − if the object is to the right of the lens (virtual object).*

Image distance

d_i is + for an image (real) formed to the right of the lens by a real object.
d_i is − for an image (virtual) formed to the left of the lens by a real object.

* This situation arises in systems containing more than one lens, where the image formed by the first lens becomes the object for the second lens. In such a case, the object of the second lens may lie to the right of that lens, in which event d_o is assigned a negative value and the object is called a virtual object.

Focal length

f is + for a converging lens.
f is − for a diverging lens.

Magnification

m is + for an image that is upright with respect to the object.
m is − for an image that is inverted with respect to the object.

Examples 9 and 10 illustrate the use of the thin-lens and magnification equations.

Example 9 The Real Image Formed by a Camera Lens

A 1.70-m-tall person is standing 2.50 m in front of a camera. The camera uses a converging lens whose focal length is 0.0500 m. (a) Find the image distance (the distance between the lens and the film) and determine whether the image is real or virtual. (b) Find the magnification and the height of the image on the film.

REASONING This optical arrangement is similar to that in Figure 26.22, where the object distance is greater than twice the focal length of the lens. Therefore, we expect the image to be real, inverted, and smaller than the object.

SOLUTION

(a) To find the image distance d_i we use the thin-lens equation with $d_o = 2.50$ m and $f = 0.0500$ m:

$$\frac{1}{d_i} = \frac{1}{f} - \frac{1}{d_o} = \frac{1}{0.0500 \text{ m}} - \frac{1}{2.50 \text{ m}} = 19.6 \text{ m}^{-1} \quad \text{or} \quad \boxed{d_i = 0.0510 \text{ m}}$$

Since the image distance is a positive number, a $\boxed{\text{real image}}$ is formed on the film.

(b) The magnification follows from Equation 26.7:

$$m = -\frac{d_i}{d_o} = -\frac{0.0510 \text{ m}}{2.50 \text{ m}} = \boxed{-0.0204}$$

The image is 0.0204 times as large as the object, and it is inverted since m is negative. Since the object height is $h_o = 1.70$ m, the image height is

$$h_i = mh_o = (-0.0204)(1.70 \text{ m}) = \boxed{-0.0347 \text{ m}}$$

PROBLEM SOLVING INSIGHT

According to the thin-lens equation, the image distance d_i has a reciprocal given by $d_i^{-1} = f^{-1} - d_o^{-1}$, where f is the focal length and d_o is the object distance. After combining the reciprocals f^{-1} and d_o^{-1}, do not forget to take the reciprocal of the result to find d_i.

Example 10 The Virtual Image Formed by a Diverging Lens

An object is placed 7.10 cm to the left of a diverging lens whose focal length is $f = -5.08$ cm (a diverging lens has a negative focal length). (a) Find the image distance and determine whether the image is real or virtual. (b) Obtain the magnification.

REASONING This situation is similar to that in Figure 26.25. The ray diagram shows that the image is virtual, erect, and smaller than the object.

SOLUTION

(a) The thin-lens equation can be used to find the image distance d_i:

$$\frac{1}{d_i} = \frac{1}{f} - \frac{1}{d_o} = \frac{1}{-5.08 \text{ cm}} - \frac{1}{7.10 \text{ cm}} = -0.338 \text{ cm}^{-1} \quad \text{or} \quad \boxed{d_i = -2.96 \text{ cm}}$$

The image distance is negative, indicating that the image is virtual and located to the left of the lens.

(b) Since d_i and d_o are known, the magnification can be determined:

$$m = -\frac{d_i}{d_o} = -\frac{(-2.96\ \text{cm})}{7.10\ \text{cm}} = \boxed{0.417} \qquad (26.7)$$

The image is upright (m is +) and smaller ($m < 1$) than the object.

The thin-lens and magnification equations can be derived by considering rays 1 and 3 in Figure 26.28*a*. Ray 1 is shown separately in part *b* of the drawing, where the angle θ is the same in each of the two pink triangles. Thus, $\tan\theta$ is the same for each triangle, so

$$\tan\theta = \frac{h_o}{f} = \frac{-h_i}{d_i - f}$$

A minus sign has been inserted in the numerator of the ratio $h_i/(d_i - f)$ for the following reason. The angle θ in Figure 26.28*b* is assumed to be positive. Since the image is inverted relative to the object, the image height h_i is a negative number. The insertion of the minus sign ensures that the term $-h_i/(d_i - f)$, and hence $\tan\theta$, is a positive quantity.

Ray 3 is shown separately in part *c* of the drawing, where the angles θ' are the same. Therefore,

$$\tan\theta' = \frac{h_o}{d_o} = \frac{-h_i}{d_i}$$

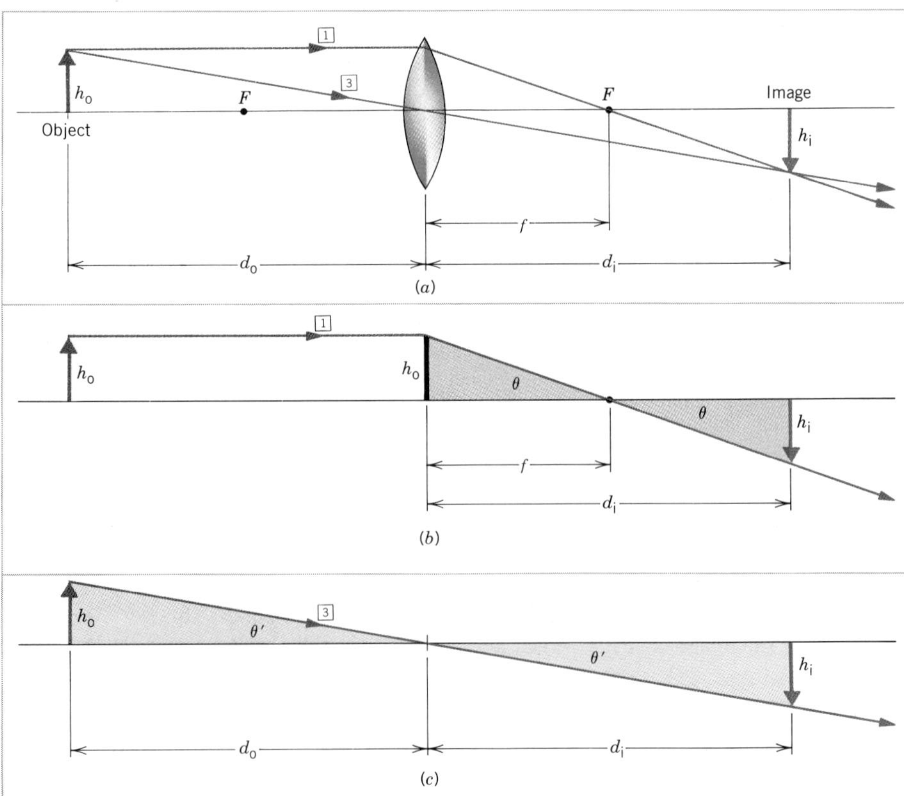

Figure 26.28 These ray diagrams are used for deriving the thin-lens and magnification equations.

A minus sign has been inserted in the numerator of the term h_i/d_i for the same reason that a minus sign was inserted earlier, namely, to ensure that tan θ' is a positive quantity. The first equation gives $h_i/h_o = -(d_i - f)/f$, while the second equation yields $h_i/h_o = -d_i/d_o$. Equating these two expressions for h_i/h_o and rearranging the result produces the thin-lens equation, $1/d_o + 1/d_i = 1/f$. The magnification equation follows directly from the equation $h_i/h_o = -d_i/d_o$, if we recognize that h_i/h_o is the magnification m of the lens.

26.9 LENSES IN COMBINATION

Many optical instruments, such as microscopes and telescopes, use a number of lenses together to produce an image. Among other things, a multiple-lens system can produce an image that is magnified more than is possible with a single lens. For instance, Figure 26.29*a* shows a two-lens system used in a microscope. The first lens, the lens closest to the object, is referred to as the *objective.* The second lens is known as the *eyepiece* (or *ocular*). The object is placed just outside the focal

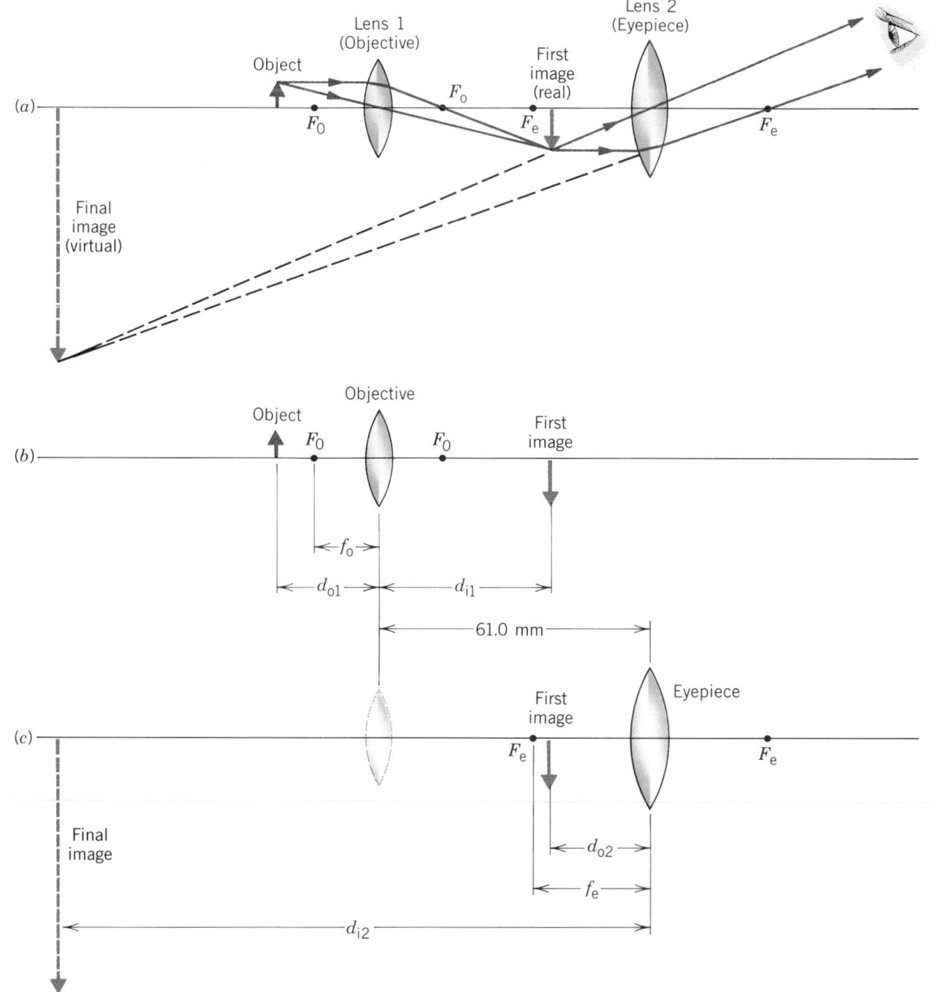

Figure 26.29 (*a*) This two-lens system can be used as a compound microscope to produce a virtual, enlarged, and inverted final image. (*b*) The objective forms the first image and (*c*) the eyepiece forms the final image.

point F_o of the objective. The image formed by the objective—called the "first image" in the drawing—is real, inverted, and enlarged compared to the object. This first image then serves as the object for the eyepiece. Since the first image falls between the eyepiece and its focal point F_e, the eyepiece forms an enlarged, virtual, final image, which is what the observer sees.

PROBLEM SOLVING INSIGHT

The location of the final image in a multiple-lens system can be determined by applying the thin-lens equation to each lens separately. The key point to remember in such situations is that ***the image produced by one lens serves as the object for the next lens,*** as the next example illustrates.

Example 11 A Microscope—Two Lenses in Combination

The objective and eyepiece of the compound microscope in Figure 26.29 are both converging lenses and have focal lengths of $f_o = 15.0$ mm and $f_e = 25.5$ mm. A distance of 61.0 mm separates the lenses. The microscope is being used to examine an object placed $d_{o1} = 24.1$ mm in front of the objective. Find the final image distance.

REASONING We begin by using the thin-lens equation to find the location of the image produced by the first lens, the objective. This image then becomes the object for the second lens, the eyepiece. The thin-lens equation can be used again to locate the final image produced by the eyepiece.

SOLUTION The "first image" distance d_{i1} can be determined using the thin-lens equation with $d_{o1} = 24.1$ mm and $f_o = 15.0$ mm (see Figure 26.29*b*):

$$\frac{1}{d_{i1}} = \frac{1}{f_o} - \frac{1}{d_{o1}} = \frac{1}{15.0 \text{ mm}} - \frac{1}{24.1 \text{ mm}} = 0.0252 \text{ mm}^{-1} \quad \text{or} \quad d_{i1} = 39.7 \text{ mm}$$

The first image now becomes the object for the eyepiece (see part *c* of the drawing). Since the distance between the lenses is 61.0 mm, the object distance for the eyepiece is $d_{o2} = 61.0 \text{ mm} - d_{i1} = 61.0 \text{ mm} - 39.7 \text{ mm} = 21.3$ mm. Noting that the focal length of the eyepiece is $f_e = 25.5$ mm, we can determine the final image distance with the aid of the thin-lens equation:

$$\frac{1}{d_{i2}} = \frac{1}{f_e} - \frac{1}{d_{o2}} = \frac{1}{25.5 \text{ mm}} - \frac{1}{21.3 \text{ mm}} = -0.0077 \text{ mm}^{-1} \quad \text{or} \quad \boxed{d_{i2} = -130 \text{ mm}}$$

The fact that d_{i2} is negative indicates that the final image is virtual and lies to the left of the eyepiece, as the drawing shows.

26.10 THE HUMAN EYE

THE ANATOMY OF THE EYE

Without doubt, the human eye is the most remarkable of all optical devices. Figure 26.30 shows some of its main anatomical features. The eyeball is approximately spherical with a diameter of about 25 mm. Light enters the eye through a transparent membrane (the *cornea*). This membrane covers a clear liquid region (the *aqueous humor*), behind which is a diaphragm (the *iris*), the *lens*, a region filled with a jelly-like substance (the *vitreous humor*), and, finally, the *retina*. The retina is the light-sensitive part of the eye, consisting of millions of structures called *rods* and *cones*. When stimulated by light, these structures send electrical impulses via the *optic nerve* to the brain, which interprets the image on the retina.

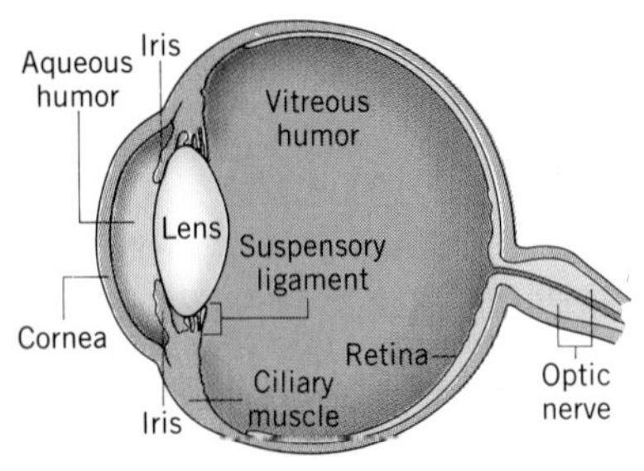

Figure 26.30 A cross-sectional view of the human eye.

The iris is the colored portion of the eye and controls the amount of light reaching the retina. The iris acts as a controller, because it is a muscular diaphragm with a variable opening at its center, through which the light passes. The opening is called the *pupil.* The diameter of the pupil varies from about 2 to 7 mm, decreasing in bright light and increasing (dilating) in dim light.

Of prime importance to the operation of the eye is the fact that the lens is flexible, and its shape can be altered by the action of the *ciliary muscle.* The lens is connected to the ciliary muscle by the *suspensory ligaments* (see the drawing). We will see shortly how the shape-changing ability of the lens affects the focusing property of the eye.

THE OPTICS OF THE EYE

Optically, the eye and the camera are similar; both have a lens system and a diaphragm with a variable opening or aperture at its center. Moreover, the retina of the eye and the film in the camera serve similar functions, for both record the image formed by the lens system. In the eye, the image formed on the retina is real, inverted, and smaller than the object, just as it is in a camera. Although the image on the retina is inverted, it is interpreted by the brain as being right-side up.

THE PHYSICS OF . . .

the human eye.

For clear vision, the eye must refract the incoming light rays, so as to form a sharp image on the retina. In reaching the retina, the light travels through five different media, each with a different index of refraction n: air ($n = 1.00$), the cornea ($n = 1.38$), the aqueous humor ($n = 1.33$), the lens ($n = 1.40$, on the average), and the vitreous humor ($n = 1.34$). Each time light passes from one medium into another, it is refracted at the boundary. Collectively, all the boundaries participate in refracting the light to form the image on the retina. However, the greatest amount of refraction, about 70% or so, occurs at the air/cornea boundary. According to Snell's law, the large refraction at this interface occurs primarily because the refractive index of air ($n = 1.00$) is so different from that of the cornea ($n = 1.38$). The refraction at all the other boundaries is relatively small, because the indices of refraction on either side of these boundaries are nearly equal. The lens itself contributes only about 20–25% of the total refraction, since the surrounding aqueous and vitreous humors have indices of refraction that are nearly the same as that of the lens.

Even though the lens contributes only a quarter of the total refraction or less, the function of the lens is an important one. The eye has a fixed image distance; that is, the distance between the lens and the retina is constant. Therefore, the only way for images to be produced on the retina for objects located at different distances is for the focal length of the lens to be adjustable. And it is the ciliary muscle that adjusts the focal length. When the eye looks at a very distant object, the ciliary muscle is not tensed. The lens has its least curvature and, consequently, its longest focal length. Under this condition the eye is said to be "fully relaxed," and the rays form a sharp image on the retina, as in Figure 26.31*a*. When the object moves closer to the eye, the ciliary muscle automatically tenses, thereby increasing the curvature of the lens, shortening the focal length, and permitting a sharp image to form again on the retina (Figure 26.31*b*). When a sharp image of an object is formed on the retina, we say the eye is "focused" on the object. The process in which the lens changes its focal length to focus on objects at different distances is called ***accommodation*** and occurs so swiftly that we are usually unaware of it.

When you hold a book too close, the print is blurred because the lens cannot adjust enough to bring the book into focus. The point nearest the eye at which an object can be placed and still produce a sharp image on the retina is called the ***near***

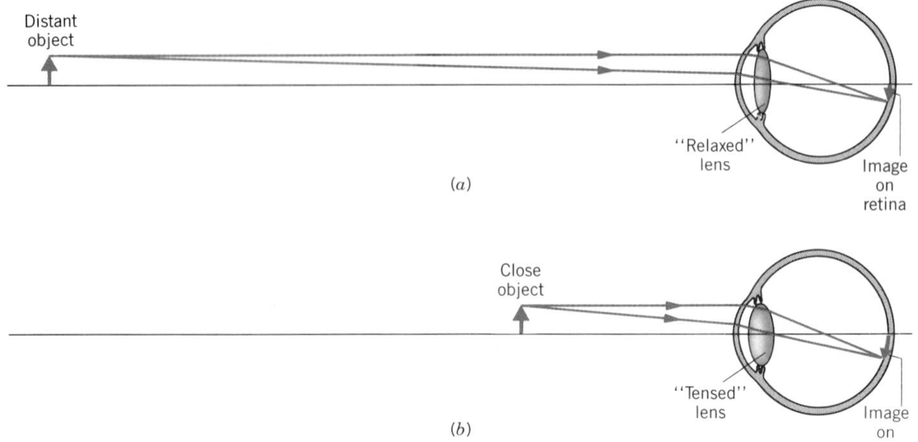

Figure 26.31 (*a*) When fully relaxed, the lens of the eye has its longest focal length, and an image of a very distant object is formed on the retina. (*b*) When the ciliary muscle is tensed, the lens is thicker and has a shorter focal length. Consequently, an image of a closer object is also formed on the retina.

point of the eye. The ciliary muscle is fully tensed when an object is placed at the near point. For people in their early twenties with normal vision, the near point is located about 25 cm from the eye. It increases to about 50 cm at age 40 and to roughly 500 cm at age 60. Since most reading material is held at a distance of 45 cm or so from the eye, older adults typically need eyeglasses to overcome the loss of accommodation. The ***far point*** of the eye is the location of the farthest object on which the fully relaxed eye can focus. A person with normal eyesight can see objects very far away, such as the planets and stars, and thus has a far point located nearly at infinity.

NEARSIGHTEDNESS

A person who is ***nearsighted (myopic)*** can focus on nearby objects but cannot clearly see objects far away. For such a person, the far point of the eye is not at infinity and may even be as close to the eye as three or four meters. When a nearsighted eye tries to focus on a distant object, the eye is fully relaxed, like a normal eye. However, the nearsighted eye has a focal length that is shorter than it should be, so rays from the distant object form a sharp image in front of the retina, as Figure 26.32*a* shows, and blurred vision results.

The nearsighted eye can be corrected with glasses or contacts that use *diverging* lenses, as Figure 26.32*b* suggests. The rays from the object diverge after leaving the eyeglass lens. Therefore, when they are subsequently refracted toward the principal axis by the eye, a sharp image is formed farther back and falls on the retina. Since the relaxed (but nearsighted) eye can focus on an object at the eye's far point—but not on objects farther away—the diverging lens is designed to transform a very distant object into an image located at the far point. Part *c* of the drawing shows this transformation, and the next example illustrates how to determine the focal length of the diverging lens that accomplishes it.

THE PHYSICS OF . . .

eyeglasses.

Example 12 Eyeglasses for the Nearsighted Person

A nearsighted person has a far point located only 521 cm from the eye. Assuming that eyeglasses are to be worn 2 cm in front of the eye, find the focal length needed for the diverging lenses of the glasses so the person can see distant objects.

REASONING In Figure 26.32*c* the far point is 521 cm away from the eye. Since the glasses are worn 2 cm from the eye, the far point is 519 cm to the left of the diverging lens. The image distance, then, is −519 cm, the negative sign indicating that the image is a virtual image formed to the left of the lens. The object is assumed to be infinitely far from the diverging lens. The thin-lens equation can be used to find the focal length of the eyeglasses.

SOLUTION With $d_i = -519$ cm and $d_o = \infty$, the focal length can be found as follows:

$$\frac{1}{f} = \frac{1}{d_o} + \frac{1}{d_i} = \frac{1}{\infty} + \frac{1}{-519 \text{ cm}} \quad \text{or} \quad \boxed{f = -519 \text{ cm}} \qquad (26.6)$$

The value for f is negative, indicating the lens is a diverging lens.

PROBLEM SOLVING INSIGHT

Eyeglasses are worn about 2 cm from the eyes. Be sure, if necessary, to take this 2 cm into account when determining the object and image distances (d_o and d_i) that are used in the thin-lens equation.

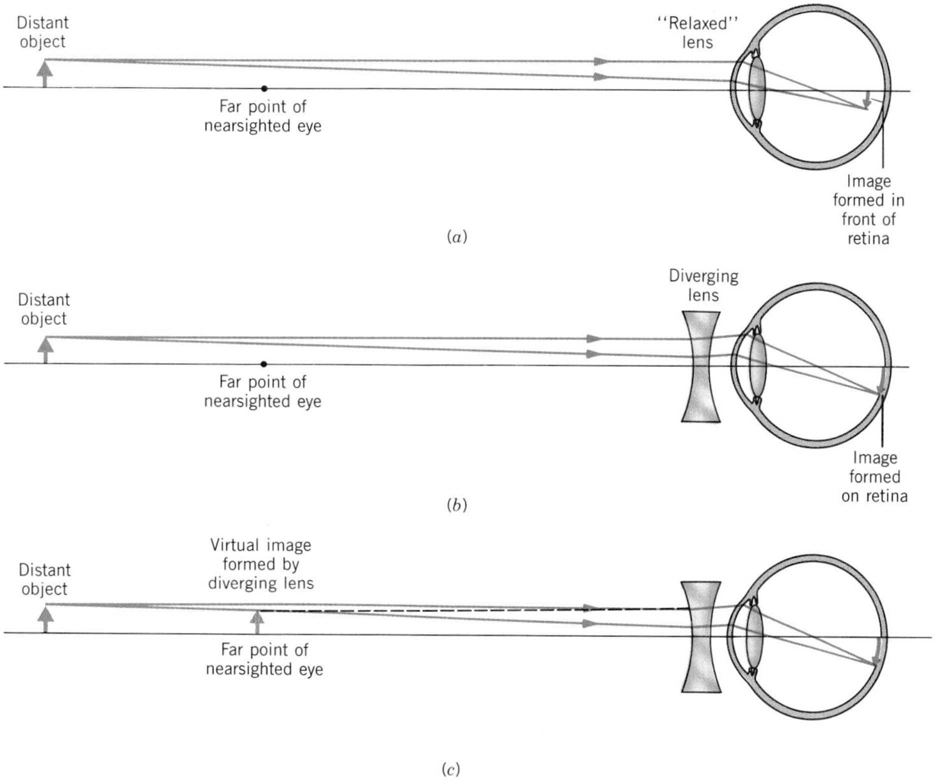

Distant object

"Relaxed" lens

Far point of nearsighted eye

Image formed in front of retina

(*a*)

Distant object

Diverging lens

Far point of nearsighted eye

Image formed on retina

(*b*)

Virtual image formed by diverging lens

Distant object

Far point of nearsighted eye

(*c*)

Figure 26.32 (*a*) When a nearsighted person views a distant object, the image is formed in front of the retina. The result is blurred vision. (*b*) With a diverging lens in front of the eye, the image is moved onto the retina and clear vision results. (*c*) The diverging lens is designed to form a virtual image at the far point of the nearsighted eye.

FARSIGHTEDNESS

A ***farsighted (hyperopic)*** person can usually see distant objects clearly, but cannot focus on those nearby. Whereas the near point of a "normal" eye is located about 25 cm from the eye, the near point of a farsighted eye may be considerably farther away than that, perhaps as far as several hundred centimeters. When a farsighted eye tries to focus on a book held closer than the near point, the eye accommodates and shortens its focal length as much as it can. However, even at its shortest, the focal length of a farsighted eye is longer than it should be. Therefore, the light rays from the book form a sharp image behind the retina, as Figure 26.33*a* indicates, leading to blurred vision.

Figure 26.33*b* shows that farsightedness can be corrected by placing a *converging* lens in front of the eye. The lens refracts the light rays more toward the

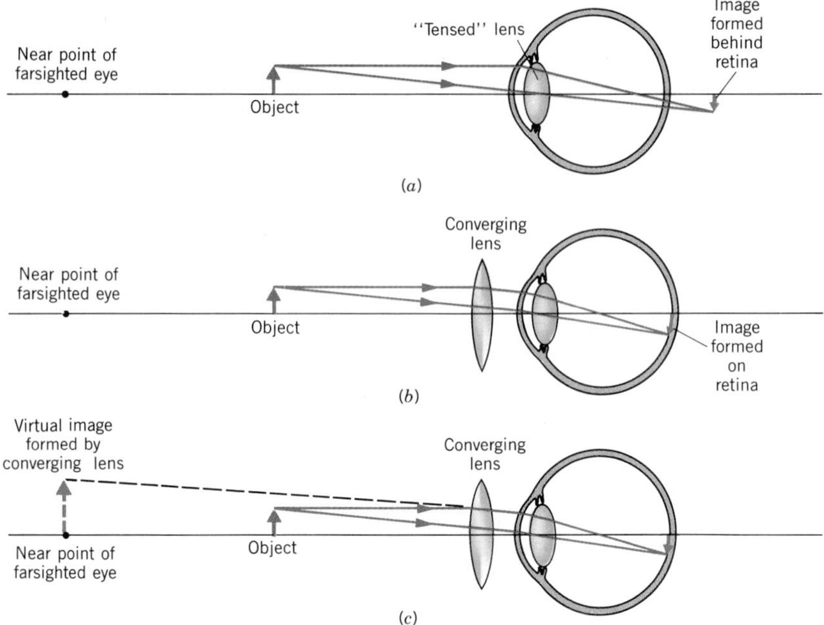

Figure 26.33 (*a*) When a farsighted person views an object located inside the near point, an image is formed behind the retina, causing blurred vision. (*b*) With a converging lens in front of the eye, the image is moved onto the retina and clear vision results. (*c*) The converging lens is designed to form a virtual image at the near point of the farsighted eye.

principal axis before they enter the eye. Consequently, when the rays are refracted even more by the eye, they converge to form an image on the retina. Part *c* of the figure illustrates what the eye sees when it looks through the converging lens. The lens is designed so that the eye perceives the light to be coming from a virtual image located at the near point. Example 13 shows how the focal length of the converging lens is determined to correct for farsightedness.

THE PHYSICS OF . . .

contact lenses.

Example 13 Contact Lenses for the Farsighted Person

A farsighted person has a near point located 210 cm from the eyes. Obtain the focal length of the converging lenses in a pair of contacts that can be used to read a book held 25.0 cm from the eyes.

REASONING A contact lens is placed directly against the eye. Thus, the object distance, which is the distance from the book to the lens, is 25.0 cm. The lens forms an image of the book at the near point of the eye, so the image distance is −210 cm. The minus sign indicates that the image is a virtual image formed to the left of the lens, as in Figure 26.33*c*. The focal length can be obtained from the thin-lens equation.

SOLUTION With $d_o = 25.0$ cm and $d_i = -210$ cm, the focal length can be determined as follows:

$$\frac{1}{f} = \frac{1}{d_o} + \frac{1}{d_i} = \frac{1}{25.0 \text{ cm}} + \frac{1}{-210 \text{ cm}} = 0.0352 \text{ cm}^{-1} \quad \text{or} \quad \boxed{f = 28.4 \text{ cm}} \qquad (26.6)$$

THE REFRACTIVE POWER OF A LENS—THE DIOPTER

The extent to which rays of light are refracted by a lens depends on its focal length. However, the optometrists who prescribe correctional lenses and the opticians who make the lenses do not specify the focal length directly in prescriptions. Instead, they use the concept of ***refractive power*** to decribe the extent to which a lens refracts light:

$$\begin{array}{c}\text{Refractive power}\\ \text{of a lens}\\ \text{(in diopters)}\end{array} = \frac{1}{f\text{ (in meters)}} \tag{26.8}$$

The refractive power is measured in units of *diopters.* One diopter is 1 m^{-1}.

Equation 26.8 shows that a converging lens has a refractive power of 1 diopter if it focuses parallel light rays to a focal point 1 m beyond the lens. If a lens refracts parallel rays even more and converges them to a focal point only 0.25 m beyond the lens, the lens has four times more refractive power, or 4 diopters. Since a converging lens has a positive focal length and a diverging lens has a negative focal length, the refractive power of a converging lens is positive while that of a diverging lens is negative. For instance, the eyeglasses in Example 12 would be described in a prescription from an optometrist in the following way: Refractive power $= 1/(-5.19\text{ m}) = -0.193$ diopters. The contact lenses in Example 13 would be described in a similar fashion: Refractive power $= 1/(0.284\text{ m}) =$ 3.52 diopters.

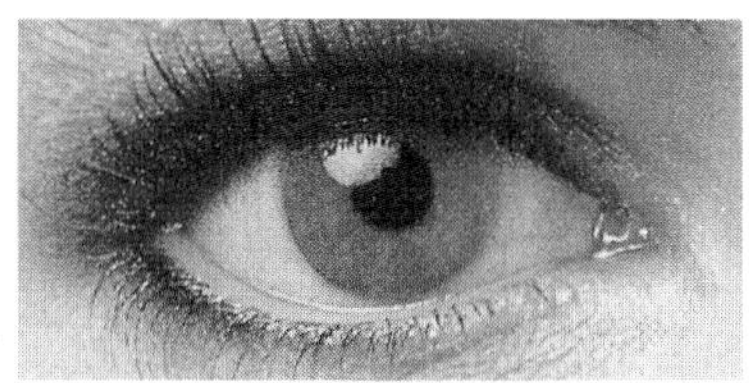

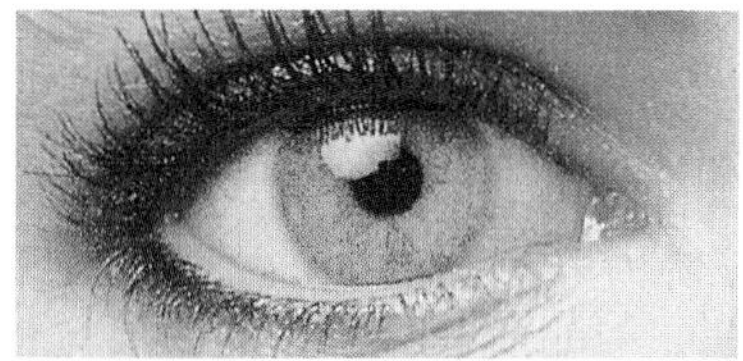

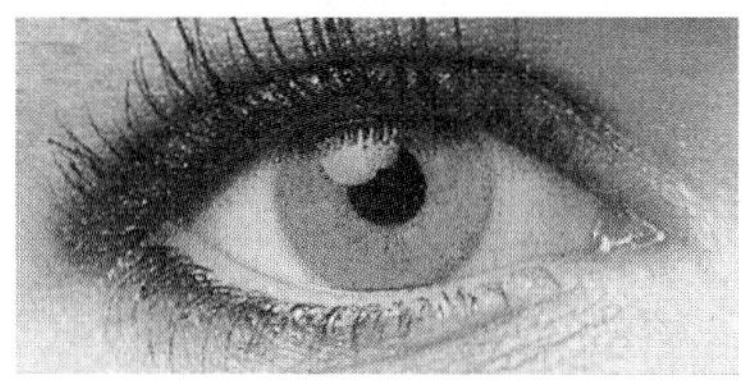

In addition to improving your vision, contact lenses can also be used to coordinate your eye color with your wardrobe.

26.11 ANGULAR MAGNIFICATION AND THE MAGNIFYING GLASS

If you hold a penny at arm's length, the penny looks larger than the moon. The reason is that the penny, being so close, forms a larger image on the retina of the eye than does the more distant moon. The brain interprets the larger image of the penny as arising from a larger object. As far as the brain is concerned, then, the size of the image on the retina determines how large an object appears to be. However, the size of the image on the retina is difficult to measure. Alternatively, the angle θ subtended by the image can be used as an indication of the image size. Figure 26.34 shows this alternative, which has the advantage that θ is also the angle subtended by the object and, hence, can be measured easily. The angle θ is

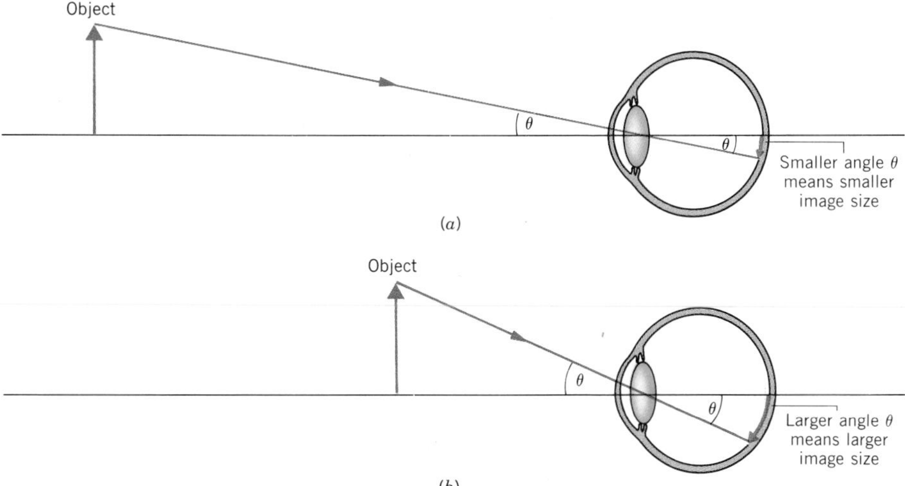

Figure 26.34 In both (*a*) and (*b*) the object is the same size, but in (*b*) the image on the retina is larger because the object is closer to the eye. The angle θ is the angular size of both the image and the object.

called the ***angular size*** of both the image and the object. The larger the angular size, the larger the image on the retina, and the larger the object appears to be.

According to Equation 8.1, the angle θ (measured in radians) is the length of the circular arc subtended by the angle divided by the radius of the arc, as Figure 26.35*a* indicates. Part *b* of the drawing shows the situation when we view an object of height h_o at a distance d_o from the eye. Comparing part *a* with part *b*, it is evident that when θ is small, h_o is approximately equal to the arc length and d_o is nearly equal to the radius. Thus, for small angles, θ can be expressed in terms of the measurable quantities h_o and d_o:

$$\theta \text{ (in radians)} = \text{Angular size} \approx \frac{h_o}{d_o}$$

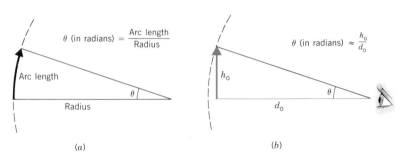

Figure 26.35 (*a*) The angle θ, measured in radians, is the arc length divided by the radius. (*b*) For small angles (less than 9°), θ is approximately equal to h_o/d_o, where h_o and d_o are the object height and distance.

This approximation is good to within one percent for angles of 9° or smaller. In the next example the angular size of the penny is compared with that of the moon.

Example 14 A Penny and the Moon

Compare the angular size of a penny (diameter $= h_o = 1.9$ cm) held at arm's length ($d_o = 71$ cm) with that of the moon (diameter $= h_o = 3.5 \times 10^6$ m, and $d_o = 3.9 \times 10^8$ m).

REASONING The angular size θ of an object is given approximately by its height h_o divided by its distance d_o from the eye, $\theta = h_o/d_o$, provided that the angle involved is less than roughly 9°. As we shall see, this approximation applies here. In this example, the "heights" of the penny and the moon are their diameters.

SOLUTION The angular sizes of the penny and moon are

[Penny] $$\theta = \frac{h_o}{d_o} = \frac{1.9 \text{ cm}}{71 \text{ cm}} = \boxed{0.027 \text{ rad } (1.5°)}$$

[Moon] $$\theta = \frac{h_o}{d_o} = \frac{3.5 \times 10^6 \text{ m}}{3.9 \times 10^8 \text{ m}} = \boxed{0.0090 \text{ rad } (0.52°)}$$

The penny thus appears to be about three times larger than the moon.

An optical instrument, such as a magnifying glass, allows us to view small or distant objects, because it produces a larger image on the retina than would be possible otherwise. In other words, an optical instrument magnifies the angular size of the object. The ***angular magnification*** (or ***magnifying power***) M is the angular size θ' of the final image produced by the instrument divided by a reference angular size θ. The reference angular size is the angular size of the object when seen without the instrument.

Angular magnification

$$M = \frac{\text{Angular size of final image produced by optical instrument}}{\text{Reference angular size of object seen without optical instrument}} = \frac{\theta'}{\theta} \quad (26.9)$$

THE PHYSICS OF . . . a magnifying glass.

A magnifying glass is the simplest device that provides angular magnification. To find its angular magnification, we first determine the reference angular size θ. In this case, θ is chosen to be the angular size of the object when placed at the near point of the eye and seen without the magnifying glass. Since an object cannot be brought closer than the near point of the eye and still produce a sharp image on the retina, θ represents the largest angular size obtainable without the magnifying glass. Figure 26.36*a* indicates that the reference angular size is $\theta \approx h_o/N$, where N is the distance from the eye to the near point. To compute θ', recall from Section 26.7 and Figure 26.24 that a magnifying glass is usually a single converging lens, with the object located inside the focal point. In this situation, Figure 26.36*b* indicates that the lens produces a virtual image that is enlarged and upright with respect to the object. Assuming the eye is next to the magnifying glass, the angular size θ' seen by the eye is $\theta' \approx h_o/d_o$, where d_o is the object distance. The angular magnification is

$$M = \frac{\theta'}{\theta} \approx \frac{h_o/d_o}{h_o/N} = \frac{N}{d_o}$$

According to the thin-lens equation, d_o is related to the image distance d_i and the focal length f of the lens by

$$\frac{1}{d_o} = \frac{1}{f} - \frac{1}{d_i}$$

Substituting this expression for $1/d_o$ into the expression above for M leads to the following result:

Angular magnification of a magnifying glass

$$M = \frac{\theta'}{\theta} \approx \left(\frac{1}{f} - \frac{1}{d_i}\right) N \quad (26.10)$$

Two special cases of this result are of interest, depending on whether the image is located as close to the eye as possible or as far away as possible. To be seen clearly, the closest the image can be relative to the eye is at the near point; for this situation, $d_i = -N$. The minus sign indicates that the image lies to the left of the lens and is virtual. Under this condition, Equation 26.10 becomes $M \approx N/f + 1$. The farthest the image can be from the eye is at infinity ($d_i = -\infty$); this occurs when the object is placed at the focal point of the lens. When the image is at infinity, Equation 26.10 simplifies to $M \approx N/f$. Clearly, the angular magnification is greater when the image is at the near point of the eye rather than at infinity. In either case, however, the greatest magnification is achieved by using a magnifying glass with the shortest possible focal length. Example 15 illustrates how to determine the angular magnification of a magnifying glass that is used in these two ways.

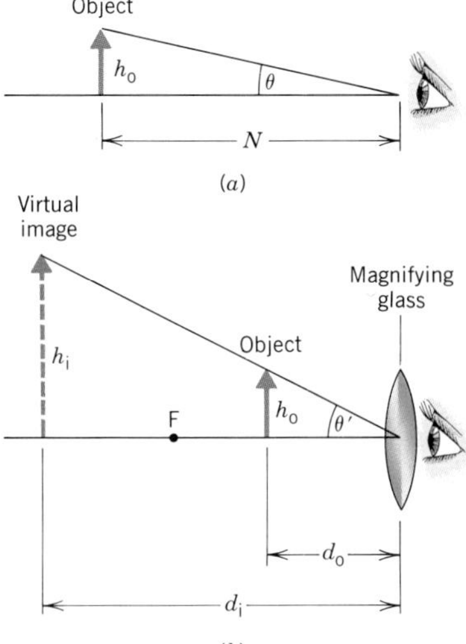

Figure 26.36 (*a*) Without a magnifying glass, the largest angular size θ occurs when the object is placed at the near point of the eye. The distance from the near point to the eye is N. (*b*) A magnifying glass produces an enlarged, virtual image of an object placed inside the focal point F of the lens. The angular size of both the image and the object is θ'.

Example 15 Examining a Diamond Ring with a Magnifying Glass

A jeweler, whose near point is 40.0 cm from his eye, is using a small magnifying glass (called a loupe) to examine a diamond ring. The lens of the magnifying glass has a focal length of 5.00 cm, and the image of the ring is −185 cm from the lens. The image distance is negative because the image is virtual and is formed on the same side of the lens as the object. (a) Determine the angular magnification of the magnifying glass. (b) Where should the image be located so the jeweler's eye is fully relaxed and has the least strain? What is the angular magnification under this "least strain" condition?

REASONING The angular magnification of the magnifying glass can be determined from Equation 26.10. In part (a) the image distance is −185 cm. In part (b) the ciliary muscle of the jeweler's eye is fully relaxed when examining the ring, so the image must be infinitely far from the eye, as Section 26.10 discusses.

SOLUTION

(a) With $f = 5.00$ cm, $d_i = -185$ cm, and $N = 40.0$ cm, the angular magnification of the magnifying glass is

$$M = \left(\frac{1}{f} - \frac{1}{d_i}\right)N = \left(\frac{1}{5.00\text{ cm}} - \frac{1}{-185\text{ cm}}\right)(40.0\text{ cm}) = \boxed{8.22}$$

(b) When the jeweler's eye is fully relaxed, the image distance is $d_i = -\infty$, so $1/d_i = 1/(-\infty) = 0$. The angular magnification is

$$M = \left(\frac{1}{f} - \frac{1}{d_i}\right)N = \left(\frac{1}{5.00\text{ cm}} - \frac{1}{-\infty}\right)(40.0\text{ cm}) = \boxed{8.00}$$

Jewelers often prefer to minimize eye strain when viewing objects, even though it means a slight reduction in angular magnification.

THE PHYSICS OF . . .

the compound microscope.

26.12 THE COMPOUND MICROSCOPE

To increase the angular magnification beyond that possible with a magnifying glass, an additional converging lens can be included to "premagnify" the object before the magnifying glass comes into play. The result is an optical instrument known as the ***compound microscope*** (Figure 26.37). As discussed in Section 26.9, the magnifying glass is called the eyepiece, and the additional lens is called the objective. The object is placed just outside the focal point F_o of the objective, as in Figure 26.38*a*.

In obtaining the angular magnification M of the compound microscope, we follow the same approach used in the last section and begin with $M = \theta'/\theta$ (Equation 26.9), where θ' is the angular size of the final image and θ is the reference angular size. As with the magnifying glass in Figure 26.36, the reference angular size is determined by the height h_o of the object when the object is located at the near point of the unaided eye: $\theta \approx h_o/N$, where N is the distance between the eye and the near point.

To find the angular size θ' of the final image produced by the microscope, note from Figure 26.38*a* that the objective produces a "first image," which then serves as the object for the eyepiece. The eyepiece, in turn, produces the final image. Part *b* of the drawing shows that the angular size θ' of the final image is equal to that of the first image, so we need only to determine θ' from the first image. The first image is inverted and its height is mh_o, where m is the magnification of the objective and h_o is the height of the object. The magnification equation (Equation 26.7) specifies m in terms of the object distance d_o and the image distance d_i, according to $m = -d_i/d_o$. If, as in part *a* of the drawing, the object is placed just outside the focal point F_o of the objective to achieve a large magnification, then $d_o \approx f_o$, where f_o is the focal length of the objective. Furthermore, if the microscope is designed so the eye is fully relaxed when viewing the final image, the final image must be very far from the eyepiece, or near "infinity." This location of the final image implies that the first image must fall just inside the focal point F_e of the eyepiece, as part *b* of the drawing indicates. Therefore, $d_i \approx L - f_e$, where L is the separation between the two lenses and f_e is the focal length of the eyepiece. With these approximations for d_o and d_i, the magnification of the objective becomes

$$m = -\frac{d_i}{d_o} \approx -\frac{(L - f_e)}{f_o}$$

Since the distance between the first image and the eyepiece is nearly equal to f_e, the angular size θ' of the first image (and also of the final image) is

$$\theta' \approx \frac{\text{Height of first image}}{\text{Distance of first image from eyepiece}} \approx \frac{mh_o}{f_e} \approx -\left(\frac{L - f_e}{f_o}\right)\frac{h_o}{f_e}$$

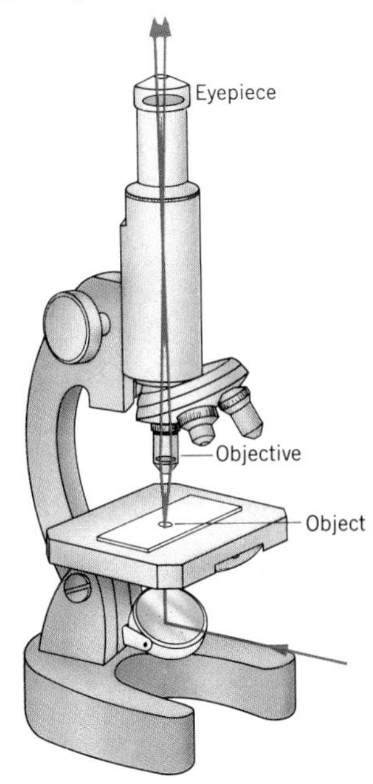

Figure 26.37 A compound microscope.

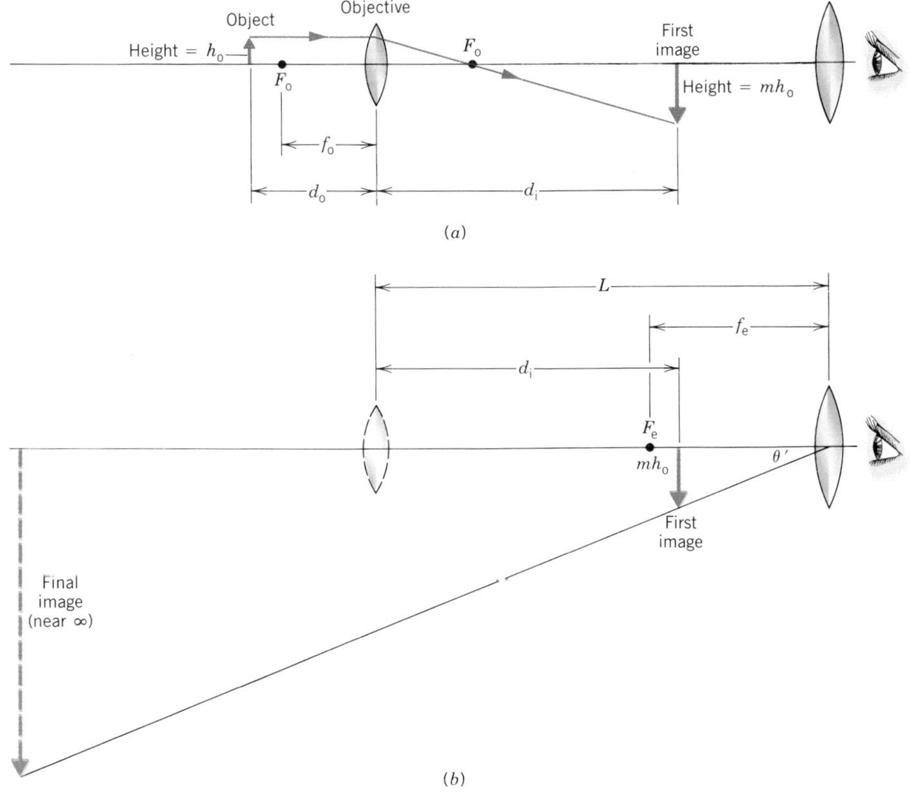

Figure 26.38 (*a*) In a two-lens compound microscope, the objective forms a "first image" of the object. This first image is enlarged, real, and inverted. (For clarity, only one ray is shown.) (*b*) The first image becomes the object for the eyepiece, which produces an enlarged, virtual, final image near infinity.

Using the expressions above for θ and θ', we find that the angular magnification of the compound microscope is

$$M = \frac{\theta'}{\theta} \approx \frac{-\left(\frac{L - f_e}{f_o}\right)\frac{h_o}{f_e}}{\frac{h_o}{N}}$$

Angular magnification of a compound microscope

$$M \approx -\frac{(L - f_e)N}{f_o f_e} \qquad (26.11)$$

Equation 26.11 shows that the angular magnification is greatest when f_o and f_e are as small as possible (since they are in the denominator) and when the distance L between the lenses is as large as possible. Furthermore, L must be greater than the sum of f_o and f_e for this equation to be valid. Example 16 deals with the angular magnification of a compound microscope.

Example 16 The Angular Magnification of a Compound Microscope

The objective of a compound microscope has a focal length of $f_o = 0.40$ cm, while that of the eyepiece is $f_e = 3.0$ cm. The two lenses are separated by a distance of $L = 20.0$ cm. A person with normal eyes ($N = 25$ cm) is using the microscope. (a) Determine the angular magnification of the microscope. (b) Compare the answer in part (a) with the largest angular magnification obtainable by using the eyepiece alone as a magnifying glass.

REASONING The angular magnification of the compound microscope can be obtained directly from Equation 26.11, since all the variables are known. When the eyepiece is used alone as a magnifying glass, as in Figure 26.36*b*, the largest angular magnification occurs when the image seen through the eyepiece is as close as possible to the eye. The image in this case is at the near point, and according to Equation 26.10, the angular magnification is $M \approx (N/f_e) + 1$.

SOLUTION

(a) The angular magnification of the compound microscope is

$$M \approx -\frac{(L - f_e)N}{f_o f_e} = -\frac{(20.0\text{ cm} - 3.0\text{ cm})(25\text{ cm})}{(0.40\text{ cm})(3.0\text{ cm})} = \boxed{-350}$$

The minus sign indicates that the final image is inverted relative to the initial object.

(b) The maximum angular magnification of the eyepiece by itself is

$$M \approx \frac{N}{f_e} + 1 = \frac{25\text{ cm}}{3.0\text{ cm}} + 1 = \boxed{9.3}$$

The effect of the objective is to increase the angular magnification of the compound microscope by a factor of $350/9.3 = 38$ compared to that of a magnifying glass.

26.13 THE TELESCOPE

THE PHYSICS OF . . .

the telescope.

A telescope is an instrument for magnifying distant objects, such as stars and planets. Like a microscope, a telescope consists of an objective and an eyepiece (also called the ocular). Since the object is usually far away, the light rays striking the telescope are nearly parallel, and the "first image" is formed just beyond the focal point F_o of the objective, as Figure 26.39*a* illustrates. The first image is real and inverted. Unlike that in the compound microscope, however, this image is *smaller* than the object. If, as in part *b* of the drawing, the telescope is constructed so the first image lies just inside the focal point F_e of the eyepiece, the eyepiece acts like a magnifying glass. It forms a final image that is greatly enlarged, virtual, and located near infinity. This final image can then be viewed with a fully relaxed eye.

The angular magnification M of a telescope, like that of a magnifying glass or a microscope, is the angular size θ' subtended by the final image of the telescope divided by the reference angular size θ of the object as seen without the telescope. For an astronomical object, such as a planet, it is convenient to use as a reference the angular size of the object seen in the sky with the unaided eye. Since the object is far away, the angular size seen by the unaided eye is nearly the same as the angle θ subtended at the objective of the telescope in Figure 26.39*a*. Moreover, θ is also the angle subtended by the first image, so $\theta \approx -h_i/f_o$, where h_i is the height of the first image and f_o is the focal length of the objective. A minus sign has been

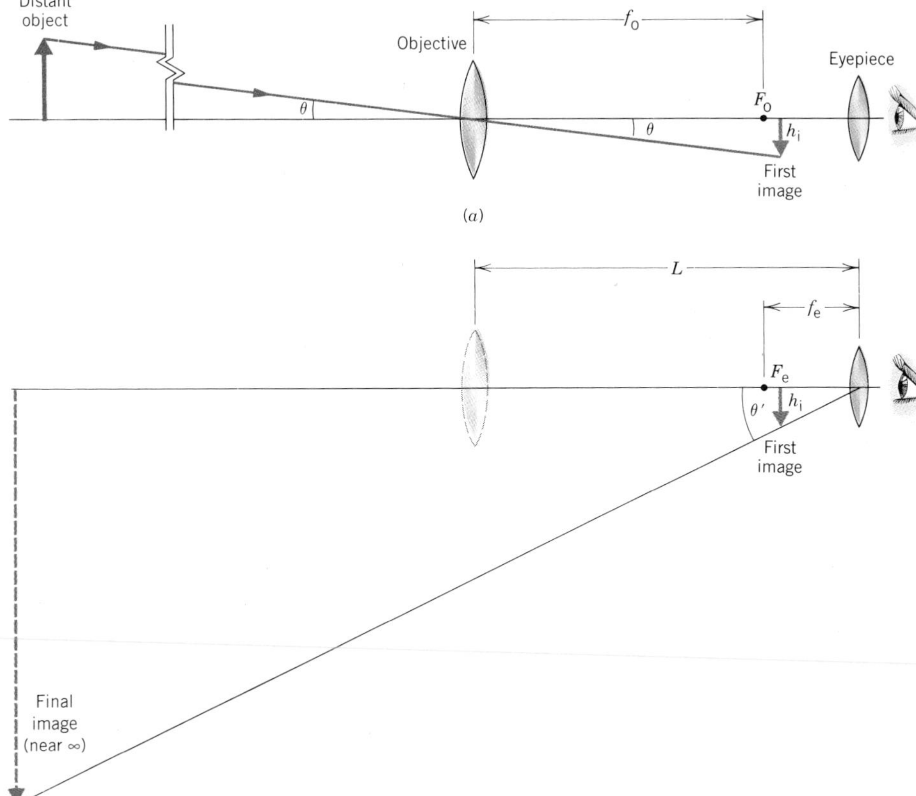

Figure 26.39 (*a*) An astronomical telescope is used to view distant objects. (Note the "break" in the principal axis, between the object and the objective.) The objective produces a real, inverted, first image. (*b*) The eyepiece magnifies the first image to produce the final image near infinity.

inserted into this equation for the following reason. The angle θ in Figure 26.39 is assumed to be positive. Since the first image is inverted relative to the object, the image height h_i is a negative number. The insertion of the minus sign ensures that the term $-h_i/f_o$, and hence θ, is a positive quantity. To obtain an expression for θ', note in part *b* of the figure that the first image is located very near the focal point F_e of the eyepiece, so $\theta' \approx h_i/f_e$, where f_e is the focal length of the eyepiece lens. The angular magnification of the telescope is, then,

Angular magnification of an astronomical telescope

$$M = \frac{\theta'}{\theta} \approx \frac{h_i/f_e}{-h_i/f_o} = -\frac{f_o}{f_e} \qquad (26.12)$$

The angular magnification is determined by the ratio of the focal length of the objective to the focal length of the eyepiece. For large angular magnifications, the objective should have a long focal length and the eyepiece a short one. Some of the design features of a telescope are the topic of the next example.

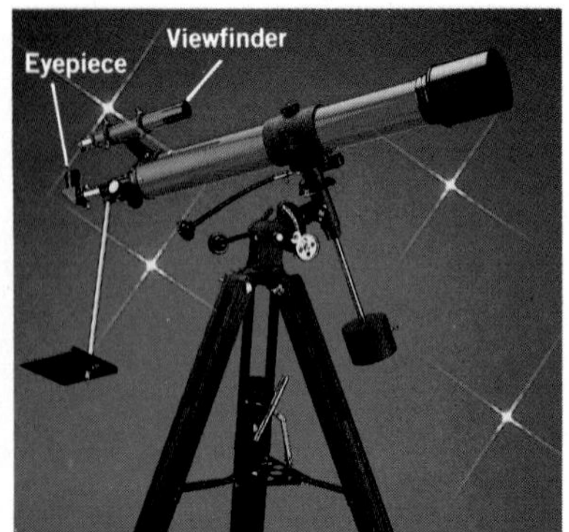

Figure 26.40 An astronomical telescope. The viewfinder is a separate small telescope with low magnification and serves as an aid in locating the object. Once the object has been found, the viewer looks through the eyepiece to obtain the full magnification of the telescope.

Example 17 The Angular Magnification of an Astronomical Telescope

The telescope shown in Figure 26.40 has the following specifications: $f_o = 985$ mm and $f_e = 5.00$ mm. From these data, find (a) the angular magnification of the telescope and (b) the approximate length of the telescope.

REASONING The angular magnification of the telescope follows directly from Equation 26.12, since the focal lengths of the objective and eyepiece are known. We can find the length of the telescope by noting that it is approximately equal to the distance L between the objective and eyepiece. Figure 26.39 shows that the first image is located just beyond the focal point F_o of the objective, and just inside the focal point F_e of the eyepiece. These two focal points are, therefore, very close together, so the distance L is approximately the sum of the two focal lengths: $L \approx f_o + f_e$.

SOLUTION

(a) The angular magnification is

$$M = -\frac{f_o}{f_e} = -\frac{985 \text{ mm}}{5.00 \text{ mm}} = \boxed{-197} \qquad (26.12)$$

(b) The approximate length of the telescope is

$$L \approx f_o + f_e = 985 \text{ mm} + 5.00 \text{ mm} = \boxed{990 \text{ mm}}$$

Two types of astronomical telescopes are used today, the ***refracting telescope*** and the ***reflecting telescope.*** Both include a lens for the eyepiece. The refracting telescope also uses a lens for the objective and is the type illustrated in Figures 26.39 and 26.40. The reflecting telescope, however, uses a concave mirror for the objective, as Figure 26.41 indicates. The incoming light rays reflect from a concave mirror at the right end. A small plane mirror intercepts the reflected, converging rays and directs them toward the eyepiece. This particular design was developed by Isaac Newton and is said to have a Newtonian focus.

Because the light from distant stars is so faint, telescopes need a great light-gathering ability, so their objectives have large diameters. Since a lens can be

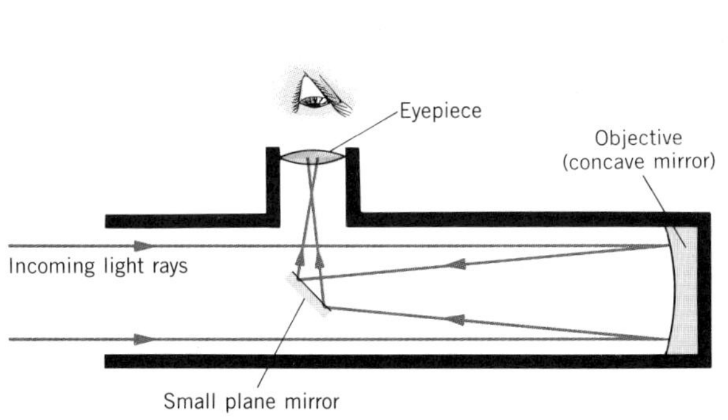

Figure 26.41 A reflecting telescope with a Newtonian focus.

supported only at its edge, a large lens begins to sag under its own weight and produce distorted images. Consequently, large concave mirrors are preferable, because they can be supported over the entire back surface. Also, a mirror has only one surface to be ground and polished rather than two, and a mirror is free from the chromatic aberration that exists in all lenses.* All the larger astronomical telescopes in the world are reflectors.

26.14 LENS ABERRATIONS

Rather than forming a sharp image, a single lens typically forms an image that is slightly out of focus. This lack of sharpness arises because the rays originating from a single point on the object are not focused to a single point on the image. As a result, each point on the image becomes a small "blur." The lack of point-to-point correspondence between object and image is called an aberration.

One common type of aberration is ***spherical aberration,*** and it occurs with converging and diverging lenses made with spherical surfaces. Figure 26.42*a* shows how spherical aberration arises with a converging lens. Ideally, all rays traveling parallel to the principal axis are refracted so they cross the axis at the same point after passing through the lens. However, rays far from the principal axis are refracted more by the lens than those closer in. Consequently, the outer rays cross the axis closer to the lens than do the inner rays, so a lens with spherical aberration does not have a unique focal point. Instead, as the drawing suggests, there is a location along the principal axis where the light converges to the smallest cross-sectional area. This area is circular and is known as the ***circle of least confusion.*** The circle of least confusion is where the most satisfactory image can be formed by the lens.

Spherical aberration can be reduced substantially by using a variable-aperture diaphragm to allow only those rays close to the principal axis to pass through the lens. Figure 26.42*b* indicates that a reasonably sharp focal point can be achieved by this method, although less light now passes through the lens. Lenses with parabolic surfaces are also used to reduce this type of aberration, but they are difficult and expensive to make.

Chromatic aberration also causes blurred images. It arises because the index of refraction of the material from which the lens is made varies with wavelength. Section 26.5 discusses how this variation leads to the phenomenon of dispersion,

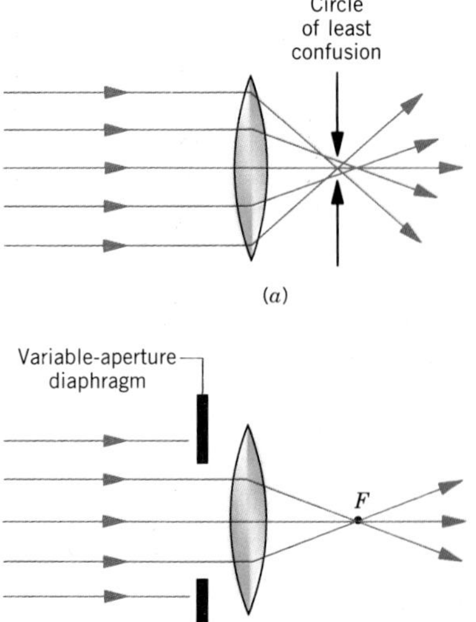

Figure 26.42 (*a*) In a converging lens, spherical aberration prevents light rays parallel to the principal axis from converging to a common point. (*b*) Spherical aberration can be reduced by allowing only rays near the principal axis to pass through the lens. The refracted rays now converge more nearly to a single focal point F.

* Section 26.14 discusses chromatic aberration.

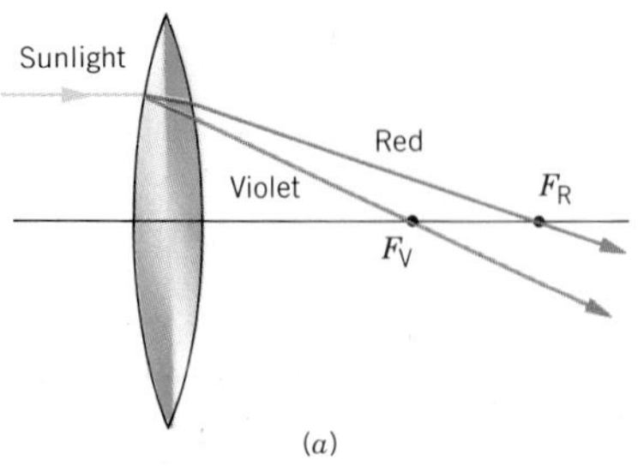

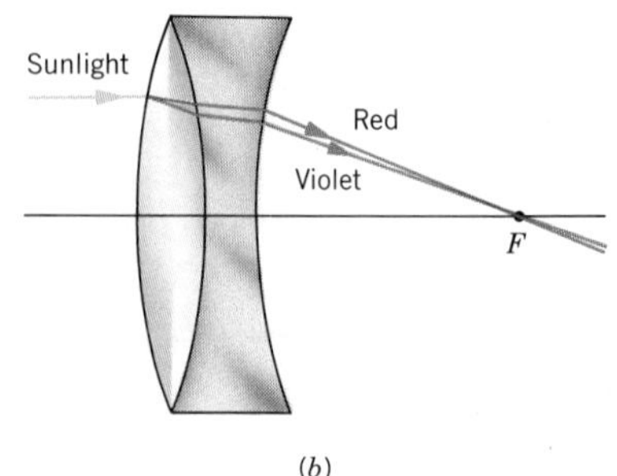

Figure 26.43 (*a*) Chromatic aberration arises when different colors are focused at different points along the principal axis: F_V = focal point for violet light, F_R = focal point for red light. (*b*) A converging and a diverging lens in tandem can be designed to bring different colors more nearly to the same focal point *F*.

whereby different colors refract by different amounts. Figure 26.43*a* shows sunlight incident on a converging lens, in which the light spreads into its color spectrum because of dispersion. For clarity, however, the picture shows only the colors at the opposite ends of the visible spectrum—red and violet. Violet is refracted more than red, so the violet ray crosses the principal axis closer to the lens than does the red ray. Thus, the focal length of the lens is shorter for violet than for red, with intermediate values of the focal length corresponding to the colors in between. As a result of chromatic aberration, an undesirable color fringe surrounds the image.

Chromatic aberration can be greatly reduced by using a compound lens, such as the combination of a converging lens and a diverging lens shown in Figure 26.43*b*. Each lens is made from a different type of glass. With this lens combination the red and violet rays almost come to a common focus and, thus, chromatic aberration is reduced. A lens combination designed to reduce chromatic aberration is called an *achromatic lens* (from the Greek "achromatos," meaning "without color"). All high-quality cameras use achromatic lenses.

INTEGRATION OF CONCEPTS

THE REFLECTION AND REFRACTION OF LIGHT AND ENERGY CONSERVATION

We first encountered reflection in Chapter 25 in association with mirrors and found that the law of reflection is obeyed. This law is also obeyed when light reflects from a surface between two transparent materials (e.g., air and water). In the current chapter, we have seen that refraction also occurs at such a surface. In refraction, light penetrates into the second material and travels in a direction that differs from the incident direction, as specified by Snell's law. Thus, reflection and refraction of light waves occur simultaneously at a surface between two transparent materials. It is important to keep in mind that light waves are electromagnetic in nature. They are composed of electric and magnetic fields, which carry energy. As a result, we can apply the principle of conservation of energy (see Chapter 6) to the phenomena of reflection and refraction. When only reflection and refraction occur at a surface, energy conservation indicates that the energy reflected plus the energy refracted must add up to equal the energy carried by the incident light. The percentage of incident energy that appears as reflected versus refracted light depends on the angle of incidence and the refractive indices of the materials on either side of the surface. For instance, when light travels from air toward water at perpendicular incidence, most of the light energy is refracted and little is reflected. But when the angle of incidence is nearly 90° and the light barely grazes the water surface, most of the light energy is reflected, with only a small amount refracted into the water. On a rainy night, you probably have experienced the annoying glare that results when light from an oncoming car just grazes the wet road. Under such conditions, most of the light energy reflects into your eyes.

LENSES AND MIRRORS

In passing through a lens, light is refracted in such a way that an image of the source of the light (the object) is formed. In this chapter, we have concentrated on lenses that are so thin that their thicknesses may be ignored. Mirrors also use light, but they depend on reflection, not refraction. In Chapter 25 we dealt with spherical mirrors. In spite of the fact that thin lenses and spherical mirrors use light in fundamentally different ways, they have much in common as optical devices. For example, they both have focal lengths. The thin-lens equation and the mirror equation show that the focal length is related to the object distance and the image distance in the same mathematical way for both kinds of devices. Both can produce real or virtual images and can produce upright or inverted images. Both can also create images that are larger or smaller than the original objects, an ability described by the same magnification equation in each case. It is convenient that thin lenses and spherical mirrors function as optical devices in so many similar ways.

SUMMARY

When light strikes the interface between two media, part of the light is reflected and the remainder is transmitted across the interface. The change in the direction of travel as light passes from one medium into another is called **refraction.** The **index of refraction** n of a material is the ratio of the speed of light c in a vacuum to the speed of light v in the material: $n = c/v$. **Snell's law of refraction** states that (1) the refracted ray, the incident ray, and the normal to the interface all lie in the same plane, and (2) the angle of refraction θ_2 is related to the angle of incidence θ_1 by $n_1 \sin \theta_1 = n_2 \sin \theta_2$, where n_1 and n_2 are the indices of refraction of the incident and refracting media, respectively. The angles are measured relative to the normal.

Because of refraction, a submerged object has an **apparent depth** that is different than its actual depth. If the observer is directly above the object, the apparent depth d' is related to the actual depth d by $d' = d(n_2/n_1)$, where n_1 and n_2 are the refractive indices of the media in which the object and the observer, respectively, are located.

When light passes from a medium of larger refractive index n_1 into one of smaller refractive index n_2, the refracted ray is bent away from the normal. If the incident ray is at the **critical angle** θ_c, the angle of refraction is $90°$. The critical angle is determined from Snell's law and is given by $\sin \theta_c = n_2/n_1$. When the angle of incidence exceeds the critical angle, all the incident light is reflected back into the medium from which it came, a phenomenon known as **total internal reflection.**

When light is incident on a nonmetallic surface at the **Brewster angle** θ_B, the reflected light is completely polarized parallel to the surface. The Brewster angle is given by $\tan \theta_B = n_2/n_1$, where n_1 and n_2 are the refractive indices of the incident and refracting media, respectively. When light is incident at the Brewster angle, the reflected and refracted rays are perpendicular to each other.

A glass prism can spread a beam of sunlight into a spectrum of colors, because the index of refraction of the glass depends on the wavelength of the light. The spreading of light into its color components is known as **dispersion.**

Converging lenses and **diverging lenses** depend on the phenomenon of refraction in forming an image. With a converging lens, paraxial rays that are parallel to the principal axis are focused to a point on the axis by the lens. This point is called the **focal point** of the lens, and its distance from the lens is the **focal length** f. Paraxial light rays that are parallel to the principal axis of a diverging lens appear to originate from its focal point after passing through the lens. The image produced by a converging or a diverging lens can be located with the help of a **ray diagram,** which can be constructed using the rays shown in Figure 26.21.

The **thin-lens equation** can be used with either converging or diverging lenses that are thin, and it relates the object distance d_o, the image distance d_i, and the focal length f of the lens: $1/d_o + 1/d_i = 1/f$. The magnification m of a lens is the ratio of the image height h_i to the object height h_o. The magnification is also related to d_i and d_o by the **magnification equation:** $m = -(d_i/d_o)$. The algebraic sign conventions for the variables appearing in the thin-lens and magnification equations are summarized in Section 26.8. When two or more lenses are used in combination, the image produced by one lens serves as the object for the next lens.

In the **human eye,** a real, inverted image is formed on a light-sensitive surface, called the retina. **Accommodation** is the process by which the focal length of the eye is automatically adjusted, so that objects at different distances produce focused images on the retina. The **near point** of the eye is the point nearest the eye at which an object can be placed and still have a sharp image produced on the retina. The **far point** of the eye is the location of the farthest object on which the fully relaxed eye can focus. For a normal eye, the near point is located 25 cm from the eye, and the far point is located at infinity.

A **nearsighted (myopic)** eye is one that can focus on nearby objects, but not on distant ones. Nearsightedness can be corrected by wearing eyeglasses or contacts made from diverging lenses. A **farsighted (hyperopic)** eye can see distant objects clearly, but not those close up. Farsightedness can be corrected by using converging lenses.

The **refractive power** of a lens is measured in diopters and is given by $1/f$, where f is the focal length of the lens in meters. A converging lens has a positive refractive power, while a diverging lens has a negative refractive power.

The **angular size** of an object is the angle that it subtends at the eye of the viewer. For small angles, the angular size in radians is $\theta \approx h_o/d_o$, where h_o is the height of the object and d_o is the object distance. The **angular magnification** M of an optical instrument is the angular size θ' of the final image produced by the instrument divided by the reference angular size θ of the object, which is that seen without the instrument: $M = \theta'/\theta$.

A **magnifying glass** is usually a single converging lens that forms an enlarged, upright, and virtual image of an object placed at or inside the focal point of the lens. For a magnifying glass held close to the eye, the angular magnification is approximately $M \approx (1/f - 1/d_i)N$, where f is the focal length of the lens, d_i is the image distance, and N is the distance of the viewer's near point from the eye.

A **compound microscope** usually consists of two lenses, an objective and an eyepiece. The final image is enlarged, inverted, and virtual. The angular magnification of such a microscope is $M \approx -(L - f_e)N/(f_o f_e)$, where f_o and f_e are, respectively, the focal lengths of the objective and eyepiece, L is the distance between the two lenses, and N is the distance of the viewer's near point from the eye.

An **astronomical telescope** magnifies distant objects with the aid of an objective and eyepiece, and it produces a final image that is inverted and virtual. The angular magnification of a telescope is $M \approx -f_o/f_e$, where f_o and f_e are the focal lengths of the objective and eyepiece.

Lens aberrations limit the formation of perfectly focused or sharp images by optical instruments. **Spherical aberration** occurs because rays that pass through the outer edge of a lens with spherical surfaces are not focused at the same point as those that pass through the center of the lens. **Chromatic aberration** arises because a lens focuses different colors at different points.

QUESTIONS

1. Light with a given frequency travels through each of the liquids listed in Table 26.1. In which liquid does this light have (a) the longest and (b) the shortest wavelength? Provide reasons for your answers.

2. Two slabs with parallel faces are made from different types of glass. A ray of light travels through air and enters each slab at the same angle of incidence, as the drawing shows. Which slab has the greater index of refraction? Why?

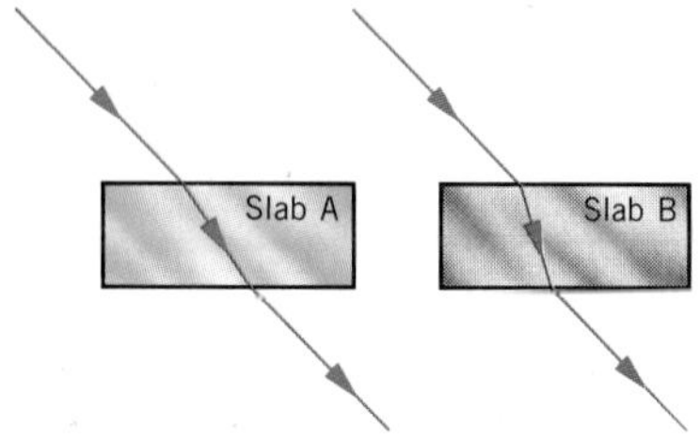

3. Two blocks, made from the same transparent material, are immersed in different liquids. A ray of light strikes each block at the same angle of incidence. From the drawing, determine which liquid has the greater index of refraction. Justify your answer.

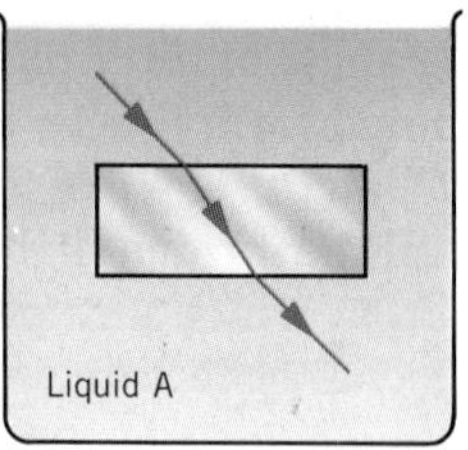

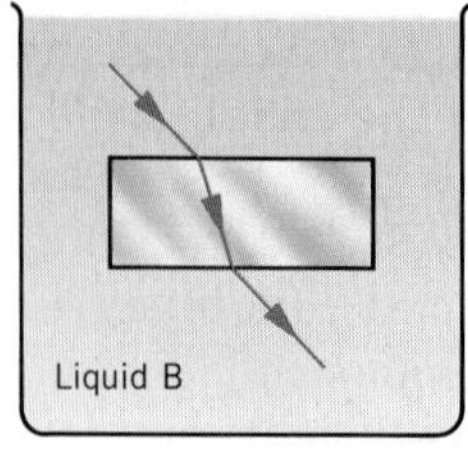

4. When an observer peers over the edge of a deep empty bowl, he does not see the entire bottom surface, so a small object lying on the bottom remains hidden from view. However, when the bowl is filled with water, the object can be seen. Explain this effect.

5. Two identical containers, one filled with water ($n = 1.33$) and the other filled with ethyl alcohol ($n = 1.36$) are viewed from directly above. Which container (if either) appears to have a greater depth of fluid? Why?

6. When you look through an aquarium window at a fish, is the fish as close as it appears? Explain.

7. At night, when it's dark outside and you are standing in a brightly lit room, it is easy to see your reflection in a window. During the day it is not so easy. Account for these facts.

8. A man is fishing from a dock. (a) If he is using a bow and arrow, should he aim above the fish, at the fish, or below the fish, to strike it? (b) How would he aim if he were using a laser gun? Give your reasoning.

9. Two rays of light converge to a point on a screen. A plane-parallel plate of glass is placed in the path of this converging light, and the glass plate is parallel to the screen. Will the point of convergence remain on the screen? If not, will the point move toward the glass or away from it? Justify your answer by drawing a diagram and showing how the rays are affected by the glass.

10. A person sitting at the beach is wearing a pair of Polaroid sunglasses and notices little discomfort due to the glare from the water on a bright sunny day. When she lies on her side, however, she notices that the glare increases. Why?

11. Linearly polarized light is incident on a surface at the Brewster's angle. The incident light is polarized such that its electric field has *no component* parallel to the surface. Is any light reflected from the surface? Explain.

12. Suppose a narrow beam of sunlight passes through a plate of glass with parallel sides, as in Figure 26.6. When the light leaves the glass, is the light dispersed into colors, as when light leaves a glass prism? Justify your answer, commenting on whether the separation of the colors (if they are separated) would be larger or smaller with thicker glass plates.

13. Refer to Figure 26.6. Note that the ray within the glass slab is traveling from a medium with a larger refractive index toward a medium with a smaller refractive index. Is it possible, for θ_1 less than 90°, that the ray within the glass will experience total internal reflection at the glass/air interface? Account for your answer.

14. Suppose you want to make a rainbow by spraying water from a garden hose into the air. (a) Where must you stand relative to the water and the sun to see the rainbow? (b) Why can't you ever walk under the rainbow?

15. A person is floating on an air mattress in the middle of a swimming pool. His friend is sitting on the side of the pool. The person on the air mattress claims that there is a light on the bottom of the pool directly beneath him. His friend insists, however, that she cannot see any light from where she sits on the side. Can both individuals be correct? Give your reasoning.

16. A beam of blue light is propagating in glass. When the light reaches the boundary between the glass and the surrounding air, the beam is totally reflected back into the glass. However, red light with the same angle of incidence is not totally reflected, and some of the light is refracted into the air. Why do these two colors behave differently?

17. A beacon in a lighthouse is to produce a parallel beam of light. The beacon consists of a bulb and a converging lens. Should the bulb be placed outside the focal point, at the focal point, or inside the focal point of the lens? State your reason.

18. Review Conceptual Example 7 as an aid in answering this question. Is it possible that a converging lens (in air) behaves as a diverging lens when surrounded by another medium? Give a reason for your answer.

19. A spherical mirror and a lens are immersed in water. Compared to the way they work in air, which one do you expect will be more affected by the water? Why?

20. A converging lens is used to project a real image onto a screen, as in Figure 26.23. A piece of black tape is then placed over the upper half of the lens. Will only the lower half of the image be visible on the screen? Justify your answer by drawing rays from various points on the object to the corresponding points on the image.

21. Review Conceptual Example 7 as an aid in answering this question. A converging lens is made from glass whose index of refraction is n. The lens is surrounded by a fluid whose index of refraction is also n. Can this lens still form an image, either real or virtual, of an object? Why?

22. In a TV mystery program, a photographic negative is introduced as evidence in a court trial. The negative shows an image of a house (now burned down) that was the scene of the crime. At the trial the defendant's acquittal depends on knowing exactly how far above the ground a window was. An expert called by the defense claims that this height can be calculated from only two pieces of information: (1) the measured height on the film, and (2) the focal length of the camera lens. Explain whether the expert is making sense, using the thin-lens and magnification equations to guide your thinking.

23. Review Conceptual Example 8 as an aid in understanding this question. Suppose two people who wear glasses are camping. One of them is nearsighted and the other is farsighted. Whose glasses may be useful in starting a fire with the sun's rays? Give your reasoning.

24. Suppose that a 21-year-old with normal vision (near point = 25 cm) is standing in front of a plane mirror. How close can he stand to the mirror and still see himself in focus?

25. If we read for a long time, our eyes become "tired." When this happens, it helps to stop reading and look at a distant object. From the point of view of the ciliary muscle, why does this refresh the eyes?

26. To a swimmer under water, objects look blurred and out of focus. However, when the swimmer wears goggles that keep the water away from the eyes, the objects appear sharp and in focus. Why do goggles improve a swimmer's underwater vision?

27. The refractive power of the lens of the eye is 15 diopters when surrounded by the aqueous and vitreous humors. If this lens is removed from the eye and surrounded by air, its refractive power increases to about 150 diopters. Why is the refractive power of the lens so much greater outside the eye?

28. Two lenses have refractive powers of 1 and 4 diopters. Draw each of the lenses and locate their focal points to scale. Parallel rays of light are incident on each lens. (a) Draw the refracted rays as they pass through the focal point of each lens. (b) From your drawings, decide which lens bends the rays to the greatest extent.

29. Jupiter is the largest planet in our solar system. Yet, to the naked eye, it looks smaller than Venus. Why?

30. By means of a ray diagram, show that the eyes of a person wearing glasses appear to be (a) smaller when the glasses use diverging lenses to correct for nearsightedness and (b) larger when the glasses use converging lenses to correct for farsightedness.

31. Can a diverging lens be used as a magnifying glass? Justify your answer with a ray diagram.

32. Who benefits more from using a magnifying glass, a person whose near point is located 25 cm away from the eyes or a person whose near point is located 75 cm away from the eyes? Provide a reason for your answer.

33. Two lenses, whose focal lengths are 3.0 and 45 cm, are used to build a telescope. Which lens should be the objective? Why?

34. Two refracting telescopes have identical eyepieces, although one telescope is twice as long as the other. Which telescope has the greater angular magnification? Provide a reason for your answer.

35. Suppose a well-designed optical instrument is composed of two converging lenses separated by 14 cm. The focal lengths of the lenses are 0.60 and 4.5 cm. Is the instrument a microscope or a telescope? Why?

36. It is often thought that virtual images are somehow less important than real images. To show that this is not true, identify which of the following instruments normally produce final images that are virtual: (a) a projector, (b) a camera, (c) a magnifying glass, (d) eyeglasses, (e) a compound microscope, and (f) an astronomical telescope.

37. Why does chromatic aberration occur in lenses, but not in mirrors?

PROBLEMS

Unless specified otherwise, use the values given in Table 26.1 for the refractive indices.

Section 26.1 The Index of Refraction

1. What is the speed of light in benzene?

2. Light travels at a speed of 2.201×10^8 m/s in a certain substance. What substance in Table 26.1 could this be? Use 2.998×10^8 m/s for the speed of light in a vacuum.

3. Find the ratio of the speed of light in diamond to the speed of light in ice.

4. A light wave has a frequency of 5.09×10^{14} Hz in water and diamond. What is the wavelength of this light in each medium?

5. A glass window ($n = 1.5$) has a thickness of 4.0×10^{-3} m. How long does it take light to pass perpendicularly through the plate?

6. The speed of light is 1.50 times larger in material *A* than it is in material *B*. Determine the ratio n_A/n_B of the refractive indices of these materials.

***7.** In a certain time, light travels 3.50 km in a vacuum. During the same time, light travels only 2.50 km in a liquid. What is the refractive index of the liquid?

***8.** When light enters a medium whose index of refraction is n, the frequency of the light does not change, but the wavelength and speed do. Derive an expression for the wavelength λ' of the light in a medium in terms of the wavelength λ in a vacuum and the index of refraction n of the medium.

Section 26.2 Snell's Law and the Refraction of Light

9. A light ray in air is incident on a water surface at a 43° angle of incidence. Find (a) the angle of reflection and (b) the angle of refraction.

10. A layer of oil ($n = 1.45$) floats on an unknown liquid. A ray of light shines from the oil into the unknown liquid. The angles of incidence and refraction are, respectively, 65.0° and 53.0°. What is the index of refraction of the unknown liquid?

11. A ray of light is propagating in water and strikes a plate of fused quartz. The angle of refraction in the quartz is measured to be 36.7°. What is the angle of incidence?

12. A beam of light is traveling in air and strikes a material. The angles of incidence and refraction are 50.0° and 30.3°, respectively. Obtain the speed of light in the material.

13. The drawing shows a coin resting on the bottom corner of a beaker filled with an unknown liquid. A ray of light from the coin travels to the surface of the liquid and is refracted as it enters into the air. A person sees the ray as it skims just above the surface of the liquid. How fast is the light traveling in the liquid?

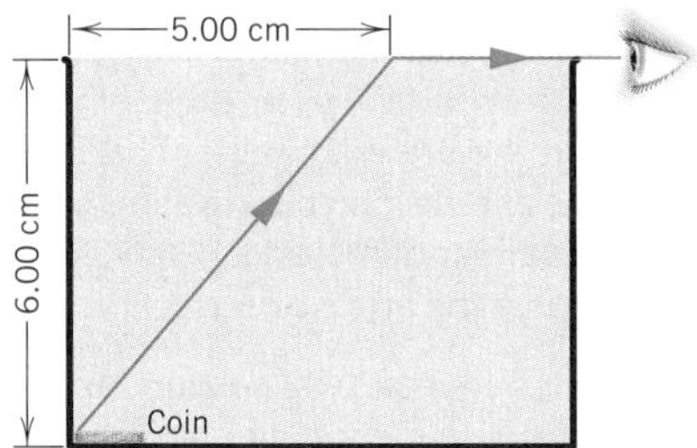

14. As an aid in understanding this problem, refer to Conceptual Example 4. A swimmer, who is looking up from under the water, sees a diving board directly above at an apparent height of 4.0 m above the water. What is the actual height of the diving board?

15. A block of crown glass is placed on top of a printed page. The block is 6.00 cm thick. When viewed directly from above, how far *above* the page does the printing appear to be?

16. A beam of light impinges from air onto a block of ice at a 60.0° angle of incidence. Assuming that this angle remains the same, find the percentage by which the angle of refraction changes when the ice turns to water, and state whether the change is an increase or a decrease.

***17.** A spotlight on a boat is 2.5 m above the water, and the light strikes the water at a point that is 8.0 m horizontally displaced from the spotlight (see the drawing). The depth of the water is 4.0 m. Determine the distance *d*, which locates the point where the light strikes the bottom.

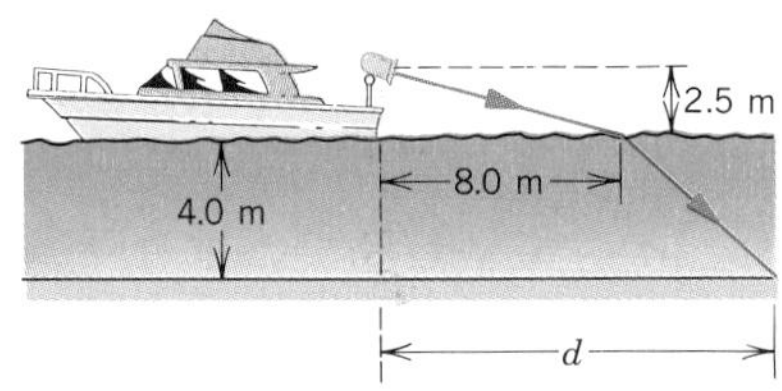

***18.** A prism is made from ice and is surrounded by air. The cross section of the prism is an isosceles right triangle. As the drawing indicates, a ray of light hits the prism. Once inside the prism, the ray travels parallel to the hypotenuse of the right triangle. Find the angle of incidence θ_1 of the entering ray and the angle of refraction θ_2 of the exiting ray.

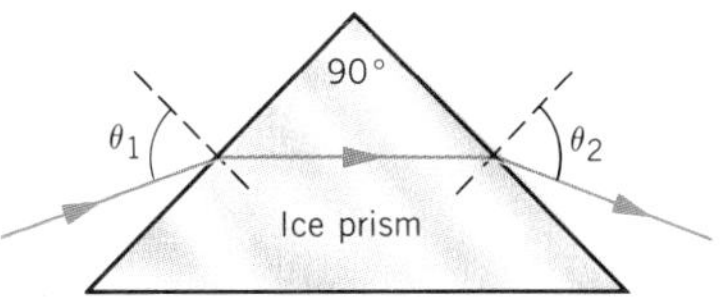

***19.** A silver medallion is sealed within a transparent block of plastic. An observer in air, viewing the medallion from directly above, sees the medallion at an apparent depth of 1.6 cm beneath the top surface of the block. How far below the top surface would the medallion appear if the observer (not wearing goggles) and the block were under water?

***20.** Refer to Figure 26.4 and assume the observer is nearly above the submerged object. For this situation, derive the expression for the apparent depth: $d' = d(n_2/n_1)$, Equation 26.3. (*Hint: Use Snell's law of refraction and the fact that the angles of incidence and refraction are small, so tan* $\theta \approx$ *sin* θ*.*)

****21.** A small logo is embedded in a thick block of crown glass ($n = 1.52$), 3.20 cm beneath the top surface of the glass. The block is put under water, so there is 1.50 cm of water above the top surface of the block. The logo is viewed from directly above by an observer in air. How far beneath the top surface of the water does the logo appear to be?

****22.** A beaker has a height of 30.0 cm. The lower half of the beaker is filled with water and the upper half is filled with oil. To a person looking down into the beaker from above, what is the apparent depth of the bottom?

****23.** The back wall of a home aquarium is a mirror that is 30.0 cm away from the front wall. The walls of the tank are negligibly thin. A fish is swimming midway between the front and back walls. (a) How far from the front wall does the fish seem to be located? (b) An image of the fish appears behind the mirror. How far is this image from the front wall of the aquarium? (c) Would the refractive index of the liquid have to be larger or smaller in order for the image of the fish to appear in *front* of the mirror, rather than behind it? Why?

Section 26.3 Total Internal Reflection

24. What is the critical angle for light emerging from ice into air?

25. One method of determining the refractive index of a transparent solid is to measure the critical angle when the solid is in air. If θ_c is found to be 40.5°, what is the index of refraction of the solid?

26. Light is propagating from diamond into crown glass. (a) Find the critical angle. (b) Is there a critical angle for light propagating from crown glass into diamond? If so, find its value.

27. The drawing shows a crown glass slab with a rectangular cross section. As illustrated, a laser beam strikes the upper surface at an angle of 60.0°. After reflecting from the upper surface, the beam reflects from the side and bottom surfaces.

(a) If the glass is surrounded by air, determine where part of the beam first exits the glass, at point *A*, *B*, or *C*. (b) Repeat part (a), assuming that the glass is surrounded by water.

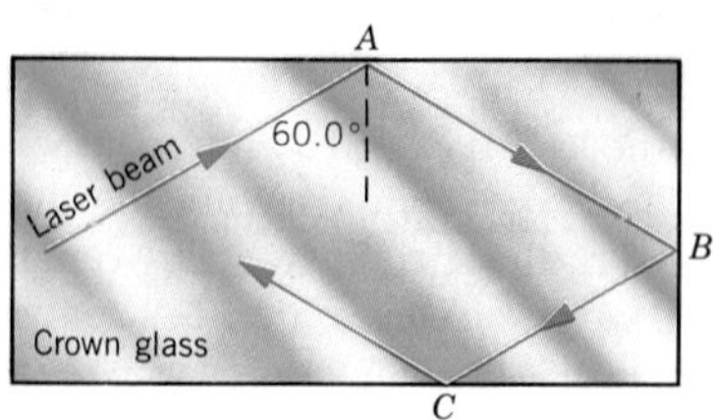

28. A person is sitting in a small boat in the ocean. A shark is swimming under water at a depth of 4.5 m. How close (measured horizontally) must the shark be from the boat before the person can see it? Assume that the person's eyes are very near the surface of the water.

29. A swimmer, under water ($n = 1.333$), looks upward toward the surface and sees a honeybee. The swimmer's line of sight makes an angle of 48.6° (the critical angle) with respect to the normal to the surface. Is the honeybee in the air above the water or floating on the surface of the water?

30. A point source of light is submerged 2.2 m below the surface of a lake and emits rays in all directions. On the surface of the lake, directly above the source, the area illuminated is a circle. What is the maximum radius that this circle could have?

31. A glass block ($n = 1.60$) is immersed in a liquid. A ray of light within the glass hits a glass–liquid surface at a 65.0° angle of incidence. Some of the light enters the liquid. What is the smallest possible refractive index for the liquid?

* **32.** Three materials, A, B, and C, have refractive indices n_A, n_B, and n_C. The materials are in the form of parallel plates and are stacked on top of one another with *A* on the bottom and *B* in the middle. A ray of light originates in material *A* and strikes the *A*–*B* surface with an angle of incidence θ_A. It is observed that the light penetrates into material *B* only when θ_A is less than 50.0° and penetrates into material C only when θ_A is less than 30.0°. Find n_B/n_A and n_B/n_C.

Section 26.4 Polarization and the Reflection and Refraction of Light

33. Light is reflected from a glass coffee table. When the angle of incidence is 56.7°, the reflected light is completely polarized parallel to the surface of the glass. What is the index of refraction of the glass?

34. Find Brewster's angle when light is reflected off a piece of glass ($n = 1.530$) submerged in ethyl alcohol ($n = 1.362$).

35. At what angle of incidence is sunlight completely polarized upon being reflected from the surface of a lake (a) in the summer and (b) in the winter when the water is frozen?

36. Light (in air) is incident at the Brewster angle on a certain piece of glass. The angle of refraction is 32.0°. Find (a) the Brewster angle and (b) the index of refraction of the glass.

37. Light is incident from air onto a beaker of benzene. If the reflected light is 100% polarized, what is the angle of refraction of the light that penetrates into the benzene?

* **38.** When light strikes the surface between two materials from above, the Brewster angle is 65.0°. What is the Brewster angle when the light encounters the same surface from below?

* **39.** In Section 26.4 it was mentioned that the reflected and refracted rays are perpendicular to each other when light strikes the surface at the Brewster angle. This is equivalent to saying that the angle of reflection plus the angle of refraction is 90°. Using Snell's law and Brewster's law, prove that the angle of reflection plus the angle of refraction is 90°.

** **40.** For a surface between two nonconducting materials, prove that the Brewster angle is never larger than the critical angle, so that light can never be 100% reflected from such a surface and simultaneously be 100% polarized.

Section 26.5 The Dispersion of Light: Prisms and Rainbows

41. A beam of sunlight encounters a plate of crown glass at a 45.00° angle of incidence. Using the data in Table 26.2, find the angle between the violet ray and the red ray in the glass.

42. Red light, traveling in air, is incident on a block of plastic. The index of refraction for the red light is 1.67 and the angle of refraction is 22.6°. Green light is also incident on the plastic with the same angle of incidence as the red light. The green light has an angle of refraction of 21.9°. What is the index of refraction for the green light?

43. Yellow light strikes a diamond at a 45.0° angle of incidence and is refracted when it enters the diamond. Blue light strikes a piece of flint glass and has the same angle of refraction as does the yellow light in the diamond. See Table 26.2 for data. What is the angle of incidence of the blue light?

44. Horizontal rays of red light ($\lambda = 660$ nm, in vacuum) and violet light ($\lambda = 410$ nm, in vacuum) are incident on the flint-glass prism shown in the drawing. See Table 26.2 for any necessary data. What is the angle of refraction for each ray as it emerges from the prism?

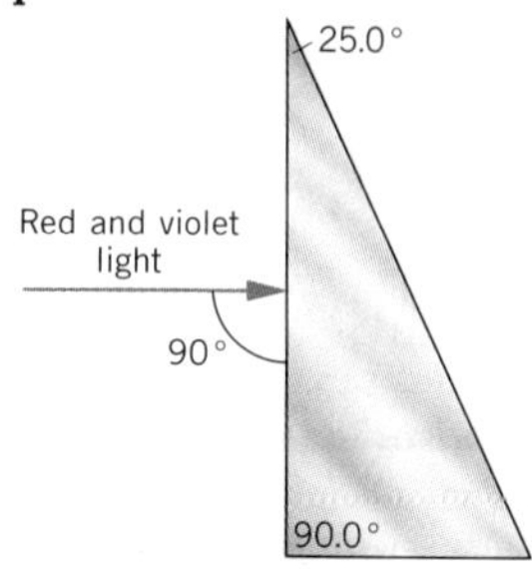

*45. Refer to Conceptual Example 7 as background material for this problem. The drawing shows a horizontal beam of light that is incident on a prism. The base of the prism is also horizontal. The prism ($n = 1.52$) is surrounded by a liquid whose index of refraction is 1.68. Determine the angle θ that the exiting light makes with the normal to the right face of the prism.

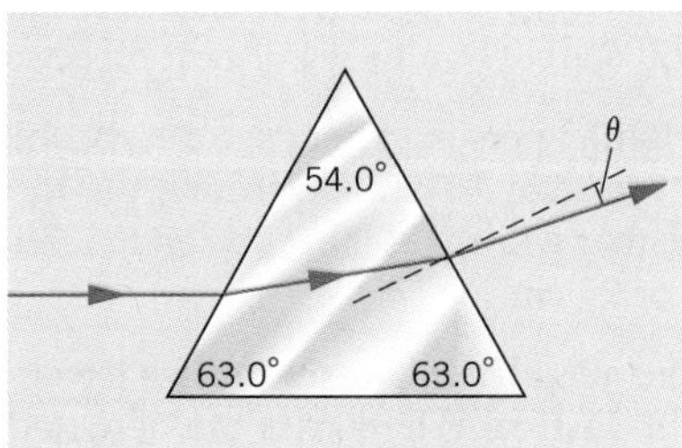

*46. This problem relates to Figure 26.14*b*, which illustrates the dispersion of light by a prism. The prism is made from flint glass (see Table 26.2), and its cross section is an equilateral triangle. The angle of incidence for both the red and violet light is 60.0°. Find the angles of refraction at which the red and violet rays emerge into the air from the prism.

Section 26.6 Lenses, Section 26.7 The Formation of Images by Lenses, Section 26.8 The Thin-Lens Equation and the Magnification Equation

(Note: When drawing ray diagrams, be sure that the object height h_o is much smaller than the focal length f of the lens or mirror. This ensures that the rays are paraxial rays.)

47. An object is located 9.0 cm in front of a converging lens ($f = 6.0$ cm). Using an accurately drawn ray diagram, determine where the image is located.

48. When a diverging lens is held 13 cm above a line of print, as in Figure 26.25, the image is 5.0 cm beneath the lens. What is the focal length of the lens?

49. A macroscopic (or macro) lens for a camera is usually a converging lens of normal focal length built into a lens barrel that can be adjusted to provide the additional lens-to-film distance needed when focusing at very close range. Suppose that a macro lens ($f = 50.0$ mm) has a maximum lens-to-film distance of 275 mm. How close can the object be located in front of the lens?

50. A person is using a converging lens ($f = 35.0$ cm) to focus an image of the moon onto a piece of paper. The diameter of the moon is 3.48×10^6 m, and its mean distance from the earth is 3.85×10^8 m. What is the diameter of the moon's image on the paper?

51. A converging lens ($f = 12.0$ cm) is held 8.00 cm in front of a newspaper. Find (a) the image distance and (b) the magnification.

52. A diverging lens has a focal length of −25 cm. (a) Find the image distance when an object is placed 38 cm from the lens. (b) Is the image real or virtual?

53. An object, 0.50 cm high, is placed 8.6 cm in front of a diverging lens whose focal length is −7.5 cm. Find the height of the image.

54. An object is located 30.0 cm to the left of a converging lens whose focal length is 50.0 cm. (a) Draw a ray diagram to scale and from it determine the image distance and the magnification. (b) Use the thin-lens and magnification equations to verify your answers to part (a).

55. A diverging lens has a focal length of −38 cm. An object is placed 28 cm in front of this lens. Calculate (a) the image distance and (b) the magnification. Is the image (c) real or virtual, (d) upright or inverted, and (e) enlarged or reduced in size?

56. A magnifying glass uses a converging lens whose focal length is 15 cm. The magnifying glass produces a virtual and upright image that is 3.0 times larger than the object. (a) How far is the object from the lens? (b) What is the image distance?

*57. See Conceptual Example 8 as an aid in understanding this problem. A camper is trying to start a fire by focusing sunlight onto a piece of paper. The diameter of the sun is 1.39×10^9 m and its mean distance from the earth is 1.50×10^{11} m. The camper is using a converging lens whose focal length is 10.0 cm. (a) What is the area of the sun's image on the paper? (b) If 0.530 W of sunlight pass through the lens, what is the intensity of the sunlight at the paper?

*58. On a roll of movie film, each picture has a width of 70.0 mm. The projector lens has a focal length of 305 mm. If the screen in a theater is 60.0 m from the projector lens, what is the width of the image projected on the screen?

*59. The moon's diameter is 3.48×10^6 m and its mean distance from the earth is 3.85×10^8 m. The moon is being photographed by a camera whose lens has a focal length of 50.0 mm. (a) Find the diameter of the moon's image on the slide film. (b) When the slide is projected onto a screen that is 15.0 m from the lens of the projector ($f = 110.0$ mm), what is the diameter of the moon's image on the screen?

*60. From a distance of sixty meters, a photographer uses a telephoto lens ($f = 500.0$ mm) to take a picture of a charging rhinoceros. How far from the rhinoceros would the photographer have to be to record an image of the same size using a lens whose focal length is 50.0 mm?

**61. An object is 20.0 cm from a converging lens, and the image falls on a screen. When the object is moved 4.00 cm closer to the lens, the screen must be moved 2.70 cm farther away from the lens to register a sharp image. Determine the focal length of the lens.

**62. A converging lens ($f = 25.0$ cm) is used to project an image of an object onto a screen. The object and the screen are 125 cm apart, and between them the lens can be placed at either of two locations. Find the two object distances.

Section 26.9 Lenses in Combination

63. A converging lens ($f = 0.65$ m) is located 1.2 m to the right of an object. A plane mirror is located 2.1 m to the right of the lens. (a) Find the final image distance, measured relative to the mirror, of this lens–mirror combination. (b) Is the final image real or virtual? Why?

64. A converging lens has a focal length of 0.080 m. An object is located 0.040 m to the left of this lens. A second converging lens has the same focal length as the first one and is located 0.120 m to the right of it. Relative to the second lens, where is the final image located?

65. A converging lens ($f = 12.0$ cm) is located 30.0 cm to the left of a diverging lens ($f = -6.00$ cm). A postage stamp is placed 36.0 cm to the left of the converging lens. (a) Locate the final image of the stamp relative to the diverging lens. (b) Find the overall magnification. (c) Is the final image real or virtual? With respect to the original object, is the final image (d) upright or inverted, and is it (e) larger or smaller?

66. An object, 0.75 cm tall, is placed 12.0 cm to the left of a diverging lens ($f = -8.00$ cm). A converging lens is placed 8.00 cm to the right of the diverging lens. The final image is virtual and is 29.0 cm to the left of the diverging lens. Determine (a) the focal length of the converging lens and (b) the height of the final image.

67. A converging lens ($f = 12.0$ cm) is 28.0 cm to the left of a diverging lens ($f = -14.0$ cm). An object is located 6.00 cm to the left of the converging lens. Draw an accurate ray diagram and from it find (a) the final image distance, measured from the diverging lens, and the overall magnification. (b) Confirm your answers to part (a) by using the thin-lens and magnification equations.

*__68.__ A coin is located 15.00 cm to the left of a converging lens ($f = 10.00$ cm). A second, identical lens is placed to the right of the first lens, such that the image formed by the combination has the same size and orientation as the original coin. Find the separation between the lenses.

*__69.__ An object is placed 20.0 cm to the left of a diverging lens ($f = -8.00$ cm). A concave mirror ($f = 12.0$ cm) is placed 30.0 cm to the right of the lens. (a) Find the final image distance, measured relative to the mirror. (b) Is the final image real or virtual? (c) Is the final image upright or inverted with respect to the original object?

**70. Two converging lenses ($f_1 = 9.00$ cm and $f_2 = 6.00$ cm) are separated by 18.0 cm. The lens on the left has the longer focal length. An object stands 12.0 cm to the left of the combination. (a) Locate the final image relative to the lens on the right. (b) Obtain the overall magnification. (c) Is the final image real or virtual? With respect to the original object, is the final image (d) upright or inverted and is it (e) larger or smaller?

Section 26.10 The Human Eye

71. A nearsighted person has a far point located only 220 cm from his eyes. Determine the focal length of contact lenses that will enable him to see distant objects clearly.

72. A farsighted person cannot see clearly objects that are closer to the eye than 73 cm. Determine the focal length of contact lenses that will enable this person to read a magazine at a distance of 25 cm.

73. Suppose your friend wears contact lenses that have a focal length of 35.1 cm. The lenses are designed so she can read a magazine held as close as 25.0 cm. Where is the near point of her unaided eyes?

74. A person has far points of 5.0 m from the right eye and 6.5 m from the left eye. Write a prescription for the refractive power of each corrective contact lens.

75. A woman can read the large print in a newspaper only when it is at a distance of 65 cm or more from her eyes. (a) Is she myopic or hyperopic? (b) What should be the refractive power of her glasses (worn 2.0 cm from the eyes), so she can read the newspaper at a distance of 25 cm from the eyes?

76. A person holds a book 25 cm in front of the effective lens of her eye; the print in the book is 2.0 mm high. If the effective lens of the eye is located 1.7 cm from the retina, what is the size (including the sign) of the print image on the retina?

*__77.__ A nearsighted person wears contacts to correct for a far point that is only 3.62 m from his eyes. The near point of his unaided eyes is 25.0 cm from his eyes. If he does not remove the lenses when reading, how close can he hold a book and see it clearly?

**78. The far point of a nearsighted person is 4.37 m from her eyes, and she wears contacts that enable her to see distant objects clearly. A tree is 12.0 m away and 3.00 m high. (a) When she looks through the contacts at the tree, what is its image distance? (b) How high is the image formed by the contacts?

**79. The contacts worn by a farsighted person allow her to see objects clearly that are as close as 25.0 cm, even though her uncorrected near point is 79.0 cm from her eyes. When she is looking at a poster, the contacts form an image of the poster at a distance of 217 cm from her eyes. (a) How far away is the poster actually located? (b) If the poster is 0.350 m tall, how tall is the image formed by the contacts?

Section 26.11 Angular Magnification and the Magnifying Glass

80. A spectator, seated in the left field stands, is watching a 1.9-m-tall baseball player who is 75 m away. On a TV screen, the same player has a 0.12-m image. To a viewer located 3.0 m from the screen, does the ball player appear to be larger or smaller than what the spectator sees? Justify your answer in both cases by calculating the angular sizes of the player.

81. A quarter (diameter = 2.4 cm) is held at arms length (70.0 cm). The sun has a diameter of 1.39×10^9 m and is 1.50×10^{11} m from the earth. What is the ratio of the angular size of the quarter to that of the sun?

82. A jeweler whose near point is 65 cm from his eye uses a small magnifying glass to examine a watch held 5.0 cm from the lens. Find the angular magnification of the magnifying glass.

83. A magnifying glass is held above a magazine such that the image is located at the near point of the eye. The near point is 0.30 m away from the eye, and the angular magnification is 3.4. Find the focal length of the magnifying glass.

84. A butterfly collector is examining a rare specimen and uses a magnifying glass with a refractive power of 10.0 diopters. The magnifying glass is held close to the eye, and the butterfly-to-lens distance is adjusted so a virtual image is formed at infinity. The angular magnification is 4.0. What is the distance between the collector's eyes and the near point?

***85.** A stamp collector is viewing a special stamp. The collector's near point is 25 cm from his eyes. (a) What is the refractive power of a magnifying glass that has an angular magnification of 6.0 when the image of the stamp is located at the near point of the eye? (b) What is the angular magnification when the image of the stamp is 45 cm from the eye?

****86.** A farsighted person can read printing as close as 25.0 cm when she wears contacts that have a focal length of 45.4 cm. One day, however, she forgets her contacts and uses a magnifying glass, which has a maximum angular magnification of 7.50 for a young person with a normal near point of 25.0 cm. What is the maximum angular magnification that the magnifying glass can provide for her?

Section 26.12 The Compound Microscope

87. An insect subtends an angle of only 4.0×10^{-3} rad at the unaided eye when placed at the near point. What is the angular size (magnitude only) when the insect is viewed through a microscope whose angular magnification has a magnitude of 160?

88. A microscope for viewing blood cells has an objective with a focal length of 0.620 cm and an eyepiece with a focal length of 4.40 cm. The distance between the objective and eyepiece is 23.0 cm. If a blood cell subtends an angle of 2.40×10^{-5} rad when viewed with the naked eye at a near point of 25.0 cm, what angle (magnitude only) does it subtend when viewed through the microscope?

89. The distance between the lenses in a microscope is 18 cm. The focal length of the objective is 1.5 cm. If the microscope is to provide an angular magnification of −83 when used by a person with a normal near point (25 cm from the eye), what must be the focal length of the eyepiece?

90. An anatomist is viewing heart muscle cells with a microscope that has two selectable objectives with refracting powers of 100 and 300 diopters. When she uses the 100-diopter objective, the image of a cell subtends an angle of 3×10^{-3} rad with the eye. What angle is subtended when she uses the 300-diopter objective?

***91.** The maximum angular magnification of a magnifying glass is 12.0 when a person uses it who has a near point that is 25.0 cm from his eyes. The same person finds that a microscope, using this magnifying glass as the eyepiece, has an angular magnification of −525. The separation between the eyepiece and the objective of the microscope is 23.0 cm. Obtain the focal length of the objective.

***92.** It is possible to interchange the eyepiece and the objective of a microscope without changing the angular magnification of the instrument, provided the separation L between the two lenses is suitably adjusted. Derive an equation that gives the new separation L' in terms of the original separation L, the focal length f_o of the original objective, and the focal length f_e of the original eyepiece.

Section 26.13 The Telescope

93. An astronomical telescope has an angular magnification of −184 and uses an objective with a focal length of 48.0 cm. What is the focal length of the eyepiece?

94. An astronomical telescope has an objective with a focal length of 96 cm and interchangeable eyepieces whose focal lengths are 3.0 and 0.80 cm. What angular magnifications are possible?

95. A refracting telescope has an objective and an eyepiece that have refractive powers of 1.25 diopters and 250 diopters, respectively. Find the angular magnification of the telescope.

96. The moon subtends an angle of 9.0×10^{-3} rad at the unaided eye. An astronomical telescope uses an eyepiece with a focal length of 0.42 m and an objective with a focal length of 1.8 m. When viewed through this telescope, what angle (magnitude only) does the moon subtend?

97. An amateur astronomer decides to build a telescope from a discarded pair of eyeglasses. One of the lenses has a refractive power of 11 diopters, while the other has a refractive power of 1.3 diopters. (a) Which lens should be the objective? (b) How far apart should the lenses be separated? (c) What is the angular magnification of the telescope?

***98.** A refracting telescope has an angular magnification of −109. The length of the barrel is 1.10 m. What are the focal lengths of the objective and the eyepiece?

***99.** The refracting telescope at Yerkes Observatory in Wisconsin has an objective whose focal length is 19.4 m. Its eyepiece has a focal length of 10.0 cm. (a) What is the angular magnification of the telescope? (b) If the telescope is used to look at a lunar crater (diameter = 1500 m), what is the size of the first image, assuming the surface of the moon is 3.77×10^8 m from the surface of the earth? (c) How close does the crater appear to be when seen through the telescope?

**100. An astronomical telescope is being used to examine a relatively close object that is only 114.00 m away from the objective of the telescope. The objective and eyepiece have focal lengths of 1.500 and 0.070 m, respectively. Noting that the expression $M \approx -f_o/f_e$ is no longer applicable because the object is so close, use the thin-lens and magnification equations to find the angular magnification of the telescope. *(Hint: See Figure 26.39 and note that the focal points F_o and F_e are so close together that the distance between them may be ignored.)*

ADDITIONAL PROBLEMS

101. A figurine is placed 15.0 cm in front of a converging lens ($f = 40.0$ cm). Using a ray diagram drawn to scale, find (a) the image distance and (b) the magnification.

102. A compound microscope has a barrel whose length is 16.0 cm and an eyepiece whose focal length is 1.4 cm. The viewer has a near point located 25 cm from his eyes. What focal length must the objective have so the angular magnification of the microscope is −320?

103. A ray of sunlight hits a frozen lake at a 45° angle of incidence. At what angle of refraction does the ray penetrate (a) the ice and (b) the water beneath the ice?

104. A ray of light is traveling in glass and strikes a glass/liquid interface. The angle of incidence is 58.0°, and the index of refraction of glass is $n = 1.50$. (a) What must be the index of refraction of the liquid such that the direction of the light entering the liquid is not changed? (b) What is the largest index of refraction that the liquid can have, such that none of the light is transmitted into the liquid and all of it is reflected back into the glass?

105. A farsighted person cannot focus clearly on objects that are less than 145 cm from his eyes. To correct this problem, the person wears eyeglasses that are located 2.0 cm in front of his eyes. Determine the focal length that will permit this person to read a newspaper at a distance of 32.0 cm from his eyes.

106. Amber ($n = 1.546$) is a transparent brown-yellow fossil resin. An insect, trapped and preserved within the amber, appears to be 2.5 cm beneath the surface, when viewed directly from above. How far below the surface is the insect actually located?

107. A person looks at a scene through a diverging lens with a focal length of −12.5 cm. The lens forms a virtual image 5.00 cm from the lens. Find the magnification.

108. A camera is supplied with two interchangeable lenses, whose focal lengths are 35.0 and 150.0 mm. A woman whose height is 1.80 m stands 8.00 m in front of the camera. What is the height (including sign) of her image that each lens produces on the film?

109. Refer to Conceptual Example 7 as an aid in understanding this problem. The drawing shows a ray of light traveling through a gas ($n = 1.00$), a solid ($n = 1.55$), and a liquid ($n = 1.55$). At what angle θ does the light enter the liquid?

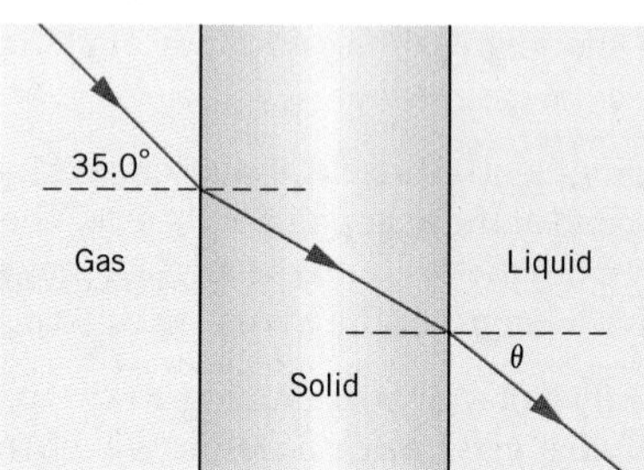

110. An engraver uses a magnifying glass ($f = 9.5$ cm) to examine some work. The image he sees is located 25 cm from his eye, which is his near point. (a) What is the distance between the work and the magnifying glass? (b) What is the angular magnification of the magnifying glass?

111. A refracting astronomical telescope for hobbyists has an angular magnification of −155. The eyepiece has a focal length of 5.00 mm. (a) Determine the focal length of the objective. (b) About how long is the telescope?

112. A ray of light in air enters a liquid at an angle of incidence of 47.00°. The angle of refraction is 29.16°. Identify the liquid on the basis of the information in Table 26.1.

113. An optometrist prescribes contact lenses that have a focal length of 55.0 cm. (a) Are the lenses converging or diverging, and (b) is the person who wears them nearsighted or farsighted? (c) Where is the unaided near point of the person located, if the lenses are designed so that objects no closer than 35.0 cm can be seen clearly?

114. To focus a camera on objects at different distances, the converging lens is moved toward or away from the film, so a sharp image always falls on the film. A camera with a telephoto lens ($f = 200.0$ mm) is to be focused on an object located first at a distance of 3.5 m and then at 50.0 m. Over what distance must the lens be movable?

115. (a) For a diverging lens ($f = -20.0$ cm), construct a ray diagram to scale and find the image distance for an object that is 20.0 cm from the lens. (b) Determine the magnification of the lens from the diagram.

***116.** A person using a magnifying glass observes that for clear vision its maximum angular magnification is 1.25 times larger than its minimum angular magnification. Assuming that the person has a near point located 25 cm from her eye, what is the focal length of the magnifying glass?

*117. An office copier uses a lens to place an image of a document onto a rotating drum. The copy is made from this image. (a) What kind of lens is used? (b) If the document and its copy are to have the same size, but are inverted with respect to one another, how far from the document is the lens located and how far from the lens is the image located? Express your answers in terms of the focal length f of the lens.

*118. A reflecting telescope uses a concave mirror whose radius of curvature is 2.4 m. If the angular magnification of the telescope is −360, what is the focal length of the eyepiece?

*119. When a converging lens is used in a camera (as in Figure 26.22), the film must be placed at a distance of 0.210 m from the lens to record an image of an object that is 4.00 m from the lens. The same lens is then used in a projector (see Figure 26.23), with the screen 0.500 m from the lens. How far from the projector lens should the film be placed?

*120. At age forty, a man requires contact lenses ($f = 65.0$ cm) to read a book held 25.0 cm from his eyes. At age forty-five, he finds that while wearing these contacts he must now hold a book 29.0 cm from his eyes. (a) By what distance has his near point changed? (b) What focal length lenses does he require at age forty-five to read a book at 25.0 cm?

**121. The angular magnification of a refracting telescope is 32 800 times larger when you look through the correct end of the telescope than when you look through the wrong end. What is the angular magnification of the telescope?

**122. The equation

$$\frac{1}{d_o} + \frac{1}{d_i} = \frac{1}{f}$$

is called the *Gaussian* form of the thin-lens equation. The drawing shows the variables d_o, d_i, and f. The drawing also shows the distances x and x', which are, respectively, the distance from the object to the focal point on the left of the lens and the distance from the focal point on the right of the lens to the image. An equivalent form of the thin-lens equation, involving x, x', and f, is called the *Newtonian* form. Show that the Newtonian form of the thin-lens equation can be written as $xx' = f^2$.

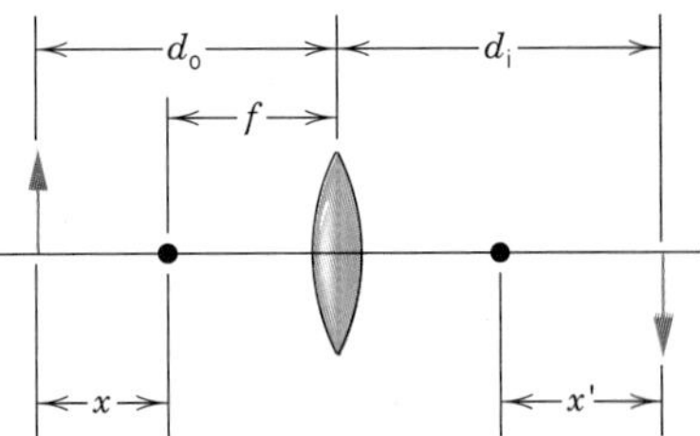

**123. Bill is farsighted and has a near point located 125 cm from his eyes. Anne is also farsighted, but her near point is 75.0 cm from her eyes. Both have glasses that correct their vision to a normal near point (25.0 cm from the eyes), and both wear the glasses 2.0 cm from the eyes. Relative to the eyes, what is the closest object that can be seen clearly (a) by Anne when she wears Bill's glasses and (b) by Bill when he wears Anne's glasses?

CHAPTER 27 INTERFERENCE AND THE WAVE NATURE OF LIGHT

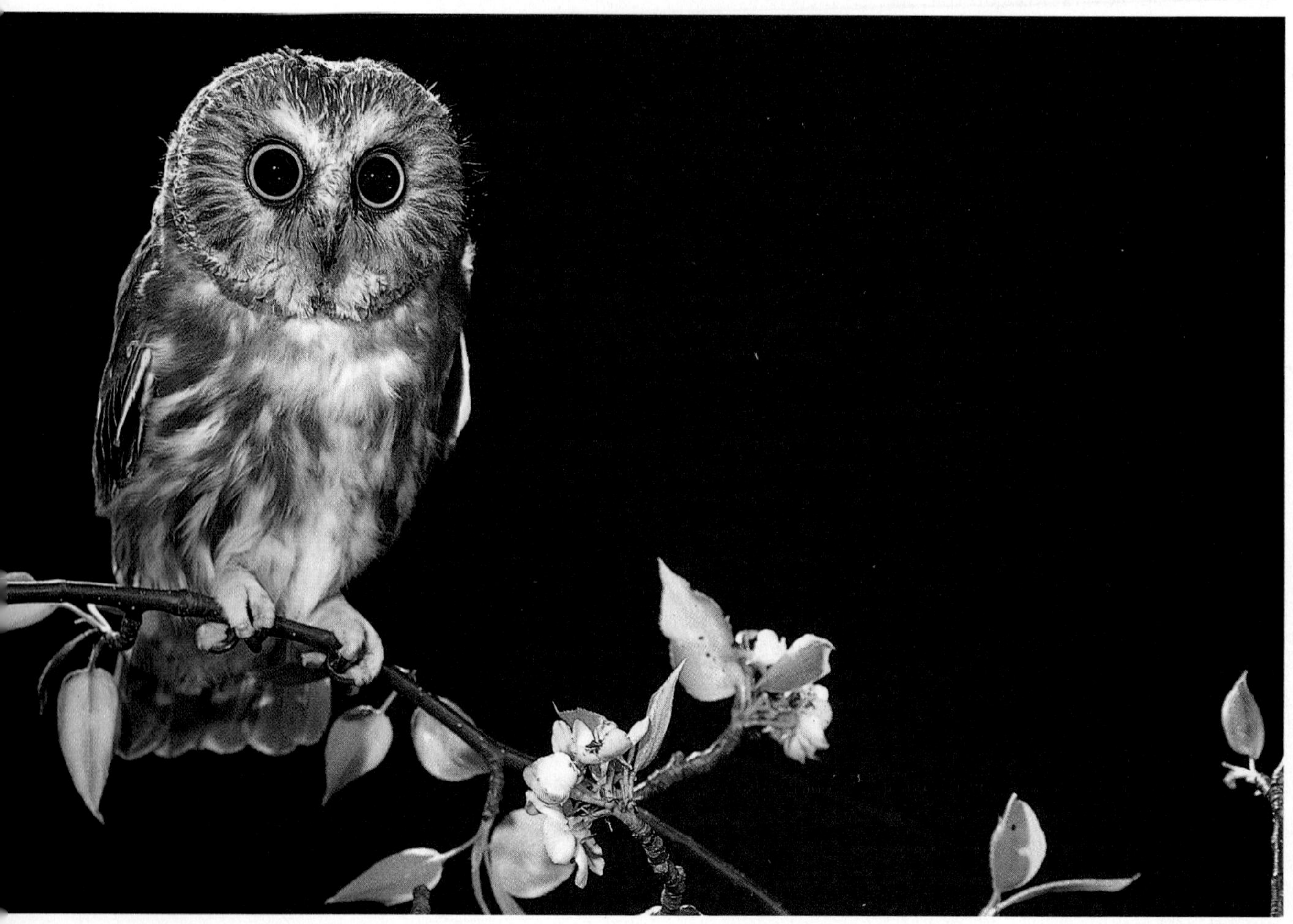

Saw-whet owls are famous for their eyes, which can distinguish between two objects that are quite close together. The ability to distinguish between closely spaced objects is related, in part, to the very large pupils and the wave nature of light, as we will discuss. Being a wave, light can exhibit effects due to interference and diffraction. The study of the interference and diffraction of light is referred to as ***wave optics*** *or* ***physical optics,*** *to distinguish it from* ***geometrical optics,*** *which deals with the straight-line motion of light and its reflection and refraction. Geometrical optics cannot explain interference and diffraction, and in this chapter we will use the principle of linear superposition to do so, a principle first presented in Chapter 17 in connection with sound waves.*

27.1 THE PRINCIPLE OF LINEAR SUPERPOSITION

The principle of linear superposition states that when two or more waves are present simultaneously at the same place, the resultant wave is the sum of the individual waves. Light is an electromagnetic wave; thus, the electric fields of two light waves passing through a given point combine to give the total electric field at that point. And the square of the electric field strength is proportional to the light intensity, which is the light energy per second per square meter. The intensity, in turn, is related to the brightness of the light.

Figure 27.1 illustrates what happens when two identical waves (same wavelength λ and amplitude) arrive at the point *P* in phase, that is, crest to crest and trough to trough. According to the principle of linear superposition, the waves reinforce each other and ***constructive interference*** occurs. The resulting total wave at *P* has an amplitude that is twice the amplitude of either individual wave, and in the case of light waves, the brightness at *P* is greater than that due to either wave alone. The waves start out in phase and are in phase at *P*, because this spot is located away from the sources of the waves at distances ℓ_1 and ℓ_2 that differ by one wavelength. In Figure 27.1, the distances are $\ell_1 = 2\frac{1}{4}$ wavelengths and $\ell_2 = 3\frac{1}{4}$ wavelengths. In general, when the waves start out in phase, constructive interference will result at *P* whenever the distances are the same or differ by any integer number of wavelengths; in other words, assuming ℓ_2 is the larger distance, whenever $\ell_2 - \ell_1 = m\lambda$, where $m = 0, 1, 2, 3, \ldots$.

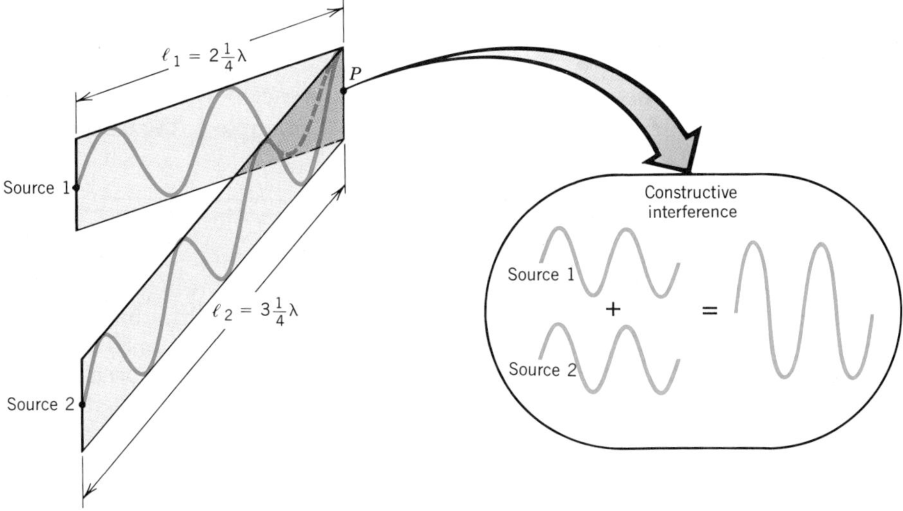

Figure 27.1 The waves emitted by source 1 and source 2 start out in phase and arrive at point *P* in phase, leading to constructive interference at that point.

Figure 27.2 shows what occurs when two identical waves arrive at the point *P* out of phase with one another, or crest to trough. Now the waves mutually cancel, according to the principle of linear superposition, and ***destructive interference*** results. With light waves this would mean that there is no brightness. The waves begin with the same phase but are out of phase at *P*, because the distances through which they travel in reaching this spot differ by one-half of a wavelength ($\ell_1 = 2\frac{3}{4}\lambda$ and $\ell_2 = 3\frac{1}{4}\lambda$ in the drawing). For waves that start out in phase, destructive interference will take place at *P* whenever the distances differ by any odd integer number of half-wavelengths, that is, whenever $\ell_2 - \ell_1 = (m + \frac{1}{2})\lambda$, where $m = 0, 1, 2, 3, \ldots$, and ℓ_2 is the larger distance.

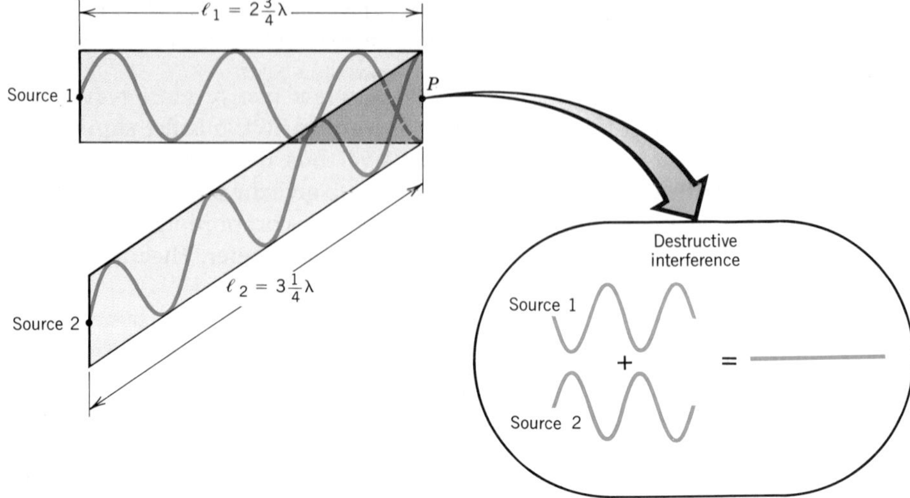

Figure 27.2 The waves emitted by the two sources have the same phase to begin with, but they arrive at point *P* out of phase. As a result, destructive interference occurs at *P*.

If constructive or destructive interference is to continue occurring at a point, the sources of the waves must be ***coherent sources.*** Two sources are coherent if the waves they emit maintain a constant phase relation. Effectively, this means that the waves do not shift relative to one another as time passes. For instance, suppose that the wave pattern of source 1 in Figure 27.2 shifted forward or backward by random amounts at random moments. Then, on average, neither constructive nor destructive interference would be observed at point *P*, because there would be no stable relation between the two wave patterns. Lasers are coherent sources of light, while incandescent light bulbs and fluorescent lamps are incoherent sources.

Many people have observed the effects of interference between two electromagnetic waves while watching TV, as Conceptual Example 1 discusses.

THE PHYSICS OF . . .

how airplanes overhead cause TV pictures to "flutter."

Conceptual Example 1 Fluttering TV Pictures

Television programming is carried by electromagnetic waves that are in the radio-frequency region of the electromagnetic spectrum. They are emitted by a transmitting antenna and detected by a receiving antenna connected to the TV set, as Figure 27.3 suggests. Sometimes, when an airplane passes overhead at a sufficiently low altitude, the TV picture flutters due to interference effects. Considering that an interference effect requires at least two sources of waves and that there is only a single transmitting antenna for the electromagnetic waves, explain how the interference arises. In addition, explain why an airplane directly over the receiving antenna causes little fluttering.

REASONING To explain how the interference arises, we recall that light can reflect from a metal surface and that light is an electromagnetic wave. The waves that carry TV programming are not visible light waves, but they also can and do reflect from airplanes. When this occurs, the receiving antenna may detect waves coming from two sources. Source 1 is the transmitting antenna of the station, which is located at a fixed distance ℓ_1 from the house. The plane acts as source 2, and its distance ℓ_2 from the house is changing as the plane moves. As a result, the difference between ℓ_1 and ℓ_2 is also changing, and the conditions of constructive and destructive interference come and go at the receiving antenna. Correspondingly, the receiving antenna delivers to the TV set a signal that increases and decreases, causing the picture to flutter as the plane passes by. Later on in this chapter, we will encounter other situations in which reflection creates the second source of waves needed for interference effects to arise.

The reason that an airplane directly above the receiving antenna causes little fluttering is a direct consequence of the law of reflection. This law states that the incident and reflected rays lie in the same plane and that the angles of incidence and reflection are equal. Consequently, most of the electromagnetic energy reflected from a plane directly above the receiving antenna travels away from the antenna, not toward it. By the time the plane is directly overhead, the fluttering effect has largely come and gone.

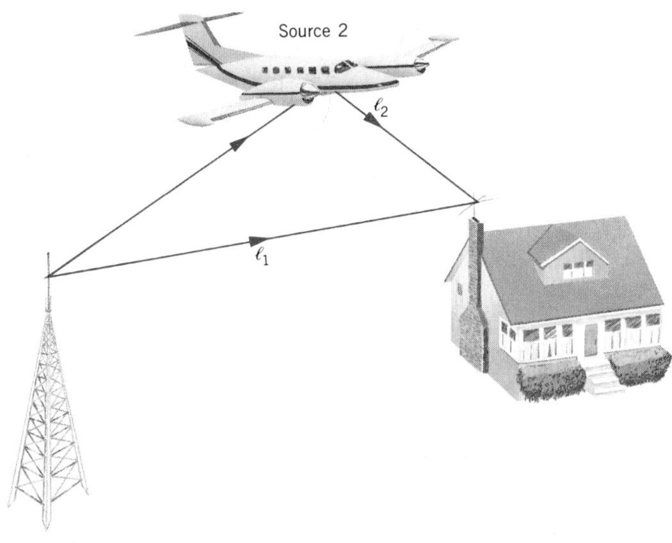

Figure 27.3 Source 1 is a TV transmitting antenna that emits electromagnetic waves. One wave reaches the house directly, while another arrives after being reflected from a passing airplane, which acts as wave source 2. As the plane flies by, conditions of constructive and destructive interference come and go at the receiving antenna, and the TV picture flutters.

27.2 YOUNG'S DOUBLE-SLIT EXPERIMENT

In 1801 the English scientist Thomas Young (1773–1829) performed a historic experiment that demonstrated the wave nature of light by showing that two overlapping light waves interfered with each other. His experiment was particularly important, because he was also able to determine the wavelength of the light from his measurements, the first such determination of this important property. Figure 27.4 shows one arrangement of Young's experiment, in which light of a single wavelength (monochromatic light) passes through a single narrow slit S_0 and falls on two closely spaced, narrow slits S_1 and S_2. These two slits act as coherent sources of light waves that interfere constructively and destructively at different points on the screen to produce a pattern of alternating bright and dark fringes. The purpose of the single slit S_0 is to ensure that only light from one direction falls on the double slit. Without it, light coming from different points on the light source would strike the double slit from different directions and cause the pattern on the screen to be washed out. The slits S_1 and S_2 act as coherent sources of light waves, because the light from each originates from the same primary source, namely, the single slit S_0.

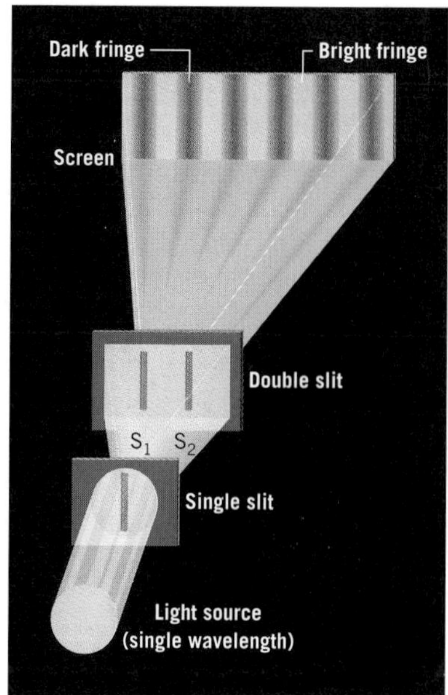

Figure 27.4 In Young's double-slit experiment, two slits S_1 and S_2 act as coherent sources of light. Light waves from these slits interfere constructively and destructively on the screen to produce, respectively, the bright and dark fringes. The slit widths and the distance between the slits have been exaggerated for clarity.

To help explain the origin of the bright and dark fringes, Figure 27.5 presents three top views of the double slit and the screen. Part *a* illustrates how a bright fringe arises directly opposite the midpoint between the two slits. At this location on the screen, the distances ℓ_1 and ℓ_2 to the slits are equal, each containing the same number of wavelengths. Therefore, constructive interference results, leading to the bright fringe. Part *b* indicates that constructive interference produces another bright fringe on one side of the midpoint when the distance ℓ_2 is larger than ℓ_1 by exactly one wavelength. A bright fringe also occurs symmetrically on

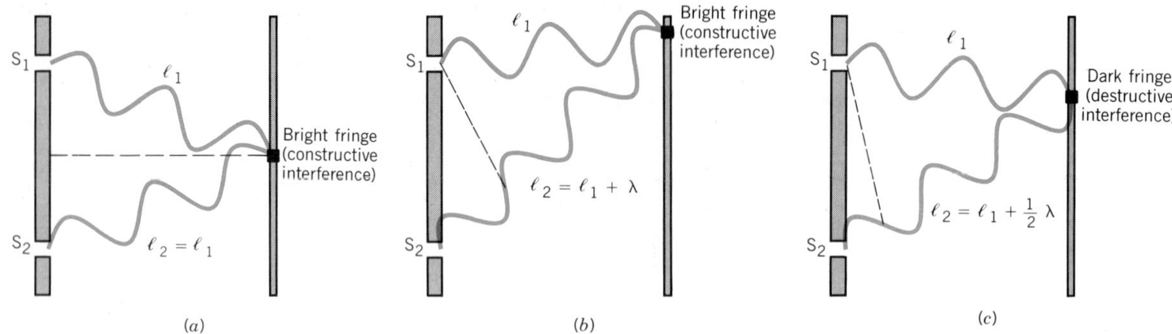

Figure 27.5 The waves from slits S_1 and S_2 interfere constructively (parts *a* and *b*) or destructively (part *c*) on the screen, depending on the difference in distances between the slits and the screen. The slit widths and the distance between the slits have been exaggerated for clarity.

the other side of the midpoint when the distance ℓ_1 exceeds ℓ_2 by one wavelength; for clarity, however, this bright fringe is not shown. Constructive interference produces additional bright fringes on both sides of the middle wherever the difference between ℓ_1 and ℓ_2 is an integer number of wavelengths. Part *c* shows how the first dark fringe arises. Here the distance ℓ_2 is larger than ℓ_1 by exactly one-half a wavelength, so the waves interfere destructively, giving rise to the dark fringe. Destructive interference creates additional dark fringes on both sides of the center wherever the difference between ℓ_1 and ℓ_2 equals an odd integer number of half-wavelengths.

The brightness of the fringes in Young's experiment varies. As an indication of the brightness, Figure 27.6 gives a graph of the light intensity for the fringe pattern. The central fringe is labeled with a zero, while the other bright fringes are numbered in ascending order on either side of the center. It can be seen that the central fringe has the greatest intensity. To either side of the center, the intensities of the other fringes decrease in a way that depends on how small the slit widths are relative to the wavelength of the light. Figure 27.7 shows a photograph of the fringe pattern observed in a typical Young's experiment.

The position of the fringes observed on the screen in Young's experiment can be calculated with the aid of Figure 27.8. If the screen is located very far away compared with the separation d of the slits, then the lines labeled ℓ_1 and ℓ_2 in part

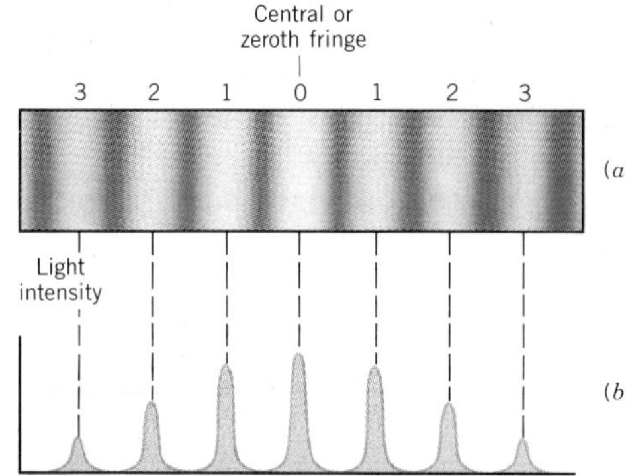

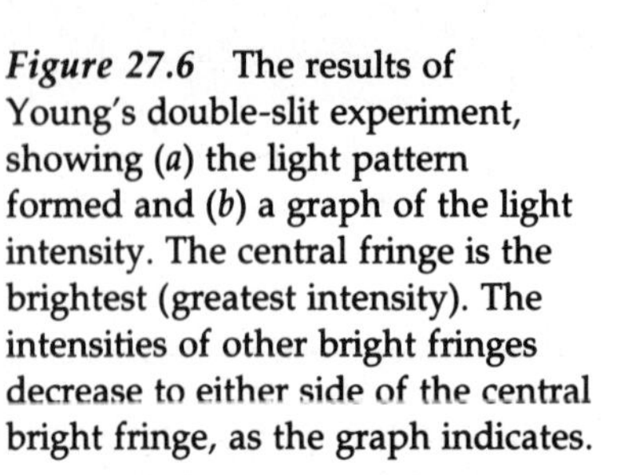

Figure 27.6 The results of Young's double-slit experiment, showing (*a*) the light pattern formed and (*b*) a graph of the light intensity. The central fringe is the brightest (greatest intensity). The intensities of other bright fringes decrease to either side of the central bright fringe, as the graph indicates.

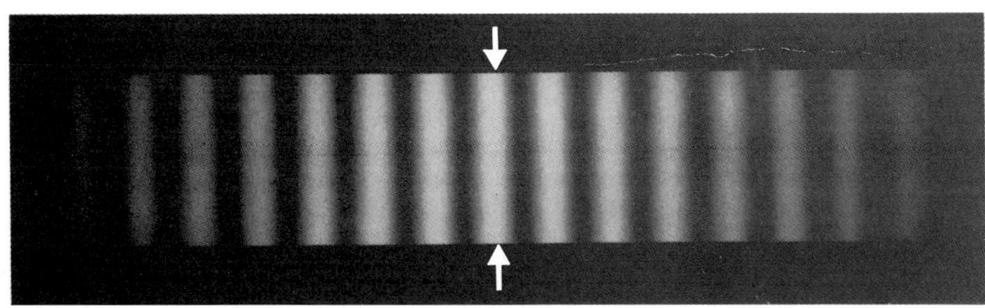

Figure 27.7 This photograph shows the fringe pattern obtained in a typical Young's double-slit experiment. The arrows indicate the central bright fringe.

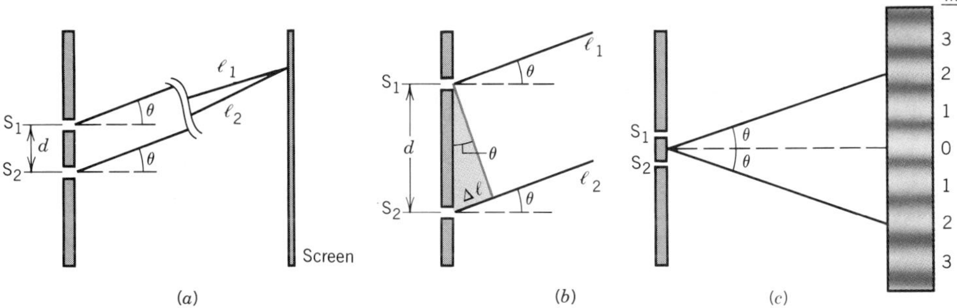

Figure 27.8 With the help of these pictures, Equation 27.1 is derived in the text. This equation gives the angles θ at which the bright fringes occur.

a are nearly parallel. Being nearly parallel, these lines make approximately equal angles θ with the horizontal. The distances ℓ_1 and ℓ_2 differ by an amount $\Delta\ell$, which is the length of the short side of the colored triangle in part *b* of the drawing. Since the triangle is a right triangle, it follows that $\Delta\ell = d \sin\theta$. Constructive interference occurs when the distances differ by an integer number m of wavelengths λ, or $\Delta\ell = d \sin\theta = m\lambda$. Therefore, the angle θ for the interference maxima can be determined from the following expression:

[Bright fringes of a double slit]

$$\sin\theta = m\frac{\lambda}{d} \qquad m = 0, 1, 2, 3, \ldots \qquad (27.1)$$

The value of m specifies the *order* of the fringe. Thus, $m = 2$ identifies the "second-order" bright fringe. Part *c* of the drawing stresses that the angle θ given by Equation 27.1 locates bright fringes on either side of the midpoint between the slits. A similar line of reasoning leads to the conclusion that the dark fringes, which lie between the bright fringes, are located according to

[Dark fringes of a double slit]

$$\sin\theta = (m + \tfrac{1}{2})\frac{\lambda}{d} \qquad m = 0, 1, 2, 3, \ldots \qquad (27.2)$$

Example 2 illustrates the application of these expressions and shows how to determine the distance of a higher-order bright fringe from the central bright fringe.

Example 2 Young's Double-Slit Experiment

Red light ($\lambda = 664$ nm in vacuum) is used in Young's experiment with the slits separated by a distance $d = 1.20 \times 10^{-4}$ m. The screen is located at a distance from the slits given by $L = 2.75$ m. Find the distance y on the screen between the central bright fringe and the third-order bright fringe (see Figure 27.9).

REASONING This problem can be solved by first using Equation 27.1 to determine the value of θ that locates the third-order bright fringe. Then trigonometry can be used to obtain the distance y.

SOLUTION According to Equation 27.1, we find

$$\sin\theta = m\frac{\lambda}{d} = 3\left(\frac{664 \times 10^{-9}\text{ m}}{1.20 \times 10^{-4}\text{ m}}\right) = 1.66 \times 10^{-2}$$

$$\theta = \sin^{-1}(1.66 \times 10^{-2}) = 0.951°$$

According to Figure 27.9, the distance y can be calculated from $\tan\theta = y/L$:

$$y = L\tan\theta = (2.75\text{ m})\tan 0.951° = \boxed{0.0456\text{ m}}$$

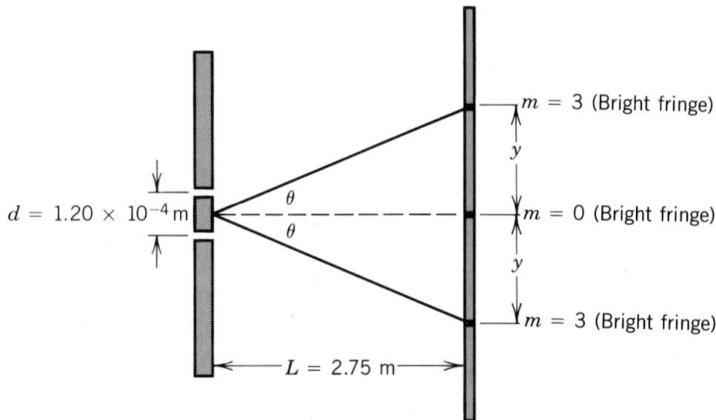

Figure 27.9 The third-order bright fringe ($m = 3$) is observed on the screen at a distance y from the central bright fringe ($m = 0$).

In the version of Young's experiment that we have discussed, monochromatic light has been used. It is possible, however, to use light that contains a mixture of wavelengths, and Conceptual Example 3 deals with some of the interesting features of the resulting interference pattern.

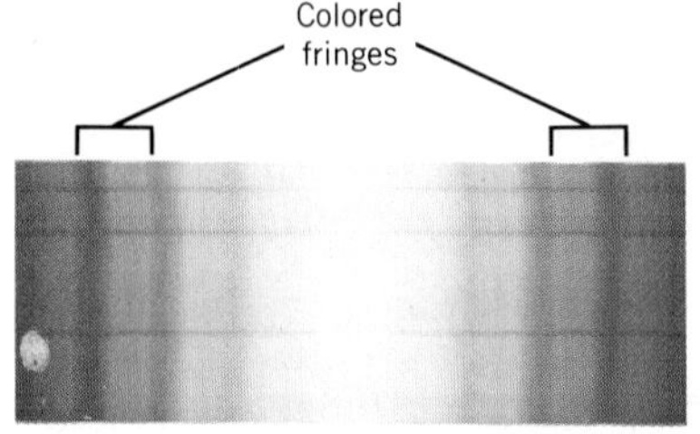

Figure 27.10 This photograph shows the results observed on the screen in one version of Young's experiment in which white light is used.

Conceptual Example 3 White Light and Young's Experiment

Figure 27.10 shows a photograph that illustrates the kind of interference fringes that can result when white light, which is a mixture of colors, is used in Young's experiment. Except for the central fringe, which is white, the bright fringes are a rainbow of colors. Why does Young's experiment separate white light into its constituent colors? In any one group of colored fringes, such as the one singled out in Figure 27.10, why is red farther out from the central fringe than green is? And finally, why is the central fringe white, rather than colored?

REASONING To understand how the color separation arises, we need to remember that each color corresponds to a different wavelength λ and that constructive and destructive interference depend on the wavelength. According to Equation 27.1

($\sin \theta = m\lambda/d$), there is a different angle that locates a bright fringe for each value of λ, or for each color. These different angles lead to the separation of colors on the observation screen. In fact, on either side of the central fringe, there is one group of colored fringes for $m = 1$, and another for each value of m.

Now, consider what it means that, within any single group of colored fringes, red is farther out from the central fringe than green is. It means that, in the equation $\sin \theta = m\lambda/d$, red light has a larger angle θ than green light does. Does this make sense? Yes, because red has the larger wavelength (see Table 26.2, where $\lambda_{red} = 660$ nm and $\lambda_{green} = 550$ nm).

In Figure 27.10, the central fringe is distinguished from all the other colored fringes by being white. In Equation 27.1, the central fringe is different from the other fringes because it is the only one for which $m = 0$. Let us see if the whiteness and the fact that $m = 0$ are related. In Equation 27.1, a value of $m = 0$ means that $\sin \theta = m\lambda/d = 0$, which reveals that $\theta = 0°$, no matter what the wavelength λ is. In other words, all wavelengths have a zeroth-order bright fringe located at the same place on the screen, so that all colors strike the screen there and mix together to produce the white central fringe.

Related Homework Material: *Problem 7*

Historically, Young's experiment provided strong evidence that light has a wavelike character. If light behaved only as a stream of "tiny particles," as others believed at the time,* then the two slits would deliver the light energy into only two bright fringes located directly opposite the slits on the screen. Instead, Young's experiment shows that wave interference redistributes the energy from the two slits into many bright fringes.

27.3 *THIN-FILM INTERFERENCE*

Young's double-slit experiment is one example of interference between light waves. Interference also occurs in more common circumstances. For instance, Figure 27.11 shows a thin film, such as gasoline floating on water. To begin with, let us assume that the film has a constant thickness. Consider what happens when monochromatic light (a single wavelength) strikes the film nearly perpendicularly. At the top surface of the film reflection occurs and produces the light wave represented by ray 1. However, refraction also occurs, and some light enters the film. Part of this light reflects from the bottom surface and passes back up through the film, eventually reentering the air. Thus, a second light wave, which is represented by ray 2, also exists. Moreover, this wave, having traversed the film twice, has traveled farther than wave 1. Because of the extra travel distance, there can be interference between the two waves. If constructive interference occurs, an observer, whose eyes detect the superposition of waves 1 and 2, would see a uniformly bright film. If destructive interference occurs, an observer would see a uniformly dark film. The controlling factor is whether the extra distance for wave 2 is an integer number of whole wavelengths or an odd integer number of half-wavelengths.

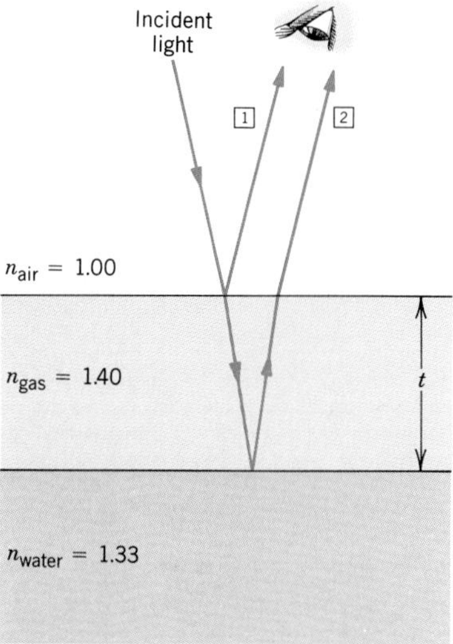

Figure 27.11 Because of reflection and refraction, two light waves, represented by rays 1 and 2, enter the eye when light shines on a thin film of gasoline floating on water. Interference occurs between these waves.

In Figure 27.11, the difference in path lengths between waves 1 and 2 occurs inside the thin film. Therefore, ***the wavelength that is important for thin-film***

* It is now known that the particle or corpuscular theory of light, which Isaac Newton promoted, does indeed explain some experiments that the wave theory cannot explain. Today, light is regarded as having both particle and wave characteristics. Chapter 29 discusses this dual nature of light.

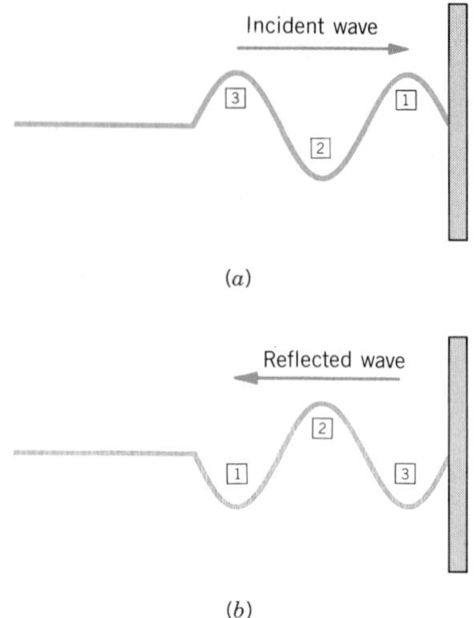

Figure 27.12 When a wave on a string reflects from a wall, the wave undergoes a phase change. Thus, after reflection an upward-pointing half-cycle of the wave becomes a downward-pointing half-cycle, and vice versa, as the numbered labels in the drawing indicate.

interference is the wavelength within the film, not the wavelength in vacuum. The wavelength within the film can be calculated from the wavelength in vacuum by using the index of refraction n for the film, since $n = c/v = (c/f)/(v/f) = \lambda_{\text{vacuum}}/\lambda_{\text{film}}$. In other words,

$$\lambda_{\text{film}} = \frac{\lambda_{\text{vacuum}}}{n} \tag{27.3}$$

In explaining the interference that can occur in Figure 27.11, we need to add one more important part to the story. Whenever waves reflect at a boundary, it is possible for them to change phase. Figure 27.12, for example, shows that a wave on a string is inverted when it reflects from the end that is tied to a wall (see also Figure 17.20). This inversion is equivalent to a half-cycle of the wave, as if the wave had traveled an additional distance of one-half of a wavelength. In contrast, a phase change does not occur when a wave on a string reflects from the end of a string that is hanging free. When light waves undergo reflection, similar phase changes occur as follows:

1. When light travels through a material with a smaller refractive index toward a material with a larger refractive index (e.g., air to gasoline), reflection at the boundary occurs along with a phase change that is equivalent to one-half of a wavelength.
2. When light travels from a larger toward a smaller refractive index, there is no phase change upon reflection at the boundary.

The next example indicates how the phase change that can accompany reflection is taken into account when dealing with thin film interference.

Example 4 Colored Thin Films of Gasoline

(a) A thin film of gasoline floats on a puddle of water. Sunlight falls almost perpendicularly on the film and reflects into your eyes. Although sunlight is white, since it contains all colors, the film has a yellow hue, because destructive interference eliminates the color of blue ($\lambda_{\text{vacuum}} = 469$ nm) from the reflected light. If the refractive indices of the blue light in gasoline and water are 1.40 and 1.33, respectively, determine the minimum nonzero thickness t of the film. (b) Repeat part (a) assuming that the gasoline is on glass ($n_{\text{glass}} = 1.52$) instead of water.

REASONING To solve this problem, we must express the condition for destructive interference in terms of the film thickness t and the wavelength in the gasoline film λ_{film}. We must also take into account any phase changes that occur upon reflection. These phase changes will be different in parts (a) and (b), because different indices of refraction are involved.

SOLUTION

(a) In Figure 27.11, the phase change for wave 1 is equivalent to one-half of a wavelength, since this light travels from a smaller refractive index ($n_{\text{air}} = 1.00$) toward a larger refractive index ($n_{\text{gas}} = 1.40$). In contrast, there is no phase change when wave 2 reflects from the bottom surface of the film, since this light travels from a larger refractive index ($n_{\text{gas}} = 1.40$) toward a smaller one ($n_{\text{water}} = 1.33$). The net phase change between waves 1 and 2 due to reflection is, thus, equivalent to one-half of a wavelength, $\frac{1}{2}\lambda_{\text{film}}$. This half-wavelength must be combined with the extra travel distance for wave 2, to determine the condition for destructive interference. For destructive interference, the combined total must be an odd integer number of half-wavelengths. Since wave 2 travels back and forth through the film and since light strikes the film nearly perpendicularly, the extra travel distance is twice the film thickness, or $2t$. Thus, the condition for destructive interference is $2t + \frac{1}{2}\lambda_{\text{film}} = \frac{1}{2}\lambda_{\text{film}}, 1\frac{1}{2}\lambda_{\text{film}}$, and so forth. This condition is satisfied when

$$2t = m\lambda_{\text{film}} \qquad m = 0, 1, 2, 3, \ldots$$

With $m = 1$, the expression above gives the minimum nonzero film thickness for which the blue color is missing in the reflected light: $t = \frac{1}{2}\lambda_{\text{film}}$. Equation 27.3 with $n = 1.40$ gives the wavelength of blue light in the film as $\lambda_{\text{film}} = (469 \text{ nm})/1.40 = 335 \text{ nm}$. Therefore, the minimum film thickness is

$$t = \tfrac{1}{2}\lambda_{\text{film}} = \tfrac{1}{2}(335 \text{ nm}) = \boxed{168 \text{ nm}}$$

(b) When the water in Figure 27.11 is replaced by glass, the phase change that accompanies the reflection of wave 2 from the bottom surface of the film is no longer zero. Instead, the phase change is equivalent to one-half of a wavelength, because the wave now travels from a smaller refractive index ($n_{\text{gas}} = 1.40$) toward a larger refractive index ($n_{\text{glass}} = 1.52$). In other words, wave 2 now behaves exactly as wave 1. Consequently, there is no net phase change between the waves due to reflection. As a result, destructive interference occurs when the extra distance traveled by wave 2 is an odd integer number of half-wavelengths, which is a different condition than that in part (a):

$$2t = (m + \tfrac{1}{2})\lambda_{\text{film}} \qquad m = 0, 1, 2, 3, \ldots$$

The minimum nonzero thickness for destructive interference corresponds to $m = 0$ in this equation, so

$$t = \frac{(m + \frac{1}{2})\lambda_{\text{film}}}{2} = \frac{\lambda_{\text{film}}}{4} = \frac{335 \text{ nm}}{4} = \boxed{83.8 \text{ nm}}$$

PROBLEM SOLVING INSIGHT

When analyzing thin-film interference effects, remember to use the wavelength of the light in the film (λ_{film}) instead of the wavelength in a vacuum (λ_{vacuum}).

Example 4 deals with a thin film that has the same yellow color everywhere. In nature, such a uniformly colored thin film would be unusual, and the next example deals with a more realistic situation.

Conceptual Example 5 Multicolored Thin Films

Under natural conditions thin films, like the soap bubble in Figure 27.13 or like gasoline on water, have a multicolored appearance that often changes while you are watching them. Why are such films multicolored, and what can be inferred from the fact that the colors change in time?

REASONING In Example 4 we have seen that a thin film can appear yellow if the uniform thickness of the film is such that destructive interference removes blue light from the reflected sunlight. The thickness of the film is the key. If the thickness were different, so that destructive interference removed green light from the reflected sunlight, the film would appear magenta. Constructive interference can also cause certain colors to appear brighter than others in the reflected light and give the film a colored appearance. The colors that are enhanced by constructive interference, like those removed by destructive interference, depend on the thickness of the film. Thus, we conclude that the different colors in a soap bubble or a thin film of gasoline on water arise because the thickness varies from place to place on the film. Moreover, the fact that the colors change as you watch them indicates that the thickness is changing. A number of factors can cause the thickness to change, including air currents, temperature fluctuations, and the pull of gravity, which tends to make a vertical film sag, leading to thicker regions at the bottom than at the top.

Related Homework Material: *Problems 13, 17, and 55*

The beautiful iridescent colors on the wings of this butterfly are due to the thin film interference of light, which occurs because there is a transparent cuticle-like material on the upper wing surfaces.

Figure 27.13 A soap bubble is multicolored when viewed in sunlight because of the effects of thin-film interference.

The colors that you see when sunlight is reflected from a thin film also depend on your viewing angle. At an oblique angle, the light corresponding to ray 2 in Figure 27.11 would travel a greater distance within the film than it does at nearly perpendicular incidence. The greater distance would lead to destructive interference for a different wavelength.

THE PHYSICS OF . . .

nonreflecting coatings for high-quality lenses.

Thin-film interference can be beneficial in optical instruments. For example, many cameras contain up to six or more lenses. Reflections from all the lens surfaces can reduce considerably the amount of light directly reaching the film. In addition, multiple reflections from the lenses often reach the film indirectly and degrade the quality of the image. To minimize such unwanted reflections, high-quality lenses are often covered with a thin nonreflective coating of magnesium fluoride ($n = 1.38$). This situation is like that in part (b) of Example 4, except the thickness is usually chosen to ensure that destructive interference eliminates the reflection of green light, which is in the middle of the visible spectrum. It should be pointed out that the absence of any reflected light does not mean that it has been destroyed by the nonreflective coating. Rather, the "missing" light has been transmitted into the coating and the lens it covers.

Another interesting illustration of thin-film interference is the air wedge. As Figure 27.14*a* shows, an air wedge is formed when two flat plates of glass are separated along one side, perhaps by a thin sheet of paper. The thickness of this film of air varies between zero, where the plates touch, and the thickness of the paper. When monochromatic light reflects from this arrangement, alternate bright and dark fringes are formed by constructive and destructive interference, as the picture indicates. Example 6 deals with the interference caused by an air wedge.

Example 6 An Air Wedge

(a) Assuming that green light ($\lambda_{\text{vacuum}} = 552$ nm) strikes the glass plates nearly perpendicularly in Figure 27.14, determine the number of bright fringes that occur between the place where the plates touch and the edge of the sheet of paper (thickness = 4.10×10^{-5} m). (b) Explain why there is a dark fringe where the plates touch.

REASONING A bright fringe occurs wherever there is constructive interference, as determined by any phase changes due to reflection and the effects of the thickness of the air wedge. We examine the phase changes and the effects of the thickness separately.

SOLUTION

(a) There is no phase change upon reflection for wave 1, since this light travels from a larger (glass) toward a smaller (air) refractive index. In contrast, there is a half-wavelength phase change for wave 2, since the ordering of the refractive indices is reversed at the lower air/glass boundary where reflection occurs. The net phase change due to reflection for waves 1 and 2, then, is equivalent to a half wavelength. Now we combine any extra distance traveled by ray 2 with this half-wavelength and determine the condition for the constructive interference that creates the bright fringes. Constructive interference occurs whenever the *combination* yields an integer number of wavelengths. Therefore, the extra travel distance alone must be an odd integer number of half-wavelengths if constructive interference is to occur. At nearly perpendicular incidence, the extra travel distance is approximately twice the thickness t of the wedge at any point, so the condition for constructive interference is

$$2t = (m + \tfrac{1}{2})\lambda_{\text{film}} \qquad m = 0, 1, 2, 3, \ldots$$

In this expression, note that the "film" is a film of air. Since the refractive index of air is nearly one, λ_{film} is virtually the same as that in vacuum, so $\lambda_{\text{film}} = 552$ nm. When t equals the thickness of the paper holding the plates apart, the corresponding value of m can be obtained from the equation above:

$$m = \frac{2t}{\lambda_{\text{film}}} - \frac{1}{2} = \frac{2(4.10 \times 10^{-5}\text{ m})}{552 \times 10^{-9}\text{ m}} - \frac{1}{2} = 148$$

Since the first bright fringe occurs when $m = 0$, the number of bright fringes is $m + 1 =$ $\boxed{149}$.

(b) Where the plates touch, there is a dark fringe because of destructive interference between the light waves represented by rays 1 and 2. Destructive interference occurs, since the thickness of the wedge is zero here and the only difference between the rays is the half-wavelength phase change due to reflection from the lower plate.

PROBLEM SOLVING INSIGHT

When light reflects from a boundary between two materials, the reflected light will experience a phase change only if the light travels from the material with a smaller refractive index toward the material with a larger refractive index. Be sure to take such a phase change into account when analyzing thin-film interference phenomena.

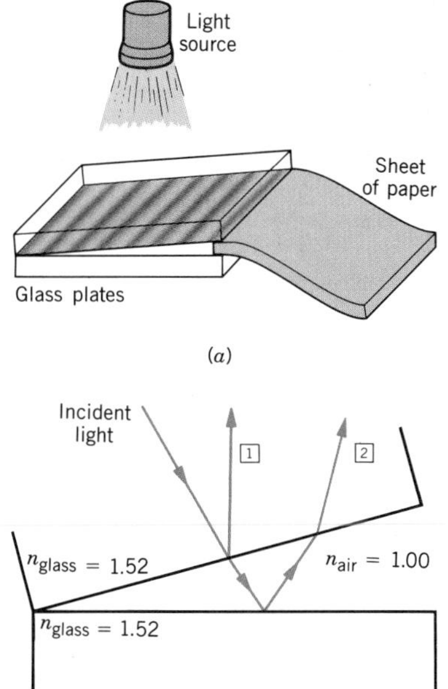

Figure 27.14 (*a*) The wedge of air formed between two flat glass plates causes an interference pattern of alternating dark and bright fringes to appear in reflected light. (*b*) A side view of the glass plates and the air wedge.

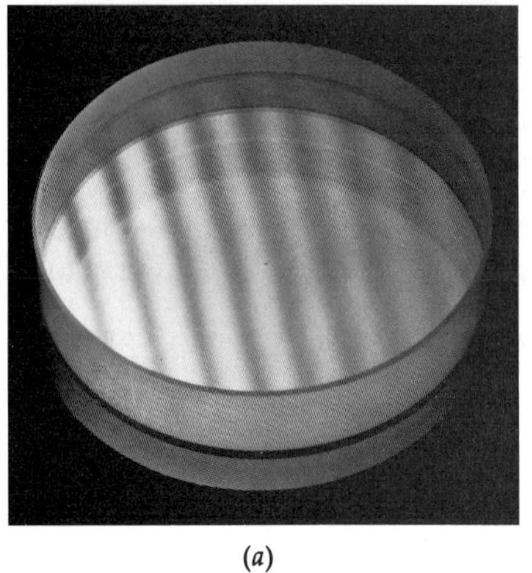

(*a*)

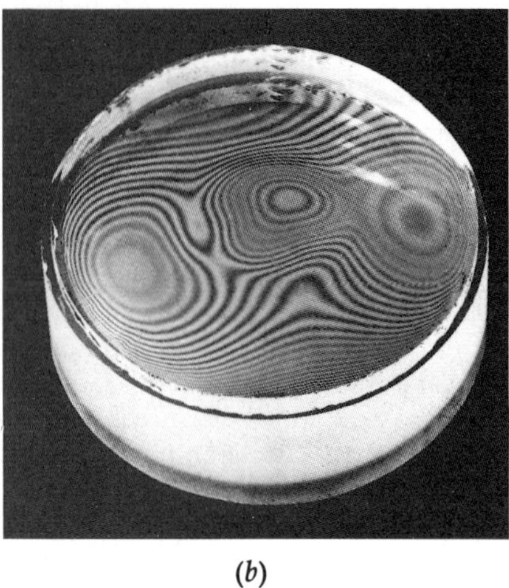

(*b*)

Figure 27.15 In reflected light a pattern of interference fringes can be observed due to the air wedge between two glass plates. (*a*) When the plates are ultraflat, or optically flat, the pattern consists of straight fringes. (*b*) When the plates are not flat, the pattern is wavy.

THE PHYSICS OF . . .

testing a surface for flatness using thin-film interference.

Figure 27.15*a* shows a photograph of the fringes observed for an air wedge between two ultraflat, or optically flat, plates (plates that are flat to within a fraction of a wavelength of the light). Part *b* shows the fringe pattern observed when the plates are not optically flat. In fact, one way to test a plate for flatness is to use it to form an air wedge with a second reference plate that is known to be flat. When the fringes are straight as in part *a*, rather than wavy as in part *b*, no further polishing is needed to flatten the plate being tested.

Another type of air wedge can also be used to determine the degree to which the surface of a lens or mirror is spherical. When an accurate spherical surface is put in contact with an optically flat plate, as in Figure 27.16*a*, the circular interference fringes shown in part *b* of the figure can be observed. The circular fringes are called *Newton's rings.* They arise in the same way that the straight fringes arise in Figure 27.14. When the curved surface is irregular, as in Figure 27.16*c*, the interference fringes are no longer circular.

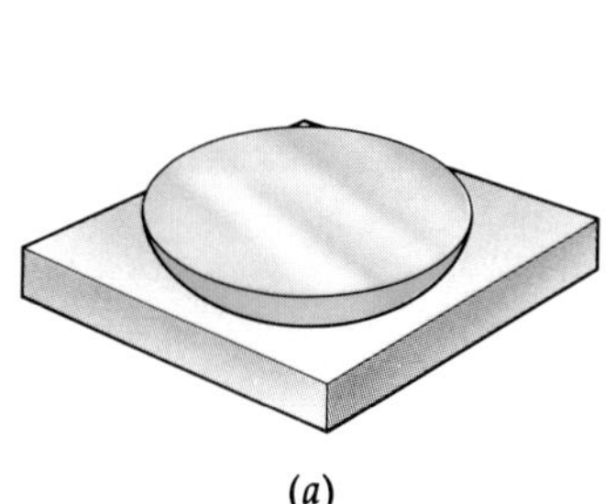

(*a*)

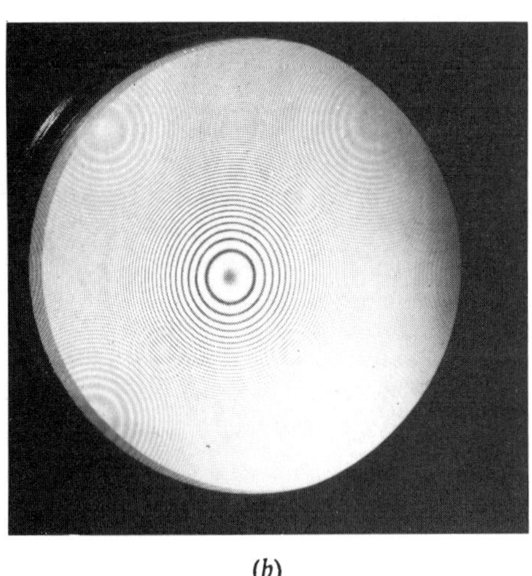

(*b*)

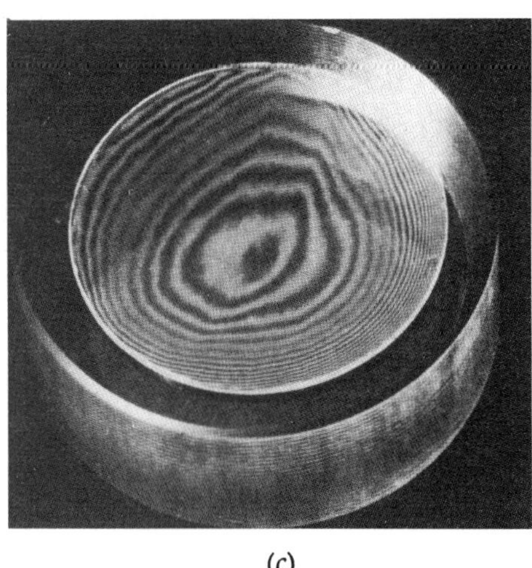

(*c*)

Figure 27.16 (*a*) The air wedge between an accurate spherical glass surface and an optically flat plate leads to (*b*) a pattern of circular interference fringes that is known as Newton's rings. (*c*) When the curved surface is irregular, the interference fringes are not circular.

27.4 THE MICHELSON INTERFEROMETER

An interferometer is an apparatus that can be used to measure the wavelength of light by utilizing interference between two light waves. One particularly famous interferometer is that developed by Albert A. Michelson (1852–1931). The Michelson interferometer uses reflection to set up conditions where two light waves interfere. Figure 27.17 presents a schematic drawing of the instrument. Waves emitted by the monochromatic light source strike a *beam splitter*, so called because it splits the beam of light into two parts. The beam splitter is a glass plate, the far side of which is coated with a thin layer of silver that reflects part of the beam upward as wave A in the drawing. The coating is so thin, however, that it also allows the remainder of the beam to pass directly through as wave F. Wave A strikes an adjustable mirror and reflects back on itself. It again crosses the beam splitter and then enters the viewing telescope. Wave F strikes a fixed mirror and returns, to be partly reflected into the viewing telescope by the beam splitter. Note that wave A passes through the glass plate of the beam splitter three times in reaching the viewing scope, while wave F passes through it only once. The compensating plate in the path of wave F has the same thickness as the beam splitter plate and ensures that wave F also passes three times through the same thickness of glass on the way to the viewing scope. Thus, an observer who views the combination of waves A and F through the telescope sees constructive or destructive interference, depending only on the difference in path lengths D_A and D_F traveled by the two waves.

THE PHYSICS OF . . .

the Michelson interferometer.

Now suppose the mirrors are perpendicular to each other, the beam splitter makes a 45° angle with each, and the distances D_A and D_F are equal. The waves A and F travel the same distance, and the field of view in the telescope is uniformly bright due to constructive interference. However, if the adjustable mirror were

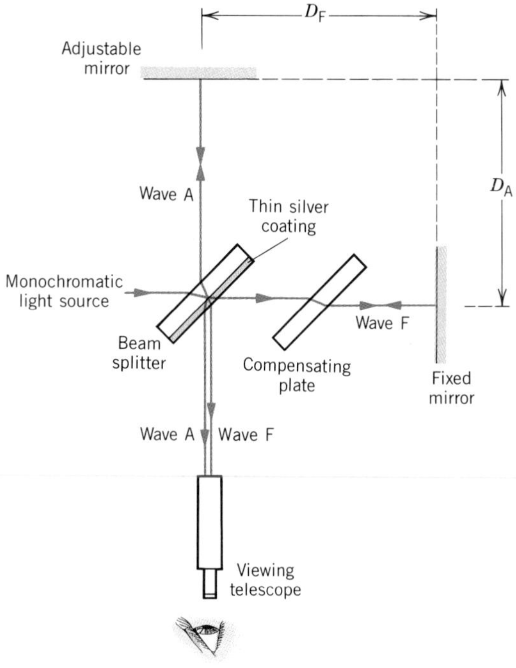

Figure 27.17 A schematic drawing of a Michelson interferometer.

moved away from the telescope by a distance of $\frac{1}{4}\lambda$, wave A would travel back and forth by an amount that is twice this value, leading to an extra distance of $\frac{1}{2}\lambda$. Then, the two waves would be out of phase when they reached the viewing scope, destructive interference would occur, and the viewer would see a dark field. If the adjustable mirror were moved further, brightness would return as soon as the waves were in phase and interfered constructively. The in-phase condition would occur when wave A travels a total extra distance of λ relative to wave F. Thus, as the mirror is continuously moved, the viewer sees the field of view change from bright to dark, then back to bright, and so on. The amount by which $D_{\mathbf{A}}$ has been changed can be measured and related to the wavelength of the light, since a bright field changes into a dark field and back again each time $D_{\mathbf{A}}$ is changed by a half wavelength. (The back-and-forth change in distance is λ.) If a sufficiently large number of wavelengths are counted in this manner, the Michelson interferometer can be used to obtain a very accurate value for the wavelength from the measured changes in $D_{\mathbf{A}}$.

27.5 DIFFRACTION

As Section 17.3 discusses, ***diffraction*** is a bending of waves around obstacles or the edges of an opening. In Figure 27.18, for example, sound waves are leaving a room through an open doorway. Because the exiting sound waves bend, or diffract, around the edges of the opening, a listener outside the room can hear the sound even when standing around the corner from the doorway.

Diffraction is an interference effect, and the Dutch scientist Christian Huygens (1629–1695) developed a principle that is useful in explaining why diffraction arises. ***Huygens' principle*** describes how a wave front that exists at one instant gives rise to the wave front that exists later on. The principle states that ***every point on a wave front acts as a source of tiny wavelets that move forward with the same speed as the wave; the wave front at a later instant is the surface that is tangent to the wavelets.***

We begin by using Huygens' principle to explain the diffraction of sound waves in Figure 27.18. The drawing shows the top view of a plane wave front of sound approaching a doorway and identifies five points on the wave front just as it is leaving the opening. According to Huygens' principle, each of these points acts as a source of wavelets, which are shown as red circular arcs at some moment after they are emitted. The tangent to the wavelets from points 2, 3, and 4 indicates that in front of the doorway the wave front is flat and is moving straight ahead at this later instant. But at the edges, points 1 and 5 are the last points that produce wavelets. Huygens' principle suggests that in conforming to the curved shape of the wavelets emitted near the edges, the new wave front moves into regions that it would not reach otherwise. The sound wave, then, bends or diffracts around the edges of the doorway.

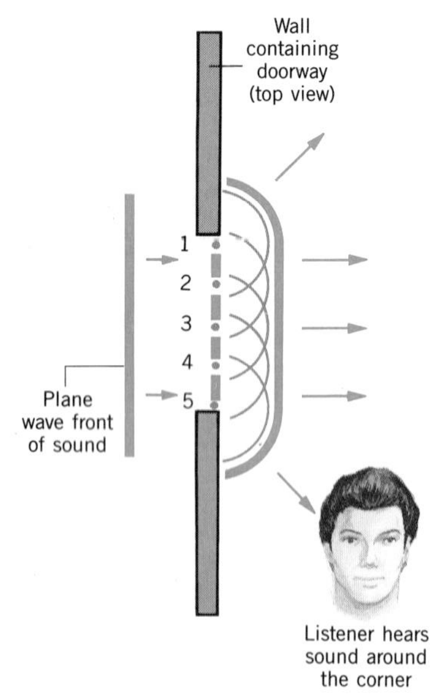

Figure 27.18 Sound bends or diffracts around the edges of a doorway, so even a person who is not standing directly in front of the opening can hear the sound. The five red points within the doorway act as sources and emit the five Huygens wavelets shown in red.

Huygens' principle applies not only to sound waves, but to all kinds of waves. For instance, light has a wavelike nature and, consequently, exhibits diffraction. Therefore, you may ask, "Since I can hear around the edges of a doorway, why can't I also see around them?" As a matter of fact, light waves do bend around the edges of a doorway. However, the degree of bending is extremely small, so the diffraction of light is not enough to allow you to see around the corner.

As we will learn, the extent to which a wave bends around the edges of an opening is determined by the ratio λ/W, where λ is the wavelength of the wave and W is the width of the opening. The photographs in Figure 27.19 illustrate the

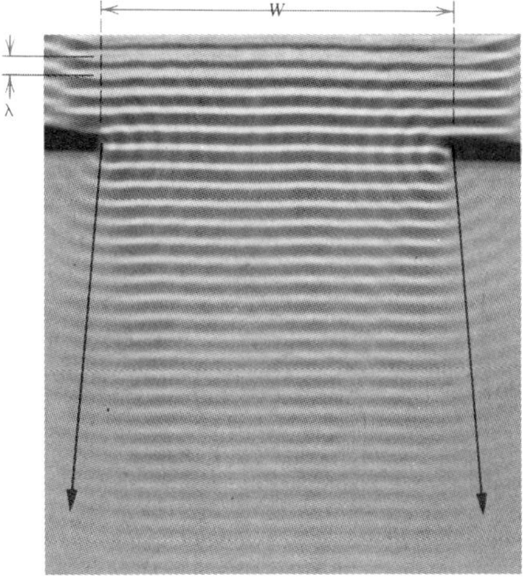

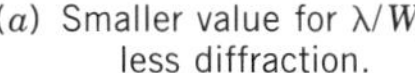
(a) Smaller value for λ/W, less diffraction.

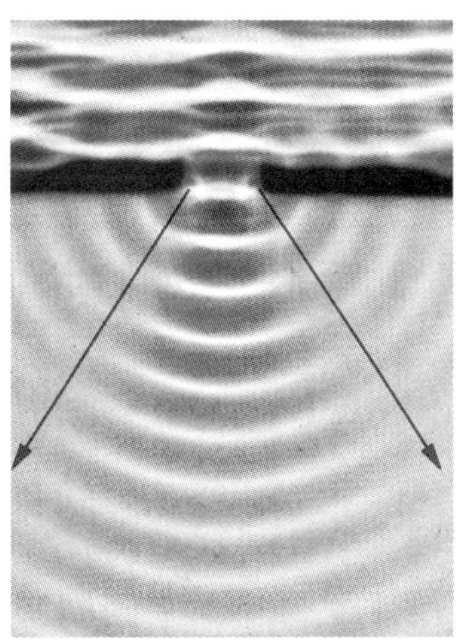
(b) Larger value for λ/W, more diffraction.

Figure 27.19 These photographs show the plane wave fronts (horizontal lines) of water waves approaching an opening whose width W is smaller in (*b*) than it is in (*a*). In addition, the wavelength λ of the waves is larger in (*b*) than in (*a*). Therefore, the ratio λ/W increases from (*a*) to (*b*) and so does the extent of the diffraction. The waves bend around the edges of the opening more in (*b*) than in (*a*).

effect of this ratio on the diffraction of water waves. In part *a*, the ratio λ/W is small, because the wavelength (as indicated by the distance between the wave fronts) is small relative to the width of the opening. The wave fronts move through the opening with little bending or diffraction into the regions around the edges of the opening. In part *b*, the wavelength is larger and the width of the opening is smaller. As a result, the ratio λ/W is larger, and the degree of bending becomes more pronounced, with the wave fronts penetrating more into the regions around the edges of the opening.

Based on the pictures in Figure 27.19, we might expect that light waves of wavelength λ will bend or diffract appreciably when they pass through an opening whose width W is small enough to make the ratio λ/W sufficiently large. This is indeed the case, as Figure 27.20 illustrates. In this picture, it is assumed that parallel rays (or plane wave fronts) of light fall on a very narrow slit and illuminate a viewing screen that is located far from the slit. Part *a* of the drawing shows what would happen if there were no diffraction. Light would pass through the slit without bending around the edges and produce an image of the slit on the screen.

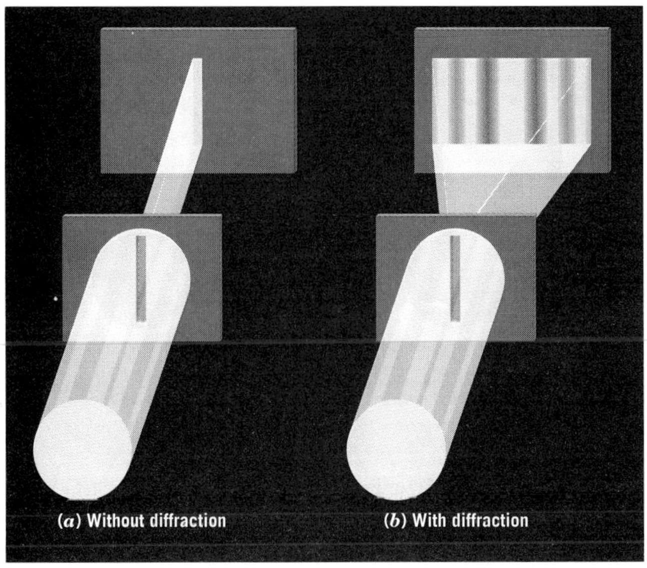
(*a*) Without diffraction (*b*) With diffraction

Figure 27.20 (*a*) If light were to pass through a very narrow slit *without* diffraction, only the region on the screen directly opposite the slit would be illuminated. (*b*) Diffraction causes the light to bend around the edges of the slit into regions it would not otherwise reach, forming a pattern of alternating bright and dark fringes on the screen. The slit width has been exaggerated for clarity.

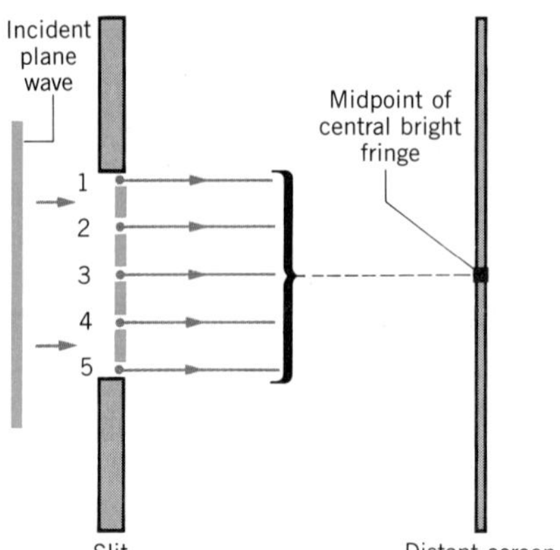

Figure 27.21 A plane wave front is incident on a single slit. This top view of the slit shows five sources of Huygens wavelets. The wavelets travel toward the midpoint of the central bright fringe on the screen, as the rays indicate. The screen is very far from the slit.

Part *b* shows what actually happens. The light diffracts around the edges of the slit and brightens regions on the screen that are not directly opposite the slit. The diffraction pattern on the screen consists of a bright central band, accompanied by a series of narrower faint fringes that are parallel to the slit itself.

To help explain how the pattern of diffraction fringes arises, Figure 27.21 shows a top view of a plane wave front approaching the slit and singles out five sources of Huygens wavelets. Consider how the light from these five sources reaches the midpoint on the screen. To simplify things, the screen is assumed to be so far from the slit that the rays from each Huygens source are nearly parallel.* Then, all the wavelets travel virtually the same distance to the midpoint, arriving there in phase. As a result, constructive interference creates a bright central fringe on the screen, directly opposite the slit.

The wavelets emitted by the Huygens sources in the slit can also interfere destructively on the screen, as Figure 27.22 illustrates. Part *a* shows the light traveling from each source toward the first dark fringe. The angle θ gives the position of this dark fringe relative to the line between the midpoint of the slit and the midpoint of the central bright fringe. Since the screen is very far from the slit, the rays of light from each Huygens source are nearly parallel and oriented at nearly the same angle θ, as in part *b* of the drawing. The wavelet from source 1

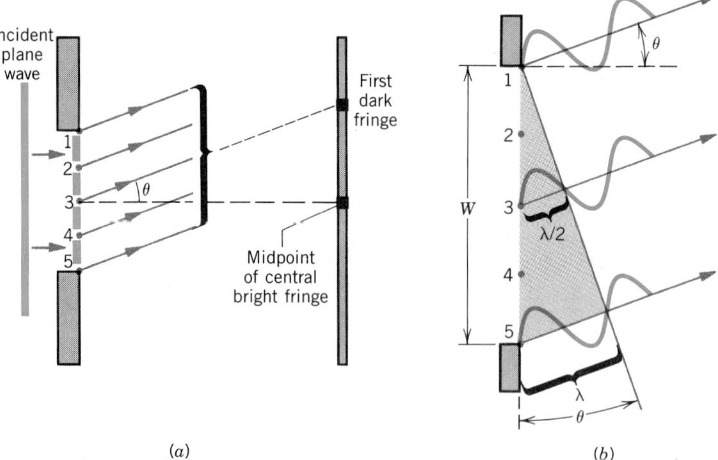

Figure 27.22 These drawings pertain to single-slit diffraction and show how destructive interference leads to the first dark fringe on either side of the central bright fringe. For clarity, only one of the dark fringes is shown. The screen is very far from the slit.

* When the rays are parallel, the diffraction is called Fraunhofer diffraction in tribute to the German optician Josef Fraunhofer (1787–1826). When the rays are not parallel, the diffraction is referred to as Fresnel diffraction, named for the French physicist Augustin Jean Fresnel (1788–1827).

travels the shortest distance to the screen, while that from source 5 travels the farthest. Destructive interference creates the first dark fringe when the extra distance traveled by the wavelet from source 5 is exactly one wavelength, as the colored right triangle in the drawing indicates. Under this condition, the extra distance traveled by the wavelet from source 3 at the center of the slit is exactly one-half of a wavelength. Therefore, wavelets from sources 1 and 3 are exactly out of phase and interfere destructively when they reach the screen. Similarly, a wavelet that originates slightly below source 1 cancels a wavelet that originates the same distance below source 3. Thus, each wavelet from the upper half of the slit cancels a corresponding wavelet from the lower half, and no light reaches the screen. As can be seen from the colored right triangle, the angle θ locating the first dark fringe is given by $\sin \theta = \lambda/W$, where W is the width of the slit.

Figure 27.23 shows the condition that leads to destructive interference at the second dark fringe on either side of the midpoint on the screen. In reaching the screen, the light from source 5 now travels a distance of two wavelengths farther than the light from source 1. Under this condition, the wavelet from source 5 travels one wavelength farther than the wavelet from source 3, and the wavelet from source 3 travels one wavelength farther than the wavelet from source 1. Therefore, each half of the slit can be treated as the entire slit was in the previous paragraph; all the wavelets from the top half interfere destructively with each other, and all the wavelets in the bottom half do likewise. As a result, no light from either half reaches the screen, and another dark fringe occurs. The colored triangle in the drawing shows that this second dark fringe occurs when $\sin \theta = 2\lambda/W$. Similar arguments hold for the third- and higher-order dark fringes, with the general result being

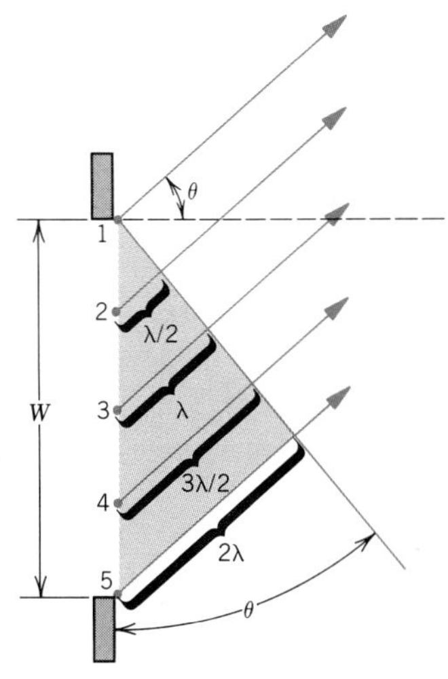

Figure 27.23 In a single-slit diffraction pattern, multiple dark fringes occur on either side of the central bright fringe. This drawing shows how destructive interference creates the second dark fringe on a very distant screen.

[Dark fringes for single-slit diffraction]

$$\sin \theta = m \frac{\lambda}{W} \qquad m = 1, 2, 3, \ldots \tag{27.4}$$

Between each pair of dark fringes there is a bright fringe due to constructive interference. The brightness of the fringes is related to the light intensity, just as loudness is related to sound intensity. The intensity of the light at any location on the screen is the amount of light energy per second per unit area that strikes the screen there. Figure 27.24 gives a graph of the light intensity along with a photograph of a single-slit diffraction pattern. The central bright fringe, which is approximately twice as wide as the other bright fringes, has by far the greatest intensity. The width of the central fringe provides one indication of the extent of diffraction, as Example 7 illustrates.

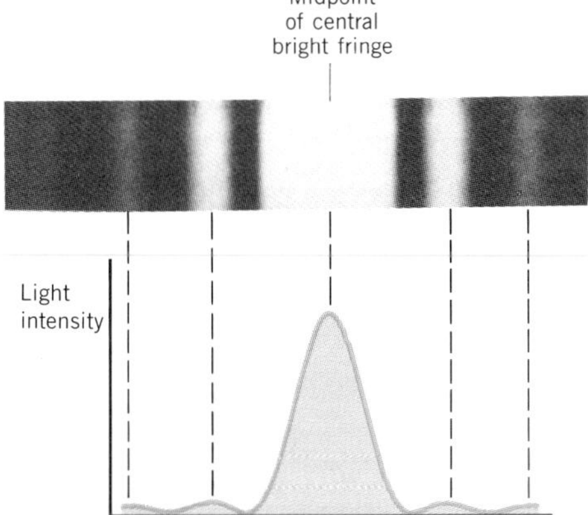

Figure 27.24 The photograph shows a single-slit diffraction pattern, with a bright and wide central fringe. The higher-order bright fringes are much less intense than the central fringe, as the graph indicates.

Example 7 Single-Slit Diffraction

Light passes through a slit and shines on a flat screen that is located $L = 0.40$ m away (see Figure 27.25). The width of the slit is $W = 4.0 \times 10^{-6}$ m. The distance between the middle of the central bright fringe and the first dark fringe is y. Determine the width $2y$ of the central bright fringe when the wavelength of the light is (a) $\lambda = 690$ nm in vacuum (red) and (b) $\lambda = 410$ nm in vacuum (violet).

REASONING The width of the central bright fringe is determined by two factors. One is the angle θ that locates the first dark fringe on either side of the midpoint. The other is the distance L between the screen and the slit. Larger values for θ and L lead to a wider central bright fringe.

SOLUTION

(a) The angle θ in Equation 27.4 locates the first dark fringe when $m = 1$; $\sin\theta = (1)\lambda/W$. Therefore,

$$\theta = \sin^{-1}\left(\frac{\lambda}{W}\right) = \sin^{-1}\left(\frac{690 \times 10^{-9}\ \text{m}}{4.0 \times 10^{-6}\ \text{m}}\right) = 9.9^\circ$$

According to Figure 27.25, $\tan\theta = y/L$, so

$$y = L\tan\theta = (0.40\ \text{m})\tan 9.9^\circ = 0.070\ \text{m}$$

The width of the central fringe, then, is $\boxed{2y = 0.14\ \text{m}}$.

(b) Repeating the calculation above with $\lambda = 410$ nm shows that $\theta = 5.9^\circ$ and $\boxed{2y = 0.083\ \text{m}}$. Notice that the central fringe is narrower when the wavelength is smaller. This occurs because diffraction always becomes less when the ratio λ/W becomes smaller.

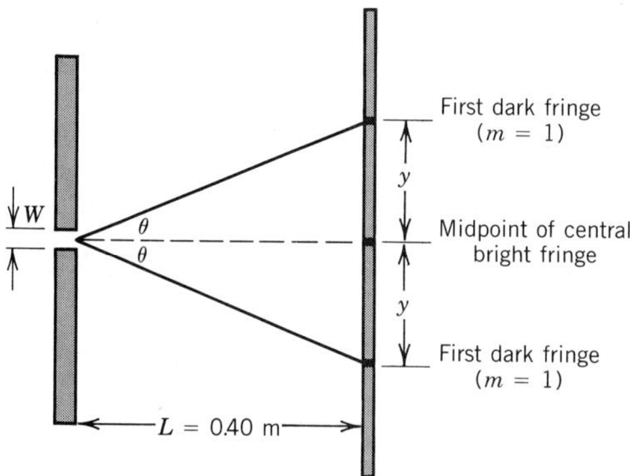

Figure 27.25 The distance $2y$ is the width of the central bright fringe.

THE PHYSICS OF . . .

producing computer chips using photolithography.

In the production of computer chips it is important to minimize the effects of diffraction. As Figure 23.32 illustrates, such chips are very small and yet contain enormous numbers of electronic components. Such miniaturization is achieved using the techniques of photolithography. The patterns on the chip are created first on a "mask," which is similar to a photographic slide. Light is then directed through the mask onto silicon wafers that have been coated with a photosensitive material. The light-activated parts of the coating can be removed chemically, to leave the ultrathin lines that form the miniature patterns on the chip. As the light passes through the narrow slit-like patterns on the mask, the light spreads out due to diffraction. If excessive diffraction occurs, the light spreads out so much that sharp patterns are not formed on the photosensitive material coating the silicon wafer. Ultraminiaturization of the patterns requires the absolute minimum of diffraction, and currently this is achieved by using ultraviolet light, which has a wavelength shorter than that of visible light. The shorter the wavelength λ, the smaller the ratio λ/W, and the less the diffraction, as illustrated in Example 7. Intensive research programs are under way to develop the technique of X-ray lithography. The wavelengths of X-rays are even shorter than those of ultraviolet light and, thus, will reduce diffraction even more, allowing further miniaturization.

Another example of diffraction can be seen when light from a point source falls on an opaque disk, such as a coin (Figure 27.26). The effects of diffraction modify the dark shadow cast by the disk in several ways. First, the light waves diffracted around the circular edge of the disk interfere constructively at the center of the shadow to produce a small bright spot. There are also circular bright fringes in the shadow area. In addition, the boundary between the circular shadow and the lighted screen is not sharply defined, but consists of concentric bright and dark fringes. The various fringes are analogous to those produced by a single slit and are due to interference between Huygens wavelets that originate from different points near the edge of the disk.

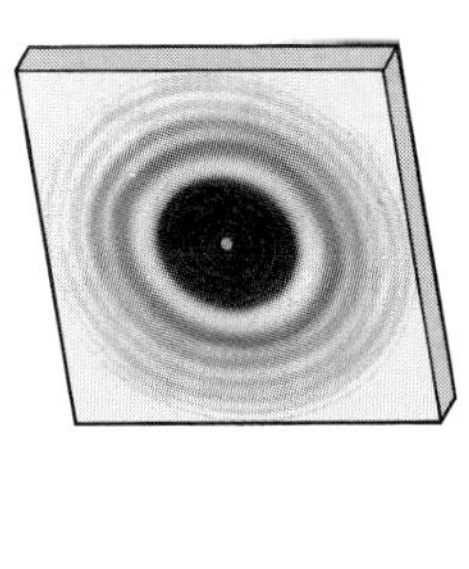

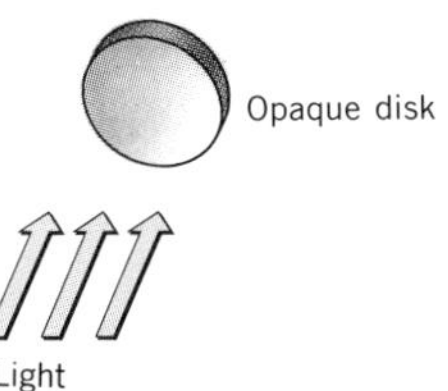

Figure 27.26 The diffraction pattern formed by an opaque disk consists of a small bright spot in the middle of the shadow, circular bright fringes within the shadow, and concentric bright and dark fringes surrounding the shadow.

27.6 RESOLVING POWER

The ***resolving power*** is the ability of an optical instrument to distinguish between two closely spaced objects. In a fashion similar to that for a single slit, diffraction also occurs when light passes through circular or nearly circular openings, such as those that admit light into cameras, microscopes, telescopes, or human eyes. The resulting diffraction pattern places a natural limit on the resolving power of such optical instruments.

Figure 27.27 shows the diffraction pattern created by a small circular opening when the viewing screen is far from the opening. The pattern consists of a central bright circular region, surrounded by alternating bright and dark circular fringes. These fringes are analogous to the rectangular fringes that a single slit produces. The angle θ in the picture locates the first circular dark fringe relative to the central bright region and is given by

$$\sin \theta = 1.22 \frac{\lambda}{D} \tag{27.5}$$

where λ is the wavelength of the light and D is the diameter of the circular opening. This expression is similar to Equation 27.4 for a slit ($\sin \theta = \lambda/D$, when $m = 1$) and is valid when the distance to the screen is much larger than the diameter of the aperture.

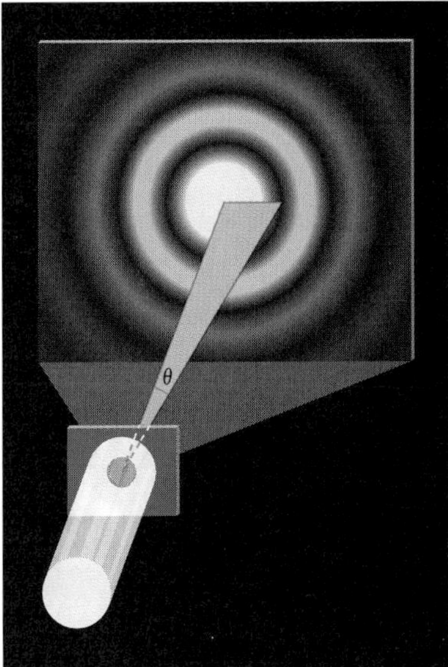

Figure 27.27 When light passes through a small circular opening, a circular diffraction pattern is observed on a viewing screen. The angle θ locates the first dark fringe relative to the central bright region. The intensities of the bright fringes, as well as the diameter of the opening, have been exaggerated in the interest of clarity.

An optical instrument with the ability to resolve two closely spaced objects can produce images of them that can be identified separately. For instance, think about the images on the film when light from two widely separated point objects passes through the circular aperture of a camera. As Figure 27.28 illustrates, each image is a circular diffraction pattern, but the two patterns do not overlap and are completely resolved. On the other hand, if the objects are sufficiently close

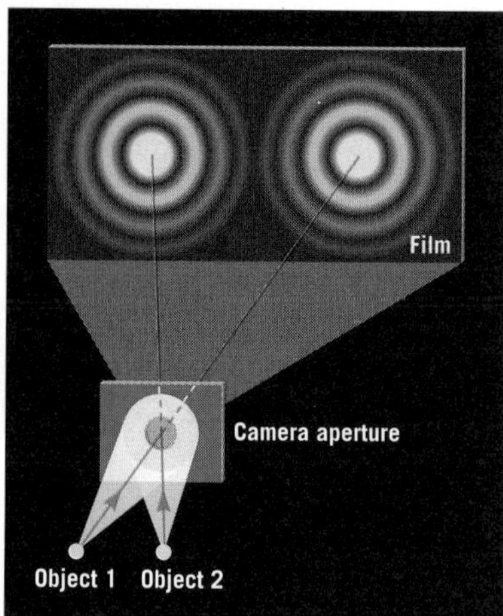

Figure 27.28 When light from two point objects passes through the circular aperture of a camera, two circular diffraction patterns are formed as images on the film. The images here are completely separated or resolved, because the objects themselves are widely separated.

together, the intensity patterns created by the diffraction overlap, as Figure 27.29*a* suggests. In fact, if the overlap is extensive, it may no longer be possible to distinguish the patterns separately. In such a case, the picture from a camera would show a single blurred object instead of two separate objects. In Figure 27.29*b* the diffraction patterns overlap, but not enough to prevent us from seeing that two objects are present. Ultimately, then, diffraction limits the ability of an optical instrument to produce distinguishable images of objects that are close together.

It is useful to have a criterion for judging whether two closely spaced objects are resolved by an optical instrument. Figure 27.29*a* presents the ***Rayleigh criterion*** for resolution, first proposed by Lord Rayleigh (1842–1919): ***Two point objects are just resolved when the first dark fringe in the diffraction pattern of one falls directly on the central bright fringe in the diffraction pattern of the other.*** The minimum angle θ_{min} between the two objects in the drawing is that given by Equation 27.5. If θ_{min} is small and is expressed in radians, $\sin\theta_{min} \approx \theta_{min}$. Then, Equation 27.5 can be rewritten as

$$\theta_{min} \approx 1.22\,\frac{\lambda}{D} \qquad (\theta_{min} \text{ in radians}) \tag{27.6}$$

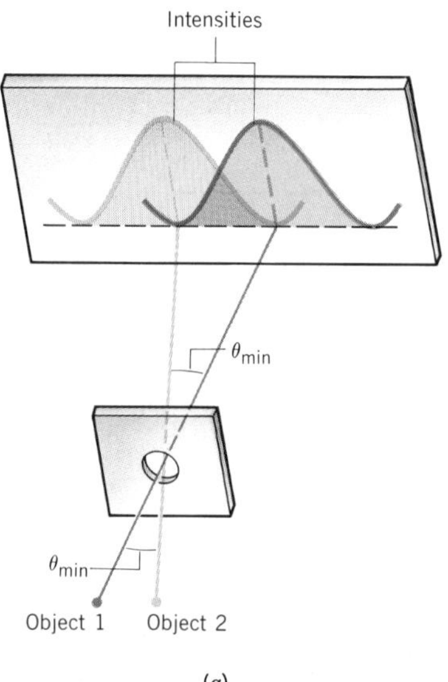

(*a*)

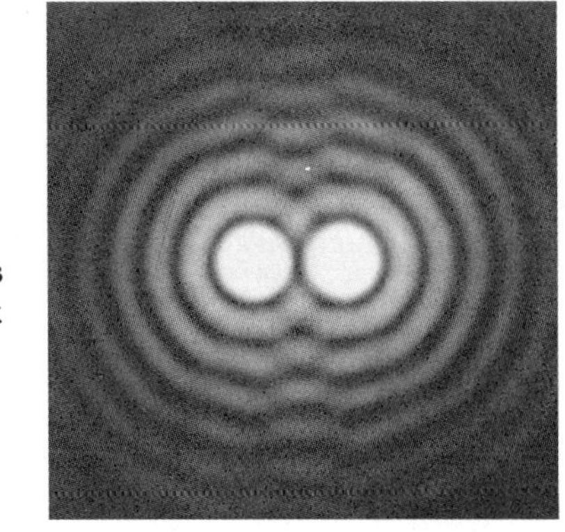

(*b*)

Figure 27.29 (*a*) According to the Rayleigh criterion, two point objects are just resolved when the first dark fringe (zero intensity) of one of the images falls on the central bright fringe (maximum intensity) of the other. (*b*) This photograph shows two overlapping but still resolvable diffraction patterns.

For a given wavelength λ and aperture diameter D, this result specifies the smallest angle that two point objects can subtend at the aperture and still be resolved. According to Equation 27.6, optical instruments designed to resolve closely spaced objects (small values of θ_{min}) must utilize the smallest possible wavelength and the largest possible aperture diameter. Examples 8 and 9 deal with the resolving power of the human eye.

THE PHYSICS OF . . .

comparing human eyes and eagle eyes.

Example 8 The Human Eye Versus the Eagle's Eye

(a) A hang glider is flying at an altitude of $H = 120$ m. Green light (wavelength = 555 nm in vacuum) enters the pilot's eye through a pupil that has a diameter $D = 2.5$ mm. The average index of refraction of the material in the eye is approximately $n = 1.36$. Determine how far apart two point objects must be on the ground if the pilot is to have any hope of distinguishing between them (see Figure 27.30). (b) An eagle's eye has a pupil with a diameter of $D = 6.2$ mm and has about the same refractive index as does a human eye. Repeat part (a) for an eagle flying at the same altitude as the glider.

REASONING According to the Rayleigh criterion, the two objects must be separated by a distance s sufficient to subtend an angle $\theta_{min} \approx 1.22\lambda/D$ at the pupil of the pilot's eye. Remembering to use radians for the angle, we will relate θ_{min} to the altitude H and the desired distance s. In applying the expression for θ_{min}, we must keep in mind that λ is the wavelength within the eye, which is where the diffraction occurs that limits the resolution.

SOLUTION

(a) The wavelength λ within the eye takes into account the refractive index of the eye and is given by Equation 27.3 as $\lambda = \lambda_{vacuum}/n = (555 \text{ nm})/1.36 = 408$ nm. Therefore,

$$\theta_{min} \approx 1.22\frac{\lambda}{D} = 1.22\left(\frac{408 \times 10^{-9}\text{ m}}{2.5 \times 10^{-3}\text{ m}}\right) = 2.0 \times 10^{-4}\text{ rad} \tag{27.6}$$

According to Equation 8.1, θ_{min} in radians is $\theta_{min} \approx s/H$, so

$$s \approx \theta_{min}H = (2.0 \times 10^{-4}\text{ rad})(120\text{ m}) = \boxed{0.024\text{ m}}$$

(b) Since the pupil of an eagle's eye is larger than that of a human eye, diffraction creates less of a limitation for the eagle. A calculation like that above, using $D = 6.2$ mm, reveals that the diffraction limit for the eagle is $\boxed{s = 0.0096\text{ m}}$.

PROBLEM SOLVING INSIGHT

The minimum angle θ_{min} between two objects that are just resolved must be expressed in radians, not degrees, when using Equation 27.6 ($\theta_{min} \approx 1.22\lambda/D$).

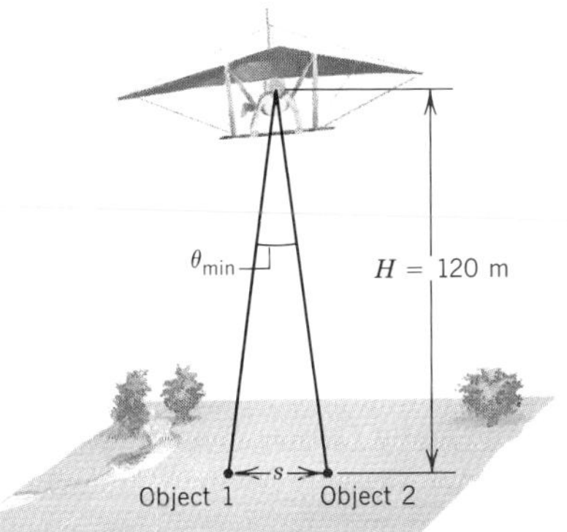

Figure 27.30 The Rayleigh criterion can be used to estimate the smallest distance s that can separate two objects on the ground, if a person on a hang glider is to have any hope of distinguishing one object from the other.

Example 9 The Great Wall of China

The Great Wall of China is an enormous structure, its width approaching 7.0 m in places. Some people even claim that it is possible to identify the separate sides of the Great Wall from the moon with the unaided eye. Evaluate this claim in the following way. Assume that green light (wavelength = 555 nm in vacuum) reflects from the Great Wall and enters the eye (pupil diameter $D = 2.5$ mm, refractive index $n = 1.36$) of an astronaut. Determine the maximum distance this astronaut could be from the earth and still resolve the two sides of the Great Wall at their widest point. Compare this maximum distance to the distance between the earth and the moon.

REASONING This problem is similar to that in Example 8, except that the astronaut takes the place of the hang glider pilot and the two sides of the Great Wall serve as object 1 and object 2 in Figure 27.30. From Example 8, we know, then, that the sides of the Great Wall must subtend an angle of at least $\theta_{min} \approx 2.0 \times 10^{-4}$ rad at the astronaut's eye if they are to be resolvable. Using this value for θ_{min} and the fact that $s = 7.0$ m, we can find H, the distance for the astronaut.

SOLUTION Referring to Figure 27.30 and using Equation 8.1, we can express θ_{min} in radians as $\theta_{min} \approx s/H$. Therefore, the maximum distance for the astronaut is

$$H \approx \frac{s}{\theta_{min}} = \frac{7.0\text{ m}}{2.0 \times 10^{-4}\text{ rad}} = \boxed{3.5 \times 10^4\text{ m}}$$

In comparison, the moon is 3.85×10^8 m from the earth. Thus, it is not possible to identify the separate sides of the Great Wall from the moon.

Many optical instruments have a resolving power that exceeds that of the human eye. The typical camera does, for instance. Conceptual Example 10 compares the abilities of the human eye and a camera to resolve two closely spaced objects.

Conceptual Example 10 What You See Is Not What You Get

The French postimpressionist artist Georges Seurat developed a technique of painting in which dots of color are placed close together on the canvas. From sufficiently far away the individual dots are not distinguishable, and the images in the picture take on a more normal appearance. Figure 27.31 shows a person in an art museum, looking at one of Seurat's famous paintings. Suppose that this person stood close to the painting and then backed up until the dots just became indistinguishable to his eyes. From this position he took a picture. When he got home and had the film developed, however, he could see the individual dots in an enlarged photograph. Why did the camera resolve the individual dots, while his eyes did not? Assume that light enters his eyes through pupils that have diameters of 2.5 mm and that the average index of refraction of the material in the eyes is 1.36. In addition, assume that light enters his camera through an aperture or opening with a diameter of 25 mm.

REASONING To answer this question, we turn to the Rayleigh criterion for resolving two point objects, namely, $\theta_{min} = 1.22\ \lambda/D$. For the moment, we ignore the wavelength λ and concentrate on the diameter D of the opening through which light enters the instrument. According to the Rayleigh criterion, a larger value of D means that the value for θ_{min} is smaller, so that an optical instrument can be farther away and still resolve the objects. In other words, a larger value of D means a greater resolving power. For the eye and the camera the diameters are $D_{eye} = 2.5$ mm and $D_{camera} = 25$ mm, so

Figure 27.31 This person is about to take a photograph of a famous painting by Georges Seurat, who developed the technique of using dots of color to construct his images. Conceptual Example 10 discusses what the person sees when the film is developed.

the diameter for the camera is ten times larger than that for the eye. Thus, at the distance at which the eye loses its ability to resolve the individual dots in the painting, the camera still has resolving power to spare.

The conclusion that we have just reached is only tentative, however, since we have yet to consider the effect of the wavelength λ. A smaller value for λ means that the value for $\theta_{min} = 1.22\ \lambda/D$ is smaller. For the eye, the pertinent wavelength is that *within the eye,* since it is within the eye that the diffraction occurs that limits the resolving power. According to Equation 27.3, the wavelength within the eye is $\lambda_{eye} = \lambda_{vacuum}/n$, where n is the index of refraction for the material inside the eye. For the camera, the wavelength is simply $\lambda = \lambda_{vacuum}$. Since the index of refraction of the eye is 1.36, the wavelength within the eye is smaller than that for the camera by a factor of 1.36. This factor is insufficient to overcome the factor of ten by which the diameter of the camera exceeds the diameter of the pupil of the eye. Therefore, the camera succeeds in resolving the dots in the painting, while the eye fails, because the camera gathers light through a much larger opening.

Related Homework Material: *Problem 35*

27.7 THE DIFFRACTION GRATING

THE PHYSICS OF . . .

the diffraction grating.

Diffraction patterns of bright and dark fringes occur when monochromatic light passes through a single or double slit. Fringe patterns also result when light passes through more than two slits, and an arrangement consisting of a large number of parallel, closely spaced slits is called a ***diffraction grating.*** Gratings with as many as 40 000 slits per centimeter can be made, depending on the production method. In one method a diamond-tipped cutting tool is used to inscribe closely spaced parallel lines on a glass plate, the spaces between the lines serving as the slits. In fact, the number of slits per centimeter is often quoted as the number of lines per centimeter.

Figure 27.32 illustrates how light travels to a distant viewing screen from each of five slits in a grating and forms the central bright fringe and the first-order bright fringes on either side. Higher-order bright fringes are also formed but are not shown in the drawing. Each bright fringe is located by an angle θ relative to the central fringe. These bright fringes are sometimes called the *principal fringes* or *principal maxima,* since they are places where the light intensity is a maximum. The term "principal" distinguishes them from other, much less bright fringes that are referred to as secondary fringes or secondary maxima.

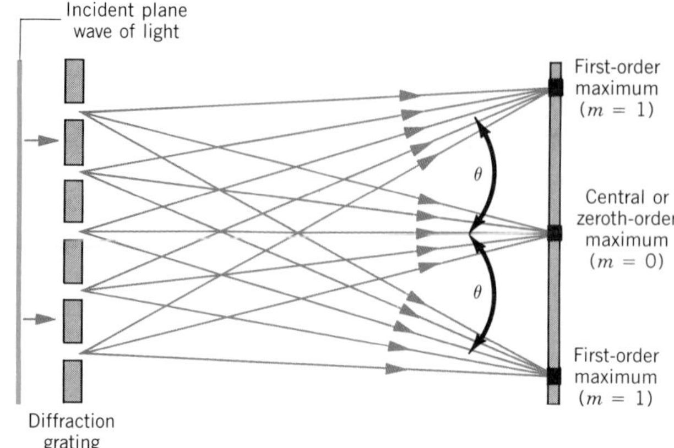

Figure 27.32 When light passes through a diffraction grating, a central bright fringe ($m = 0$) and higher-order bright fringes ($m = 1, 2, \ldots$) form when the light falls on a distant viewing screen.

Constructive interference creates the principal fringes. To show how, we assume the screen is far from the grating, so that the rays remain nearly parallel while the light travels toward the screen, as in Figure 27.33. In reaching the place on the screen where the first-order maximum is located, light from slit 2 travels a distance of one wavelength farther than light from slit 1. Similarly, light from slit 3 travels one wavelength farther than light from slit 2, and so forth, as emphasized by the colored right triangles. For the first-order maximum, the enlarged view of one of these right triangles shows that constructive interference occurs when $\sin\theta = \lambda/d$, where d is the separation between the slits. The second-order maximum forms when the extra distance traveled by light from adjacent slits is two wavelengths, so that $\sin\theta = 2\lambda/d$. The general result is

Principal maxima of a diffraction grating

$$\sin\theta = m\frac{\lambda}{d} \qquad m = 0, 1, 2, 3, \ldots \tag{27.7}$$

The separation d between the slits can be calculated from the number of slits per centimeter of grating; for instance, a grating with 2500 slits per centimeter has a slit separation of $d = 1/2500 \text{ cm} = 4.0 \times 10^{-4}$ cm.

Equation 27.7 is identical to Equation 27.1 for the double slit. A grating, however, produces bright fringes that are much *narrower* or *sharper* than those from a double slit, as the intensity patterns in Figure 27.34 reveal. Consider a grating with 100 slits/cm, for instance. The extra distance traveled by light from

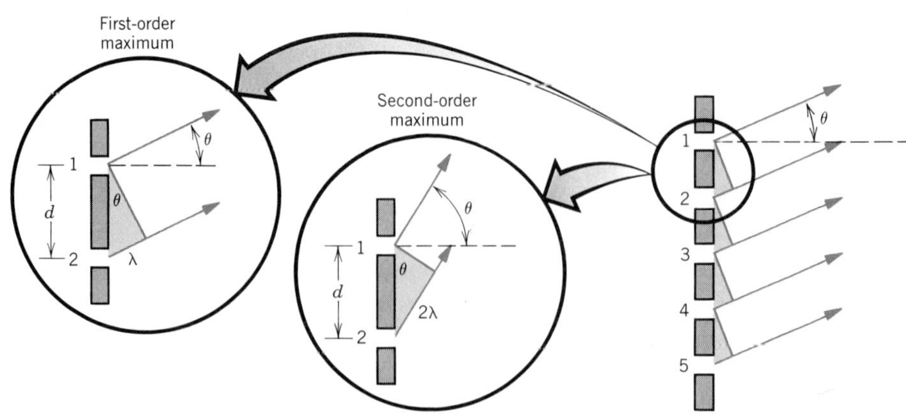

Figure 27.33 The conditions shown here lead to the first- and second-order intensity maxima in the diffraction pattern produced by a diffraction grating.

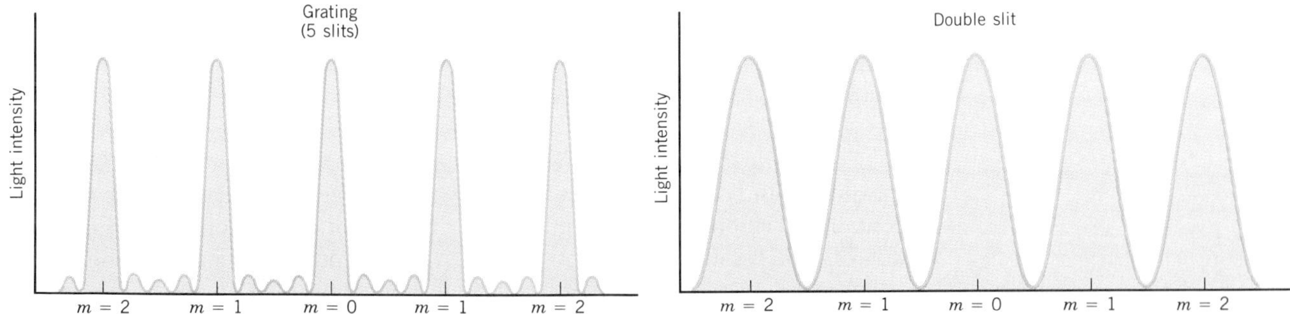

Figure 27.34 The bright fringes produced by a diffraction grating are much narrower than those produced by a double slit. Note the three small secondary bright fringes between the principal bright fringes of the grating. For a large number of slits, these secondary fringes become very small.

adjacent slits in forming a first-order maximum is exactly one wavelength λ. For an extra distance slightly larger than one wavelength, the light reaches the screen at a point slightly displaced from the maximum. If the extra distance is $\lambda + \lambda/100$, this slightly displaced point is already a place where complete destructive interference occurs, according to the following reasoning. The extra distance traveled by light from slit 51 compared to that from slit 1 is $50(\lambda + \lambda/100) = 50.5\lambda$. The additional half wavelength means that crests and troughs from these two slits combine to create destructive interference. The same result applies to light from slits 52 and 2, 53 and 3, and so on—in other words, to all the light from the grating. If a grating contains 10 000 slits/cm instead of 100 slits/cm, then destructive interference occurs when the extra distance traveled by light from adjacent slits is $\lambda + \lambda/10\,000$, instead of $\lambda + \lambda/100$. Thus, for a greater number of slits per centimeter, a smaller displacement from the maximum is required to produce the adjacent point of destructive interference on the screen, with the result that the principal fringes are even narrower. Note in Figure 27.34 that between the principal fringes there are secondary maxima with much smaller intensities.

The next example illustrates the ability of a grating to separate the components in a mixture of colors.

Example 11 Separating Colors with a Diffraction Grating

A mixture of violet light ($\lambda = 410$ nm in vacuum) and red light ($\lambda = 660$ nm in vacuum) falls on a grating that contains 1.0×10^4 lines/cm. For each wavelength, find the angle θ that locates the first-order maxima.

REASONING Before Equation 27.7 can be used here, a value for the separation d between the slits is needed: $d = 1/(1.0 \times 10^4 \text{ lines/cm}) = 1.0 \times 10^{-4}$ cm, or 1.0×10^{-6} m. For violet light, the angle θ_{violet} for the first-order maxima ($m = 1$) is given by $\sin \theta_{\text{violet}} = m\lambda_{\text{violet}}/d$, with an analogous equation applying for the red light.

SOLUTION For violet light, the angle locating the first-order maxima is

$$\theta_{\text{violet}} = \sin^{-1} \frac{\lambda_{\text{violet}}}{d} = \sin^{-1}\left(\frac{410 \times 10^{-9} \text{ m}}{1.0 \times 10^{-6} \text{ m}}\right) = \boxed{24^\circ}$$

For red light, a similar calculation with $\lambda_{red} = 660 \times 10^{-9}$ m shows that $\boxed{\theta_{red} = 41°}$.
Because θ_{violet} and θ_{red} are different, separate first-order bright fringes are seen for violet and red light on a viewing screen.

If the light in Example 11 had been sunlight, the angles for the first-order maxima would cover all values in the range between 24° and 41°, since sunlight contains all colors or wavelengths between violet and red. Consequently, a rainbow-like dispersion of the colors would be observed to either side of the central fringe on a screen. The central bright fringe would be white, however, since all the colors overlap there.

THE PHYSICS OF . . .

a grating spectroscope.

An instrument designed to measure the angles at which the principal maxima of a grating occur is called a grating spectroscope. With a measured value of the angle, calculations such as those in Example 11 can be turned around to provide the corresponding value of the wavelength. As we will point out in Chapter 30, the atoms in a hot gas emit discrete wavelengths, and determining the values of these wavelengths is one important technique used to identify the atoms. Figure 27.35 contains a sketch of a grating spectroscope. The slit that admits light from the source (e.g., a hot gas) is located at the focal point of the collimating lens, so the light rays striking the grating are parallel. The telescope is used to detect the bright fringes and, hence, to measure the angle θ.

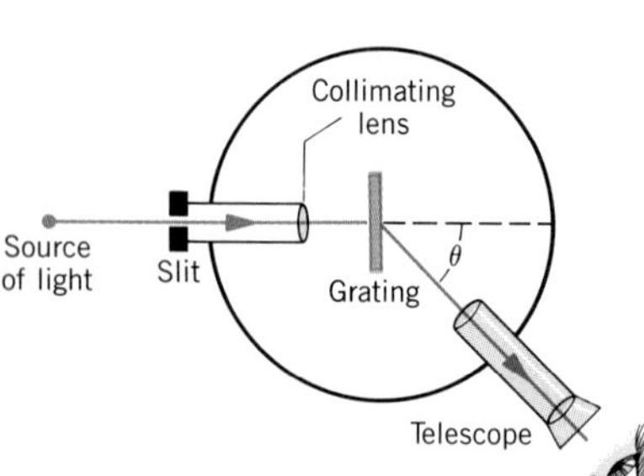

Figure 27.35 A grating spectroscope.

27.8 COMPACT DISCS AND THE USE OF INTERFERENCE

The compact disc (CD) has revolutionized stereo sound reproduction and uses interference effects in some interesting ways. A CD contains a spiral track that holds the audio information and is analogous to the spiral groove on an LP record. The audio information on the CD track, however, is detected using a laser beam that reflects from the bottom of the disc, as Figure 27.36 illustrates. The information is encoded in the form of raised areas on the bottom of the disc. These raised areas appear as "pits" when viewed from the *top* or labeled side of the CD. They are separated by flat areas called "land." The pits and land are covered with a transparent plastic coating, which has been omitted from the drawing for simplicity.

THE PHYSICS OF . . .

retrieving information from compact discs.

As the CD rotates, the laser beam reflects off the disc and into a detector. The reflected light intensity fluctuates as the pits and land areas pass by, and the fluctuations convey the audio information as a series of binary numbers (zeros and ones). To make the fluctuations easier to detect, the pit thickness t (see Figure 27.36) is chosen with destructive interference in mind. As the laser beam overlaps the edges of a pit, part of the beam is reflected from the raised pit surface and part from the land. The part that reflects from the land travels an additional distance of $2t$. The thickness of the pits is chosen so that $2t$ is one-half of a wavelength of the laser beam in the plastic coating. With this choice, destructive interference occurs when the two parts of the reflected beam combine. As a result, there is markedly less reflected intensity when the laser beam passes over a pit edge than when it passes over land alone. Thus, the fluctuations in reflected light that occur while the disc rotates are large enough to detect because of the effects of destructive interference. Example 12 determines the theoretical thickness of the pits on a compact disc. In reality, a value slightly less than that obtained in the example is used for technical reasons that are not pertinent here.

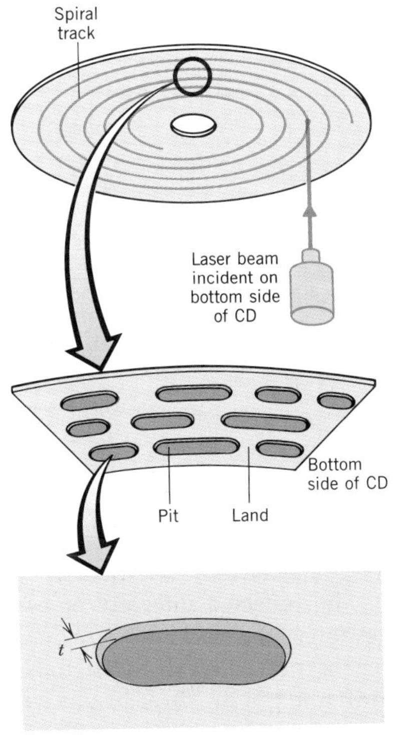

Figure 27.36 The bottom surface of a compact disc (CD) carries the audio information in the form of raised areas ("pits") and land along a spiral track. A CD is played by using a laser beam that strikes the bottom surface of the disc and reflects from it.

Example 12 Pit Thickness on a Compact Disc

The laser in a CD player has a wavelength of 790 nm in a vacuum. The plastic coating over the pits has an index of refraction of $n = 1.5$. Find the thickness of the pits on a CD.

REASONING As we have discussed, the thickness t is chosen so that $2t = \frac{1}{2}\lambda_{\text{coating}}$ in order to achieve destructive interference. Equation 27.3 gives the wavelength in the plastic coating as $\lambda_{\text{coating}} = \lambda_{\text{vacuum}}/n$.

SOLUTION The thickness of the pit is

$$t = \frac{\lambda_{\text{coating}}}{4} = \frac{\lambda_{\text{vacuum}}}{4n} = \frac{790 \times 10^{-9}\ \text{m}}{4(1.5)} = \boxed{1.3 \times 10^{-7}\ \text{m}}$$

THE PHYSICS OF . . .

the three-beam tracking method for compact discs.

As a CD rotates, the laser beam must accurately follow the pits and land along the spiral track. One type of CD player utilizes a three-beam tracking method to ensure that the laser beam follows the spiral properly. Figure 27.37 shows that a diffraction grating is the key element in this method. Before the laser beam strikes the CD, the beam passes through a grating that produces a central maximum and two first-order maxima, one on either side. As the picture indicates, the central maximum beam falls directly on the spiral track. This beam reflects into a detector, and the reflected light intensity fluctuates as the pits and land areas pass by, the fluctuations conveying the audio information. The two first-order maxima beams are called tracking beams. They hit the CD between the arms of the spiral and also reflect into detectors of their own. Under perfect conditions, the intensities of the two reflected tracking beams do not fluctuate, since they originate from the smooth surface between the arms of the spiral where there are no pits. As a

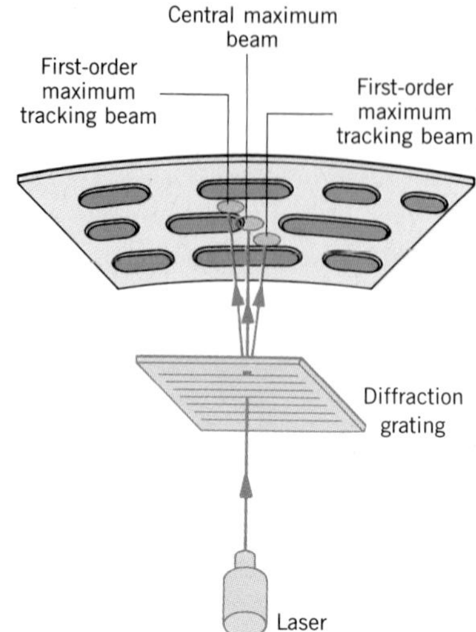

Figure 27.37 A three-beam tracking method is often used in CD players to ensure that the laser follows the spiral track correctly. The three beams are derived from a single laser beam by using a diffraction grating.

result, each tracking beam detector puts out the same constant electrical signal. However, if the tracking drifts to either side, the reflected intensity of each tracking beam changes because of the pits. In response, the tracking beam detectors produce different electrical signals. The difference between the signals is used in a "feedback" circuit to correct for the drift and put the three beams back into their proper positions.

THE PHYSICS OF . . .

X-ray diffraction.

27.9 X-RAY DIFFRACTION

Not all diffraction gratings are commercially made. Nature also creates diffraction gratings, although these gratings do not look like an array of closely spaced slits. Instead, nature's gratings are the arrays of regularly spaced atoms that exist in crystalline solids. For example, Figure 27.38 shows a crystal of ordinary salt (NaCl). Typically, the atoms in a crystalline solid are separated by distances of about 1.0×10^{-10} m, so we might expect a crystalline array of atoms to act like a grating with roughly this "slit" spacing for electromagnetic waves of the appropriate wavelength. Assuming that $\sin \theta = 0.5$ and that $m = 1$ in Equation 27.7, it follows that $0.5 = \lambda/d$. A value of $d = 1.0 \times 10^{-10}$ m in this equation gives a wavelength of $\lambda = 0.5 \times 10^{-10}$ m. This wavelength is much shorter than that of

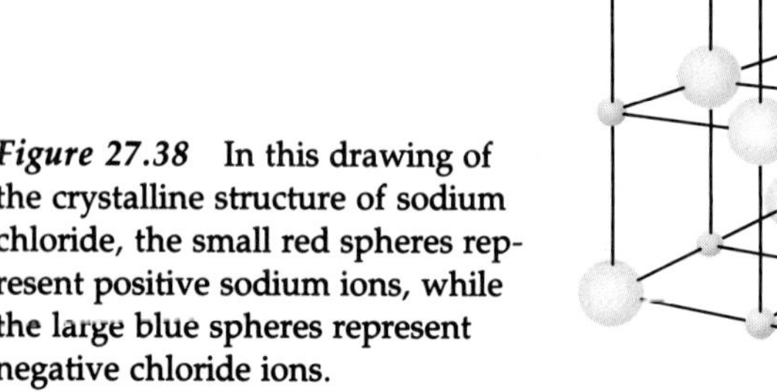

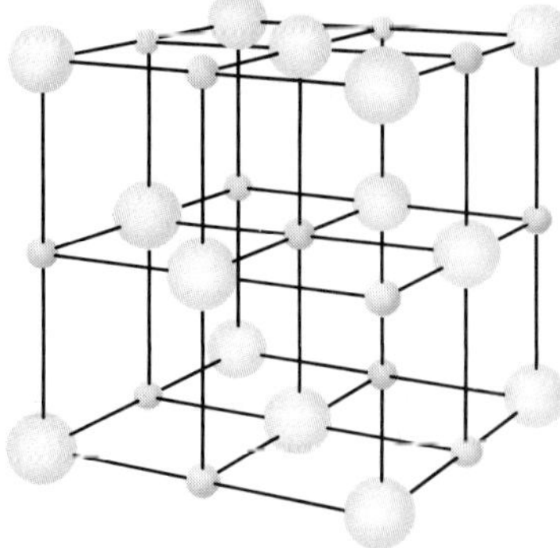

Figure 27.38 In this drawing of the crystalline structure of sodium chloride, the small red spheres represent positive sodium ions, while the large blue spheres represent negative chloride ions.

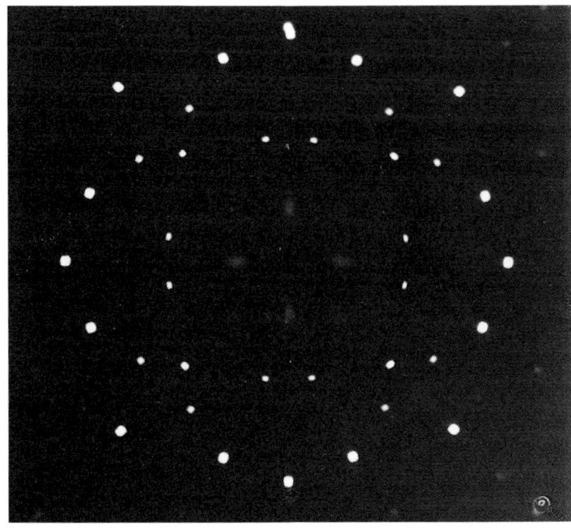

Figure 27.39 The X-ray diffraction pattern from crystalline NaCl.

visible light and falls in the X-ray region of the electromagnetic spectrum. (See Figure 24.6.)

A diffraction pattern does indeed result when X-rays are directed onto a crystalline material, as Figure 27.39 illustrates for a crystal of NaCl. The pattern consists of a complicated arrangement of spots, because a crystal has a complex three-dimensional structure. It is from patterns such as these that the spacing between atoms and the nature of the crystal structure can be determined.

SUMMARY

The **principle of linear superposition** states that when two or more waves are present simultaneously in the same region of space, the resultant wave is the sum of the individual waves. According to this principle, two or more light waves can interfere constructively or destructively when they exist at the same place at the same time, provided they originate from **coherent sources.** Two sources are coherent if they emit waves that have a constant phase relationship.

In **Young's double-slit experiment,** light passes through a pair of closely spaced narrow slits and produces a pattern of alternating bright and dark fringes on a viewing screen. The fringes arise because of constructive and destructive interference. The angle θ for the mth higher-order bright fringe is given by $\sin\theta = m\lambda/d$, where d is the spacing between the narrow slits, λ is the wavelength of the light, and $m = 0, 1, 2, 3, \ldots$. Similarly, the angle for the dark fringes is given by $\sin\theta = (m + \frac{1}{2})\lambda/d$.

Constructive and destructive interference of light waves can occur with **thin films** of transparent materials. The interference occurs between light waves that reflect from the top and bottom surfaces of the film. One important factor in thin-film interference is the thickness of the film relative to the wavelength of the light within the film. The wavelength within the film is $\lambda_{\text{film}} = \lambda_{\text{vacuum}}/n$, where n is the refractive index of the film. A second important factor is the phase change that can occur when light undergoes reflection at each surface of the film; the exact nature of the phase change is discussed in Section 27.3.

Diffraction is a bending of waves around obstacles or the edges of an opening. Diffraction is an interference effect that can be explained with the aid of **Huygens' principle.** This principle states that every point on a wave front acts as a source of tiny wavelets that move forward with the same speed as the wave; the wave front at a later instant is the surface that is tangent to the wavelets. When light passes through a single narrow slit and falls on a viewing screen, a pattern of bright and dark fringes is formed because of the superposition of such wavelets. The angle θ for the mth dark fringe on either side of the central bright fringe is given by $\sin\theta = m\lambda/W$, where W is the slit width and $m = 1, 2, 3, \ldots$.

The **resolving power** of an optical instrument is the ability of the instrument to distinguish between two closely spaced objects. Resolving power is limited by the diffraction that occurs when light waves enter an instrument, often through a circular opening. Consideration of the diffraction fringes for a circular opening leads to the **Rayleigh criterion** for resolution. This criterion specifies that two point objects are just resolved when the first dark fringe in the diffraction pattern of one falls directly on the central bright fringe in the diffraction pattern of the other. According to this specification, the minimum angle (in radians) that two point objects can subtend at an aperture of diameter D and still be resolved as separate objects is $\theta_{min} \approx 1.22\lambda/D$, where λ is the wavelength of the light.

A **diffraction grating** is a device consisting of a large number of parallel, closely spaced slits. When light passes through a diffraction grating and falls on a viewing screen, the light forms a pattern of bright and dark fringes. The bright fringes are referred to as principal maxima and are found at an angle θ, such that $\sin\theta = m\lambda/d$, where d is the separation between two successive slits and $m = 0, 1, 2, 3, \ldots$.

QUESTIONS

1. Suppose that a radio station broadcasts simultaneously from two transmitting antennas at two different locations. Is it clear that your radio will have better reception with two transmitting antennas, rather than one? Justify your answer.

2. (a) How would the pattern of bright and dark fringes produced in a Young's double-slit experiment change if the light waves coming from *both* of the slits had their phases shifted by an amount equivalent to a half wavelength? (b) How would the pattern change, if the light coming from *only one* of the slits had its phase shifted by an amount equivalent to a half wavelength?

3. Replace slits S_1 and S_2 in Figure 27.4 with identical in-phase loudspeakers and use the same ac electrical signal to drive them. The two sound waves produced will then be identical, and you will have the audio equivalent of Young's double-slit experiment. In terms of loudness and softness, describe what you would hear as you walked along the screen, starting from the center and going to either end.

4. Consider Young's experiment in Figure 27.4 with one of the slits, S_1 or S_2, covered. The screen is bright everywhere, indicating that light energy is reaching every point on it. However, when the covered slit is uncovered, a certain spot on the screen becomes dark, indicating that light energy is no longer reaching that spot. Where did the energy go, and is energy conservation violated here? Explain.

5. A camera lens is covered with a nonreflective coating that eliminates the reflection of perpendicularly incident green light. Recalling Snell's law of refraction, would you expect the reflected green light to be eliminated if it were incident on the nonreflective coating at an angle of 45° rather than perpendicularly? Justify your answer.

6. When sunlight reflects from a thin film of soapy water, the film appears multicolored, because destructive interference removes different wavelengths from the light reflected at different places, depending on the thickness of the film. As the film becomes thinner and thinner, it looks darker and darker in reflected light, appearing black just before it breaks. The blackness means that destructive interference removes *all* wavelengths from the reflected light when the film is very thin. Explain why.

7. Two pieces of the same glass are covered with thin films of different materials. The thickness of each film is the same. In reflected sunlight, however, the films have different colors. Give two reasons that account for the difference in colors.

8. In Figure 27.16*b* there is a dark spot at the center of the pattern of Newton's rings. By considering the phase changes that occur when light reflects from the upper curved surface and the lower flat surface, account for the dark spot.

9. A thin film of a material is floating on water ($n = 1.33$). When the material has a refractive index of $n = 1.20$, the film looks bright in reflected light as its thickness approaches zero. But when the material has a refractive index of $n = 1.45$, the film looks black in reflected light as its thickness approaches zero. Explain these observations in terms of constructive and destructive interference and the phase changes that occur when light waves undergo reflection.

10. A transparent coating is deposited on a glass plate and has a refractive index that is *larger than that of the glass,* not smaller, as it is for a typical nonreflective coating. For a certain wavelength within the coating, the thickness of the coating is a quarter wavelength. The coating *enhances* the reflection of the light corresponding to this wavelength. Explain why, referring to part (a) of Example 4 in the text to guide your thinking.

11. On most cameras one can select the *f*-number setting, or *f*-stop. The *f*-number gives the ratio of the focal length of the camera lens to the diameter of the aperture through which light enters the camera. If one wishes to resolve two closely spaced objects in a picture, should a small or a large *f*-number setting be used? Account for your answer.

12. Do any dark fringes appear in a single-slit diffraction pattern (see Figure 27.20*b*) when the wavelength of the light is greater than the width of the slit? Why? *(Hint: See Equation 27.4.)*

13. Account for the fact that a sound wave diffracts much more than a light wave does when the two pass through the same doorway.

14. Light enters the eye through an opening called the pupil. On the inner side of the pupil is the material from which the eye is made (average index of refraction = 1.36). Suppose, for comparison, that a sheet of opaque material has a hole cut into it. Light passes through this hole into the air on the other side. Assuming that the pupil and the hole have the same diameter, in which case does green light ($\lambda_{\text{vacuum}} = 550$ nm) diffract more, when it enters the eye or when it passes through the hole in the opaque material? Give your reasoning.

15. Four light bulbs are arranged at the corners of a rectangle that is three times longer than it is wide. You look at this arrangement perpendicular to the plane of the rectangle. From very far away, your eyes cannot resolve the individual bulbs and you see a single "smear" of light. From close in, you see the individual bulbs. Between these two extremes, what do you see? Draw two pictures to illustrate the possibilities that exist, depending on how far away you are. Explain your drawings.

16. Suppose the pupil of your eye were elliptical instead of circular in shape, with the long axis of the ellipse oriented in the vertical direction. (a) Would the resolving power of your eye be the same in the horizontal and vertical directions? (b) In which direction would the resolving power be greatest? Justify your answers by discussing how the diffraction of light waves would differ in the two directions.

17. Suppose you were designing an eye and could select the size of the pupil and the wavelength of the electromagnetic waves to which the eye is sensitive. As far as the limitation created by diffraction is concerned, rank the following design choices in order of decreasing resolving power (greatest first): (a) large pupil and ultraviolet wavelengths, (b) small pupil and infrared wavelengths, and (c) small pupil and ultraviolet wavelengths.

18. In our discussion of single-slit diffraction, we considered the ratio of the wavelength λ to the width W of the slit. We ignored the height of the slit, in effect assuming that the height was much larger than the width. Suppose the height and width were the same size, so that diffraction in both dimensions occurred. How would the diffraction pattern in Figure 27.20*b* be altered? Give your reasoning.

19. What would happen to the distance between the bright fringes produced by a diffraction grating if the entire interference apparatus (light source, grating, and screen) were immersed in water? Why?

PROBLEMS

Section 27.1 The Principle of Linear Superposition, Section 27.2 Young's Double-Slit Experiment

1. The transmitting antenna for a radio station is 7.00 km from your house. The frequency of the electromagnetic wave broadcast by this station is 536 kHz. The station builds a second transmitting antenna that broadcasts the same electromagnetic wave in phase with the original one. The new antenna is 8.12 km from your house. Does constructive or destructive interference occur at the receiving antenna of your radio? Show your calculations.

2. Two sources are in phase and emit waves that have a wavelength of 0.44 m. Determine whether constructive or destructive interference occurs at a point whose distances from the two sources are as follows: (a) 1.32 and 3.08 m; (b) 2.67 and 3.33 m; (c) 2.20 and 3.74 m; (d) 1.10 and 4.18 m.

3. In a Young's double-slit experiment, the angle that locates the second-order bright fringe is 2.0°. The slit separation is 3.8×10^{-5} m. What is the wavelength of the light?

4. In a Young's double-slit experiment, the angle that locates the first dark fringe on either side of the central bright fringe is 1.6°. Find the ratio of the slit separation d to the wavelength λ of the light.

5. A flat observation screen is placed at a distance of 4.5 m from a pair of slits. The separation on the screen between the central bright fringe and the first-order bright fringe is 0.037 m. The light illuminating the slits has a wavelength of 490 nm. Determine the slit separation.

6. In a Young's double-slit experiment, the seventh dark fringe is located 0.025 m to the side of the central bright fringe on a flat screen, which is 1.1 m away from the slits. The slits are 1.4×10^{-4} m apart. What is the wavelength of the light being used?

*7. Review Conceptual Example 3 before attempting this problem. Two slits are 0.158 mm apart. A mixture of red light (wavelength = 665 nm) and yellow-green light (wavelength = 565 nm) falls on the slits. A flat observation screen is located 2.24 m away. What is the distance on the screen between the third-order red fringe and the third-order yellow-green fringe?

*8. At most, how many bright fringes can be formed on either side of the central bright fringe when light of wavelength 625 nm falls on a double slit whose slit separation is 3.76×10^{-6} m?

*9. In a Young's double-slit experiment the separation y

between the first-order bright fringe and the central bright fringe on a flat screen is 0.0240 m, when light is used that has a wavelength of 475 nm. Assume that the angles that locate the fringes on the screen are small enough so that $\sin\theta \approx \tan\theta$. Find the separation y when the light has a wavelength of 611 nm.

****10.** In Young's experiment a mixture of orange light (611 nm) and blue light (471 nm) shines on the double slit. The centers of the first-order bright blue fringes lie at the outer edges of a screen that is located 0.500 m away from the slits. However, the first-order bright orange fringes fall off the screen. By how much and in which direction (toward or away from the slits) should the screen be moved, so that the centers of the first-order bright orange fringes just appear on the screen? It may be assumed that θ is small, so that $\sin\theta \approx \tan\theta$.

Section 27.3 Thin-Film Interference

11. A mixture of red light ($\lambda_{\text{vacuum}} = 661$ nm) and green light ($\lambda_{\text{vacuum}} = 551$ nm) shines perpendicularly on a soap film ($n = 1.33$) that has air on either side. What is the minimum nonzero thickness of the film, so that destructive interference causes it to look red in reflected light?

12. A layer of transparent plastic ($n = 1.61$) on glass ($n = 1.52$) looks dark when viewed in reflected light whose wavelength is 589 nm in vacuum. Find the two smallest possible nonzero values for the thickness of the layer.

13. In preparation for this problem review Conceptual Example 5. A mixture of red light ($\lambda_{\text{vacuum}} = 661$ nm) and blue light ($\lambda_{\text{vacuum}} = 472$ nm) shines perpendicularly on a thin layer of gasoline ($n_{\text{gas}} = 1.40$) lying on water ($n_{\text{water}} = 1.33$). The gasoline layer has a uniform thickness of 2.36×10^{-7} m. Destructive interference removes one of the colors from the reflected light. By means of suitable calculations decide whether the gasoline looks red or blue in reflected light.

14. A nonreflective coating of magnesium fluoride ($n = 1.38$) covers the glass ($n = 1.52$) of a camera lens. Assuming that the coating prevents reflection of yellow-green light (wavelength in vacuum = 565 nm), determine the minimum nonzero thickness that the coating can have.

15. A transparent film ($n = 1.43$) is deposited on a glass plate ($n = 1.52$) to form a nonreflective coating. The film has a thickness of 1.07×10^{-7} m. What is the longest possible wavelength of light (in vacuum) for which this film has been designed?

16. Example 6(a) in the text deals with the air wedge formed between two plates of glass ($n = 1.52$). Repeat this example, assuming that the wedge of air is replaced by water ($n = 1.33$).

***17.** Review Conceptual Example 5 before attempting this problem. A film of gasoline ($n = 1.40$) floats on water ($n = 1.33$). Yellow light (wavelength = 580 nm in vacuum) shines perpendicularly on this film. (a) Determine the minimum nonzero thickness of the film, such that the film appears bright yellow due to constructive interference. (b) Repeat part (a), assuming that the gasoline film is on glass ($n = 1.52$) instead of water.

***18.** A layer of glycerol ($n = 1.47$) on glass ($n = 1.52$) has a thickness of 1.02×10^{-7} m. This is the minimum nonzero thickness for which the layer will look dark in a certain monochromatic light. What is the next largest thickness that the layer could have and still look dark in the same light?

***19.** A film of oil lies on wet pavement. The refractive index of the oil exceeds that of the water. The film has the minimum nonzero thickness such that it appears dark due to destructive interference when viewed in red light (wavelength = 660 nm in vacuum). Assuming that the visible spectrum extends from 380 to 750 nm, what are the visible wavelength(s) (in vacuum) for which the film will appear bright due to constructive interference?

****20.** A uniform layer of water ($n = 1.33$) lies on a glass plate ($n = 1.52$). Light shines perpendicularly on the layer. Because of constructive interference, the layer looks maximally bright when the wavelength of the light is 432 nm in vacuum and *also* when it is 648 nm in vacuum. (a) Obtain the minimum thickness of the film. (b) Assuming that the film has the minimum thickness and that the visible spectrum extends from 380 to 750 nm, determine the visible wavelength(s) (in vacuum) for which the film appears completely dark.

Section 27.5 Diffraction

21. A diffraction pattern forms when light passes through a single slit. The wavelength of the light is 675 nm. Determine the angle that locates the first dark fringe when the width of the slit is (a) 1.8×10^{-4} m and (b) 1.8×10^{-6} m.

22. How wide does a single slit have to be before the angle locating the first-order dark fringe is less than 0.10°, when the wavelength of the light is 660 nm?

23. A doorway is 0.91 m wide. (a) Obtain the angle that locates the first dark fringe in the Fraunhofer diffraction pattern formed when red light (wavelength = 660 nm) passes through the doorway. (b) Repeat part (a) for a 440-Hz tone (concert A), assuming that the speed of sound is 343 m/s.

24. A slit whose width is 4.50×10^{-5} m is located 1.23 m from a flat screen. Light shines through the slit and falls on the screen. Find the width of the central fringe of the diffraction pattern when the wavelength of the light is 580 nm.

25. Light shines through a single slit whose width is 5.6×10^{-4} m. A diffraction pattern is formed on a flat screen located 4.0 m away. The distance between the middle of the central bright fringe and the first dark fringe is 3.5 mm. What is the wavelength of the light?

26. A loudspeaker produces an 1100-Hz tone and a 3100-Hz tone. The sound waves are emitted through a vertically oriented slit. The angle locating the first diffraction minimum, or audio "dark fringe," is 15° for the 3100-Hz tone. What is the corresponding angle for the 1100-Hz tone?

***27.** The width of a slit is 2.0×10^{-5} m. Light with a wavelength of 480 nm passes through this slit and falls on a screen that is located 0.50 m away. In the diffraction pattern, find the width of the bright fringe that is next to the central bright fringe.

***28.** A single slit (width $= 4.85 \times 10^{-7}$ m) produces *no* dark fringes on an observation screen, even though light shines on the slit. Determine a minimum value for the wavelength of the light being used.

***29.** The central bright fringe in a single-slit diffraction pattern has a width that equals the distance between the screen and the slit. Find the ratio λ/W of the wavelength of the light to the width of the slit.

****30.** In a single-slit diffraction pattern, the central fringe is 450 times as wide as the slit. The screen is 18 000 times as far from the slit as the slit is wide. What is the ratio λ/W, where λ is the wavelength of the light shining through the slit and W is the width of the slit? Assume that the angle that locates a dark fringe on the screen is small, so that $\sin \theta \approx \tan \theta$.

Section 27.6 Resolving Power

31. You are looking down at the earth from inside a commercial jetliner flying at an altitude of 8690 m, and the pupil of your eye has a diameter of 2.00 mm. The average refractive index of the material in the eye is 1.36. Determine how far apart two cars must be on the ground if you are to have any hope of distinguishing between them in (a) red light (wavelength = 665 nm in vacuum) and (b) violet light (wavelength = 405 nm in vacuum).

32. In a dot matrix printer, an array of dots is used to form the printed characters. If the dots are close enough together, they cannot be resolved individually by the eye and, therefore, appear to form solid lines. Suppose that the pupil of the eye has a diameter of 2.0 mm in bright yellow-green light (wavelength = 563 nm in vacuum), that the material in the eye has an average refractive index of $n = 1.36$, and that the printed page is to be read at a distance of 0.31 m. Considering the limit created by diffraction, find the smallest separation between the dots that the eye can see.

33. Late one night on a highway, a car speeds by you and fades into the distance. Under these conditions the pupils of your eyes (average refractive index = 1.36) have diameters of about 7.0 mm. The taillights of this car are separated by a distance of 1.2 m and emit red light (wavelength = 660 nm in vacuum). How far away from you is this car when its taillights appear to merge into a single spot of light because of the effects of diffraction?

34. Two stars are 3.7×10^{11} m apart and are equally distant from the earth. A telescope has an objective lens with a diameter of 1.02 m and just detects these stars as separate objects. Assume that light of wavelength 550 nm is being observed. Diffraction effects, rather than atmospheric turbulence, limit the resolving power of the telescope. Find the maximum distance that these stars could be from the earth.

35. Review Conceptual Example 10 as background for this problem. In addition to the data given there, assume that the dots in the painting are separated by 1.5 mm and that the wavelength of the light is $\lambda_{\text{vacuum}} = 550$ nm. Find the distance at which the dots merge for (a) the eye and (b) the camera.

36. The largest refracting telescope in the world is at the Yerkes Observatory in Williams Bay, Wisconsin. The objective of the telescope has a diameter of 1.02 m. With light whose wavelength is 565 nm, it is said that this telescope can resolve two objects at a distance of 3.75×10^{4} m that are separated by 0.0254 m. Verify this statement with a calculation of your own.

***37.** In an experiment, red light from a ruby laser (wavelength = 694.3 nm) is passed through a telescope in reverse and is sent on its way to the moon. At the surface of the moon, which is 3.77×10^{8} m away, the light strikes a reflector left there by astronauts. The reflected light returns to the earth, where it is detected. When it leaves the telescope, the circular beam of light has a diameter of about 0.20 m, and diffraction causes the beam to spread as the light travels to the moon. In effect, the first circular dark fringe in the diffraction pattern defines the size of the central bright spot on the moon. Determine the diameter (not the radius) of the central bright spot on the moon.

***38.** You are using a microscope to examine a blood sample. Recall from Section 26.12 that the sample should be placed just outside the focal point of the objective lens of the microscope. (a) If the specimen is being illuminated with light of wavelength λ and the diameter of the objective equals its focal length, determine the closest distance between two blood cells that can just be resolved. Express your answer in terms of λ. (b) Based on your answer to (a), should you use light with a longer wavelength or a shorter wavelength if you wish to resolve two blood cells that are even closer together?

***39.** The pupil of an eagle's eye has a diameter of 6.0 mm. The refractive index in the eye is 1.36. Two field mice are separated by 0.010 m. From a distance of 175 m, the eagle sees them as one unresolved object and dives toward them at a speed of 17 m/s. Assume that the eagle's eye detects light that has a wavelength of 550 nm in a vacuum. How much time passes until the eagle sees the mice as separate objects?

****40.** Two concentric circles of light emit light whose wavelength is 555 nm. The larger circle has a radius of 4.0 cm, while the smaller circle has a radius of 1.0 cm. When taking a picture of these lighted circles, a camera admits light through an aperture of 12.5 mm. What is the maximum distance at which the camera can (a) distinguish one circle from the other and (b) reveal that the inner circle is a circle of light rather than a solid disk of light?

Section 27.7 The Diffraction Grating, Section 27.8 Compact Discs and the Use of Interference

41. The diffraction gratings discussed in the text are transmission gratings, because light *passes through* them. There are also gratings in which the light *reflects from* the grating to form a pattern of fringes. Equation 27.7 also applies to a reflection grating with straight parallel lines when the incident light shines perpendicularly on the grating. The surface of a compact disc (CD) has a multicolored appearance because it acts like a reflection grating and spreads sunlight into its colors. The arms of the spiral track on the CD are separated by 1.1×10^{-6} m. Using Equation 27.7, estimate the angle that corresponds to the first-order maximum for a wavelength of (a) 660 nm (red) and (b) 410 nm (violet).

42. Light of wavelength 490 nm falls on a diffraction grating. The third-order maximum occurs at an angle of 25°. Find the separation between the slits of the grating.

43. A diffraction grating produces a first-order bright fringe that is 0.0894 m away from the central bright fringe on a flat screen. The separation between the slits of the grating is 4.17×10^{-6} m, and the distance between the grating and the screen is 0.625 m. What is the wavelength of the light shining on the grating?

44. When a grating is used with light that has a wavelength of 575 nm, a second-order maximum is formed at an angle of 11.2°. How many lines per centimeter does this grating have?

45. The wavelength of the laser beam used in a compact disc player is 790 nm. Suppose that a diffraction grating produces first-order tracking beams that are 1.2 mm apart at a distance of 3.0 mm from the grating. Estimate the spacing between the slits of the grating.

46. A grating and an observation screen are fixed in position. For a wavelength λ_1, adjacent principal maxima are separated on the screen by 2.7 cm. For another wavelength λ_2, the same principal maxima are separated by 3.2 cm. Find the ratio λ_2/λ_1, assuming that the screen is far enough from the grating that $\sin\theta \approx \tan\theta$.

***47.** Monochromatic light shines on a diffraction grating. When the light source and the grating are in air, the first-order maximum occurs at an angle of 33°. At what angle does the first order maximum occur when the source and the grating are immersed in water ($n = 1.33$)?

***48.** A diffraction grating contains 4820 lines/cm and is used with blue light (wavelength = 470 nm). What is the highest-order bright fringe that can be seen with this grating?

***49.** The separation between the slits of a grating is 2.2×10^{-6} m. This grating is used with light that contains all wavelengths between 410 and 660 nm. Rainbow-like spectra form on a screen 3.2 m away. How wide (in meters) is (a) the first-order spectrum and (b) the second-order spectrum?

****50.** There are 5620 lines per centimeter in a grating that is used with light whose wavelength is 471 nm. A flat observation screen is located at a distance of 0.750 m from the grating. What is the minimum width that the screen must have so the *centers* of all the principal maxima formed on either side of the central maximum fall on the screen?

ADDITIONAL PROBLEMS

51. A single slit has a width of 2.1×10^{-6} m and is used to form a diffraction pattern. Find the angle that locates the second dark fringe when the wavelength of the light is (a) 430 nm and (b) 660 nm.

52. A rock concert is being held in an open field. Two loudspeakers are separated by 7.00 m. As an aid in arranging the seating, a test is conducted in which both speakers vibrate in phase and produce an 80.0-Hz bass tone simultaneously. The speed of sound is 343 m/s. A reference line is marked out in front of the speakers, perpendicular to the midpoint of the line between the speakers. Relative to either side of this reference line, what is the smallest angle that locates the places where destructive interference occurs? People seated in these places would have trouble hearing the 80.0-Hz bass tone.

53. A Young's double-slit experiment is performed using light that has a wavelength of 630 nm. The separation between the slits is 5.3×10^{-5} m. Find the angles that locate the first-, second-, and third-order bright fringes on the screen.

54. The first dark fringe in the diffraction pattern of a single slit is located at an angle of $\theta_A = 34°$. With the same light, the first dark fringe formed with another single slit is at an angle of $\theta_B = 56°$. Find the ratio W_A/W_B of the widths of the two slits.

55. For background on this problem, review Conceptual Example 5. A mixture of yellow light (wavelength = 580 nm in vacuum) and violet light (wavelength = 410 nm in vacuum) falls perpendicularly on a film of gasoline that is floating on a puddle of water. For both wavelengths, the refractive index of gasoline is $n = 1.40$ and of water is $n = 1.33$. What is the minimum nonzero thickness of the film in a spot that looks (a) yellow and (b) violet because of destructive interference?

56. It is claimed that some professional baseball players can see which way the ball is spinning as it travels toward home plate. One way to judge this claim is to estimate the distance at which a batter can first hope to resolve two points on opposite sides of a baseball, which has a diameter of 0.0738 m. (a) Estimate this distance, assuming that the pupil of the eye has a diameter of 2.0 mm, the material within the eye has a refractive index of 1.36, and the wavelength of the light is 550 nm in vacuum. (b) Considering that the distance between the pitcher's mound and home plate is 18.4 m, can you rule out the claim based on your answer to part (a)?

57. A telescope is being used to view two objects that are separated by 480 m on the moon's surface. The surface of the moon is 3.77×10^8 m away from the surface of the earth. Assume that diffraction effects, rather than atmospheric turbulence, limits the resolving power of the telescope and that the wavelength of the light being used is 550 nm. Determine the diameter that the objective lens of this telescope must have if the two objects on the moon are to be resolved.

***58.** The first-order maximum produced by a grating is located by an angle of $\theta = 28°$. What is the angle for the second-order maximum with the same light?

***59.** The same diffraction grating is used with two different wavelengths of light, λ_A and λ_B. The fourth-order principal maximum of light *A* exactly overlaps the third-order principal maximum of light *B*. Find the ratio λ_A/λ_B.

***60.** Orange light ($\lambda_{vacuum} = 611$ nm) shines on a soap film ($n = 1.33$) that has air on either side of it. The light strikes the film perpendicularly. What is the minimum thickness of the film for which constructive interference causes it to look bright in reflected light?

***61.** Violet light (wavelength = 410 nm) and red light (wavelength = 660 nm) lie at opposite ends of the visible spectrum. (a) For each wavelength, find the angle θ that locates the first-order maximum produced by a grating with 3300 lines/cm. This grating converts a mixture of all colors between violet and red into a rainbow-like dispersion between the two angles. Repeat the calculation above for (b) the second-order maximum and (c) the third-order maximum. (d) From your results, decide whether there is any overlap between any of the "rainbows" and specify which orders overlap.

***62.** Two parallel slits are illuminated by light composed of two wavelengths, one of which is 645 nm. On a viewing screen, the light whose wavelength is known produces its third dark fringe at the same place where the light whose wavelength is unknown produces its fourth-order bright fringe. The fringes are counted relative to the central or zeroth-order bright fringe. What is the unknown wavelength?

****63.** Two gratings *A* and *B* have slit separations d_A and d_B, respectively. They are used with the same light and the same observation screen. When grating *A* is replaced with grating *B*, it is observed that the first-order maximum of *A* is exactly replaced by the second-order maximum of *B*. (a) Determine the ratio d_B/d_A of the spacings between the slits of the gratings. (b) Find the next two principal maxima of grating *A* and the principal maxima of *B* that exactly replace them when the gratings are switched. Identify these maxima by their order numbers.

****64.** A piece of curved glass has a radius of curvature of 10.0 m and is used to form Newton's rings, as in Figure 27.16. Not counting the dark spot at the center of the pattern, there are one hundred dark fringes, the last one being at the outer edge of the curved piece of glass. The light being used has a wavelength of 654 nm in vacuum. What is the radius of the outermost dark ring in the pattern?

INTERSECTION

Physics & Space Science

Our ability to put probes into deep space and satellites into earth orbit provides the human race with a great opportunity to understand both the earth and our astronomical environment. By using orbiting telescopes and cameras that are sensitive to visible light and to infrared and ultraviolet radiation, scientists are able to gather tremendous amounts of information about the earth's surface. By using other telescopes in orbit that are sensitive to X-rays and gamma rays, astronomers are able to learn many things about the stars and interstellar matter that are otherwise invisible to us on earth.

Imagine that you are a Space Shuttle astronaut. Consider the physics involved in putting you in orbit, and the physics that you use once you are there. The Space Shuttle rocket engines exert 6.3 million pounds of thrust. By firing tremendous streams of hot gases out the bottom of the Shuttle's rockets, you are forced upward into space. The Shuttle has only enough fuel to put you into an orbit 300 miles above the earth's surface; it cannot carry enough fuel to take you to the moon.

The acceleration you feel getting into space is determined by Newton's second law, $\Sigma F = ma$. To maintain the height you reach, the Shuttle is put in orbit; otherwise, like a rock thrown straight up in the air, it would fall right back to the surface. As you move around the earth while in orbit, your potential and kinetic energies change. However, the sum of the two energies remains constant throughout the orbit. (Actually, your Shuttle continually loses energy owing to its colliding with air molecules even at that altitude, but the effect is negligible over the length of time you will be up there.)

The crew has three assignments: To examine the effects on agriculture of the prolonged drought that has occurred in the western United States, to study the sun's X-ray emissions, and to salvage a spy satellite.

Your primary diagnostic instrument for studying crops is an infrared camera. Everything stores and re-emits energy received from the sun, and living things also generate and emit their own energy. Some of this energy is emitted as infrared radiation. The health of trees and crops can be determined by the infrared radiation they are generating and emitting. When the infrared-sensitive film is developed on earth, scientists will be able to see which forests and crop regions are not thriving. With that information in hand, they will be able to provide vital information on how to reallocate water.

As your Shuttle phases over the Caribbean, you observe the characteristic spiral shape of a developing tropical storm. The spiral shape results from effects of the earth's rotation on regions of different pressure in the atmosphere, an application of basic physics principles. You photograph the storm and radio news of it to Houston.

Now one of your colleagues uses the Shuttle's mechanical arm to pull an X-ray telescope out of the cargo bay and to aim it at the sun. Although it appears yellow to our eyes, the sun not only emits all visible colors of light but also emits all kinds of nonvisible radiation, including radio waves, infrared radiation, ultraviolet radiation, X-rays, and gamma rays. To a telescope sensitive to X-rays, the sun does not appear at all like the uniform disk we normally see. It is splotchy as seen by means of X-rays, because this radiation leaks out through holes created in the sun's surface by magnetic fields. The distribution of X-rays you record will give astronomers important insights into the sun's internal activity.

A piece of space debris, pulled by the earth's gravity, flashes past the Shuttle. The earth's atmosphere creates enormous friction on the surface of the space rock (called a meteoroid) as it is pulled earthward. The friction generates heat, which causes the meteoroid to begin melting and shedding its outer layer. Staring out a window, you are treated to the sight of the burning meteor, as the meteoroid is called while it glows in our atmosphere. Without the protection of the heat tiles on the bottom of the Shuttle, the same thing would happen to your ship as you reentered the atmosphere. Material scientists developed the heat tiles by using the principles of basic physics and chemistry, and the perfection of the tiles was essential to reusable spacecraft.

The Hubble space telescope is one of the many artificial satellites that orbit the earth. Initially its utility was very limited due to construction flaws. These flaws were repaired in orbit in December, 1993, during a highly successful mission of the Space Shuttle. The astronauts are shown here working on the telescope in the Shuttle's bay. The repaired Hubble now provides astronomers with views of the universe that are unsurpassed in clarity of detail and richness of information.

Ahead is the rapidly rotating satellite you are to retrieve. Its gyroscopes have failed, and now it is useless because its antennae don't remain aimed at the earth; it cannot reliably receive commands sent to it. Retrieving and repairing it will save millions of dollars, compared with replacing it. You prepare for a space walk. You and a colleague will go out and spin the satellite down so that it can be safely retrieved in the Shuttle's cargo bay. The conservation of angular momentum makes this challenging, both in grabbing the satellite and in spinning it down. Any ideas how you'll do it?

Returning to earth, the Shuttle becomes one of the most expensive gliders in history. There is no room for mistakes, since the Shuttle's engines cannot be used to bring your ship around for a second try. The Shuttle's aerodynamics, state of the art navigation and communication, and outstanding piloting bring you down safely.

FOR INFORMATION ABOUT CAREERS IN SPACE SCIENCE

Professional Careers Sourcebook, K. M. Savage and C. A. Dorgan, eds., 1990, Gale Research, Inc., publisher, provides addresses and phone numbers for dozens of guides and other resources about careers in many space and related earth science fields.

Encyclopedia of Careers and Vocational Guidance, Vol. 2, 8th edition, William E. Hopke, ed., 1990, J. G. Ferguson, publisher, has an in-depth description of the working conditions astronauts, and space and related earth scientists encounter.

You can also write to the following, among many others listed in the books above:

- American Institute of Physics, 335 East 45th St., New York, NY 10017
- American Astronomical Society, 2000 Florida Ave., NW, No. 300, Washington, DC 20009
- NASA, Office of Educational Programs and Services, 400 Maryland Ave., Washington, DC 20025
- American Institute of Aeronautics and Astronautics, 370 L'Enfant Promenade, SW, Washington, DC 20024
- American Society of Civil Engineers, 345 East 47th St., New York, NY 10017

PART SIX

MODERN PHYSICS

Near the turn of the century, scientists believed they had a firm understanding of the physical aspect of our world. For more than two hundred years, Newton's laws had accurately described the motion of all types of objects, from projectiles to orbiting planets. An understanding of heat and sound had been established by using a kinetic theory of matter based on Newton's laws. And physicists had attained a clear picture of electricity and magnetism, as well as considerable insight into the nature of electromagnetic waves.

Yet all was not well, for there remained important phenomena and new experimental data that could not be explained with these so-called classical theories. The shortcomings of the classical theories turned out to be so serious that, during the first part of the twentieth century, physicists were forced to develop entirely new views of the universe. First came Einstein's theory of special relativity, which describes how the world appears when viewed at speeds near the speed of light. Then came quantum mechanics, which offers extraordinary insight into the microscopic realm of the atom, a world that consists of electrons, protons, neutrons, and molecules. The special theory of relativity and quantum mechanics form the foundation of what is now called "modern physics."

The first chapter in our discussion of modern physics deals with Einstein's theory of special relativity. We will see that our common-sense notions about space and time are radically altered. And special relativity predicts a surprising union between mass and energy, as expressed in Einstein's famous equation, $E_0 = mc^2$, where E_0 is the rest energy, m is the mass, and c is the speed of light. In the following two chapters, we will encounter some of the experiments that led up to the development of quantum mechanics. Then the "wave-particle duality" of nature will be discussed, according to which a wave can exhibit particle-like characteristics and a particle can exhibit wave-like characteristics. In addition, we will see how quantum mechanics describes the nature of the atom, how X-rays are produced, and how lasers operate. Finally, the remaining two chapters treat some of the more recent developments in modern physics, including nuclear physics and radioactivity, the biological effects of ionizing radiation, and elementary particles.

CHAPTER 28 SPECIAL RELATIVITY

Ask anyone to name the most famous scientist of the twentieth century and you are likely to get Albert Einstein (1879–1955) for an answer. Few individuals have become as well known as he, in recognition for the remarkable number, diversity, and fundamental significance of his scientific accomplishments. He is best known for the theory of special relativity, published in 1905 when he was 26 years old. Even today, this theory is a rich source of amazement, for it alters in fundamental ways many of our basic ideas about the physical universe. We will see that time no longer has a unique meaning in the aftermath of Einstein's theory. And neither does length. For example, due to relativistic effects, space travelers would make a high-speed journey to a distant planet in a much shorter time than that registered on the clocks left behind on earth. In addition, we will find that mass (inertia) and energy are not independent ideas as classical physics assumes. Instead, they are equivalent, with the result that the energy E_0 of a stationary object is related to its mass m by Einstein's famous equation, $E_0 = mc^2$, where c is the speed of light in a vacuum. Another surprising result of special relativity is that an object with mass cannot be accelerated to speeds at or beyond the speed of light, no matter how much force is applied. Thus, the speed of light represents the ultimate speed for any moving object that has mass. We will find that relativistic effects become significant when the speed of an object is an appreciable fraction of the speed of light. Special relativity also modifies our notions about how relative velocities are measured by different observers, ideas that were first presented in Section 3.4. All of these features of special relativity will be discussed in this chapter. But first, we need to review the idea of an inertial reference frame, since this idea plays such a fundamental role in the theory.

28.1 EVENTS AND INERTIAL REFERENCE FRAMES

The theory of special relativity deals with the way that an event is measured by observers who are moving relative to the event. An ***event,*** such as the launching of the space shuttle in Figure 28.1, is a physical "happening" that occurs at a certain place and time. In this drawing two observers are watching the lift-off, one standing on the earth and one seated in an airplane that is flying at a constant velocity relative to the earth. To record the event, each observer uses a ***reference frame*** that consists of a set of x, y, z axes (called a *coordinate system*) and a clock. The coordinate systems are used to establish where the event occurs, and the clocks are used to specify when it happens. Each observer is at rest relative to his own reference frame. However, the earth-based observer and the airborne observer are moving relative to each other and so are their respective reference frames.

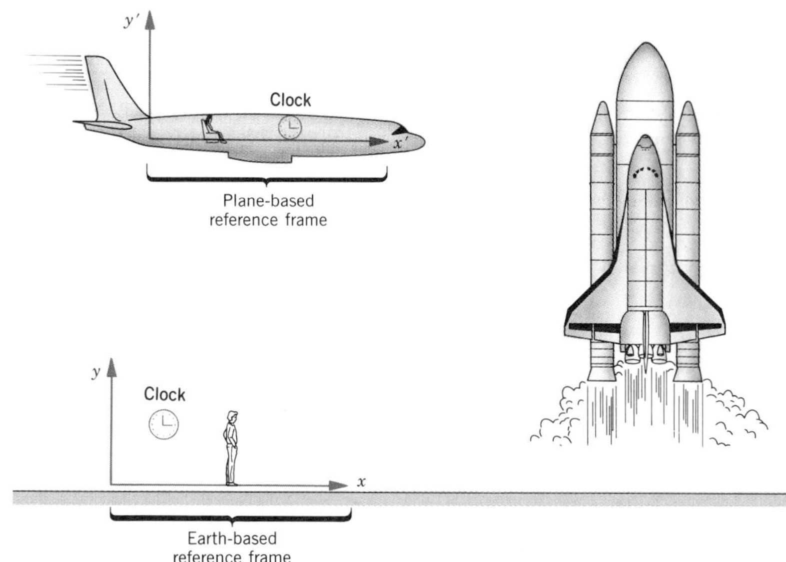

Figure 28.1 Using an earth-based reference frame, an observer standing on the earth records the location and time of an event (the lift-off). Likewise, an observer in the airplane uses a plane-based reference frame to describe the event.

The theory of special relativity deals with a "special" kind of reference frame, called an ***inertial reference frame.*** As Section 4.2 discusses, an inertial reference frame is one in which Newton's law of inertia is valid. That is, if the net force acting on a body is zero, the body either remains at rest or moves at a constant velocity. In other words, the acceleration of such a body is zero when measured in an inertial reference frame. Rotating and otherwise accelerating reference frames are not inertial reference frames. The earth-based reference frame in Figure 28.1 is not quite an inertial frame, because it is subjected to centripetal accelerations as the earth spins on its axis and revolves around the sun. In most situations, however, the effects of these accelerations are small, so we can consider an earth-based reference frame to be an inertial one. To the extent that the earth-based reference frame is an inertial frame, so is the plane-based reference frame, for the plane moves at a constant velocity relative to the earth. The next section discusses why inertial reference frames are important in relativity.

28.2 THE POSTULATES OF SPECIAL RELATIVITY

Einstein built his theory of special relativity on two fundamental assumptions or postulates about the way nature behaves.

> ***The Postulates of Special Relativity***
>
> 1. **The Relativity Postulate.** The laws of physics are the same in every inertial reference frame.
> 2. **The Speed of Light Postulate.** The speed of light in a vacuum, measured in any inertial reference frame, always has the same value of c, no matter how fast the source of light and the observer are moving relative to each other.

It is not too difficult to accept the relativity postulate. For instance, in Figure 28.1 each observer, using his own inertial reference frame, can make measurements on the motion of the space shuttle. The relativity postulate asserts that both observers find their data to be consistent with Newton's laws of motion. Similarly, both observers find that the behavior of the electronics on board the space shuttle is described by the laws of electromagnetism, such as Faraday's law of electromagnetic induction. According to the relativity postulate, ***any inertial reference frame is as good as any other for expressing the laws of physics, because the laws are the same in all such frames.*** In other words, with regard to inertial reference frames, nature does not play favorites.

Since the laws of physics are the same in all inertial reference frames, there is no experiment that can distinguish between an inertial frame that is at rest and one that is moving at a constant velocity. When you are seated on the aircraft in Figure 28.1, for instance, it is just as valid to say that you are at rest and the earth is moving as it is to say the reverse. It is not possible to single out one particular inertial reference frame as being at "absolute rest." Consequently, it is meaningless to talk about the "absolute velocity" of an object—that is, its velocity measured relative to a reference frame at "absolute rest." Thus, the earth moves relative to the sun, which itself moves relative to the center of our galaxy. And the galaxy moves relative to other galaxies, and so on. According to Einstein, only the relative velocity between objects, not their absolute velocities, can be measured and is physically meaningful.

While the relativity postulate seems plausible, the speed of light postulate defies common sense. For instance, Figure 28.2 illustrates a person standing on the bed of a truck that is moving at a constant speed of 15 m/s relative to the ground. Now, suppose you are standing on the ground and the person on the truck shines a flashlight at you. The person on the truck observes the speed of light to be c. What do you measure for the speed of light? You might guess that the

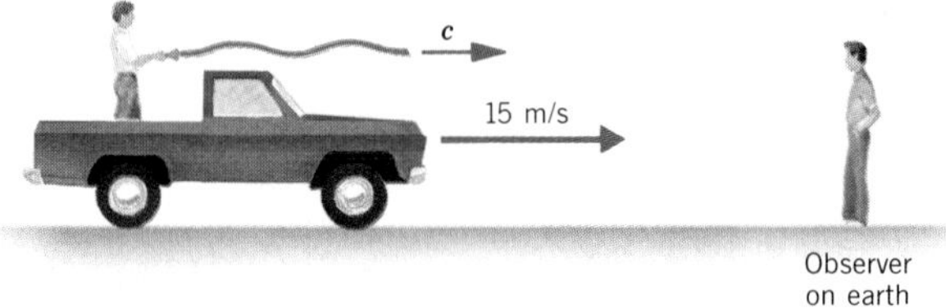

Figure 28.2 Both the person on the truck and the observer on the earth measure the speed of the light to be c, regardless of the speed of the truck.

speed of light would be $c + 15$ m/s. However, this guess is inconsistent with the speed of light postulate, which states that all observers in inertial reference frames measure the speed of light to be c—nothing more, nothing less. Therefore, you must also measure the speed of light to be c, the same as that measured by the person on the truck. According to the speed of light postulate, the fact that the flashlight is moving toward you has no influence whatsoever on the speed of the light approaching you. This property of light, although surprising, has been verified many times by experiment.

Since waves, such as water waves and sound waves, require a medium through which to propagate, it was natural for scientists before Einstein to assume that light did too. This hypothetical medium was called the *luminiferous ether* and was assumed to fill all of space. Furthermore, it was believed that light traveled at the speed c only when measured with respect to the ether. According to this view, an observer moving relative to the ether would measure a speed for light that was slower or faster than c, depending on whether the observer moved with or against the light. During the years 1883–1887, however, the American scientists A. A. Michelson and E. W. Morley carried out a series of famous experiments whose results were not consistent with the ether theory. Their results indicated that the speed of light is indeed the same in all inertial reference frames and does not depend on the motion of the observer relative to the source of the light. These experiments, and others, led eventually to the demise of the ether theory and the acceptance of the theory of special relativity.

28.3 THE RELATIVITY OF TIME: TIME DILATION

TIME DILATION

Common experience indicates that time passes just as quickly for a person standing on the ground as it does for an astronaut in a spacecraft. In contrast, the theory of special relativity reveals that the person on the ground measures time passing more slowly for the astronaut than for himself. We can see how this curious effect arises with the help of the clock illustrated in Figure 28.3. This clock uses a pulse of light to mark time. A short pulse of light is emitted by a light source, reflects from a mirror, and then strikes a detector that is situated next to the source. Each time a pulse reaches the detector, a "tick" registers on the chart recorder, another short pulse of light is emitted, and the cycle repeats. Thus, the time interval between successive "ticks" is marked by a beginning event (the firing of the light source) and an ending event (the pulse striking the detector). The source and detector are so close to each other that the two events can be considered to occur at the same location.

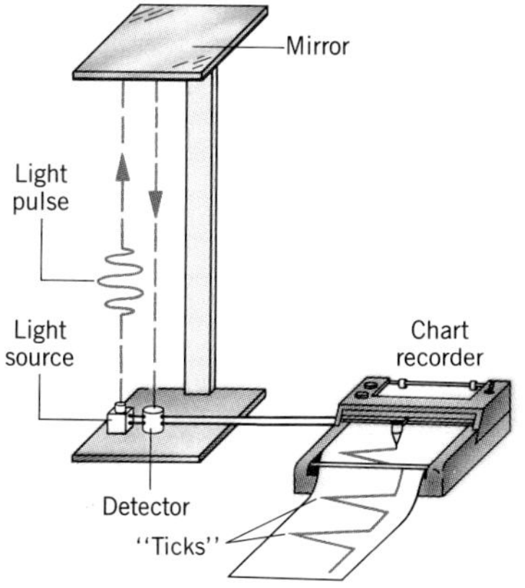

Figure 28.3 A light clock.

Suppose two identical clocks are built. One is kept on earth, and the other is placed aboard a spacecraft that travels at a constant velocity relative to the earth. The astronaut is at rest with respect to the clock on the spacecraft and, therefore, sees the light pulse move on the up/down path shown in Figure 28.4*a*. According to the astronaut, the time interval Δt_0 required for the light to follow this path is the distance $2D$ divided by the speed of light c; $\Delta t_0 = 2D/c$. To the astronaut, Δt_0 is the time interval between the "ticks" of the spacecraft clock, that is, the time interval between the beginning and ending events of the clock. An earth-based observer, however, does *not* measure Δt_0 as the time interval between these two events. Since the spacecraft is moving, the earth-based observer sees the light

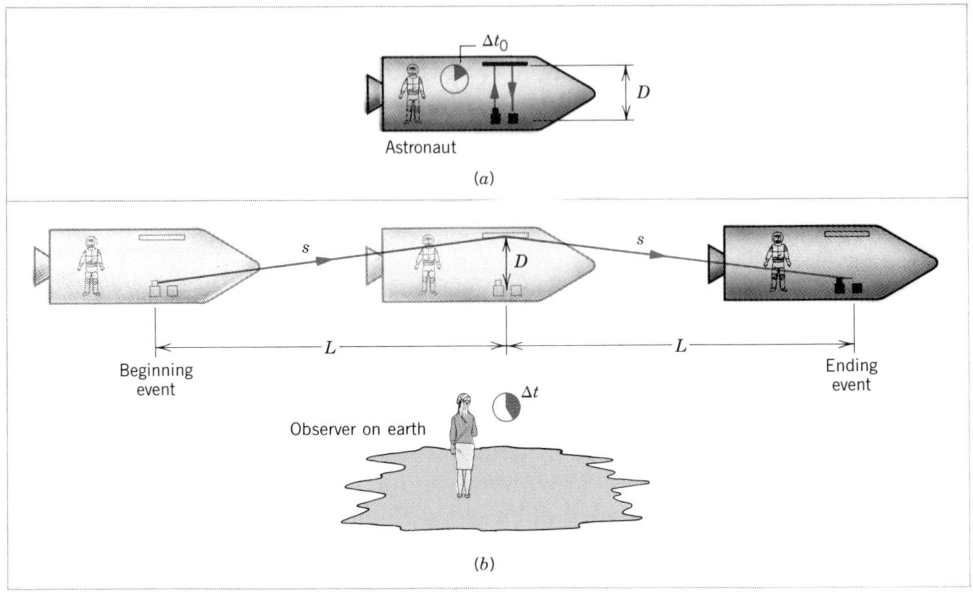

Figure 28.4 (*a*) The astronaut measures the time interval Δt_0 between successive "ticks" of his light clock. (*b*) An observer on earth watches the astronaut's clock and sees the light pulse travel a greater distance between "ticks" than it does in part *a*. Consequently, the earth-based observer measures a time interval Δt between "ticks" that is greater than Δt_0.

pulse follow the diagonal path shown in red in part *b* of the drawing. This diagonal path is longer than the up/down path seen by the astronaut. But light travels at the *same speed c* for both observers, in accord with the speed of light postulate. Therefore, the earth-based observer measures a time interval Δt between the two events that is *greater* than the time interval Δt_0 measured by the astronaut. In other words, the earth-based observer, using his own earth-based clock to measure the performance of the astronaut's clock, finds that the astronaut's clock runs slowly. This result of the theory of special relativity is known as ***time dilation.*** (To *dilate* means to expand, and the time interval Δt is "expanded" relative to Δt_0.)

The time interval Δt that the earth-based observer measures in Figure 28.4*b* can be determined as follows. While the light pulse travels from the source to the detector, the spacecraft moves a distance $2L = v\,\Delta t$ to the right, where v is the speed of the spacecraft relative to the earth. From the drawing it can be seen that the light pulse travels a total diagonal distance of $2s$ during the time interval Δt. Applying the Pythagorean theorem, we find that

$$2s = 2\sqrt{D^2 + L^2} = 2\sqrt{D^2 + \left(\frac{v\,\Delta t}{2}\right)^2}$$

But the distance $2s$ is also equal to the speed of light times the time interval Δt, so $2s = c\,\Delta t$. Therefore,

$$c\,\Delta t = 2\sqrt{D^2 + \left(\frac{v\,\Delta t}{2}\right)^2}$$

Squaring this result and solving for Δt gives

$$\Delta t = \frac{2D}{c} \frac{1}{\sqrt{1 - \frac{v^2}{c^2}}}$$

But $2D/c = \Delta t_0$, the time interval between successive "ticks" of the spacecraft's clock as measured by the astronaut. With this substitution, the equation above can be expressed as

[Time dilation]

$$\Delta t = \frac{\Delta t_0}{\sqrt{1 - \frac{v^2}{c^2}}} \tag{28.1}$$

The symbols in this formula are summarized below:

Δt_0 = proper time interval between two events, as measured by an observer who is at rest with respect to the events and who views the events as occurring *at the same place*

Δt = time interval measured by an observer who is in motion with respect to the events and who views the events as occurring at *different places*

v = relative speed between the two observers

c = speed of light in a vacuum

For a speed v that is less than c, the term $\sqrt{1 - v^2/c^2}$ in Equation 28.1 is less than 1, and the dilated time interval Δt is greater than Δt_0. Example 1 illustrates this time dilation effect.

Example 1 Time Dilation

The spacecraft in Figure 28.4 is moving past the earth at a constant speed v that is 0.92 times the speed of light. Thus, $v = (0.92)(3.0 \times 10^8 \text{ m/s})$, which is often written as $v = 0.92c$. The astronaut measures the time interval between successive "ticks" of the spacecraft clock to be $\Delta t_0 = 1.0$ s. What is the time interval Δt that an earth observer measures between "ticks" of the astronaut's clock?

REASONING Since the clock on the spacecraft is moving relative to the earth observer, the earth observer measures a greater time interval Δt between "ticks" than does the astronaut, who is at rest relative to the clock. The dilated time interval Δt can be determined from the time dilation relation, Equation 28.1.

SOLUTION The dilated time interval is

$$\Delta t = \frac{\Delta t_0}{\sqrt{1 - \frac{v^2}{c^2}}} = \frac{1.0 \text{ s}}{\sqrt{1 - \left(\frac{0.92c}{c}\right)^2}} = \boxed{2.6 \text{ s}}$$

From the point of view of the earth-based observer, the astronaut is using a clock that is running slowly, for the earth-based observer measures a time between "ticks" that is longer (2.6 s) than what the astronaut measures (1.0 s).

The fact that each of these clocks has a different time is allowed by special relativity, but only if each one moves at a different speed relative to the others.

Example 1 shows that time dilation is appreciable when the speed v is comparable to the speed of light. The speeds we experience in everyday life, however, are far too small for time dilation to be noticeable. For instance, for a clock aboard a jetliner traveling at $v = 0.000\ 000\ 75c$ (about 500 miles per hour), the time intervals Δt and Δt_0 in Example 1 would differ by only 2.8×10^{-13} s. This small difference in time means that 110 000 years would have to pass on an earth-based clock before the clock on the jetliner would lose 1 second.

PROPER TIME INTERVAL

In Figure 28.4 both the astronaut and the person standing on the earth are measuring the time interval between a beginning event (the firing of the light source) and an ending event (the light pulse striking the detector). For the astronaut, who is at rest with respect to the light clock, the two events occur at the same location. (Remember, we are assuming that the light source and detector are so close together that they are considered to be at the same place.) Being at rest with respect to a clock is the usual or "proper" situation, so the time interval Δt_0 measured by the astronaut is called the ***proper time interval.*** In general, the proper time interval Δt_0 between two events is the time interval measured by an observer who is at rest relative to the events and sees the events at the *same location* in space. On the other hand, the earth-based observer does not see the two events occurring at the same location in space, since the spacecraft is in motion. The time interval Δt that the earth-based observer measures is, therefore, not a proper time interval in the sense that we have defined it.

To understand situations involving time dilation, it is essential to distinguish between Δt_0 and Δt. In such situations it is helpful if one first identifies the two events that define the time interval. These events may be something other than the firing of a light source and the light pulse striking a detector. Then determine the reference frame in which the two events occur at the same place. An observer at rest in this reference frame measures the proper time interval Δt_0.

THE PHYSICS OF . . .

space travel and special relativity.

SPACE TRAVEL

One of the intriguing aspects of time dilation occurs in conjunction with space travel. Since enormous distances are involved, travel to even the closest star outside our solar system would take a long time. However, as the following example shows, the time for such a trip can be considerably less for the passengers than one might guess.

Example 2 Space Travel

Alpha Centauri, a nearby star in our galaxy, is 4.3 light-years away. This means that, as measured by a person on earth, it would take light 4.3 years to reach this star. If a rocket leaves for Alpha Centauri at a speed of $v = 0.95c$ relative to the earth, by how much will the passengers have aged, according to their own clock, when they reach their destination?

REASONING The two events in this problem are the departure from earth and the arrival at Alpha Centauri. At departure, earth is just outside the spaceship. Upon arrival at the destination, Alpha Centauri is just outside. Therefore, relative to the passengers, the two events occur at the same place, namely, just outside the spaceship. Thus, the passengers measure the proper time interval Δt_0 on their clock, and it is this interval that we must find. For a person left behind on earth, the events occur at *different places,* so such a person measures the dilated time interval Δt rather than the proper time interval. Since it takes 4.3 years for light to traverse the distance and the rocket is moving at $v = 0.95c$ relative to the earth, a person on earth measures the dilated time interval to be $\Delta t = (4.3 \text{ years})/0.95 = 4.5$ years. This value can be used with the time-dilation equation to find the proper time interval Δt_0.

SOLUTION Using the time-dilation equation, we find that the proper time interval by which the passengers judge their own aging is

$$\Delta t_0 = \Delta t \sqrt{1 - \frac{v^2}{c^2}} = (4.5 \text{ years}) \sqrt{1 - \left(\frac{0.95c}{c}\right)^2} = \boxed{1.4 \text{ years}}$$

Thus, the people aboard the rocket have aged by only 1.4 years when they reach Alpha Centauri, and not the 4.5 years an earthbound observer has calculated.

PROBLEM SOLVING INSIGHT

In dealing with time dilation, decide which time interval is the proper time interval according to the following two steps: (1) Identify the two events that define the time interval. (2) Determine the reference frame in which the two events occur at the same place; an observer at rest in this frame measures the proper time interval Δt_0.

VERIFICATION OF TIME DILATION

A striking confirmation of time dilation was achieved in 1971 by an experiment carried out by J. C. Hafele and R. E. Keating.* They transported very precise cesium-beam atomic clocks around the world on commercial jets. Since the speed of a jet plane is considerably less than c, the time-dilation effect is extremely small. However, the atomic clocks were accurate to about $\pm 10^{-9}$ s, so the effect could be measured. The clocks were in the air for 45 hours, and their times were compared to reference atomic clocks kept on earth. The experimental results revealed that, within experimental error, the readings on the clocks on board the planes were different from those on earth by an amount that agreed with the prediction of relativity.

Time dilation has also been confirmed with experiments using subatomic particles called *muons.* These particles are created high in the atmosphere, at altitudes of about 10 000 m. When at rest, muons are short-lived, existing for a time of about 2.2×10^{-6} s before disintegrating into other particles. With such a short lifetime, these particles could never make it down to the earth's surface, even if they traveled close to the speed of light. However, *a large number of muons do reach the earth.* The only way they can do so is to live longer because of time dilation, as Example 3 illustrates.

Mission specialist Mae Jemison working aboard the space shuttle Endeavour.

* J. C. Hafele and R. E. Keating, Around the world atomic clocks: relativistic time gains observed. *Science,* Vol. 168, 14 July 1972.

Example 3 The Lifetime of a Muon

The average lifetime of a muon at rest is 2.2×10^{-6} s. A muon created in the upper atmosphere, thousands of meters above sea level, travels toward the earth at a speed of $v = 0.998c$. Find, on the average, (a) how long a muon lives according to an observer on earth, and (b) how far the muon travels before disintegrating.

REASONING The two events of interest are the generation and subsequent disintegration of the muon. When the muon is at rest, these events occur at the same place, so the muon's average (at rest) lifetime of 2.2×10^{-6} s is a proper time interval Δt_0. When the muon moves at a speed $v = 0.998c$ relative to the earth, an observer on the earth measures a dilated lifetime Δt that is given by Equation 28.1. The average distance x traveled by a muon, as measured by an earth observer, is equal to the muon's speed times the dilated time interval.

SOLUTION

(a) The observer on earth measures a dilated lifetime given by

$$\Delta t = \frac{\Delta t_0}{\sqrt{1 - \frac{v^2}{c^2}}} = \frac{2.2 \times 10^{-6}\ \text{s}}{\sqrt{1 - \left(\frac{0.998c}{c}\right)^2}} = \boxed{35 \times 10^{-6}\ \text{s}} \qquad (28.1)$$

PROBLEM SOLVING INSIGHT

The proper time interval Δt_0 is always shorter than the dilated time interval Δt.

(b) The distance traveled by the muon before it disintegrates is

$$x = v\ \Delta t = (0.998)(3.00 \times 10^8\ \text{m/s})(35 \times 10^{-6}\ \text{s}) = \boxed{1.0 \times 10^4\ \text{m}}$$

Thus, the dilated, or extended, lifetime provides sufficient time for the muon to reach the surface of the earth. If its lifetime were only 2.2×10^{-6} s, a muon would travel only 660 m before disintegrating and could never reach the earth.

28.4 THE RELATIVITY OF LENGTH: LENGTH CONTRACTION

Because of time dilation, observers moving at a constant velocity relative to each other measure different time intervals between two events. For instance, Example 2 in the previous section illustrates that a trip from earth to Alpha Centauri at a speed of $v = 0.95c$ takes 4.5 years according to a clock on earth, but only 1.4 years according to a clock in the rocket. These two times differ by the factor $\sqrt{1 - v^2/c^2}$. Since the times for the trip are different, one might ask if the observers measure different distances between earth and Alpha Centauri. The answer, according to

To space travelers heading toward a distant galaxy, such as the Spiral Galaxy shown here, the distance is not as great as observers on earth measure it to be, because of the effects of length contraction in special relativity.

special relativity, is yes. After all, both the earth-based observer and the rocket passenger agree that the relative speed between the rocket and earth is $v = 0.95c$. Since speed is distance divided by time and the time is different for the two observers, it follows that the distances must also be different, if the relative speed is to be the same for both individuals. Thus, the earth observer determines the distance to Alpha Centauri to be $L_0 = v\ \Delta t = (0.95c)(4.5\ \text{years}) = 4.3$ light-years. On the other hand, a passenger aboard the rocket finds the distance is only $L = v\ \Delta t_0 = (0.95c)(1.4\ \text{years}) = 1.3$ light-years. The passenger, measuring the shorter time, also measures the shorter distance. This shortening of the distance between two points is one example of a phenomenon known as ***length contraction.***

The relation between the distances measured by two observers in relative motion at a constant velocity can be obtained with the aid of Figure 28.5. Part *a* of the drawing shows the situation from the point of view of the earth-based observer. This person measures the time of the trip to be Δt, the distance to be L_0, and the relative speed of the rocket to be $v = L_0/\Delta t$. Part *b* of the drawing presents the point of view of the passenger, for whom the rocket is at rest, and the earth and Alpha Centauri appear to move by at a speed v. The passenger determines the distance of the trip to be L, the time to be Δt_0, and the relative speed to be $v = L/\Delta t_0$. Since the relative speed computed by the passenger equals that computed by the earth-based observer, it follows that $v = L/\Delta t_0 = L_0/\Delta t$. Using this result and the time-dilation equation, Equation 28.1, we obtain the following relation between L and L_0:

[**Length contraction**]

$$L = L_0 \sqrt{1 - \frac{v^2}{c^2}} \qquad (28.2)$$

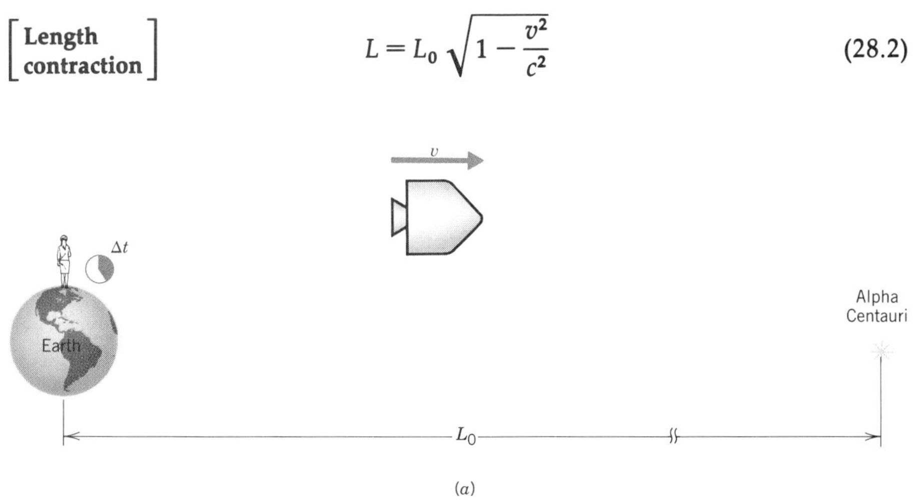

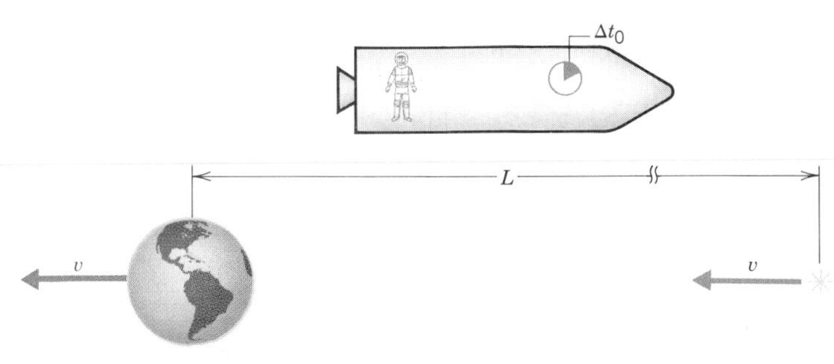

Figure 28.5 (*a*) As measured by an observer on the earth, the distance to Alpha Centauri is L_0 and the time required to make the trip is Δt. (*b*) According to the passenger on the spacecraft, the earth and Alpha Centauri move with speed v relative to the craft. The passenger measures the distance and time of the trip to be L and Δt_0, respectively, both quantities less than those in part *a*.

The length L_0 is called the ***proper length;*** it is the length (or distance) between two points *as measured by an observer at rest with respect to them.* Since v is less than c, the term $\sqrt{1 - v^2/c^2}$ is less than 1, and L is less than L_0. It is important to note that this length contraction occurs only along the direction of the motion. Those dimensions that are perpendicular to the motion are not shortened, as the next example discusses.

Example 4 The Contraction of a Meter Stick

An astronaut, using a meter stick that is at rest relative to a cylindrical spacecraft, measures the length and diameter of the spacecraft to be 82 and 21 m, respectively. The spacecraft moves with a constant speed of $v = 0.95c$ relative to the earth, as in Figure 28.5. What are the dimensions of the spacecraft, as measured by an observer on earth?

REASONING The length of 82 m is a proper length L_0, since it is measured using a meter stick that is at rest relative to the spacecraft. The length L measured by the observer on earth can be determined from the length-contraction formula, Equation 28.2. On the other hand, the diameter of the spacecraft is perpendicular to the motion, so the earth observer does not measure any change in the diameter.

PROBLEM SOLVING INSIGHT

The proper length L_0 is always larger than the contracted length L.

SOLUTION The length L of the spacecraft, as measured by the observer on earth, is

$$L = L_0\sqrt{1 - \frac{v^2}{c^2}} = (82\text{ m})\sqrt{1 - \left(\frac{0.95c}{c}\right)^2} = \boxed{26\text{ m}}$$

Both the astronaut and the observer on earth measure the same value for the diameter of the spacecraft: $\boxed{\text{Diameter} = 21\text{ m}}$. Figure 28.5*a* shows the size of the spacecraft as measured by the earth observer, while part *b* shows the size measured by the astronaut.

When dealing with relativistic effects we need to distinguish carefully between the criteria for the proper time interval and the proper length. The proper time interval Δt_0 between two events is the time interval measured by an observer who is at rest relative to the events and who sees them occurring at the *same place.* All other moving inertial observers will measure a larger value for this time interval. The proper length L_0 of an object is the length measured by an observer who is *at rest* with respect to the object. All other moving inertial observers will measure a shorter value for this length. The observer who measures the proper time interval may not be the same one who measures the proper length. For instance, Figure 28.5 shows that the astronaut measures the proper time interval Δt_0 for the trip between earth and Alpha Centauri, while the earth-based observer measures the proper length (or distance) ΔL_0 for the trip.

It should be emphasized that the word "proper" in the phrases proper time and proper length does *not* mean that these quantities are the correct or preferred quantities in any absolute sense. If this were so, the observer measuring these quantities would be using a preferred reference frame for making the measurement, a situation that is prohibited by the relativity postulate. According to this postulate, all inertial reference frames are equivalent. Therefore, when two observers are moving relative to each other at a constant velocity, each measures the other person's clock to run more slowly than his own, and each measures the other person's length to be contracted.

28.5 RELATIVISTIC MOMENTUM

Thus far we have discussed how time intervals and distances between two events are measured by observers moving at a constant velocity relative to each other. The theory of special relativity also alters our ideas about momentum and energy.

Recall from Chapter 7 that when two or more objects interact, the principle of conservation of linear momentum applies if the system of objects is isolated. (An isolated system is one in which the sum of the external forces acting on the objects is zero.) This principle states that the total linear momentum of an isolated system remains constant at all times. The conservation of linear momentum is a law of physics and, in accord with the relativity postulate, is valid in all inertial reference frames. That is, when the total linear momentum is conserved in one inertial reference frame, it is conserved in all inertial reference frames.

As an example of momentum conservation, suppose several people are watching two billiard balls collide on a frictionless pool table. One person is standing next to the pool table, and the other is moving past the table with a constant velocity. Since the two balls constitute an isolated system, the relativity postulate requires that both observers must find the total linear momentum of the two-ball system to be the same before, during, and after the collision. For this kind of situation in Chapter 7 (see Example 8 there), we defined the linear momentum $\mathbf{p}$ of an object to be the product of its mass m and velocity $\mathbf{v}$. As a result, the magnitude of the momentum was $p = mv$. As long as the speed of an object is considerably smaller than the speed of light, this definition is adequate. However, when the speed approaches the speed of light, an analysis of the collision shows that the total linear momentum is not conserved in all inertial reference frames if one defines linear momentum as the product of mass and velocity. In order to preserve the conservation of linear momentum, it is necessary to modify this definition as shown below in Equation 28.3. The momentum given by this expression, which is valid for all speeds, is called the ***relativistic momentum*** and is conserved in all inertial reference frames.

[Relativistic momentum]

$$p = \frac{mv}{\sqrt{1 - \frac{v^2}{c^2}}} \tag{28.3}$$

We see that the relativistic momentum differs from the nonrelativistic momentum (mv) by the same factor $\sqrt{1 - v^2/c^2}$ that occurs in the time-dilation and length-contraction equations. Since this factor is always less than 1 and occurs in the denominator of Equation 28.3, the relativistic momentum is always larger than the nonrelativistic momentum. To illustrate how the two momenta differ as the speed of an object increases, Figure 28.6 shows a plot of the ratio of the relativistic momentum to the nonrelativistic momentum as a function of speed. According to Equation 28.3, this momentum ratio is just $1/\sqrt{1 - v^2/c^2}$. The graph shows that for speeds attained by ordinary objects, such as cars and planes, the relativistic and nonrelativistic momenta are almost equal because their ratio is nearly 1. Thus, at speeds much less than the speed of light, either the nonrelativistic momentum or the relativistic momentum can be used to describe collisions. On the other hand, when the speed of the object becomes comparable to the speed of light, the relativistic momentum becomes significantly greater than the nonrelativistic momentum, and the relativistic momentum must be used. Example 5 deals with the relativistic momentum of an electron traveling close to the speed of light.

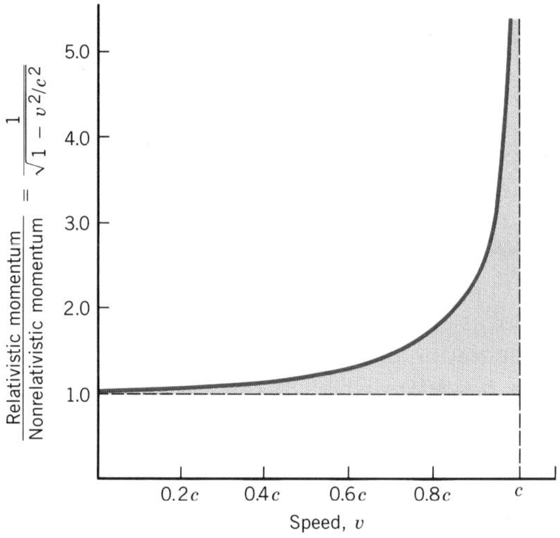

Figure 28.6 This graph shows how the ratio of the relativistic momentum to the nonrelativistic momentum increases as the speed of an object approaches the speed of light.

Figure 28.7 The Stanford three-kilometer linear accelerator accelerates electrons to nearly the speed of light.

Example 5 The Relativistic Momentum of a High-Speed Electron

The particle accelerator at Stanford University (Figure 28.7) is three kilometers long and accelerates electrons to a speed of 0.999 999 999 7c, which is very nearly equal to the speed of light. Find the relativistic momentum of an electron emerging from the accelerator and compare its value with the nonrelativistic value.

REASONING AND SOLUTION The relativistic momentum of the electron can be obtained from Equation 28.3 if we recall that the mass of an electron is $m = 9.11 \times 10^{-31}$ kg:

$$p = \frac{mv}{\sqrt{1 - \frac{v^2}{c^2}}} = \frac{(9.11 \times 10^{-31}\text{ kg})(0.999\,999\,999\,7c)}{\sqrt{1 - \frac{(0.999\,999\,999\,7c)^2}{c^2}}} = \boxed{1 \times 10^{-17}\text{ kg}\cdot\text{m/s}}$$

This value for the momentum agrees with that measured experimentally when the electrons are deflected by a magnetic field as they emerge from the accelerator. The relativistic momentum is greater than the nonrelativistic momentum by a factor of

$$\frac{1}{\sqrt{1 - \frac{v^2}{c^2}}} = \frac{1}{\sqrt{1 - \frac{(0.999\,999\,999\,7c)^2}{c^2}}} = \boxed{4 \times 10^4}$$

28.6 THE EQUIVALENCE OF MASS AND ENERGY

THE TOTAL ENERGY OF AN OBJECT

One of the most astonishing results of special relativity is that mass and energy are equivalent, in the sense that a gain or loss of mass can be regarded equally well as a gain or loss of energy. Consider, for example, an object of mass m traveling at a speed v. Einstein showed that the ***total energy*** E of the moving object is related to its mass and speed by the following relation:

[Total energy of an object]

$$E = \frac{mc^2}{\sqrt{1 - \frac{v^2}{c^2}}} \tag{28.4}$$

To gain some understanding of Equation 28.4, consider the special case when the object is at rest. When $v = 0$, the total energy is called the *rest energy* E_0, and Equation 28.4 reduces to Einstein's now-famous equation:

[Rest energy of an object when $v = 0$]

$$E_0 = mc^2 \tag{28.5}$$

The rest energy represents the energy equivalent of the mass of an object at rest. As Example 6 shows, even a small mass is equivalent to an enormous amount of energy.

Example 6 The Energy Equivalent of a Golf Ball

A 0.046-kg golf ball is lying on the green. (a) Find the rest energy of the golf ball. (b) If this rest energy were used to operate a 75-W light bulb, for how many years could the bulb stay on?

REASONING The rest energy E_0 that is equivalent to the mass m of the golf ball is found from the relation $E_0 = mc^2$. The 75-W light bulb consumes 75 J of energy per second. If the entire rest energy of the golf ball is available to keep the light bulb burning, the bulb could stay on for a time equal to the rest energy divided by the power of the light bulb.

SOLUTION

(a) The rest energy of the golf ball is

$$E_0 = mc^2 = (0.046 \text{ kg})(3.0 \times 10^8 \text{ m/s})^2 = \boxed{4.1 \times 10^{15} \text{ J}} \tag{28.5}$$

(b) This rest energy can keep the light bulb burning for a time t given by

$$t = \frac{\text{Rest energy}}{\text{Power}} = \frac{4.1 \times 10^{15} \text{ J}}{75 \text{ W}} = 5.5 \times 10^{13} \text{ s}$$

Since one year contains 3.2×10^7 s, we find $\boxed{t = 1.7 \times 10^6 \text{ yr}}$, or 1.7 million years!

When an object is accelerated from rest to a speed v, the object acquires kinetic energy in addition to its rest energy. The total energy E is the sum of the rest energy E_0 and the kinetic energy KE, or $E = E_0 + \text{KE}$. Using Equations 28.4 and 28.5, we can write the kinetic energy as

$$\text{KE} = E - E_0 = mc^2 \left(\frac{1}{\sqrt{1 - \frac{v^2}{c^2}}} - 1 \right) \tag{28.6}$$

This equation is the relativistically correct expression for the kinetic energy of an object of mass m moving at speed v; the kinetic energy is the difference between the object's total energy E and its rest energy E_0.

Equation 28.6 looks nothing like the kinetic energy expression introduced in Chapter 6, namely, $\text{KE} = \frac{1}{2}mv^2$. However, for speeds much less than the speed of light ($v \ll c$), the relativistic equation for the kinetic energy reduces to $\text{KE} = \frac{1}{2}mv^2$, as can be seen by using the binomial expansion* to represent the square root term in Equation 28.6:

$$\frac{1}{\sqrt{1-\dfrac{v^2}{c^2}}} = 1 + \frac{1}{2}\left(\frac{v^2}{c^2}\right) + \frac{3}{8}\left(\frac{v^2}{c^2}\right)^2 + \cdots$$

Suppose v is much smaller than c, say $v = 0.01c$. The second term in the binomial expansion has the value $\frac{1}{2}(v^2/c^2) = 5.0 \times 10^{-5}$, while the third term has the much smaller value $\frac{3}{8}(v^2/c^2)^2 = 3.8 \times 10^{-9}$. The additional terms are even smaller than the third term, so if $v \ll c$, we can neglect the third and additional terms in comparison with the first and second terms. Substituting the first two terms of the binomial expansion into Equation 28.6 gives

$$\text{KE} \approx mc^2\left(1 + \frac{1}{2}\frac{v^2}{c^2} - 1\right) = \tfrac{1}{2}mv^2$$

which is the familiar form for the kinetic energy. However, Equation 28.6 gives the correct kinetic energy for all speeds and must be used for speeds near the speed of light, as in Example 7.

* The binomial expansion states that $(1-x)^n = 1 - nx + n(n-1)x^2/2 + \cdots$. In our case, $x = v^2/c^2$ and $n = -1/2$.

Example 7 A High-Speed Electron

An electron ($m = 9.109 \times 10^{-31}$ kg) is accelerated from rest to a speed of $v = 0.9995c$ in a particle accelerator. Determine the electron's (a) rest energy, (b) total energy, and (c) kinetic energy.

REASONING AND SOLUTION

(a) The electron's rest energy is

$$E_0 = mc^2 = (9.109 \times 10^{-31}\ \text{kg})(2.998 \times 10^8\ \text{m/s})^2 = \boxed{8.187 \times 10^{-14}\ \text{J}} \quad (28.5)$$

Energy is often expressed in units of electron volts (eV). Since $1\ \text{eV} = 1.602 \times 10^{-19}$ J, the electron's rest energy is

$$8.187 \times 10^{-14}\ \text{J}\left(\frac{1\ \text{eV}}{1.602 \times 10^{-19}\ \text{J}}\right) = \boxed{5.11 \times 10^5\ \text{eV} \quad \text{or} \quad 0.511\ \text{MeV}}$$

(b) The total energy of an electron traveling at a speed of $v = 0.9995c$ is

$$E = \frac{mc^2}{\sqrt{1-\dfrac{v^2}{c^2}}} = \frac{(9.109 \times 10^{-31}\ \text{kg})(2.998 \times 10^8\ \text{m/s})^2}{\sqrt{1-\left(\dfrac{0.9995c}{c}\right)^2}}$$

$$= \boxed{2.59 \times 10^{-12}\ \text{J} \quad \text{or} \quad 16.2\ \text{MeV}} \quad (28.4)$$

(c) The kinetic energy is the difference between the total energy and the rest energy:

$$\text{KE} = E - E_0 = 2.59 \times 10^{-12}\ \text{J} - 8.2 \times 10^{-14}\ \text{J}$$

$$= \boxed{2.51 \times 10^{-12}\ \text{J} \quad \text{or} \quad 15.7\ \text{MeV}} \qquad (28.6)$$

For comparison, if the kinetic energy of the electron had been calculated from $\frac{1}{2}mv^2$, a value of 0.26 MeV would have been obtained, which is much smaller than the actual value of 15.7 MeV.

Since mass and energy are equivalent, any change in one is accompanied by a corresponding change in the other. For instance, life on earth is dependent on electromagnetic energy (light) from the sun. Because this energy is leaving the sun, there is a decrease in the sun's mass. Example 8 illustrates how to determine this decrease.

Example 8 The Sun Is Losing Mass

The sun radiates electromagnetic energy at the rate of 3.92×10^{26} W. (a) What is the change in the sun's mass during each second that it is radiating energy? (b) The mass of the sun is 1.99×10^{30} kg. What fraction of the sun's mass is lost during a human lifetime of 75 years?

REASONING Since power is energy per unit time, the amount of electromagnetic energy radiated during each second is 3.92×10^{26} J. Thus, during each second, the sun's rest energy decreases by this amount. The change ΔE_0 in the sun's rest energy is related to the change Δm in its mass by $\Delta E_0 = (\Delta m)c^2$, according to Equation 28.5.

SOLUTION

(a) For each second that the sun radiates energy, the change in its mass is

$$\Delta m = \frac{\Delta E_0}{c^2} = \frac{3.92 \times 10^{26}\ \text{J}}{(3.00 \times 10^8\ \text{m/s})^2} = \boxed{4.36 \times 10^9\ \text{kg}}$$

Over 4 billion kilograms of mass are lost by the sun during each second.

(b) The amount of mass lost by the sun in 75 years is

$$\Delta m = (4.36 \times 10^9\ \text{kg/s}) \left(\frac{3.16 \times 10^7\ \text{s}}{1\ \text{year}} \right) (75\ \text{years}) = 1.0 \times 10^{19}\ \text{kg}$$

While this is an enormous amount of mass, it represents only a tiny fraction of the sun's mass:

$$\frac{\Delta m}{m_{\text{sun}}} = \frac{1.0 \times 10^{19}\ \text{kg}}{1.99 \times 10^{30}\ \text{kg}} = \boxed{5.0 \times 10^{-12}}$$

Any change in the energy of a system causes a change in the mass of the system according to $\Delta E_0 = (\Delta m)c^2$. It does not matter whether the change in energy is due to a change in electromagnetic energy, potential energy, thermal energy, or so on. While any change in energy gives rise to a change in mass, in most instances the change in mass is too small to be detected. For instance, when 4186 J of heat is used to raise the temperature of 1 kg of water by 1 C°, the mass changes by only $\Delta m = \Delta E_0/c^2 = (4186\ \text{J})/(3.00 \times 10^8\ \text{m/s})^2 = 4.7 \times 10^{-14}$ kg. Conceptual Example 9 further illustrates how a change in the energy of an object leads to an equivalent change in its mass.

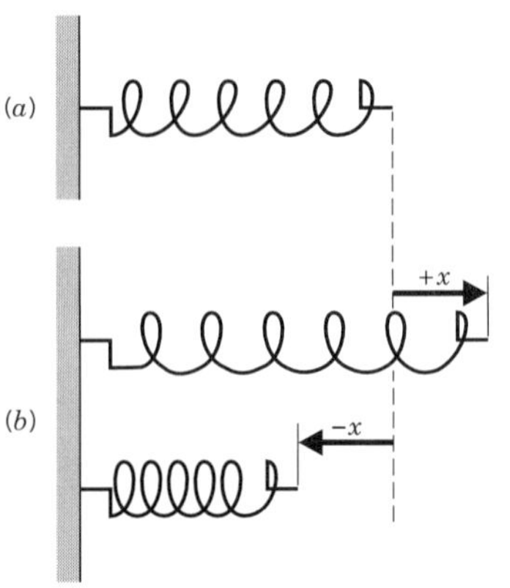

Figure 28.8 (*a*) This spring is unstrained and assumed to have no mass. (*b*) When the spring is either stretched or compressed by an amount x, it gains elastic potential energy and, hence, mass.

Conceptual Example 9 When Is a Massless Spring Not Massless?

Figure 28.8*a* shows a top view of a spring lying on a horizontal table. The spring is initially unstrained and assumed to be massless. Now, suppose that the spring is either stretched or compressed by an amount x from its unstrained length, as part *b* of the drawing shows. Is the mass of the spring still zero, or has it changed? And, if the mass has changed, does it change more when the spring is stretched or when it is compressed?

REASONING Whenever a spring is stretched or compressed, its elastic potential energy changes. In Section 10.5 we showed that the elastic potential energy of an ideal spring is equal to $\frac{1}{2}kx^2$, where k is the spring constant and x is the amount of stretch or compression. Consistent with the theory of special relativity, any change in the total energy of a system, including a change in the elastic potential energy, is equivalent to a change in the mass of the system. Thus, the mass of a strained spring is greater than that of an unstrained (and massless) spring. Furthermore, since the elastic potential energy depends on x^2, the increase in mass of the spring is the same whether it is compressed or stretched, provided x is the same in both cases.

While a strained spring does, in principle, have mass, the mass is exceedingly small. This is because the mass is equal to the elastic potential energy divided by c^2, and since c^2 is so large, the mass is correspondingly very small.

It is also possible to transform matter itself into other forms of energy, just as potential energy can be transformed into kinetic energy and vice versa. For example, the positron (see Section 31.4), created in high-energy accelerators, has the same mass as an electron, but an opposite electrical charge. If these two particles of matter collide, they are completely annihilated, and a burst of high-energy electromagnetic waves is produced. Thus, matter is transformed into electromagnetic waves, the energy of the electromagnetic waves being equal to the total energies of the two colliding particles.

The transformation of electromagnetic waves into matter also happens. In one experiment, an extremely high-energy electromagnetic wave, called a gamma ray (see Section 31.4), passes close to the nucleus of an atom. If the gamma ray has sufficient energy, it can create an electron and a positron. The gamma ray disappears, and the two particles of matter appear in its place. Except for picking up some momentum, the nearby nucleus remains unchanged. The process in which the gamma ray is transformed into two antiparticles is known as *pair production.*

THE SPEED OF LIGHT IS THE ULTIMATE SPEED

One of the important consequences of the theory of special relativity is that objects with mass cannot reach the speed of light. Thus, the speed of light represents the ultimate speed. To see that this speed limitation is a consequence of special relativity, consider Equation 28.6, which gives the kinetic energy of a moving object. As v approaches the speed of light c, the $\sqrt{1 - v^2/c^2}$ term in the denominator approaches zero. Hence, the kinetic energy becomes infinitely large. However, the work–energy theorem (Chapter 6) tells us that an infinite amount of work would have to be done to give the object an infinite kinetic energy. Since an infinite amount of work is not available, we are left with the conclusion that particles with mass cannot attain the speed of light.

28.7 THE RELATIVISTIC ADDITION OF VELOCITIES

The velocity of an object relative to an observer plays a central role in special relativity, for the effects on time, length, momentum, and energy depend on how fast the relative motion is, compared to the speed of light. To determine the velocity of an object relative to an observer, it is sometimes necessary to add two or more velocities together. For instance, Figure 28.9 illustrates a truck moving at a constant velocity of $v = 15$ m/s toward an observer standing on the earth. Suppose someone on the truck throws a baseball toward the observer at a velocity of $u' = 8.0$ m/s relative to the truck. We might conclude that the observer on earth sees the ball approaching at a velocity of $u = u' + v = 23$ m/s. Although this conclusion seems reasonable, careful measurements would show that this is not quite right. The equation $u = u' + v$ is not valid, because if the velocity of the truck were sufficiently close to the speed of light, the equation would predict that the observer on the earth could see the baseball moving at a velocity greater than the speed of light. This is an impossibility, since no object with a finite mass can move faster than the speed of light.

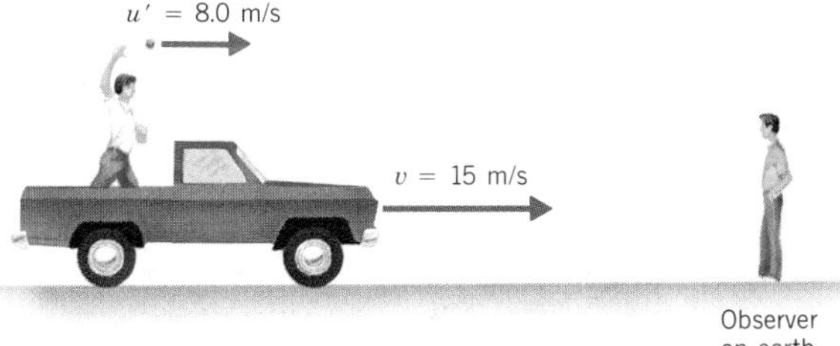

Figure 28.9 The truck is approaching the earth-based observer at a relative velocity of $v = 15$ m/s. The velocity of the baseball relative to the truck is $u' = 8.0$ m/s.

For the case where the truck and ball are moving along the same direction, the theory of special relativity states that the velocities are related according to the ***velocity-addition formula:***

[Velocity addition]

$$u = \frac{u' + v}{1 + \frac{u'v}{c^2}} \tag{28.7}$$

In this equation the symbols have the following meanings:

u = the velocity of the object as measured by the observer on the earth

u' = the velocity of the object measured by the person on the truck, which itself is moving at a velocity v relative to the earth

When the motion occurs along a straight line, the velocities in Equation 28.7 can have either positive or negative values, depending on whether they are directed along the positive or negative direction. For instance, in Figure 28.9, $u' = 8.0$ m/s and $v = 15$ m/s, assuming that the direction to the right is positive. Equation 28.7 differs from the nonrelativistic formula ($u = u' + v$) by the presence of the $u'v/c^2$ term in the denominator. When u' and v are small compared to c, the $u'v/c^2$ term is small compared to 1, so the velocity-addition formula reduces to $u \approx u' + v$. However, when either u' or v is comparable to c, the results can be quite different, as Example 10 illustrates.

Example 10 The Relativistic Addition of Velocities

Imagine a hypothetical situation in which the truck in Figure 28.9 is approaching the observer on the earth at a relative velocity of $v = 0.8c$. A person riding on the truck throws a baseball toward the observer at a velocity of $u' = 0.5c$ relative to the truck. At what velocity does the observer on earth see the ball approaching?

REASONING The observer on earth does *not* see the baseball approaching at $u = 0.5c + 0.8c = 1.3c$. This cannot be, because the velocity of the ball would then exceed the speed of light. The velocity-addition formula gives the correct velocity, which is less than the speed of light.

SOLUTION The earth-based observer sees the ball approaching with a velocity of

$$u = \frac{u' + v}{1 + \dfrac{u'v}{c^2}} = \frac{0.5c + 0.8c}{1 + \dfrac{(0.5c)(0.8c)}{c^2}} = \frac{1.3c}{1 + 0.4} = \boxed{0.93c}$$

Example 10 has discussed how the speed of a baseball is viewed by observers in different inertial reference frames. The next example deals with a similar situation, except that the baseball is replaced by the light of a laser beam.

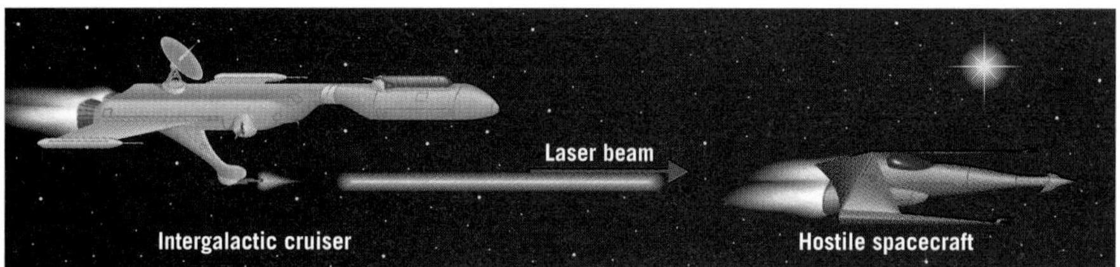

Figure 28.10 An intergalactic cruiser, closing in on a hostile spacecraft, fires a beam of laser light.

Conceptual Example 11 The Speed of a Laser Beam

An intergalactic cruiser is approaching a hostile spacecraft at a relative speed of $0.7c$, as Figure 28.10 shows. Both vehicles are moving at a constant velocity, and therefore each constitutes an inertial reference frame. The cruiser fires a beam of laser light at the enemy. At what speed do the renegades aboard the spacecraft see the laser beam (a) approach them and (b) move away from the cruiser?

REASONING

(a) According to the speed of light postulate, *all* observers in inertial reference frames measure the speed of light in a vacuum to be c. Thus, the renegades aboard the hostile spacecraft see the laser beam travel toward them at the speed of light, even though the beam is emitted from a source that itself is moving at seven-tenths the speed of light.

(b) The renegades aboard the spacecraft see the cruiser approach them at a relative speed of $0.7c$, and they also see the laser beam approach them at a relative speed of c. Since both these speeds are measured relative to the *same* inertial reference frame, namely, that of the renegades, they see the laser beam leaving the cruiser at a relative speed that is the difference between the two speeds, or $c - 0.7c = 0.3c$. The velocity-addition formula, Equation 28.7, is not applicable here, because both velocities are measured relative to the *same* inertial reference frame (the renegade's reference frame).

The velocity-addition formula is used only when the velocities are measured relative to different inertial reference frames.

Related Homework Material: *Question 12, Problem 36*

It is a straightforward matter to show that the velocity-addition formula is consistent with the speed of light postulate. Consider Figure 28.11, which shows a person riding on a truck and holding a flashlight. The speed of the light, as measured by this person, is $u' = c$. According to the observer standing on the earth, the speed of this light is given by the velocity-addition formula as

$$u = \frac{u' + v}{1 + \frac{u'v}{c^2}} = \frac{c + v}{1 + \frac{cv}{c^2}} = \frac{(c + v)c}{(c + v)} = c$$

Thus, the velocity-addition formula indicates that the observer on earth and the person on the truck both measure the speed of light to be c, independent of the relative velocity v between them. This is exactly what the speed of light postulate states.

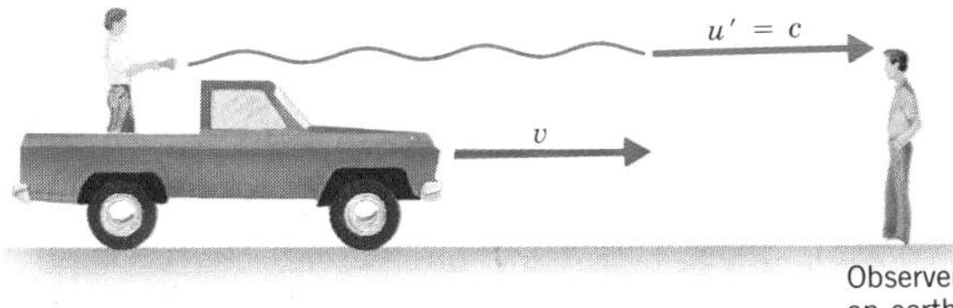

Figure 28.11 The speed of the light emitted by the flashlight is c relative to both the truck and the observer on earth.

INTEGRATION OF CONCEPTS

SPECIAL RELATIVITY AND NEWTONIAN MECHANICS

Einstein's theory of special relativity reveals a number of startling results that conflict with our traditional ideas about space and time. Among its most famous revelations are that moving clocks run slow, moving objects appear shortened, mass and energy are equivalent, and the speed of light is the ultimate speed for an object with mass. However surprising these predictions are, a large number of experiments are in complete agreement with the special theory of relativity, and today scientists accept it.

But what about our traditional ideas concerning space and time? We should remember that these ideas come from centuries of experience and experiments that support the concepts developed by Galileo, Newton, and others. These traditional concepts have been discussed in Part I of this text under the title of "Mechanics," or "Newtonian Mechanics," as it is often called. Newtonian mechanics, in contrast to the special theory of relativity, presumes that time and length are the same in different inertial reference frames. And Newtonian mechanics does not remotely hint that mass and energy are equivalent, or that the speed of light is the ultimate speed.

Is the theory provided by Newton and others wrong, then, because it fails to predict those things for which special relativity has become famous? Newtonian mechanics is certainly not wrong. It is just more limited in scope than is special relativity, which applies to all speeds between zero and the speed of light. The Newtonian view, in contrast, is valid only for speeds much smaller than the speed of light. In fact, in the limit of small speeds, both the Newtonian and the relativistic pictures of reality are in agreement. Thus, special relativity does not contradict the results of Newtonian mechanics, but generalizes them. Suppose, for example, that your car could attain speeds that are near the speed of light. As you accelerated toward the speed of light, you would observe the world around you change gradually. At the start, you would see the familiar Newtonian picture, where, for example, length is not contracted and time is not dilated. As you accelerated, this picture would change, until you saw a relativistic picture in which the lengths of objects speeding by the car are contracted and the moving clocks run slow.

SUMMARY

The special theory of relativity is based on two postulates. The **relativity postulate** states that the laws of physics are the same in every inertial reference frame. The **speed of light postulate** says that the speed of light in a vacuum, measured in any inertial reference frame, always has the same value of c, no matter how fast the source of light and the observer are moving relative to each other.

The **proper time interval** Δt_0 between two events is the time interval measured by an observer who is at rest relative to the events and views them occurring at the same location. A moving observer who does *not* see the two events occurring at the same location measures a dilated time interval Δt. The dilated time interval is greater than the proper time interval, according to the **time-dilation equation:** $\Delta t = \Delta t_0/\sqrt{1 - v^2/c^2}$. In this expression, v is the relative speed between the observer who measures Δt_0 and the observer who measures Δt.

The **proper length** L_0 between two points is the length measured by an observer who is at rest relative to the points. An observer moving with a relative speed v parallel to the line between the two points does not measure the proper length. Instead, such an observer measures a contracted length L given by the **length-contraction formula:** $L = L_0\sqrt{1 - v^2/c^2}$.

An object of mass m, moving with speed v, has a **relativistic momentum** given by $p = mv/\sqrt{1 - v^2/c^2}$.

Energy and mass are equivalent. The total energy E of an object of mass m, moving at speed v, is $E = mc^2/\sqrt{1 - v^2/c^2}$. The total energy of an object is the sum of its rest energy, $E_0 = mc^2$, and its kinetic energy KE: $E = E_0 + \text{KE}$. The kinetic energy is, therefore, $\text{KE} = E - E_0$. The speed of an object with mass cannot equal the speed of light, which is the **ultimate speed** for such an object.

When an object is moving with respect to a reference frame that itself is moving relative to an observer, the **velocity-addition formula** (Equation 28.7) gives the velocity of the object as measured by the observer.

QUESTIONS

1. The speed of light in water is c/n, where $n = 1.33$ is the index of refraction of water. Thus, the speed of light in water is less than c. Why doesn't this violate the speed of light postulate?

2. A baseball player at home plate hits a pop fly straight up (the beginning event) that is caught by the catcher at home plate (the ending event). Which of the following observers record the proper time interval between the two events: (a) a spectator sitting in the stands, (b) a spectator sitting on the couch and watching the game on TV, and (c) the third baseman running in to cover the play? Explain your answers.

3. The earth spins on its axis once each day. To a person viewing the earth from an inertial reference frame in space, which clock runs slower, a clock at the north pole or one at the equator? Why? (Ignore the orbital motion of the earth about the sun.)

4. Suppose you are standing at a railroad crossing, watching a train go by. (a) Both you and a passenger in the train are looking at a clock on the train. Which of you measures the proper time interval? (b) Who measures the proper length of the train car? (c) Who measures the proper distance between the railroad ties under the track? Justify your answers.

5. There are tables that list data for the various particles of matter that physicists have discovered. Often, such tables list the masses of the particles in units of energy, such as in MeV (million electron volts), rather than in kilograms. Why is this possible?

6. Why is it easier to accelerate an electron to a speed that is close to the speed of light, compared to accelerating a proton to the same speed?

7. Light travels in water at a speed of 2.26×10^8 m/s. Is it possible, in principle, for a particle that has mass to travel through water at a faster speed than does light? If so, why doesn't this violate the speed of light postulate?

8. Do two positive, electric charges separated by a finite distance have more mass than when they are infinitely far apart (assume that the charges remain stationary)? Provide a reason for your answer.

9. One system consists of two stationary charges, $+q$ and $-q$, separated by a distance r. Another system consists of two stationary charges, $+q$ and $+q$, that are also separated by a distance r. Which system of charges, if either, has the greater mass? Provide a reason for your answer.

10. Suppose a parallel plate capacitor is initially uncharged. The capacitor is charged up by removing electrons from one plate and placing them on the other plate. Does the uncharged or the charged capacitor, if either, have the greater mass? Why?

11. The speed limit on many interstate highways is 65 miles per hour. If the speed of light were 65 miles per hour, would you be able to drive at the speed limit? Give your reasoning.

12. Review Conceptual Example 11 as an aid in answering this question. A person is approaching you in a truck that is traveling very close to the speed of light. This person throws a baseball toward you. Relative to the truck, the ball is thrown with a speed nearly equal to the speed of light, so the person on the truck sees the baseball move away from the truck at a very high speed. Yet you see the baseball move away from the truck very slowly. Why? Use the velocity-addition formula to guide your thinking.

13. Which of the following quantities will two observers always measure to be the *same*, regardless of the relative velocity between the observers: (a) the time interval between two events; (b) the length of an object; (c) the speed of light; (d) the relative speed between the observers. In each case, give a reason for your answer.

14. If the speed of light were infinitely large instead of 3.0×10^8 m/s, would the effects of time dilation and length contraction be observable? Explain, using the equations presented in the text to support your reasoning.

PROBLEMS

Before doing any calculations involving time dilation or length contraction, it is useful to identify which observer measures the proper time interval Δt_0 or the proper length L_0.

Section 28.3 The Relativity of Time: Time Dilation

1. A law enforcement officer in an intergalactic "police car" turns on a red flashing light and sees it generate a flash every 1.5 s. A person on earth measures that the time between flashes is 2.5 s. How fast is the "police car" moving relative to the earth?

2. A particle known as a pion lives, on average, for a proper time of 2.6×10^{-8} s before breaking apart into other particles. How long does this particle live according to a laboratory observer if the particle moves past the observer at a speed of $0.67c$?

3. In 1986, Kristin Otto set a world's record for the 100-m freestyle. Suppose that this race had been monitored from a spaceship traveling at a speed of $0.900c$ relative to the earth and that the space travelers measured the time interval of the race to be 125.6 s. What was the time recorded on earth?

4. A spacecraft is passing through the solar system at a speed of $0.850c$ relative to the earth. What does the captain measure for the number of hours in an earth day if the spacecraft is (a) moving toward the earth or (b) away from the earth?

5. A radar antenna is rotating at an angular speed of 0.25 rad/s, as measured on earth. To an observer moving past the antenna at a speed of $0.80c$, what is its angular speed?

***6.** An astronaut travels at a speed of 7800 m/s relative to the earth, a speed that is very small compared to c. According to a clock on the earth, the trip lasts 15 days. Determine the *difference* (in seconds) between the time recorded by the earth clock and the astronaut's clock. (*Hint: When $v \ll c$, the following approximation is valid:* $\sqrt{1 - v^2/c^2} \approx 1 - \frac{1}{2}(v^2/c^2)$.)

***7.** A 5.00-kg object oscillates back and forth at the end of a spring whose spring constant is 49.3 N/m. An observer is traveling at a speed of 2.80×10^8 m/s relative to the fixed end of the spring. What does this observer measure for the period of oscillation?

****8.** A certain type of bacteria is known to double in number every 24.0 hours. Two cultures of these bacteria are prepared, each consisting initially of one bacterium. One culture is left on earth and the other placed on a rocket that travels at a speed of $0.866c$ relative to the earth. At a time when the earthbound culture has grown to 256 bacteria, how many bacteria are in the culture on the rocket?

Section 28.4 The Relativity of Length: Length Contraction

9. A UFO streaks across the sky at a speed of $0.90c$ relative to the earth. A person on earth determines the length of the UFO to be 230 m along the direction of its motion. What length does the person measure for the UFO when it lands?

10. The mean distance between earth and Jupiter is 6.29×10^{11} m. How fast would you have to travel in a spacecraft so the distance has decreased to 2.00×10^{11} m?

11. Our galaxy, the Milky Way, has a diameter (proper length) of about 1.2×10^5 light-years. As far as an astronaut is concerned, how long (in years) would it take to cross the Milky Way, if the speed of the spacecraft is $0.999\ 99c$?

12. A world record for the fastest jet-engine car was set by Richard Noble when his car attained an average speed of 283 m/s (633 mi/h) over a distance of 604 m. If the speed of light were 355 m/s, what distance would Noble have measured while driving the car?

13. Suppose the straight-line distance between New York and San Francisco is 4.2×10^6 m (neglecting the curvature of the earth). A UFO is flying between these two cities at a speed of $0.70c$ relative to the earth. What do the voyagers aboard the UFO measure for this distance?

14. Two spaceships A and B are exploring a new planet. Relative to this planet, spaceship A has a speed of $0.60c$, while spaceship B has a speed of $0.80c$. What is the ratio D_A/D_B of the values for the planet's diameter that each spaceship measures in a direction that is parallel to its motion?

15. Suppose you are traveling in space and pass a rectangular landing pad on a planet. Your spacecraft has a speed of $0.85c$ relative to the planet and moves in a direction parallel to the length of the pad. While moving, you measure the length to be 1800 m and the width to be 1500 m. What are the dimensions of the landing pad according to the engineer who built it?

***16.** As the drawing shows, a carpenter on a space station has constructed a 30.0° ramp. A rocket moves past the space station with a relative speed of $0.850c$ in a direction parallel to side x. What does a person aboard the rocket measure for the angle of the ramp?

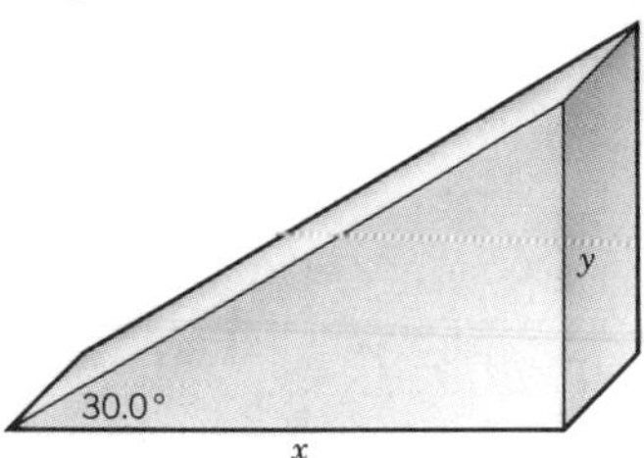

****17.** A rectangle has the dimensions of 3.0 m × 2.0 m when viewed by someone at rest with respect to it. When you move past the rectangle along one of its sides, the rectangle looks like a square. What dimensions do you observe when you move at the same speed along the adjacent side of the rectangle?

****18.** A woman and a man are on separate rockets, which are flying parallel to each other and have a relative speed of $0.940c$. The woman measures the same value for the length of her own rocket and for the length of the man's rocket. What is the ratio of the value that the man measures for the length of his own rocket to the value he measures for the length of the woman's rocket?

Section 28.5 Relativistic Momentum

19. A small meteor, moving through the solar system at a speed of $0.70c$ relative to the earth, has a mass of 10.2 kg. What is the relativistic momentum of the meteor?

20. A car, whose mass is 1550 kg, is traveling at 15.0 m/s. If the speed of light were 25.0 m/s, what would be the momentum of the car as measured by a person standing on the ground?

21. A rocket of mass 1.40×10^5 kg has a relativistic momentum of 3.15×10^{13} kg·m/s. How fast is the rocket traveling?

22. A woman is 1.7 m tall and has a mass of 49 kg. She moves past an observer with the direction of the motion parallel to her height. The observer measures her relativistic momentum to be 3.0×10^{10} kg·m/s. What does the observer measure for her height?

***23.** Starting from rest, two skaters "push off" against each other on smooth level ice, where friction is negligible. One is a woman and one is a man. The woman moves away with a velocity of +2.5 m/s relative to the ice. The mass of the woman is 54 kg, and the mass of the man is 88 kg. Assuming that the speed of light is 3.0 m/s, so that the relativistic momentum must be used, find the recoil velocity of the man relative to the ice. *(Hint: This problem is similar to Example 6 in Chapter 7.)*

Section 28.6 The Equivalence of Mass and Energy

24. Radium is a radioactive element whose nucleus emits an α particle (a helium nucleus) that has a kinetic energy of about 7.8×10^{-13} J (4.9 MeV). To what amount of mass is this energy equivalent?

25. The total amount of energy consumed in the United States during 1990 is estimated to have been about 8.6×10^{19} J. One penny has a mass of 2.9×10^{-3} kg. How many pennies have the equivalent of this amount of energy?

26. How fast is a proton traveling when its total energy is twice its rest energy?

27. The amount of heat required to melt 1 kg of ice at 0 °C is 3.35×10^5 J. What is the difference between the mass of the water and that of the ice? Which has the greater mass?

28. An electron and a positron each have a mass of 9.11×10^{-31} kg. They collide and annihilate each other, with only electromagnetic radiation appearing after the collision. If each particle is moving at a speed of $0.20c$ relative to the laboratory before the collision, determine the energy of the electromagnetic radiation. *(Hint: For each particle, use the expression for its total energy.)*

29. Determine the ratio of the relativistic kinetic energy to the nonrelativistic kinetic energy ($\frac{1}{2}mv^2$) when a particle has a speed of (a) $1.00 \times 10^{-3}c$ and (b) $0.970c$.

30. An elementary particle called a pion has been observed to decay completely into electromagnetic radiation. The pion has a mass of 2.4×10^{-28} kg. (a) What is the kinetic energy of the pion at a speed of $0.850c$? (b) How much energy in the form of electromagnetic radiation is released when the high-speed pion decays? *(Hint: Use the expression for its total energy.)*

***31.** An electron is accelerated from rest through a potential difference of 2.40×10^7 V. (a) What is the kinetic energy (in joules) of the electron? (b) What is the speed of the electron?

***32.** How close would two electrons have to be positioned so that their total mass doubles?

Section 28.7 The Relativistic Addition of Velocities

33. A rocket ship is moving directly toward the earth with a velocity of $0.80c$ relative to the earth. The ship sends out a pulse of light that is aimed at the earth. What is the velocity that a person on earth measures for the approaching pulse?

34. An observer on the earth sees a spaceship approaching at a velocity of $0.50c$. The spacecraft then launches an exploration vehicle that, according to the earth observer, approaches at $0.70c$. What is the velocity of the exploration vehicle relative to the spaceship?

35. It has been proposed that spaceships of the future will be powered by ion propulsion engines. In one such engine the ions are to be ejected with a speed of $0.80c$ relative to the engine. If the ship were traveling away from the earth with a velocity of $0.70c$, what would be the velocity of the ions relative to the earth? (Be sure to assign the correct plus or minus signs to the velocities, assuming that the direction away from earth is positive.)

***36.** Refer to Conceptual Example 11 as an aid in solving this problem. An intergalactic cruiser has two types of guns: a photon cannon that fires a beam of laser light, and an ion gun that shoots atomic ions at a velocity of $0.950c$ relative to the cruiser. The cruiser closes in on an alien spacecraft at a velocity of $0.800c$ relative to this spacecraft. The captain fires both types of guns. At what velocity do the aliens see (a) the laser light and (b) the ions approach them? At what velocity do the aliens see (c) the laser light and (d) the ions move away from the cruiser?

***37.** A person on earth notices a rocket approaching from the right at a speed of $0.75c$ and another rocket approaching from the left at $0.65c$. What is the relative velocity between the two rockets, as measured by a passenger on one of them?

****38.** Two atomic particles approach each other in a head-on collision. Each particle has a mass of 2.16×10^{-25} kg. The speed of each particle is 2.40×10^8 m/s when measured by an observer standing in the laboratory. (a) What is the speed of one particle as seen by the other particle? (b) Determine the relativistic momentum of one particle, as would be observed by the other.

ADDITIONAL PROBLEMS

39. A nuclear power reactor generates 3.0×10^9 W of power. In one year, what is the change in the mass of the nuclear fuel due to the energy being taken from the reactor?

40. Suppose that you are on board a spacecraft moving toward the earth at a speed of $0.960c$. You have just finished exercising, and your heart is beating at a rate of 155 beats per minute. Your pulse rate is also being monitored on earth. What is your pulse rate according to the clock on earth?

41. How fast must a meter stick be moving if its length is observed to shrink to one-half a meter?

42. At what speed is the relativistic momentum of a particle three times its nonrelativistic momentum?

43. How much work must be done on an electron to accelerate it from rest to a speed of $0.990c$?

***44.** A rocket is moving away from the earth with a speed of $0.75c$. An escape pod of length 45 m (as measured by the rocket crew) is launched from the rocket toward the earth with a speed of $0.55c$ relative to the rocket. What is the length of the escape pod as determined by an observer on earth?

***45.** An unstable particle is at rest and suddenly breaks up into two fragments. No external forces act on the particle or its fragments. One of the fragments has a velocity of $+0.800c$ and a mass of 1.67×10^{-27} kg, while the other has a mass of 5.01×10^{-27} kg. What is the velocity of the more massive fragment? *(Hint: This problem is similar to Example 6 in Chapter 7.)*

***46.** Four kilograms of water are heated from 20.0 °C to 60.0 °C. (a) How much heat is required to produce this change in temperature? [The specific heat capacity of water is 4186 J/(kg·C°).] (b) By how much does the mass of the water increase?

****47.** Twins who are 19.0 years of age leave the earth and travel to a distant planet 12.0 light-years away. Assume that the planet and earth are at rest with respect to each other. The twins depart at the same time on different spaceships. One twin travels at a speed of $0.900c$, while the other twin travels at $0.500c$. (a) According to the theory of special relativity, what is the difference between their ages when they meet again at the earliest possible time? (b) Which twin is older?

CHAPTER 29 PARTICLES AND WAVES

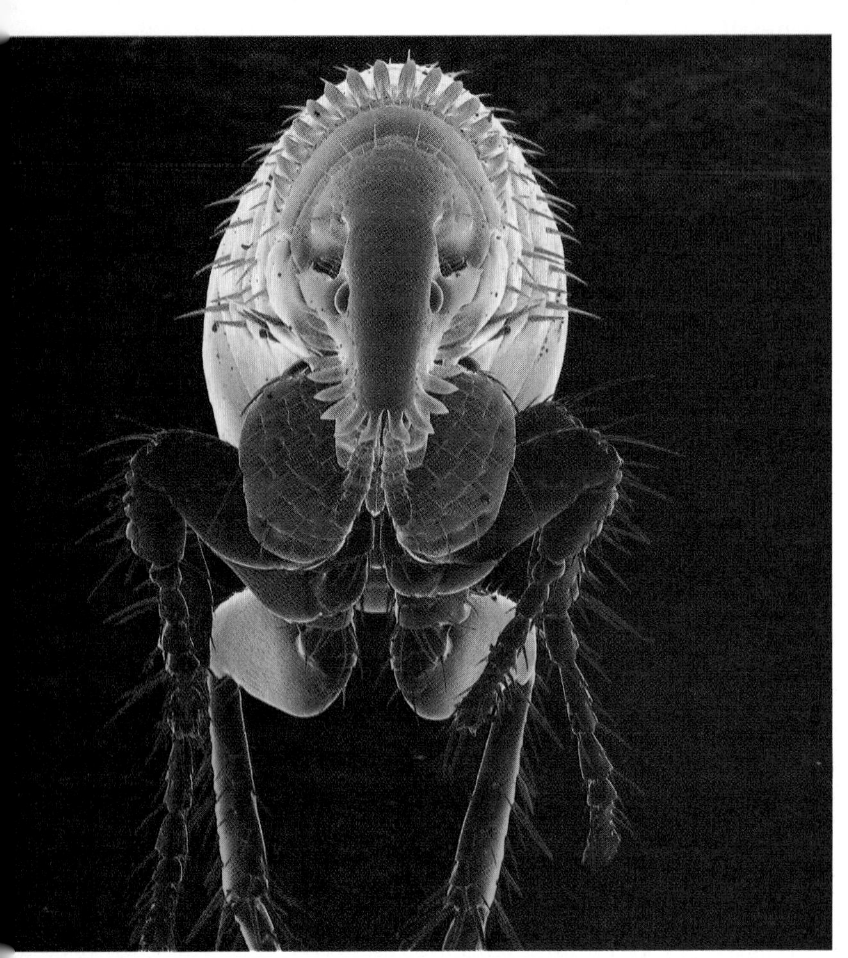

This highly magnified view of a cat flea was made with a scanning electron microscope (SEM). Unlike an optical microscope, which operates with light waves, an electron microscope uses electrons. The exceptional resolution of fine detail that can be obtained in a SEM image is a result of the fact that particles of matter, such as the electron, can behave as waves. The experimental and theoretical basis for this remarkable behavior will be discussed in this chapter. We will find that it is possible to determine the wavelength for a moving particle. In Chapter 27 the Rayleigh criterion for resolution revealed that smaller wavelengths lead to higher resolution. For the electrons in an electron microscope, the wavelength is very small indeed and is directly responsible for the exceptional resolution of SEM images. The wave-like nature of a particle is only one of several surprises that this chapter contains. We will also see that light and other electromagnetic waves can exhibit some of the same characteristics that are normally associated with particles. In particular, we will find that electromagnetic waves can be regarded as being composed of discrete packets of energy, called photons, and that a photon has momentum, just as a particle does. We will present the chain of experimental evidence that supports the photon picture of electromagnetic radiation. The chapter concludes with an introduction to one of the most unusual scientific principles, the Heisenberg uncertainty principle, which places a limit on our knowledge of certain aspects of the physical world. The basic physics that you will read about here was discovered during the first quarter of this century. As a result, that period was one of the most exciting and remarkable in the entire history of science.

29.1 THE WAVE-PARTICLE DUALITY

The ability to exhibit interference effects is an essential characteristic of waves. For instance, Section 27.2 discusses Young's famous experiment in which light passes through two closely spaced slits and produces a pattern of bright and dark fringes on a screen (see Figure 27.4). The fringe pattern is a direct indication that interference is occurring between the light waves coming from each slit.

One of the most incredible discoveries of twentieth-century physics is that particles can also behave like waves and exhibit interference effects. For instance, Figure 29.1 shows a version of Young's experiment performed by directing *a beam of electrons* onto a double slit. In this experiment, the screen is like a television screen and glows wherever an electron strikes it. Part *a* of the picture indicates the pattern that would be seen on the screen if each electron, behaving strictly as a particle, were to pass through one slit or the other and strike the screen. The pattern would consist of an image of each slit. Part *b* shows the pattern actually observed, which consists of bright and dark fringes, reminiscent of that obtained when light waves pass through a double slit. The fringe pattern indicates that the electrons are exhibiting interference effects that are associated with waves.

But how can electrons behave like waves in the experiment shown in Figure 29.1*b*? And what kind of waves are they? The answers to these profound questions will be discussed later in this chapter. For the moment, we intend only to emphasize that the picture of an electron as a tiny discrete particle of matter does not account for the fact that the electron can behave as a wave in some circumstances. In other words, the electron exhibits a dual nature, with both particle-like characteristics and wave-like characteristics.

There is another interesting question: If a particle can exhibit wave-like properties, can waves exhibit particle-like behavior? As the next three sections reveal, the answer is yes. In fact, experiments that demonstrated the particle-like behavior of waves were performed near the beginning of the twentieth century, before the experiments that demonstrated the wave-like properties of the electron. Scientists now accept the ***wave-particle duality*** as an essential part of nature: ***Waves can exhibit particle-like characteristics, and particles can exhibit wave-like characteristics.***

Section 29.2 begins the remarkable story of the wave-particle duality by discussing the electromagnetic waves that are radiated by a perfect blackbody. It is appropriate to begin with blackbody radiation, because it provided the first link in the chain of experimental evidence leading to our present understanding of the wave-particle duality.

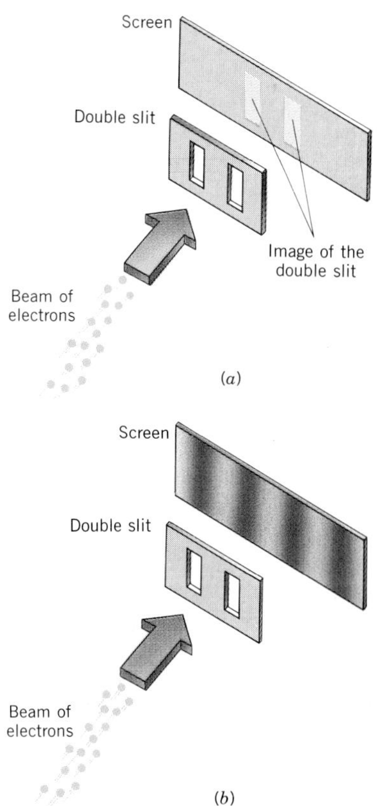

Figure 29.1 **(*a*) If electrons behaved as discrete particles with no wave properties, they would pass through one or the other of the two slits and strike the screen, causing it to glow and produce exact images of the slits. (*b*) In reality, the screen reveals a pattern of bright and dark fringes, similar to the pattern produced when a beam of light is used and interference occurs between the light waves coming from each slit.**

29.2 BLACKBODY RADIATION AND PLANCK'S CONSTANT

All bodies, no matter how hot or cold, continuously radiate electromagnetic waves. For instance, we see the glow of very hot objects, because they emit electromagnetic waves in the visible region of the spectrum. A temperature of about 1700 K produces the white-hot appearance of the filament in an incandescent light bulb, while a temperature near 1000 K creates the characteristic cherry red color of burning charcoal. However, at relatively low temperatures, cooler objects emit visible light waves only weakly and, as a result, do not appear to be

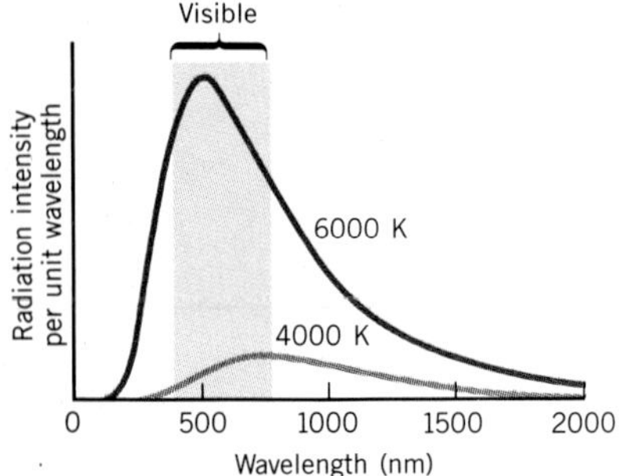

Figure 29.2 **The electromagnetic radiation emitted by a perfect blackbody has an intensity per unit wavelength that varies from wavelength to wavelength, as each curve indicates. At the higher temperature, the intensity per unit wavelength is greater, and the maximum occurs at a shorter wavelength.**

glowing. Certainly the human body, at only 310 K, does not emit enough visible light to be seen in the dark with the unaided eye. But the body does emit electromagnetic waves in the infrared region of the spectrum, and these can be detected with infrared sensitive devices.

At a given temperature, the intensities of the electromagnetic waves emitted by an object vary from wavelength to wavelength throughout the visible, infrared, and other regions of the spectrum. Figure 29.2 illustrates how the intensity per unit wavelength depends on wavelength for a perfect blackbody emitter. As Section 13.3 discusses, a perfect blackbody at a constant temperature absorbs and reemits all the electromagnetic radiation that falls on it. The two curves in the drawing show that at a higher temperature the maximum emitted intensity per unit wavelength increases and shifts toward shorter wavelengths. In accounting for the shape of these curves, the German physicist Max Planck (1858–1947) took the first step toward our present understanding of the wave–particle duality.

In 1900 Planck calculated the blackbody radiation curves, using a model that represents a blackbody as a large number of atomic oscillators, each of which emits and absorbs electromagnetic waves. To obtain agreement between the theoretical and experimental curves, Planck assumed that the energy E of an atomic oscillator could have only the discrete values of $E = 0, hf, 2hf, 3hf$, and so on. In other words, he assumed that

$$E = nhf \qquad n = 0, 1, 2, 3, \ldots \tag{29.1}$$

where n is a positive integer, f is the frequency of vibration (in hertz), and h is a constant now called ***Planck's constant.**** Experiment has shown that Planck's constant has a value of

$$h = 6.626\,0755 \times 10^{-34}\ \mathrm{J \cdot s}$$

The radical feature of Planck's assumption was that the energy of an atomic oscillator could have only discrete values (hf, $2hf$, $3hf$, etc.), with energies in between these values being forbidden. Whenever the energy of a system can have only certain definite values, and nothing in between, the energy is said to be *quantized.* This quantization of the energy was unexpected on the basis of the traditional physics of the time. However, it was soon realized that energy quantization had wide-ranging implications.

Conservation of energy requires that the energy carried off by the electromagnetic waves must equal the energy lost by the atomic oscillators. Suppose, for example, that an oscillator with an energy of $3hf$ emits an electromagnetic wave. According to Equation 29.1, the next smallest allowed value for the energy of the oscillator is $2hf$. In such a case, the energy carried off by the electromagnetic wave would have the value of hf, equaling the amount of energy lost by the oscillator. Thus, Planck's model for blackbody radiation sets the stage for the idea that electromagnetic energy occurs as a collection of discrete amounts or packets of energy, the energy of a packet being equal to hf. As the next section discusses, it was Einstein who made the specific proposal that light consists of such energy packets.

* It is now known that the energy of a harmonic oscillator is $E = (n + \frac{1}{2})hf$, the extra term of $\frac{1}{2}$ being unimportant to the present discussion.

29.3 PHOTONS AND THE PHOTOELECTRIC EFFECT

Einstein proposed that light consists of energy packets in connection with a phenomenon called the ***photoelectric effect.*** Figure 29.3 illustrates the effect. If light with a sufficiently high frequency shines on a metal plate, electrons are ejected from the plate. The ejected electrons move toward a positive electrode called the collector and cause a current to register on the ammeter. Because the electrons are ejected with the aid of light, they are called ***photoelectrons.*** As will be discussed shortly, a number of features of the photoelectric effect could not be explained solely with the ideas of classical physics.

In 1905 Einstein presented an explanation of the photoelectric effect that took advantage of Planck's work concerning blackbody radiation. It was primarily for his theory of the photoelectric effect that he was awarded the Nobel prize in physics in 1921. In his photoelectric theory, Einstein proposed that light of frequency f could be regarded as a collection of discrete packets of energy, each packet containing an amount of energy E given by

[Energy of a photon]

$$E = hf \tag{29.2}$$

where h is Planck's constant. Today these energy packets are called ***photons.*** The light energy given off by a light bulb, for instance, is carried by photons. The brighter the light shining on a given area, the greater is the number of photons per second that strike the area. Example 1 estimates the number of photons emitted per second by a typical light bulb.

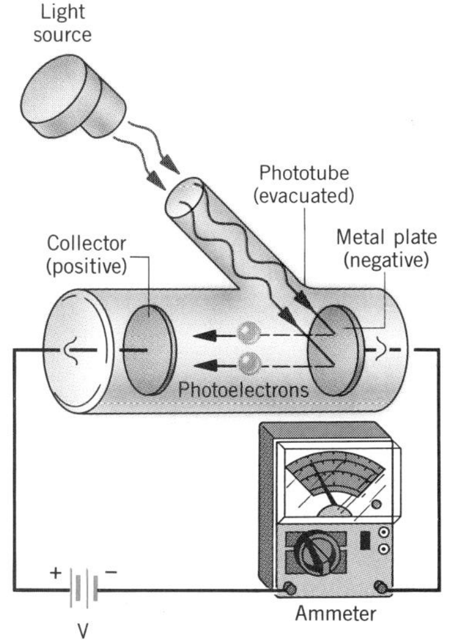

Figure 29.3 In the photoelectric effect, light shines on a metal surface, and if the frequency of the light is sufficiently high, electrons are ejected from the surface. These photoelectrons, as they are called, are drawn to the positive collector, thus producing a current.

Example 1 Photons from a Light Bulb

In converting electrical energy into light energy, a sixty-watt incandescent light bulb operates at about 2.1% efficiency. Assuming that all the light is green light (vacuum wavelength = 555 nm), determine the number of photons per second given off by the bulb.

REASONING The number of photons emitted per second can be found by dividing the amount of light energy emitted per second by the energy E of one photon. The energy of a single photon is $E = hf$, according to Equation 29.2. The frequency f of the photon is related to its wavelength λ by Equation 16.1 as $f = c/\lambda$.

SOLUTION At an efficiency of 2.1%, the light energy emitted per second by a sixty-watt bulb is (0.021)(60.0 J/s) = 1.3 J/s. The energy of a single photon is

$$E = hf = \frac{hc}{\lambda} = \frac{(6.63 \times 10^{-34}\ \text{J}\cdot\text{s})(3.00 \times 10^{8}\ \text{m/s})}{555 \times 10^{-9}\ \text{m}} = 3.58 \times 10^{-19}\ \text{J}$$

Therefore,

$$\begin{array}{l}\text{Number of}\\ \text{photons emitted}\\ \text{per second}\end{array} = \frac{1.3\ \text{J/s}}{3.58 \times 10^{-19}\ \text{J/photon}} = \boxed{3.6 \times 10^{18}\ \text{photons/s}}$$

According to Einstein, when light shines on a metal, a photon can give up its energy to an electron in the metal. If the photon has enough energy to do the work

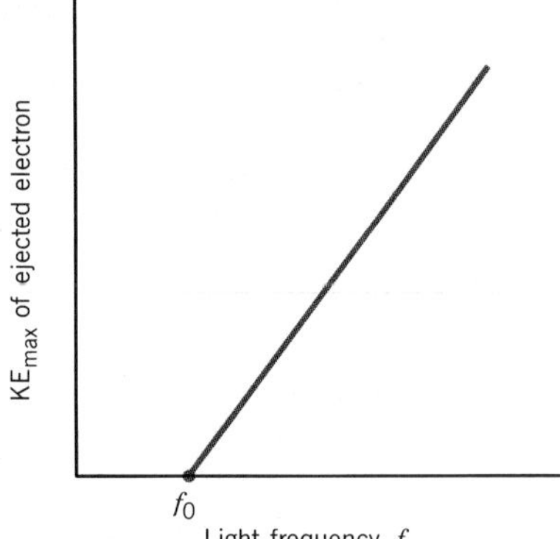

Figure 29.4 Photons of light can eject electrons from a metal when the light frequency is above a minimum value f_0. For frequencies above this minimum value, the ejected electrons have a maximum kinetic energy KE_{max} that is linearly related to the frequency of the light, as the graph shows.

of removing the electron from the metal, the electron can be ejected. The work required depends on how strongly the electron is held. For the *least strongly* held electrons, the necessary work has a minimum value W_0 and is called the ***work function*** of the metal. If a photon has energy in excess of the work needed to remove an electron, the excess energy appears as kinetic energy of the ejected electron. Thus, the least strongly held electrons are ejected with the maximum kinetic energy KE_{max}. Einstein applied the conservation of energy principle and proposed the following relation to describe the photoelectric effect:

$$\underbrace{hf}_{\substack{\text{Photon}\\\text{energy}}} = \underbrace{KE_{max}}_{\substack{\text{Maximum}\\\text{kinetic energy}\\\text{of ejected}\\\text{electron}}} + \underbrace{W_0}_{\substack{\text{Minimum}\\\text{work needed to}\\\text{eject electron}}} \qquad (29.3)$$

According to this equation, $KE_{max} = hf - W_0$, which is plotted in Figure 29.4, with KE_{max} along the ordinate and f along the abscissa. The graph is a straight line that crosses the abscissa at $f = f_0$. At this frequency, the electron departs from the metal with no kinetic energy ($KE_{max} = 0$). According to Equation 29.3, when $KE_{max} = 0$ the energy hf_0 of the incident photon is equal to the work function W_0 of the metal: $hf_0 = W_0$.

The photon picture provides an explanation for a number of features of the photoelectric experiment that are difficult to explain without photons. It is known, for instance, that only light with a frequency above a certain minimum value f_0 will eject electrons. If the frequency of the light is below this value, no electrons are ejected, regardless of how intense the light is. The next example illustrates how Einstein's theory accounts for this minimum value of the frequency.

Example 2 The Photoelectric Effect for a Silver Surface

The work function for a silver surface is $W_0 = 4.73$ eV. Find the minimum frequency that light must have to eject electrons from this surface.

REASONING The minimum frequency f_0 is that frequency at which the photon energy equals the work function W_0 of the metal, so the electron is ejected with zero kinetic energy. Since 1 eV $= 1.60 \times 10^{-19}$ J, the work function expressed in joules is $W_0 = (4.73 \text{ eV})(1.60 \times 10^{-19} \text{ J/1 eV}) = 7.57 \times 10^{-19}$ J. Using Equation 29.3, we find that the minimum frequency needed to eject an electron from the metal is

$$hf_0 - \underbrace{KE_{max}}_{=0} + W_0 \quad \text{or} \quad f_0 - \frac{W_0}{h}$$

SOLUTION The minimum frequency f_0 is

$$f_0 = \frac{W_0}{h} = \frac{7.57 \times 10^{-19} \text{ J}}{6.63 \times 10^{-34} \text{ J} \cdot \text{s}} = \boxed{1.14 \times 10^{15} \text{ Hz}}$$

Photons with frequencies less than f_0 do not have enough energy to eject electrons from a silver surface. Since $\lambda_0 = c/f_0$, the wavelength of this light is $\lambda_0 = 263$ nm, which is in the ultraviolet region of the electromagnetic spectrum.

PROBLEM SOLVING INSIGHT

The work function of a metal is the minimum energy needed to eject an electron from the metal. An electron that has received this minimum energy has no kinetic energy once outside the metal.

Another significant feature of the photoelectric effect is that the maximum kinetic energy of the ejected electrons remains the same when the intensity of the light increases, provided the light frequency remains the same. As the light intensity increases, more photons per second strike the metal, and consequently more electrons per second are ejected. However, since the frequency is the same for each photon, the energy of each photon is also the same. Thus, the ejected electrons always have the same maximum kinetic energy.

Whereas the photon model of light explains the photoelectric effect satisfactorily, the electromagnetic wave picture of light does not. Certainly, it is possible to imagine that the electric field of an electromagnetic wave would cause electrons in the metal to oscillate and tear free from the surface when the amplitude of oscillation becomes large enough. However, were this the case, higher intensity light would eject electrons with a greater maximum kinetic energy, a fact that experiment does not confirm. Moreover, in the electromagnetic wave picture, a relatively long time would be required with low-intensity light before the electrons would build up a sufficiently large oscillation amplitude to tear free. Instead, experiment shows that even the weakest light intensity causes electrons to be ejected almost instantaneously, provided the frequency of the light is above the minimum value f_0. The failure of the electromagnetic wave picture to explain the photoelectric effect does not mean that the wave model should be abandoned. But we must recognize that the wave picture does not account for all the characteristics of light. The photon model also makes an important contribution to our understanding of the way light behaves when it interacts with matter.

Because a photon has energy, the photon can eject an electron from a metal surface when it interacts with the electron. However, a photon is different from a normal particle. A normal particle has a mass and can travel at speeds up to, but not equal to, the speed of light. A photon, on the other hand, travels at the speed of light in a vacuum and does not exist as an object at rest. The energy of a photon is entirely kinetic in nature, for it has no rest energy and no mass. To show that a photon has no mass, we rewrite Equation 28.4 for the total energy E as

$$E\sqrt{1-\frac{v^2}{c^2}} = mc^2$$

The term $\sqrt{1-(v^2/c^2)}$ is zero because a photon travels at the speed of light, $v = c$. Since the energy E of the photon is finite, the left side of the equation above is zero. Thus, the right side must also be zero, so $m = 0$ and the photon has no mass.

THE PHYSICS OF . . .

a photoelectric burglar alarm.

There are a number of interesting applications of the photoelectric effect. These applications depend on the fact that the moving photoelectrons in Figure 29.3 constitute a current, a current that changes as the intensity of the light changes. For example, one type of burglar alarm uses a beam of light that passes across a room before striking the metal surface within the phototube. Ultraviolet light is often used because it is invisible to the naked eye. When an intruder passes through the beam, the light intensity drops momentarily. The corresponding drop in the current of photoelectrons is sensed by electronic circuitry and activates an alarm.

THE PHYSICS OF . . .

an optical soundtrack in a motion picture.

Another application of the photoelectric effect occurs in motion pictures. The sound produced by most motion pictures is contained in the optical soundtrack of the film. The soundtrack is printed alongside the picture frames and consists of a pattern of light and dark regions, as Figure 29.5 illustrates. During the showing of

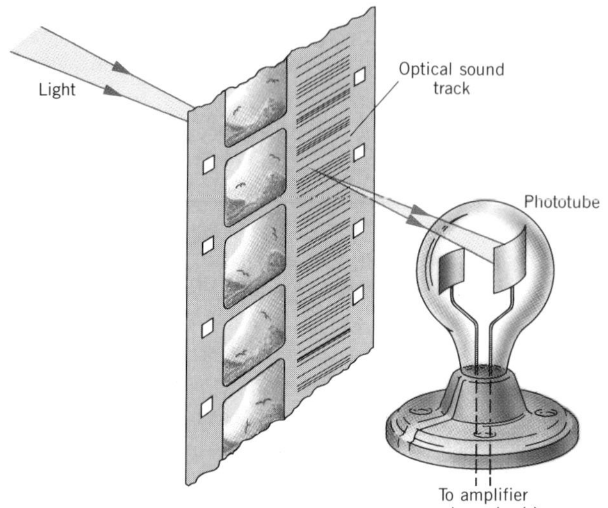

Figure 29.5 The optical soundtrack is adjacent to the picture frames, and it varies the intensity of the light reaching the phototube.

a film, a beam of light in the projector passes through the soundtrack and onto a phototube located behind the film. As the film moves past the beam, the light and dark regions of the soundtrack vary the intensity of the light reaching the phototube, producing a fluctuating current. The fluctuations in the current are a replica of the sound encoded on the soundtrack. The current is then sent to an amplifier, which drives the speakers.

29.4 THE MOMENTUM OF A PHOTON AND THE COMPTON EFFECT

Although Einstein presented his photon model for the photoelectric effect in 1905, it was not until 1923 that the photon picture began to achieve widespread acceptance. It was then that the American physicist Arthur H. Compton (1892–1962) used the photon model to explain his research on the scattering of X-rays by the electrons in graphite. X-rays are high-frequency electromagnetic waves and, like light, they are composed of photons.

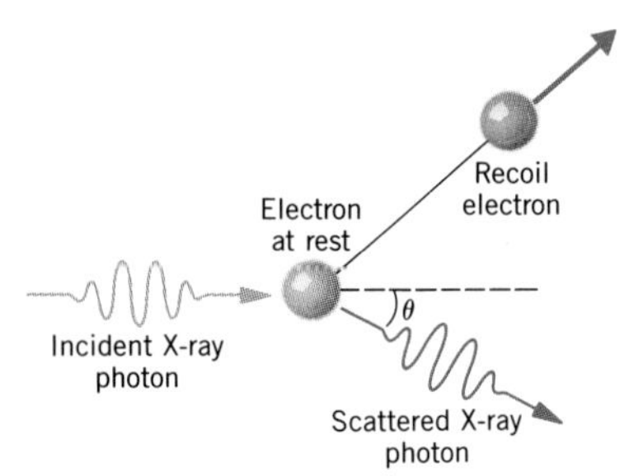

Figure 29.6 In an experiment performed by Arthur H. Compton, an X-ray photon collides with a stationary electron. The scattered photon and the recoil electron depart the collision in different directions.

Figure 29.6 illustrates what happens when an X-ray photon strikes an electron in a piece of graphite. Like two billiard balls colliding on a pool table, the X-ray photon scatters in one direction, and the electron recoils in another direction after the collision. Compton observed that the scattered photon has a frequency f' that is smaller than the frequency f of the incident photon, indicating that the photon loses energy during the collision. In addition, he found that the difference between the two frequencies depends on the angle θ at which the scattered photon leaves the collision. The phenomenon in which an X-ray photon is scattered from an electron, the scattered photon having a smaller frequency than the incident photon, is called the ***Compton effect.***

In Section 7.3 the collision between two objects is analyzed using the fact that the total kinetic energy and the total linear momentum of the objects are the same before and after the collision. Similar analysis can be applied to the collision between a photon and an electron. The electron is assumed to be initially at rest and essentially free, that is, not bound to the atoms of the material. According to the principle of conservation of energy,

$$\underbrace{hf}_{\substack{\text{Energy of}\\\text{incident}\\\text{photon}}} = \underbrace{hf'}_{\substack{\text{Energy of}\\\text{scattered}\\\text{photon}}} + \underbrace{\text{KE}}_{\substack{\text{Kinetic energy}\\\text{of recoil}\\\text{electron}}} \tag{29.4}$$

where the relation $E = hf$ has been used for the photon energies. It follows, then, that $hf' = hf - \text{KE}$, which shows that the energy and corresponding frequency f' of the scattered photon are less than the energy and frequency of the incident photon, just as Compton observed. Since $\lambda' = c/f'$, the wavelength of the scattered X-rays is larger than that of the incident X-rays.

For an initially stationary electron, conservation of total linear momentum requires that

$$\begin{matrix}\text{Momentum of}\\\text{incident photon}\end{matrix} = \begin{matrix}\text{Momentum of}\\\text{scattered photon}\end{matrix} + \begin{matrix}\text{Momentum of}\\\text{recoil electron}\end{matrix} \tag{29.5}$$

To find an expression for the magnitude p of the photon's momentum, we use Equations 28.3 and 28.4. According to these equations, the momentum of any particle is $p = mv/\sqrt{1 - (v^2/c^2)}$ and its total energy is $E = mc^2/\sqrt{1 - (v^2/c^2)}$. Dividing these two equations, we find that $p/E = v/c^2$. Since a photon travels at the speed of light, $v = c$ and $p/E = 1/c$. Therefore, the momentum of a photon is $p = E/c$. But the energy of a photon is $E = hf$, while the wavelength is $\lambda = c/f$. Therefore, the magnitude of the momentum is

$$p = \frac{hf}{c} = \frac{h}{\lambda} \tag{29.6}$$

Using Equations 29.4, 29.5, and 29.6, Compton showed that the difference between the wavelength λ' of the scattered photon and the wavelength λ of the incident photon is related to the scattering angle θ by

$$\lambda' - \lambda = \frac{h}{mc}(1 - \cos\theta) \tag{29.7}$$

In this equation m is the mass of the electron. The quantity h/mc is referred to as the ***Compton wavelength of the electron,*** and has the value $h/mc = 2.43 \times 10^{-12}$ m. Since $\cos\theta$ varies between $+1$ and -1, the shift $\lambda' - \lambda$ in the wavelength can vary between zero and $2h/mc$, depending on the value of θ, a fact observed by Compton. (The reason why photons are more affected by scattering from electrons than from more massive particles, such as atoms or molecules, is the subject of question 8 at the end of this chapter.) The photoelectric effect and the Compton effect provided compelling evidence that electromagnetic waves can exhibit particle-like characteristics attributable to energy packets called photons.

In the Compton effect the electron recoils because it gains some of the photon's momentum. In principle, then, the momentum that photons have can be used to make other objects move. Conceptual Example 3 considers a propulsion system for an interstellar spaceship that is based on the momentum of a photon.

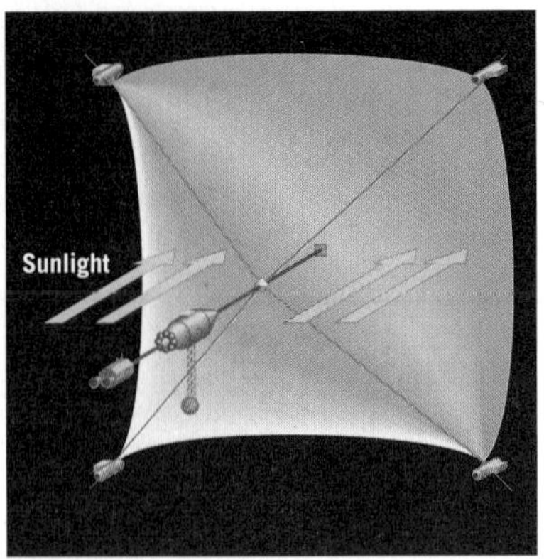

Figure 29.7 A solar sail provides the propulsion for this interstellar spaceship.

Conceptual Example 3 Solar Sails and Interstellar Spaceships

One propulsion method for interstellar spaceships found in science fiction novels uses a large sail. The intent is that sunlight striking the sail creates a force that pushes the ship away from the sun (Figure 29.7), much as the wind propels a sailboat. Does such a design have any hope of working and, if so, should the surface of the sail facing the sun be shiny like a mirror or black, in order to produce the greatest possible force?

REASONING There is certainly reason to believe that the design might work. In Conceptual Example 3 in Chapter 7, we found that hailstones striking the roof of a car exert a force on it because they have momentum. Photons also have momentum, so, like the hailstones, they can apply a force to the sail.

In Chapter 7 we were guided by the impulse-momentum theorem in assessing the force, and it will also be useful here. This theorem (Equation 7.4) states that when a net force acts on an object, the impulse of the net force is equal to the change in momentum of the object. Greater impulses lead to greater forces for a given time interval. Thus, when a photon collides with the sail, the photon's momentum changes because of the force applied by the sail to the photon. Newton's action–reaction law indicates that the photon simultaneously applies a force of equal magnitude, but opposite direction to the sail. It is this reaction force that propels the spaceship, and it will be greater when the momentum change experienced by the photon is greater. The surface of the sail facing the sun, then, should be such that it causes the largest possible momentum change for the impinging photons.

As we found in Section 13.3, radiation reflects from a shiny mirror-like surface and is absorbed by a black surface. Now, consider a photon that strikes the sail perpendicularly. In reflecting from a mirror-like surface, its momentum changes from its value in the forward direction to a value of the same magnitude in the reverse direction. This is a greater change than what occurs when the photon is absorbed by a black surface. Then, the momentum changes only from its value in the forward direction to a value of zero. Consequently, the surface of the sail facing the sun should be shiny in order to produce the greatest possible propulsion force. A shiny surface causes the photons to bounce like the hailstones on the roof of the car and, in so doing, apply a greater force to the sail.

Related Homework Material: *Question 10, Problem 48*

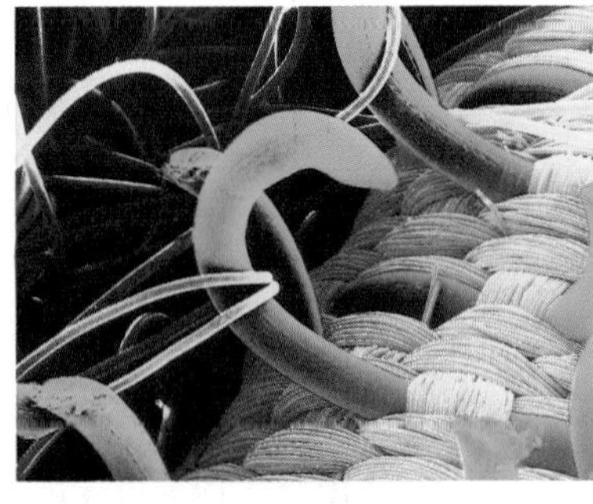

The scanning electron microscope uses the wave nature of the electron to produce highly magnified and detailed images, such as this one of the common fastener Velcro. Velcro is manufactured in two separate pieces, one with a surface containing hooks (right) and the other containing loops (left). When the two are pressed together, the hooks catch in the loops to form a bond.

29.5 THE DE BROGLIE WAVELENGTH AND THE WAVE NATURE OF MATTER

As a graduate student in 1923, Louis de Broglie (1892–1987) made the astounding suggestion that since light waves could exhibit particle-like behavior, particles of matter should exhibit wave-like behavior. De Broglie proposed that the wavelength λ of a particle is given by the same relation (Equation 29.6) that applies to a photon:

[De Broglie wavelength]

$$\lambda = \frac{h}{p} \tag{29.8}$$

where h is Planck's constant and p is the magnitude of the relativistic momentum of the particle. Today, λ is known as the ***de Broglie wavelength*** of the particle.

Confirmation of de Broglie's suggestion came in 1927 from the experiments of the American physicists Clinton J. Davisson (1881–1958) and Lester H. Germer (1896–1971) and, independently, the English physicist George P. Thomson

(1882–1975). Davisson and Germer directed a beam of electrons onto a crystal of nickel and observed that the electrons exhibited a diffraction behavior, analogous to that seen when X-rays are diffracted by a crystal (see Section 27.9 for a discussion of X-ray diffraction). The wavelength of the electrons revealed by the diffraction pattern matched that predicted by de Broglie's hypothesis, $\lambda = h/p$. More recently, Young's double-slit experiment has been performed with electrons and reveals the effects of wave interference illustrated in Figure 29.1.

Particles other than electrons can also exhibit wave-like properties. For instance, neutrons are sometimes used in diffraction studies of crystal structure. Figure 29.8 compares the neutron diffraction pattern and the X-ray diffraction pattern caused by a crystal of rock salt (NaCl).

Although all moving particles have a de Broglie wavelength, the effects of this wavelength are observable only for particles whose masses are very small, on the order of the mass of an electron or a neutron, for instance. Example 4 illustrates why.

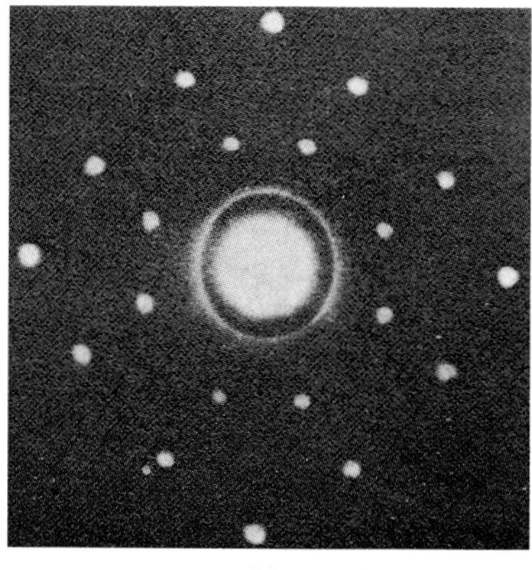

(*a*)

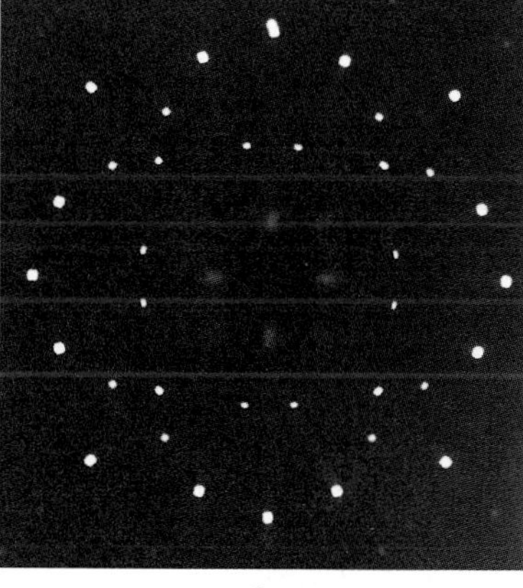

(*b*)

Figure 29.8 (*a*) The neutron diffraction pattern and (*b*) the X-ray diffraction pattern for a crystal of sodium chloride (NaCl).

Example 4 The de Broglie Wavelength of an Electron and a Baseball

Determine the de Broglie wavelength for (a) an electron (mass $= 9.1 \times 10^{-31}$ kg) moving at a speed of 6.0×10^6 m/s and (b) a baseball (mass $= 0.15$ kg) moving at a speed of 13 m/s.

REASONING In each case, the de Broglie wavelength is given by Equation 29.8 as Planck's constant divided by the magnitude of the momentum. Since the speeds are small compared to the speed of light, we can ignore relativistic effects and express the magnitude of the momentum as the product of the mass and the speed.

SOLUTION
(a) For the momentum of the electron, we have

$$p = mv = (9.1 \times 10^{-31}\ \text{kg})(6.0 \times 10^6\ \text{m/s}) = 5.5 \times 10^{-24}\ \text{kg}\cdot\text{m/s}$$

The de Broglie wavelength of the electron is

$$\lambda = \frac{h}{p} = \frac{6.63 \times 10^{-34}\ \text{J}\cdot\text{s}}{5.5 \times 10^{-24}\ \text{kg}\cdot\text{m/s}} = \boxed{1.2 \times 10^{-10}\ \text{m}} \qquad (29.8)$$

A de Broglie wavelength of 1.2×10^{-10} m is about the size of the interatomic spacing in a solid, such as the nickel crystal used by Davisson and Germer, and, therefore, leads to the observed diffraction effects.

(b) Calculations similar to those in part (a) show that the magnitude of the baseball's momentum and its de Broglie wavelength are $p = 2.0$ kg·m/s and $\boxed{\lambda = 3.3 \times 10^{-34}\ \text{m}}$. This wavelength is incredibly small, even by comparison with the size of an atom (10^{-10} m) or a nucleus (10^{-14} m). A wavelength of 3.3×10^{-34} m is so small that the wave characteristics of a baseball cannot be observed.

The de Broglie equation for particle wavelength provides no hint as to what kind of wave is associated with a particle of matter. To gain some insight into the nature of this wave, we now turn our attention to Figure 29.9. This picture shows how the fringe pattern emerges on the screen when electrons are used in a version of Young's double-slit experiment. The bright fringes occur in places on the screen where particle waves coming from each slit interfere constructively, while the dark fringes occur in places where the waves interfere destructively.

When an electron passes through the double-slit arrangement and strikes a spot on the screen, the screen glows at that spot, and Figure 29.9 illustrates how the spots accumulate in time. As more and more electrons strike the screen, the spots eventually form the fringe pattern that is evident in part *d* of the drawing. Bright fringes occur where there is a high probability of electrons striking the screen, and dark fringes occur where there is a low probability. Here lies the key to understanding particle waves. ***Particle waves are waves of probability,*** waves whose magnitude at a point in space gives an indication of the probability that the particle will be found at that point. At the place where the screen is located, the pattern of probabilities conveyed by the particle waves causes the fringe pattern to emerge. The fact that no fringe pattern is apparent in part *b* of the picture does not mean that there are no probability waves present; it just means that too few electrons have struck the screen for the fringe pattern to be recognizable.

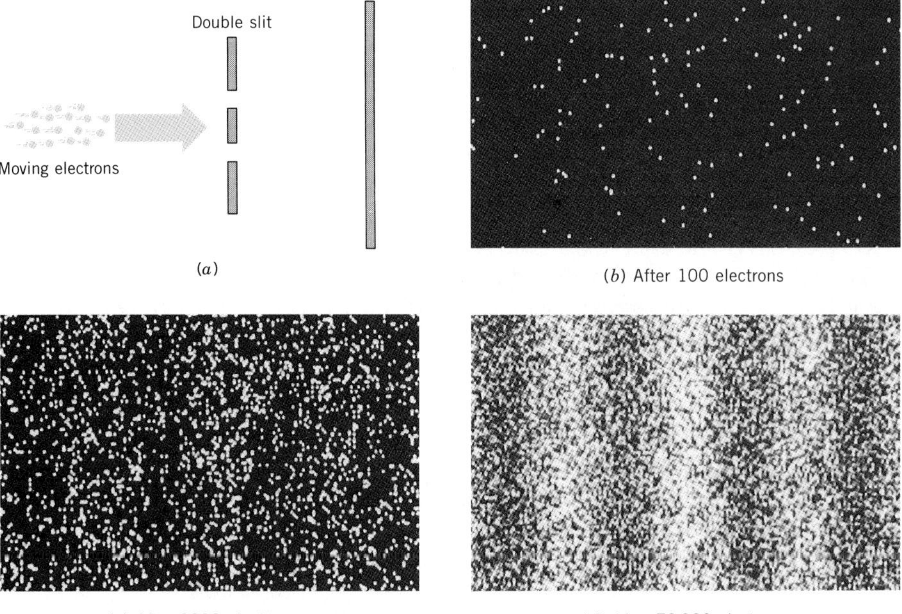

Figure 29.9 This electron version of Young's double-slit experiment was produced by Tonomura et al. The characteristic fringe pattern becomes recognizable only after a sufficient number of electrons have struck the screen. (*Source:* A. Tonomura, J. Endo, T. Matsuda, and T. Kawasaki, *Am. J. Phys.* 57(2): 117, Feb. 1989.)

The pattern of probabilities that leads to the fringes in Figure 29.9 is analogous to the pattern of light intensities that is responsible for the fringes in Young's original experiment with light waves (see Figure 27.4). Section 24.4 discusses the fact that the intensity of the light is proportional to either the square of the electric field strength or the square of the magnetic field strength of the wave. In an analogous fashion in the case of particle waves, the probability is proportional to the square of the magnitude Ψ (Greek letter psi) of the wave. Ψ is referred to as the ***wave function*** of the particle.

In 1925 the Austrian physicist Erwin Schrödinger (1887–1961) and the German physicist Werner Heisenberg (1901–1976) independently developed theoretical frameworks for determining the wave function. In so doing, they established a new branch of physics called ***quantum mechanics.*** The word "quantum" refers to the fact that in the world of the atom, where particle waves must be considered, the particle energy is quantized, so only certain energies are allowed. To understand the structure of the atom and the phenomena related to it, quantum mechanics is essential, and the Schrödinger equation for calculating the wave function is now widely used. In the next chapter, we will explore the structure of the atom based on the ideas of quantum mechanics.

29.6 THE HEISENBERG UNCERTAINTY PRINCIPLE

As the previous section discusses, the bright fringes in Figure 29.9 indicate the places where there is a high probability of an electron striking the screen. And since there are a number of bright fringes, there is more than one place where each electron has some probability of hitting. Yet, any given electron can strike the screen in only one place after passing through the double slit. As a result, it is not possible to specify in advance exactly where on the screen an individual electron will hit. All we can do is speak of the probability that the electron may end up in a number of different places. No longer is it possible to say, as Newton's laws would suggest, that a single electron, fired through the double slit, will travel directly forward in a straight line and strike the screen. This simple picture just does not apply when a particle as small as an electron passes through a pair of closely spaced narrow slits. Because the wave nature of particles is important in such circumstances, we lose the ability to predict with 100% certainty the path that a single particle will follow. Instead, only the average behavior of large numbers of particles is predictable, and the behavior of any individual particle is uncertain.

To see more clearly into the nature of the uncertainty, consider electrons passing through a single slit, as in Figure 29.10. After a sufficient number of electrons strike the screen, a diffraction pattern emerges. The electron diffraction pattern consists of alternating bright and dark fringes and is analogous to that for light waves shown in Figure 27.24. Figure 29.10 shows the slit and locates the first dark fringe on either side of the central bright fringe. The central fringe is bright because electrons strike the screen over the entire region between the dark fringes. If the electrons striking the screen outside the central bright fringe are neglected, the extent to which the electrons are diffracted is given by the angle θ in the drawing. To reach locations within the central bright fringe, some electrons must have acquired momentum in the y direction, despite the fact that they enter the slit traveling along the x direction and have no momentum in the y direction to

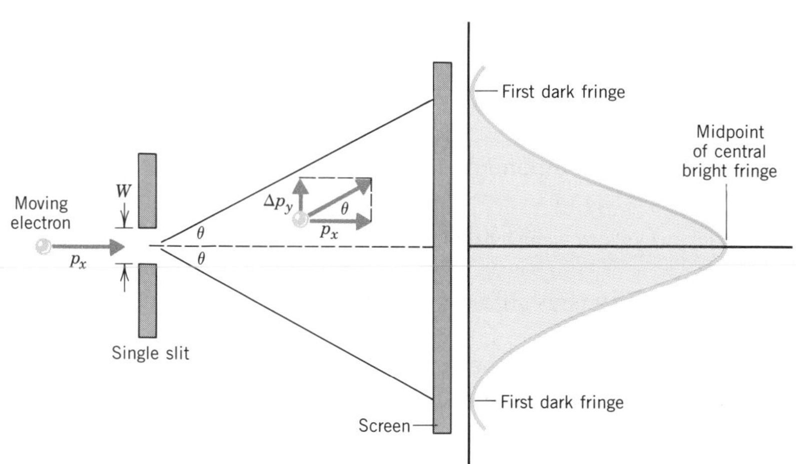

Figure 29.10 When a sufficient number of electrons pass through a single slit and strike the screen, a diffraction pattern of bright and dark fringes emerges. (Only the central bright fringe is shown.) This pattern is due to the wave nature of the electrons and is analogous to that produced by light waves.

start with. The figure illustrates that the y component of the momentum may be as large as Δp_y. The notation Δp_y indicates the difference between the maximum value of the y component of the momentum after the electron passes through the slit and its value of zero before the electron passes through the slit. Δp_y represents the *uncertainty* in the y component of the momentum, in that a diffracted electron may have any value from zero to Δp_y.

It is possible to relate Δp_y to the width W of the slit. To do this, we assume that Equation 27.4, which applies to light waves, also applies to particle waves whose de Broglie wavelength is λ. This equation, $\sin\theta = \lambda/W$, specifies the angle θ that locates the first dark fringe. If θ is small, $\sin\theta \approx \tan\theta$. Moreover, Figure 29.10 indicates that $\tan\theta = \Delta p_y/p_x$, where p_x is the x component of the momentum of the electron. Therefore, $\Delta p_y/p_x \approx \lambda/W$. But $p_x = h/\lambda$ according to de Broglie's equation, so that

$$\frac{\Delta p_y}{h/\lambda} \approx \frac{\lambda}{W}$$

As a result,

$$\Delta p_y \approx \frac{h}{W} \tag{29.9}$$

which indicates that a smaller slit width leads to a larger uncertainty in the y component of the electron's momentum.

It was Heisenberg who first suggested that the uncertainty Δp_y in the y component of the momentum is related to the uncertainty in the y position of the electron as the electron passes through the slit. Since the electron can pass through anywhere over the width W, the uncertainty in the y position of the electron is $\Delta y = W$. Substituting Δy for W in Equation 29.9 shows that $\Delta p_y \approx h/\Delta y$ or $(\Delta p_y)(\Delta y) \approx h$. The result of Heisenberg's more complete analysis is given below in Equation 29.10 and is known as the ***Heisenberg uncertainty principle.*** Note that the Heisenberg principle is a general principle with wide applicability. It does not just apply to the case of single slit diffraction, which we have used here for the sake of convenience.

The Heisenberg Uncertainty Principle

$$(\Delta p_y)(\Delta y) \geq \frac{h}{2\pi} \tag{29.10}$$

Δy = uncertainty in a particle's position along the y direction
Δp_y = uncertainty in the y component of the linear momentum of the particle

$$(\Delta E)(\Delta t) \geq \frac{h}{2\pi} \tag{29.11}$$

ΔE = uncertainty in the energy of a particle when the particle is in a certain state
Δt = time interval during which the particle is in the state

The Heisenberg uncertainty principle places limits on the accuracy with which the momentum and position of a particle can be specified simultaneously. These limits are not just limits due to faulty measuring techniques. They are fundamental limits imposed by nature, and there are no ways to circumvent such limits. Equation 29.10 indicates that Δp_y and Δy cannot both be arbitrarily small at the same time. If one is small, then the other must be large, so that their product equals or exceeds Planck's constant divided by 2π. For example, if the position of a particle is known exactly, so that Δy is zero, then Δp_y is an infinitely large number, and the momentum of the particle is completely uncertain. Conversely, if we assume that Δp_y is zero, then Δy is an infinitely large number, and the position of the particle is completely uncertain. In other words, the Heisenberg uncertainty principle states that it is impossible to specify precisely both the momentum and position of a particle at the same time.

There is also an uncertainty principle that deals with energy and time, as expressed by Equation 29.11. The product of the uncertainty ΔE in the energy of a particle and the time interval Δt during which the particle remains in a given energy state is greater than or equal to Planck's constant divided by 2π. Therefore, the shorter the lifetime of a particle in a given state, the greater is the uncertainty in the energy of that state.

Example 5 shows that the uncertainty principle has significant consequences for the motion of tiny particles such as electrons but has little effect on the motion of macroscopic objects, even those with as little mass as a Ping-Pong ball.

Example 5 The Heisenberg Uncertainty Principle

Assume that the position of an object is known so precisely that the uncertainty in the position is only $\Delta y = 1.5 \times 10^{-11}$ m. (a) Determine the minimum uncertainty in the momentum of the object. Find the corresponding minimum uncertainty in the speed of the object, if the object is (b) an electron (mass $= 9.1 \times 10^{-31}$ kg) and (c) a Ping-Pong ball (mass $= 2.2 \times 10^{-3}$ kg).

REASONING The minimum uncertainty Δp_y in the y component of the momentum is given by the Heisenberg uncertainty principle as $\Delta p_y = h/(2\pi\,\Delta y)$, where Δy is the uncertainty in the position of the object. Both the electron and the Ping-Pong ball have the same uncertainty in their momenta, because they have the same uncertainty in their positions. However, these objects have very different masses. As a result, we will find that the uncertainty in the speeds of these objects is very different.

SOLUTION

(a) The minimum uncertainty in the y component of the momentum is

$$\Delta p_y = \frac{h}{2\pi\,\Delta y} = \frac{6.63 \times 10^{-34}\ \text{J}\cdot\text{s}}{2\pi(1.5 \times 10^{-11}\ \text{m})} = \boxed{7.0 \times 10^{-24}\ \text{kg}\cdot\text{m/s}} \qquad (29.10)$$

(b) Since $\Delta p_y = m\,\Delta v_y$, the minimum uncertainty in the speed of the electron is

$$\Delta v_y = \frac{\Delta p_y}{m} = \frac{7.0 \times 10^{-24}\ \text{kg}\cdot\text{m/s}}{9.1 \times 10^{-31}\ \text{kg}} = \boxed{7.7 \times 10^{6}\ \text{m/s}}$$

PROBLEM SOLVING INSIGHT

The Heisenberg uncertainty principle states that the product of Δp_y and Δy is greater than or equal to $h/2\pi$. The minimum uncertainty occurs when the product is equal to $h/2\pi$.

Thus, the small uncertainty in the y position of the electron gives rise to a large uncertainty in the speed of the electron.

(c) The uncertainty in the speed of the Ping-Pong ball is

$$\Delta v_y = \frac{\Delta p_y}{m} = \frac{7.0 \times 10^{-24}\ \text{kg}\cdot\text{m/s}}{2.2 \times 10^{-3}\ \text{kg}} = \boxed{3.2 \times 10^{-21}\ \text{m/s}}$$

Because the mass of the Ping-Pong ball is relatively large, the small uncertainty in its y position gives rise to an uncertainty in its speed that is much smaller than that for the electron. Thus, in contrast to the electron, we can know simultaneously where the ball is and how fast it is moving, to a very high degree of certainty.

Example 5 emphasizes how the uncertainty principle imposes different uncertainties on the speeds of an electron (small mass) and a Ping-Pong ball (large mass). For objects like the ball, which have relatively large masses, the uncertainties in position and speed are so small that they have no effect on our ability to determine simultaneously where such objects are and how fast they are moving. The uncertainties calculated in Example 5 depend on more than just the mass, however. They also depend on Planck's constant, which is a very small number. It is interesting to speculate about what life would be like if Planck's constant were much larger than 6.63×10^{-34} J · s. Conceptual Example 6 deals with just such speculation.

Conceptual Example 6 What If Planck's Constant Were Large?

A bullet leaving the barrel of a gun is analogous to an electron passing through the single slit in Figure 29.10. With this analogy in mind, what would hunting be like if Planck's constant had a relatively large value instead of its small value of 6.63×10^{-34} J · s?

REASONING It takes two to hunt, the hunter and the hunted. Consider the hunter first, and what it would mean to take aim if Planck's constant were large. The bullet moving down the barrel of the gun has a de Broglie wavelength given by Equation 29.8 as $\lambda = h/p$. This wavelength is large, since Planck's constant h is now large. (We are assuming that the magnitude p of the bullet's momentum is not abnormally large.) Remember from Section 27.5 that the diffraction of light waves through an opening is greater when the wavelength is greater, other things being equal. We expect, therefore, that with λ being comparable to the opening in the barrel, the hunter will have to contend with appreciable diffraction when the bullet leaves the barrel. This means that, in spite of being aimed directly at the target, the bullet may come nowhere near it. In Figure 29.10, for comparison, diffraction allows the electron to strike the screen on either side of the maximum intensity point of the central bright fringe. Under such circumstances, the hunter might as well not bother aiming. He'll have just as much success by scattering his shots around randomly, hoping to hit something.

Now consider the target and what it means that the hunter "aimed directly at the target." This must mean that the target's location is known, at least approximately. In other words, the uncertainty Δy in the target's position is reasonably small. But the uncertainty principle indicates that the minimum uncertainty in the target's momentum is $\Delta p = h/(2\pi\,\Delta y)$. Since Planck's constant is now large, Δp is also large. The target, therefore, may have a large momentum and be moving rapidly. Even if the bullet happens to reach the place where the hunter aimed, the target may no longer be there. Hunting would be vastly different indeed, if Planck's constant had a very large value.

Related Homework Material: *Problem 37*

INTEGRATION OF CONCEPTS

PHOTONS AND THE CONSERVATION PRINCIPLES FOR ENERGY AND MOMENTUM

The conservation of energy and the conservation of momentum are two of the most firmly established principles in physics. A large body of experimental evidence, accumulated over many years, indicates that these principles are fundamental. The word "fundamental" means that other phenomena and concepts must obey these principles. If a newly observed phenomenon or proposed idea seems to conflict with energy or momentum conservation, scientists immediately suspect that an important feature of the story has been left out. In this sense, then, these conservation principles serve as a test that new ideas must pass. The photon picture of light, for instance, indicates that light consists of packets of energy. Each photon has a discrete energy given by the product of Planck's constant and the frequency of the light. And each photon has a momentum given by Planck's constant divided by the wavelength of the light. This picture was not widely accepted until Einstein and Compton showed that these relationships for photon energy and momentum could be used with the principles of energy and momentum conservation to explain the photoelectric effect and the Compton effect. Scientists now routinely incorporate photon energy and momentum when they apply the conservation principles for energy and momentum. Thus, the newer physics of the twentieth century combines with the older classical physics to give a more complete view of the physical world.

SUMMARY

The **wave–particle duality** refers to the fact that a wave can exhibit particle-like characteristics and a particle can exhibit wave-like characteristics.

At a constant temperature, a perfect blackbody absorbs and reemits all the electromagnetic radiation that falls on it. Max Planck calculated the emitted radiation intensity per unit wavelength as a function of wavelength. In his theory, Planck assumed that a blackbody consists of atomic oscillators that can have only quantized energies. Planck's quantized energies are given by $E = nhf$, where $n = 0, 1, 2, 3, \ldots$, h is **Planck's constant** (6.63×10^{-34} J·s), and f is the vibration frequency.

All electromagnetic radiation consists of **photons,** which are packets of energy. The energy of a photon is $E = hf$, where h is Planck's constant and f is the frequency of the light. A photon in a vacuum always travels at the speed of light c and has no mass. The **photoelectric effect** is the phenomenon in which light shining on a metal surface causes electrons to be ejected from the surface. The **work function** W_0 of a metal is the minimum work that must be done to eject an electron from the metal. In accordance with the conservation of energy, the electrons ejected from a metal have a maximum kinetic energy KE_{max} that is related to the energy hf of the incident photon by $hf = KE_{max} + W_0$.

The **Compton effect** is the scattering of a photon by an electron in a material, the scattered photon having a smaller frequency than the incident photon. The magnitude p of the photon's momentum is $p = h/\lambda$, and in the Compton effect, part of it is transferred to the recoiling electron. The difference between the wavelength λ' of the scattered photon and the wavelength λ of the incident photon is related to the scattering angle θ by $\lambda' - \lambda = (h/mc)(1 - \cos\theta)$, where m is the mass of the electron and the quantity h/mc is known as the Compton wavelength of the electron.

The **de Broglie wavelength** of a particle is $\lambda = h/p$, where p is the magnitude of the relativistic momentum of the particle. Because of its de Broglie wavelength, a particle can exhibit wave-like characteristics. The wave associated with a particle is a wave of probability.

The **Heisenberg uncertainty principle** places limits on our knowledge about the behavior of a particle. The uncertainty principle indicates that $(\Delta p_y)(\Delta y) \geq h/2\pi$, where Δy and Δp_y are, respectively, the uncertainties in the position and momentum of the particle. The uncertainty principle also states that $(\Delta E)(\Delta t) \geq h/2\pi$, where ΔE is the uncertainty in the energy of a particle when the particle is in a certain state and Δt is the time interval during which the particle is in the state.

QUESTIONS

1. The photons emitted by a source of light do *not* all have the same energy. Is the source monochromatic? Give your reasoning.

2. Which of the colored lights (red, orange, yellow, green, or blue) on a Christmas tree emits photons with (a) the least energy and (b) the greatest energy? Account for your answers.

3. Does a photon emitted by a higher-wattage red light bulb have more energy than a photon emitted by a lower-wattage red bulb? Justify your answer.

4. When a sufficient number of visible light photons strike a piece of photographic film, the film becomes exposed. An X-ray photon is more energetic than a visible light photon. Yet, most photographic films are not exposed by the X-ray machines used at airport security checkpoints. Explain what these observations imply about the number of photons emitted by the X-ray machines.

5. Radiation of a given wavelength causes electrons to be emitted from the surface of one metal but not from the surface of another metal. Explain why this could be.

6. In the photoelectric effect, light of a given frequency ejects two electrons A and B from the surface of a metal. The kinetic energies of the electrons are KE_A and KE_B, and experiment reveals that KE_A is greater than KE_B. The work needed to eject electron A is W_A, while that needed to eject electron B is W_B. Based on the conservation of energy principle, is the ratio W_A/W_B greater than, equal to, or less than one? Defend your answer.

7. In a Compton scattering experiment, an electron is accelerated straight ahead in the same direction as that of the incident X-ray photon. Which way does the scattered photon move? Explain your reasoning, using the principle of conservation of momentum.

8. Photons can undergo Compton scattering from a molecule such as nitrogen, just as they do from an electron. However, the change in photon wavelength is much less than when an electron is scattered. Explain why, using Equation 29.7 for a nitrogen molecule instead of an electron.

9. Photons have momentum, as well as energy. If two photons in a vacuum have the same energy, do they necessarily have the same momentum? Justify your answer.

10. Review Conceptual Example 3 as background for this question. The photograph shows a device called a radiometer. The four rectangular panels are black on one side and shiny like a mirror on the other side. In bright light the panel arrangement spins around, in a direction from the black side of a panel toward the shiny side. Do photon collisions with both sides of the panels (collisions such as those discussed in Conceptual Example 3) cause the observed spinning? Give your reasoning.

11. A stone is dropped from the top of a building. As the stone falls, does its de Broglie wavelength increase, decrease, or remain the same? Provide a reason for your answer.

12. An electron and a neutron have different masses. Is it possible, according to Equation 29.8, that they can have the same de Broglie wavelength? Account for your answer.

13. In Figure 29.1 replace the electrons with protons that have the same speed. With the aid of Equation 27.1 for the bright fringes in Young's double-slit experiment and Equation 29.8, decide whether the angular separation between the fringes would increase, decrease, or remain the same when compared to that produced by the electrons.

PROBLEMS

In working these problems, ignore relativistic effects.

Section 29.3 Photons and the Photoelectric Effect

1. Ultraviolet light is responsible for sun tanning. Find the wavelength of an ultraviolet photon whose energy is 6.4×10^{-19} J.

2. Two photons have energies of 3.3×10^{-16} J and 1.3×10^{-20} J. Using Figure 24.6, identify the appropriate region in the electromagnetic spectrum for each of these photons.

3. An FM radio station broadcasts at a frequency of 98.1 MHz. The power radiated from the antenna is 5.0×10^4 W. How many photons per second does the antenna emit?

4. The work function for a sodium surface is 2.28 eV. What is the maximum wavelength that an electromagnetic wave can have and still eject electrons from this surface?

5. Ultraviolet light with a frequency of 3.00×10^{15} Hz strikes a metal surface and ejects electrons that have a maximum kinetic energy of 6.1 eV. What is the work function (in eV) of the metal?

6. Light is shining perpendicularly on the surface of the earth with an intensity of 680 W/m^2. Assuming all the photons in the light have a wavelength of 730 nm, determine the number of photons per second per square meter that reach the earth.

7. Radiation of a certain wavelength causes electrons with a maximum kinetic energy of 0.68 eV to be ejected from a metal whose work function is 2.75 eV. What will be the maximum kinetic energy (in eV) with which this same radiation ejects electrons from another metal whose work function is 2.17 eV?

8. An AM radio station broadcasts an electromagnetic wave at a frequency of 665 kHz, while an FM station broadcasts at 91.9 MHz. How many AM photons are needed to have a total energy equal to that of one FM photon?

*9. An owl has good night vision because its eyes can detect a light intensity as small as 5.0×10^{-13} W/m^2. What is the minimum number of photons per second that an owl eye can detect if its pupil has a diameter of 8.5 mm and the light has a wavelength of 510 nm?

*10. Radiation with a wavelength of 281 nm shines on a metal surface and ejects electrons that have a maximum speed of 3.48×10^5 m/s. Which one of the following metals is it, the values in parentheses being the work functions: potassium (2.24 eV), calcium (2.71 eV), uranium (3.63 eV), aluminum (4.08 eV), and gold (4.82 eV)?

*11. The maximum wavelength for which an electromagnetic wave can eject electrons from a platinum surface is 196 nm. When radiation with a wavelength of 141 nm shines on the surface, what is the maximum speed of the ejected electrons?

*12. Example 1 in the text calculates the number of photons per second given off by a sixty-watt incandescent light bulb. The photons are emitted uniformly in all directions. From a distance of 3.1 m you glance at this bulb for 0.10 s. The light from the bulb travels directly to your eye and does not reflect from anything. The pupil of the eye has a diameter of 2.0 mm. How many photons enter your eye?

**13. (a) How many photons (wavelength = 620 nm) must be absorbed to melt a 2.0-kg block of ice at 0 °C into water at 0 °C? (b) On the average, how many H_2O molecules does one photon convert from the ice phase to the water phase?

**14. A laser emits 1.30×10^{18} photons per second in a beam of light that has a diameter of 2.00 mm and a wavelength of 514.5 nm. Determine (a) the average electric field strength and (b) the average magnetic field strength for the electromagnetic wave that constitutes the beam.

Section 29.4 The Momentum of a Photon and the Compton Effect

15. The microwaves used in a microwave oven have a wavelength of about 0.13 m. What is the momentum of a microwave photon?

16. A photon of red light has a wavelength of 750 nm, while a photon of violet light has a wavelength of 380 nm. Find the ratio $p_{\text{violet}}/p_{\text{red}}$ of the photon momenta.

17. A photon has the same momentum as an electron moving with a speed of 2.0×10^5 m/s. What is the wavelength of the photon?

18. Determine the *change* in the photon's wavelength that occurs when an electron scatters an X-ray photon (a) straight back at an angle of $\theta = 180.0°$ and (b) at an angle of $\theta = 30.0°$. All angles are measured as in Figure 29.6.

19. In a Compton scattering experiment, the incident X-rays have a wavelength of 0.2685 nm, while the scattered X-rays have a wavelength of 0.2702 nm. At what angle θ in Figure 29.6 are the X-rays scattered?

*20. The X-rays detected at a scattering angle of $\theta = 163°$ in Figure 29.6 have a wavelength of 0.1867 nm. Find (a) the wavelength of an incident photon, (b) the energy of an incident photon, (c) the energy of a scattered photon, and (d) the kinetic energy of the recoil electron. (For accuracy, use $h = 6.626 \times 10^{-34}$ J·s and $c = 2.998 \times 10^8$ m/s.)

*21. What is the maximum amount by which the wavelength of an incident photon could change when it undergoes Compton scattering from a nitrogen molecule (N_2)?

*22. Suppose that the electron in Figure 29.6 is accelerated straight ahead along the dashed line, instead of upward and to the right, as shown. What is the change in wavelength of the photon?

Section 29.5 The de Broglie Wavelength and the Wave Nature of Matter

23. A honeybee (mass = 1.3×10^{-4} kg) is crawling at a speed of 0.020 m/s. What is the de Broglie wavelength of the bee?

24. A particle has a speed of 1.2×10^6 m/s. Its de Broglie wavelength is 8.4×10^{-14} m. What is the mass of the particle?

25. The interatomic spacing in a crystal of table salt is 0.282 nm. This crystal is being studied in a neutron diffraction experiment, similar to the one that produced the photograph in Figure 29.8*a*. How fast must a neutron (mass = 1.67×10^{-27} kg) be moving to have a de Broglie wavelength of 0.282 nm?

26. The de Broglie wavelength of a proton in a particle accelerator is 1.30×10^{-14} m. Determine the kinetic energy (in joules) of the proton.

27. How fast does a proton have to be moving to have the same de Broglie wavelength as an electron does when the electron moves at 4.5×10^6 m/s?

28. Recall from Section 14.3 that the average kinetic energy of an atom in a monatomic ideal gas is given by $\overline{KE} = \frac{3}{2}kT$, where $k = 1.38 \times 10^{-23}$ J/K and T is the Kelvin temperature of the gas. Determine the de Broglie wavelength of a helium atom (mass $= 6.65 \times 10^{-27}$ kg) that has the average kinetic energy at room temperature (293 K).

***29.** From a cliff 9.5 m above a lake, a young woman (mass = 41 kg) jumps from rest, straight down into the water. At the instant she strikes the water, what is her de Broglie wavelength?

***30.** In a Young's double-slit experiment performed with electrons, the two slits are separated by a distance of 2.0×10^{-6} m. The first-order bright fringes are located on the observation screen at a position given by $\theta = 1.6 \times 10^{-4}$ degrees in Equation 27.1. Find (a) the wavelength, (b) the momentum, and (c) the kinetic energy of the electrons.

***31.** In a television picture tube, electrons are accelerated from rest through a potential difference of 21 000 V. What is the de Broglie wavelength of the electrons?

****32.** The kinetic energy of a particle is equal to the energy of a photon. The particle moves at 5.0% of the speed of light. Find the ratio of the photon wavelength to the de Broglie wavelength of the particle.

Section 29.6 The Heisenberg Uncertainty Principle

33. In the lungs there are tiny sacs of air, which are called alveoli. The average diameter of one of these sacs is 0.25 mm. Consider an oxygen molecule (mass $= 5.3 \times 10^{-26}$ kg) trapped within a sac. What is the minimum uncertainty in the velocity of this oxygen molecule?

34. The speed of a golf ball (mass = 0.045 kg) and of an electron is 71 m/s. If the uncertainty in the speed is 1.0%, estimate the minimum uncertainty in the position of each object.

35. An electron is trapped within a sphere whose diameter is 2.0×10^{-15} m. What is the minimum uncertainty in the electron's momentum?

36. A prisoner (mass = 75 kg) paces back and forth in his cell, and the uncertainty in his speed is 0.10 m/s. (a) Use the uncertainty principle to estimate the minimum uncertainty in his position. (b) Repeat part (a), assuming that Planck's constant has a value of 663 J·s instead of 6.63×10^{-34} J·s.

37. Review Conceptual Example 6 as background for this problem. When electrons pass through a single slit, as in Figure 29.10, they form a diffraction pattern. As Section 29.6 discusses, the central bright fringe extends to either side of the midpoint, according to an angle θ given by $\sin\theta = \lambda/W$, where λ is the de Broglie wavelength of the electron and W is the width of the slit. When λ is the same size as W, $\theta = 90°$, and the central fringe fills the entire observation screen. In this case, an electron passing through the slit has roughly the same probability of hitting the screen either straight ahead or anywhere off to one side or the other. Now, imagine yourself in a world where Planck's constant is large enough so you exhibit similar effects when you walk through a 0.90-m-wide doorway. If your mass is 82 kg and you walk at a speed of 0.50 m/s, how large would Planck's constant have to be in this hypothetical world?

***38.** Suppose the minimum uncertainity in the position of a particle is equal to its de Broglie wavelength. If the particle has an average speed of 4.5×10^5 m/s, what is the minimum uncertainty in its speed?

ADDITIONAL PROBLEMS

39. What is (a) the wavelength of a 1.0-eV photon and (b) the de Broglie wavelength of a 1.0-eV electron?

40. The wavelengths (in vacuum) of visible light occur between 380 and 750 nm. Determine the range of photon energies (in joules) to which this range of wavelengths corresponds.

41. As Section 27.5 discusses, sound waves diffract or bend around the edges of a doorway. Larger wavelengths diffract more than smaller wavelengths. (a) The speed of sound is 343 m/s. With what speed would a 55.0-kg person have to move through a doorway to diffract to the same extent as a 128-Hz bass tone? (b) At the speed calculated in part (a), how long (in years) would it take the person to move a distance of one meter?

42. An electron and a proton have the same speed. What is the ratio $\lambda_{electron}/\lambda_{proton}$ of their de Broglie wavelengths?

43. Incident X-rays have a wavelength of 0.3365 nm and are scattered by the "free" electrons in graphite. The scattering angle in Figure 29.6 is $\theta = 125°$. What is the magnitude of the momentum of (a) the incident photon and (b) the scattered photon? (For accuracy, use $h = 6.626 \times 10^{-34}$ J·s and $c = 2.998 \times 10^8$ m/s.)

44. A magnesium surface has a work function of 3.68 eV. Electromagnetic waves with a wavelength of 215 nm strike the surface and eject electrons. Find the maximum kinetic energy of the ejected electrons. Express your answer in electron volts.

***45.** The width of the central bright fringe in a diffraction pattern on a screen is identical when either electrons or red light (vacuum wavelength = 661 nm) pass through a single slit. The distance between the screen and the slit is the same in each case and is large compared to the slit width. (a) How fast are the electrons moving? (b) To judge whether the speed in (a) is fast or slow for an electron, determine the speed acquired by an electron in accelerating from rest through a potential difference of only one volt.

***46.** At night, approximately 530 photons per second must enter an unaided human eye for an object to be seen, assuming the light is green. The light bulb in Example 1 in the text emits green light uniformly in all directions, and the diameter of the pupil of the eye is 7.0 mm. What is the maximum distance from which the bulb could be seen?

***47.** An electron and a proton have the same kinetic energy. Determine the ratio of the de Broglie wavelength of the electron to that of the proton.

****48.** Review Conceptual Example 3 before attempting this problem. A beam of visible light has a wavelength of 395 nm and shines perpendicularly on a surface. As a result, there are 3.0×10^{18} photons per second striking the surface. By using the impulse-momentum theorem (Section 7.1), obtain the average force that this beam applies to the surface when (a) the surface is a mirror, so the momentum of each photon is reversed after reflection, and (b) the surface is black, so each photon is absorbed and the momentum of the photon is reduced to zero in the process.

CHAPTER 30 THE NATURE OF THE ATOM

The idea that all matter is composed of atoms is fundamental to our modern view of the world. It has given us a firm basis for understanding the properties of solids, liquids, and gases. This understanding has led to a host of useful devices, one of the most famous being the laser. In the photograph above, laser beams are being used to measure the atmospheric effects that cause stars to twinkle. With such measurements, computers can adjust telescopes to obtain images of incredible clarity. The laser beams arise because atoms generate light as they undergo transitions from a higher to a lower atomic energy level. Our venture into the atomic world begins with the concept of the nuclear atom and continues with the Bohr model of the hydrogen atom. This model introduces many basic atomic features, including the notions of discrete energy states and quantum numbers. We will see, however, that quantum mechanics has displaced the Bohr model and provides a more complete description of the atom. Following an overview of quantum mechanics, the Pauli exclusion principle will be discussed. This principle is important for explaining how electrons are arranged in complex atoms and how the elements are ordered in the periodic table. We will then apply our knowledge of atomic structure to describe how X-rays and laser light are produced.

30.1 RUTHERFORD SCATTERING AND THE NUCLEAR ATOM

An atom contains a small, positively charged nucleus (radius $\approx 10^{-15}$ m), which is surrounded at relatively large distances (radius $\approx 10^{-10}$ m) by a number of electrons, as Figure 30.1*a* illustrates. In the natural state, an atom is electrically neutral, because the nucleus contains a number of protons (each with a charge of $+e$) that equals the number of electrons (each with a charge of $-e$). This model of the atom is universally accepted now and is referred to as the "nuclear atom."

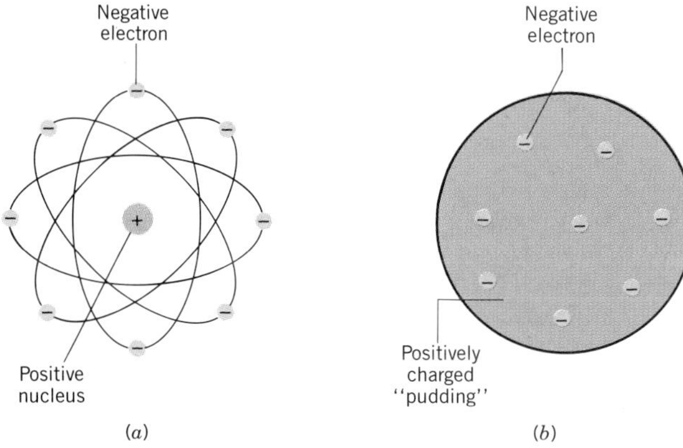

Figure 30.1 (*a*) The nuclear atom. (*b*) The "plum pudding" model of the atom (now discredited).

The nuclear atom is a relatively recent idea. In the early part of the twentieth century a widely accepted model, due to the English physicist Joseph J. Thomson (1856–1940), pictured the atom very differently. In Thomson's view there was no nucleus at the center of an atom. Instead, the positive charge was assumed to be spread throughout the atom, forming a kind of "paste" or "pudding," in which the negative electrons were suspended like "plums." Figure 30.1 compares this "plum-pudding" model with the currently accepted view of the atom.

The "plum-pudding" model was discredited in 1911 when the New Zealand physicist Ernest Rutherford (1871–1937) published experimental results that the model could not explain. As Figure 30.2 indicates, Rutherford and his co-workers directed a beam of alpha particles (α particles) at a thin metal foil made of gold. Alpha particles are positively charged particles (the nuclei of helium atoms, although this was not recognized at the time) emitted by some radioactive materials. If the "plum-pudding" model were correct, the α particles would be expected to pass nearly straight through the foil. After all, there is nothing in this model to deflect the relatively massive α particles, since the electrons have a comparatively small mass and the positive charge is spread out in a diluted "pudding." Using a zinc sulfide screen, which flashed briefly when struck by an α particle, Rutherford and co-workers were able to determine that not all the α particles passed straight through the foil. Instead, some were deflected at large angles, even backward. Rutherford himself said, "It was almost as incredible as if you had fired a fifteen inch shell at a piece of tissue and it came back and hit you." Rutherford concluded that the positive charge, instead of being distributed thinly and uniformly throughout the atom, was concentrated in a small region called the nucleus.

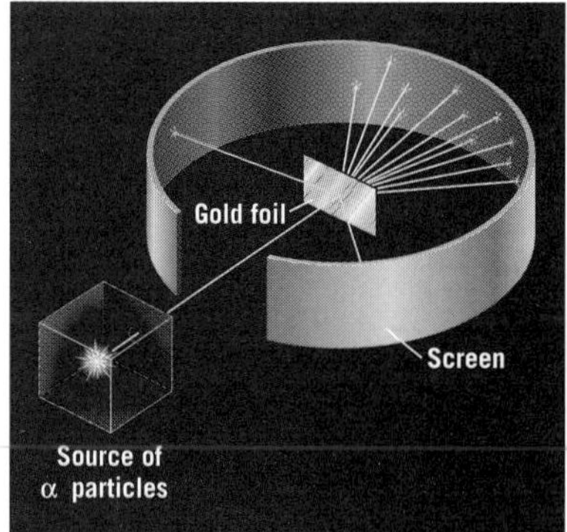

Figure 30.2 A Rutherford scattering experiment in which α particles are scattered by a thin metal foil. The entire apparatus is placed in a vacuum chamber (not shown).

But how could the electrons in a nuclear atom remain separated from the positively charged nucleus? If the electrons were stationary, they would be pulled inward by the attractive electric force of the nuclear charge. Therefore, it was realized that the electrons had to be moving around the nucleus in some fashion, like planets revolving around the sun. However, the dimensions of the atom are such that it contains a large fraction of empty space. If the dimensions of our solar system had the same proportions as those of the atom, there would be much more empty space around the sun than there actually is, as Conceptual Example 1 discusses.

Conceptual Example 1 Atoms Are Mostly Empty Space

In the planetary model of the atom, the nucleus (radius $\approx 1 \times 10^{-15}$ m) is analogous to the sun (radius $\approx 7 \times 10^{8}$ m). Electrons orbit (radius $\approx 1 \times 10^{-10}$ m) the nucleus like the earth orbits (radius $\approx 1.5 \times 10^{11}$ m) the sun. If the dimensions of the solar system had the same proportions as those of the atom, would the earth be closer to or farther away from the sun than it actually is?

REASONING The radius of an electron orbit is one hundred thousand times larger than the radius of the nucleus: $(1 \times 10^{-10}\text{ m})/(1 \times 10^{-15}\text{ m}) = 10^{5}$. If the radius of the earth's orbit were one hundred thousand times larger than the radius of the sun, the earth's orbit would have a radius of $10^{5} \times (7 \times 10^{8}\text{ m}) = 7 \times 10^{13}$ m. This is more than four hundred times greater than the actual orbital radius of 1.5×10^{11} m, so the earth would be much farther away from the sun. In fact, it would be more than ten times farther from the sun than Pluto is, which is our most distant planet and has an orbital radius of only about 6×10^{12} m. An atom, then, contains a much greater fraction of empty space than our solar system does.

Related Homework Material: *Problem 1*

While the planetary model of the atom is easy to visualize, it too is fraught with difficulties. For instance, an electron moving on a curved path has a centripetal acceleration, as Section 5.2 discusses. And when an electron is accelerating, it radiates electromagnetic waves. The difficulty is that the waves carry away energy, which decreases the energy of the electrons. With their energy constantly being depleted, the electrons would spiral inward and eventually collapse into the nucleus. Since matter is stable, such a collapse does not occur. Thus, the planetary model, while providing a more realistic picture of the atom than the "plum-pudding" model, must be telling only part of the story. The full story of atomic structure is fascinating, and the next section describes another aspect of it.

30.2 LINE SPECTRA

We have seen in Sections 13.3 and 29.2 that all objects emit electromagnetic waves, and we will see in Section 30.3 how this radiation arises. For a solid object, such as the hot filament of a light bulb, these waves have a continuous range of wavelengths, some of which are in the visible region of the spectrum. The continuous range of wavelengths is characteristic of the entire collection of atoms that make up the solid. In contrast, individual atoms, free of the strong interactions that are present in a solid, emit only certain specific wavelengths, rather than a continuous range. These wavelengths are characteristic of the atom and provide important clues about its structure. To study the behavior of individual atoms, low-pressure gases are used in which the atoms are relatively far apart.

A low-pressure gas in a sealed tube can be made to emit electromagnetic waves by applying a sufficiently large potential difference between two electrodes located within the tube. With a grating spectroscope like that in Figure 27.35, the individual wavelengths emitted by the gas can be separated and identified as a series of bright fringes or lines. The series of lines is called a ***line spectrum.*** The simplest line spectrum is that of the hydrogen atom (H).* Figure 30.3 shows the visible part of the line spectrum of atomic hydrogen, along with the visible parts

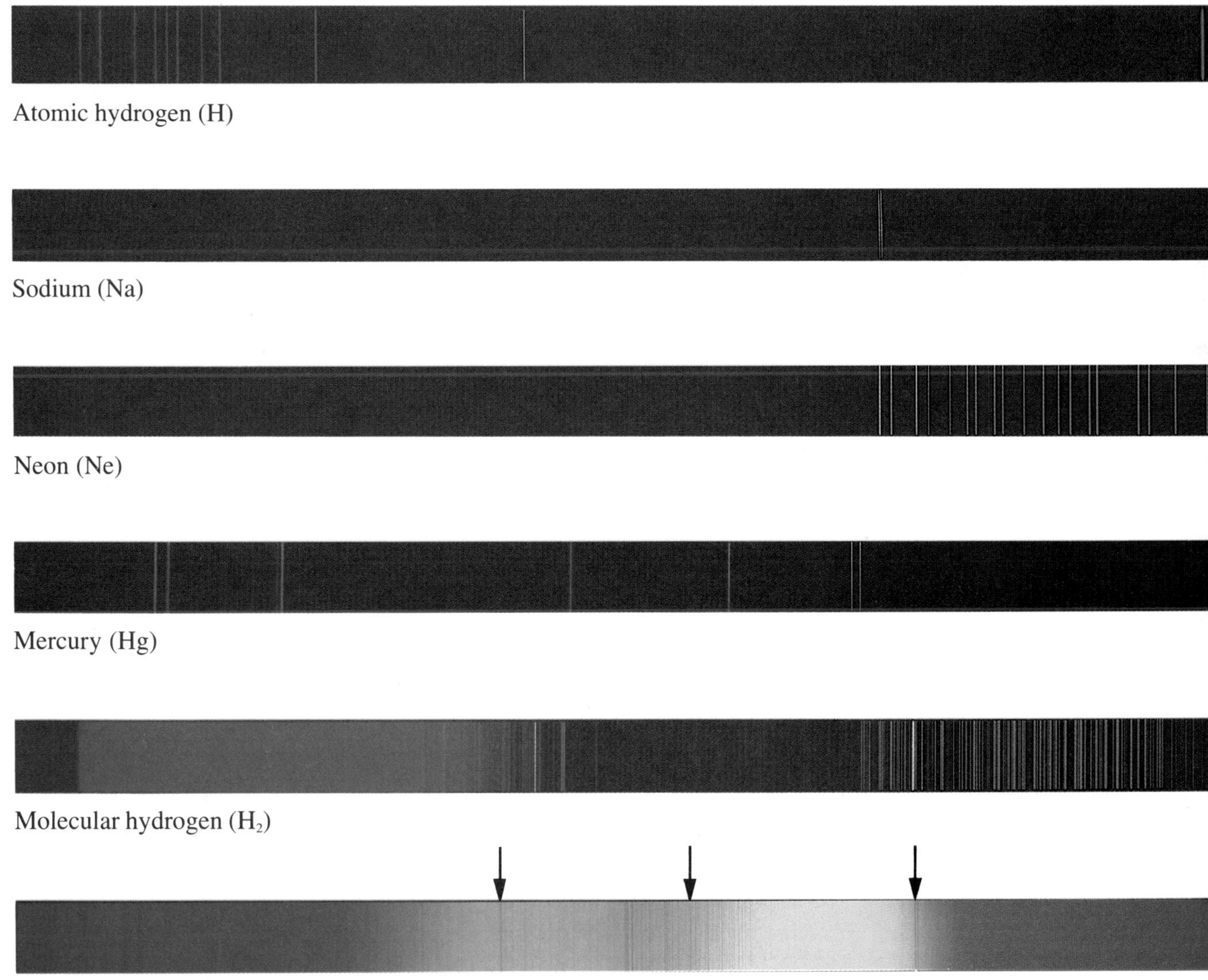

Figure 30.3 Line spectra for various atoms and molecules, along with the continuous spectrum of the sun. The dark lines in the sun's spectrum are called Fraunhofer lines, three of which are marked by arrows.

* Molecular hydrogen (H_2) consists of two hydrogen atoms bonded together and has a more complicated line spectrum than that of atomic hydrogen.

THE PHYSICS OF . . .

neon signs and mercury vapor street lamps.

of the line spectra of more complicated atoms such as neon and mercury. The specific visible wavelengths emitted by neon and mercury are familiar, because they give neon signs and mercury vapor street lamps their characteristic colors.

Much effort has been devoted to understanding the pattern of wavelengths observed in the line spectrum of atomic hydrogen. In addition to the series of lines found in the visible region, analogous series have been found in the nonvisible regions of the electromagnetic spectrum, at both shorter and longer wavelengths. In schematic form, Figure 30.4 illustrates some of the series of lines for atomic hydrogen. The group of lines in the visible region is known as the ***Balmer series,*** in recognition of Johann J. Balmer (1825–1898), a Swiss schoolteacher who found an empirical equation that gave the values for the observed wavelengths. This equation is given below, along with similar equations that apply to the ***Lyman series*** and ***Paschen series,*** which are also shown in the drawing:

[Lyman series] $$\frac{1}{\lambda} = R\left(\frac{1}{1^2} - \frac{1}{n^2}\right) \qquad n = 2, 3, 4, \ldots \qquad (30.1)$$

[Balmer series] $$\frac{1}{\lambda} = R\left(\frac{1}{2^2} - \frac{1}{n^2}\right) \qquad n = 3, 4, 5, \ldots \qquad (30.2)$$

[Paschen series] $$\frac{1}{\lambda} = R\left(\frac{1}{3^2} - \frac{1}{n^2}\right) \qquad n = 4, 5, 6, \ldots \qquad (30.3)$$

In these equations, the constant term R has the value of $R = 1.097 \times 10^7\ \text{m}^{-1}$ and is called the *Rydberg constant.* An essential feature of each group of lines is that there is a long and a short wavelength limit, with the lines being increasingly crowded together toward the short wavelength limit. Figure 30.4 also gives the wavelength limits for each of the three series, and Example 2 determines them for the Balmer series.

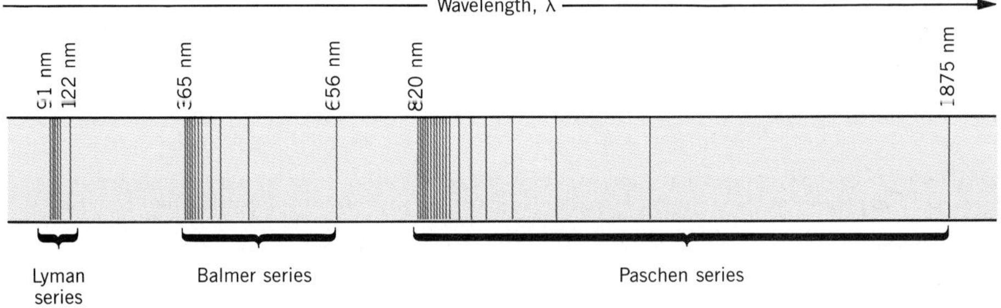

Figure 30.4 Line spectrum of atomic hydrogen. Only the Balmer series lies in the visible region of the electromagnetic spectrum.

Example 2 The Balmer Series

Find (a) the longest and (b) the shortest wavelengths of the Balmer series.

REASONING Each wavelength in the series corresponds to one value for the integer n in Equation 30.2. Longer wavelengths are associated with smaller values of n. The longest wavelength of the series occurs when n has its smallest value of $n = 3$. The shortest wavelength arises when n has a very large value, so that $1/n^2$ is essentially zero.

SOLUTION
(a) With $n = 3$, Equation 30.2 reveals that for the longest wavelength

$$\frac{1}{\lambda} = R\left(\frac{1}{2^2} - \frac{1}{n^2}\right) = (1.097 \times 10^7\ \text{m}^{-1})\left(\frac{1}{2^2} - \frac{1}{3^2}\right)$$

$$= 1.524 \times 10^6\ \text{m}^{-1} \quad \text{or} \quad \boxed{\lambda = 656\ \text{nm}}$$

(b) With $1/n^2 = 0$, Equation 30.2 reveals that for the shortest wavelength

$$\frac{1}{\lambda} = (1.097 \times 10^7\ \text{m}^{-1})\left(\frac{1}{2^2} - 0\right) = 2.743 \times 10^6\ \text{m}^{-1} \quad \text{or} \quad \boxed{\lambda = 365\ \text{nm}}$$

Mercury vapor street lamps produce a characteristic pale green light because of the line spectrum emitted by the mercury atoms in the lamps.

Equations 30.1–30.3 are useful, because they reproduce the wavelengths that hydrogen atoms radiate. However, these equations are empirical, and they provide no insight as to *why* certain wavelengths are radiated and others are not. It was the Danish physicist Niels Bohr (1885–1962) who provided the first model of the atom that predicted the discrete wavelengths emitted by atomic hydrogen. Bohr's model started us on the way toward understanding how the structure of the atom restricts the radiated wavelengths to certain values.

30.3 THE BOHR MODEL OF THE HYDROGEN ATOM

THE MODEL

In 1913 Bohr presented a model that led to equations such as Balmer's for predicting the specific wavelengths that the hydrogen atom radiates. Bohr's theory begins with Rutherford's picture of an atom as a nucleus surrounded by electrons moving in circular orbits. In analyzing this picture, Bohr made a number of assumptions in order to combine the new quantum ideas of Planck and Einstein with the traditional description of a particle in uniform circular motion.

Adopting Planck's idea of quantized energy levels, Bohr hypothesized that in a hydrogen atom there can be only certain values of the total energy (electron kinetic energy plus potential energy). These allowed energy levels correspond to different orbits for the electron as it moves around the nucleus, the larger orbits being associated with larger total energies. Figure 30.5 illustrates two of the orbits. In addition, Bohr assumed that an electron in one of these orbits *does not* radiate electromagnetic waves. For this reason, the orbits are called ***stationary orbits*** or ***stationary states.*** Bohr recognized that radiationless orbits violated the laws of physics, as they were then known. But the assumption of such orbits was necessary, because the traditional laws indicated that an electron radiates electromagnetic waves as it accelerates around a circular path, and the loss of the energy carried by the waves would lead to the collapse of the orbit.

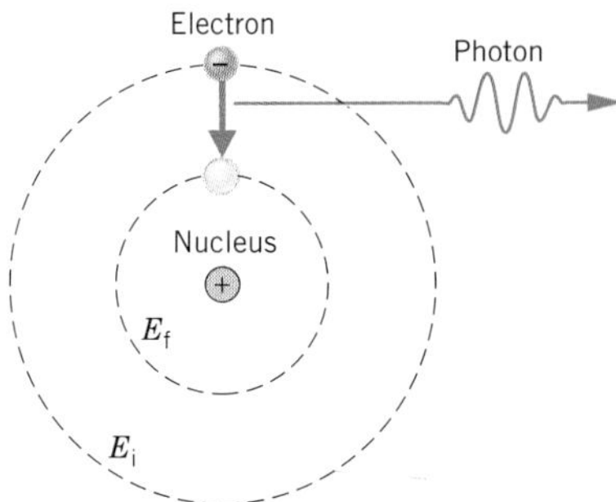

Figure 30.5 In the Bohr model of the hydrogen atom, a photon is emitted when the electron drops from a larger, higher energy orbit (energy $= E_i$) to a smaller, lower energy orbit (energy $= E_f$).

To incorporate Einstein's photon concept, Bohr theorized that a photon is emitted only when the electron *changes* orbits from a larger one with a higher energy to a smaller one with a lower energy, as Figure 30.5 indicates. But how do electrons get into the higher-energy orbits in the first place? They get there by picking up energy when atoms collide, which happens more often when a gas is heated, or by acquiring energy when a high voltage is applied to a gas.

When an electron in an initial orbit with a larger energy E_i changes to a final orbit with a smaller energy E_f, the emitted photon has an energy of $E_i - E_f$, consistent with the law of conservation of energy. But according to Einstein, the energy of a photon is hf, where f is its frequency and h is Planck's constant. Thus,

$$E_i - E_f = hf \tag{30.4}$$

Since the frequency of an electromagnetic wave is related to the wavelength by $f = c/\lambda$, Bohr could use Equation 30.4 to determine the wavelengths radiated by a hydrogen atom. First, however, he had to derive expressions for the energies E_i and E_f.

THE ENERGIES AND RADII OF THE BOHR ORBITS

For an electron of mass m and speed v in an orbit of radius r, the total energy is the kinetic energy (KE $= \frac{1}{2}mv^2$) of the electron plus the electric potential energy. The potential energy is the product of the charge ($-e$) on the electron and the electric potential produced by the positive nuclear charge, in accord with Equation 19.3. We assume that the nucleus contains Z protons,* for a total nuclear charge of $+Ze$. The electric potential at a distance r from a point charge of $+Ze$ is given as $+kZe/r$ by Equation 19.6, where $k = 8.988 \times 10^9\ \mathrm{N \cdot m^2/C^2}$. The electric potential energy is, then, EPE $= (-e)(+kZe/r)$. Consequently, the total energy E of the atom is

$$E = \mathrm{KE} + \mathrm{EPE} = \tfrac{1}{2}mv^2 - \frac{kZe^2}{r} \tag{30.5}$$

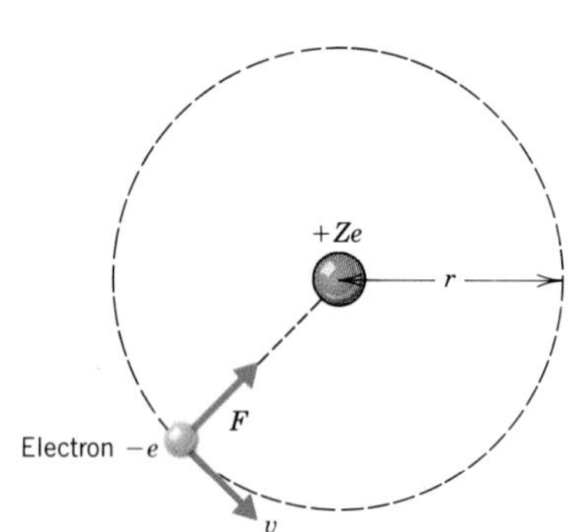

Figure 30.6 In the Bohr model, the electron is in uniform circular motion around the nucleus. The centripetal force F is the electrostatic force of attraction that the positive nuclear charge exerts on the electron.

But a centripetal force of magnitude mv^2/r (Equation 5.3) must act on a particle in uniform circular motion. As Figure 30.6 indicates, the centripetal force is provided by the electrostatic force of attraction that the protons in the nucleus exert on the electron. According to Coulomb's law, the magnitude of the electrostatic force is $F = kZe^2/r^2$. Therefore, $mv^2/r = kZe^2/r^2$, or

$$mv^2 = \frac{kZe^2}{r} \tag{30.6}$$

We can use this relation to eliminate the term mv^2 from Equation 30.5, with the result that

$$E = \frac{1}{2}\left(\frac{kZe^2}{r}\right) - \frac{kZe^2}{r} = -\frac{kZe^2}{2r} \tag{30.7}$$

The total energy of the atom is negative, because the negative electric potential energy is larger in magnitude than the positive kinetic energy.

A value for the radius r is needed, if Equation 30.7 is to be useful. To determine r, Bohr made an assumption about the angular momentum of the electron. The angular momentum L is given by Equation 9.10 as $L = I\omega$, where $I = mr^2$ is the moment of inertia of the electron moving on its circular path and $\omega = v/r$ is the angular speed of the electron in radians per second. Thus, the angular momentum is $L = (mr^2)(v/r) = mvr$. Bohr conjectured that the angular momentum of the electron can assume only certain discrete values; in other words, L is quantized. He postulated that the allowed values are integer multiples of Planck's constant divided by 2π:

* For hydrogen, Z = 1, but we also wish to consider situations in which Z is greater than 1.

$$L_n = mv_n r_n = n\frac{h}{2\pi} \qquad n = 1, 2, 3, \ldots \tag{30.8}$$

Solving this equation for v_n and substituting the result into Equation 30.6 leads to the following expression for the radius r_n of the nth Bohr orbit:

$$r_n = \left(\frac{h^2}{4\pi^2 mke^2}\right)\frac{n^2}{Z} \qquad n = 1, 2, 3, \ldots \tag{30.9}$$

With $h = 6.626 \times 10^{-34}$ J·s, $m = 9.109 \times 10^{-31}$ kg, $k = 8.988 \times 10^9$ N·m²/C², and $e = 1.602 \times 10^{-19}$ C, this expression reveals that

Radii for Bohr orbits (in meters)

$$r_n = (5.29 \times 10^{-11})\frac{n^2}{Z} \qquad n = 1, 2, 3, \ldots \tag{30.10}$$

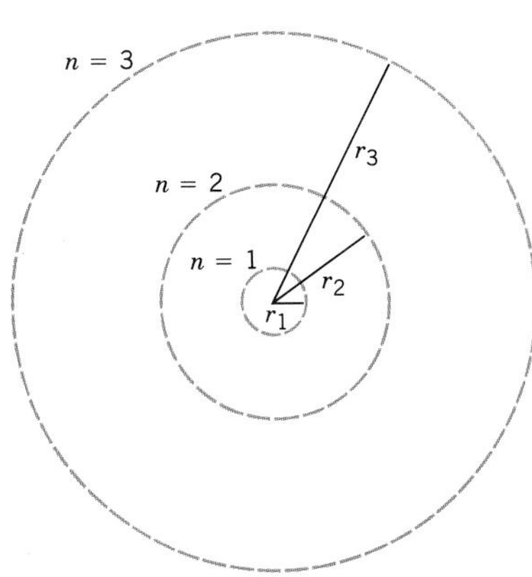

Figure 30.7 **The first Bohr orbit in the hydrogen atom has a radius $r_1 = 5.29 \times 10^{-11}$ m. The second and third Bohr orbits have radii $r_2 = 4r_1$ and $r_3 = 9r_1$, respectively.**

Therefore, in the hydrogen atom ($Z = 1$) the smallest Bohr orbit ($n = 1$) has a radius of $r_1 = 5.29 \times 10^{-11}$ m. This particular value is called the ***Bohr radius.*** Figure 30.7 shows the first three Bohr orbits for the hydrogen atom.

The expression for the radius of a Bohr orbit can be substituted into Equation 30.7 to show that the corresponding total energy for the nth orbit is

$$E_n = -\left(\frac{2\pi^2 mk^2e^4}{h^2}\right)\frac{Z^2}{n^2} \qquad n = 1, 2, 3, \ldots \tag{30.11}$$

Substituting values for h, m, k, and e into this expression yields

Bohr energy levels in joules

$$E_n = -(2.18 \times 10^{-18}\ \text{J})\frac{Z^2}{n^2} \qquad n = 1, 2, 3, \ldots \tag{30.12}$$

Often, atomic energies are expressed in electron volts rather than joules. Since 1.60×10^{-19} J = 1 eV, the result above can be rewritten as

Bohr energy levels in electron volts

$$E_n = -(13.6\ \text{eV})\frac{Z^2}{n^2} \qquad n = 1, 2, 3, \ldots \tag{30.13}$$

ENERGY LEVEL DIAGRAMS

It is useful to represent the energy values given by Equation 30.13 on an *energy level diagram,* as in Figure 30.8. In this diagram, which applies to the hydrogen atom ($Z = 1$), the highest energy level corresponds to $n = \infty$ in Equation 30.13 and has an energy of 0 eV. This is the energy of the atom when the electron is completely removed ($r = \infty$) from the nucleus and is at rest. In contrast, the lowest energy level corresponds to $n = 1$ and has a value of -13.6 eV. The lowest energy level is called the ***ground state,*** to distinguish it from the higher levels, which are called ***excited states.*** Observe how the energies of the excited states come closer and closer together as n increases.

The electron in a hydrogen atom at room temperature spends most of its time in the ground state. To raise the electron from the ground state ($n = 1$) to the highest possible excited state ($n = \infty$), 13.6 eV of energy must be supplied. Supplying this amount of energy removes the electron from the atom, producing the positive

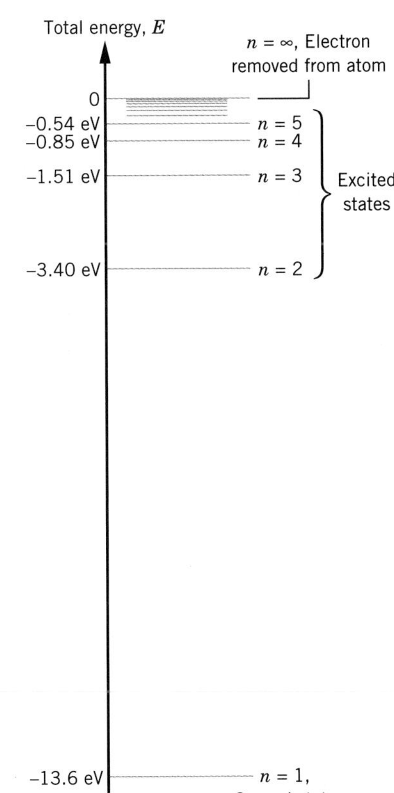

Figure 30.8 **Energy level diagram for the hydrogen atom.**

hydrogen ion H^+. The energy needed to remove the electron is called the ***ionization energy.*** Thus, the Bohr model predicts that the ionization energy of atomic hydrogen is 13.6 eV, in excellent agreement with the experimental value. In Example 3 the Bohr model is applied to doubly ionized lithium.

Example 3 The Ionization Energy of Li^{2+}

The Bohr model does not apply when more than one electron orbits the nucleus, because the model does not account for the electrostatic force that one electron exerts on another. For instance, an electrically neutral lithium atom (Li) contains three electrons in orbit around a nucleus that includes three protons ($Z = 3$), and Bohr's analysis is not applicable. However, the Bohr model can be used for the doubly charged positive ion of lithium (Li^{2+}) that results when two electrons are removed from the neutral atom, leaving only one electron to orbit the nucleus. Obtain the ionization energy that is needed to remove the remaining electron from Li^{2+}.

REASONING The lithium ion Li^{2+} contains three times the positive nuclear charge as that of the hydrogen atom. Therefore, the orbiting electron is attracted more strongly to the nucleus in Li^{2+} than in the hydrogen atom. As a result, we expect that more energy is required to ionize Li^{2+} than the 13.6 eV required for atomic hydrogen.

SOLUTION The Bohr energy levels for Li^{2+} are given by Equation 30.13 with $Z = 3$; $E_n = -(13.6\text{ eV})(3^2/n^2)$. Therefore, the ground state ($n = 1$) energy is

$$E_1 = -(13.6\text{ eV})\frac{3^2}{1^2} = -122\text{ eV}$$

To remove the electron from Li^{2+}, 122 eV of energy must be supplied: $\boxed{\text{Ionization energy} = 122\text{ eV}}$. This value for the ionization energy agrees well with the experimental value of 122.4 eV.

THE LINE SPECTRA OF THE HYDROGEN ATOM

To determine the wavelengths emitted by the hydrogen atom, Bohr substituted Equation 30.11 for the energies into Equation 30.4 and used $f = c/\lambda$. He obtained the following result:

$$\frac{1}{\lambda} = \frac{2\pi^2 mk^2e^4}{h^3c}(Z^2)\left(\frac{1}{n_f^2} - \frac{1}{n_i^2}\right) \tag{30.14}$$

$$n_i, n_f = 1, 2, 3, \ldots \quad \text{and} \quad n_i > n_f$$

With the known values for h, m, k, e, and c, it can be seen that $2\pi^2mk^2e^4/(h^3c) = 1.097 \times 10^7\text{ m}^{-1}$, in agreement with the Rydberg constant R that appears in Equations 30.1–30.3. The agreement between the theoretical and experimental values of the Rydberg constant was a major accomplishment of Bohr's theory.

With $Z = 1$ and $n_f = 1$, Equation 30.14 reproduces Equation 30.1 for the Lyman series. Thus, Bohr's model shows that the Lyman series of lines occurs when electrons make transitions from higher energy levels with $n_i = 2, 3, 4, \ldots$ to the first energy level where $n_f = 1$. Figure 30.9 shows these transitions. Notice that when an electron makes a transition from $n_i = 2$ to $n_f = 1$, the longest wavelength photon in the Lyman series is emitted, since the energy change is the smallest possible. When an electron makes a transition from the highest level where $n_i = \infty$ to the lowest level where $n_f = 1$, the shortest wavelength is emitted,

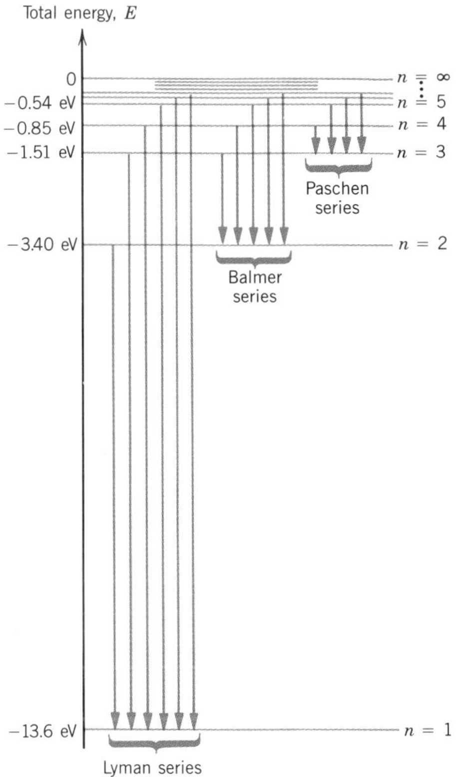

Figure 30.9 The Lyman, Balmer, and Paschen series of lines in the hydrogen atom spectrum correspond to transitions that the electron makes between higher and lower energy levels, as indicated here.

since the energy change is the largest possible. Since the higher energy levels are increasingly close together, the lines in the series become more and more crowded together toward the short wavelength limit, as observed. Figure 30.9 also shows the energy level transitions for the Balmer and Paschen series. In the Balmer series $n_i = 3, 4, 5, \ldots$, while $n_f = 2$. In the Paschen series $n_i = 4, 5, 6, \ldots$, while $n_f = 3$. The next example deals further with the line spectrum of the hydrogen atom.

Example 4 The Brackett Series for Atomic Hydrogen

In the line spectrum of atomic hydrogen there is also a group of lines known as the Brackett series. These lines are produced when electrons, excited to high energy levels, make transitions to the $n = 4$ level. Determine (a) the longest wavelength in this series and (b) the wavelength that corresponds to the transition from $n_i = 6$ to $n_f = 4$. (c) Refer to Figure 24.6 and identify the spectral region in which these lines are found.

REASONING The longest wavelength corresponds to the transition that has the smallest energy change. This would be between the $n_i = 5$ and $n_f = 4$ levels in Figure 30.9.

SOLUTION

(a) Using Equation 30.14 with $Z = 1$, $n_i = 5$, and $n_f = 4$, we find that

$$\frac{1}{\lambda} = (1.097 \times 10^7\ \text{m}^{-1})(1^2)\left(\frac{1}{4^2} - \frac{1}{5^2}\right)$$

$$= 2.468 \times 10^5\ \text{m}^{-1} \quad \text{or} \quad \boxed{\lambda = 4051\ \text{nm}}$$

PROBLEM SOLVING INSIGHT

In the line spectrum of atomic hydrogen, all lines in a given series (e.g., the Lyman series) are identified by a single value of the quantum number n_f, for the *lower energy level into which an electron falls.* Each line in a given series, however, corresponds to a different value of the quantum number n_i, for the *higher energy level where an electron originates.*

(b) The calculation here is similar to that above:

$$\frac{1}{\lambda} = (1.097 \times 10^7\ \mathrm{m}^{-1})(1^2)\left(\frac{1}{4^2} - \frac{1}{6^2}\right)$$

$$= 3.809 \times 10^5\ \mathrm{m}^{-1} \quad \text{or} \quad \boxed{\lambda = 2625\ \mathrm{nm}}$$

(c) According to Figure 24.6, these lines lie in the infrared region of the spectrum.

The various lines in the hydrogen atom spectrum are produced when electrons change from higher to lower energy levels. During these energy-level transitions photons are emitted, and, consequently, the spectral lines are called ***emission lines.*** Electrons can also make transitions from lower energy levels to higher energy levels, in a process known as absorption. In this case, an atom absorbs a photon that has precisely the energy needed to produce a transition from a lower energy level to a higher energy level. Thus, if photons with a continuous range of wavelengths pass through a gas and then are analyzed with a grating spectroscope, a series of dark ***absorption lines*** appear in the continuous spectrum. The dark lines indicate the wavelengths that have been removed by the absorption process. Such absorption lines can be seen in Figure 30.3 in the spectrum of the sun, where they are called Fraunhofer lines, after their discoverer. They are due to atoms, located in the outer and cooler layers of the sun, that absorb radiation coming from within the sun. The inner and hotter portion of the sun emits a continuous spectrum of wavelengths, since it is too hot for individual atoms to retain their structures.

absorption lines in the sun's spectrum.

The Bohr model provides a great deal of insight into atomic structure. However, this model is now known to be oversimplified and has been superseded by a more detailed picture provided by quantum mechanics and the Schrödinger equation (see Section 30.5).

30.4 DE BROGLIE'S EXPLANATION OF BOHR'S ASSUMPTION ABOUT ANGULAR MOMENTUM

Of all the assumptions Bohr made in his model of the hydrogen atom, perhaps the most puzzling is the one about the angular momentum of the electron ($L_n = mv_n r_n = nh/2\pi$; $n = 1, 2, 3, \ldots$). Why should the angular momentum have only those values that are integer multiples of Planck's constant divided by 2π? In 1923, ten years after Bohr's work, de Broglie pointed out that his own theory for the wavelength of a moving particle could provide an answer to this question.

In de Broglie's way of thinking, the electron in its circular Bohr orbit must be pictured as a particle wave. And like waves traveling on a string, particle waves can lead to standing waves under resonance conditions. Section 17.5 discusses these conditions for a string. Standing waves form when the total distance traveled by a wave down the string and back is one wavelength, two wavelengths, or any integer number of wavelengths. The total distance around a Bohr orbit of radius r is the circumference of the orbit or $2\pi r$. By the same reasoning, then, the condition for standing particle waves for the electron in a Bohr orbit would be

$$2\pi r = n\lambda \qquad n = 1, 2, 3, \ldots$$

where n is the number of whole wavelengths that fit into the circumference of the circle. But according to Equation 29.8 the de Broglie wavelength of the electron is $\lambda = h/p$, where p is the magnitude of the electron's momentum. If the speed of the electron is much less than the speed of light, the momentum is $p = mv$, and the condition for standing particle waves becomes $2\pi r = nh/(mv)$. A rearrangement of this result gives

$$mvr = n\frac{h}{2\pi} \qquad n = 1, 2, 3, \ldots$$

which is just what Bohr assumed for the angular momentum of the electron. As an example, Figure 30.10 illustrates the standing particle wave on a Bohr orbit for which $2\pi r = 4\lambda$.

De Broglie's explanation of Bohr's assumption about angular momentum emphasizes an important fact, namely, that particle waves play a central role in the structure of the atom. Moreover, the theoretical framework of quantum mechanics provides the basis for determining the wave function Ψ (Greek letter *psi*) that represents a particle wave. The next section deals with the picture that quantum mechanics gives for atomic structure, a picture that supersedes the Bohr model. In any case, the Bohr model can be applied when a single electron orbits the nucleus, while the theoretical framework of quantum mechanics can be applied, in principle, to atoms that contain an arbitrary number of electrons.

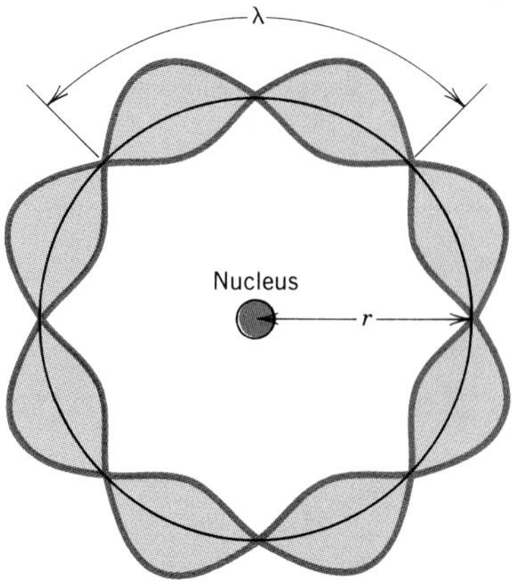

Figure 30.10 De Broglie suggested standing particle waves as an explanation for Bohr's angular momentum assumption. Here, a standing particle wave is illustrated on a Bohr orbit where four de Broglie wavelengths fit into the circumference of the orbit.

30.5 THE QUANTUM MECHANICAL PICTURE OF THE HYDROGEN ATOM

QUANTUM NUMBERS

The picture of the hydrogen atom that quantum mechanics and the Schrödinger equation provide differs in a number of ways from the Bohr model. The Bohr model uses a single integer number n to identify the various electron orbits and the associated energies. Because this number can have only discrete values, rather than a continuous range of values, n is called a ***quantum number.*** In contrast, quantum mechanics reveals that four different quantum numbers are needed to describe each state of the hydrogen atom. These four are described below.

1. **The principal quantum number n.** As in the Bohr model, this number determines the total energy of the atom and can have only integer values: $n = 1, 2, 3, \ldots$. In fact, the Schrödinger equation predicts* that the energy of the hydrogen atom is identical to that obtained from the Bohr model: $E_n = -(13.6\text{ eV})\,Z^2/n^2$.
2. **The orbital quantum number ℓ.** This number determines the angular momentum of the electron due to its orbital motion. The values that ℓ can have depend on the value of n, and only the following integers are allowed:

$$\ell = 0, 1, 2, \ldots, (n-1)$$

For instance, if $n = 1$, the orbital quantum number can have only the value

* This prediction requires that small relativistic effects and small interactions within the atom be ignored, and assumes that the hydrogen atom is not located in an external magnetic field.

$\ell = 0$, but if $n = 4$, the values $\ell = 0$, 1, 2, and 3 are possible. The magnitude L of the angular momentum of the electron is

$$L = \sqrt{\ell(\ell + 1)}\,\frac{h}{2\pi} \tag{30.15}$$

3. **The magnetic quantum number m_ℓ.** The word "magnetic" is used here because an externally applied magnetic field influences the energy of the atom, and this quantum number is used in describing the effect. The effect was discovered by the Dutch physicist Pieter Zeeman (1865–1943) and, hence, is known as the *Zeeman effect.* When there is no external magnetic field, m_ℓ plays no role in determining the energy. In either event, the magnetic quantum number determines the component of the angular momentum along a specific direction, which is called the z direction by convention. The values that m_ℓ can have depend on the value of ℓ, with only the following positive and negative integers being permitted:

$$m_\ell = -\ell, \ldots, -2, -1, 0, +1, +2, \ldots, +\ell$$

For example, if the orbital quantum number is $\ell = 2$, then the magnetic quantum number can have the values $m_\ell = -2, -1, 0, +1$, and $+2$. The component L_z of the angular momentum in the z direction is

$$L_z = m_\ell \frac{h}{2\pi} \tag{30.16}$$

4. **The spin quantum number m_s.** This number is needed because the electron has an intrinsic property called spin angular momentum. Loosely speaking, we can view the electron as spinning while it orbits the nucleus, analogous to the way the earth spins as it moves around the sun. There are two possible values for the spin quantum number of the electron:

$$m_s = +\tfrac{1}{2} \quad \text{or} \quad m_s = -\tfrac{1}{2}$$

Sometimes the phrases "spin up" and "spin down" are used to refer to the directions of the spin angular momentum associated with the values for m_s.

Table 30.1 summarizes the four quantum numbers that are needed to describe each state of the hydrogen atom. One set of values for n, ℓ, m_ℓ, and m_s corresponds to one state. As the principal quantum number n increases, the number of possible combinations of the four quantum numbers rises rapidly, as Example 5 illustrates.

Table 30.1 Quantum Numbers for the Hydrogen Atom

Name	Symbol	Allowed Values
Principal quantum number	n	1, 2, 3, . . .
Orbital quantum number	ℓ	0, 1, 2, . . . , $(n - 1)$
Magnetic quantum number	m_ℓ	$-\ell, \ldots, -2, -1, 0, +1, +2, \ldots, +\ell$
Spin quantum number	m_s	$-\tfrac{1}{2}, +\tfrac{1}{2}$

Example 5 Quantum Mechanical States of the Hydrogen Atom

Determine the number of possible states for the hydrogen atom when the principal quantum number is (a) $n = 1$ and (b) $n = 2$.

REASONING Each different combination of the four quantum numbers summarized in Table 30.1 corresponds to a different state. We begin with the value for n and find the allowed values for ℓ. Then, for each ℓ value we find the possibilities for m_ℓ. Finally, m_s may be $+\frac{1}{2}$ or $-\frac{1}{2}$ for each group of values for n, ℓ, and m_ℓ.

SOLUTION

(a) The diagram below shows the possibilities for ℓ, m_ℓ, and m_s when $n = 1$:

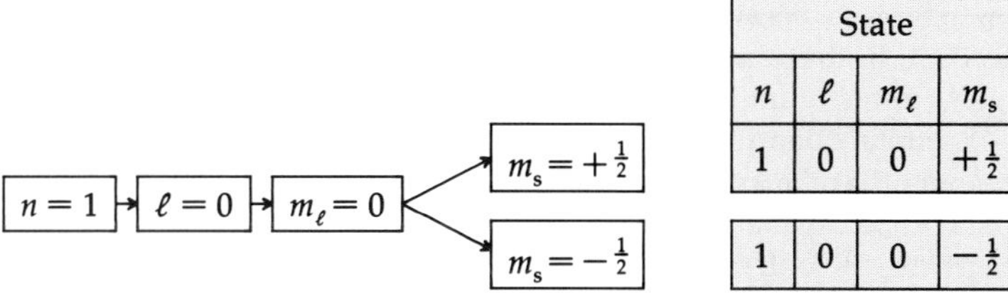

State			
n	ℓ	m_ℓ	m_s
1	0	0	$+\frac{1}{2}$
1	0	0	$-\frac{1}{2}$

Thus, there are two different states for the hydrogen atom. These two states have the same energy, since they have the same value of n.

(b) When $n = 2$, there are eight possible combinations for the values of n, ℓ, m_ℓ, and m_s, as the diagram below indicates:

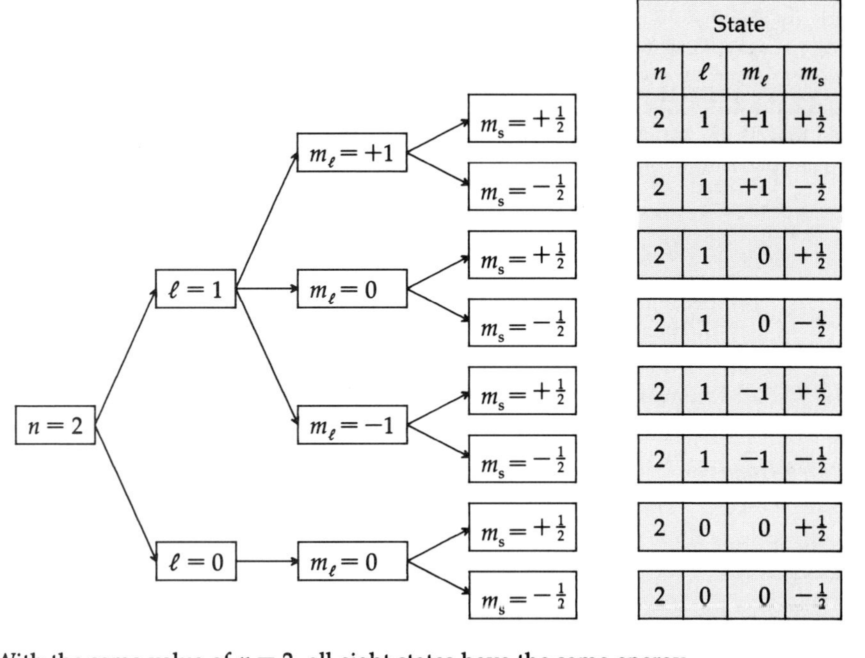

State			
n	ℓ	m_ℓ	m_s
2	1	+1	$+\frac{1}{2}$
2	1	+1	$-\frac{1}{2}$
2	1	0	$+\frac{1}{2}$
2	1	0	$-\frac{1}{2}$
2	1	−1	$+\frac{1}{2}$
2	1	−1	$-\frac{1}{2}$
2	0	0	$+\frac{1}{2}$
2	0	0	$-\frac{1}{2}$

With the same value of $n = 2$, all eight states have the same energy.

Quantum mechanics provides a more accurate description of reality than does the Bohr model. It is important to realize that the two pictures of atomic structure differ substantially, as Conceptual Example 6 illustrates.

Conceptual Example 6 The Bohr Model Versus Quantum Mechanics

Consider two hydrogen atoms. There are no external magnetic fields present, and the electron in each atom has the same energy. Is it possible for the electrons in these atoms (a) to have zero orbital angular momentum and (b) to have different orbital angular momenta? Evaluate these questions according to the Bohr model and to quantum mechanics.

REASONING

(a) In both the Bohr model and in quantum mechanics, the energy is proportional to $1/n^2$, according to Equation 30.13, where n is the principal quantum number. Moreover, the value of n may be $n = 1, 2, 3, \ldots$, and may not be zero. In the Bohr model, the fact that n may not be zero means that it is not possible for the orbital angular momentum to be zero, because the angular momentum is proportional to n, according to Equation 30.8. In the quantum mechanical picture, however, the magnitude of the orbital angular momentum is proportional to $\sqrt{\ell(\ell+1)}$, as given by Equation 30.15. Here, ℓ is the orbital quantum number and may take on the values $\ell = 0, 1, 2, \ldots, (n-1)$. We note that ℓ may be zero, no matter what the value for n. Consequently, $\sqrt{\ell(\ell+1)}$, and therefore the orbital angular momentum, may be zero for the electron in each of the atoms, in contrast to the case for the Bohr model.

(b) If the electrons have the same energy, they have the same value for the principal quantum number n. In the Bohr model, this means that they cannot have different values for the orbital angular momentum L_n, since $L_n = nh/(2\pi)$, according to Equation 30.8. Whereas the Bohr model uses the same quantum number n to specify both energy and orbital angular momentum, quantum mechanics uses different quantum numbers n and ℓ. In quantum mechanics, the energy is determined by n when external magnetic fields are absent, and the orbital angular momentum is determined by ℓ. Since $\ell = 0, 1, 2, \ldots, (n-1)$, different values of ℓ are compatible with the same value of n. For instance, if $n = 2$ for both of the electrons, one of them could have $\ell = 0$ while the other could have $\ell = 1$. As a result, the electrons would have different angular momenta, even though they have the same energy.

The following table summarizes the discussion from parts (a) and (b):

	Bohr Model	Quantum Mechanics
(a) For a given n, can the angular momentum ever be zero?	No	Yes
(b) For a given n, can the angular momentum have different values?	No	Yes

Related Homework Material: *Question 5, Problem 27*

ELECTRON PROBABILITY CLOUDS

According to the Bohr model, the nth orbit is a circle of radius r_n, and every time the position of the electron in this orbit is measured, the electron is found exactly a distance r_n away from the nucleus. This simplistic picture is now known to be incorrect. In its place, the quantum mechanical picture of the atom is now accepted as the correct one. Suppose the electron is in a quantum mechanical state for which $n = 1$, and we imagine making a number of measurements of the electron's position with respect to the nucleus. We would find that the position of the electron is uncertain, in the sense that there is a probability of finding the electron sometimes very near the nucleus, sometimes very far from the nucleus, and sometimes at intermediate locations. The probability is determined by the wave function Ψ, as Section 29.5 discusses. We can make a three-dimensional picture of our findings by marking a dot at each location where the electron is found. After a sufficient number of measurements is made, a picture of the quantum mechanical state emerges. A greater number of dots occur at places where the probability of finding the electron is higher. Figure 30.11 shows the spatial distribution for an electron in a state for which $n = 1$, $\ell = 0$, and $m_\ell = 0$. This picture is constructed from so many measurements that the individual dots are no longer visible, but have merged to form a kind of probability "cloud" whose density changes gradually from place to place. The dense regions indicate places where the probability of finding the electron is higher, while the less dense regions indicate places where the probability is lower. Also indicated in Figure 30.11 is the radius where quantum mechanics predicts the greatest probability per unit radial distance of finding the electron in the $n = 1$ state. This radius matches exactly the radius of 5.29×10^{-11} m found for the first Bohr orbit.

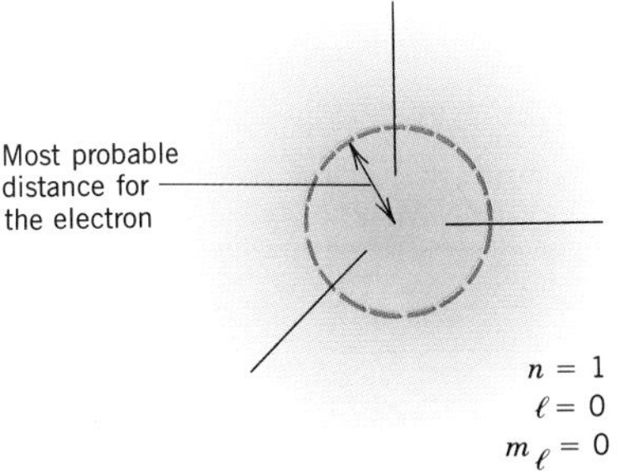

Figure 30.11 The electron probability cloud for the ground state ($n = 1$, $\ell = 0$, $m_\ell = 0$) of the hydrogen atom.

For a principal quantum number of $n = 2$, the probability clouds are different than for $n = 1$. In fact, more than one cloud shape is possible, because with $n = 2$ the orbital quantum number can be either $\ell = 0$ or $\ell = 1$. While the value of ℓ does not affect the energy of the hydrogen atom, the value does have a significant effect on the shape of the probability clouds. Figure 30.12*a* shows the cloud for $n = 2$, $\ell = 0$, and $m_\ell = 0$. Part *b* of the drawing shows that when $n = 2$, $\ell = 1$, and $m_\ell = 0$, the cloud has a two-lobe shape with the nucleus at the center between the lobes. For larger values of n, the probability clouds become increasingly complex and are spread out over larger volumes of space.

The probability cloud picture of the electron in a hydrogen atom is very different from the well-defined orbit of the Bohr model. The fundamental reason for this difference is to be found in the Heisenberg uncertainty principle, as Conceptual Example 7 discusses.

Conceptual Example 7 The Uncertainty Principle and the Hydrogen Atom

In the Bohr model of the hydrogen atom, the electron in the ground state is in an orbit that has a radius of exactly 5.29×10^{-11} m. Furthermore, as calculated in Example 4 in Chapter 18, the speed of the electron in this orbit is exactly 2.18×10^6 m/s. Considering the Heisenberg uncertainty principle, is this a realistic picture of atomic structure?

REASONING The Bohr model indicates that the electron is located exactly at a radius of 5.29×10^{-11} m, so the uncertainty Δy in its radial position is zero. We will now show that an uncertainty of $\Delta y = 0$ is not consistent with the Heisenberg principle, so that the Bohr picture is not realistic. We begin by choosing a value for the uncertainty in the momentum Δp_y and then use the uncertainty principle, $(\Delta p_y)(\Delta y) \geq h/(2\pi)$, to determine the corresponding value for Δy.

The magnitude of the electron's momentum is the product of its mass m and speed v, so that $\Delta p = \Delta(mv) = m\Delta v$. It seems reasonable to assume that the uncertainty Δv in the speed should be less than 2.18×10^6 m/s. This means that the speed is somewhere between zero and twice the value known to apply to the ground state in the Bohr model. A much larger uncertainty would mean that the electron could have so much energy that it would not likely remain in orbit. With $\Delta v = 2.18 \times 10^6$ m/s and $m = 9.11 \times 10^{-31}$ kg for the mass of the electron, we find that the minimum uncertainty in the radial position of the electron is $\Delta y = h/(2\pi\, m\Delta v) = 5.3 \times 10^{-11}$ m. This uncertainty is quite

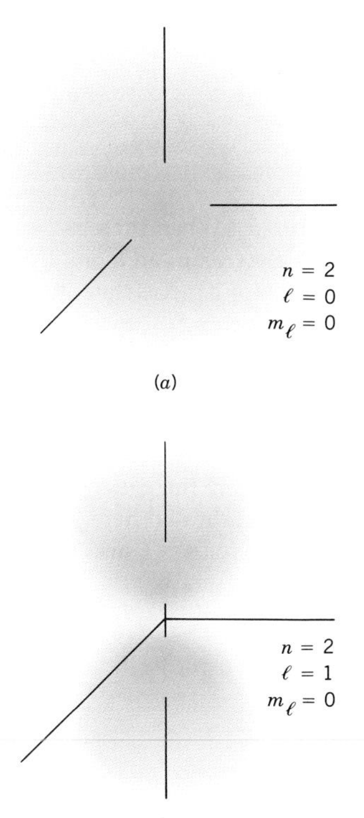

Figure 30.12 The electron probability clouds for the hydrogen atom when (*a*) $n = 2$, $\ell = 0$, $m_\ell = 0$ and (*b*) $n = 2$, $\ell = 1$, $m_\ell = 0$.

large, for it is virtually equal to the Bohr radius. According to the uncertainty principle, then, the radial position of the electron can be anywhere from approximately zero to twice the Bohr radius. The single-radius orbit of the Bohr model does not correctly represent this aspect of reality at the atomic level. Quantum mechanics, however, does correctly represent it in terms of a probability cloud picture of atomic structure.

30.6 THE PAULI EXCLUSION PRINCIPLE AND THE PERIODIC TABLE OF THE ELEMENTS

MULTIPLE-ELECTRON ATOMS

Except for hydrogen, all electrically neutral atoms contain more than one electron, the number being given by the atomic number Z of the element. In addition to being attracted by the nucleus, the electrons repel each other. This repulsion contributes to the total energy of a multiple-electron atom. As a result, the one-electron energy expression for hydrogen, $E_n = -(13.6\text{ eV})\, Z^2/n^2$, does not apply to other neutral atoms. However, the simplest approach for dealing with a multiple-electron atom still uses the four quantum numbers n, ℓ, m_ℓ, and m_s.

Detailed quantum mechanical calculations reveal that the energy level of each state of a multiple-electron atom depends on both the principal quantum number n and the orbital quantum number ℓ. Figure 30.13 illustrates that the energy generally increases as n increases. Furthermore, for a given n, the energy also increases as ℓ increases, but there are some exceptions, as the drawing indicates.

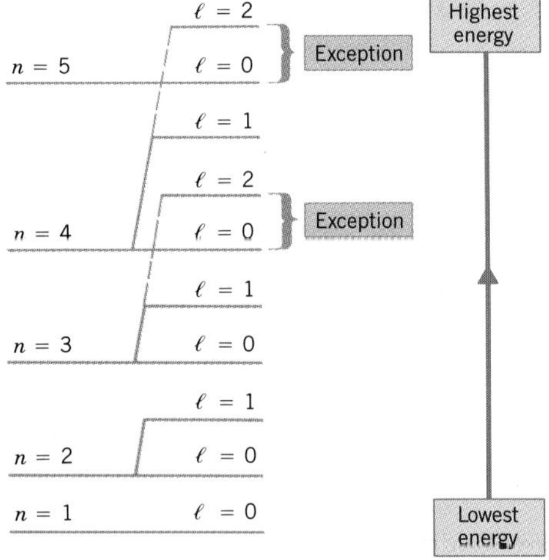

Figure 30.13 When there is more than one electron in an atom, the total energy of a given state depends on the principal quantum number n and the orbital quantum number ℓ. Generally, the energy increases with increasing n and, for a fixed n, with increasing ℓ. There are exceptions to the general rule, however, as indicated here. For clarity, levels for $n = 6$ and higher are not shown.

In a multiple-electron atom, all electrons with the same value of n are said to be in the same ***shell.*** Electrons with $n = 1$ are in a single shell (sometimes called the K shell), electrons with $n = 2$ are in another shell (the L shell), those with $n = 3$ are in a third shell (the M shell), and so on. Those electrons with the same values for both n and ℓ are often referred to as being in the same ***subshell.*** The $n = 1$ shell consists of a single $\ell = 0$ subshell. The $n = 2$ shell has two subshells, one with $\ell = 0$ and one with $\ell = 1$. Similarly, the $n = 3$ shell has three subshells.

In the hydrogen atom near room temperature, the electron spends most of its time in the lowest energy level or ground state, namely, in the $n = 1$ shell.

Similarly, when an atom contains more than one electron and is near room temperature, the electrons spend most of their time in the lowest energy levels possible. The lowest energy state for an atom is called the ***ground state.*** However, when a multiple-electron atom is in its ground state, not every electron is crowded into the $n = 1$ shell. The reason the electrons are not all in the same shell is that they obey a principle discovered by the Austrian physicist Wolfgang Pauli (1900–1958).

The Pauli Exclusion Principle

No two electrons in an atom can have the same set of values for the four quantum numbers n, ℓ, m_ℓ, and m_s.

For instance, suppose two electrons in an atom have three quantum numbers that are identical: $n = 3$, $m_\ell = 1$, and $m_s = -\frac{1}{2}$. According to the exclusion principle, it is not possible for each to have $\ell = 2$, for example, since each would then have the same four quantum numbers. Each electron must have a different value for ℓ ($\ell = 1$ and $\ell = 2$, for instance) and, consequently, be in a different subshell. With the aid of the Pauli exclusion principle, we can determine which energy levels are occupied by the electrons in an atom in its ground state, as the next example demonstrates.

Example 8 Ground States of Atoms

Determine which of the energy levels in Figure 30.13 are occupied by the electrons in the ground state of hydrogen (1 electron), helium (2 electrons), lithium (3 electrons), beryllium (4 electrons), and boron (5 electrons).

REASONING In the ground state of an atom the electrons are in the lowest available energy levels. Consistent with the Pauli exclusion principle, they fill those levels "from the bottom up," that is, from the lowest to the highest energy.

SOLUTION As the colored dot ( ) in Figure 30.14 indicates, the electron in the hydrogen atom (H) is in the $n = 1$, $\ell = 0$ subshell, which has the lowest possible energy. A second electron is present in the helium atom (He), and both electrons can have the quantum numbers $n = 1$, $\ell = 0$, and $m_\ell = 0$. However, in accord with the Pauli exclusion principle, each electron must have a different spin quantum number, $m_s = +\frac{1}{2}$ for one electron and $m_s = -\frac{1}{2}$ for the other. Thus, the drawing shows both electrons in the lowest energy level.

The third electron that is present in the lithium atom (Li) would violate the exclusion principle if it were also in the $n = 1$, $\ell = 0$ subshell, no matter what the value for m_s. Thus, the $n = 1$, $\ell = 0$ subshell is filled when occupied by two electrons. With this level filled, the $n = 2$, $\ell = 0$ subshell becomes the next lowest energy level available and is

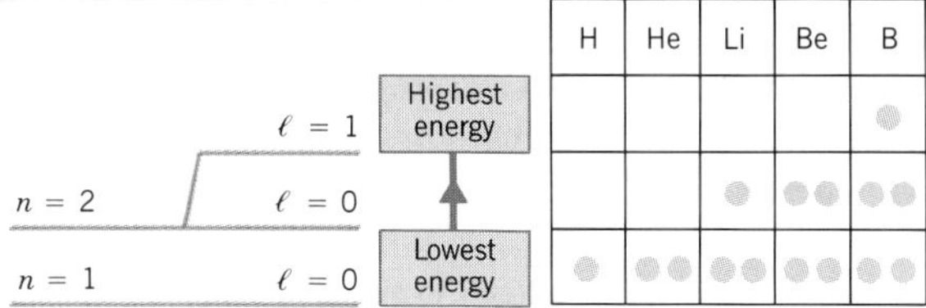

H	He	Li	Be	B
				●
		●	●●	●●
●	●●	●●	●●	●●

Figure 30.14 The electrons ( ) in the ground state of an atom fill the available energy levels "from the bottom up," that is, from the lowest to the highest energy, consistent with the Pauli exclusion principle. The ranking of the energy levels in this figure is meant to apply for a given atom, not between one atom and another.

where the third electron of lithium is found (see Figure 30.14). In the beryllium atom (Be), the fourth electron is in the $n = 2$, $\ell = 0$ subshell, along with the third electron. This is possible, since the third and fourth electrons can have different values for m_s.

With the first four electrons in place as discussed above, the fifth electron in the boron atom (B) cannot fit into the $n = 1$, $\ell = 0$ or the $n = 2$, $\ell = 0$ subshell without violating the exclusion principle. Therefore, the fifth electron is found in the $n = 2$, $\ell = 1$ subshell, which is the next available energy level with the lowest energy, as Figure 30.14 indicates. For this electron, m_ℓ can be $-1, 0,$ or $+1$, and m_s can be $+\frac{1}{2}$ or $-\frac{1}{2}$ in each case. However, in the absence of an external magnetic field, all six of these possibilities correspond to the same energy.

Because of the Pauli exclusion principle, there is a maximum number of electrons that can fit into an energy level or subshell. Example 8 shows that the $n = 1$, $\ell = 0$ subshell can hold at most two electrons. The $n = 2$, $\ell = 1$ subshell, however, can hold six electrons, because with $\ell = 1$, there are three possibilities for m_ℓ ($-1, 0,$ and $+1$), and for each of these, the value of m_s can be $+\frac{1}{2}$ or $-\frac{1}{2}$. In general, m_ℓ can have the values $0, \pm 1, \pm 2, \ldots, \pm \ell$, for $2\ell + 1$ possibilities. Since each of these can be combined with two possibilities for m_s, the total number of different combinations for m_ℓ and m_s is $2(2\ell + 1)$. This, then, is the maximum number of electrons the ℓth subshell can hold, as Figure 30.15 summarizes.

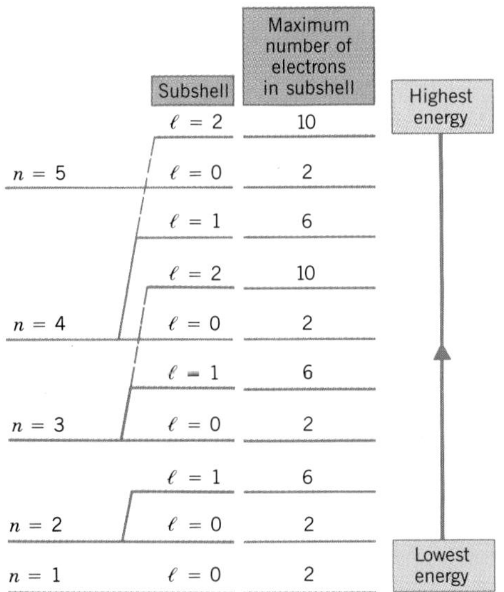

Figure 30.15 The maximum number of electrons that the ℓth subshell can hold is $2(2\ell + 1)$.

SHORTHAND NOTATION FOR THE ELECTRONIC CONFIGURATION OF THE ATOM

For historical reasons, there is a widely used convention in which each subshell of an atom is referred to by a letter, rather than by the value of its orbital quantum number ℓ. For instance, an $\ell = 0$ subshell is called an s subshell. An $\ell = 1$ subshell and an $\ell = 2$ subshell are known as p and d subshells, respectively. The higher values of $\ell = 3, 4,$ etc., are referred to as f, g, etc., in alphabetical sequence, as Table 30.2 indicates.

This convention of letters is used in a shorthand notation that is convenient for simultaneously indicating the principal quantum number n, the orbital quantum number ℓ, and the number of electrons in the n, ℓ subshell. The notation is as follows:

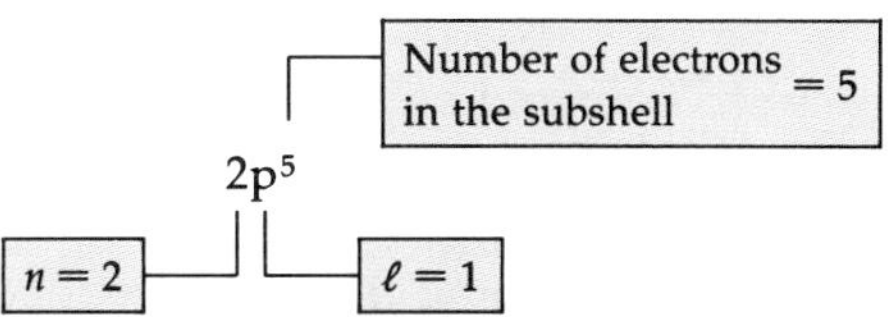

With this notation, the arrangement or configuration of the electrons in an atom can be specified efficiently. For instance, in Example 8, we found that the electron configuration for boron has two electrons in the $n = 1, \ell = 0$ subshell, two in the $n = 2, \ell = 0$ subshell, and one in the $n = 2, \ell = 1$ subshell. In shorthand notation this arrangement is expressed as $1s^2\ 2s^2\ 2p^1$. Table 30.3 gives the ground state electron configurations written in this fashion for elements containing up to eighteen electrons. The first five entries are those worked out in Example 8.

Table 30.2 The Convention of Letters Used to Refer to the Orbital Quantum Number

Orbital Quantum Number ℓ	Letter
0	s
1	p
2	d
3	f
4	g
5	h

Table 30.3 Ground State Electronic Configurations of Atoms

Element	Number of Electrons	Configuration of the Electrons
Hydrogen (H)	1	$1s^1$
Helium (He)	2	$1s^2$
Lithium (Li)	3	$1s^2\ 2s^1$
Beryllium (Be)	4	$1s^2\ 2s^2$
Boron (B)	5	$1s^2\ 2s^2\ 2p^1$
Carbon (C)	6	$1s^2\ 2s^2\ 2p^2$
Nitrogen (N)	7	$1s^2\ 2s^2\ 2p^3$
Oxygen (O)	8	$1s^2\ 2s^2\ 2p^4$
Fluorine (F)	9	$1s^2\ 2s^2\ 2p^5$
Neon (Ne)	10	$1s^2\ 2s^2\ 2p^6$
Sodium (Na)	11	$1s^2\ 2s^2\ 2p^6\ 3s^1$
Magnesium (Mg)	12	$1s^2\ 2s^2\ 2p^6\ 3s^2$
Aluminum (Al)	13	$1s^2\ 2s^2\ 2p^6\ 3s^2\ 3p^1$
Silicon (Si)	14	$1s^2\ 2s^2\ 2p^6\ 3s^2\ 3p^2$
Phosphorus (P)	15	$1s^2\ 2s^2\ 2p^6\ 3s^2\ 3p^3$
Sulfur (S)	16	$1s^2\ 2s^2\ 2p^6\ 3s^2\ 3p^4$
Chlorine (Cl)	17	$1s^2\ 2s^2\ 2p^6\ 3s^2\ 3p^5$
Argon (Ar)	18	$1s^2\ 2s^2\ 2p^6\ 3s^2\ 3p^6$

THE PERIODIC TABLE

THE PHYSICS OF . . .
the periodic table of the elements.

Each entry in the periodic table of the elements often includes the ground state electronic configuration, as Figure 30.16 illustrates for argon. To save space, only the configuration of the outermost electrons and unfilled subshells is specified.

Originally the periodic table was developed by the Russian chemist Dmitri Mendeleev (1834–1907) on the basis that certain groups of elements exhibit similar chemical properties. There are eight of these groups, plus the transition elements in the middle of the table, the lanthanide series, and the actinide series. The similar chemical properties within a group can be explained on the basis of

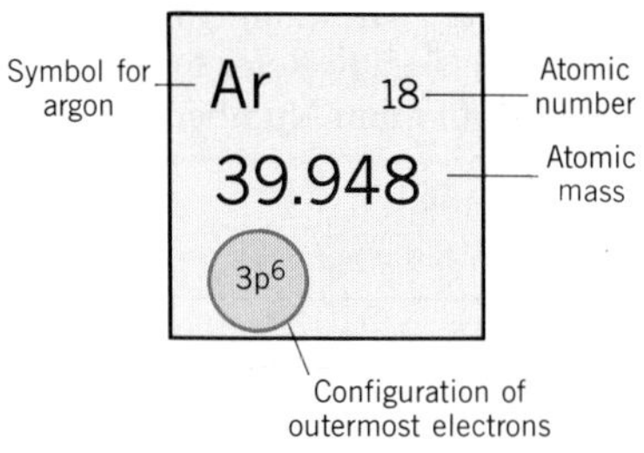

Figure 30.16 The entries in the periodic table of the elements often include the ground state configuration of the outermost electrons.

the configurations of the outer electrons of the elements in the group. Thus, quantum mechanics and the Pauli exclusion principle offer an explanation for the chemical behavior of the atoms.

The full periodic table can be found on the inside of the back cover. Group 0, the last column of elements on the right side of the table, consists of the noble gases, such as helium (He), neon (Ne), and argon (Ar). Chemically, these elements are relatively inert, because their outermost electrons form a shell or subshell that is completely full. Being full, the shell or subshell is very stable, not readily forming chemical bonds by accepting electrons from or donating electrons to other elements.

Group I is made up of the alkali metals. Sodium (Na) and potassium (K) are familiar members of this group. These elements are chemically reactive, because they have only a single electron in an outermost s subshell. This electron can be easily lost to other elements in a chemical reaction. Thus, elements in group I often form singly charged positive ions, such as the sodium ion Na^+.

Group VII consists of the halogens and includes fluorine (F), chlorine (Cl), and bromine (Br). These elements have outermost electrons in a p subshell that is only one electron shy of being full. The halogens are highly reactive. Their chemistry is characterized by reactions in which they accept a single electron from other elements to form a stable, filled subshell. For this reason, the halogens readily form ions such as the chloride ion Cl^-, which carry a single negative charge.

By looking at the other groups (II–VI) in the periodic table, you can see that elements within a group have outermost electrons in either s or p subshells. Within a group, the subshells are filled to the same extent. The transition elements are elements formed when electrons fill out primarily 3d, 4d, 5d, and 6d subshells. The lanthanide series involves completing mainly the 5d and 4f subshells. And finally, the actinide series corresponds to filling out primarily the 5f and 6d subshells.

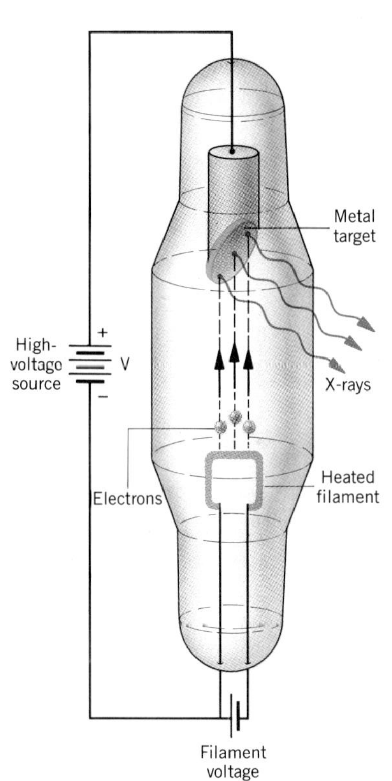

Figure 30.17 In an X-ray tube, electrons are emitted by a heated filament, accelerate through a large potential difference V, and strike a metal target. The X-rays originate from the interaction between the electrons and the metal target.

30.7 X-RAYS

X-rays were discovered by Wilhelm K. Roentgen (1845–1923), a Dutch physicist who performed much of his work in Germany. X-rays can be produced when electrons, accelerated through a large potential difference, collide with a metal target made from molybdenum or platinum, for example. The target is contained within an evacuated glass tube, as Figure 30.17 shows. A plot of X-ray intensity per unit wavelength versus the wavelength looks similar to Figure 30.18 and consists of sharp peaks or lines superimposed on a broad continuous spectrum. The sharp peaks are called characteristic lines or ***characteristic X-rays,*** because they are characteristic of the target material. The broad continuous spectrum is referred to as ***Bremsstrahlung*** (German for "braking radiation"). Bremsstrahlung X-rays are emitted when the electrons decelerate or "brake" upon hitting the target.

In Figure 30.18 the characteristic lines are marked K_α and K_β, because they involve the $n = 1$ or K shell of a metal atom. If an electron with enough energy strikes the target, one of the K-shell electrons can be knocked entirely out of a target atom. An electron in one of the outer shells can then fall into the K shell, and an X-ray photon is emitted in the process. Example 9 shows that a large potential difference is needed to operate an X-ray tube, so the electrons impinging on the metal target have sufficient energy to generate the characteristic X-rays.

Example 9 The Voltage Needed to Operate an X-Ray Tube

Strictly speaking, the Bohr model does not apply to multiple-electron atoms, but it can be used to make estimates. Use the Bohr model to estimate the minimum energy that an incoming electron must have to knock a K-shell electron entirely out of an atom in a platinum (Z = 78) target in an X-ray tube.

REASONING According to the Bohr model, the energy of a K-shell electron is given by Equation 30.13 with $n = 1$: $E_n = -(13.6 \text{ eV}) Z^2/n^2$. When striking a platinum target, an incoming electron must have at least enough energy to raise the K-shell electron from this low energy level up to the 0-eV level that corresponds to a very large distance from the nucleus. Only then will the incoming electron knock the K-shell electron out of the platinum atom.

SOLUTION The energy of the Bohr $n = 1$ level is

$$E_1 = -(13.6 \text{ eV}) \frac{Z^2}{n^2} = -(13.6 \text{ eV}) \frac{77^2}{1^2} = -8.1 \times 10^4 \text{ eV}$$

In this calculation we have used 77 rather than 78 for the value of Z. In so doing, we account approximately for the fact that each of the two K-shell electrons applies a repulsive force to the other. This repulsive force tends to balance the attractive force of one nuclear proton. In effect, one electron shields the other from the force of that proton. Therefore, to raise the K-shell electron up to the 0-eV level, the minimum energy for an incoming electron is $\boxed{8.1 \times 10^4 \text{ eV}}$. One electron volt is the kinetic energy acquired when an electron accelerates from rest through a potential difference of one volt. Thus, a potential difference of 81 000 V must be applied to the X-ray tube.

PROBLEM SOLVING INSIGHT

Equation 30.13 for the Bohr energy levels [$E_n = -(13.6 \text{ eV})Z^2/n^2$, $n = 1$] can be used in rough calculations of the energy levels involved in the production of K_α X-rays. In such calculations, the value of the atomic number Z must be reduced by one, to account approximately for the shielding of one K-shell electron by the other K-shell electron.

The K_α line in Figure 30.18 arises when an electron in the $n = 2$ level falls into the vacancy that the impinging electron has created in the $n = 1$ level. Similarly, the K_β line arises when an electron in the $n = 3$ level falls to the $n = 1$ level. Example 10 determines an estimate for the K_α wavelength of platinum.

Example 10 The K_α Characteristic X-Ray for Platinum

Use the Bohr model to estimate the wavelength of the K_α line in the X-ray spectrum of platinum.

REASONING This example is very similar to Example 4, which deals with the emission line spectrum of the hydrogen atom. As in that example, we use Equation 30.14, this time with the initial value of n being $n_i = 2$ and the final value being $n_f = 1$. As in Example 9, a value of 77 rather than 78 is used for Z, to account approximately for the shielding effect of the single K-shell electron in canceling out the attraction of one nuclear proton.

SOLUTION Using Equation 30.14, we find that

$$\frac{1}{\lambda} = (1.097 \times 10^7 \text{ m}^{-1})(77^2) \left(\frac{1}{1^2} - \frac{1}{2^2} \right)$$

$$= 4.9 \times 10^{10} \text{ m}^{-1} \quad \text{or} \quad \boxed{\lambda = 2.0 \times 10^{-11} \text{ m}}$$

This answer is close to an experimental value of 1.9×10^{-11} m.

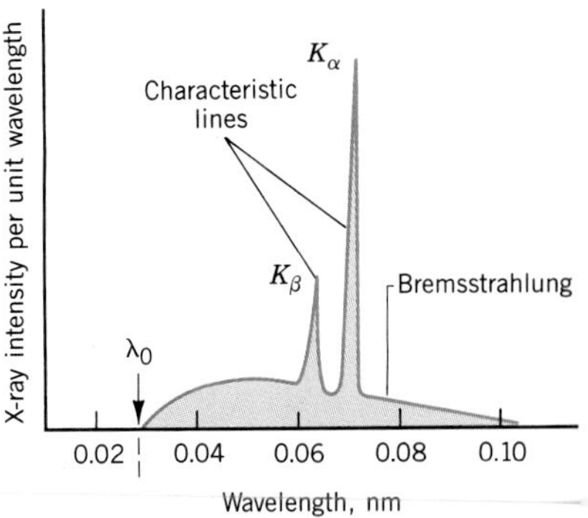

Figure 30.18 When a molybdenum target is bombarded with electrons that have been accelerated from rest through a potential difference of 45 000 V, the X-ray spectrum shown here is produced. The vertical axis is not drawn to scale.

THE PHYSICS OF . . .
X-rays.

Another interesting feature of the X-ray spectrum in Figure 30.18 is the sharp cutoff that occurs at a wavelength of λ_0 on the short wavelength side of the Bremsstrahlung. This cutoff wavelength is independent of the target material, but depends on the energy of the impinging electrons. An impinging electron cannot give up any more than all of its kinetic energy when decelerated by the metal target in an X-ray tube. Thus, at most, an emitted X-ray photon can have an energy equal to the kinetic energy KE of the electron and a frequency given by Equation 29.2 as $f = (\text{KE})/h$, where h is Planck's constant. But the kinetic energy acquired by an electron in accelerating from rest through a potential difference V is eV, according to earlier discussions in Section 19.2; V is the potential difference applied across the X-ray tube. Thus, the maximum photon frequency is $f_0 = (eV)/h$. Since $f_0 = c/\lambda_0$, a maximum frequency corresponds to a minimum wavelength, which is the cutoff wavelength λ_0:

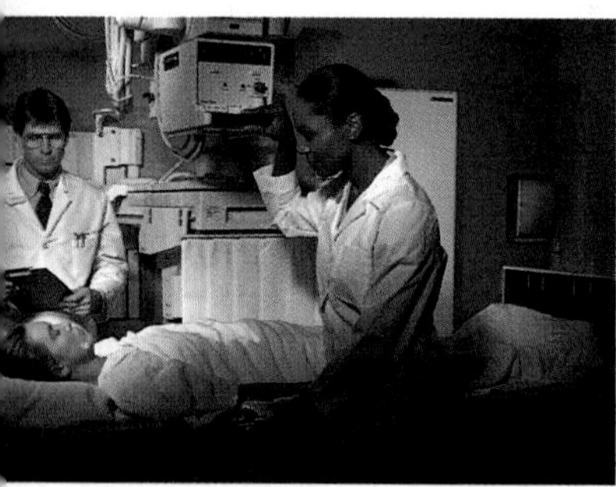

Technicians are preparing to take an abdominal X-ray of this patient.

$$\lambda_0 = \frac{hc}{eV} \tag{30.17}$$

Figure 30.18, for instance, assumes a potential difference of 45 000 V, which corresponds to a cutoff wavelength of

$$\lambda_0 = \frac{(6.63 \times 10^{-34}\ \text{J}\cdot\text{s})(3.00 \times 10^{8}\ \text{m/s})}{(1.60 \times 10^{-19}\ \text{C})(45\ 000\ \text{V})} = 2.8 \times 10^{-11}\ \text{m}$$

The medical profession began using X-rays for diagnostic purposes almost immediately after their discovery. When a conventional X-ray is obtained, the patient is typically positioned in front of a piece of photographic film, and a single burst of radiation is directed through the patient and onto the film. Since the dense structure of bone absorbs X-rays much more than soft tissue does, a shadow-like picture is recorded on the film. As useful as such pictures are, they have an inherent limitation. The image on the film is a superposition of all the "shadows" that result as the radiation passes through one layer of body material after another. Interpreting which part of a conventional X-ray corresponds to which layer of body material is very difficult.

THE PHYSICS OF . . .
CAT scanning.

The technique known as CAT scanning or CT scanning has greatly extended the ability of X-rays to provide images from specific locations within the body. The acronym CAT stands for computerized axial tomography or computer-assisted tomography, while the shorter version CT stands for computerized tomography. In this technique a series of X-ray images are obtained as indicated in Figure 30.19. A number of X-ray beams form a "fanned out" array of radiation and pass simultaneously through the patient. Each of the beams is detected on the

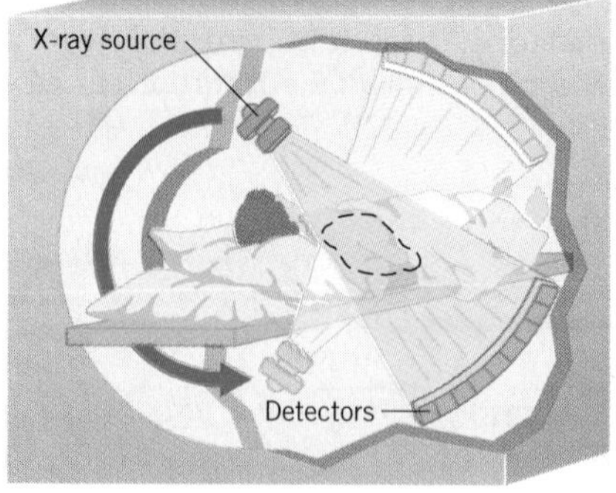

(*a*)

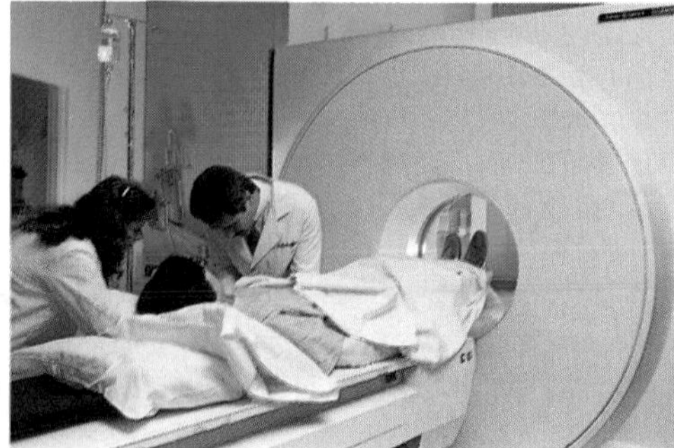

(*b*)

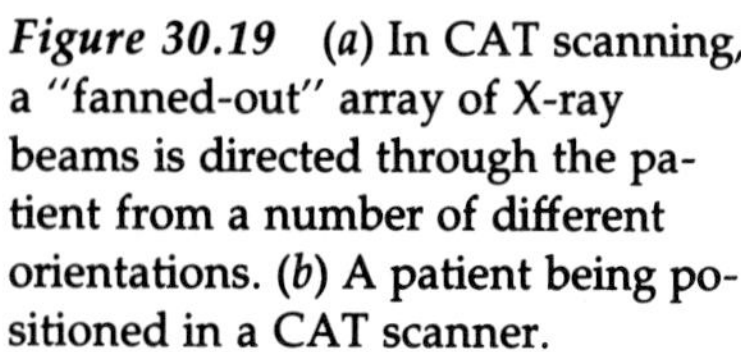

Figure 30.19 (*a*) In CAT scanning, a "fanned-out" array of X-ray beams is directed through the patient from a number of different orientations. (*b*) A patient being positioned in a CAT scanner.

other side by its own detector, which records the beam intensity. The various intensities are slightly different, depending on the nature of the body material through which the beams have passed. At this stage, there is little difference from a conventional X-ray technique, except that more than one beam is used.

The new feature of CAT scanning that leads to dramatic improvements over the conventional technique is that the X-ray source and its detectors can be rotated to a new orientation. Thus, the "fanned-out" array of beams can be sent through the patient from various directions. Figure 30.19*a* singles out two directions for illustration. In reality many different orientations are used, and the intensity of each beam in the array is recorded as a function of orientation. The way in which the intensity of a beam changes from one orientation to another is used as input to a computer. The computer then constructs a highly resolved image of the cross-sectional slice of body material through which the "fan" of radiation has passed. In effect, the CAT scanning technique makes it possible to take an X-ray picture of a cross-sectional "slice" that is perpendicular to the body's long axis. In fact, the word "axial" in the phrase "computerized axial tomography" refers to the body's long axis. Figure 30.20*a* shows a two-dimensional CAT scan of the type discussed here, while part *b* of the figure shows a three-dimensional version.

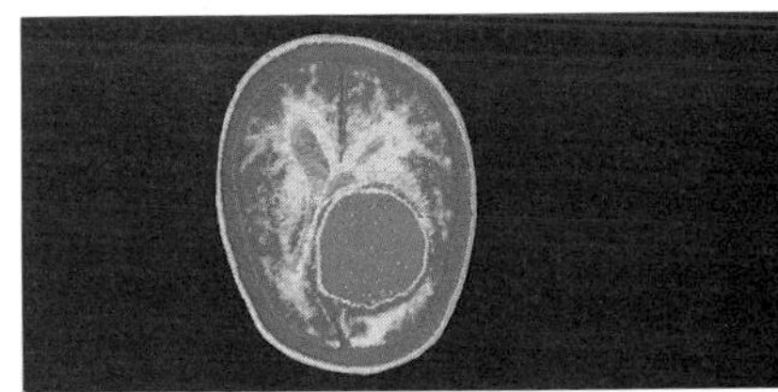

(a)

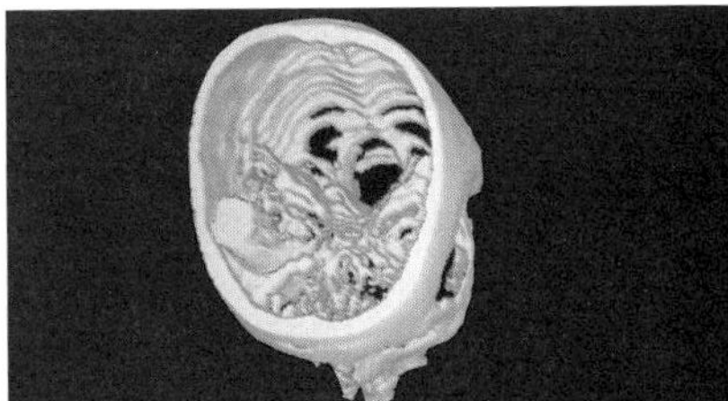

(b)

Figure 30.20 (*a*) This two-dimensional CAT scan of a brain reveals a large intracranial tumor (colored purple). (*b*) Three-dimensional CAT scans are now available, and this example reveals an arachnoid cyst (colored yellow) within a skull. In both photographs, the colors are artificial, having been computer generated to aid in distinguishing anatomical features.

30.8 THE LASER

The laser is certainly one of the most useful inventions of the twentieth century. Today, there are many types of lasers, and most of them work in a way that depends directly on the quantum mechanical structure of the atom.

THE PHYSICS OF . . .

the laser.

When an electron makes a transition from a higher energy state to a lower energy state, a photon is emitted. The emission process can be one of two types, spontaneous emission or stimulated emission. In ***spontaneous emission*** (see Figure 30.21*a*), the photon is emitted spontaneously, in a random direction, without external provocation. In ***stimulated emission*** (see Figure 30.21*b*), an incoming photon induces or stimulates the electron to change energy levels. To produce stimulated emission, however, the incoming photon must have an energy that exactly matches the difference between the energies of the two levels, namely, $E_i - E_f$. Stimulated emission is similar to a resonance process, in which the incoming photon "jiggles" the electron at just the frequency to which it is particularly sensitive and, thus, causes the change between energy levels. This frequency is given by Equation 30.4 as $f = (E_i - E_f)/h$. The operation of lasers depends on stimulated emission.

Stimulated emission has three important features. First, one photon goes in and two photons come out. In this sense, the process amplifies the number of photons. In fact, this is the origin of the word "laser," which is an acronym for light amplification by the stimulated emission of radiation. Second, the emitted photon travels in the same direction as the incoming photon. And third, the emitted photon is exactly in step with or has the same phase as the incoming photon. In other words, the two electromagnetic waves that these two photons represent are coherent and are locked in step with one another. In contrast, two photons emitted by the filament of an incandescent light bulb are emitted independently. They are not coherent, since one does not stimulate the emission of the other.

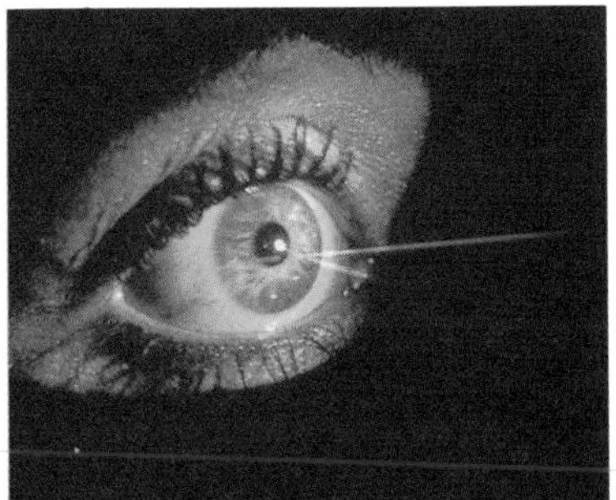

This helium-neon laser beam is being used to diagnose diseases of the eye. The use of laser technology in the field of ophthalmology is widespread.

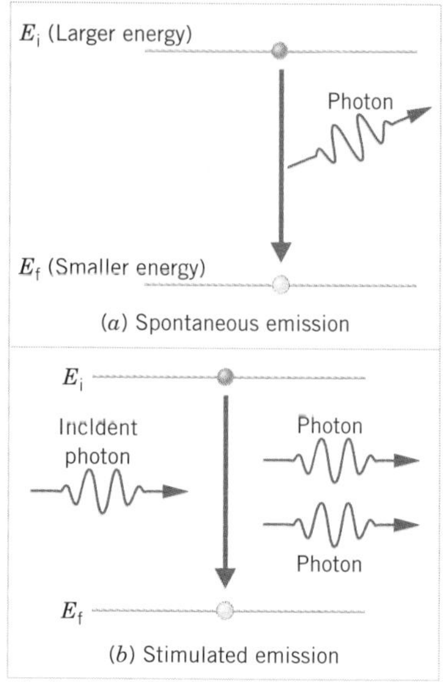

Figure 30.21 (*a*) Spontaneous emission of a photon occurs when the electron (●) makes an unprovoked transition from a higher to a lower energy level, the photon departing in a random direction. (*b*) Stimulated emission of a photon occurs when an incoming photon with the correct energy induces an electron to change energy levels, the emitted photon traveling in the same direction as the incoming photon.

While stimulated emission plays a pivotal role in a laser, other factors are also important. For instance, an external source of energy must be available to excite electrons into higher energy levels. The energy can be provided in a number of ways, including intense flashes of ordinary light and high-voltage discharges. If sufficient energy is delivered to the atoms, more electrons will be excited to a higher energy level than remain in a lower energy level, a condition known as a ***population inversion.*** Figure 30.22 compares a normal energy level population with a population inversion. The population inversions used in lasers involve a higher energy state that is ***metastable,*** in the sense that electrons remain in it for a much longer period of time than they do in an ordinary excited state (10^{-3} s versus 10^{-8} s, for example). The requirement of a metastable higher energy state is essential, so that there is more time to enhance the population inversion.

Figure 30.23 shows the widely used helium/neon laser. To sustain the necessary population inversion a high voltage is discharged across a low-pressure mixture of 15% helium and 85% neon contained in a glass tube. The laser process begins when an atom, via spontaneous emission, emits a photon parallel to the axis of the tube. This photon, via stimulated emission, causes another atom to emit two photons parallel to the tube axis. These two photons, in turn, stimulate two more atoms, yielding four photons. Four yield eight, and so on, in a kind of avalanche effect. To ensure that more and more photons are created by stimulated emission, both ends of the tube are silvered to form mirrors that reflect the photons back and forth through the helium/neon mixture. One end is only partially silvered, however, so that some of the photons can escape from the tube to form the laser beam. When the stimulated emission in a laser involves only a single pair of energy levels, the output beam has a single frequency or wavelength; that is, the radiation in the laser beam is monochromatic.

A laser beam is also exceptionally narrow. The width is determined by the size of the opening through which the beam exits, and very little spreading-out occurs, except that due to diffraction around the edges of the opening. A laser beam does not spread much, because any photons emitted at an angle with respect to the tube axis are quickly reflected out the sides of the tube by the silvered ends (see Figure 30.23). These ends are carefully arranged to be perpendicular to the tube axis. Since all the power in a laser beam can be confined to a narrow region, the intensity, or power per unit area, can be quite large.

Figure 30.24 shows the pertinent energy levels for a helium/neon laser. By coincidence, helium and neon have nearly identical metastable higher energy states, respectively located 20.61 and 20.66 eV above the ground state. The high-voltage discharge across the gaseous mixture excites electrons in helium atoms to

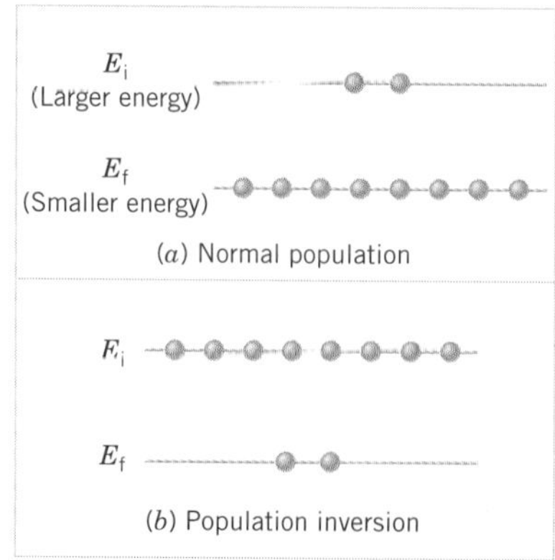

Figure 30.22 (*a*) In a normal situation at room temperature, most of the electrons in atoms are found in a lower or ground state energy level. (*b*) If an external energy source is provided to excite electrons into a higher energy level, a population inversion can be created, in which more electrons are in the higher level than in the lower level.

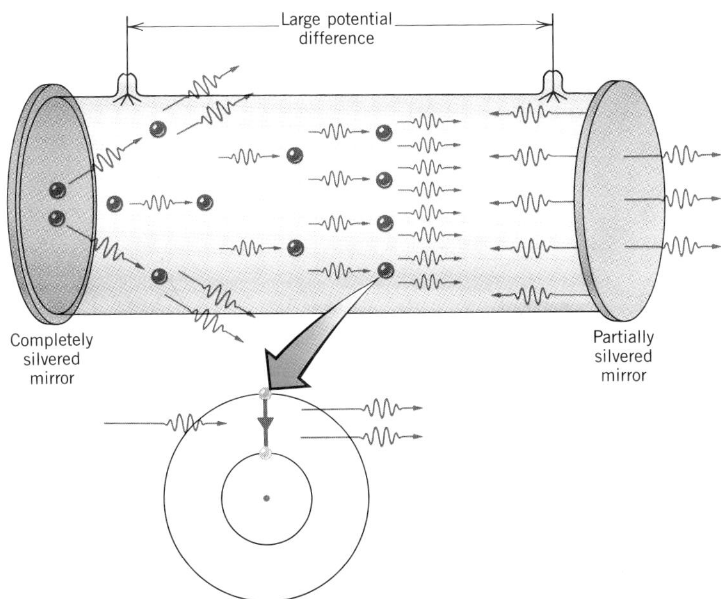

Figure 30.23 A schematic drawing of a helium/neon laser.

the 20.61-eV state. Then, when an excited helium atom collides inelastically with a neon atom, the 20.61 eV of energy is given to an electron in the neon atom, along with 0.05 eV of kinetic energy from the moving atoms. As a result, the electron in the neon atom is raised to the 20.66-eV state. In this fashion, a population inversion is sustained in the neon, relative to an energy level that is 18.70 eV above the ground state. In producing the laser beam, stimulated emission causes electrons in neon to drop from the 20.66-eV level to the 18.70-eV level. The energy change of 1.96 eV corresponds to a wavelength of 633 nm, which is in the red region of the visible spectrum.

The helium/neon laser is not the only kind of laser. There are many different types, including the ruby laser, the argon-ion laser, the carbon dioxide laser, the gallium arsenide solid-state laser, and chemical dye lasers. Depending on the type and whether the laser operates continuously or in pulses, the available beam power ranges from milliwatts to megawatts. Since lasers provide coherent monochromatic electromagnetic radiation that can be confined to an intense narrow beam, they are useful in a wide variety of situations. Today they are used to reproduce music in compact disc players, to weld parts of automobile frames together, to measure distances accurately in surveying, to transmit telephone conversations and other forms of communication over long distances, and to

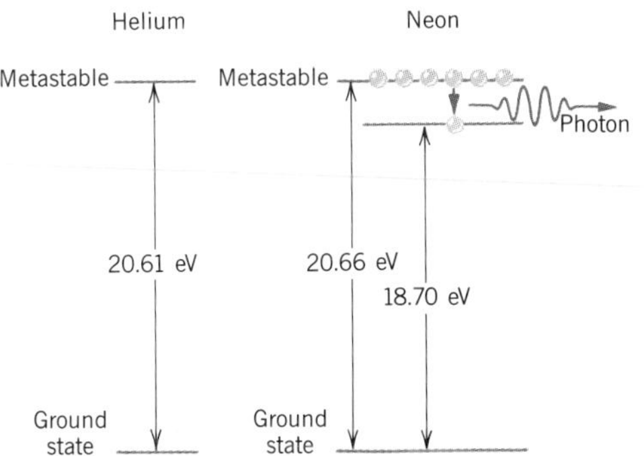

Figure 30.24 These energy levels are involved in the operation of a helium/neon laser.

study molecular structure. Medical applications of lasers include delicate eye surgery, removal of kidney stones, and removal of tooth and gum decay. These are only a few of the uses for lasers. Many other uses have been found since the laser was invented in 1960.

THE PHYSICS OF . . .

holography.

*30.9 HOLOGRAPHY

One of the most familiar applications of lasers is in holography, which is a process for producing three-dimensional images. The information used to produce a holographic image is captured on photographic film, which is referred to as a hologram. Figure 30.25 illustrates how a hologram is made. Laser light strikes a half-silvered mirror, or beamsplitter, which reflects part and transmits part of the light. In the drawing, the reflected part is called the object beam, because it illuminates the object (a chess piece). The transmitted part is called the reference beam. The object beam reflects from the chess piece at points such as A and B and, together with the reference beam, falls on the film. One of the main characteristics of laser light is that it is coherent. Thus, the light from the two beams has a stable phase relationship, like the light from the two slits in Young's double slit experiment (see Section 27.2). Because of the stable phase relationship and because the two beams travel different distances, an interference pattern is formed on the film. This pattern is the hologram and, although much more complex, is analogous to the pattern of bright and dark fringes formed in the double-slit experiment.

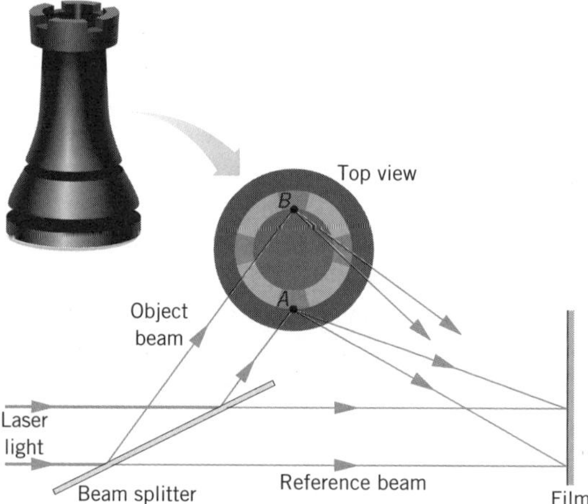

Figure 30.25 An arrangement used to produce a hologram.

Figure 30.26 shows in greater detail how a holographic interference pattern arises. This drawing considers only the reference beam and the light (wavelength $= \lambda$) coming from point A on the chess piece. As we know from Section 27.1, constructive interference between the two light waves leads to a bright fringe; constructive interference occurs when the waves, in reaching the film, travel distances that differ by an integer number m of wavelengths. As Figure 30.26 indicates, ℓ_m is the distance between point A and the place on the film where the mth order bright fringe occurs. ℓ_0 is the perpendicular distance that the reference beam would travel from point A to the $m = 0$ bright fringe. In addition, r_m is the vertical distance that locates the bright fringe. In terms of these distances, we know that

$$\ell_m - \ell_0 = m\lambda \qquad \text{(condition for constructive interference)}$$

$$\ell_0^2 + r_m^2 = \ell_m^2 \qquad \text{(Pythagorean theorem)}$$

The first of these equations indicates that $\ell_m = m\lambda + \ell_0$, which can be substituted into the second equation. The result can be rearranged to show that

$$r_m^2 = m\lambda(m\lambda + 2\ell_0)$$

Since ℓ_0 is typically much larger than λ ($\ell_0 \approx 10^{-1}$ m and $\lambda \approx 10^{-6}$ m), it follows that $r_m \approx \sqrt{m\lambda 2\ell_0}$. In other words, r_m is roughly proportional to $\sqrt{m}$. Therefore, the fringes are farther apart near the top of the film than they are near the bottom. For example, for the $m = 1$ and $m = 2$ fringes, we have $r_2 - r_1 \propto \sqrt{2} - \sqrt{1} = 0.41$, while for the $m = 2$ and $m = 3$ fringes, we have $r_3 - r_2 \propto \sqrt{3} - \sqrt{2} = 0.32$.

In addition to the fringe pattern just discussed, the total interference pattern on the hologram includes interference effects that are related to light coming from point B and other locations on the object in Figure 30.25. The total pattern is very complicated. Nevertheless, the fringe pattern for point A alone is sufficient to illustrate the fact that a hologram can be used to produce both a virtual image and a real image of the object, as we will now see.

To produce the holographic images, the laser light is directed through the interference pattern on the film, as in Figure 30.27. The pattern can be thought of as a kind of diffraction grating, with the bright fringes analogous to the space between the slits of the grating. Section 27.7 discusses how light passing through a grating is split into higher-order bright fringes that are symmetrically located on either side of a central bright fringe. Figure 30.27 shows the three rays corresponding to the central and first-order bright fringes, as they originate from a spot near the top and a spot near the bottom of the film. The angle θ, which locates the

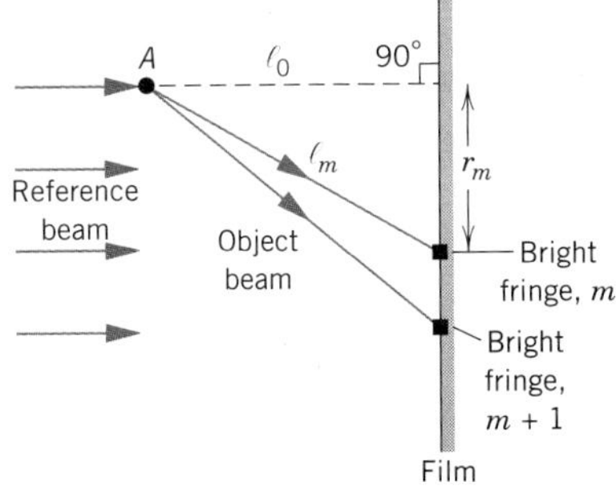

Figure 30.26 This drawing helps to explain how the interference pattern arises on the film when light from point A (see Figure 30.25) and light from the reference beam combine there.

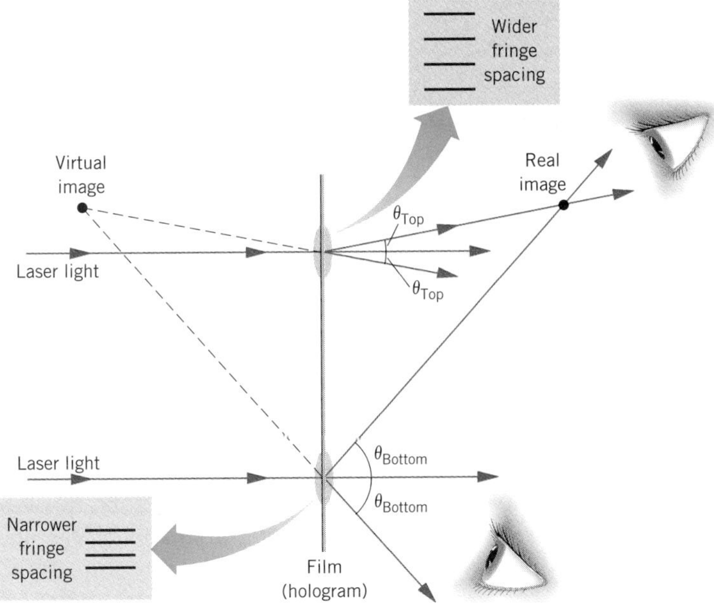

Figure 30.27 When the laser light used to produce a hologram is shone through it, both a real and a virtual image of the object are produced.

first-order fringes relative to the central fringe, is given by Equation 27.7 as $\sin\theta = \lambda/d$, where d is the separation between the slits of the grating. When the slit separation is greater, as it is near the top of the film, the angle is smaller. When the slit separation is smaller, as it is near the bottom, the angle is larger. Thus, Figure 30.27 has been drawn with θ_{TOP} smaller than θ_{BOTTOM}. Of the three rays emerging from the film at the top and the three at the bottom, we use the uppermost one in each case to locate the real image of point A on the chess piece. The real image is located where these two rays intersect, when extended to the right. To locate the virtual image, we use the lower ray in each of the three-ray bundles at the top and bottom of the film. When projected to the left, they appear to be originating from the spot where the projections intersect, that is, from the virtual image.

A holographic image differs greatly from a photographic image. The most obvious difference is that a hologram provides a three-dimensional image, while photographs are two dimensional. The reason that holographic images are three dimensional is inherent in the interference pattern formed on the film. In Figure 30.25 part of this pattern arises because the light emitted by point A travels different distances in reaching different spots on the film than does the light in the reference beam. The same can be said about the light emitted from point B and other places as well. As a result, the total interference pattern contains information about how much farther from the film the various points on the object are. It is because of this information that holographic images are three dimensional. Furthermore, as Figure 30.28 illustrates, it is possible to "walk around" a holographic image and view it from different angles, as you would view the original object.

(a)

(b)

Figure 30.28 This hologram of an award-winning architectural model depicts in the central background an Omnisphere theater in front of a museum of Science and Technology. Two views of the same hologram are shown: (*a*) looking perpendicularly at the hologram and (*b*) looking at an angle from the perpendicular.

There is also a vast difference between the methods used to produce holograms and photographs. As Section 26.7 discusses, a converging lens is used in a camera to produce a photograph. The lens focuses the light rays originating from a point on the object to a corresponding point on the film. To produce a hologram, lenses are not used in such a fashion, and a point on the object *does not* correspond to a single point on the film. Notice in Figure 30.25 that light from point A diverges on its way to the film. There is no lens to make it converge to a single corresponding point on the film, as there is in a camera. The light falls over the entire exposed region of the film and contributes everywhere to the interference pattern that is formed. Therefore, a hologram may be cut into smaller pieces, and each piece will contain some information about the light originating from point A. For this reason, each smaller piece can be used to produce a three-dimensional image of the object. On the other hand, it is not possible to reconstruct the entire image in a photograph from only a small piece of the original film.

The holograms discussed here are typically viewed with the aid of the laser light used to produce them. There are also other kinds of holograms. Credit cards, for example, sometimes use holograms for identification purposes. This kind of hologram is called a rainbow hologram and is designed to be viewed in white light that is reflected from it. Other applications of holography are becoming increasingly widespread and include head-up displays for the instrument panels in high-performance fighter planes, laser scanners at supermarket checkout counters, computerized data storage and retrieval systems, methods for high-precision biomedical measurements, and others.

INTEGRATION OF CONCEPTS

THE BOHR MODEL AND PHYSICS PRINCIPLES

Physicists often use models to help them understand how nature works. Most models, like the Bohr model of the hydrogen atom, are theoretical and exist as a collection of interconnected relations that express fundamental principles of nature. The principles are put together in a way that, hopefully, allows the physicist to predict the results of experiments. To the extent that the predictions of the model match experimental results, the model is a successful one. The Bohr model was the first to succeed in predicting the measured line spectra of the hydrogen atom.

The Bohr model brings together a remarkably wide array of physics principles to create a picture of atomic structure. The model is based on the ideas of uniform circular motion and centripetal force, as expressed by Newton's second law of motion. It incorporates the concepts of kinetic energy and potential energy. It uses Coulomb's law for the electrostatic force that one charged particle exerts on another. It uses the electrostatic potential energy that is related to the conservative nature of the electrostatic force. Bohr's model stresses the importance of angular momentum. And it takes advantage of the relation between frequency, wavelength, and the speed of an electromagnetic wave. All of the ideas above are the older and traditional ones that have become known as classical physics.

The Bohr model not only draws on the ideas of classical physics. It also combines them with many of the newer concepts of physics that emerged during the early twentieth century and are known as modern physics. It uses Planck's idea of quantized energy levels and Einstein's idea of photons. And it introduces the ideas of radiationless orbits and quantized angular momentum.

Today, the Schrödinger equation provides us with the quantum mechanical picture of the atom, which supersedes the Bohr model. Nevertheless, we should not lose sight of the fact that the Bohr model is a *tour de force* of integrating the principles of classical and modern physics.

SUMMARY

The idea of a **nuclear atom** originated in 1911, as a result of experiments by Ernest Rutherford, in which α particles were scattered by a thin metal foil. The phrase "nuclear atom" refers to the fact that an atom consists of a small, positively charged nucleus surrounded at relatively large distances by a number of electrons, whose total negative charge equals the positive nuclear charge when the atom is electrically neutral.

A **line spectrum** is a series of discrete electromagnetic wavelengths emitted by the atoms of a low-pressure gas that is subjected to a sufficiently high potential difference. Certain groups of discrete wavelengths are referred to as "series." The line spectrum of atomic hydrogen includes, among others, the **Lyman series,** the **Balmer series,** and the **Paschen series** of wavelengths.

The Bohr model applies to atoms or ions that have

only a single electron orbiting a nucleus containing Z protons. This model assumes that the electron exists in circular orbits that are called **stationary orbits,** because the electron does not radiate electromagnetic waves while in them. According to this model, a photon is emitted only when an electron changes from a higher energy orbit to a lower energy orbit. The model also assumes that the orbital angular momentum L_n of the electron can only have values that are integer multiples of Planck's constant divided by 2π: $L_n = n(h/2\pi)$; $n = 1, 2, 3, \ldots$. With the assumptions above, it can be shown that the nth Bohr orbit has a radius (in meters) of $r_n = (5.29 \times 10^{-11})(n^2/Z)$ and that the total energy associated with this orbit is $E_n = -(13.6 \text{ eV})(Z^2/n^2)$. The **ionization energy** is the energy needed to remove an electron completely from an atom. The Bohr model predicts that the wavelengths comprising the line spectrum emitted by a hydrogen atom are given by Equation 30.14.

Quantum mechanics describes the hydrogen atom in terms of four quantum numbers: (1) **the principal quantum number** n, which can have the integer values $n = 1, 2, 3, \ldots$; (2) **the orbital quantum number** ℓ, which can have the integer values $\ell = 0, 1, 2, \ldots, (n-1)$; (3) **the magnetic quantum number** m_ℓ, which can have the positive and negative integer values $m_\ell = -\ell, \ldots, -2, -1, 0, +1, +2, \ldots, +\ell$, and (4) **the spin quantum number** m_s, which, for an electron, can be either $m_s = +\frac{1}{2}$ or $m_s = -\frac{1}{2}$. According to quantum mechanics, an electron does not reside in a circular orbit but, rather, has some probability of being found at various distances from the nucleus.

The Pauli exclusion principle states that no two electrons in an atom can have the same set of values for the four quantum numbers n, ℓ, m_ℓ, and m_s. This principle determines the way in which the electrons in multiple-electron atoms are distributed into shells (defined by the value of n) and subshells (defined by the values of n and ℓ). Table 30.2 summarizes the conventional notation for atomic subshells. The arrangement of the periodic table of the elements is related to the exclusion principle.

X-rays are electromagnetic waves emitted when high-energy electrons strike a metal target contained within an evacuated glass tube. The emitted X-ray spectrum of wavelengths consists of sharp "peaks" or "lines," called **characteristic X-rays,** superimposed on a broad continuous range of wavelengths called **Bremsstrahlung.** The minimum wavelength, or cutoff wavelength, of the Bremsstrahlung is determined by the kinetic energy of the electrons striking the target in the X-ray tube.

A **laser** is a device that generates electromagnetic waves via a process known as **stimulated emission.** In this process, one photon stimulates the production of another photon by causing an electron in an atom to fall from a higher energy level to a lower energy level. Because of this mechanism of photon production, the electromagnetic waves generated by a laser are coherent and may be confined to a very narrow beam.

QUESTIONS

1. At room temperature, most of the atoms of atomic hydrogen contain electrons that are in the ground state or $n = 1$ energy level. A tube is filled with atomic hydrogen. Electromagnetic radiation with a continuous spectrum of wavelengths, including those in the Lyman, Balmer, and Paschen series, enters one end of this tube and leaves the other end. The exiting radiation is found to contain absorption lines. To which one (or more) of the series do the wavelengths of these absorption lines correspond? Explain.

2. Refer to Question 1. Suppose the electrons in the atoms are mostly in excited states. Do you expect there to be a greater or lesser number of absorption lines in the exiting radiation, compared to the situation when the electrons are in the ground state? Account for your answer.

3. When the outermost electron in an atom is in an excited state, the atom is more easily ionized than when the outermost electron is in the ground state. Why?

4. In the Bohr model for the hydrogen atom, the closer the electron is to the nucleus, the smaller is the total energy of the atom. Is this also true in the quantum mechanical picture of the hydrogen atom? Justify your answer.

5. Conceptual Example 6 provides background for this question. Consider two different hydrogen atoms. The electron in each atom is in an excited state. Is it possible for the electrons to have different energies but the same orbital angular momentum L, according to (a) the Bohr model and (b) quantum mechanics? In each case, give a reason for your answer.

6. In the quantum mechanical picture of the hydrogen atom, the orbital angular momentum of the electron *may* be zero in any of the possible energy states. For which energy state *must* the orbital angular momentum be zero? Give your reasoning.

7. Can a 5g subshell contain (a) 22 electrons and (b) 17 electrons? Why?

8. An electronic configuration for manganese (Z = 25) is written as $1s^2\ 2s^2\ 2p^6\ 3s^2\ 3p^6\ 3d^4\ 4s^2\ 4p^1$. Does this configuration of the electrons represent the ground state or an excited state? Consult Figure 30.15 for the order in which the subshells fill, and account for your answer.

9. Explain why you would not expect hydrogen and helium atoms in their ground state to emit characteristic X-rays.

10. The drawing shows the X-ray spectra produced by an X-ray tube when the tube is operated at two different potential differences. Explain why the characteristic lines occur at the same wavelengths in the two spectra, while the cutoff wavelength λ_0 shifts to the right when a smaller potential difference is used to operate the tube.

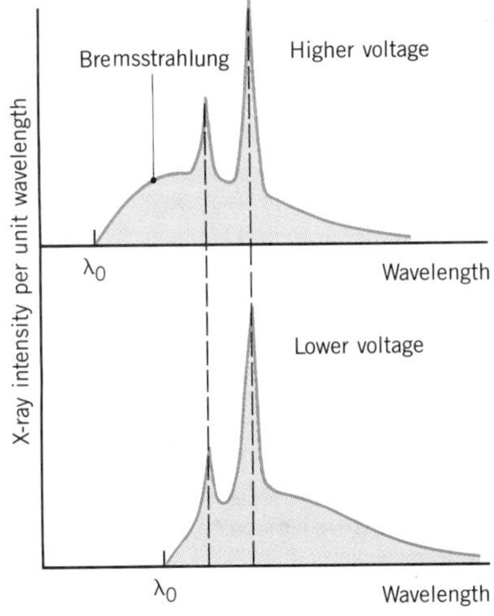

Question 10

11. In the production of X-rays, it is possible to create Bremsstrahlung X-rays without producing the characteristic X-rays. Explain how this can be accomplished by adjusting the electric potential difference used to operate the X-ray tube.

12. The short wavelength side of X-ray spectra ends abruptly at a cutoff wavelength λ_0 (see Figure 30.18). Does this cutoff wavelength depend on the target material used in the X-ray tube? Give your reasoning.

13. Explain why a laser beam focused to a small spot can cut through a piece of metal.

PROBLEMS

In working these problems, ignore relativistic effects.

Section 30.1 Rutherford Scattering and the Nuclear Atom

1. Review Conceptual Example 1 in preparation for this problem. At a shooting range, the bull's-eye of a target has a radius of 3 cm. Assume that the target shooter is analogous to an electron in orbit (radius $\approx 1 \times 10^{-10}$ m) about the nucleus (radius $\approx 1 \times 10^{-15}$ m) of an atom. How far (in kilometers) from the target is the shooter standing?

2. The nucleus of the hydrogen atom has a radius of about 1×10^{-15} m. The electron is normally at a distance of about 5.3×10^{-11} m from the nucleus. Assuming the hydrogen atom is a sphere with a radius of 5.3×10^{-11} m, find (a) the volume of the atom, (b) the volume of the nucleus, and (c) the percentage of the volume of the atom that is occupied by the nucleus.

3. The mass of an α particle is 6.64×10^{-27} kg. An α particle used in a scattering experiment has a kinetic energy of 7.00×10^{-13} J. What is the de Broglie wavelength of the particle?

4. The nucleus of a hydrogen atom is a single proton, which has a radius of about 1.0×10^{-15} m. The single electron in a hydrogen atom normally orbits the nucleus at a distance of 5.3×10^{-11} m. What is the ratio of the density of the hydrogen nucleus to the density of the complete hydrogen atom?

*5. There are Z protons in the nucleus of an atom, where Z is the atomic number of the element. An α particle carries a charge of $+2e$. In a scattering experiment, an α particle, heading directly toward a nucleus in a metal foil, will come to a halt when all the particle's kinetic energy is converted to electric potential energy. In such a situation, how close will an α particle with a kinetic energy of 5.0×10^{-13} J come to a gold nucleus (Z = 79)?

*6. The nucleus of an aluminum atom contains 13 protons and has a radius of 3.6×10^{-15} m. How much work (in electron volts) must be done to bring a proton from infinity, where it is at rest, to the "surface" of an aluminum nucleus?

Section 30.2 Line Spectra, Section 30.3 The Bohr Model of the Hydrogen Atom

7. If the line with the longest wavelength in the Balmer series for atomic hydrogen is counted as the first line, what is the wavelength of the third line?

8. On a piece of graph paper, make a copy of Figure 30.8. This figure shows the energy level diagram for the hydrogen atom (Z = 1), as determined from Equation 30.13. On the same piece of graph paper, alongside the hydrogen atom diagram, draw to scale the energy level diagram that Equation 30.13 predicts for singly ionized helium He^+ (Z = 2). In these drawings include only the first two energy levels for hydrogen

and the first four levels for He^+. Which levels have the same energy in both diagrams?

9. In the line spectrum of atomic hydrogen there is also a group of lines known as the Pfund series. These lines are produced when electrons, excited to high energy levels, make transitions to the $n = 5$ level. Determine (a) the longest wavelength and (b) the shortest wavelength in this series. (c) Refer to Figure 24.6 and identify the region of the electromagnetic spectrum in which these lines are found.

10. It is possible to use electromagnetic radiation to ionize atoms. To do so, the atoms must absorb the radiation, the photons of which must have enough energy to remove an electron from an atom. What is the longest radiation wavelength that can be used to ionize the ground state hydrogen atom?

11. The electron in a hydrogen atom is in the first excited state, when the electron acquires an additional 2.86 eV of energy. What is the quantum number of the state into which the electron moves?

12. Find the energy (in joules) of the photon that is emitted when the electron in a hydrogen atom undergoes a transition from the $n = 7$ energy level to produce a line in the Paschen series.

13. What is the radius for the $n = 5$ Bohr orbit in a doubly ionized lithium atom Li^{2+} ($Z = 3$)?

14. In the hydrogen atom, what is the total energy (in electron volts) of an electron that is in an orbit with a radius of 4.761×10^{-10} m?

15. Using the Bohr model, compare the nth orbit of a triply ionized beryllium atom Be^{3+} ($Z = 4$) to the nth orbit of a hydrogen atom (H) by calculating the ratio (Be^{3+}/H) of the following quantities: (a) the energies and (b) the radii.

***16.** For atomic hydrogen, the Paschen series of lines occurs when $n_f = 3$, while the Brackett series occurs when $n_f = 4$ in Equation 30.14. Using this equation, show that the ranges of wavelengths in these two series overlap.

***17.** In the Bohr model, Equation 30.12 or 30.13 gives the total energy (kinetic plus potential). In a certain Bohr orbit, the total energy is -4.90 eV. For this orbit, determine the kinetic energy and the electric potential energy of the electron.

***18.** In an unidentified ionized atom, only one electron moves about the nucleus. The radius of the $n = 3$ Bohr orbit is 2.38×10^{-10} m. What is the energy (in electron volts) of the $n = 7$ orbit for this atom?

***19.** A wavelength of 410.2 nm is emitted by the hydrogen atoms in a high-voltage discharge tube. What are the initial and final values of the quantum number n for the energy level transition that produces this wavelength? *(Hint: Identify the region of the electromagnetic spectrum in which the given wavelength is to be found.)*

****20.** (a) Derive an expression for the speed of the electron in the nth Bohr orbit, in terms of Z, n, and the constants k, e, and h. For the hydrogen atom, determine the speed in (b) the $n = 1$ orbit and (c) the $n = 2$ orbit. (d) Generally, when speeds are less than one-tenth the speed of light, the effects of special relativity can be ignored. Are the speeds found in (b) and (c) consistent with ignoring relativistic effects in the Bohr model?

****21.** A diffraction grating is used in the first order to separate the wavelengths in the Balmer series of atomic hydrogen. (Section 27.7 discusses diffraction gratings.) The grating and an observation screen are separated by a distance of 75.0 cm, as Figure 27.32 illustrates. You may assume that θ is small, so $\sin\theta \approx \theta$ when radian measure is used for θ. How many lines per centimeter should the grating have, so the longest and the next-to-the-longest wavelengths in the series are separated by 5.00 cm on the screen?

Section 30.5 The Quantum Mechanical Picture of the Hydrogen Atom

22. An electron is in the excited state for which $n = 3$. What is the maximum possible magnitude for the orbital angular momentum for this electron (in J·s)?

23. Suppose the value of the principal quantum number is $n = 4$. What are the possible values for the magnetic quantum number m_ℓ?

24. Write down the eighteen possible sets of the four quantum numbers when the principal quantum number is $n = 3$.

25. The principal quantum number for an electron in an atom is $n = 6$, while the magnetic quantum number is $m_\ell = 2$. What possible values for the orbital quantum number ℓ could this electron have?

26. An electron in an atom has a value for the magnetic quantum number of $m_\ell = 4$. What are the *minimum* values that (a) the orbital quantum number and (b) the principal quantum number can have?

***27.** Review Conceptual Example 6 as background for this problem. For the hydrogen atom, the Bohr model and quantum mechanics both give the same value for the energy of the nth state. However, they do not give the same value for the orbital angular momentum L. (a) For $n = 1$, determine the values of L (in units of $h/2\pi$) predicted by the Bohr model and quantum mechanics. (b) Repeat part (a) for $n = 2$, noting that quantum mechanics permits more than one value of ℓ when the electron is in the $n = 2$ state.

***28.** For an electron in a hydrogen atom, the z component of the angular momentum has a *maximum* value of $L_z = 2.11 \times 10^{-34}$ J·s. Find the three smallest possible values (algebraically) for the total energy (in electron volts) that this atom could have.

****29.** An electron is in the $n = 5$ state. What is the smallest possible value for the angle between the z component of the orbital angular momentum and the orbital angular momentum?

Section 30.6 The Pauli Exclusion Principle and the Periodic Table of the Elements

30. In the style indicated in Table 30.3, write down the ground state electronic configuration of calcium Ca (Z = 20). Refer to Figure 30.15 for the order in which the subshells fill.

31. Following the style used in Table 30.3, determine the electronic configuration of the ground state for yttrium Y (Z = 39). Refer to Figure 30.15 to see the order in which the subshells fill.

32. Figure 30.15 was constructed using the Pauli exclusion principle and indicates that the $n = 1$ shell holds 2 electrons, the $n = 2$ shell holds 8 electrons, and the $n = 3$ shell holds 18 electrons. These numbers can be obtained by adding the numbers given in the figure for the subshells contained within a given shell. How many electrons can be put into (a) the $n = 4$ shell and (b) the $n = 5$ shell, neither of which is completely shown in the figure?

33. When an electron makes a transition between energy levels of an atom, there are no restrictions on the initial and final values of the principal quantum number n. According to quantum mechanics, however, there is a rule that restricts the initial and final values of the orbital quantum number ℓ. This rule is called a *selection rule* and states that $\Delta\ell = \pm 1$. In other words, when an electron makes a transition between energy levels, the value of ℓ can only increase or decrease by one. The value of ℓ may not remain the same or increase or decrease by more than one. According to this rule, which of the following energy level transitions are allowed: (a) 2s → 1s, (b) 2p → 1s, (c) 4p → 2p, (d) 4s → 2p, and (e) 3d → 3s?

***34.** What is the atom with the smallest atomic number that contains the same number of electrons in its s subshells as it does in its d subshell? Refer to Figure 30.15 for the order in which the subshells fill.

Section 30.7 X-Rays

35. Suppose an X-ray machine in a doctor's office uses a potential difference of 61 kV to operate the X-ray tube. What is the shortest X-ray wavelength emitted by this machine?

36. By using the Bohr model decide which element is likely to emit a K_α X-ray with a wavelength of 4.5×10^{-9} m.

37. Molybdenum has an atomic number of Z = 42. Using the Bohr model, estimate the wavelength of the K_α X-ray.

38. The atomic number of tungsten is Z = 74. According to the Bohr model, what is the energy (in joules) of a K_α X-ray photon?

39. An X-ray tube is being operated at a potential difference of 45.0 kV. What is the Bremsstrahlung wavelength that corresponds to 25.0% of the kinetic energy with which an electron collides with the metal target in the tube?

***40.** Metal A emits a K_α X-ray that has a wavelength 2.63 times longer than that of the K_α X-ray emitted by metal B. The atomic number of metal A is $Z_A = 51$. Use the Bohr model to find an estimate for Z_B, the atomic number of metal B.

***41.** An X-ray tube contains a silver (Z = 47) target. The high voltage in this tube is increased from zero. Using the Bohr model, find the value of the voltage at which the K_α X-ray just appears in the X-ray spectrum.

Section 30.8 The Laser

42. A laser used in the removal of tooth decay has a wavelength of 193 nm. What is the energy (in joules) of a photon that this laser produces?

43. In the helium/neon laser, there is an energy difference of 1.96 eV between the levels that participate in stimulated emission. Verify, by performing a calculation, that the laser produces a wavelength of 633 nm.

44. A carbon dioxide laser produces a wavelength of 1.06×10^{-5} m. A semiconductor laser produces a wavelength of 7.90×10^{-7} m. (a) Which laser produces the more energetic photons? Explain. (b) By what factor are they more energetic?

45. A laser is used in eye surgery to weld a detached retina back into place. The wavelength of the laser beam is 514 nm, while the power is 2.0 W. During surgery, the laser beam is turned on for 0.10 s. During this time, how many photons are emitted by the laser?

***46.** Fusion is the process by which the sun produces energy. One experimental technique for creating controlled fusion utilizes a solid-state laser that emits a wavelength of 1060 nm and can produce a power of 1.0×10^{14} W for a pulse duration of 1.1×10^{-11} s. In contrast, the helium/neon laser used at the checkout counter in a bar-code scanner emits a wavelength of 633 nm and produces a power of about 1.0×10^{-3} W. How long (in days) would the helium/neon laser have to operate to produce the same number of photons that the solid-state laser produces in 1.1×10^{-11} s?

ADDITIONAL PROBLEMS

47. In the style shown in Table 30.3, write down the ground state electronic configuration for arsenic As (Z = 33). Note from Figure 30.15 that the 4s subshell fills before the 3d subshell.

48. What is the minimum potential difference that must be applied to an X-ray tube to knock a K-shell electron completely out of an atom in a copper (Z = 29) target? Use the Bohr model as needed.

49. Determine the ionization energy (in electron volts) that is needed to remove the remaining electron from a singly ionized helium atom He^+ ($Z = 2$).

50. Write down the fourteen sets of the four quantum numbers that correspond to the electrons in a completely filled 4f subshell.

51. The K_β characteristic X-ray line for tungsten has a wavelength of 1.84×10^{-11} m. What is the difference in energy between the two energy levels that give rise to this line? Express the answer in joules and in electron volts.

52. A 790-kg synchronous communications satellite has a period of one day. The radius of the orbit is 4.23×10^7 m. Suppose that Bohr's assumption about the angular momentum of the electron in orbit about the nucleus applies to this satellite. What is the quantum number n for the orbit of the satellite?

53. For a doubly ionized lithium atom Li^{2+} ($Z = 3$), what is the principal quantum number of the state in which the electron has the same total energy as a ground state electron has in the hydrogen atom?

***54.** The total orbital angular momentum of the electron in a hydrogen atom has a magnitude of $L = 3.66 \times 10^{-34}$ J·s. In the quantum mechanical picture of the atom, what values can the angular momentum component L_z have? *(Hint: There are seven possible values.)*

***55.** In the line spectrum emitted by doubly ionized lithium atoms Li^{2+} ($Z = 3$), the shortest wavelength in one series of lines is 162.1 nm. This line is produced when an electron makes a transition to a final energy level whose principal quantum number is n_f. What is the value of n_f?

***56.** Consider a particle of mass m that can exist only between $x = 0$ and $x = +L$ on the x axis. We could say that this particle is confined to a "box" of length L. In this situation, imagine the standing de Broglie waves that can fit into the box. For example, the drawing shows the first three possibilities. Note in this picture that there are either one, two, or three half-wavelengths that fit into the distance L. Use Equation 29.8 for the de Broglie wavelength of a particle and derive an expression for the allowed energies (only kinetic energy) that the particle can have. This expression involves m, L, Planck's constant, and a quantum number n that can have only the values 1, 2, 3, ...

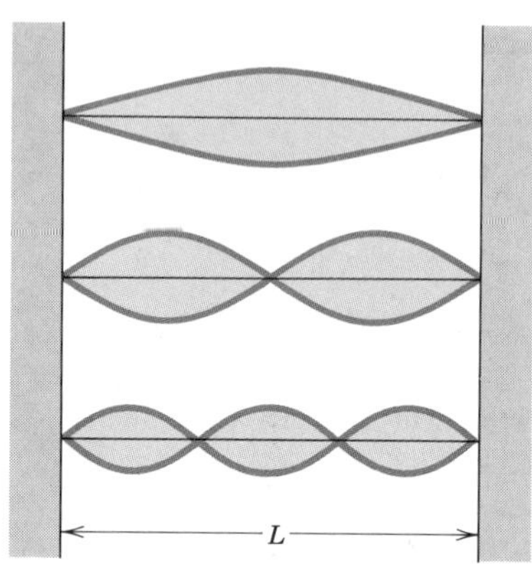

***57.** The Bohr model can be applied to singly ionized helium He^+ ($Z = 2$). Using this model, consider the series of lines that is produced when the electron makes a transition from higher energy levels into the $n_f = 4$ level. Some of the lines in this series lie in the visible region of the spectrum (380–750 nm). What are the values of n_i for the energy levels from which the electron makes the transitions corresponding to these lines?

****58.** (a) Derive an expression for the time it takes the electron in the nth Bohr orbit to make one complete revolution around the nucleus. Express your answer in terms of Z, n, and the constants k, e, h, and m. For a hydrogen atom, determine this time for (b) the $n = 1$ orbit and (c) the $n = 2$ orbit.

****59.** A certain species of ionized atoms produces an emission line spectrum according to the Bohr model, but the number of protons Z in the nucleus is unknown. A group of lines in the spectrum forms a series in which the shortest wavelength is 22.79 nm and the longest wavelength is 41.02 nm. Find the next-to-the-longest wavelength in the series of lines.

NUCLEAR PHYSICS AND RADIOACTIVITY

CHAPTER 31

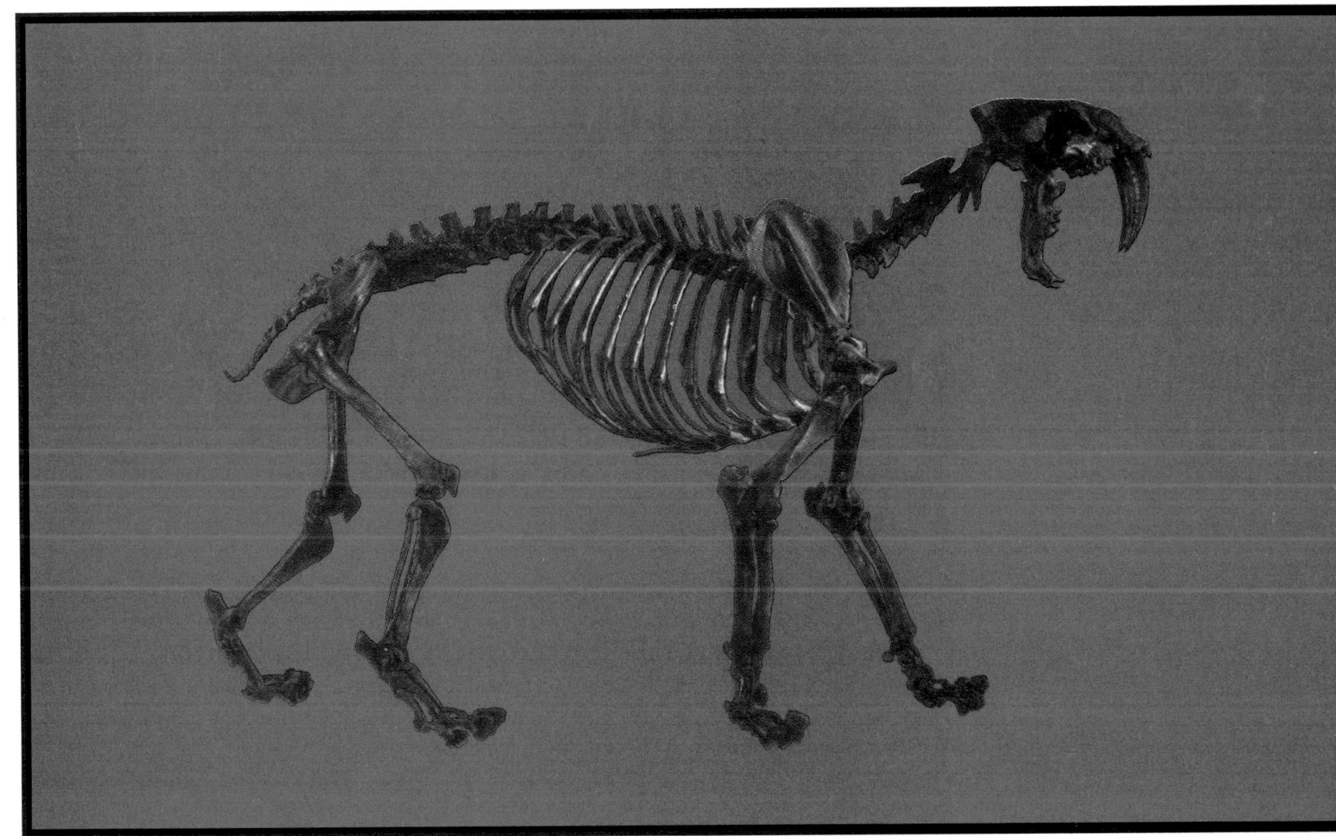

The nucleus of an atom is incredibly small, for if the diameter of an atom were enlarged to the size of a football field, the nucleus would only be the size of a small BB. Although small, the nucleus contains more than 99.9% of the mass of an atom, because each of its constituents is about 1800 times more massive than an electron. We will see that many nuclei, particularly those of the heavier elements, are unstable and spontaneously break apart, or disintegrate, into other nuclei. This spontaneous disintegration is called radioactive decay, and it is accompanied by the emission of certain types of particles and high-energy photons. While these particles and photons can be very dangerous to individuals and to the environment, radioactive decay also has many beneficial uses. For example, paleontologists use the decay of radioactive nuclei to date the fossil remains of extinct species, such as the saber-toothed cat in the photograph above. In medicine, there are many beneficial applications of radioactive decay, including the diagnosis and treatment of various kinds of cancer and other diseases. In this chapter, we will study the common types of radioactive decay and what they imply about nuclear structure.

31.1 NUCLEAR STRUCTURE

Atoms consist of electrons in orbit about a central nucleus. As we have seen in Chapter 30, the electron orbits are quantum mechanical in nature and have interesting characteristics. In our previous discussion, however, little has been said about the nucleus itself. But the nucleus is fascinating in its own right, and now we consider it in greater detail.

The nucleus of an atom consists of neutrons and protons, collectively referred to as ***nucleons.*** The ***neutron,*** discovered in 1932 by the English physicist James Chadwick (1891 – 1974), carries no electrical charge and has a mass slightly larger than that of a proton (see Table 31.1).

Table 31.1 Properties of Particles in the Atom

Particle	Electric Charge (C)	Mass: Kilograms (kg)	Mass: Atomic Mass Units (u)
Electron	-1.60×10^{-19}	$9.109\ 390 \times 10^{-31}$	$5.485\ 799 \times 10^{-4}$
Proton	$+1.60 \times 10^{-19}$	$1.672\ 623 \times 10^{-27}$	1.007 276
Neutron	0	$1.674\ 929 \times 10^{-27}$	1.008 665

The number of protons in the nucleus is different in different elements and is given by the ***atomic number*** Z. In an electrically neutral atom, the number of nuclear protons equals the number of electrons in orbit around the nucleus. The number of neutrons in the nucleus is N. The total number of protons and neutrons is referred to as the ***atomic mass number*** A, because the total nuclear mass is *approximately* equal to A times the mass of a single nucleon:

$$A = Z + N \tag{31.1}$$

Sometimes A is also called the ***nucleon number.*** A shorthand notation is often used to specify Z and A along with the chemical symbol for the element. For instance, the nuclei of all naturally occurring aluminum atoms have $A = 27$, and the atomic number for aluminum is $Z = 13$. In shorthand notation, then, the aluminum nucleus is specified as ${}^{27}_{13}\mathrm{Al}$. The number of neutrons in an aluminum nucleus is $N = A - Z = 14$. In general, for an element whose chemical symbol is X, the symbol for the nucleus is

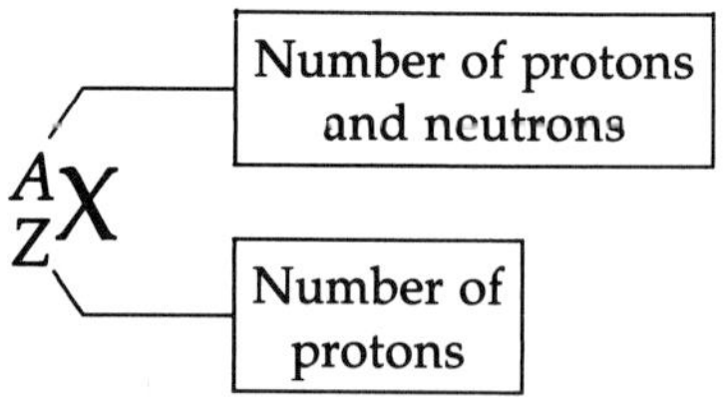

For a proton the symbol is ${}^{1}_{1}\mathrm{H}$, since the proton is the nucleus of a hydrogen atom. A neutron is denoted by ${}^{1}_{0}\mathrm{n}$. In the case of an electron we use ${}^{0}_{-1}\mathrm{e}$, where $A = 0$ because an electron has no nucleus and $Z = -1$ because the electron has a negative charge.

Nuclei that contain the same number of protons, but a different number of neutrons, are known as ***isotopes.*** Carbon, for example, occurs in nature in two stable forms. In most carbon atoms (98.90%), the nucleus is the $^{12}_{6}C$ isotope and consists of six protons and six neutrons. A small fraction (1.10%), however, contains nuclei that have six protons and seven neutrons, namely, the $^{13}_{6}C$ isotope. The percentages given above are the natural abundances of the isotopes. The atomic masses in the periodic table are average atomic masses, taking into account the abundances of the various isotopes.

The protons and neutrons in the nucleus are clustered together to form an approximately spherical region, as Figure 31.1 illustrates. Experiment shows that the radius r of the nucleus depends on the atomic mass number A and is given approximately in meters by

$$r \approx (1.2 \times 10^{-15}\text{ m})A^{1/3} \tag{31.2}$$

The radius of the aluminum nucleus ($A = 27$), for example, is $r \approx (1.2 \times 10^{-15}\text{ m})27^{1/3} = 3.6 \times 10^{-15}$ m. Equation 31.2 leads to an important conclusion concerning the nuclear density of different atoms, as Conceptual Example 1 discusses.

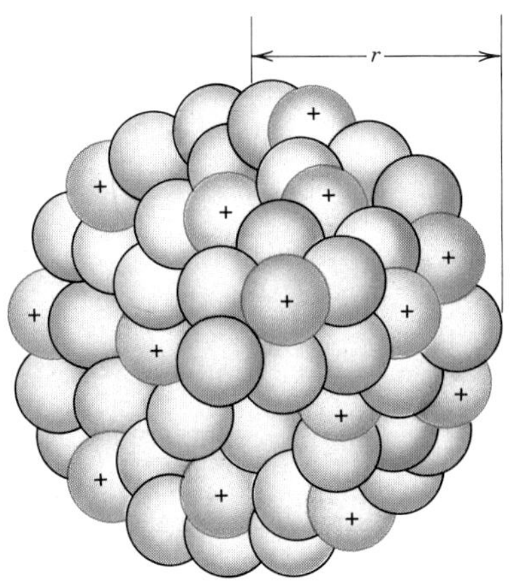

Figure 31.1 The nucleus is approximately spherical (radius = r) and contains protons (⊕) clustered closely together with neutrons (○).

Conceptual Example 1 Nuclear Density

It is well known that solid lead and gaseous oxygen contain different atoms and that the density of lead is much greater than that of oxygen. Using Equation 31.2, decide whether the density of the *nucleus* in a lead atom is greater than, approximately equal to, or less than that in an oxygen atom.

REASONING We know from Equation 11.1 that density is mass divided by volume. The total mass M of a nucleus is approximately equal to the number of nucleons A times the mass m of a single nucleon, since the masses of a proton and a neutron are nearly the same: $M \approx Am$. The values of A are different for lead and oxygen atoms, however. The volume V of the nucleus is approximately spherical and has a radius r, so that $V = \frac{4}{3}\pi r^3$. But r^3 is proportional to A, as we can see by cubing each side of Equation 31.2. Therefore, the volume V is also proportional to A. Thus, when the total mass M is divided by the volume V, the factor of A appears in both the numerator and denominator and is eliminated algebraically from the result, no matter what the value of A is. We find, then, that the density of the nucleus in a lead atom is approximately the same as it is in an oxygen atom. In general, Equation 31.2 implies that the nuclear density has nearly the same value in all atoms. The difference in densities between solid lead and gaseous oxygen arises mainly because of the difference in how closely the atoms themselves are packed together in the solid and gaseous phases.

Related Homework Material: Problem 10

31.2 THE STRONG NUCLEAR FORCE AND THE STABILITY OF THE NUCLEUS

Two positive charges that are as close together as they are in a nucleus repel one another with a very strong electrostatic force. What, then, keeps the nucleus from flying apart? Clearly, some kind of attractive force must hold the nucleus together, since many kinds of naturally occurring atoms contain stable nuclei. The gravitational force of attraction between nucleons is too weak to counteract the

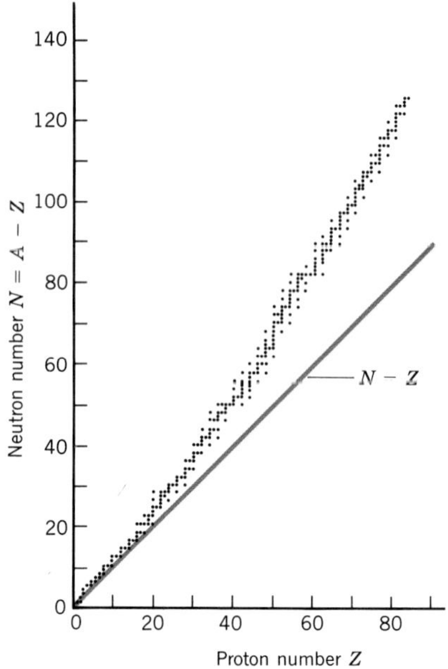

Figure 31.2 With few exceptions, the naturally occurring stable nuclei have a number N of neutrons that equals or exceeds the number Z of protons.

repulsive electric force, so it must be that a different type of force holds the nucleus together. This force is the ***strong nuclear force*** and is one of only four fundamental forces that have been discovered, fundamental in the sense that all forces in nature can be explained in terms of these four. We have already encountered two other fundamental forces, the gravitational force and the electromagnetic force. The remaining one will be mentioned in Section 31.5.

Many features of the strong nuclear force are well known. The strong nuclear force is almost independent of electric charge. At a given separation distance, almost the same nuclear force of attraction exists between two protons, two neutrons, or between a proton and a neutron. The range of action of the strong nuclear force is extremely short, with the force of attraction being very strong when two nucleons are as close as 10^{-15} m and essentially zero at larger distances. In contrast, the electric force between two protons decreases to zero only gradually as the separation distance increases to large values and, therefore, has a relatively long range of action.

The limited range of action of the strong nuclear force plays an important role in the stability of the nucleus. For a nucleus to be stable, the electrostatic repulsion between the protons must be balanced by the attraction between the nucleons due to the strong nuclear force. But one proton repels all other protons within the nucleus, since the electrostatic force has such a long range of action. In contrast, a proton or a neutron attracts only its nearest neighbors via the strong nuclear force. As the number Z of protons in the nucleus increases under these conditions, the number N of neutrons has to increase even more, if stability is to be maintained. Figure 31.2 shows a plot of N versus Z for naturally occurring elements that have stable nuclei. For reference, the plot also includes the straight line that represents the condition $N = Z$. With few exceptions, the points representing stable nuclei fall above this reference line, reflecting the fact that the number of neutrons becomes greater than the number of protons as the atomic number Z increases.

As more and more protons occur in a nucleus, there comes a point when a balance of repulsive and attractive forces cannot be achieved by an increased number of neutrons. Eventually, the limited range of action of the strong nuclear force prevents extra neutrons from balancing the long-range electric repulsion of extra protons. The stable nucleus with the largest number of protons ($Z = 83$) is that of bismuth, ${}^{209}_{83}\text{Bi}$, which contains 126 neutrons. All nuclei with more than 83 protons (e.g., uranium, $Z = 92$) are unstable and spontaneously break apart or rearrange their internal structures as time passes. This spontaneous disintegration or rearrangement of internal structure is called ***radioactivity,*** first discovered in 1896 by the French physicist Henri Becquerel (1852–1908). Section 31.4 discusses radioactivity in greater detail.

31.3 THE MASS DEFECT OF THE NUCLEUS AND NUCLEAR BINDING ENERGY

Because of the strong nuclear force, the nucleons in a stable nucleus are held tightly together. Therefore, energy is required to separate a stable nucleus into its constituent protons and neutrons, as Figure 31.3 illustrates. The more stable the nucleus is, the greater the amount of energy needed to break it apart. The required energy is called the ***binding energy*** of the nucleus.

In Einstein's theory of special relativity, energy and mass are equivalent. A change Δm in the mass of a system is equivalent to a change ΔE_0 in the rest energy of the system by an amount $\Delta E_0 = (\Delta m)c^2$, where c is the speed of light in a

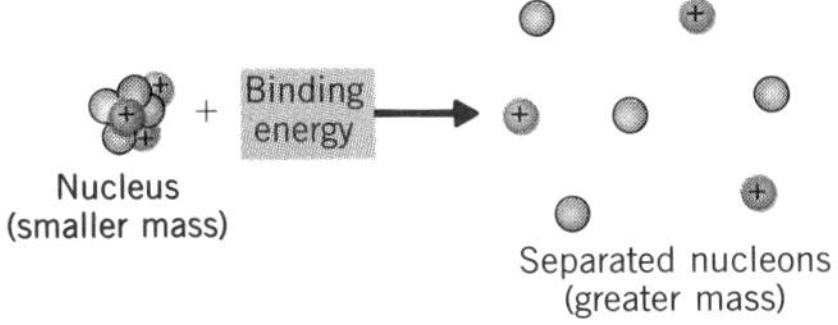

Figure 31.3 Energy must be supplied to break the nucleus apart into its constituent protons and neutrons. Each of the separated nucleons is at rest and out of the range of the forces of the other nucleons.

vacuum. Thus, in Figure 31.3, the binding energy used to disassemble the nucleus appears as extra mass of the separated nucleons. In other words, the sum of the individual masses of the separated protons and neutrons is greater by an amount Δm than the mass of the stable nucleus. The difference in mass Δm is known as the ***mass defect*** of the nucleus. As Example 2 shows, the binding energy of a nucleus can be determined from the mass defect according to Equation 31.3:

$$\text{Binding energy} = (\text{Mass defect})c^2 = (\Delta m)c^2 \tag{31.3}$$

Example 2 The Binding Energy of the Helium Nucleus

The most abundant isotope of helium has a ${}^4_2\text{He}$ nucleus whose mass is 6.6447×10^{-27} kg. For this nucleus, find (a) the mass defect and (b) the binding energy.

REASONING The symbol ${}^4_2\text{He}$ indicates that the helium nucleus contains $Z = 2$ protons and $N = 4 - 2 = 2$ neutrons. To obtain the mass defect Δm, we first determine the sum of the individual masses of the separated protons and neutrons. Then we subtract from this sum the mass of the ${}^4_2\text{He}$ nucleus. Finally, we use Equation 31.3 to calculate the binding energy from the value for Δm.

SOLUTION

(a) Using data from Table 31.1, we find that the sum of the individual masses of the nucleons is

$$\underbrace{2(1.6726 \times 10^{-27}\ \text{kg})}_{\text{Two protons}} + \underbrace{2(1.6749 \times 10^{-27}\ \text{kg})}_{\text{Two neutrons}} = 6.6950 \times 10^{-27}\ \text{kg}$$

This value is greater than the mass of the intact ${}^4_2\text{He}$ nucleus, and the mass defect is

$$\Delta m = 6.6950 \times 10^{-27}\ \text{kg} - 6.6447 \times 10^{-27}\ \text{kg} = \boxed{0.0503 \times 10^{-27}\ \text{kg}}$$

(b) According to Equation 31.3, the binding energy is

$$\begin{matrix}\text{Binding} \\ \text{energy}\end{matrix} = (\Delta m)c^2 = (0.0503 \times 10^{-27}\ \text{kg})(3.00 \times 10^8\ \text{m/s})^2 = 4.53 \times 10^{-12}\ \text{J}$$

Usually, binding energies are expressed in energy units of electron volts instead of joules ($1\ \text{eV} = 1.60 \times 10^{-19}$ J):

$$\begin{matrix}\text{Binding} \\ \text{energy}\end{matrix} = (4.53 \times 10^{-12}\ \text{J})\left(\frac{1\ \text{eV}}{1.60 \times 10^{-19}\ \text{J}}\right) = 2.83 \times 10^7\ \text{eV} = \boxed{28.3\ \text{MeV}}$$

This value is more than two million times greater than the energy required to remove an orbital electron from an atom.

In calculations such as that in Example 2, it is customary to use the ***atomic mass unit*** (u) instead of the kilogram. As introduced in Section 14.1, the atomic mass unit is one-twelfth of the mass of a ${}^{12}_6\text{C}$ atom of carbon. In terms of this unit, the mass of a $^{12}_{6}$C atom is exactly 12 u. Table 31.1 also gives the masses of the electron, the proton, and the neutron in atomic mass units. For future use, the energy equivalent of one atomic mass unit can be calculated by observing that the mass of a proton is 1.6726×10^{-27} kg or 1.0073 u, so that

$$1 \text{ u} = (1 \text{ u}) \left(\frac{1.6726 \times 10^{-27} \text{ kg}}{1.0073 \text{ u}} \right) = 1.6605 \times 10^{-27} \text{ kg}$$

and

$$\Delta E_0 = (\Delta m)c^2 = (1.6605 \times 10^{-27} \text{ kg})(2.9979 \times 10^8 \text{ m/s})^2 = 1.4924 \times 10^{-10} \text{ J}$$

In electron volts, therefore, one atomic mass unit is equivalent to

$$1 \text{ u} = (1.4924 \times 10^{-10} \text{ J}) \left(\frac{1 \text{ eV}}{1.6022 \times 10^{-19} \text{ J}} \right) = 9.315 \times 10^8 \text{ eV} = 931.5 \text{ MeV}$$

A table of the isotopes, such as that in Appendix F, gives masses in atomic mass units. Typically, however, the given masses are not nuclear masses. They are *atomic masses,* that is, the masses of neutral atoms, including the mass of the orbital electrons. Example 3 deals again with the $^{4}_{2}$He nucleus and shows how to take into account the effect of the orbital electrons when using data from a table of isotopes to determine binding energies.

Example 3 The Binding Energy of the Helium Nucleus, Revisited

Using atomic mass units instead of kilograms, obtain the binding energy of the $^{4}_{2}$He nucleus.

REASONING To determine the binding energy, we calculate the mass defect in atomic mass units and then use the fact that one atomic mass unit is equivalent to 931.5 MeV of energy. The table in Appendix F gives a mass of 4.0026 u for $^{4}_{2}$He, *which includes the mass of the two electrons in the neutral helium atom.* To calculate the mass defect, we must subtract 4.0026 u from the sum of the individual masses of the nucleons, including the mass of the electrons. As Figure 31.4 illustrates, the electron mass will be included if the masses of two hydrogen atoms are used in the calculation instead of the masses of two protons. The mass of a $^{1}_{1}$H hydrogen atom is 1.0078 u according to Appendix F, and the mass of a neutron is given in Table 31.1 as 1.0087 u.

SOLUTION The sum of the individual masses is

$$\underbrace{2(1.0078 \text{ u})}_{\text{Two hydrogen atoms}} + \underbrace{2(1.0087 \text{ u})}_{\text{Two neutrons}} = 4.0330 \text{ u}$$

The mass defect is $\Delta m = 4.0330 \text{ u} - 4.0026 \text{ u} = 0.0304$ u. Since 1 u is equivalent to 931.5 MeV, the binding energy is $\boxed{\text{Binding energy} = 28.3 \text{ MeV}}$, which matches that obtained in Example 2.

To see how the nuclear binding energy varies from nucleus to nucleus, it is necessary to compare the binding energy for each nucleus on a per-nucleon basis. Figure 31.5 shows a graph in which the binding energy divided by the nucleon number A is plotted against the nucleon number itself. In the graph, the peak for the $^{4}_{2}$He isotope of helium indicates that the $^{4}_{2}$He nucleus is particularly stable. The

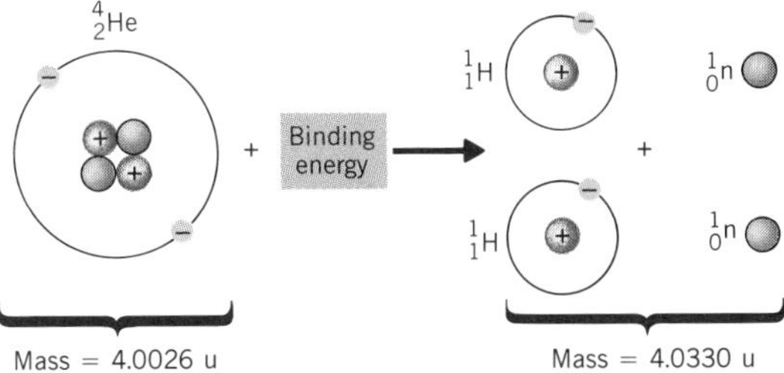

Figure 31.4 Tables of isotopes usually give the mass of the neutral atom (including the orbital electrons), rather than the mass of the nucleus. When using data from such tables to determine the mass defect of a nucleus, the mass of the orbital electrons must be taken into account, as this drawing illustrates for the $^4_2$He isotope of helium. See Example 3.

binding energy per nucleon increases rapidly for nuclei with small masses and reaches a maximum of approximately 8.7 MeV/nucleon for a nucleon number of about $A = 60$. For greater nucleon numbers, the binding energy per nucleon decreases gradually. Eventually, the binding energy per nucleon decreases enough so there is insufficient binding energy to hold the nucleus together. Nuclei more massive than the $^{209}_{83}$Bi nucleus of bismuth are unstable and hence radioactive.

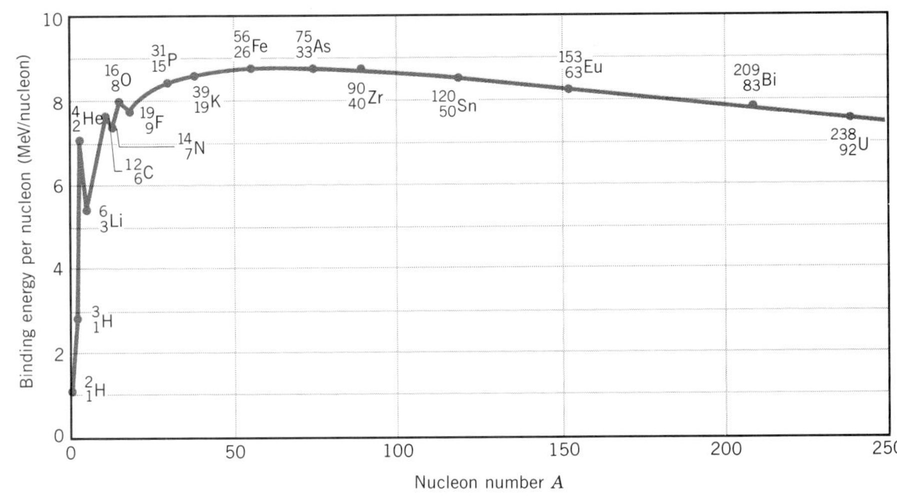

Figure 31.5 A plot of binding energy per nucleon versus the nucleon number A.

31.4 RADIOACTIVITY

When an unstable or radioactive nucleus disintegrates spontaneously, certain kinds of particles and/or high-energy photons are released. These particles and photons are collectively called "rays." Three kinds of rays are produced by naturally occurring radioactivity: **α rays, β rays,** and **γ rays.** They are named according to the first three letters of the Greek alphabet alpha (α), beta (β), and gamma (γ), to indicate the extent of their ability to penetrate matter. α rays are the least penetrating, being blocked by a thin (≈ 0.01 mm) sheet of lead, while β rays penetrate lead to a much greater distance (≈ 0.1 mm). γ rays are the most penetrating and can pass through an appreciable thickness (≈ 100 mm) of lead. The disintegration process that produces α, β, and γ rays must obey the laws of physics that are summarized below:

1. The conservation of ***mass/energy*** (see Sections 6.8 and 28.6)
2. The conservation of ***electric charge*** (see Section 18.2)
3. The conservation of ***linear momentum*** (see Section 7.2)
4. The conservation of ***angular momentum*** (see Section 9.6)
5. The conservation of ***nucleon number***

Except for the conservation of nucleon number, these laws have been discussed earlier. No nuclear process has ever been observed in which the number of nucleons present before the process has differed from the number of nucleons after the process. Therefore, the number of nucleons is conserved during a nuclear disintegration. As applied to the disintegration of a nucleus, the laws require that the energy, electric charge, linear momentum, angular momentum, and nucleon number that a nucleus possesses must remain unchanged when the nucleus disintegrates into nuclear fragments and accompanying α, β, or γ rays.

The three types of radioactivity that occur naturally can be observed in a relatively simple experiment. A piece of radioactive material is placed at the bottom of a narrow hole in a lead cylinder. The cylinder is located within an evacuated chamber, as Figure 31.6 illustrates. A magnetic field is directed perpendicular to the plane of the paper, and a photographic plate is positioned above the hole. Three spots appear on the developed plate, which are associated with the radioactivity of the nuclei in the material. Since moving particles are deflected by a magnetic field only when they are electrically charged, an analysis of this experiment reveals that two types of radioactivity (α and β rays, as it turns out) consist of charged particles, while the third type (γ rays) does not.

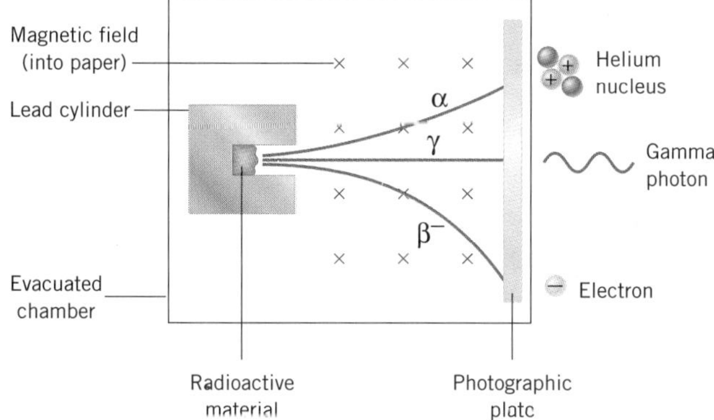

Figure 31.6 α and β rays are deflected by a magnetic field and, therefore, consist of moving charged particles. γ rays are not deflected by a magnetic field and, consequently, must be uncharged.

α DECAY

When a nucleus disintegrates and produces α rays, it is said to undergo ***α decay.*** Experimental evidence shows that α rays consist of positively charged particles, each one being the ^4_2He nucleus of helium. Thus, an α particle has a charge of $+2e$ and a nucleon number of $A = 4$. Since the grouping of 2 protons and 2 neutrons in a ^4_2He nucleus is particularly stable, as we have seen in connection with Figure 31.5, it is not surprising that an α particle can be ejected as a unit from a more massive unstable nucleus.

Figure 31.7 shows the disintegration process for one example of α decay:

$$\underset{\substack{\text{Parent}\\ \text{nucleus}\\ \text{(uranium)}}}{{}^{238}_{92}\text{U}} \rightarrow \underset{\substack{\text{Daughter}\\ \text{nucleus}\\ \text{(thorium)}}}{{}^{234}_{90}\text{Th}} + \underset{\substack{\alpha\text{ particle}\\ \text{(helium}\\ \text{nucleus)}}}{{}^{4}_{2}\text{He}}$$

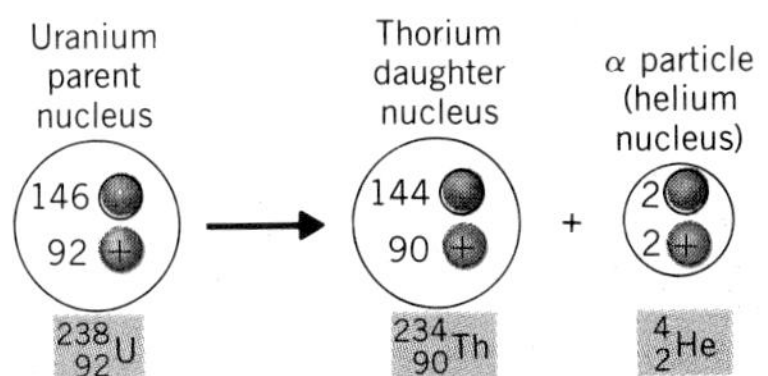

Figure 31.7 **α decay occurs when an unstable parent nucleus emits an α particle and in the process is converted into a different or daughter nucleus.**

The original nucleus is referred to as the parent nucleus (P), and the nucleus remaining after disintegration is called the daughter nucleus (D). Upon emission of an α particle, the uranium ${}^{238}_{92}\text{U}$ parent is converted into the ${}^{234}_{90}\text{Th}$ daughter, which is an isotope of thorium. The parent and daughter nuclei are different, so α decay converts one element into another, a process known as ***transmutation.***

Electric charge is conserved during α decay. In Figure 31.7, for instance, 90 of the 92 protons in the uranium nucleus end up in the thorium nucleus, and the remaining 2 protons are carried off by the α particle. The total number of 92, however, is the same before and after disintegration. α decay also conserves the number of nucleons, for the number is the same before (238) and after (234 + 4) disintegration. Consistent with the conservation of electric charge and nucleon number, the general form for α decay is

[α decay]

$$\underset{\substack{\text{Parent}\\ \text{nucleus}}}{{}^{A}_{Z}\text{P}} \rightarrow \underset{\substack{\text{Daughter}\\ \text{nucleus}}}{{}^{A-4}_{Z-2}\text{D}} + \underset{\substack{\alpha\text{ particle}\\ \text{(helium nucleus)}}}{{}^{4}_{2}\text{He}}$$

When a nucleus releases an α particle, the nucleus also releases energy. In fact, the energy released by radioactive decay is responsible, in part, for keeping the interior of the earth hot and, in some places, even molten. The following example shows how the conservation of mass/energy can be used to determine the amount of energy released in α decay.

Example 4 α Decay and the Release of Energy

Determine the energy released when α decay converts ${}^{238}_{92}\text{U}$ into ${}^{234}_{90}\text{Th}$.

REASONING Energy is released during the α decay. As a result, the combined mass of the ${}^{234}_{90}\text{Th}$ daughter nucleus and the α particle is less than the mass of the ${}^{238}_{92}\text{U}$ parent nucleus. The difference in mass is equivalent to the energy released. We determine the difference in mass in atomic mass units and then use the fact that 1 u is equivalent to 931.5 MeV.

SOLUTION Appendix F gives the masses shown below:

$$\underset{238.0508\text{ u}}{{}^{238}_{92}\text{U}} \longrightarrow \underbrace{\underset{234.0436\text{ u}}{{}^{234}_{90}\text{Th}} + \underset{4.0026\text{ u}}{{}^{4}_{2}\text{He}}}_{238.0462\text{ u}}$$

The decrease in mass is 238.0508 u − 238.0462 u = 0.0046 u. As usual, the masses from Appendix F are atomic masses and include the mass of the orbital electrons. But this causes no error here, because the same total number of electrons is included for ${}^{238}_{92}\text{U}$, on the one hand, and for ${}^{234}_{90}\text{Th}$ plus ${}^{4}_{2}\text{He}$, on the other. Since 1 u is equivalent to 931.5 MeV, the released energy is $\boxed{4.3\text{ MeV}}$.

When α decay occurs as in Example 4, the energy released appears as kinetic energy of the recoiling $^{234}_{90}$Th nucleus and the α particle, except for a small portion carried away as a γ ray. Conceptual Example 5 discusses how the $^{234}_{90}$Th nucleus and the α particle share in the released energy.

Conceptual Example 5 How Energy Is Shared in the α Decay of $^{238}_{92}$U

In Example 4, the energy released by the α decay of $^{238}_{92}$U was found to be 4.3 MeV. Since this energy is carried away as kinetic energy of the recoiling $^{234}_{90}$Th nucleus and the α particle, it follows that $KE_{Th} + KE_{\alpha} = 4.3$ MeV. However, KE_{Th} and KE_{α} are not equal, so that the $^{234}_{90}$Th nucleus and the α particle do not share equally in the released energy. Which particle carries away more kinetic energy, the $^{234}_{90}$Th nucleus or the α particle?

REASONING Kinetic energy depends on the mass m and speed v of a particle, since $KE = \frac{1}{2}mv^2$. The $^{234}_{90}$Th nucleus has a much greater mass than the α particle, and since the kinetic energy is proportional to the mass, it is tempting to conclude that the $^{234}_{90}$Th nucleus has the greater kinetic energy. This conclusion is not correct, however, since it does not take into account the fact that the $^{234}_{90}$Th nucleus and the α particle have different speeds after the decay. In fact, we expect the thorium nucleus to recoil with the smaller speed precisely *because* it has the greater mass. The decaying $^{238}_{92}$U is like a father and his young daughter on ice skates, pushing off against one another. The more massive father recoils with much less speed than the daughter. We can use the principle of conservation of linear momentum to verify our expectation.

As Section 7.2 discusses, the conservation principle states that the total linear momentum of an isolated system remains constant. An isolated system is one for which the vector sum of the external forces acting on the system is zero, and the decaying $^{238}_{92}$U nucleus fits this description. It is stationary initially, and since momentum is mass times velocity, its initial momentum is zero. In its final form, the system consists of the $^{234}_{90}$Th nucleus and the α particle and has a final total momentum of $m_{Th}v_{Th} + m_{\alpha}v_{\alpha}$. According to momentum conservation, the initial and final values of the total momentum of the system must be the same, so that $m_{Th}v_{Th} + m_{\alpha}v_{\alpha} = 0$. Solving this equation for the velocity of the thorium nucleus, we find that $v_{Th} = -m_{\alpha}v_{\alpha}/m_{Th}$. Thus, since m_{Th} is much greater than m_{α}, we can see that the speed of the thorium nucleus is less than the speed of the α particle. Moreover, the kinetic energy depends on the square of the speed and only the first power of the mass. As a result, the much greater speed of the α particle ensures that the α particle has the greater kinetic energy.

Related Homework Material: *Problem 28*

THE PHYSICS OF . . .

radioactivity and smoke detectors.

One widely used application of α decay is in smoke detectors. Figure 31.8 illustrates how a smoke detector operates. Two small and parallel metal plates are separated by a distance of about one centimeter. A tiny amount of radioactive material at the center of one of the plates emits α particles, which collide with air molecules. During the collisions, the air molecules are ionized to form positive and negative ions. The voltage from a battery causes one plate to be positive and the other negative, so that each plate attracts ions of opposite charges. As a result there is a current in the circuit attached to the plates. The presence of smoke particles between the plates reduces the current, since the ions that collide with a smoke particle are usually neutralized. The drop in current that smoke particles cause is used to trigger an alarm.

β DECAY

The β rays in Figure 31.6 are deflected by the magnetic field in a direction opposite to that of the positively charged α rays. Consequently, these β rays, which are the most common kind, consist of negatively charged particles or β^- particles. Experiment shows that β^- particles are electrons. As an illustration of β^- decay, consider the thorium $^{234}_{90}\text{Th}$ nucleus, which decays by emitting a β^- particle, as in Figure 31.9:

$$\underset{\substack{\text{Parent}\\\text{nucleus}\\\text{(thorium)}}}{^{234}_{90}\text{Th}} \rightarrow \underset{\substack{\text{Daughter}\\\text{nucleus}\\\text{(protactinium)}}}{^{234}_{91}\text{Pa}} + \underset{\substack{\beta^-\text{ particle}\\\text{(electron)}}}{^{0}_{-1}\text{e}}$$

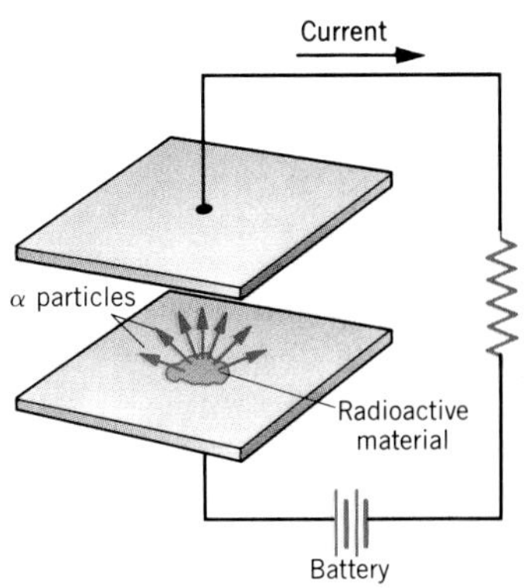

Figure 31.8 A smoke detector.

β^- decay, like α decay, causes a transmutation of one element into another. In this case, thorium $^{234}_{90}\text{Th}$ is converted into protactinium $^{234}_{91}\text{Pa}$. The law of conservation of charge is obeyed, since the net number of positive charges is the same before (90) and after (91 − 1) the β^- emission. The law of conservation of nucleon number is obeyed, since the nucleon number remains at $A = 234$. The general form for β^- decay is

[β^- decay]
$$\underset{\substack{\text{Parent}\\\text{nucleus}}}{^{A}_{Z}\text{P}} \rightarrow \underset{\substack{\text{Daughter}\\\text{nucleus}}}{_{Z+1}^{\;\;\;A}\text{D}} + \underset{\substack{\beta^-\text{ particle}\\\text{(electron)}}}{^{0}_{-1}\text{e}}$$

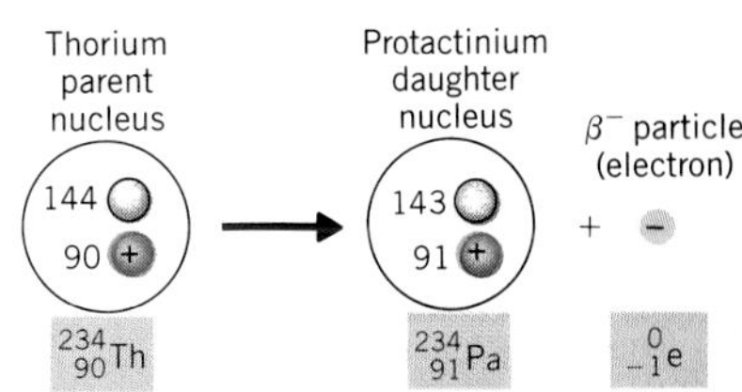

Figure 31.9 β decay occurs when a neutron in an unstable parent nucleus decays into a proton and an electron, the electron being emitted as the β^- particle. In the process, the parent nucleus is transformed into the daughter nucleus.

The electron emitted in β^- decay does *not* actually exist within the parent nucleus and is *not* one of the orbital electrons. Instead, the electron is created when a neutron decays into a proton and an electron; when this occurs, the proton number of the parent nucleus increases from Z to Z + 1 and the nucleon number remains unchanged. The electron is usually fast-moving and escapes from the atom, leaving behind a positively charged atom.

Example 6 illustrates that energy is released during β^- decay, just as it is during α decay, and that the conservation of mass/energy applies.

Example 6 β^- Decay and the Release of Energy

Find the energy released when β^- decay changes $^{234}_{90}\text{Th}$ into $^{234}_{91}\text{Pa}$.

REASONING AND SOLUTION To find the energy released, we follow the usual procedure of determining how much the mass has decreased because of the decay and then calculating the equivalent energy. The masses (see Appendix F) are shown below:

$$\underset{234.043\ 59\ \text{u}}{^{234}_{90}\text{Th}} \longrightarrow \underbrace{^{234}_{91}\text{Pa} + {}^{0}_{-1}\text{e}}_{234.043\ 30\ \text{u}}$$

When the $^{234}_{90}\text{Th}$ nucleus of a thorium atom is converted into a $^{234}_{91}\text{Pa}$ nucleus, the number of orbital electrons remains the same, so the resulting protactinium atom is missing one orbital electron. However, the mass taken from Appendix F includes all 91 electrons of a neutral protactinium atom. In effect, then, the value of 234.043 30 u for $^{234}_{91}\text{Pa}$ already includes the mass of the β^- particle. The mass decrease that accompanies the β^- decay is 234.043 59 u − 234.043 30 u = 0.000 29 u. The equivalent energy (1 u = 931.5 MeV) is $\boxed{0.27\text{ MeV}}$. This is the maximum kinetic energy that the emitted electron can have.

PROBLEM SOLVING INSIGHT

In β^- decay, be careful not to include the mass of the electron ($^{0}_{-1}$e) twice. As discussed here for the daughter atom ($^{234}_{91}$Pa), the mass given in Appendix F already includes the mass of the emitted electron.

A second kind of β decay sometimes occurs.† In this process the particle emitted by the nucleus is a ***positron,*** rather than an electron. A positron, also called a β^+ particle, has the same mass as an electron, but carries a charge of $+e$ instead of $-e$. The disintegration process for β^+ decay is

[β^+ decay]
$$\underset{\text{Parent nucleus}}{{}^{A}_{Z}\text{P}} \rightarrow \underset{\text{Daughter nucleus}}{{}_{Z-1}^{\;\;\;A}\text{D}} + \underset{\beta^+\text{ particle (positron)}}{{}^{0}_{1}\text{e}}$$

The emitted positron does *not* exist within the nucleus but, rather, is created when a nuclear proton is transformed into a neutron. In the process, the proton number of the parent nucleus decreases from Z to Z − 1, and the nucleon number remains the same. As with β^- decay, the laws of conservation of charge and nucleon number are obeyed, and there is a transmutation of one element into another.

γ DECAY

The nucleus, like the orbital electrons, exists only in discrete energy states or levels. When a nucleus changes from an excited energy state (denoted by an asterisk *) to a lower energy state, a photon is emitted. The process is similar to the one discussed in Section 30.3 for the photon emission that leads to the hydrogen atom line spectrum. With nuclear energy levels, however, the photon has a much greater energy and is called a γ ray. The γ decay process is written as follows:

[γ decay]
$$\underset{\text{Excited energy state}}{{}^{A}_{Z}\text{P}^*} \rightarrow \underset{\text{Lower energy state}}{{}^{A}_{Z}\text{P}} + \underset{\gamma\text{ ray}}{\gamma}$$

γ decay does *not* cause a transmutation of one element into another. In the next example the wavelength of one particular γ ray photon is determined.

Example 7 The Wavelength of a Photon Emitted in γ Decay

What is the wavelength of the 0.186 MeV γ ray photon emitted by radium ${}^{226}_{88}\text{Ra}$?

REASONING The photon energy is the difference between two nuclear energy levels. Equation 30.4 gives the relation between the energy level separation ΔE and the frequency f of the photon as $\Delta E = hf$. Since $f\lambda = c$, the wavelength of the photon is $\lambda = hc/\Delta E$.

SOLUTION First we must convert the photon energy into joules:

$$\Delta E = (0.186 \times 10^6 \text{ eV})\left(\frac{1.60 \times 10^{-19}\text{ J}}{1 \text{ eV}}\right) = 2.98 \times 10^{-14}\text{ J}$$

The wavelength of the photon is

$$\lambda = \frac{hc}{\Delta E} = \frac{(6.63 \times 10^{-34}\text{ J}\cdot\text{s})(3.00 \times 10^8\text{ m/s})}{2.98 \times 10^{-14}\text{ J}} = \boxed{6.67 \times 10^{-12}\text{ m}}$$

PROBLEM SOLVING INSIGHT

The energy ΔE of a γ ray photon, like that of photons in other regions of the electromagnetic spectrum (visible, infrared, microwave, etc.), is equal to the product of Planck's constant h and the frequency f of the photon: $\Delta E = hf$.

† A third kind of β decay also occurs, in which a nucleus pulls in or captures one of the orbital electrons from outside the nucleus. The process is called ***electron capture,*** or ***K capture,*** since the electron normally comes from the innermost or K shell.

31.5 THE NEUTRINO

When a β particle is emitted by a radioactive nucleus, energy is simultaneously released, as Example 6 illustrates. Experimentally, however, it is found that most β particles do not have enough kinetic energy to account for all the energy released. If a β particle carries away only part of the energy, where does the remainder go? The question puzzled physicists until 1930, when Wolfgang Pauli proposed that part of the energy is carried away by another particle that is emitted along with the β particle. This additional particle is called the ***neutrino,*** and its existence was verified experimentally in 1956. The Greek letter nu (ν) is used to symbolize the neutrino. For instance, the β^- decay of thorium $^{234}_{90}Th$ (see Section 31.4) is more correctly written as

$$^{234}_{90}\text{Th} \rightarrow {}^{234}_{91}\text{Pa} + {}^{0}_{-1}\text{e} + \bar{\nu}$$

The bar above the ν is included, because the neutrino emitted in this particular decay process is an antimatter neutrino or antineutrino. A normal neutrino (ν without the bar) is emitted when β^+ decay occurs.

The neutrino has zero electrical charge. Moreover, at present there is no convincing experimental evidence to indicate that the neutrino has any mass. A particle with zero mass, like a photon, travels at the speed of light. The neutrino, therefore, travels near or at the speed of light. The emission of β particles and neutrinos involves a fundamental force that has not been mentioned before in this text. This force is much weaker than the strong nuclear force and weaker than the electromagnetic force; hence, it is referred to as the ***weak nuclear force.***

31.6 RADIOACTIVE DECAY AND ACTIVITY

The question of which radioactive nucleus in a group disintegrates at a given instant is decided like the drawing of numbers in a state lottery; individual disintegrations occur randomly. As time passes, the number N of parent nuclei decreases, as Figure 31.10 shows. This graph of N versus time indicates that the decrease occurs in a smooth fashion, with N approaching zero after enough time has passed. To help describe the graph, it is useful to define the ***half-life*** $T_{1/2}$ of a radioactive isotope as the time required for one-half of the nuclei present to disintegrate. For example, radium $^{226}_{88}Ra$ has a half-life of 1600 years, for it takes this amount of time for one-half of a given quantity of this isotope to disintegrate into radon $^{222}_{86}Rn$. In another 1600 years, one-half of the remaining radium atoms will disintegrate, leaving only one-fourth of the original number intact. In Figure 31.10, the number of nuclei present at time $t = 0$ is $N = N_0$, while the number present at $t = T_{1/2}$ is $N = \frac{1}{2}N_0$. The number present at $t = 2T_{1/2}$ is $N = \frac{1}{4}N_0$, and so forth. The value of the half-life depends on the nature of the radioactive nucleus. Values ranging from a fraction of a second to billions of years have been found, as Table 31.2 indicates. Example 8 deals with the half-life of radon $^{222}_{86}Rn$.

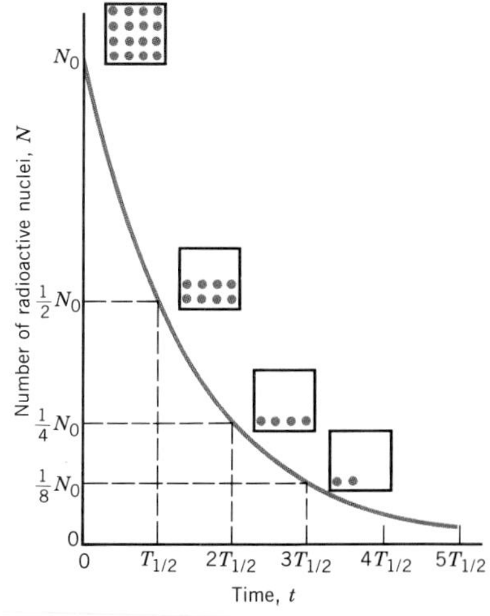

Figure 31.10 The half-life $T_{1/2}$ of a radioactive decay is the time in which one-half of the radioactive nuclei disintegrate.

Table 31.2 Some Half-lives for Radioactive Decay

Isotope		Half-life	Decay Mode
Polonium	$^{214}_{84}Po$	1.64×10^{-4} s	α, γ
Krypton	$^{89}_{36}Kr$	3.16 min	β^-, γ
Radon	$^{222}_{86}Rn$	3.83 days	α, γ
Strontium	$^{90}_{38}Sr$	28.5 yr	β^-
Radium	$^{226}_{88}Ra$	1.6×10^3 yr	α, γ
Carbon	$^{14}_{6}C$	5.73×10^3 yr	β^-
Uranium	$^{238}_{92}U$	4.47×10^9 yr	α, γ
Indium	$^{115}_{49}In$	4.41×10^{14} yr	β^-

THE PHYSICS OF . . .

radioactive radon gas in houses.

Example 8 The Radioactive Decay of Radon Gas

Radon $^{222}_{86}Rn$ is a radioactive gas produced when radium $^{226}_{88}Ra$ undergoes α decay. There is growing concern about radon as a health hazard, because it can become trapped in houses, entering primarily through cracks in walls and floors and in the drinking water. Suppose 3.0×10^7 radon atoms are trapped in a basement at the time the basement is sealed against further entry of the gas. The half-life of radon is 3.83 days. How many radon atoms remain after 31 days?

REASONING During each half-life, the number of radon atoms is reduced by a factor of two. Thus, we determine the number of half-lives there are in a period of 31 days and reduce the number of radon atoms by a factor of two for each one.

SOLUTION In a period of 31 days there are 31 days/3.83 days = 8.1 half-lives. In 8 half-lives the number of radon atoms is reduced by a factor of $2^8 = 256$. Ignoring the difference between 8 and 8.1 half-lives, we find that the number of atoms remaining is $3.0 \times 10^7/256 = \boxed{1.2 \times 10^5}$.

The ***activity*** of a radioactive sample is the number of disintegrations per second that occur. Each time a disintegration occurs, the number N of radioactive nuclei decreases. As a result, the activity can be obtained by dividing ΔN, the change in the number of nuclei, by Δt, the time interval during which the change takes place; the average activity over the time interval Δt is the magnitude of $\Delta N/\Delta t$. Since the decay of any individual nucleus is completely random, the number of disintegrations per second that occur in a sample is proportional to the number of radioactive nuclei present, so that

$$\frac{\Delta N}{\Delta t} = -\lambda N \tag{31.4}$$

where λ is a proportionality constant referred to as the ***decay constant.*** The minus sign is present in this equation because each disintegration decreases the number N of nuclei originally present.

The SI unit for activity is the *becquerel* (Bq); one becquerel equals one disintegration per second. Activity is also measured in terms of a unit called the *curie* (Ci), in honor of Marie (1867–1934) and Pierre (1859–1906) Curie, the discoverers of radium and polonium. Historically, the curie was chosen as a unit because it is roughly the activity of one gram of pure radium. In terms of becquerels,

$$1 \text{ Ci} = 3.70 \times 10^{10} \text{ Bq}$$

The activity of the radium put into the dial of a watch to make it glow in the dark is about 4×10^4 Bq, and the activity used in radiation therapy for cancer treatment is approximately a billion times greater, or 4×10^{13} Bq.

The mathematical expression for the graph of N versus t shown in Figure 31.10 can be obtained from Equation 31.4 with the aid of calculus. The result for the number N of radioactive nuclei present at time t is

$$N = N_0 \, e^{-\lambda t} \qquad (31.5)$$

In cases of severe radon contamination, the soil itself must be removed, which is what the workers here are doing.

assuming that the number at $t = 0$ is N_0. The exponential e has the value $e = 2.718 \ldots$, and many calculators provide the value of e^x. We can relate the half-life $T_{1/2}$ of a radioactive nucleus to its decay constant λ in the following manner. By substituting $N = \frac{1}{2}N_0$ and $t = T_{1/2}$ into Equation 31.5, we find that $\frac{1}{2} = e^{-\lambda T_{1/2}}$. Taking the natural logarithm of both sides of this equation reveals that $\ln 2 = \lambda T_{1/2}$ or

$$T_{1/2} = \frac{\ln 2}{\lambda} = \frac{0.693}{\lambda} \qquad (31.6)$$

The following example illustrates the use of Equations 31.5 and 31.6.

Example 9 The Activity of Radon $^{222}_{86}Rn$

As in Example 8, suppose there are 3.0×10^7 radon atoms ($T_{1/2} = 3.83$ days or 3.31×10^5 s) trapped in a basement. (a) How many radon atoms remain after 31 days? Find the ctivity (b) just after the basement is sealed against further entry of radon and (c) 31 days later.

REASONING AND SOLUTION

(a) The answer can be obtained directly from Equation 31.5, provided the decay constant is first determined from the half-life:

$$\lambda = \frac{0.693}{T_{1/2}} = \frac{0.693}{3.83 \text{ days}} = 0.181 \text{ days}^{-1}$$

$$N = N_0 \, e^{-\lambda t} = (3.0 \times 10^7)e^{-(0.181 \text{ days}^{-1})(31 \text{ days})} = \boxed{1.1 \times 10^5}$$

This value is slightly less than that found in Example 8, because there we ignored the difference between 8.0 and 8.1 half-lives.

(b) The activity can be obtained from Equation 31.4, provided the decay constant is expressed in reciprocal seconds: $\lambda = 0.693/(3.31 \times 10^5 \text{ s}) = 2.09 \times 10^{-6} \text{ s}^{-1}$. According to Equation 31.4,

$$\frac{\Delta N}{\Delta t} = -\lambda N = -(2.09 \times 10^{-6} \text{ s}^{-1})(3.0 \times 10^7) = -63 \text{ disintegrations/s}$$

The activity is the magnitude of $\Delta N/\Delta t$, so initially $\boxed{\text{Activity} = 63 \text{ Bq}}$.

(c) From part (a), the number of radioactive nuclei remaining at the end of 31 days is $N = 1.1 \times 10^5$, and reasoning similar to that in part (b) reveals that $\boxed{\text{Activity} = 0.23 \text{ Bq}}$.

THE PHYSICS OF . . .

radioactive dating.

31.7 RADIOACTIVE DATING

One important application of radioactivity is the determination of the age of archeological or geological samples. If an object contains radioactive nuclei when it is formed, then the decay of these nuclei marks the passage of time like a clock, half of the nuclei disintegrating during each half-life. If the half-life is known, a measurement of the number of nuclei present today relative to the number present initially can give the age of the sample. According to Equation 31.4, the activity of a sample is proportional to the number of radioactive nuclei, so one way to obtain the age is to compare present activity with initial activity. A more accurate way is to determine the present number of radioactive nuclei with the aid of a mass spectrometer.

The present activity of a sample can be measured, but how is it possible to know what the original activity was, perhaps thousands of years ago? Radioactive dating methods entail certain assumptions that make it possible to estimate the original activity. For instance, the radiocarbon technique utilizes the ${}^{14}_{6}C$ isotope of carbon, which undergoes β^- decay with a half-life of 5730 yr. This isotope is present in the earth's atmosphere at an equilibrium concentration of about one atom for every 8.3×10^{11} atoms of normal carbon ${}^{12}_{6}C$. It is often assumed* that this value has remained constant over the years, because ${}^{14}_{6}C$ is created when cosmic rays interact with the earth's upper atmosphere, a production method that offsets the loss via β^- decay. Moreover, nearly all living organisms ingest the equilibrium concentration of ${}^{14}_{6}C$. However, once an organism dies, metabolism no longer sustains the input of ${}^{14}_{6}C$, and β^- decay causes half of the ${}^{14}_{6}C$ nuclei to disintegrate every 5730 yr.

It is possible to calculate the ${}^{14}_{6}C$ activity of one gram of carbon in a living organism. One gram of carbon (atomic mass = 12 u) is 1.0/12 mol, and since there are 6.02×10^{23} atoms per mole (Avogadro's number), the number of ${}^{14}_{6}C$ atoms present is

$$\left(\frac{1.0}{12}\ \text{mol}\right)\left(6.02 \times 10^{23}\ \frac{\text{atoms}}{\text{mol}}\right)\left(\frac{1}{8.3 \times 10^{11}}\right) = 6.0 \times 10^{10}\ \text{atoms}$$

Since the half-life is 5730 yr (1.81×10^{11} s), the decay constant of ${}^{14}_{6}C$ is $\lambda = 0.693/T_{1/2} = 0.693/(1.81 \times 10^{11}\ \text{s}) = 3.83 \times 10^{-12}\ \text{s}^{-1}$. Therefore, Equation 31.4 indicates that the activity, or the magnitude of $\Delta N/\Delta t$, is

$$\begin{array}{l}\text{Activity of one}\\ \text{gram of carbon in}\\ \text{a living organism}\end{array} = \lambda N = (3.83 \times 10^{-12}\ \text{s}^{-1})(6.0 \times 10^{10}) = 0.23\ \text{Bq}$$

Radioactive dating is used to determine the age of fossils such as this one. It is a head of a *Tyrannosaurus rex* that was found embedded in South Dakota sandstone.

An organism that lived thousands of years ago presumably had an activity of about 0.23 Bq per gram of carbon. When the organism died, the activity began decreasing. From a sample of the remains, the current activity per gram of carbon

* The assumption that the ${}^{14}_{6}C$ concentration has always been at its present equilibrium value has been evaluated by comparing ${}^{14}_{6}C$ ages with ages determined by counting tree rings. More recently, ages determined using the radioactive decay of uranium ${}^{238}_{92}U$ have been used for comparison. These comparisons indicate that the equilibrium value of the ${}^{14}_{6}C$ concentration has indeed remained constant for the past 1000 years. However, from there back about 30 000 years, it appears that the ${}^{14}_{6}C$ concentration in the atmosphere was larger than its present value by up to 40%. In this text, as a first approximation we ignore such discrepancies.

can be measured and compared to the value of 0.23 Bq to determine the time that has transpired since death. This procedure is illustrated in Example 10.

Example 10 The Iceman

On September 19, 1991, a German tourist made a startling discovery during a walking trip in the Italian Alps. He found the Iceman, a Stone Age traveler whose body had become trapped in the ice of a glacier. Figure 31.11 shows the remarkably well-preserved remains, which were dated using the radiocarbon method. Material found with the body had a ${}^{14}_{6}C$ activity of about 0.121 Bq per gram of carbon. Determine the age of the Iceman's remains.

REASONING According to Equation 31.5, the number of nuclei remaining at time t is $N = N_0 e^{-\lambda t}$. Multiplying both sides of this expression by the decay constant λ and recognizing that the product of λ and N is the activity A, we find that $A = A_0 e^{-\lambda t}$, where $A_0 = 0.23$ Bq is the activity at time $t = 0$ for one gram of carbon. The decay constant λ can be determined from the value of 5730 yr for the half-life of ${}^{14}_{6}C$, using Equation 31.6. With known values for A_0 and λ, the given activity of $A = 0.121$ Bq per gram of carbon can be used to find the age t of the Iceman's remains.

SOLUTION For ${}^{14}_{6}C$, the decay constant is $\lambda = 0.693/T_{1/2} = 0.693/(5730\ \text{yr}) = 1.21 \times 10^{-4}\ \text{yr}^{-1}$. Since $A = 0.121$ Bq and $A_0 = 0.23$ Bq, the age can be determined from

$$A = 0.121\ \text{Bq} = (0.23\ \text{Bq})e^{-(1.21\times10^{-4}\ \text{yr}^{-1})t}$$

Taking the natural logarithm of both sides of this result gives

$$\ln\left(\frac{0.121\ \text{Bq}}{0.23\ \text{Bq}}\right) = -\ (1.21 \times 10^{-4}\ \text{yr}^{-1})t$$

The age of the sample is $\boxed{t = 5.3 \times 10^3\ \text{yr}}$.

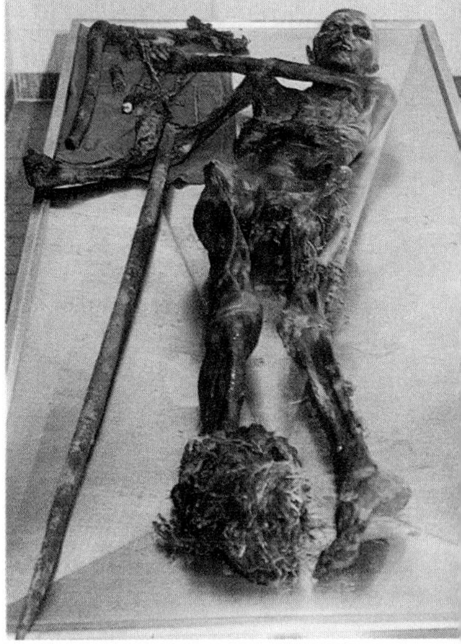

Figure 31.11 These remains of the Iceman were discovered in the ice of a glacier in the Italian Alps in 1991. Radiocarbon dating reveals that they are 5300 years old.

Radiocarbon dating is not the only radioactive dating method. For example, other methods utilize uranium ${}^{238}_{92}U$, potassium ${}^{40}_{19}K$, and lead ${}^{210}_{82}Pb$. For such methods to be useful, the half-life of the radioactive species must be neither too short nor too long relative to the age of the sample to be dated, as Conceptual Example 11 discusses.

Conceptual Example 11 Dating a Bottle of Wine

A bottle of red wine is thought to have been sealed about 5 years ago. The wine contains a number of different kinds of atoms, including carbon, oxygen, and hydrogen. Each of these has a radioactive isotope. The radioactive isotope of carbon is the familiar ${}^{14}_{6}C$, with a half-life of 5730 yr. The radioactive isotope of oxygen is ${}^{15}_{8}O$ and has a half-life of 122.2 s. The radioactive isotope of hydrogen is ${}^{3}_{1}H$ and is called tritium; its half-life is 12.33 yr. The activity of each of these isotopes is known at the time the bottle was sealed. However, only one of the isotopes is useful for determining the age of the wine accurately. Which is it?

REASONING In a dating method that measures the activity of a radioactive isotope, the age of the sample is determined from the change in the activity during the time period in question. Here the expected age is about 5 years. This period is only a small fraction of the 5730-yr half-life of ${}^{14}_{6}C$. As a result, relatively few of the ${}^{14}_{6}C$ nuclei would decay during the wine's life, and the measured activity would change little from its initial value. To obtain an accurate age from such a small change would require prohibitively precise measurements. Nor is the ${}^{15}_{8}O$ isotope very useful. The difficulty is its relatively short half-life of 122.2 s. During a 5-year period, so many half-lives of 122.2 s would occur that the activity would decrease to a vanishingly small level. It would not be even possible to measure it. The only remaining option is the tritium isotope of hydrogen. The expected age of 5 yr is long enough relative to the half-life of 12.33 yr that a measurable change in activity will occur, but not so long that the activity will have completely vanished for all practical purposes.

Related Homework Material: *Question 14, Problem 45*

31.8 RADIOACTIVE DECAY SERIES

When an unstable parent nucleus decays, the resulting daughter nucleus is sometimes also unstable. If so, the daughter then decays and produces its own daughter, and so on, until a completely stable nucleus is produced. This sequential decay of one nucleus after another is called a ***radioactive decay series.*** Examples 4–6 discuss the first two steps of a series that begins with uranium ${}^{238}_{92}U$:

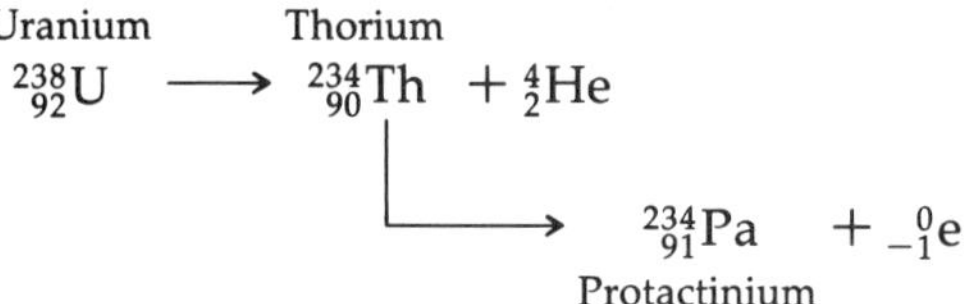

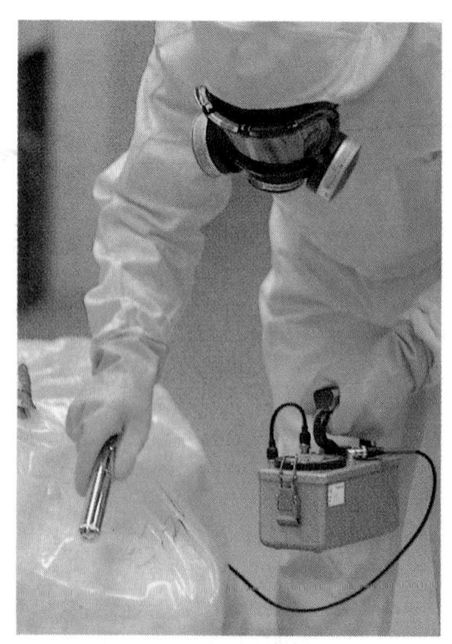

Workers in nuclear industries monitor levels of radioactivity as part of routine safety inspections.

Furthermore, Example 8 deals with radon ${}^{222}_{86}Rn$, which is formed down the line in the ${}^{238}_{92}U$ radioactive decay series. Figure 31.12 shows the entire series. At several points in the series, branches occur, because more than one kind of decay is possible for an intermediate species. Ultimately, however, the series ends with lead ${}^{206}_{82}Pb$, which is stable.

The ${}^{238}_{92}U$ series and other such series are the only sources of some of the radioactive elements found in nature. Radium ${}^{226}_{88}Ra$, for instance, has a half-life of 1600 yr, which is short enough that all the ${}^{226}_{88}Ra$ created when the earth was formed billions of years ago has now disappeared. The ${}^{238}_{92}U$ series provides a continuing supply of ${}^{226}_{88}Ra$, however.

THE PHYSICS OF . . .

detectors of radiation.

31.9 DETECTORS OF RADIATION

There are a number of devices that can be used to detect the effects of the particles and photons (γ rays) emitted when a radioactive nucleus decays. Such devices detect the ionization that these particles and photons cause as they pass through matter.

The most familiar detector is the ***Geiger counter,*** which Figure 31.13 illustrates. The Geiger counter consists of a gas-filled metal cylinder. The α, β, or γ rays enter the cylinder through a thin window at one end. γ rays can also penetrate directly

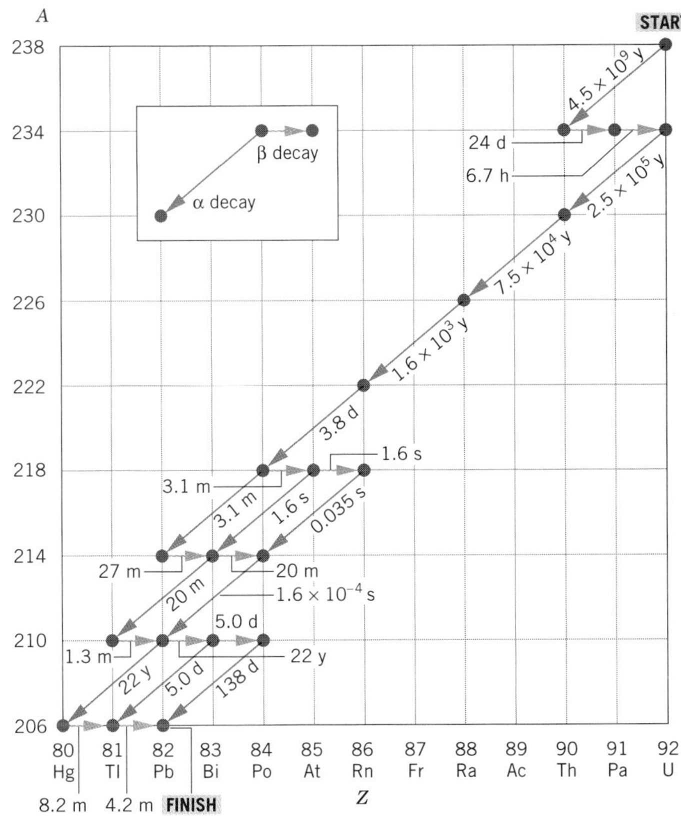

Figure 31.12 A radioactive decay series that begins with uranium $^{238}_{92}U$ and ends with lead $^{206}_{82}Pb$. The half-lives are given in seconds (s), minutes (m), hours (h), days (d), or years (y). The insert in the upper left corner of the graph identifies the type of decay that each nucleus undergoes.

through the metal. A wire electrode runs along the center of the tube and is kept at a high positive voltage (1000–3000 V) relative to the outer cylinder. When a high-energy particle or photon enters the cylinder, it collides with and ionizes a gas molecule. The electron produced from the gas molecule accelerates toward the positive wire, ionizing other molecules in its path. Additional electrons are formed, and an avalanche of electrons rushes toward the wire, leading to a pulse of current through the resistor R. This pulse can be counted or made to produce a "click" in a loudspeaker. The number of counts or clicks is related to the number of high-energy particles or photons present, or equivalently, to the number of disintegrations that produced the particles or photons.

The ***scintillation counter*** is another important radiation detector. As Figure 31.14 indicates, this device consists of a scintillator mounted on a photomultiplier tube. Often the scintillator is a crystal (e.g., cesium iodide) containing a small amount of impurity (thallium), but plastic, liquid, and gaseous scintillators are also used. In response to ionizing radiation, the scintillator emits a flash of visible light. The photons of the flash then strike the photocathode of the photomultiplier tube. The photocathode is made of a material that emits electrons because of the photoelectric effect. These photoelectrons are then attracted to a special electrode kept at a voltage of about +100 V relative to the photocathode. The electrode is coated with a substance that emits several additional electrons for every electron striking it. The additional electrons are attracted to a second similar electrode (voltage = +200 V) where they generate even more electrons. Commercial photomultiplier tubes contain as many as 15 of these special electrodes, so photoelectrons resulting from the light flash of the scintillator lead to a cascade of

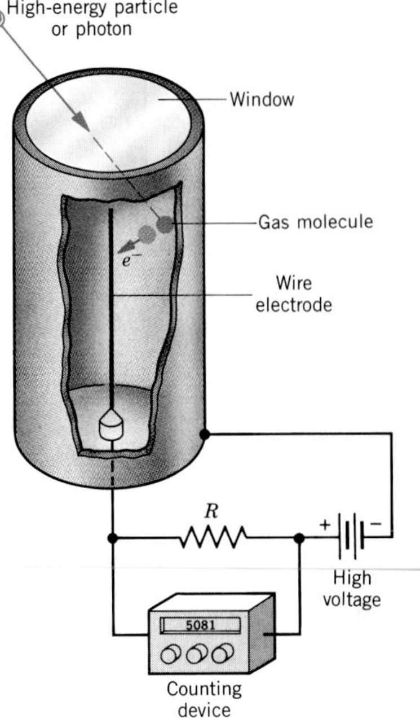

Figure 31.13 A Geiger counter.

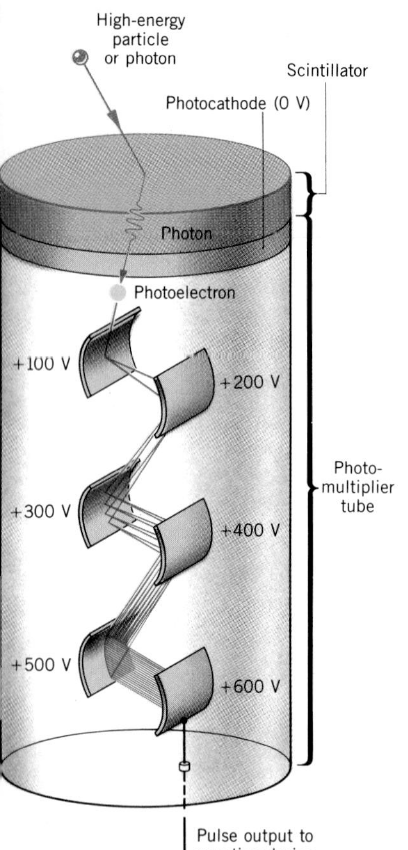

Figure 31.14 A scintillation counter.

Figure 31.15 A photograph of particle tracks in a bubble chamber.

electrons and a pulse of current. As in a Geiger tube, the current pulses can be counted.

Ionizing radiation can also be detected with several types of ***semiconductor detectors.*** Such devices utilize n- and p-type materials, and their operation depends on the electrons and holes formed in the materials as a result of the radiation. One of the main advantages of semiconductor detectors is their ability to discriminate between two particles with only slightly different energies.

A number of instruments provide a pictorial representation of the path that high-energy particles follow after they are emitted from unstable nuclei. In a ***cloud chamber,*** a gas is cooled just to the point where it will condense into droplets, provided nucleating agents are available on which the droplets can form. When a high-energy particle, such as an α particle or a β particle, passes through the gas, the ions it leaves behind serve as nucleating agents, and droplets form along the path of the particle. A ***bubble chamber*** works in a similar fashion, except it contains a liquid that is just at the point of boiling. Tiny bubbles form along the trail of a high-energy particle passing through the liquid. The paths revealed in a cloud or bubble chamber can be photographed to provide a permanent record of the event. Figure 31.15 shows a photograph of the tracks in a bubble chamber. A ***photographic emulsion*** also can be used directly to produce a record of the path taken by a particle of ionizing radiation. The ions formed as the particle passes through the emulsion cause silver to be deposited along the track when the emulsion is developed.

SUMMARY

The nucleus of an atom consists of protons and neutrons, which are collectively referred to as **nucleons.** A **neutron** is an electrically neutral particle whose mass is slightly larger than that of a proton. The **atomic number** Z is the number of protons in the nucleus. The **atomic mass number** or **nucleon number** A is the total number of protons and neutrons in the nucleus: $A = Z + N$, where N is the number of neutrons. Nuclei that contain the same number of protons, but a different number of neutrons, are called **isotopes.**

The **strong nuclear force** is the force of attraction between nucleons and is one of the four fundamental forces of nature. This force balances the electrostatic force of repulsion between protons and holds the nucleus together. The strong nuclear force has a very short range of action and is almost independent of electric charge.

The **binding energy** of a nucleus is the energy required to separate the nucleus into its constituent protons and neutrons. The binding energy is equal to $(\Delta m)c^2$, where c is the speed of light in a vacuum and Δm is the mass defect of the nucleus. The **mass defect** is the amount by which the sum of the individual masses of the protons and neutrons exceeds the mass of the parent nucleus.

When specifying nuclear masses, it is customary to use the **atomic mass unit** (u), which is one-twelfth of the mass of a $^{12}_{6}C$ atom. One atomic mass unit is equivalent to an energy of 931.5 MeV.

Unstable nuclei spontaneously decay by breaking apart or rearranging their internal structures in a process called **radioactivity.** Naturally occurring radioactivity produces α rays, β rays, and γ rays. **α rays** consist of positively charged particles, each particle being the $^{4}_{2}He$ nucleus of helium. The most common kind of **β ray** consists of negatively charged particles, or β^- particles, which are electrons. Another kind of β ray consists of positively charged particles, or β^+ particles. A β^+ particle, also called a **positron,** has the same mass as an electron but carries a charge of $+e$ instead of $-e$. **γ rays** are high-energy photons. If a radioactive parent nucleus disintegrates into a daughter nucleus that has a different atomic number, one element has been converted into another element, the conversion being referred to as a **transmutation.**

The **neutrino** is an electrically neutral particle that has near zero or zero mass. The neutrino travels near or at the speed of light and is emitted along with β particles.

The **half-life** $T_{1/2}$ of a radioactive isotope is the time required for one-half of the nuclei present to disintegrate or decay. The **activity** is the number of disintegrations per second that occur. In other words, the activity is the magnitude of $\Delta N/\Delta t$, where ΔN is the change in the number N of radioactive nuclei and Δt is the time interval during which the change occurs. The SI unit for activity is the becquerel (Bq), one becquerel being one disintegration per second. Radioactive decay obeys the following relation: $\Delta N/\Delta t = -\lambda N$, where λ is the **decay constant.** This equation can be solved to show that $N = N_0 e^{-\lambda t}$, where N_0 is the number of nuclei present initially. The decay constant is related to the half-life according to $T_{1/2} = 0.693/\lambda$.

The sequential decay of one nucleus after another is called a **radioactive decay series,** and Figure 31.12 illustrates one such series.

QUESTIONS

1. A material is known to be an isotope of lead, although the particular isotope is not known. From such limited information, which of the following quantities can you specify: (a) its atomic number, (b) its neutron number, and (c) its atomic mass number? Explain.

2. Two nuclei have different nucleon numbers A_1 and A_2. Are the two nuclei necessarily isotopes of the same element? Give your reasoning and support it with an example from Appendix F.

3. Two nuclei have the same radius, even though they contain different numbers of protons and different numbers of neutrons. Explain how this is possible.

4. Using Figure 31.5, rank the following nuclei in ascending order according to the binding energy per nucleon (smallest first): phosphorus $^{31}_{15}P$, cobalt $^{59}_{27}Co$, tungsten $^{184}_{74}W$, and thorium $^{232}_{90}Th$.

5. Describe qualitatively how the radius of the daughter nucleus compares to that of the parent nucleus for (a) α decay and (b) β decay. Justify your answers in terms of Equation 31.2.

6. Suppose that the range of action of the strong nuclear force were longer than it actually is. Compared to what is actually required, would a greater or smaller number of neutrons be needed to hold together a nucleus that contains a given number of protons? Account for your answer.

7. Uranium $^{238}_{92}U$ decays into thorium $^{234}_{90}Th$ by means of α decay, as Example 4 in the text discusses. A reasonable question to ask is, Why doesn't the $^{238}_{92}U$ nucleus just emit a single proton, instead of an α particle? This hypothetical decay scheme is shown below, along with the pertinent atomic masses:

$$^{238}_{92}U \rightarrow {}^{237}_{91}Pa + {}^{1}_{1}H$$

Uranium	Protactinium	Proton
238.050 78 u	237.051 14 u	1.007 83 u

For a decay to be possible, it must bring the parent nucleus toward a more stable state by allowing the release of energy.

Compare the total mass of the products of this hypothetical decay with the mass of $^{238}_{92}U$ and decide whether the emission of a single proton is possible for $^{238}_{92}U$. Explain.

8. Explain why unstable nuclei with short half-lives typically have only a small or zero natural abundance.

9. The half-life of indium $^{115}_{49}In$ is 4.41×10^{14} yr. Thus, one-half of the nuclei in a sample of this isotope will decay in this time, which is very long. Is it possible for any single nucleus in the sample to decay after only one second has passed? Justify your answer.

10. (a) Is it possible for two different samples that contain different radioactive elements to have the same activity? (b) Is it possible for two different samples of the same radioactive element to have different activities? In each case, defend your answer.

11. To which of the following objects, each about 1000 yr old, can the radiocarbon dating technique *not* be applied: a wooden box, a gold statue, and a skeleton? Explain.

12. Two isotopes have half-lives that are short, relative to the known age of the earth. Today, one of these isotopes is found in nature and one is not. How is this possible, considering that both should have long since decayed to zero? Explain.

13. Suppose there were a greater number of carbon $^{14}_{6}C$ atoms in an animal living 5000 yr ago than is currently believed. When the bones of this animal are tested today using radiocarbon dating, is the age obtained too small or too large? Give your reasoning.

14. Review Conceptual Example 11 as an aid in answering this question. Tritium is an isotope of hydrogen and undergoes β^- decay with a half-life of 12.33 yr. Like carbon $^{14}_{6}C$, tritium is produced in the atmosphere because of cosmic rays and can be used in a radioactive dating technique. Can tritium dating be used to determine a reliable date for a sample that is about 700 yr old? Account for your answer.

15. Because of radioactive decay, one element can be transmuted into another. Thus, a container of uranium $^{238}_{92}U$ ultimately becomes a container of lead $^{206}_{82}Pb$, as Figure 31.12 indicates. How long does the transmutation of $^{238}_{92}U$ entirely into $^{206}_{82}Pb$ require—several decades, several centuries, thousands of years, millions of years, or billions of years? State your reasoning.

PROBLEMS

The data given for atomic masses in these problems include the mass of the electrons orbiting the nucleus of the electrically neutral atom.

Section 31.1 Nuclear Structure, Section 31.2 The Strong Nuclear Force and the Stability of the Nucleus

1. In each of the following cases, what does the symbol X represent and how many neutrons are in the nucleus: (a) $^{195}_{78}X$, (b) $^{32}_{16}X$, (c) $^{63}_{29}X$, (d) $^{11}_{5}X$, and (e) $^{239}_{94}X$? Use the periodic table on the inside of the back cover as needed.

2. What is the radius of a nucleus of uranium $^{238}_{92}U$?

3. Two isotopes of chlorine occur in nature. The $^{35}_{17}Cl$ isotope has an atomic mass of 34.968 85 u and a natural abundance of 75.77%. The $^{37}_{17}Cl$ isotope has an atomic mass of 36.965 90 u and a natural abundance of 24.23%. By a calculation of your own, verify that the value of 35.45 u listed in the periodic table is a weighted average of the individual atomic masses.

4. By what factor does the nucleon number of a nucleus have to increase in order for the nuclear radius to double?

5. The largest stable nucleus has a nucleon number of 209, while the smallest has a nucleon number of 1. If each nucleus is assumed to be a sphere, what is the ratio (largest/smallest) of the surface areas of these spheres?

6. In the nucleus of gold $^{197}_{79}Au$, what is the magnitude of the electrostatic force of repulsion that one proton exerts on another, assuming that the centers of the protons are located at opposite ends of a diameter of the gold nucleus?

*7. Two naturally occurring isotopes of carbon are $^{12}_{6}C$ (atomic mass = 12.000 000 u) and $^{13}_{6}C$ (atomic mass = 13.003 355 u). In one gram of each of these isotopes there are different numbers of atoms. Which contains more atoms, and how many more?

*8. The nucleus of an atom has a volume of 9.4×10^{-43} m^3. This nucleus contains 78 neutrons. Identify the nucleus in the form $^{A}_{Z}X$. Use the periodic table on the inside of the back cover as needed.

*9. One isotope (X) contains an equal number of protons and neutrons. Another isotope (Y) of the same element has twice the number of neutrons as the first isotope does. Determine the ratio r_Y/r_X of the nuclear radii of the isotopes.

**10. Conceptual Example 1 provides some useful background for this problem. (a) Determine an approximate value for the density (in kg/m^3) of the nucleus. (b) If a BB (radius = 2.3 mm) from an air rifle had a density equal to the nuclear density, what mass would the BB have? (c) Assuming the mass of a supertanker is about 1.5×10^8 kg, how many "supertankers" of mass would this hypothetical BB have?

Section 31.3 The Mass Defect of the Nucleus and Nuclear Binding Energy

11. Determine the mass defect of the nucleus for cobalt

$^{59}_{27}$Co, which has an atomic mass of 58.933 198 u. Express your answer in (a) atomic mass units and (b) kilograms.

12. Find the binding energy (in MeV) for aluminum $^{27}_{13}$Al (atomic mass = 26.981 539 u).

13. Find the binding energy (in MeV) for lithium $^{7}_{3}$Li (atomic mass = 7.016 003 u).

14. What is the binding energy (in MeV) for oxygen $^{16}_{8}$O (atomic mass = 15.994 915 u)?

15. For radium $^{226}_{88}$Ra (atomic mass = 226.025 402 u) obtain (a) the mass defect in atomic mass units, (b) the binding energy in MeV, and (c) the binding energy per nucleon.

***16.** (a) Energy is required to separate a nucleus into its constituent nucleons, as Figure 31.3 indicates; this energy is the *total* binding energy of the nucleus. In a similar way one can speak of the energy that binds a single nucleon to the remainder of the nucleus. For example, separating nitrogen $^{14}_{7}$N into nitrogen $^{13}_{7}$N and a neutron takes energy equal to the binding energy of the neutron, as shown below:

$$^{14}_{7}\text{N} + \text{Energy} \rightarrow {}^{13}_{7}\text{N} + {}^{1}_{0}\text{n}$$

Find the energy (in MeV) that binds the neutron to the $^{14}_{7}$N nucleus by considering the mass of $^{13}_{7}$N and the mass of $^{1}_{0}$n, as compared to the mass of $^{14}_{7}$N (see Appendix F for masses). (b) Similarly, one can speak of the energy that binds a single proton to the $^{14}_{7}$N nucleus:

$$^{14}_{7}\text{N} + \text{Energy} \rightarrow {}^{13}_{6}\text{C} + {}^{1}_{1}\text{H}$$

Following the procedure outlined in part (a), determine the energy (in MeV) that binds the proton to the $^{14}_{7}$N nucleus. (c) Which nucleon is more tightly bound, the neutron or the proton?

***17.** Two isotopes of the same element have the same binding energy. One isotope contains two more neutrons than the other. What is the difference between the atomic masses (in atomic mass units) of these isotopes?

Section 31.4 Radioactivity

18. α decay occurs for each of the nuclei given below. Write the decay process for each, including the chemical symbols and values for Z and *A* for the daughter nuclei: (a) $^{228}_{90}$Th and (b) $^{231}_{91}$Pa.

19. Write the α decay process for each of the nuclei given below, including the chemical symbols and values for Z and *A* for the daughter nuclei: (a) $^{235}_{92}$U and (b) $^{239}_{94}$Pu.

20. For the following nuclei, each undergoing β^- decay, write the decay process, identifying each daughter nucleus with its chemical symbol and values for Z and *A*: (a) $^{14}_{6}$C and (b) $^{212}_{82}$Pb.

21. Write the β^- decay process for $^{60}_{27}$Co, including the chemical symbol and values for Z and *A*.

22. Find the energy (in MeV) released when α decay converts radium $^{226}_{88}$Ra (atomic mass = 226.025 40 u) into radon $^{222}_{86}$Rn (atomic mass = 222.017 57 u).

23. Tritium $^{3}_{1}$H (atomic mass = 3.016 050 u) undergoes β^- decay and produces $^{3}_{2}$He (atomic mass = 3.016 030 u). Determine the energy (in MeV) released.

24. What is the wavelength of the 0.510-MeV γ ray that is emitted by radon $^{222}_{86}$Rn?

25. Write the β^+ decay process for each of the following nuclei, being careful to include Z and *A* and the proper chemical symbol for each daughter nucleus: (a) $^{18}_{9}$F and (b) $^{15}_{8}$O.

***26.** Thorium $^{232}_{90}$Th undergoes α decay to produce a daughter nucleus that itself undergoes β^- decay. In the form $^{A}_{Z}$X, identify the nucleus that ultimately results.

***27.** Determine the symbol $^{A}_{Z}$X for the parent nucleus whose α decay produces the same daughter as the β^- decay of thallium $^{208}_{81}$Tl.

***28.** Review Conceptual Example 5 as background for this problem. The α decay of uranium $^{238}_{92}$U produces thorium $^{234}_{90}$Th (atomic mass = 234.0436 u). In Example 4, the energy released in this decay is determined to be 4.3 MeV. Determine how much of this energy is carried away by the recoiling $^{234}_{90}$Th daughter nucleus and how much by the α particle. Assume that the energy of each particle is kinetic energy, and ignore the small amount of energy carried away by the γ ray that is also emitted. In addition, ignore relativistic effects.

****29.** Find the energy (in MeV) released when β^+ decay converts sodium $^{22}_{11}$Na (atomic mass = 21.994 434 u) into neon $^{22}_{10}$Ne (atomic mass = 21.991 383 u). Notice that the atomic mass for $^{22}_{11}$Na includes the mass of 11 electrons, whereas the atomic mass for $^{22}_{10}$Ne includes the mass of only 10 electrons.

****30.** Sodium $^{24}_{11}$Na emits a γ ray that has an energy of 0.423 MeV. Assuming that the $^{24}_{11}$Na nucleus is initially at rest, use the conservation of linear momentum to find the speed with which the nucleus recoils. Ignore relativistic effects.

Section 31.6 Radioactive Decay and Activity

31. In 9.0 days the number of radioactive nuclei decreases to one-eighth the number present initially. What is the half-life (in days) of the material?

32. The $^{3}_{1}$H isotope of hydrogen is called tritium and has a half-life of 12.33 yr. What is its decay constant in units of s^{-1}?

33. The number of radioactive nuclei present at the start of an experiment is 4.60×10^{15}. The number present twenty days later is 8.14×10^{14}. What is the half-life (in days) of the nuclei?

34. Iodine $^{131}_{53}$I is used in diagnostic and therapeutic techniques in the treatment of thyroid disorders. This isotope has a half-life of 8.04 days. What percentage of an initial sample of $^{131}_{53}$I remains after 20.0 days?

35. Strontium $^{90}_{38}Sr$ has a half-life of 28.5 yr. It is chemically similar to calcium, enters the body through the food chain, and collects in the bones. Consequently, $^{90}_{38}Sr$ is a particularly serious health hazard. How long (in years) will it take for 99.9900% of the $^{90}_{38}Sr$ released in a nuclear reactor accident to disappear?

36. To make the dial of a watch glow in the dark, 1.000×10^{-9} kg of radium $^{226}_{88}Ra$ is used. The half-life of this isotope is 1.60×10^3 yr. How many kilograms of radium *disappear* while the watch is in use for fifty years?

37. If the activity of a radioactive substance is initially 398 disintegrations/min and two days later it is 285 disintegrations/min, what is the activity four days later still, or six days after the start? Give your answer in disintegrations/min.

***38.** To see why one curie of activity was chosen to be 3.7×10^{10} Bq, determine the activity (in disintegrations per second) of one gram of radium $^{226}_{88}Ra$ ($T_{1/2} = 1.6 \times 10^3$ yr).

***39.** Two waste products from nuclear reactors are strontium $^{90}_{38}Sr$ and cesium $^{134}_{55}Cs$. The half-life of $^{90}_{38}Sr$ is 28.5 yr, while that of $^{134}_{55}Cs$ is 2.06 yr. If these two species are initially present in a ratio of $Sr/Cs = 7.80 \times 10^{-3}$, what is this ratio fifteen years later?

***40.** A sample of ore containing a radioactive element has an activity of 4.0×10^4 Bq. How many grams of the element are in the sample, assuming the element is (a) radium $^{226}_{88}Ra$ ($T_{1/2} = 1.6 \times 10^3$ yr) and (b) uranium $^{238}_{92}U$ ($T_{1/2} = 4.47 \times 10^9$ yr)?

***41.** Two radioactive nuclei A and B are present in equal numbers to begin with. Three days later, there are three times as many A nuclei as there are B nuclei. The half-life of species B is 1.50 days. Find the half-life of species A.

****42.** Outside the nucleus, the neutron decays into a proton, an electron, and an antineutrino. The half-life for the neutron is 10.4 min. Over what distance will a beam of 5.00-eV neutrons travel before the number of neutrons per unit volume of the beam decreases to 75.0% of its initial value? Ignore relativistic effects.

Section 31.7 Radioactive Dating, Section 31.8 Radioactive Decay Series

43. Bones of the woolly mammoth have been found in North America. The youngest of these bones has a $^{14}_{6}C$ activity per gram of carbon that is about 21% of what was present in the live animal. How long ago (in years) did this animal disappear from North America?

44. The practical limit to ages that can be determined by radiocarbon dating is about 41 000 yr. In a 41 000-yr-old sample, what percentage of the original $^{14}_{6}C$ atoms remains?

45. Review Conceptual Example 11 before starting to solve this problem. The number of unstable nuclei remaining after a time $t = 5.00$ yr is N, and the number present initially is N_0. Find the ratio N/N_0 for (a) $^{14}_{6}C$ (half-life = 5730 yr), (b) $^{15}_{8}O$ (half-life = 122.2 s; use $t = 1.00$ h, since otherwise the answer is out of the range of your calculator), and (c) $^{3}_{1}H$ (half-life = 12.33 yr). Verify that your answers are consistent with the reasoning in Conceptual Example 11.

46. The half-life for the α decay of uranium $^{238}_{92}U$ is 4.47×10^9 yr. Determine the age (in years) of a rock that contains sixty percent of its original $^{238}_{92}U$ atoms.

47. A sample of fossilized bones has a $^{14}_{6}C$ activity of 0.0061 Bq per gram of carbon. (a) Find the age of the sample, assuming that the activity per gram of carbon in a living organism has been constant at a value of 0.23 Bq. (b) Evidence suggests that the value of 0.23 Bq might have been as much as 40% larger. Repeat part (a), taking into account this 40% increase.

***48.** Using the isotope table in Appendix F, construct a plot like that in Figure 31.12, showing the radioactive series that begins with thorium $^{232}_{90}Th$ and ends with lead $^{208}_{82}Pb$. You need not include half-lives.

****49.** When any radioactive dating method is used, experimental error in the measurement of the sample's activity leads to error in the estimated age. In an application of the radiocarbon dating technique to certain fossils, an activity of 0.10 Bq per gram of carbon is measured to within an accuracy of $\pm$ten percent. Find the age of the fossils and the maximum error (in years) in the value obtained. Assume that there is no error in the 5730-year half-life of $^{14}_{6}C$ nor in the value of 0.23 Bq per gram of carbon in a living organism.

ADDITIONAL PROBLEMS

50. The $^{208}_{82}Pb$ isotope of lead has an atomic mass of 207.976 627 u. Obtain the binding energy per nucleon (in MeV).

51. How many protons and neutrons are there in the nucleus of (a) oxygen $^{18}_{8}O$ and (b) tin $^{120}_{50}Sn$?

52. Carbon $^{14}_{6}C$ (atomic mass = 14.003 241 u) is converted into nitrogen $^{14}_{7}N$ (atomic mass = 14.003 074 u) via β^- decay. (a) Write this process in symbolic form, giving Z and A for the parent and daughter nuclei and the β^- particle. (b) Determine the energy (in MeV) released.

53. How many half-lives are required for the number of radioactive nuclei to decrease to one one-millionth of the initial number?

54. In the form $^{A}_{Z}X$, identify the daughter nucleus that results when (a) plutonium $^{242}_{94}Pu$ undergoes α decay, (b) sodium $^{24}_{11}Na$ undergoes β^- decay, and (c) nitrogen $^{13}_{7}N$ undergoes β^+ decay.

55. The shroud of Turin is a religious artifact known since the Middle Ages. In 1988 its age was measured using the radiocarbon dating technique, which revealed that the shroud could not have been made before 1200 AD. Of the $^{14}_{6}C$ nuclei that were present in the living matter from which the shroud was made, what percentage remained in 1988?

***56.** Two isotopes of a certain element have binding energies that differ by 5.03 MeV. The isotope with the larger binding energy contains one more neutron than the other isotope. Find the difference in atomic mass between the two isotopes.

***57.** The photomultiplier tube in a commercial scintillator counter contains 15 of the special electrodes or dynodes. Each dynode produces 3 electrons for every electron that strikes it. One photoelectron strikes the first dynode. What is the maximum number of electrons that strike the 15th dynode?

***58.** A device used in radiation therapy for cancer contains 0.50 g of cobalt $^{60}_{27}Co$ (59.933 819 u). The half-life of $^{60}_{27}Co$ is 5.27 yr. Determine the activity of the radioactive material.

***59.** Plutonium $^{239}_{94}Pu$ (atomic mass = 239.052 16 u) undergoes α decay. Assuming that all the released energy is in the form of kinetic energy of the α particle and ignoring the recoil of the daughter nucleus, find the speed of the α particle. Ignore relativistic effects.

****60.** Both gold $^{198}_{79}Au$ ($T_{1/2} = 2.69$ days) and iodine $^{131}_{53}I$ ($T_{1/2} = 8.04$ days) are used in diagnostic medicine related to the liver. At the time laboratory supplies are monitored, the activity of the gold is observed to be five times greater than the activity of the iodine. How many days later will the two activities be equal?

Detecting Art Forgery

One of the most bizarre episodes of art forgery occurred in 1945, when the painter Hans van Meegeren was captured by the Dutch police. Van Meegeren was accused of selling a Vermeer painting, *The Woman Taken in Adultery,* to Hermann Goering in 1942. The sale was considered to be an act of collaboration with the Germans during World War II, an offense punishable by death. Facing this possibility, van Meegeren asserted that not only was the Vermeer a forgery of his own doing, but also that he had done other "Vermeers."

Chemical analysis of the alleged forgeries was inconclusive, for van Meegeren had been careful to use paints with chemical compositions consistent with the paints used by the artists whose works he was forging. He had even re-used canvases of old paintings. He also used an artificial formaldehyde resin medium instead of oil, so that the paint would harden (or age) like old paints. Furthermore, he had rolled his paintings to create cracks in the now-brittle paint, like those found in old works. Such cracks are called crackle.

So confident was van Meegeren in the quality of his forgeries that he felt the need to produce one "in his own defense," as it were. This he did, creating another Vermeer called *Jesus Among the Doctors.* The work was good enough to get him acquitted of charges of collaboration with the enemy. Nevertheless, he still faced charges of forgery.

After two years of trying to prove his guilt as a forger in one trial (so as to avoid being executed) van Meegeren found himself in 1947 trying to *avoid* being proven guilty as a forger in a second trial! The prosecution used physics to convict him. X-ray images were made of the alleged forgeries and revealed details of the painting below its surface. While van Meegeren had prepared the canvases well, he had made one mistake. He had rolled each canvas to create crackle on its surface. But he had not taken care to assure that the crackle he created coincided with the crackle on the underlying painting that he had so carefully scraped off. The crackle below the surface showed up on the X-ray images. Van Meegeren was convicted of forgery and sentenced to a year in prison.

Hans van Meergeren painting a Vermeer-style canvas at the request of authorities.

So convincing was van Meegeren's work that for twenty years after his death, art scholars debated whether at least one of his alleged works, *Christ and His Disciples at Emmaus,* was actually done by Vermeer as had been believed before 1945. In 1968 a technique was developed to determine the time when lead-based white paints were made, based on the decay of uranium $^{238}_{92}U$ in them. When the paint is made, some of the daughter elements in the radioactive decay series of the uranium are removed. It takes about 270 years for the equilibrium of decay elements to return, so any paint younger than 270 years shows different quantities of the different daughters. The painting in question was analyzed and shown to be made from paint less than a hundred years old. The painting could not have been done by Vermeer, who painted in the seventeenth century.

For Further Reading See: Fleming, S., *Authenticity in Art: The Scientific Detection of Forgery.* New York: Crane, Russak, 1975.

IONIZING RADIATION, NUCLEAR ENERGY, AND ELEMENTARY PARTICLES

CHAPTER 32

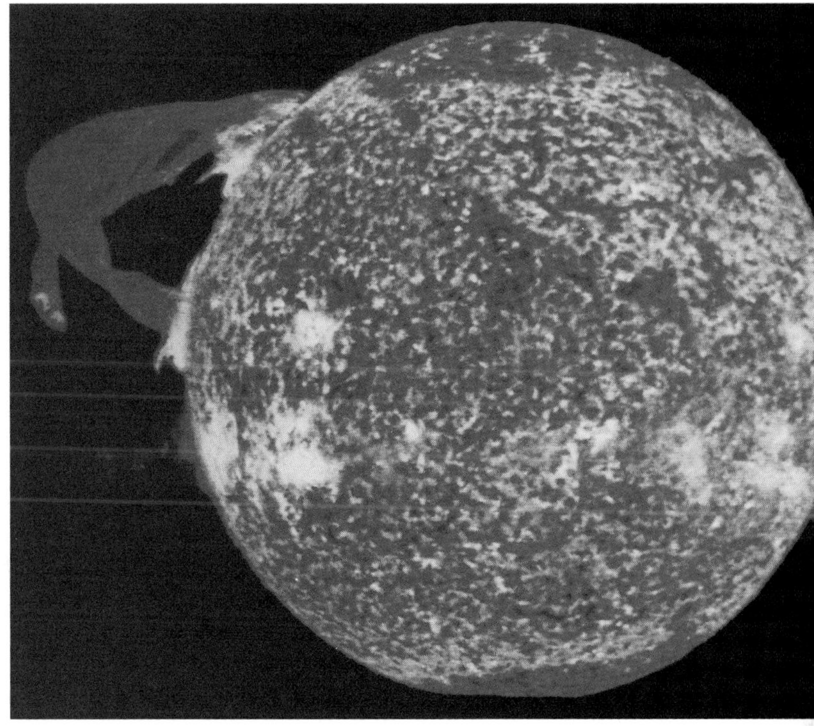

Ionizing radiation originates from many places. It comes from sources beyond the earth, such as the sun, particularly during solar flares like the one in the photograph at the right. There are natural earth-bound sources too, one example being radon. And man-made sources contribute significantly, as they do in diagnostic and therapeutic medical applications. While these applications are beneficial, ionizing radiation always has the potential for harm. Therefore, we will examine the ways in which ionizing radiation is measured. Then, starting with Section 32.2, we will discuss two ways in which energy is obtained from nuclear reactions. Fission is one of these, and in this method, a heavy nucleus is split into smaller fragments, with the release of a relatively large amount of energy. Controlled fission is the means by which nuclear reactors ultimately generate electrical energy. The other energy-releasing process, fusion, occurs when lighter nuclei combine to form a heavier nucleus. Stars such as the sun produce energy by this method. Finally, we will turn to the elementary particles that have been found within the nucleus. We will discuss the current theory of particle physics that states that protons and neutrons, as well as other particles, are made of smaller, indivisible particles called quarks.

32.1 BIOLOGICAL EFFECTS OF IONIZING RADIATION

IONIZING RADIATION

Ionizing radiation consists of photons and/or moving particles that have sufficient energy to knock an electron out of an atom or molecule, thus forming an ion. The photons usually lie in the ultraviolet, X-ray, or γ-ray regions of the electromagnetic spectrum, while the moving particles can be the α and β particles emitted during radioactive decay. An energy of roughly 1 to 35 eV is needed to ionize an atom or molecule, and the particles and γ rays emitted during nuclear disintegration often have energies of several million eV. Therefore, a single α particle, β particle, or γ ray can ionize thousands of molecules.

Nuclear radiation is potentially harmful to humans, because the ionization it produces can significantly alter the structure of molecules within a living cell. The alterations cause the cell to malfunction and, if severe enough, can lead to the death of the cell and even the organism itself. Despite the potential hazards, ionizing radiation is used in medicine for diagnostic and therapeutic purposes, such as locating bone fractures and treating cancer. The hazards can be minimized only if the fundamentals of radiation exposure, including dose units and the biological effects of radiation, are understood.

Exposure is a measure of the ionization produced in air by X-rays or γ rays, and it is defined in the following manner. A beam of X-rays or γ rays is sent through a mass m of dry air at standard temperature and pressure (STP: 0 °C, 1 atm pressure). In passing through the air, the beam produces positive ions whose total charge is q. Exposure is defined as the total charge per unit mass of air: exposure $= q/m$. The SI unit for exposure is coulombs per kilogram (C/kg). However, the first radiation unit to be defined was the *roentgen* (R), and it is still used today. With q expressed in coulombs (C) and m in kilograms (kg), the exposure in roentgens is given by

$$\text{Exposure (in roentgens)} = \left(\frac{1}{2.58 \times 10^{-4}}\right)\frac{q}{m} \tag{32.1}$$

Thus, when X-rays or γ rays produce an exposure of one roentgen, $q = 2.58 \times 10^{-4}$ C of positive charge are produced in $m = 1$ kg of dry air:

$$1\ \text{R} = 2.58 \times 10^{-4}\ \text{C/kg} \qquad \text{(dry air, at STP)}$$

Since the concept of exposure is defined in terms of the ionizing abilities of X-rays and γ rays in air, it does not specify the effect of radiation on living tissue. For biological purposes, the ***absorbed dose*** is a more suitable quantity, because it is the energy absorbed from the radiation per unit mass of absorbing material:

$$\text{Absorbed dose} = \frac{\text{Energy absorbed}}{\text{Mass of absorbing material}} \tag{32.2}$$

The SI unit of absorbed dose is the *gray* (Gy), which is a unit of energy divided by a unit of mass: 1 Gy = 1 J/kg. Equation 32.2 is applicable to all types of radiation and absorbing media. Another unit is often used for absorbed dose, namely, the *rad* (rd). The word rad is an acronym for radiation absorbed dose. The rad and the gray are related by

$$1\ \text{rad} = 0.01\ \text{gray}$$

The amount of biological damage produced by ionizing radiation is different for different kinds of radiation. For instance, a 1-rad dose of neutrons is far more likely to produce eye cataracts than a 1-rad dose of X-rays. To compare the damage caused by different types of radiation, the ***relative biological effectiveness*** (RBE) is used.* The relative biological effectiveness of a particular type of radiation compares the dose of that radiation needed to produce a certain biological effect to the dose of 200-keV X-rays needed to produce the same biological effect:

$$\text{Relative biological effectiveness (RBE)} = \frac{\text{The dose of 200-keV X-rays that produces a certain biological effect}}{\text{The dose of radiation that produces the same biological effect}} \quad (32.3)$$

The RBE depends on the nature of the ionizing radiation and its energy, as well as the type of tissue being irradiated. Table 32.1 lists some typical RBE values for different kinds of radiation, assuming that an "average" biological tissue is being irradiated. The values of RBE = 1 indicate that γ rays and β^- particles produce the same biological damage as do 200-keV X-rays. The larger RBE values indicate that protons, α particles, and fast neutrons cause substantially more damage. The RBE is often used in conjunction with the absorbed dose to reflect the damage-producing character of the radiation. The product of the absorbed dose in rads (not in grays) and the RBE is the ***biologically equivalent dose:***

$$\text{Biologically equivalent dose (in rem)} = \text{Absorbed dose (in rad)} \times \text{RBE} \quad (32.4)$$

The unit for the biologically equivalent dose is the *rem,* short for **r**oentgen **e**quivalent, **m**an. Example 1 illustrates the use of the biologically equivalent dose.

Table 32.1 Relative Biological Effectiveness (RBE) for Various Types of Radiation

Type of Radiation	RBE
200-keV X-rays	1
γ rays	1
β^- particles (electrons)	1
Protons	10
α particles	10–20
Neutrons	
Slow	2
Fast	10

Example 1 Comparing Absorbed Doses of γ Rays and Neutrons

A biological tissue is irradiated with γ rays that have an RBE of 0.70. The absorbed dose of γ rays is 850 rd. The tissue is then exposed to neutrons whose RBE is 3.5. The biologically equivalent dose of the neutrons is the same as that of the γ rays. What is the absorbed dose of neutrons?

REASONING The biologically equivalent doses of the neutrons and the γ rays are the same. Therefore, the tissue damage produced in each case is the same. However, the RBE of the neutrons is larger than the RBE of the γ rays by a factor of $3.5/0.70 = 5.0$. Consequently, we will find that the absorbed dose of the neutrons is only one-fifth as great as that of the γ rays.

SOLUTION The biologically equivalent dose of the γ rays is the product of the absorbed dose (in rads) and the RBE:

$$\text{Biologically equivalent dose of } \gamma \text{ rays} = (850 \text{ rad})(0.70) = 6.0 \times 10^2 \text{ rem} \quad (32.4)$$

For the neutrons (RBE = 3.5), the biologically equivalent dose is the same. Therefore, 6.0×10^2 rem = (Absorbed dose of neutrons)(3.5) and

$$\text{Absorbed dose of neutrons} = \frac{6.0 \times 10^2 \text{ rem}}{3.5} = \boxed{170 \text{ rd}}$$

* The RBE is sometimes called the *quality factor* (QF).

THE EFFECTS OF IONIZING RADIATION ON HUMANS

Everyone is continually exposed to background radiation from natural sources, such as cosmic rays (high-energy particles that come from outside the solar system), radioactive materials in the environment, radioactive nuclei—primarily carbon $^{14}_{6}C$ and potassium $^{40}_{19}K$—within our own bodies, and radon. Table 32.2 lists the average biologically equivalent doses received from these sources by a person in the United States. According to this table, radon is a major contributor to the natural background radiation. Radon is an odorless radioactive gas and poses a health hazard because, when inhaled, it can damage the lungs and cause cancer. Radon is found in soil and rocks and enters houses through cracks and crevices in the foundation. The amount of radon in the soil varies greatly throughout the country, with some localities having significant amounts and others having virtually none. Accordingly, the dose that any individual receives can vary widely from the average value of 200 mrem/yr given in Table 32.2 (1 mrem $= 10^{-3}$ rem). In many houses, the entry of radon can be reduced significantly by sealing the foundation against entry of the gas and providing good ventilation so it does not accumulate.

Table 32.2 Average Biologically Equivalent Doses of Radiation Received by a U. S. Resident[a]

Source of Radiation	Biologically Equivalent Dose (mrem/yr)[b]
Natural background radiation	
Cosmic rays	28
Radioactive earth and air	28
Internal radioactive nuclei	39
Inhaled radon	≈200
Man-made radiation	
Consumer products	10
Medical/dental diagnostics	39
Nuclear medicine	14
Rounded total:	360

[a] National Council on Radiation Protection and Measurement, Report No. 93, "Ionizing Radiation Exposure of the Population of the United States," 1987.
[b] 1 mrem $= 10^{-3}$ rem.

To the natural background of radiation, a significant amount of man-made radiation has been added, mostly from medical/dental diagnostic X-rays. Table 32.2 indicates an average total dose of 360 mrem/yr from all sources.

The effects of radiation on humans can be grouped into two categories, according to the time span between initial exposure and the appearance of physiological effects: (1) short-term or acute effects that appear within a matter of minutes, days, or weeks, and (2) long-term or latent effects that appear years, decades, or even generations later.

Radiation sickness is the general term applied to the acute effects of radiation. Depending on the severity of the dose, a person with radiation sickness can exhibit nausea, vomiting, fever, diarrhea, and loss of hair. Ultimately, death can occur. The severity of radiation sickness is related to the dose received, and in the following discussion the biologically equivalent doses quoted are whole-body, single doses. A dose less than 50 rem causes no short-term, ill effects. A dose between 50 and 300 rem brings on radiation sickness, the severity increasing with increasing dosage. A whole-body dose in the range of 400–500 rem is classified as an LD_{50} dose, meaning that it is a lethal dose (LD) for about 50% of the people so exposed; death occurs within a few months. Whole-body doses greater than 600 rem result in death for almost all individuals.

Long-term or latent effects of radiation may appear as a result of high-level, brief exposure or low-level exposure over a long period of time. Some long-term effects are hair loss, eye cataracts, and various kinds of cancer. In addition, genetic defects caused by mutated genes may be passed on from one generation to the next.

Because of the hazards of radiation, the federal government has established dose limits. The permissible dose for an individual is defined as the dose, accumulated over a long period of time or resulting from a single exposure, that carries negligible probability of a severe health hazard. Current federal standards (1991) state that an individual in the general population should not receive more than 500 mrem of man-made radiation each year, *exclusive* of medical sources. A person exposed to radiation in the workplace (e.g., a radiation therapist) should not receive more than 5 rem per year from work-related sources.

32.2 INDUCED NUCLEAR REACTIONS

Section 31.4 discusses how a radioactive parent nucleus disintegrates spontaneously into a daughter nucleus. It is also possible to bring about or "induce" the disintegration of a stable nucleus by striking it with another nucleus, an atomic or subatomic particle, or a γ-ray photon. A ***nuclear reaction*** is said to occur whenever the incident nucleus, particle, or photon causes a change to occur in a target nucleus.

In 1919 Ernest Rutherford observed that when an α particle strikes a nitrogen nucleus, an oxygen nucleus and a proton are produced. This nuclear reaction is written as

$$\underbrace{{}^{4}_{2}\text{He}}_{\substack{\text{Incident}\\ \alpha\text{ particle}}} + \underbrace{{}^{14}_{7}\text{N}}_{\substack{\text{Nitrogen}\\ \text{(target)}}} \longrightarrow \underbrace{{}^{17}_{8}\text{O}}_{\text{Oxygen}} + \underbrace{{}^{1}_{1}\text{H}}_{\text{Proton}}$$

Because the incident α particle induces the transmutation of nitrogen into oxygen, this reaction is an example of an ***induced nuclear transmutation.***

Nuclear reactions are often written in a shorthand form. For example, the reaction above is designated by ${}^{14}_{7}\text{N}\,(\alpha, p)\,{}^{17}_{8}\text{O}$. The first and last symbols represent the initial and final nuclei, respectively. The symbols inside the parentheses denote the incident α particle (on the left) and the small emitted particle or proton (on the right). Some other induced nuclear transmutations are listed below, together with the corresponding shorthand notations:

Nuclear Reaction	Notation
${}^{1}_{0}n + {}^{10}_{5}B \rightarrow {}^{7}_{3}Li + {}^{4}_{2}He$	${}^{10}_{5}B\ (n, \alpha)\ {}^{7}_{3}Li$
$\gamma + {}^{25}_{12}Mg \rightarrow {}^{24}_{11}Na + {}^{1}_{1}H$	${}^{25}_{12}Mg\ (\gamma, p)\ {}^{24}_{11}Na$
${}^{1}_{1}H + {}^{13}_{6}C \rightarrow {}^{14}_{7}N + \gamma$	${}^{13}_{6}C\ (p, \gamma)\ {}^{14}_{7}N$

In any nuclear reaction, both the total electric charge of the nucleons and the total number of nucleons are conserved during the process, as discussed in Section 31.4. The fact that these quantities are conserved makes it possible to identify the nucleus produced in a nuclear reaction, as the next example illustrates.

Example 2 An Induced Nuclear Transmutation

An α particle strikes an aluminum ${}^{27}_{13}Al$ nucleus, and a nucleus ${}^{A}_{Z}X$ and a neutron are produced:

$$ {}^{4}_{2}He + {}^{27}_{13}Al \rightarrow {}^{A}_{Z}X + {}^{1}_{0}n $$

Identify the nucleus produced, including its atomic number Z (the number of protons) and its atomic mass number A (the number of nucleons).

REASONING The total electric charge of the nucleons is conserved, so that we can set the total charge before the reaction equal to the total charge after the reaction. Thus, the total number of protons is the same before the reaction as it is afterwards. The total number of nucleons is also conserved, so that we can set the total number before the reaction equal to the total number after the reaction. These two conserved quantities will allow us to identify the nucleus produced.

SOLUTION The conservation of total electric charge and total number of nucleons leads to the equations listed below:

Conserved Quantity	Before Reaction		After Reaction
Total electric charge (number of protons)	$2 + 13$	$=$	$Z + 0$
Total number of nucleons	$4 + 27$	$=$	$A + 1$

Solving these equations for Z and A gives $Z = 15$ and $A = 30$. Since $Z = 15$ identifies the element as phosphorus, the nucleus produced is $\boxed{{}^{30}_{15}P}$.

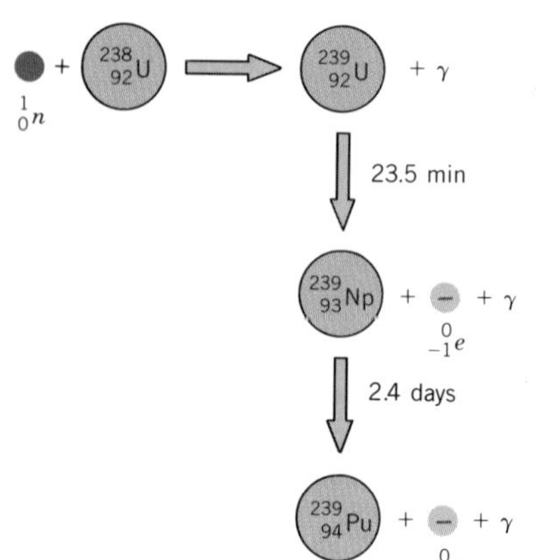

Figure 32.1 An induced nuclear reaction in which ${}^{238}_{92}U$ is transmuted into the transuranium element plutonium ${}^{239}_{94}Pu$.

Induced nuclear transmutations can be used to produce isotopes that are not found naturally. In 1934, Enrico Fermi suggested a method for producing elements with a higher atomic number than uranium ($Z = 92$). These elements—neptunium ($Z = 93$), plutonium ($Z = 94$), americium ($Z = 95$), and so on—are known as *transuranium elements.* None of the transuranium elements occurs naturally. They are created in a nuclear reaction between a suitably chosen lighter element and a small incident particle, usually a neutron or an α particle. For example, Figure 32.1 shows a reaction that produces plutonium from uranium. A neutron is captured by a uranium $^{238}_{92}U$ nucleus, producing $^{239}_{92}U$ and a γ ray. The $^{239}_{92}U$ nucleus is radioactive and decays with a half-life of 23.5 min into neptunium $^{239}_{93}Np$. Neptunium is also radioactive and disintegrates with a half-life of 2.4 days into plutonium $^{239}_{94}Pu$. Plutonium is the final product and has a half-life of 24 100 yr.

The neutrons that participate in nuclear reactions can have kinetic energies that cover a wide range. In particular, those that have a kinetic energy of about 0.04 eV or less are called ***thermal neutrons.*** The name derives from the fact that such a relatively small kinetic energy is comparable to the average translational kinetic energy of a molecule at room temperature. Thermal neutrons are used in one type of bomb detection system that can expose hidden explosives. As Figure 32.2 illustrates, the system bathes luggage suspected of containing a bomb in low doses of thermal neutrons. Some of the neutrons are captured by the nuclei of the luggage and its contents, including explosives. These nuclei subsequently emit γ rays, the energies of which are unique to the nuclei that emit them. By analyzing the γ rays, it is possible to determine the chemical natures of the materials in the luggage. In particular, the system looks for certain nitrogen compounds that signal the presence of a bomb.

THE PHYSICS OF . . .

a bomb detection system.

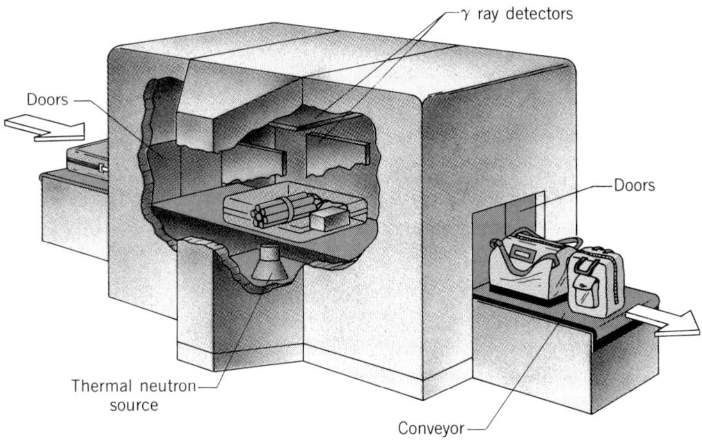

Figure 32.2 A bomb detection system that uses thermal neutrons.

32.3 NUCLEAR FISSION

THE FISSION PROCESS

In 1939 four German scientists, Otto Hahn, Lise Meitner, Fritz Strassmann, and Otto Frisch, made an important discovery that ushered in the atomic age. They found that a uranium nucleus, after absorbing a neutron, splits into two fragments, each with a smaller mass than the original nucleus. The splitting of a massive nucleus into two less-massive fragments is known as ***nuclear fission.***

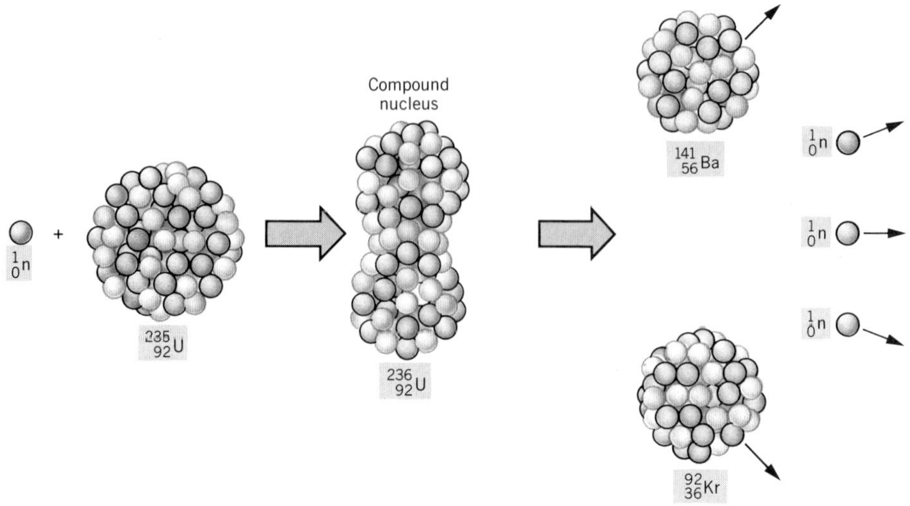

Figure 32.3 A slowly moving neutron causes the uranium nucleus $^{235}_{92}\text{U}$ to fission into barium $^{141}_{56}\text{Ba}$, krypton $^{92}_{36}\text{Kr}$, and three neutrons.

Figure 32.3 shows a fission reaction in which a uranium $^{235}_{92}\text{U}$ nucleus is split into barium $^{141}_{56}\text{Ba}$ and krypton $^{92}_{36}\text{Kr}$ nuclei. The reaction begins when $^{235}_{92}\text{U}$ absorbs a slowly moving neutron, creating a "compound nucleus," $^{236}_{92}\text{U}$. The compound nucleus disintegrates quickly into $^{141}_{56}\text{Ba}$, $^{92}_{36}\text{Kr}$, and three neutrons according to the following reaction:

$$^{1}_{0}\text{n} + {}^{235}_{92}\text{U} \longrightarrow \underbrace{^{236}_{92}\text{U}}_{\substack{\text{Compound}\\ \text{nucleus}\\ \text{(unstable)}}} \longrightarrow \underbrace{^{141}_{56}\text{Ba}}_{\text{Barium}} + \underbrace{^{92}_{36}\text{Kr}}_{\text{Krypton}} + \underbrace{3\,^{1}_{0}\text{n}}_{\text{3 neutrons}}$$

This reaction is only one of the many possible reactions that can occur when uranium fissions. For example, another reaction is

$$^{1}_{0}\text{n} + {}^{235}_{92}\text{U} \longrightarrow \underbrace{^{236}_{92}\text{U}}_{\substack{\text{Compound}\\ \text{nucleus}\\ \text{(unstable)}}} \longrightarrow \underbrace{^{140}_{54}\text{Xe}}_{\text{Xenon}} + \underbrace{^{94}_{38}\text{Sr}}_{\text{Strontium}} + \underbrace{2\,^{1}_{0}\text{n}}_{\text{2 neutrons}}$$

Some reactions produce as many as 5 neutrons; however, the average number produced per fission is 2.5.

When a neutron collides with and is absorbed by a uranium nucleus, the uranium nucleus begins to vibrate and becomes distorted. The vibration continues until the distortion becomes so severe that the attractive strong nuclear force can no longer balance the electrostatic repulsion between the nuclear protons. At this point, the nucleus bursts apart into fragments, which carry off energy, primarily in the form of kinetic energy. The energy carried off by the fragments is enormous and was stored in the original nucleus primarily in the form of electric potential energy. An average of roughly 200 MeV of energy is released per fission. This energy is approximately 10^8 times greater than the energy released per molecule in an ordinary chemical reaction, such as the combustion of gasoline or coal. Example 3 demonstrates how to estimate the energy released during the fission of a nucleus.

Example 3 The Energy Released During Nuclear Fission

Estimate the amount of energy released when a massive nucleus ($A = 240$) fissions.

REASONING Figure 31.5 shows that the binding energy of a nucleus with $A = 240$ is about 7.6 MeV per nucleon. We assume that this nucleus fissions into two fragments, each with $A \approx 120$. According to Figure 31.5, the binding energy of the fragments increases to about 8.5 MeV per nucleon. Consequently, when a massive nucleus fissions, there is a release of about 8.5 MeV − 7.6 MeV = 0.9 MeV of energy per nucleon.

SOLUTION Since there are 240 nucleons involved in the fission process, the total energy released per fission is approximately (0.9 MeV/nucleon)(240 nucleons) ≈ $\boxed{200 \text{ MeV}}$.

Virtually all naturally occurring uranium is composed of two isotopes. These isotopes and their natural abundances are ${}^{238}_{92}U$ (99.275%) and ${}^{235}_{92}U$ (0.720%). Although ${}^{238}_{92}U$ is by far the most abundant isotope, the probability that it will capture a neutron and fission is very small. For this reason, ${}^{238}_{92}U$ is not the isotope of choice for generating nuclear energy. In contrast, the isotope ${}^{235}_{92}U$ readily captures a neutron and fissions, *provided the neutron is a thermal neutron* (kinetic energy ≈ 0.04 eV or less). The probability of a thermal neutron causing ${}^{235}_{92}U$ to fission is about five hundred times greater than a neutron whose energy is relatively high, say 1 MeV. Thermal neutrons can also be used to fission other nuclei, such as plutonium ${}^{239}_{94}Pu$. Conceptual Example 4 deals with one of the reasons why thermal neutrons are useful for inducing nuclear fission.

Conceptual Example 4 Thermal Neutrons Versus Thermal Protons or Alpha Particles

Why is it possible for a thermal neutron (i.e., one with a relatively small kinetic energy) to penetrate a nucleus, whereas a proton or an α particle would need a much larger amount of energy to penetrate the same nucleus?

REASONING To penetrate a nucleus, a particle such as a neutron, a proton, or an α particle must have enough kinetic energy to do the work of overcoming any repulsive force that it encounters along the way. A repulsive force can arise, because protons in the nucleus are electrically charged. A neutron is electrically neutral, however, so it encounters no electrostatic force of repulsion from the nuclear protons as it approaches. Hence, a neutron needs relatively little energy to reach the nucleus. In contrast, a proton and an α particle both carry a positive charge. In approaching a target nucleus, either one encounters an electrostatic force of repulsion from the nuclear protons. It is true that they also experience the attractive strong nuclear force from both the nuclear protons and neutrons. But the strong nuclear force has an extremely short range of action and, therefore, comes into play only when an impinging particle reaches the nucleus. In comparison, the electrostatic force of repulsion has a long range of action and is encountered by an incoming charged particle throughout the entire journey to the target nucleus. Consequently, an impinging proton or α particle needs a relatively large amount of kinetic energy to do the work of overcoming the repulsive electrostatic force.

***Related Homework Material:** Problem 42*

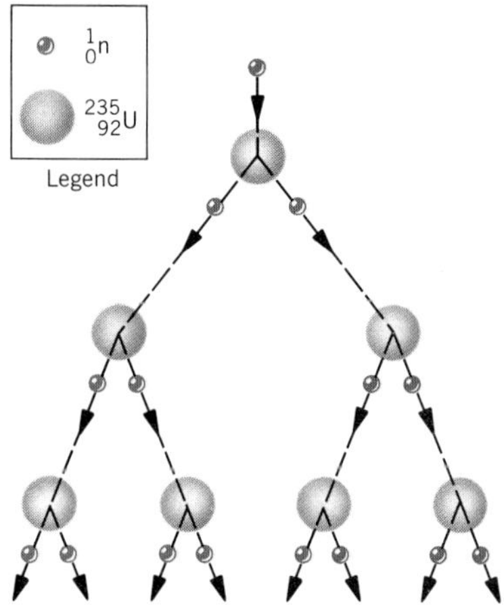

Figure 32.4 A chain reaction. For clarity, it is assumed that each fission generates two neutrons (2.5 neutrons are actually liberated on the average). The fission fragments are not shown.

CHAIN REACTION

The fact that the uranium fission reaction releases 2.5 neutrons, on the average, makes it possible for a self-sustaining series of fissions to occur. As Figure 32.4 illustrates, each neutron released can initiate another fission process, resulting in the emission of still more neutrons, followed by more fissions, and so on. A ***chain reaction*** is a series of nuclear fissions whereby some of the neutrons produced by each fission cause additional fissions. During an uncontrolled chain reaction, it would not be unusual for the number of fissions to increase a thousandfold within a few millionths of a second. With an average energy of about 200 MeV being released per fission, an uncontrolled chain reaction can generate an incredible amount of energy in a very short time, as happens in an atomic bomb (which is actually a *nuclear* bomb).

By limiting the number of neutrons in the environment of the fissile nuclei, it is possible to establish a condition whereby each fission event contributes, on average, only *one neutron* that fissions another nucleus (see Figure 32.5). In this manner, the chain reaction and the rate of energy production are *controlled.* The controlled fission chain reaction is the principle behind the nuclear reactors used in the commercial generation of electric power.

32.4 NUCLEAR REACTORS

BASIC COMPONENTS

THE PHYSICS OF . . .

nuclear reactors.

A nuclear reactor is a type of furnace in which energy is generated by a controlled fission chain reaction. The first nuclear reactor was built by Enrico Fermi in 1942, on the floor of a squash court under the west stands of Stagg Field at the University of Chicago. Today, there are many kinds and sizes of reactors, but they all have three basic components: fuel elements, a neutron moderator, and control rods. Figure 32.6 illustrates these components.

The ***fuel elements*** contain the fissile fuel and, for example, may be in the shape of thin rods about 1 cm in diameter. In a large power reactor there may be thousands of fuel elements placed close together, the entire region of fuel elements being known as the ***reactor core.***

Uranium $^{235}_{92}U$ is a common reactor fuel. Since the natural abundance of this isotope is only about 0.7%, there are special uranium-enrichment plants to increase the percentage. Most commercial reactors use uranium in which the amount of $^{235}_{92}U$ has been enriched to about 3%.

While neutrons with energies of about 0.04 eV (or less) readily fission $^{235}_{92}U$, the neutrons released during the fission process have significantly greater energies of several MeV or so. Consequently, a nuclear reactor must contain some type of material that will decrease or moderate the speed of such energetic neutrons so they can readily fission additional $^{235}_{92}U$ nuclei. The material that slows down the neutrons is called a ***moderator.*** One commonly used moderator is water. When an energetic neutron leaves a fuel element, the neutron enters the surrounding water and collides with water molecules. With each collision, the neutron loses an appreciable fraction of its energy and slows down. Once slowed down to thermal energy by the moderator, a process that takes less than 10^{-3} s, the neutron is capable of initiating a fission event upon reentering a fuel element.

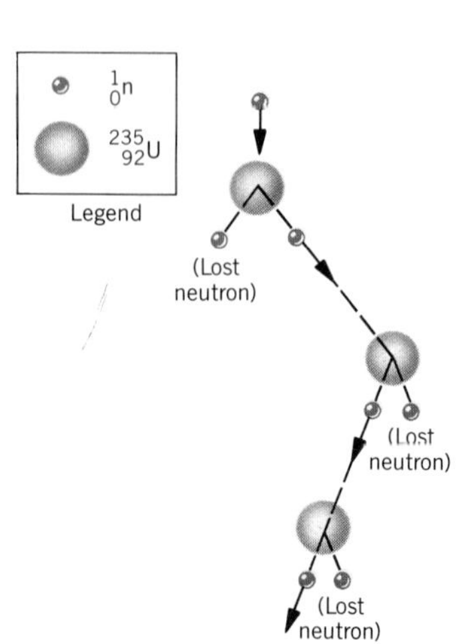

Figure 32.5 In a controlled chain reaction, only one neutron, on average, from each fission event causes another nucleus to fission. As a result, energy is released at a steady or controlled rate.

If the output power from a reactor is to remain constant, only one neutron from each fission event must trigger a new fission, as Figure 32.5 suggests. When each

The decrease in mass, or the mass defect, is $\Delta m = (5.030\ \text{u} - 5.012\ \text{u}) = 0.018\ \text{u}$. Since 1 u is equivalent to 931.5 MeV, the energy released is $\boxed{17\ \text{MeV}}$.

There are five nucleons that participate in the fusion, so the energy released per nucleon is about 3.4 MeV. This energy per nucleon is greater than that released in a fission process (≈ 0.9 MeV per nucleon). Thus, for a given mass of fuel, a fusion reaction yields more energy than a fission reaction.

Because fusion reactions release substantial amounts of energy, there is considerable interest in fusion reactors, although to date no commercial units have been constructed. The difficulties in building a fusion reactor arise mainly because the two low-mass nuclei must be brought sufficiently near each other so that the short-range strong nuclear force can pull them together, leading to fusion. But each nucleus has a positive charge and repels the other electrically. For the nuclei to get sufficiently close in the presence of the repulsive electric force, they must have large kinetic energies to start with. For example, to start a deuterium–deuterium fusion reaction, it has been estimated that each nucleus needs an initial kinetic energy of about 0.25 MeV.

In Chapter 14, we saw that the average translational kinetic energy $\overline{\text{KE}}$ of an atom in an ideal gas is directly proportional to the Kelvin temperature T of the gas according to $\overline{\text{KE}} = \frac{3}{2}kT$ (Equation 14.6), where $k = 1.38 \times 10^{-23}$ J/K is the Boltzmann constant. The kinetic energy of 0.25 MeV (4.0×10^{-14} J) needed to start a fusion reaction corresponds to a gas temperature of

$$T = \frac{2(\overline{\text{KE}})}{3k} = \frac{2(4.0 \times 10^{-14}\ \text{J})}{3(1.38 \times 10^{-23}\ \text{J/K})} = 2 \times 10^{9}\ \text{K}$$

This is two billion kelvins! Actually, the temperature of the deuterium gas need not be quite this high, because this temperature corresponds to the average kinetic energy of the nuclei. Some nuclei within the gas have energies substantially greater than the average energy, and these higher-energy nuclei can fuse together to produce a net outflow of energy. Typically, the temperature needed to start a deuterium–deuterium fusion reaction is about 4×10^{8} K. Nevertheless, it is no trivial task to create such a temperature in the laboratory. Reactions that require such extremely high temperatures are called ***thermonuclear reactions.*** The most important thermonuclear reactions occur in stars, such as our own sun. The energy radiated by the sun comes from such reactions deep within its core, where the temperature is high enough to initiate the fusion process. One group of reactions thought to occur in the sun is the *proton–proton* cycle. This cycle is a series of reactions whereby six protons form a helium nucleus, two positrons, two γ rays, two protons, and two neutrinos. The energy released by the proton–proton cycle is about 25 MeV.

Man-made fusion reactions have been carried out in a fusion-type nuclear bomb—commonly called a hydrogen bomb. In a hydrogen bomb, the fusion reaction is ignited by a fission bomb using uranium or plutonium. The temperature produced by the fission bomb is sufficiently high to initiate a thermonuclear reaction where, for example, hydrogen isotopes are fused into helium, releasing even more energy.

For fusion to be useful as a commercial energy source, the energy must be released in a steady, controlled manner—unlike the uncontrolled energy released by a hydrogen bomb. To date, scientists have not succeeded in constructing a fusion device that produces more energy on a continual basis than is expended in operating the device. A fusion device uses a high temperature to start a reaction, and under such a condition, all the atoms are completely ionized to form a *plasma* (a gas composed of charged particles, like $^2_1H^+$ and e^-). The problem is to confine the hot plasma for a long enough time so that collisions among the ions can lead to fusion.

THE PHYSICS OF . . .

nuclear fusion using magnetic confinement.

One ingenious method of confining the plasma, called *magnetic confinement,* uses a magnetic field. Charges moving in a magnetic field are subject to magnetic forces, and it is hoped that the plasma can be confined to a region of space by these forces. The problem of magnetic confinement is a difficult one, although steady progress is being made. Figure 32.9 shows the Tokamak Fusion Test Reactor, which uses magnetic confinement and has succeeded in producing a one-second burst of energy at a power level of more than 5 megawatts.

THE PHYSICS OF . . .

nuclear fusion using inertial confinement.

Another type of confinement scheme, known as *inertial confinement,* is also being developed. Tiny, solid pellets of fuel are dropped into a container. As each pellet reaches the center of the container, a number of high-intensity lasers or electron beams strike the pellet simultaneously. The heating causes the exterior of the pellet to vaporize almost instantaneously. However, the inertia of the vaporized atoms keeps them from expanding outward as fast as the vapor is being formed. As a result, high pressures, high densities, and high temperatures are achieved at the center of the pellet, thus causing fusion. Figure 32.10 shows the inertial confinement facility at the Lawrence Livermore National Laboratory.

When compared to fission, fusion has some attractive features as an energy source. As we have seen in Example 5, fusion yields more energy than fission, for a given mass of fuel. Moreover, one type of fuel, ^2_1H (deuterium), is found in the waters of the oceans and is plentiful, cheap, and easy to separate from the common ^1_1H isotope of hydrogen. Fissile materials like naturally occurring uranium $^{235}_{92}U$ are much less available, and supplies could be depleted within a century or two. However, the commercial use of fusion to provide cheap energy remains in the future.

Figure 32.9 The Tokamak Fusion Test Reactor at Princeton University uses the method of magnetic confinement to contain the hot plasma during nuclear fusion.

32.6 ELEMENTARY PARTICLES

SETTING THE STAGE

By 1932 the electron, the proton, and the neutron had been discovered and were thought to be nature's three ***elementary particles,*** in the sense that they were the

Figure 32.10 The NOVA Laser Facility at Lawrence Livermore National Laboratory at Lawrence, California. Short pulses of laser light travel simultaneously through each of the cylindrical blue tubes. Inside the spherical chamber, the pulses strike a small pellet containing deuterium and tritium. The laser pulses heat the pellet to an extremely high temperature, initiating fusion.

basic building blocks from which all matter is constructed. Experimental evidence obtained since then, however, shows that several hundred additional particles exist, and scientists no longer believe that the proton and the neutron are elementary particles.

Most of these new particles have masses greater than the electron's mass, and many are more massive than protons or neutrons. Virtually all the new particles are unstable and decay with times between about 10^{-6} and 10^{-23} s.

Often, new particles are produced by accelerating protons or electrons to high energies and letting them collide with a target nucleus. For example, Figure 32.11 shows a collision between an energetic proton and a stationary proton. If the incoming proton has sufficient energy, the collision produces an entirely new particle, the *neutral pion* (π^0). The π^0 particle lives for only about 0.8×10^{-16} s before it decays into two γ-ray photons. Since the pion did not exist before the collision, the pion was created from part of the incident proton's energy. Because a new particle such as the neutral pion is often created from energy, it is customary to report the mass of the particle in terms of its equivalent *rest energy* (see Equation 28.5). Often energy units of MeV are used. For instance, detailed analyses of experiments reveal that the mass of the π^0 particle is equivalent to a rest energy of 135.0 MeV. For comparison, the more massive proton has a rest energy of 938.3 MeV. Analyses of experiments also provide the electric charge and other properties of particles created in high-energy collisions.

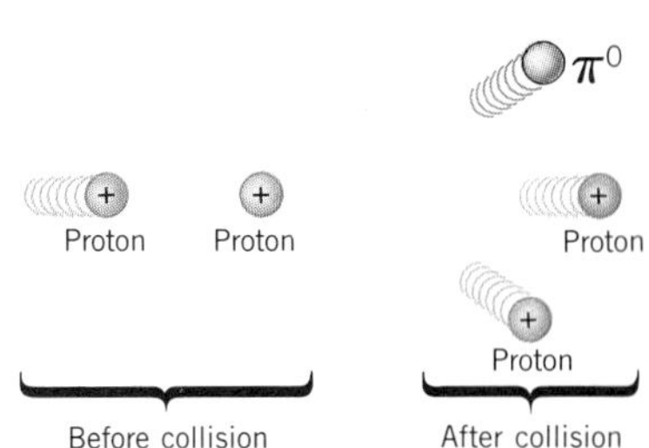

Figure 32.11 When an energetic proton collides with a stationary proton, a neutral pion (π^0) is produced. Part of the energy of the incident proton goes into creating the pion.

In the limited space available here, it is not possible to describe all the new particles that have been found. However, we will highlight some of the more significant discoveries.

NEUTRINOS

In 1930 Wolfgang Pauli suggested that a particle called the ***neutrino*** (now known as the electron neutrino) should accompany the β decay of a radioactive nucleus. As Section 31.5 discusses, the neutrino has no electric charge, has a very small (possibly zero) mass, and travels near or at the speed of light. Neutrinos were finally discovered in 1956. Today, neutrinos are created in abundance in nuclear reactors and particle accelerators and are thought to be plentiful in the universe.

POSITRONS AND ANTIPARTICLES

The year 1932 saw the discovery of the *positron* (a contraction for "positive electron"). The positron has the same mass as the electron, but carries an opposite charge of $+e$. A collision between a positron and an electron is likely to annihilate both particles, converting them into electromagnetic energy in the form of γ rays. For this reason, positrons never coexist with ordinary matter for any appreciable length of time. The mutual annihilation of a positron and an electron lies at the heart of an important medical diagnostic technique, as Conceptual Example 6 discusses.

THE PHYSICS OF . . .

PET scanning.

Conceptual Example 6 Positron Emission Tomography

Positron emission tomography, or PET scanning, as it is known, utilizes positrons in the following way. Certain radioactive isotopes decay by positron emission, for example, oxygen ${}^{15}_{8}\text{O}$. Such isotopes can be injected into the body, where they collect at specific sites. The positron (${}^{0}_{1}\text{e}$) emitted during the decay of the isotope encounters an electron (${}^{0}_{-1}\text{e}$) in the body tissue almost at once. The resulting mutual annihilation produces two γ-ray photons (${}^{0}_{1}\text{e} + {}^{0}_{-1}\text{e} \rightarrow \gamma + \gamma$), which are detected by devices mounted on a ring around the patient, as Figure 32.12 shows. The two photons strike oppositely positioned detectors on the ring and, in doing so, reveal the line along which the annihilation occurred. Such information leads to a computer-generated image that can be useful in diagnosing abnormalities at the site where the radioactive isotope collected. How does the principle of conservation of linear momentum explain the fact that the photons strike detectors located opposite one another?

REASONING The momentum conservation principle states that the total linear momentum of an isolated system remains constant (see Section 7.2). Therefore, the total momentum of the γ-ray photons will be equal to the total momentum of the positron and electron before annihilation, provided that the system is isolated. An isolated system is a system on which no net external force acts. The positron and the electron do exert electrostatic forces on one another, however, since they carry electric charges. But these are internal, not external, forces and cannot change the total linear momentum of the two-particle system. The total photon momentum, then, must equal the total momentum of the positron and the electron, which is nearly zero, to the extent that the two particles have much less momentum than the photons do. With a total linear momentum of zero, the momentum vector of one photon must point opposite to the momentum vector of the other photon. Thus, the two photons depart from the annihilation site traveling in opposite directions and strike oppositely located detectors.

Related Homework Material: *Problem 38*

PET-scanning is being used here to evaluate a brain disease.

The positron is an example of an antiparticle, and after its discovery, scientists came to realize that for every type of particle there is a corresponding type of antiparticle. The antiparticle is a form of matter that has the same mass as the particle, but carries an opposite electric charge (e.g., the electron–positron pair) or a magnetic moment that is oriented in an opposite direction relative to the spin (e.g., the neutrino–antineutrino pair). A few electrically neutral particles, like the photon and the neutral pion (π^0), are their own antiparticles.

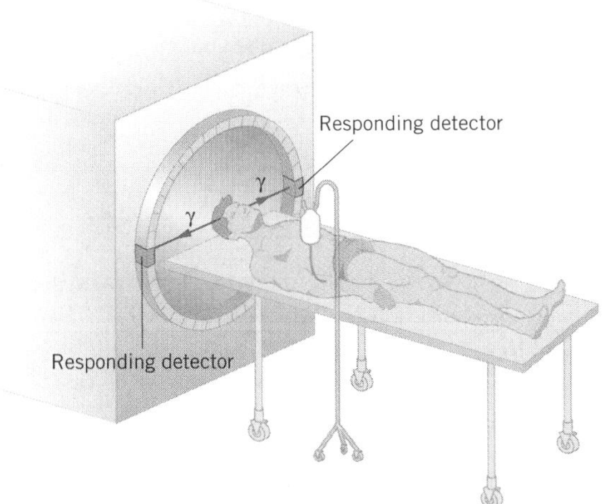

Figure 32.12 Positron emission tomography, or PET scanning, uses a radioactive isotope that has been injected into the body. The isotope decays by emitting a positron, which annihilates an electron in the body tissue, producing two γ-ray photons. These photons strike detectors mounted on opposite sides of a ring that surrounds the patient.

MUONS AND PIONS

In 1937 the American physicists S. H. Neddermeyer and C. D. Anderson discovered a new charged particle whose mass was about 207 times greater than the mass of the electron. The particle is designated by the Greek letter μ (mu) and is known as a *muon.* There are two muons that have the same mass but opposite charge: the particle μ^- and its antiparticle μ^+. The μ^- muon has the same charge as the electron, while the μ^+ muon has the same charge as the positron. Both muons are unstable, with a lifetime of 2.2×10^{-6} s. The μ^- muon decays into an electron (β^-), a muon neutrino (ν_μ), and an electron antineutrino ($\bar{\nu}_e$), according to the following reaction:

$$\mu^- \rightarrow \beta^- + \nu_\mu + \bar{\nu}_e$$

The μ^+ muon decays into a positron (β^+), a muon antineutrino ($\bar{\nu}_\mu$), and an electron neutrino (ν_e):

$$\mu^+ \rightarrow \beta^+ + \bar{\nu}_\mu + \nu_e$$

Muons interact with protons and neutrons via the weak nuclear force.

The Japanese physicist Hidekei Yukawa (1907–1981) predicted in 1935 that ***pions*** exist, but they were not discovered until 1947. Pions come in three varieties: one that is positively charged, the negatively charged antiparticle with the same mass, and the neutral pion, mentioned earlier, which is its own antiparticle. The symbols for these pions are, respectively, π^+, π^-, and π^0. The charged pions are unstable and have a lifetime of 2.6×10^{-8} s. The decay of a charged pion almost always produces a muon:

$$\pi^- \rightarrow \mu^- + \bar{\nu}_\mu$$
$$\pi^+ \rightarrow \mu^+ + \nu_\mu$$

As mentioned earlier, the neutral pion π^0 is also unstable and decays into two γ-ray photons, the lifetime being 0.8×10^{-16} s. The pions are of great interest because, unlike the muons, the pions interact with protons and neutrons via the strong nuclear force.

CLASSIFICATION OF PARTICLES

It is useful to group the known particles into three families, the photons, the leptons, and the hadrons, as Table 32.3 summarizes. This grouping is made according to the nature of the force by which a particle interacts with other particles. The ***photon family,*** for instance, has only one member, the photon. The photon interacts only with charged particles, and the interaction is only via the *electromagnetic force.* No other particle behaves in this manner.

Table 32.3 Some Particles and Their Properties

Family	Particle	Particle Symbol	Antiparticle Symbol	Rest Energy (MeV)	Lifetime (s)
Photon	Photon	γ	Self[a]	0	Stable
Lepton	Electron	e^- (or β^-)	e^+ (or β^+)	0.511	Stable
	Muon	μ^-	μ^+	105.7	2.2×10^{-6}
	Tau	τ^-	τ^+	1784	10^{-13}
	Electron neutrino	ν_e	$\bar{\nu}_e$	≈ 0	Stable
	Muon neutrino	ν_μ	$\bar{\nu}_\mu$	≈ 0	Stable
	Tau neutrino	ν_τ	$\bar{\nu}_\tau$	≈ 0	Stable
Hadron					
Mesons					
	Pion	π^+	π^-	139.6	2.6×10^{-8}
		π^0	Self[a]	135.0	0.8×10^{-16}
	Kaon	K^+	K^-	493.7	1.2×10^{-8}
		K^0_S	$\bar{K}^0_S$	497.7	0.9×10^{-10}
		K^0_L	$\bar{K}^0_L$	497.7	5.2×10^{-8}
	Eta	η^0	Self[a]	548.8	$< 10^{-18}$
	⋮				
	Plus other mesons				
Baryons					
	Proton	p	$\bar{p}$	938.3	Stable
	Neutron	n	$\bar{n}$	939.6	900
	Lambda	Λ^0	$\bar{\Lambda}^0$	1116	2.6×10^{-10}
	Sigma	Σ^+	$\bar{\Sigma}^-$	1189	0.8×10^{-10}
		Σ^0	$\bar{\Sigma}^0$	1192	6×10^{-20}
		Σ^-	$\bar{\Sigma}^+$	1197	1.5×10^{-10}
	Omega	Ω^-	Ω^+	1672	0.82×10^{-10}
	⋮				
	Plus other baryons				

[a] The particle is its own antiparticle.

The ***lepton family*** consists of particles that interact by means of the *weak nuclear force.* Leptons can also exert gravitational and electromagnetic (if the leptons are charged) forces on other particles. The four better-known leptons are the electron, the muon, the electron neutrino ν_e, and the muon neutrino ν_μ. Table 32.3 lists these particles together with their antiparticles. Recently, two other leptons have been discovered, the tau particle (τ) and its neutrino (ν_τ), bringing the number of particles in the lepton family to six.

The ***hadron family*** contains the particles that interact by means of the *strong nuclear force and the weak nuclear force.* Hadrons can also interact by gravitational and electromagnetic forces, but at short distances ($\leq 10^{-15}$ m) the strong nuclear force dominates. Among the hadrons are the proton, the neutron, and the pions. As Table 32.3 indicates, most hadrons are short-lived. The hadrons are subdivided into two groups, the ***mesons*** and the ***baryons,*** for a reason that will be discussed in connection with the idea of quarks.

QUARKS

As more and more hadrons were discovered, it became clear that they were not all elementary particles. The suggestion was made that the hadrons are made up of smaller, more elementary particles called ***quarks.*** In 1963 a quark theory was advanced independently by M. Gell-Mann (1929–) and G. Zweig (1937–). The theory proposed that there are three quarks and three corresponding antiquarks, and that hadrons are constructed from combinations of these. Thus, the quarks are elevated to the status of elementary particles for the hadron family. The particles in the photon and lepton families are considered to be elementary, and as such they are not composed of quarks.

The three quarks were named *up* (u), *down* (d), and *strange* (s), and were assumed to have, respectively, fractional charges of $+\frac{2}{3}e$, $-\frac{1}{3}e$, and $-\frac{1}{3}e$. In other words, a quark possesses a charge smaller than the charge of an electron. Table 32.4 lists the symbols and electric charges of these quarks. Experimentally, quarks should be recognizable by their fractional charges, but in spite of an extensive search for them, free quarks have never been found.

Table 32.4 Quarks and Antiquarks

	Quarks		Antiquarks	
Name	Symbol	Charge	Symbol	Charge
Up	u	$+\frac{2}{3}e$	$\bar{u}$	$-\frac{2}{3}e$
Down	d	$-\frac{1}{3}e$	$\bar{d}$	$+\frac{1}{3}e$
Strange	s	$-\frac{1}{3}e$	$\bar{s}$	$+\frac{1}{3}e$
Charm	c	$+\frac{2}{3}e$	$\bar{c}$	$-\frac{2}{3}e$
Top	t	$+\frac{2}{3}e$	$\bar{t}$	$-\frac{2}{3}e$
Bottom	b	$-\frac{1}{3}e$	$\bar{b}$	$+\frac{1}{3}e$

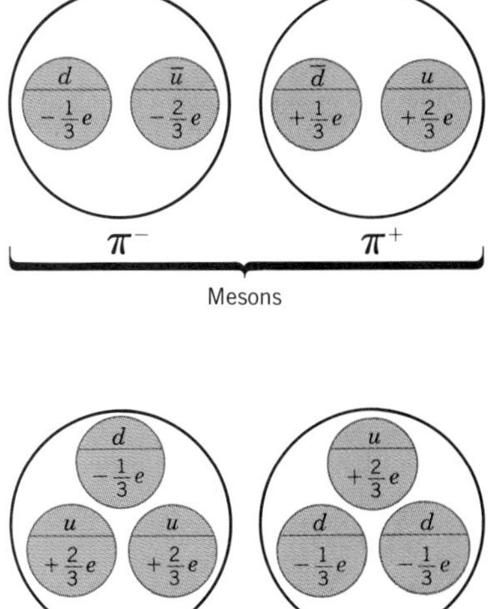

Figure 32.13 According to the original quark model of hadrons, all mesons consist of a quark and an antiquark, while baryons contain three quarks.

According to the original quark theory, the mesons are different from the baryons, for each meson consists of only two quarks—a quark and an antiquark—while a baryon contains three quarks. For instance, the π^- pion (a meson) is composed of a d quark and a $\bar{u}$ antiquark, $\pi^- = d + \bar{u}$, as Figure 32.13 shows. These two quarks combine to give the π^- pion a net charge of $-e$. Similarly, the

π^+ pion is a combination of the $\bar{d}$ and u quarks, $\pi^+ = \bar{d} + u$. In contrast, protons and neutrons, being baryons, consist of three quarks. A proton contains the combination $d + u + u$, while a neutron contains the combination $d + d + u$ (see Figure 32.13). These groups of three quarks give the correct charges for the proton and neutron.

The original quark model was extremely successful in predicting not only the correct charges for the hadrons, but other properties as well. Except for the fact that no one had succeeded in isolating a quark, the quark theory was a phenomenal success. However, in 1974 a new particle, the J/ψ meson, was discovered. This meson has a rest energy of 3100 MeV, much larger than the rest energies of other known mesons. The existence of the J/ψ meson could be explained only if a new quark–antiquark pair existed; this new quark was named *charm* (*c*). With the discovery of more and more particles, it has been necessary to postulate a fifth and a sixth quark; their names are *top* (*t*) and *bottom* (*b*), although some scientists prefer to call these quarks *truth* and *beauty*. Today, there is firm evidence for the five quarks called up, down, strange, charm, and bottom, and there is provisional evidence for the sixth or top quark. Each quark has a corresponding antiquark. All of the hundreds of the known hadrons can be accounted for in terms of these six quarks and their antiquarks. Whether the story is complete, however, remains to be seen.

In addition to their electric charges, quarks also have other properties. For example, each quark is believed to possess a characteristic called ***color***, for which there are three possibilities: blue, green, or red. The corresponding possibilities for the antiquarks are antiblue, antigreen, and antired. The use of the term "color" and the specific choices of blue, green, and red is arbitrary and pure whimsy, for the visible colors of the electromagnetic spectrum have nothing to do with quark properties. The quark property called color, however, is an important one, for it brings the quark model into agreement with the Pauli exclusion principle and enables the model to account for experimental observations that are otherwise difficult to explain.

THE STANDARD MODEL

The various elementary particles that have been discovered can interact via one or more of the four fundamental forces, which are the gravitational force, the strong nuclear force, the weak nuclear force, and the electromagnetic force. In particle physics, the phrase "***the standard model***" refers to the currently accepted explanation for the strong nuclear force, the weak nuclear force, and the electromagnetic force. In this model, the strong nuclear force between quarks is described in terms of the concept of color, the theory being referred to as quantum chromodynamics. And according to the standard model, the weak nuclear force and the electromagnetic force are separate manifestations of a single even more fundamental interaction, referred to as the electroweak interaction. The theory for the electroweak interaction was developed by Sheldon Glashow (1932–), Abdus Salam (1926–), and Steven Weinberg (1933–), who received a Nobel prize for their achievement in 1979.

In the standard model, our understanding of the building blocks of matter follows the hierarchical pattern illustrated in Figure 32.14. Molecules, such as

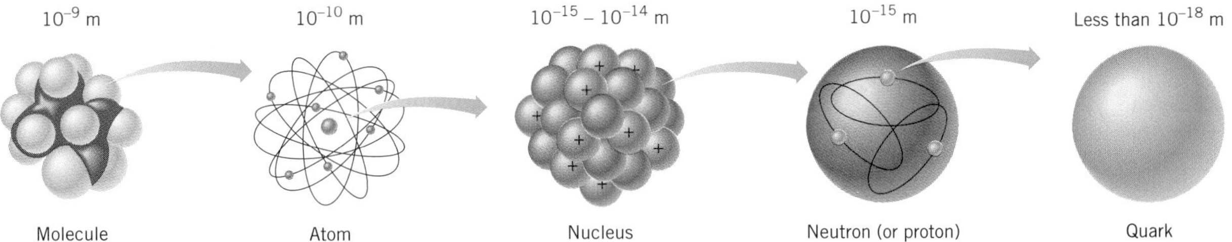

Figure 32.14 The current view of how matter is composed of basic units, starting with a molecule and ending with a quark. The approximate sizes of each unit are also shown.

water (H_2O) and glucose ($C_6H_{12}O_6$), are composed of atoms. Each atom consists of a nucleus that is surrounded by a cloud of electrons. The nucleus, in turn, is made up of protons and neutrons, which are composed of quarks.

SUMMARY

Ionizing radiation consists of photons and/or moving particles that have enough energy to ionize an atom or molecule. **Exposure** is a measure of the ionization produced in air by X-rays or γ rays. An exposure of one roentgen (R) produces 2.58×10^{-4} coulombs of positive charge in 1 kg of dry air at STP conditions.

The **absorbed dose** is the amount of energy absorbed from the radiation per unit mass of absorbing material. The SI unit of absorbed dose is the gray (Gy); 1 Gy = J/kg. However, the rad (rd) is another unit that is often used; 1 rd = 0.01 Gy.

The amount of biological damage produced by ionizing radiation is different for different types of radiation. The **relative biological effectiveness** (RBE) is the dose of 200-keV X-rays required to produce a certain biological effect divided by the dose of a given type of radiation to produce the same biological effect. The **biologically equivalent dose** is the product of the absorbed dose (in rads) and the RBE. The unit for the biologically equivalent dose is the rem.

An **induced nuclear transmutation** is the process whereby an incident particle or photon strikes a nucleus and causes the production of a new element.

Nuclear fission occurs when a massive nucleus splits into two less-massive fragments. Fission can be induced by the absorption of a thermal neutron. When a massive nucleus fissions, energy is released, because the binding energy per nucleon is greater for the fragments than for the original nucleus. Neutrons are also released during nuclear fission. These neutrons can, in turn, induce other nuclei to fission and lead to a process known as a **chain reaction.** A **fission reactor** is a device that generates energy by a controlled chain reaction.

In a **fusion** process, two nuclei with smaller masses combine to form a single nucleus with a larger mass. Energy is released by fusion when the binding energy per nucleon is greater for the larger nucleus than for the smaller nuclei.

Subatomic particles are divided into three families: the **photon** family, the **lepton** family, and the **hadron** family. **Elementary particles** are the basic building blocks of matter. All members of the photon and lepton families are elementary particles. The quark theory proposes that the hadrons are not elementary particles, but are composed of elementary particles called **quarks.** Currently, the hundreds of hadron particles can be accounted for in terms of six quarks and their antiquarks. The **standard model** consists of two parts: (1) the currently accepted explanation for the strong nuclear force in terms of the quark concept of color and (2) the theory of the electroweak interaction.

QUESTIONS

1. Two different types of radiation have the same RBE. Does this mean that each type delivers the same amount of energy to the tissue that it irradiates? Justify your answer.

2. When a dentist X-rays your teeth, a lead apron is placed over your chest and lower body. What is the purpose of this apron?

3. State whether the two quantities in each of the following cases are related and, if so, give the relation between them: (a) rads and grays, and (b) rads and roentgens.

4. Explain why the following reactions are *not* allowed: (a) ${}^{60}_{28}\text{Ni}\ (\alpha, p)\ {}^{62}_{29}\text{Cu}$, (b) ${}^{27}_{13}\text{Al}\ (n, n)\ {}^{28}_{13}\text{Al}$, (c) ${}^{39}_{19}\text{K}\ (p, \alpha)\ {}^{36}_{17}\text{Cl}$.

5. Thermal neutrons and thermal protons have nearly the same speed. But thermal neutrons and thermal electrons have very different speeds. Account for this observation, and explain whether the neutrons or the electrons have the greater speed.

6. Would a release of energy accompany the fission of a nucleus of nucleon number 25 into two fragments of about equal mass? Using the curve in Figure 32.8, account for your answer.

7. In the fission of ${}^{235}_{92}\text{U}$ there are, on the average, 2.5 neutrons released per fission. Suppose a *different* element is being fissioned and, on the average, only 1.0 neutron is released per fission. If a small fraction of the thermal neutrons absorbed by the nuclei does *not* produce a fission, can a self-sustaining chain reaction be produced using this element? Explain.

8. The mass of coal consumed in a coal-burning electric power plant is about two million times greater than the mass of ${}^{235}_{92}\text{U}$ used to fuel a comparable nuclear power plant. Why?

9. Refer to Figure 32.8 and decide if the fusion of two nuclei, each with a nucleon number of 60, would release energy. Give your reasoning.

10. Explain the difference between fission and fusion and why each process produces energy.

PROBLEMS

Section 32.1 Biological Effects of Ionizing Radiation

1. A beam of γ rays passes through 4.0×10^{-3} kg of dry air and generates 1.7×10^{12} ions, each with a charge of $+e$. What is the exposure (in roentgens)?

2. A film badge worn by a radiologist indicates that she has received an absorbed dose of 2.5×10^{-5} Gy. The mass of the radiologist is 65 kg. How much energy has she absorbed?

3. A person who receives a 500-rem dose of proton radiation (RBE = 10) has a 50% chance of dying within a few months or so. What is the absorbed dose (in rads) of this radiation?

4. What absorbed dose (in rads) of α particles (RBE = 20) causes as much biological damage as a 60-rad dose of protons (RBE = 10)?

5. A 75-kg person is exposed to 45 mrem of α particles (RBE = 12). How much energy has this person absorbed?

6. Someone stands near a radioactive source and receives doses of the following types of radiation: γ rays (20 mrad, RBE = 1), electrons (30 mrad, RBE = 1), protons (4 mrad, RBE = 10), and slow neutrons (5 mrad, RBE = 2). What is the total biologically equivalent dose (in mrem) received?

7. A beam of particles is directed at a 0.015-kg tumor. There are 1.6×10^{10} particles per second reaching the tumor, and the energy of each particle is 4.0 MeV. The RBE for the radiation is 14. Find the biologically equivalent dose given to the tumor in 25 s.

***8.** A water sample receives a 750-rad dose of radiation. Find the rise in the water temperature.

***9.** What absorbed dose (in rad) of γ rays is required to change a block of ice at 0.0° C into steam at 100.0° C?

***10.** A 2.0-kg tumor is being irradiated by a radioactive source. The tumor receives an absorbed dose of 12 Gy in a time of 850 s. Each disintegration of the radioactive source produces a particle that enters the tumor and delivers an energy of 0.40 MeV. What is the activity $\Delta N/\Delta t$ (see Section 31.6) of the radioactive source?

Section 32.2 Induced Nuclear Reactions

11. What is the nucleon number A in the reaction ${}^{27}_{13}\text{Al}\ (\alpha, n)\ {}^{A}_{15}\text{P}$?

12. What is the atomic number Z and the element X in the reaction ${}^{10}_{5}\text{B}\ (\alpha, \text{p})\ {}^{13}_{Z}\text{X}$?

13. Write the reactions below in the shorthand form discussed in the text.

$$\text{(a)}\ {}^{27}_{13}\text{Al} + {}^{1}_{0}\text{n} \rightarrow {}^{27}_{12}\text{Mg} + {}^{1}_{1}\text{H}$$
$$\text{(b)}\ {}^{40}_{18}\text{Ar} + {}^{4}_{2}\text{He} \rightarrow {}^{43}_{19}\text{K} + {}^{1}_{1}\text{H}$$

14. Write the equation for the reaction ${}^{17}_{8}\text{O}\ (\gamma, \alpha n)\ {}^{12}_{6}\text{C}$. The notation "$\alpha n$" means that an α particle and a neutron are produced by the reaction.

15. Complete the following nuclear reactions, assuming that the unknown quantity signified by the question mark is a single entity:

(a) ${}^{43}_{20}\text{Ca}\ (\alpha, ?)\ {}^{46}_{21}\text{Sc}$
(b) ${}^{9}_{4}\text{Be}\ (?, n)\ {}^{12}_{6}\text{C}$
(c) ${}^{9}_{4}\text{Be}\ (p, \alpha)\ ?$
(d) $?\ (\alpha, p)\ {}^{17}_{8}\text{O}$
(e) ${}^{55}_{25}\text{Mn}\ (n, \gamma)\ ?$

***16.** During a nuclear reaction, an unknown particle is absorbed by a copper ${}^{63}_{29}\text{Cu}$ nucleus, and the reaction products are ${}^{62}_{29}\text{Cu}$, a neutron, and a proton. What is the name, atomic number, and nucleon number of the *compound nucleus?*

***17.** Consider the induced nuclear reaction ${}^{2}_{1}\text{H} + {}^{14}_{7}\text{N} \rightarrow {}^{12}_{6}\text{C} + {}^{4}_{2}\text{He}$. Using the atomic masses given in Appendix F, determine the energy (in MeV) released when the ${}^{12}_{6}\text{C}$ and ${}^{4}_{2}\text{He}$ nuclei are formed in this manner.

Section 32.3 Nuclear Fission, Section 32.4 Nuclear Reactors

18. Determine the number of neutrons released during the following fission reaction: ${}^{1}_{0}\text{n} + {}^{235}_{92}\text{U} \rightarrow {}^{133}_{51}\text{Sb} + {}^{99}_{41}\text{Nb} +$ neutrons.

19. ${}^{235}_{92}\text{U}$ absorbs a thermal neutron and fissions into rubidium ${}^{93}_{37}\text{Rb}$ and cesium ${}^{141}_{55}\text{Cs}$. What nucleons are produced by the fission, and how many are there?

20. During an underground nuclear test, an atomic bomb is detonated. The bomb produces an amount of energy equivalent to 36 kilotons of TNT (1.0 kiloton of TNT releases about 5.0×10^{12} J of energy). To what amount of mass is this energy equivalent?

21. When a ${}^{235}_{92}\text{U}$ nucleus fissions, about 200 MeV of energy is released. What is the ratio of this energy to the rest energy of the uranium nucleus?

22. What energy (in MeV) is liberated by the following fission reaction?

$$\underbrace{{}^{1}_{0}\text{n}}_{1.009\ \text{u}} + \underbrace{{}^{235}_{92}\text{U}}_{235.044\ \text{u}} \longrightarrow \underbrace{{}^{141}_{56}\text{Ba}}_{140.914\ \text{u}} + \underbrace{{}^{92}_{36}\text{Kr}}_{91.926\ \text{u}} + \underbrace{3\,{}^{1}_{0}\text{n}}_{3(1.009\ \text{u})}$$

23. A particular fission reaction produces an energy of 210 MeV per fission. How many fissions per second occur if a reactor is generating 130 MW of power?

24. Neutrons released by a fission reaction must be slowed by collisions with the moderator nuclei before the neutrons can cause further fissions. Suppose a 1.5-MeV neutron leaves each collision with 65% of its incident energy. How many collisions are required to reduce the neutron's energy to at least 0.040 eV, which is the energy of a thermal neutron?

***25.** When 1.0 kg of coal is burned, about 3.0×10^{7} J of energy is released. If the energy released per ${}^{235}_{92}\text{U}$ fission is 2.0×10^{2} MeV, how many kilograms of coal must be burned to produce the same energy as 1.0 kg of ${}^{235}_{92}\text{U}$?

***26.** The energy consumed in one year in the United States is about 8.6×10^{19} J. When each ${}^{235}_{92}\text{U}$ nucleus fissions, about 2.0×10^{2} MeV of energy is released. How many kilograms of ${}^{235}_{92}\text{U}$ would be needed to generate this energy if all the nuclei fissioned?

***27.** The water that cools a reactor core enters the reactor at 216 °C and leaves at 287 °C. (The water is pressurized, so it does not turn to steam.) The core is generating 5.6×10^{9} W of power. Assume that the specific heat of water is 4420 J/(kg · C°) over the temperature range stated above, and find the mass of water that passes through the core each second.

****28.** A 20.0 kiloton atomic bomb releases as much energy as 20.0 kilotons of TNT (1.0 kiloton of TNT releases about 5.0×10^{12} J of energy). Recall that about 2.0×10^{2} MeV of energy is released when each ${}^{235}_{92}\text{U}$ nucleus fissions. (a) How many ${}^{235}_{92}\text{U}$ nuclei are fissioned to produce the bomb's energy? (b) How many grams of uranium are fissioned? (c) What is the equivalent mass (in grams) of the bomb's energy?

****29.** A nuclear power plant is 25% efficient, meaning that 25% of the power it generates goes into producing electricity. The remaining 75% is wasted as heat. The plant generates 8.0×10^{8} watts of electric power. If each fission releases 2.0×10^{2} MeV of energy, how many kilograms of ${}^{235}_{92}\text{U}$ are fissioned per year?

Section 32.5 Nuclear Fusion

30. The fusion of two deuterium nuclei (${}^{2}_{1}\text{H}$, mass = 2.0141 u) can yield a tritium nucleus (${}^{3}_{1}\text{H}$, mass = 3.0161 u) and a proton (${}^{1}_{1}\text{H}$, mass = 1.0078 u). What is the energy (in MeV) released in this reaction?

31. Two deuterium (${}^{2}_{1}\text{H}$) nuclei can fuse and form ${}^{3}_{2}\text{He}$ and a neutron. The atomic masses are ${}^{2}_{1}\text{H}$ (2.0141 u), ${}^{3}_{2}\text{He}$ (3.0160 u), and ${}^{1}_{0}\text{n}$ (1.0087 u). Find the energy (in MeV) released.

32. In one type of fusion reaction a proton fuses with a neutron to form a deuterium nucleus: ${}^{1}_{1}\text{H} + {}^{1}_{0}\text{n} \rightarrow {}^{2}_{1}\text{H}$. The masses are ${}^{1}_{1}\text{H}$ (1.0078 u), ${}^{1}_{0}\text{n}$ (1.0087 u), and ${}^{2}_{1}\text{H}$ (2.0141 u). How much energy (in MeV) is released by this reaction?

***33.** Imagine your car is powered by a fusion engine in which the following reaction occurs: $3\,{}^{2}_{1}\text{H} \rightarrow {}^{4}_{2}\text{He} + {}^{1}_{1}\text{H} + {}^{1}_{0}\text{n}$.

The masses are ${}^{2}_{1}\text{H}$ (2.0141 u), ${}^{4}_{2}\text{He}$ (4.0026 u), ${}^{1}_{1}\text{H}$ (1.0078 u), and ${}^{1}_{0}\text{n}$ (1.0087 u). The engine uses 6.1×10^{-6} kg of deuterium ${}^{2}_{1}\text{H}$ fuel. If one gallon of gasoline produces 2.1×10^{9} J of energy, how many gallons of gasoline would have to be burned to equal the energy released by all the deuterium fuel?

***34.** One proposed fusion reaction combines lithium ${}^{6}_{3}\text{Li}$ (6.015 u) with deuterium ${}^{2}_{1}\text{H}$ (2.014 u) to give helium ${}^{4}_{2}\text{He}$ (4.003 u): ${}^{2}_{1}\text{H} + {}^{6}_{3}\text{Li} \rightarrow 2{}^{4}_{2}\text{He}$. How many kilograms of lithium would be needed to supply the energy needs of one household for a year, estimated to be 3.8×10^{10} J?

****35.** The proton–proton cycle thought to occur in the sun consists of the following sequence of reactions:

$$\begin{aligned}
&(1)\ {}^{1}_{1}\text{H} + {}^{1}_{1}\text{H} \longrightarrow {}^{2}_{1}\text{H} + {}^{0}_{1}\text{e} + \nu \\
&(2)\ {}^{1}_{1}\text{H} + {}^{2}_{1}\text{H} \longrightarrow {}^{3}_{2}\text{He} + \gamma \\
&(3)\ {}^{3}_{2}\text{He} + {}^{3}_{2}\text{He} \longrightarrow {}^{4}_{2}\text{He} + {}^{1}_{1}\text{H} + {}^{1}_{1}\text{H}
\end{aligned}$$

In these reactions ${}^{0}_{1}\text{e}$ is a positron (mass = 0.000 549 u), ν is a neutrino (mass $\approx$ 0), and γ is a gamma ray photon (mass = 0). Note that reaction (3) uses two ${}^{3}_{2}\text{He}$ nuclei, which are formed by *two* reactions of type (1) and *two* reactions of type (2). Verify that the proton–proton cycle generates about 25 MeV of energy. When using atomic masses from Appendix F, be sure to account for the fact that there are two electrons in two hydrogen atoms, while there is one electron in a single deuterium (${}^{2}_{1}\text{H}$) atom. The mass of one electron is 0.000 549 u.

Section 32.6 Elementary Particles

36. What are the quarks from which an antiproton is constructed?

37. The lambda particle Λ^0 has an electric charge of zero. It is a baryon and, hence, is composed of three quarks. They are all different. One of these quarks is the up quark u, and there are no antiquarks present. Make a list of the three possibilities for the quarks contained in Λ^0. (Other information is needed to decide which one of these possibilities is the Λ^0 particle.)

38. Review Conceptual Example 6 as background for this problem. An electron and its antiparticle annihilate each other, producing two γ-ray photons. The kinetic energies of the particles are negligible. For each photon, determine (a) its energy (in MeV), (b) its wavelength, and (c) the magnitude of its momentum.

39. A high-energy proton collides with a stationary proton, and the reaction $p + p \rightarrow n + p + \pi^+$ occurs. The rest energy of the π^+ pion is 139.6 MeV. Ignore momentum conservation and find the minimum energy (in MeV) the incident proton must have.

40. A collision between two protons produces three new particles: $p + p \rightarrow p + \pi^+ + \Lambda^0 + K^0$. The rest energies of the new particles are π^+ (139.6 MeV), Λ^0 (1116 MeV), and K^0 (497.7 MeV). Note that one proton disappears during the reaction. How much of the protons' incident energy (in MeV) is transformed into matter during this reaction?

***41.** Suppose a neutrino is created and has an energy of 35 MeV. (a) If the neutrino, like the photon, has no mass and travels at the speed of light, find the momentum of the neutrino. (b) Determine the de Broglie wavelength of the neutrino.

***42.** Review Conceptual Example 4 as background for this problem. An energetic proton is fired at a stationary proton. For the reaction to produce new particles, the two protons must approach each other to within a distance of about 8.0×10^{-15} m. The moving proton must have a sufficient speed to overcome the repulsive Coulomb force. What must be the minimum initial kinetic energy (in MeV) of the proton?

ADDITIONAL PROBLEMS

43. During an X-ray examination, a person is exposed to radiation at a rate of 3.1×10^{-5} grays per second. The exposure time is 0.10 s, and the mass of the exposed tissue is 1.2 kg. Determine the energy absorbed.

44. A Σ^+ particle (see Table 32.3) decays into a π^0 particle and a proton: $\Sigma^+ \rightarrow \pi^0 + p$. Ignore the kinetic energy of the Σ^+ particle, and determine how much energy (in MeV) is released in the process.

45. A nitrogen ${}^{14}_{7}\text{N}$ nucleus absorbs a deuterium ${}^{2}_{1}\text{H}$ nucleus during a nuclear reaction. What is the name, atomic number, and nucleon number of the compound nucleus?

46. Uranium ${}^{235}_{92}\text{U}$ fissions into two fragments plus three neutrons: ${}^{1}_{0}\text{n} + {}^{235}_{92}\text{U} \rightarrow (2\text{ fragments}) + 3{}^{1}_{0}\text{n}$. The mass of a neutron is 1.008 665 u and that of ${}^{235}_{92}\text{U}$ is 235.043 924 u. If 225.0 MeV of energy is released during the fission, what is the combined mass of the two fragments?

47. Within the core of a nuclear reactor there are 3.0×10^{19} nuclei fissioning each second. The energy released by each fission is about 2.0×10^{2} MeV. Determine the power (in watts) being generated.

***48.** During an X-ray examination of the chest, a person receives an exposure of 0.015 R. How many singly charged ions would be produced if the X-rays passed through 2.0 m^3 of dry air at STP conditions (density of air = 1.29 kg/m^3)?

***49.** Deuterium (${}^{2}_{1}\text{H}$) is an attractive fuel for fusion reactions because it is abundant in the waters of the oceans. In the oceans, deuterium makes up about 0.015% of the hydrogen in the water (H_2O). (a) How many deuterium atoms are there in one kilogram of water? (b) If each deuterium nucleus produces about 7.2 MeV in a fusion reaction, how many kilograms of ocean water would be needed to supply the energy needs of the United States for one year, estimated to be 8.6×10^{19} J?

***50.** (a) If each fission of a $^{235}_{92}U$ nucleus releases about 2.0×10^2 MeV of energy, determine the energy (in joules) released by the complete fissioning of 1.0 gram of $^{235}_{92}U$. (b) How many grams of $^{235}_{92}U$ are consumed in one year, in order to supply the energy needs of a household that uses 30.0 kWh of energy per day, on the average?

***51.** Suppose the $^{239}_{94}Pu$ nucleus fissions into two fragments whose mass ratio is 0.32:0.68. With the aid of Figure 32.8, estimate the energy (in MeV) released during this fission.

****52.** One kilogram of dry air at STP conditions is exposed to 1.0 R of X-rays. One roentgen is defined by Equation 32.1. An equivalent definition can be based on the fact that an exposure of one roentgen deposits 8.3×10^{-3} J of energy per kilogram of dry air. Using the two definitions and assuming that all ions produced are singly charged, determine the average energy (in eV) needed to produce one ion in air.

INTERSECTION

Physics & Medicine

Medicine has made more advances in the last one hundred years than it had during the previous 10 thousand years. Are the phenomenal achievements in healing occurring on their own, or are they accompanied, perhaps driven, by other scientific progress? In particular, just how much does medicine benefit from the insights physicists have made into the nature of fundamental physical processes over the past century?

Consider the physics that is involved in helping you in a medical emergency. Suppose, for example, that you were involved in a serious accident in which you received head and abdominal wounds. You're transported to the hospital in an ambulance powered by an internal combustion engine moving on nearly frictionless bearings. Your vital signs are called into the hospital over a static-free FM radio. Once in the bright, clean, emergency room your wounds are X-rayed, and the films are developed in a matter of minutes. The doctor examines the negative through a screen that is back-lit by fluorescent lights. Meanwhile, you are connected to a cardiac monitor, and its output is displayed on a cathode-ray tube (CRT) monitor.

The skull X-rays are inconclusive, but the abdominal X-rays show clear signs of internal damage, and the doctor determines that immediate surgery is necessary. She operates on you with a laser, as well as knives of surgical stainless steel. Your blood pressure is monitored continuously and automatically. Your internal wounds are repaired, the incisions sutured, and you are rolled down the hall to the magnetic resonance imager (MRI) for a brain scan to see if your brain was injured. Happily, it was not.

With all the exposure television and personal experience have given you, you are probably not surprised by any of the details of this incident. But if you lived in the middle of the last century, say, you'd consider virtually every step nothing short of miraculous. In fact, the fundamental physics discovered over the past century plays crucial roles in every step of this scenario. Consider: The engine in the ambulance has been optimized by using the physics of combustion and thermodynamics; the vehicle's bearings are created from special, scientifically derived compounds for smooth running and long life; the FM (frequency modulation) radio is a result of decades of refinement of electromagnetic theory; the air in the emergency room is electrostatically cleaned to prevent infection; the X-ray machine and X-ray sensitive, rapid-development film are results of fundamental research in physics done since the end of the last century. The heart monitor measures changes in your body's internal resistance, and the CRT, a simplified television monitor, is based on the atomic theory of the atom, Ohm's law of electrical circuits, and electromagnetism. The automatic blood pressure monitor uses a piezoelectric crystal that converts changes in pressure that the blood pressure cuff senses as blood pulses through your veins into electrical impulses.

One of the crowning applications of physics in medicine is the series of internal imaging devices (Computerized Axial Tomography—CAT, Magnetic Resonance Imaging—MRI, and Positron Emission Tomography—PET). These apply the theories of electromagnetism, atomic and nuclear physics, and computer technology to allow physicians to see details of the inner workings of the human body.

If you become a medical professional, virtually every test you make, and every procedure you do, will be steeped in applications of fundamental physics. And new advances in physics are continually being applied to medicine. For example, during the past decade, surgeons have moved beyond metal scalpels to using lasers and ultrasonics in some applications. The laser was first built in 1960, based on a detailed study of atomic physics. Ultrasonics, the generation, detection, and application of ultrahigh-frequency sound waves, evolved in the 1940s and was refined over the next several decades. By understanding the basic physics of compressional waves (sound), physicists have learned how to focus them for use in pinpoint surgery.

Another interesting application of physics to medicine is radioactivity. Exposure to unrestricted X-radiation is lethal to life. Among other things, it modifies DNA, allowing for unrestricted cell growth, which we call cancer. The impact of nuclear weapons use and tests, and the radiation leak from the Chernobyl nuclear reactor, clearly demonstrate this. On the other hand, restricted, focused radiation is successfully used in medicine to combat cancer. A well-collimated (focused) beam of X-rays can be aimed to kill just cancer cells, leaving nearby healthy cells intact.

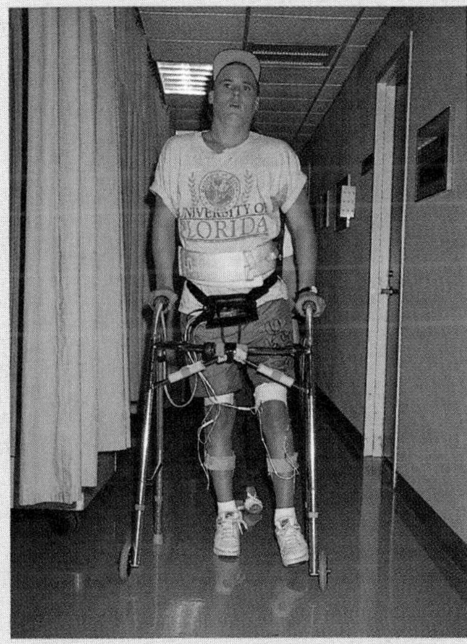

Microcomputer-controlled neuromuscular stimulation is a new technique that may some day enable paralyzed people to walk. With the aid of electrodes applied to the surface of the body, this technique uses electricity to cause contraction of the muscles that control walking.

The profound advances in our understanding of physics during this century have enabled medical researchers to understand many of the human body's functions, right down to the molecular level. As a result, medicine now provides us with cures for many of the most crippling and lethal diseases that, until now, have claimed thousands, and even millions of lives, every year. Even as recently as the early 1950s, before the vaccine for polio was developed, most people were petrified to be out and about during the summer, when polio spread like wildfire. And it is at the level of basic physics and chemistry that researchers are learning about, and searching for the cure and prevention of cancer, AIDS, and other viral diseases.

If the advances in physics made during this past century had not come along until now, most of the devices described in this essay would *still* not exist. This process of applying basic physics to medicine is not over yet. To this very day, new discoveries in science are finding important and immediate applications in medicine. And it's fair to say that without the present understanding of the basic principles of physics, medicine would still be a medieval art, and the quality of our lives would be much lower.

FOR INFORMATION ABOUT CAREERS IN MEDICINE

Professional Careers Sourcebook, K. M. Savage and C. A. Dorgan, eds., 1990, Gale Research, Inc., publisher, provides addresses and phone numbers for dozens of guides and other resources about careers in many medically related fields.

Encyclopedia of Careers and Vocational Guidance, Vol. 2, 8th edition, William E. Hopke, ed., 1990, J. G. Ferguson, publisher, has an in-depth description of the working conditions physicians encounter.

You can also write to the following, among many others listed in the books above:

- American Institute of Physics, 335 East 45th St., New York, NY 10017
- American Medical Association, 535 North Dearborn St., Chicago, IL 60610

POWERS OF TEN AND SCIENTIFIC NOTATION

APPENDIX A

In science, very large and very small decimal numbers are conveniently expressed in terms of powers of ten, some of which are listed below:

$$10^3 = 10 \times 10 \times 10 = 1000 \qquad 10^{-3} = \frac{1}{10 \times 10 \times 10} = 0.001$$

$$10^2 = 10 \times 10 = 100 \qquad 10^{-2} = \frac{1}{10 \times 10} = 0.01$$

$$10^1 = 10 \qquad 10^{-1} = \frac{1}{10} = 0.1$$

$$10^0 = 1$$

Using powers of ten, we can write the radius of the earth in the following way, for example:

$$\text{Earth radius} = 6\ 380\ 000 \text{ m} = 6.38 \times 10^6 \text{ m}$$

The factor of ten raised to the sixth power is ten multiplied by itself six times, or one million, so the earth's radius is 6.38 million meters. Alternatively, the factor of ten raised to the sixth power indicates that the decimal point in the term 6.38 is to be moved six places *to the right* to obtain the radius as a number without powers of ten.

For numbers less than one, negative powers of ten are used. For instance, the Bohr radius of the hydrogen atom is

$$\text{Bohr radius} = 0.000\ 000\ 000\ 0529 \text{ m} = 5.29 \times 10^{-11} \text{ m}$$

The factor of ten raised to the minus eleventh power indicates that the decimal point in the term 5.29 is to be moved eleven places *to the left* to obtain the radius as a number without powers of ten. Numbers expressed with the aid of powers of ten are said to be in ***scientific notation.***

Calculations that involve the multiplication and division of powers of ten are carried out as in the following examples:

$$(2.0 \times 10^6)(3.5 \times 10^3) = (2.0 \times 3.5) \times 10^{6+3} = 7.0 \times 10^9$$

$$\frac{9.0 \times 10^7}{2.0 \times 10^4} = \left(\frac{9.0}{2.0}\right) \times 10^7 \times 10^{-4} = \left(\frac{9.0}{2.0}\right) \times 10^{7-4} = 4.5 \times 10^3$$

The general rules for such calculations are:

$$\frac{1}{10^n} = 10^{-n} \qquad \text{(A-1)}$$

$$10^n \times 10^m = 10^{n+m} \quad \text{(Exponents added)} \qquad \text{(A-2)}$$

$$\frac{10^n}{10^m} = 10^{n-m} \quad \text{(Exponents subtracted)} \qquad \text{(A-3)}$$

where n and m are any positive or negative number.

Scientific notation is convenient because of the ease with which it can be used in calculations. Moreover, scientific notation provides a convenient way to express the significant figures in a number, as Appendix B discusses.

APPENDIX B SIGNIFICANT FIGURES

The number of ***significant figures*** in a number is the number of digits whose values are known with certainty. For instance, a person's height is measured to be 1.78 m, with the measurement error being in the third decimal place. All three digits are known with certainty, so that the number contains three significant figures. If a zero is given as the last digit to the right of the decimal point, the zero is presumed to be significant. Thus, the number 1.780 m contains four significant figures. As another example, consider a distance of 1500 m. This number contains only two significant figures, the one and the five. The zeros immediately to the left of the unexpressed decimal point are not counted as significant figures. However, zeros located between significant figures are significant, so a distance of 1502 m contains four significant figures.

Scientific notation is particularly convenient from the point of view of significant figures. Suppose it is known that a certain distance is fifteen hundred meters, to four significant figures. Writing the number as 1500 m presents a problem, because it implies that only two significant figures are known. In contrast, the scientific notation of 1.500×10^3 m has the advantage of indicating that the distance is known to four significant figures.

When two or more numbers are used in a calculation, the number of significant figures in the answer is limited by the number of significant figures in the original data. For instance, a rectangular garden with sides of 9.8 m and 17.1 m has an area of (9.8 m)(17.1 m). A calculator gives 167.58 m^2 for this product. However, one of the original lengths is known only to two significant figures, so the final answer is limited to only two significant figures and should be rounded off to 170 m^2. In general, ***when numbers are multiplied or divided, the final answer has a number of significant figures that equals the smallest number of significant figures in any of the original factors.***

The number of significant figures in the answer to an addition or a subtraction is also limited by the original data. Consider the total distance along a biker's trail that consists of three segments with the distances shown below:

	2.5 km
	11 km
	5.26 km
Total	18.76 km

The distance of 11 km contains no significant figures to the right of the decimal point. Therefore, neither does the sum of the three distances, and the total distance should not be reported as 18.76 km. Instead, the answer is rounded off to 19 km. In general, ***when numbers are added or subtracted, the last significant figure in the answer occurs in the last column (counting from left to right) containing a number that results from a combination of digits that are all significant.*** In the answer of 18.76 km, the eight is the sum of 2 + 1 + 5, each digit being significant. However, the seven is the sum of 5 + 0 + 2, and the zero is not significant, since it comes from the 11-km distance, which contains no significant figures to the right of the decimal point.

ALGEBRA APPENDIX C

C1 PROPORTIONS AND EQUATIONS

Physics deals with physical variables and the relations between them. Typically, variables are represented by the letters of the English and Greek alphabets. Sometimes, the relation between variables is expressed as a proportion or inverse proportion. Other times, however, it is more convenient or necessary to express the relation by means of an equation, which is governed by the rules of algebra.

If two variables are ***directly proportional*** and one of them doubles, then the other variable also doubles. Similarly, if one variable is reduced to one-half its original value, then the other is also reduced to one-half its original value. In general, if x is directly proportional to y, then increasing or decreasing one variable by a given factor causes the other variable to change in the same way by the same factor. This kind of relation is expressed as $x \propto y$, where the symbol $\propto$ means "is proportional to."

Since the proportional variables x and y always increase and decrease by the same factor, the ratio of x to y must have a constant value, or $x/y = k$, where k is a constant, independent of the values for x and y. Consequently, a proportionality such as $x \propto y$ can also be expressed in the form of an equation: $x = ky$. The constant k is referred to as a ***proportionality constant.***

If two variables are ***inversely proportional*** and one of them increases by a given factor, then the other decreases by the same factor. An inverse proportion is written as $x \propto 1/y$. This kind of proportionality is equivalent to the following equation: $xy = k$, where k is a proportionality constant, independent of x and y.

C2 SOLVING EQUATIONS

Some of the variables in an equation typically have known values, and some do not. It is often necessary to solve the equation so that a variable whose value is unknown is expressed in terms of the known quantities. ***In the process of solving an equation, it is permissible to manipulate the equation in any way, as long as a change made on one side of the equals sign is also made on the other side.*** For example, consider the equation $v = v_0 + at$. Suppose values for v, v_0, and a are available, and the value of t is required. To solve the equation for t, we begin by subtracting v_0 from *both* sides:

$$\begin{array}{rcl} v & = & v_0 + at \\ -v_0 & & -v_0 \\ \hline v - v_0 & = & at \end{array}$$

Next, we divide *both* sides of $v - v_0 = at$ by the quantity a:

$$\frac{v - v_0}{a} = \frac{at}{a} = (1)t$$

On the right side, the a in the numerator divided by the a in the denominator equals one, so that

$$t = \frac{v - v_0}{a}$$

It is always possible to check the correctness of the algebraic manipulations performed in solving an equation by substituting the answer back into the original equation. In the previous example, we substitute the answer for t into $v = v_0 + at$:

$$v = v_0 + a\left(\frac{v - v_0}{a}\right) = v_0 + (v - v_0) = v$$

The result $v = v$ implies that our algebraic manipulations were done correctly.

Algebraic manipulations other than addition, subtraction, multiplication, and division may play a role in solving an equation. The same basic rule applies, however: Whatever is done to the left side of an equation must also be done to the right side. As another example, suppose it is necessary to express v_0 in terms of v, a, and x, where $v^2 = v_0^2 + 2ax$. By subtracting $2ax$ from both sides, we isolate v_0^2 on the right:

$$\begin{array}{rcl} v^2 & = & v_0^2 + 2ax \\ -2ax & & -2ax \\ \hline v^2 - 2ax & = & v_0^2 \end{array}$$

To solve for v_0, we take the positive or negative square root of *both* sides of $v^2 - 2ax = v_0^2$:

$$v_0 = \pm\sqrt{v^2 - 2ax}$$

C3 SIMULTANEOUS EQUATIONS

When more than one variable in a single equation is unknown, additional equations are needed if solutions are to be found for all of the unknown quantities. Thus, the equation $3x + 2y = 7$ cannot be solved by itself to give unique values for both x and y. However, if x and y simultaneously obey the equation $x - 3y = 6$, then both unknowns can be found.

There are a number of methods by which such simultaneous equations can be solved. One method is to solve one equation for x in terms of y and substitute the result into the other equation to obtain an expression containing only the single unknown variable y. The equation $x - 3y = 6$, for instance, can be solved for x by adding $3y$ to each side, with the result that $x = 6 + 3y$. The substitution of this expression for x into the equation $3x + 2y = 7$ is shown below:

$$3x + 2y = 7$$

$$3(6 + 3y) + 2y = 7$$

$$18 + 9y + 2y = 7$$

We find, then, that $18 + 11y = 7$, a result that can be solved for y:

$$\begin{array}{rcr} 18 + 11y & = & 7 \\ -18 & & -18 \\ \hline 11y & = & -11 \end{array}$$

Dividing both sides of this result by 11 shows that $y = -1$. The value of $y = -1$ can be substituted in either of the original equations to obtain a value for x:

$$
\begin{aligned}
x - 3y &= 6 \\
x - 3(-1) &= 6 \\
x + 3 &= 6 \\
-3 \quad & \quad -3 \\
\hline
x &= 3
\end{aligned}
$$

C4 THE QUADRATIC FORMULA

Equations occur in physics that include the square of a variable. Such equations are said to be *quadratic* in that variable and often can be put into the following form:

$$ax^2 + bx + c = 0 \tag{C-1}$$

where a, b, and c are constants independent of x. This equation can be solved to give the ***quadratic formula,*** which is

$$x = \frac{-b \pm \sqrt{b^2 - 4ac}}{2a} \tag{C-2}$$

The $\pm$ in the quadratic formula indicates that there are two solutions. For instance, if $2x^2 - 5x + 3 = 0$, then $a = 2$, $b = -5$, and $c = 3$. The quadratic formula gives the two solutions as follows:

[Solution 1: Plus sign]

$$
\begin{aligned}
x &= \frac{-b + \sqrt{b^2 - 4ac}}{2a} \\
&= \frac{-(-5) + \sqrt{(-5)^2 - 4(2)(3)}}{2(2)} \\
&= \frac{+5 + \sqrt{1}}{4} = \frac{3}{2}
\end{aligned}
$$

[Solution 2: Minus sign]

$$
\begin{aligned}
x &= \frac{-b - \sqrt{b^2 - 4ac}}{2a} \\
&= \frac{-(-5) - \sqrt{(-5)^2 - 4(2)(3)}}{2(2)} \\
&= \frac{+5 - \sqrt{1}}{4} = 1
\end{aligned}
$$

APPENDIX D EXPONENTS AND LOGARITHMS

Appendix A discusses powers of ten such as 10^3, which means ten multiplied by itself three times, or $10 \times 10 \times 10$. The three is referred to as an ***exponent.*** The use of exponents extends beyond powers of ten. In general, the term y^n means the factor y is multiplied by itself n times. For example, y^2, or y squared, is familiar and means $y \times y$. Similarly, y^5 means $y \times y \times y \times y \times y$.

The rules that govern algebraic manipulations of exponents are the same as those given in Appendix A (see Equations A-1, A-2, and A-3) for powers of ten:

$$\frac{1}{y^n} = y^{-n} \tag{D-1}$$

$$y^n y^m = y^{n+m} \qquad \text{(Exponents added)} \tag{D-2}$$

$$\frac{y^n}{y^m} = y^{n-m} \qquad \text{(Exponents subtracted)} \tag{D-3}$$

To the three rules above we add two more that are useful. One of these is

$$y^n z^n = (yz)^n \tag{D-4}$$

The following example helps to clarify the reasoning behind this rule:

$$3^2 5^2 = (3 \times 3)(5 \times 5) = (3 \times 5)(3 \times 5) = (3 \times 5)^2$$

The other additional rule is

$$(y^n)^m = y^{nm} \qquad \text{(Exponents multiplied)} \tag{D-5}$$

To see why this rule applies, consider the following example:

$$(5^2)^3 = (5^2)(5^2)(5^2) = 5^{2+2+2} = 5^{2\times 3}$$

Roots, such as a square root or a cube root, can be represented with fractional exponents. For instance,

$$\sqrt{y} = y^{1/2} \quad \text{and} \quad \sqrt[3]{y} = y^{1/3}$$

In general, the nth root of y is given by

$$\sqrt[n]{y} = y^{1/n} \tag{D-6}$$

The rationale for Equation D-6 can be explained using the fact that $(y^n)^m = y^{nm}$. For instance, the fifth root of y is the number that, when multiplied by itself five times, gives back y. As shown below, the term $y^{1/5}$ satisfies this definition:

$$(y^{1/5})(y^{1/5})(y^{1/5})(y^{1/5})(y^{1/5}) = (y^{1/5})^5 = y^{(1/5)\times 5} = y$$

Logarithms are closely related to exponents. To see the connection between the two, note that it is possible to express any number y as another number B raised to the exponent x. In other words,

$$y = B^x \tag{D-7}$$

The exponent x is called the ***logarithm*** of the number y. The number B is called the ***base number.*** One of two choices for the base number is usually used. If $B = 10$, the logarithm is known as the *common logarithm,* for which the notation "log" applies:

[Common logarithm] $$y = 10^x \quad \text{or} \quad x = \log y \tag{D-8}$$

If $B = e = 2.718 \ldots$, the logarithm is referred to as the *natural logarithm,* and the notation "ln" is used:

[Natural logarithm] $$y = e^z \quad \text{or} \quad z = \ln y \tag{D-9}$$

The two kinds of logarithms are related by

$$\ln y = 2.3026 \log y \tag{D-10}$$

Both kinds of logarithms are often given on calculators.

The logarithm of the product or quotient of two numbers A and C can be obtained from the logarithms of the individual numbers according to the rules below. These rules are illustrated here for natural logarithms, but they are the same for any kind of logarithm.

$$\ln (AC) = \ln A + \ln C \tag{D-11}$$

$$\ln \left(\frac{A}{C}\right) = \ln A - \ln C \tag{D-12}$$

Thus, the logarithm of the product of two numbers is the sum of the individual logarithms, and the logarithm of the quotient of two numbers is the difference between the individual logarithms. Another useful rule concerns the logarithm of a number A raised to an exponent n:

$$\ln A^n = n \ln A \tag{D-13}$$

Rules D-11, D-12, and D-13 can be derived from the definition of the logarithm and the rules governing exponents.

APPENDIX E GEOMETRY AND TRIGONOMETRY

E1 GEOMETRY

ANGLES

Two angles are equal if

1. They are vertical angles (see Figure E1).
2. Their sides are parallel (see Figure E2).
3. Their sides are mutually perpendicular (see Figure E3).

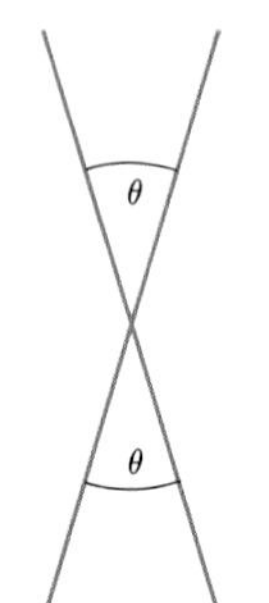
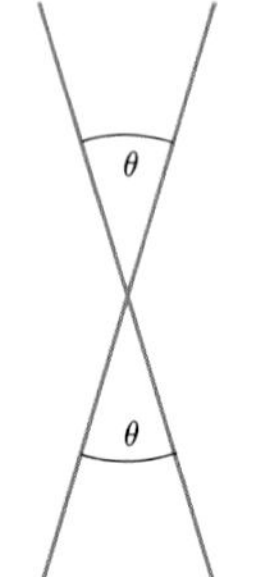

Figure E1

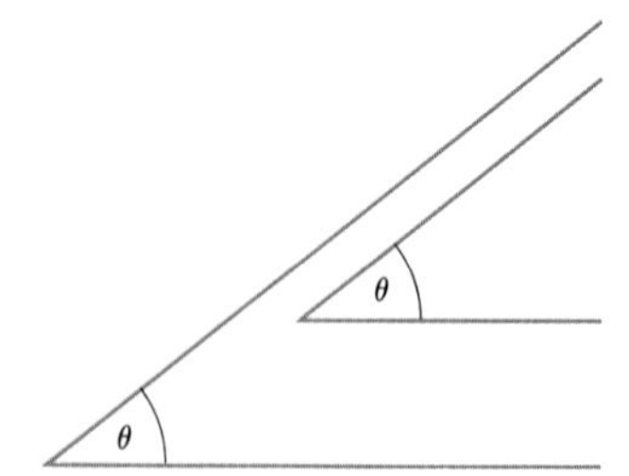

Figure E2

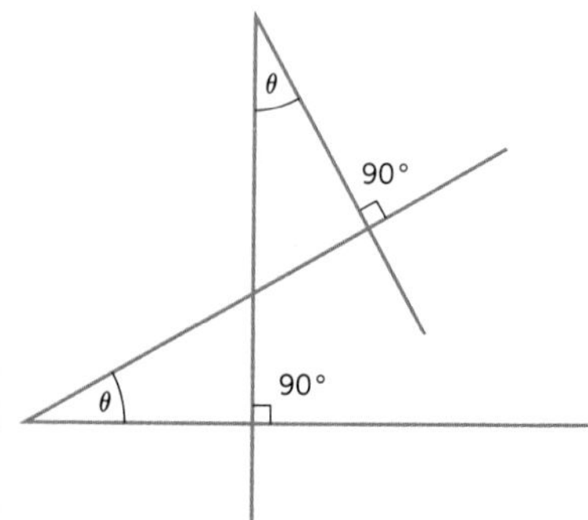

Figure E3

TRIANGLES

1. The ***sum of the angles*** of any triangle is 180° (see Figure E4).
2. A ***right triangle*** has one angle that is 90°.
3. An ***isosceles triangle*** has two sides that are equal.
4. An ***equilateral triangle*** has three sides that are equal. Each angle of an equilateral triangle is 60°.
5. Two triangles are ***similar*** if two of their angles are equal (see Figure E5). The corresponding sides of similar triangles are proportional to each other:

$$\frac{a_1}{a_2} = \frac{b_1}{b_2} = \frac{c_1}{c_2}$$

6. Two similar triangles are ***congruent*** if they can be placed on top of one another to make an exact fit.

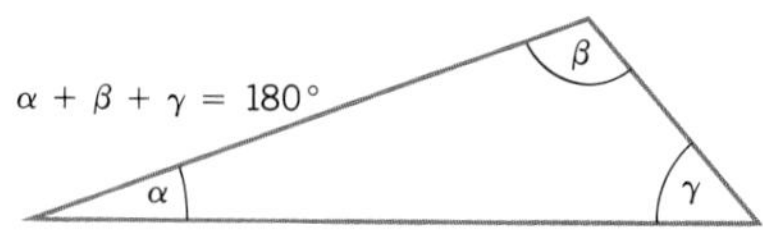

Figure E4

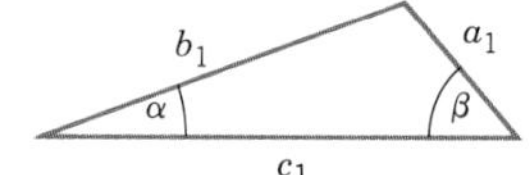

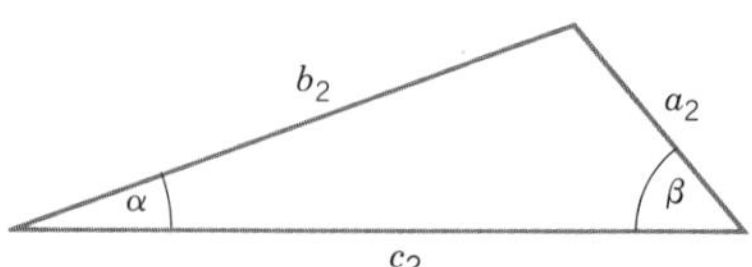

Figure E5

CIRCUMFERENCES, AREAS, AND VOLUMES OF SOME COMMON SHAPES

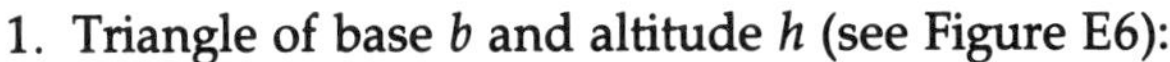

1. Triangle of base b and altitude h (see Figure E6):

$$\text{Area} = \tfrac{1}{2}bh$$

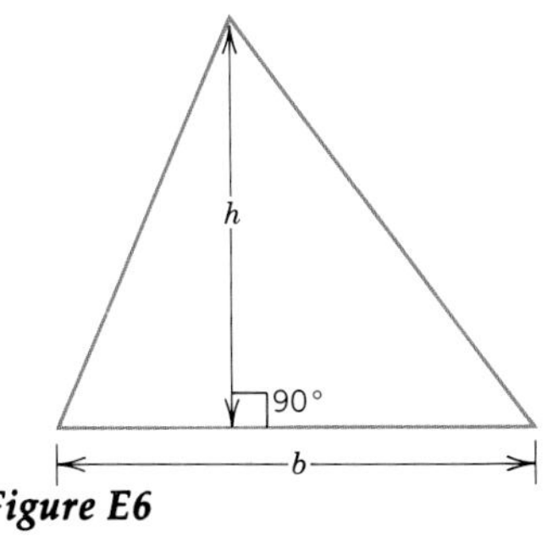

Figure E6

2. Circle of radius r:

$$\text{Circumference} = 2\pi r$$
$$\text{Area} = \pi r^2$$

3. Sphere of radius r:

$$\text{Surface area} = 4\pi r^2$$
$$\text{Volume} = \tfrac{4}{3}\pi r^3$$

4. Right circular cylinder of radius r and height h (see Figure E7):

$$\text{Surface area} = 2\pi r^2 + 2\pi rh, \text{ Volume} = \pi r^2 h.$$

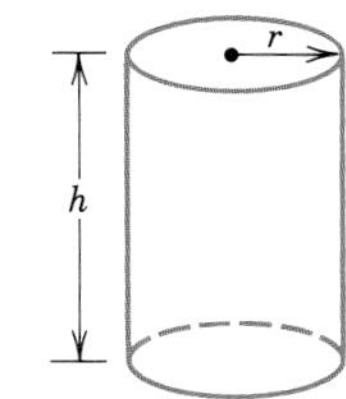

Figure E7

E2 TRIGONOMETRY

BASIC TRIGONOMETRIC FUNCTIONS

1. For a right triangle, the sine, cosine, and tangent of an angle θ are defined as follows (see Figure E8):

$$\sin\theta = \frac{\text{Side opposite }\theta}{\text{Hypotenuse}} = \frac{h_o}{h}$$

$$\cos\theta = \frac{\text{Side adjacent to }\theta}{\text{Hypotenuse}} = \frac{h_a}{h}$$

$$\tan\theta = \frac{\text{Side opposite }\theta}{\text{Side adjacent to }\theta} = \frac{h_o}{h_a}$$

2. The secant ($\sec\theta$), cosecant ($\csc\theta$), and cotangent ($\cot\theta$) of an angle θ are defined as follows:

$$\sec\theta = \frac{1}{\cos\theta} \qquad \csc\theta = \frac{1}{\sin\theta} \qquad \cot\theta = \frac{1}{\tan\theta}$$

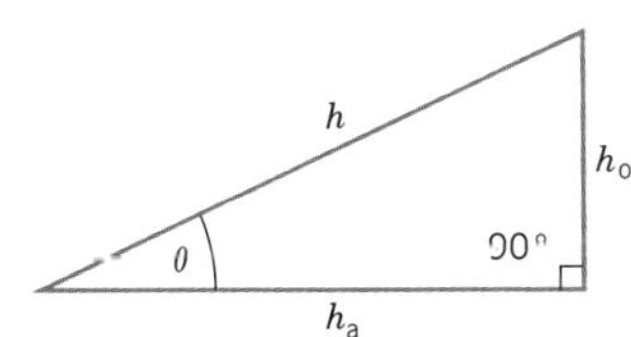

Figure E8

TRIANGLES AND TRIGONOMETRY

1. The ***Pythagorean theorem*** states that the square of the hypotenuse of a right triangle is equal to the sum of the squares of the other two sides (see Figure E8):

$$h^2 = h_o^2 + h_a^2$$

2. The ***law of cosines*** and the ***law of sines*** apply to any triangle, not just a right triangle, and they relate the angles and the lengths of the sides (see Figure E9):

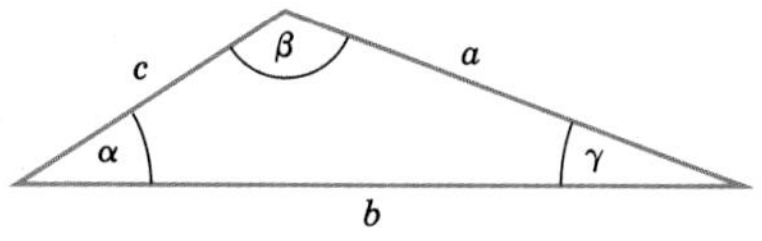

Figure E9

[Law of cosines] $c^2 = a^2 + b^2 - 2ab \cos \gamma$

[Law of sines] $\dfrac{a}{\sin \alpha} = \dfrac{b}{\sin \beta} = \dfrac{c}{\sin \gamma}$

OTHER TRIGONOMETRIC IDENTITIES

1. $\sin(-\theta) = -\sin \theta$
2. $\cos(-\theta) = \cos \theta$
3. $\tan(-\theta) = -\tan \theta$
4. $\sin \theta / \cos \theta = \tan \theta$
5. $\sin^2 \theta + \cos^2 \theta = 1$
6. $\sin(\alpha \pm \beta) = \sin \alpha \cos \beta \pm \cos \alpha \sin \beta$

$$\text{If } \alpha = 90°,\ \sin(90° \pm \beta) = \cos \beta$$
$$\text{If } \alpha = \beta,\ \sin 2\beta = 2 \sin \beta \cos \beta$$

7. $\cos(\alpha \pm \beta) = \cos \alpha \cos \beta \mp \sin \alpha \sin \beta$

$$\text{If } \alpha = 90°,\ \cos(90° \pm \beta) = \mp \sin \beta$$
$$\text{If } \alpha = \beta,\ \cos 2\beta = \cos^2 \beta - \sin^2 \beta = 1 - 2 \sin^2 \beta$$

SELECTED ISOTOPES[a] APPENDIX F

Atomic Number Z	Element	Symbol	Atomic Mass Number A	Atomic Mass u	Percent Abundance, or Decay Mode If Radioactive	Half-life (If Radioactive)
0	(Neutron)	n	1	1.008 665	β^-	10.37 min
1	Hydrogen	H	1	1.007 825	99.985	
	Deuterium	D	2	2.014 102	0.015	
	Tritium	T	3	3.016 050	β^-	12.33 yr
2	Helium	He	3	3.016 030	0.000 138	
			4	4.002 603	$\approx$100	
3	Lithium	Li	6	6.015 121	7.5	
			7	7.016 003	92.5	
4	Beryllium	Be	7	7.016 928	EC, γ	53.29 days
			9	9.012 182	100	
5	Boron	B	10	10.012 937	19.9	
			11	11.009 305	80.1	
6	Carbon	C	11	11.011 432	β^+, EC	20.39 min
			12	12.000 000	98.90	
			13	13.003 355	1.10	
			14	14.003 241	β^-	5730 yr
7	Nitrogen	N	13	13.005 738	β^+, EC	9.965 min
			14	14.003 074	99.634	
			15	15.000 108	0.366	
8	Oxygen	O	15	15.003 065	β^+, EC	122.2 s
			16	15.994 915	99.762	
			18	17.999 160	0.200	
9	Fluorine	F	18	18.000 937	EC, β^+	1.8295 h
			19	18.998 403	100	
10	Neon	Ne	20	19.992 435	90.51	
			22	21.991 383	9.22	
11	Sodium	Na	22	21.994 434	β^+, EC, γ	2.602 yr
			23	22.989 767	100	
			24	23.990 961	β^-, γ	14.659 h
12	Magnesium	Mg	24	23.985 042	78.99	
13	Aluminum	Al	27	26.981 539	100	
14	Silicon	Si	28	27.976 927	92.23	
			31	30.975 362	β^-, γ	2.622 h
15	Phosphorus	P	31	30.973 762	100	
			32	31.973 907	β^-	14.282 days

[a] Data for atomic masses are taken from *Handbook of Chemistry and Physics,* 66th ed., CRC Press, Boca Raton, FL. The masses are those for the neutral atom, including the Z electrons. Data for percent abundance, decay mode, and half-life are taken from E. Browne and R. Firestone, *Table of Radioactive Isotopes,* V. Shirley, Ed., Wiley, New York, 1986. α = alpha particle emission, β^- = negative beta emission, β^+ = positron emission, γ = γ-ray emission, EC = electron capture.

APPENDIX F: Selected Isotopes—*Continued*

Atomic Number Z	Element	Symbol	Atomic Mass Number A	Atomic Mass u	Percent Abundance, or Decay Mode If Radioactive	Half-life (If Radioactive)
16	Sulfur	S	32	31.972 070	95.02	
			35	34.969 031	β^-	87.51 days
17	Chlorine	Cl	35	34.968 852	75.77	
			37	36.965 903	24.23	
18	Argon	Ar	40	39.962 384	99.600	
19	Potassium	K	39	38.963 707	93.2581	
			40	39.963 999	β^-, EC, γ	1.277×10^9 yr
20	Calcium	Ca	40	39.962 591	96.941	
21	Scandium	Sc	45	44.955 910	100	
22	Titanium	Ti	48	47.947 947	73.8	
23	Vanadium	V	51	50.943 962	99.750	
24	Chromium	Cr	52	51.940 509	83.789	
25	Manganese	Mn	55	54.938 047	100	
26	Iron	Fe	56	55.934 939	91.72	
27	Cobalt	Co	59	58.933 198	100	
			60	59.933 819	β^-, γ	5.271 yr
28	Nickel	Ni	58	57.935 346	68.27	
			60	59.930 788	26.10	
29	Copper	Cu	63	62.939 598	69.17	
			65	64.927 793	30.83	
30	Zinc	Zn	64	63.929 145	48.6	
			66	65.926 034	27.9	
31	Gallium	Ga	69	68.925 580	60.1	
32	Germanium	Ge	72	71.922 079	27.4	
			74	73.921 177	36.5	
33	Arsenic	As	75	74.921 594	100	
34	Selenium	Se	80	79.916 520	49.7	
35	Bromine	Br	79	78.918 336	50.69	
36	Krypton	Kr	84	83.911 507	57.0	
			89	88.917 640	β^-, γ	3.16 min
			92	91.926 270	β^-, γ	1.840 s
37	Rubidium	Rb	85	84.911 794	72.165	
38	Strontium	Sr	86	85.909 267	9.86	
			88	87.905 619	82.58	
			90	89.907 738	β^-	28.5 yr
			94	93.915 367	β^-, γ	1.235 s
39	Yttrium	Y	89	88.905 849	100	
40	Zirconium	Zr	90	89.904 703	51.45	
41	Niobium	Nb	93	92.906 377	100	
42	Molybdenum	Mo	98	97.905 406	24.13	
43	Technecium	Tc	98	97.907 215	β^-, γ	4.2×10^6 yr
44	Ruthenium	Ru	102	101.904 348	31.6	
45	Rhodium	Rh	103	102.905 500	100	

APPENDIX F: Selected Isotopes—*Continued*

Atomic Number Z	Element	Symbol	Atomic Mass Number *A*	Atomic Mass *u*	Percent Abundance, or Decay Mode If Radioactive	Half-life (If Radioactive)
46	Palladium	Pd	106	105.903 478	27.33	
47	Silver	Ag	107	106.905 092	51.839	
			109	108.904 757	48.161	
48	Cadmium	Cd	114	113.903 357	28.73	
49	Indium	In	115	114.903 880	95.7; β^-	4.41×10^{14} yr
50	Tin	Sn	120	119.902 200	32.59	
51	Antimony	Sb	121	120.903 821	57.3	
52	Tellurium	Te	130	129.906 229	33.8; β^-	2.5×10^{21} yr
53	Iodine	I	127	126.904 473	100	
			131	130.906 114	β^-, γ	8.040 days
54	Xenon	Xe	132	131.904 144	26.9	
			136	135.907 214	8.9	
			140	139.921 620	β^-, γ	13.6 s
55	Cesium	Cs	133	132.905 429	100	
			134	133.906 696	β^-, EC, γ	2.062 yr
56	Barium	Ba	137	136.905 812	11.23	
			138	137.905 232	71.70	
			141	140.914 363	β^-, γ	18.27 min
57	Lanthanum	La	139	138.906 346	99.91	
58	Cerium	Ce	140	139.905 433	88.48	
59	Praseodymium	Pr	141	140.907 647	100	
60	Neodymium	Nd	142	141.907 719	27.13	
61	Promethium	Pm	145	144.912 743	EC, α, γ	17.7 yr
62	Samarium	Sm	152	151.919 729	26.7	
63	Europium	Eu	153	152.921 225	52.2	
64	Gadolinium	Gd	158	157.924 099	24.84	
65	Terbium	Tb	159	158.925 342	100	
66	Dysprosium	Dy	164	163.929 171	28.2	
67	Holmium	Ho	165	164.930 319	100	
68	Erbium	Er	166	165.930 290	33.6	
69	Thulium	Tm	169	168.934 212	100	
70	Ytterbium	Yb	174	173.938 859	31.8	
71	Lutetium	Lu	175	174.940 770	97.41	
72	Hafnium	Hf	180	179.946 545	35.100	
73	Tantalum	Ta	181	180.947 992	99.988	
74	Tungsten (wolfram)	W	184	183.950 928	30.67	
75	Rhenium	Re	187	186.955 744	62.60; β^-	4.6×10^{10} yr
76	Osmium	Os	191	190.960 920	β^-, γ	15.4 days
			192	191.961 467	41.0	
77	Iridium	Ir	191	190.960 584	37.3	
			193	192.962 917	62.7	

APPENDIX F: Selected Isotopes—*Continued*

Atomic Number Z	Element	Symbol	Atomic Mass Number A	Atomic Mass u	Percent Abundance, or Decay Mode If Radioactive	Half-life (If Radioactive)
78	Platinum	Pt	195	194.964 766	33.8	
79	Gold	Au	197	196.966 543	100	
			198	197.968 217	β^-, γ	2.6935 days
80	Mercury	Hg	202	201.970 617	29.80	
81	Thallium	Tl	205	204.974 401	70.476	
			208	207.981 988	β^-, γ	3.053 min
82	Lead	Pb	206	205.974 440	24.1	
			207	206.975 872	22.1	
			208	207.976 627	52.4	
			210	209.984 163	α, β^-, γ	22.3 yr
			211	210.988 735	β^-, γ	36.1 min
			212	211.991 871	β^-, γ	10.64 h
			214	213.999 798	β^-, γ	26.8 min
83	Bismuth	Bi	209	208.980 374	100	
			211	210.987 255	α, β^-, γ	2.14 min
			212	211.991 255	β^-, α, γ	1.0092 h
84	Polonium	Po	210	209.982 848	α, γ	138.376 days
			212	211.988 842	α, γ	45.1 s
			214	213.995 176	α, γ	163.69 μs
			216	216.001 889	α, γ	150 ms
85	Astatine	At	218	218.008 684	α, β^-	1.6 s
86	Radon	Rn	220	220.011 368	α, γ	55.6 s
			222	222.017 570	α, γ	3.825 days
87	Francium	Fr	223	223.019 733	α, β^-, γ	21.8 min
88	Radium	Ra	224	224.020 186	α, γ	3.66 days
			226	226.025 402	α, γ	1.6×10^3 yr
			228	228.031 064	β^-, γ	5.75 yr
89	Actinium	Ac	227	227.027 750	α, β^-, γ	21.77 yr
			228	228.031 015	β^-, γ	6.13 h
90	Thorium	Th	228	228.028 715	α, γ	1.913 yr
			231	231.036 298	β^-, γ	1.0633 days
			232	232.038 054	100; α, γ	1.405×10^{10} yr
			234	234.043 593	β^-, γ	24.10 days
91	Protactinium	Pa	231	231.035 880	α, γ	3.276×10^4 yr
			234	234.043 303	β^-, γ	6.70 h
			237	237.051 140	β^-, γ	8.7 min
92	Uranium	U	232	232.037 130	α, γ	68.9 yr
			233	233.039 628	α, γ	1.592×10^5 yr
			235	235.043 924	0.7200; α, γ	7.037×10^8 yr
			236	236.045 562	α, γ	2.342×10^7 yr
			238	238.050 784	99.2745; α, γ	4.468×10^9 yr
			239	239.054 289	β^-, γ	23.47 min
93	Neptunium	Np	239	239.052 933	β^-, γ	2.355 days
94	Plutonium	Pu	239	239.052 157	α, γ	2.411×10^4 yr
			242	242.058 737	α, γ	3.763×10^5 yr

APPENDIX F: Selected Isotopes—*Continued*

Atomic Number Z	Element	Symbol	Atomic Mass Number A	Atomic Mass u	Percent Abundance, or Decay Mode If Radioactive	Half-life (If Radioactive)
95	Americium	Am	243	243.061 375	α, γ	7.380×10^3 yr
96	Curium	Cm	245	245.065 483	α, γ	8.5×10^3 yr
97	Berkelium	Bk	247	247.070 300	α, γ	1.38×10^3 yr
98	Californium	Cf	249	249.074 844	α, γ	350.6 yr
99	Einsteinium	Es	254	254.088 019	α, γ, β^-	275.7 days
100	Fermium	Fm	253	253.085 173	EC, α, γ	3.00 days
101	Mendelevium	Md	255	255.091 081	EC, α	27 min
102	Nobelium	No	255	255.093 260	EC, α	3.1 min
103	Lawrencium	Lr	257	257.099 480	α, EC	646 ms
104	Rutherfordium	Rf	261	261.108 690	α	1.08 min
105	Hahnium	Ha	262	262.113 760	α	34 s

ANSWERS TO ODD-NUMBERED PROBLEMS

CHAPTER 1

1. 27.4 km

3. (*a*) 5700 s (*b*) 86 400 s

5. 115 m^3

7. (*a*) correct (*b*) incorrect (*c*) incorrect (*d*) correct (*e*) correct

9. 3.13×10^8 m^3 **11.** 713 m

13. 80.1 km, at 25.9° south of west **15.** 0.707 m

17. 340 m **19.** 35.3°

21. 7.80 km, at 35.0° north of west

23. 200 N, due east or 600 N, due west

25. 5.70×10^2 N, at 33.6° south of west

27. (*a*) 67.9 units, at 45.0° south of west (*b*) 67.9 units, at 45.0° north of west

29. 35 N, at 30° north of west

31. (*a*) 6.00 units (*b*) 36.9° north of west (*c*) 6.00 units (*d*) 36.9° south of west

33. (*a*) −128 m (*b*) 68.1 m

35. (*a*) 93 N (*b*) 36° above $+x$ axis

37. (*a*) 41.2 m/s (*b*) 52.8 m/s

39. (*a*) 58° above $+x$ axis (*b*) 240 N

41. $F_x = 322$ N, $F_y = 209$ N, $F_z = 279$ N

43. (*a*) The magnitude of **A** + **B** is larger than the magnitude of **A** or the magnitude of **B**. The angle made by **A** + **B** with the $+x$ axis is larger than θ. (*b*) 105 m, at 68.2° above the $+x$ axis.

45. 4.80 km, at 24.0° north of east

47. 268 km, at 38.5° north of east

49. (*a*) 10.4 units (*b*) 12.0 units

51. (*a*) 167 units (*b*) 89 units **53.** 6.88 km, 26.9°

55. 3.73 m

57. 222 m, 235.8° counterclockwise from the $+x$ axis

59. 3.00 m, at 42.8° above the negative x axis

61. (*a*) 2.2 km (*b*) 36° east of north

63. (*a*) 6.43 m, at 70.0° above the negative x axis (*b*) 7.66 m, at 20.0° below the negative x axis

65. 1975 m, at 11.3° north of west

67. x component = −288 units, where the minus sign denotes the direction due west. y component = +156 units, where the plus sign denotes due north.

CHAPTER 2

1. 7.2 m/s **3.** 54 m **5.** 60 s **7.** 28 m

9. 34 km/h, due north **11.** 5.60 m/s^2, due east

13. 3.0 m/s^2 **15.** 8.0 m/s **17.** 6.0×10^1 m/s

19. 13 m/s **21.** 4.5 m **23.** 2.0×10^1 m

25. 5.0×10^1 m **27.** (*a*) 2.0×10^1 s (*b*) 580 m

29. 96.9 m/s **31.** (*a*) 5.26 m/s (*b*) 233 m

33. The answer is a derivation. **35.** 1.7 s

37. 5.7 m **39.** 15.4 m/s **41.** 91.5 m/s

43. 6.12 s **45.** 47 m/s **47.** 39 m/s, upward

49. 5.0 m **51.** 0.40 s **53.** 0.932 m/s

55. 2.5 m **57.** 21.8 m/s

59. $v_A = 24$ km/h, $v_B = 8$ km/h, $v_C = -5$ km/h

61. (*a*) v_A is positive, v_B is zero, v_C is negative, v_D is positive (*b*) $v_A = 5.0$ km/h, $v_B = 0$, $v_C = -3.8$ km/h, $v_D = 2.5$ km/h

63. The answer is in graphical form.

65. $d_1 = 0.018$ m, $d_2 = 0.071$ m, $d_3 = 0.16$ m

67. 2 m **69.** (*a*) 4.0 s (*b*) 4.0 s **71.** 6.6

73. (*a*) 26 700 m (*b*) 6.74 m/s **75.** 35 m/s

77. 2.0×10^1 m **79.** 0.81 km **81.** 3.3 m/s^2

CHAPTER 3

1. 41 m **3.** 8600 m **5.** 16.9 m/s **7.** 4.7 m/s

9. 6.0×10^1 m/s

11. (*a*) 2.99×10^4 m/s (*b*) 2.69×10^4 m/s

13. 5.92×10^3 m/s **15.** 4.42 s

17. (*a*) 6.0×10^1 m (*b*) 290 m

19. The ball arrives first by 0.70 s.

21. (*a*) 1.78 s (*b*) 20.8 m/s **23.** −0.82 m

25. (*a*) 7.82 s (*b*) 135 m/s **27.** 2.40 m

29. 48 m **31.** 14.9 m **33.** 8.8 m/s

35. (*a*) 5.5 s (*b*) 42 m/s

37. (*a*) 1380 m (*b*) 66.0° below the horizontal

39. 23 m/s at 31° above the horizontal **41.** 4

43. 5.8 m/s **45.** 0.141° and 89.859°

47. The answer is a proof. **49.** 24 m/s, due south

51. 32 s **53.** (*a*) 2.0×10^3 s (*b*) 1.8 km

55. 8.92° south of west **57.** 7.50 m/s, 23.1° north of east

59. 0.82 **61.** 27.4 m **63.** 515 m/s **65.** 30.0 m

67. (*a*) 23.5 m/s (*b*) 53.5° above the horizontal

69. 12 km. Air resistance slows down the bullet.

71. 0.15 m/s **73.** The answer is a proof.

75. 2.44×10^6 m **77.** 4.2 m/s

CHAPTER 4

1. 6500 N, due north **3.** 36 500 N **5.** 3560 N

7. 2.61×10^5 N **9.** 35.4 m/s

11. −1.20 m/s^2 (to the left)

13. $a_{left} = 14.4\ m/s^2$ at 56.3° above the $+x$ axis, $a_{center} = 18.5\ m/s^2$ at 27.2° above the $+x$ axis, $a_{right} = 20.0\ m/s^2$ along the $+x$ axis
15. $1.3\ m/s^2$, due south
17. 51.8 N at 88° above the $+x$ axis **19.** $1.20\ m/s^2$, Up
21. 4.77×10^{20} N
23. (*a*) 1130 N, His mass is 115 kg.
(*b*) 0 N, His mass is 115 kg.
25. 22.2 N **27.** 130 N **29.** 0.997 **31.** 241 hours
33. 3.3×10^5 m/s **35.** 2.83 **37.** 730 N **39.** 645 N
41. The block will move, $3.72\ m/s^2$ **43.** $1.57\ m/s^2$
45. (*a*) $-0.980\ m/s^2$ (*b*) 29.5 m **47.** 0.235
49. 58.8 N
51. (*a*) 1400 N (*b*) 2400 N **53.** 9.70 N
55. (*a*) 57 600 N (*b*) 20 600 N **57.** 0.444
59. 251 N **61.** 7260 N **63.** 286 N
65. (*a*) 79.0 N (*b*) 219 N **67.** 220 N, 64° north of east
69. (*a*) 1.0×10^2 N (*b*) 38 N
71. (*a*) 914 N (*b*) 822 N **73.** 612 N
75. (*a*) 1610 N (*b*) 2640 N
77. 36 200 N and 9570 N
79. (*a*) $5.97\ m/s^2$ (*b*) 101 N **81.** 1850 N
83. $7.1\ m/s^2$ **85.** $3.9\ m/s^2$ **87.** 49.1 m **89.** 0.50
91. (*a*) 1.00×10^2 N (*b*) 41.6 N
93. (*a*) 1710 N (*b*) No net force is required.
95. (*a*) 7.40×10^5 N (*b*) 1.67×10^9 N
97. 1.00×10^2 N at 53.1° south of east
99. (*a*) 45 N (*b*) 37 N **101.** 2.64×10^6 m
103. (*a*) $0.0640\ m/s^2$ (*b*) 1.58×10^4 km
105. (*a*) $3.56\ m/s^2$ (*b*) 281 N **107.** 0.200
109. 8.7 s **111.** 0.265 m **113.** 16.3 N
115. 38 m **117.** 0.665

CHAPTER 5

1. 0.016 s **3.** 160 s
5. (*a*) $5.0 \times 10^1\ m/s^2$ (*b*) 0 (*c*) $2.0 \times 10^1\ m/s^2$
7. $0.79\ m/s^2$ **9.** 2.2 **11.** 0.68 m/s **13.** 606 N
15. 3.3 m **17.** 2.6 m
19. (*a*) 5.1×10^3 N (*b*) 15 m/s **21.** 23° **23.** 39°
25. 2.12×10^6 N **27.** 4.20×10^4 m/s **29.** 262 m
31. 1.43×10^4 s **33.** 7910 m/s **35.** 0.125
37. 4720 m **39.** (*a*) 5.7 m/s (*b*) 4.8 m/s
41. 2.9×10^4 N **43.** 9.65 m/s **45.** 332 m
47. (*a*) 3.00×10^4 m/s (*b*) 2.02×10^{30} kg
49. 27.5 days **51.** 66 m **53.** $3.22\ m/s^2$
55. (*a*) 247 m/s (*b*) $(8.30 \times 10^2)g$
57. (*a*) The centripetal force is provided by the normal force exerted by the wall on the riders. (*b*) 1670 N (*c*) 0.323

CHAPTER 6

1. 42.8° **3.** 2.2×10^2 N
5. (*a*) 2980 J (*b*) 3290 J
7. 1.88×10^7 J **9.** −23 J **11.** 203 N **13.** 3180 J
15. (*a*) 913 J (*b*) 913 J
17. (*a*) 2.35×10^{11} J (*b*) -2.35×10^{11} J
19. 9×10^3 m/s **21.** 1.0×10^3 N
23. (*a*) 25 m (*b*) 5.6 s **25.** 7.47×10^4 J
27. 2.39×10^5 J **29.** (*a*) −44 J (*b*) 44 J
31. 5.24×10^5 J **33.** 6.6 m/s **35.** 1.4 m
37. -1.0×10^3 J, Mechanical energy is not conserved, because of the frictional force that acts on the sled.
39. 126 m/s **41.** 42 m **43.** 18 m **45.** 40.8 kg
47. −4.4 J **49.** -4.51×10^4 J
51. (*a*) −1086 J
(*b*) −2.01 m, The skater is below the starting point.
53. 4130 N **55.** 31° **57.** 127 m
59. (*a*) 1.1×10^2 W (*b*) 5.3×10^2 W **61.** 73.5 s
63. (*a*) 990 W (*b*) 240 W
65. (*a*) 4.00×10^3 W (*b*) 3.35×10^4 W **67.** 43.2 J
69. (*a*) 3.0 J (*b*) 0 J (*c*) −9.0 J
71. (*a*) 1.50×10^2 J (*b*) 7.07 m/s **73.** 2.2×10^3 J
75. 6.4×10^5 J **77.** 55.8 m/s
79. The work done by the force **P** and the kinetic frictional force are, respectively, 1.0×10^3 J and −940 J. The normal force and the weight of the box do no work.
81. 3.1 m/s **83.** 0.327 m

CHAPTER 7

1. 14 N·s
3. (*a*) 5.80×10^2 kg·m/s, 1.16×10^3 J
(*b*) 1.40×10^2 kg·m/s, 1.16×10^3 J
5. 4.24 kg·m/s, at 45.0° south of east
7. (*a*) 1.3 kg·m/s, parallel to the velocity of the ball
(*b*) 220 N, parallel to the velocity of the ball
9. (*a*) 6.58×10^4 kg·m/s, at −31.4° or 31.4° south of east.
(*b*) 1.04×10^6 J
11. 9.00×10^2 N **13.** $v_2 = 9.28$ m/s, $v_3 = 19.4$ m/s
15. -1.5×10^{-4} m/s **17.** $+7.1 \times 10^5$ m/s
19. (*a*) Bonzo, since he has the smaller recoil velocity.
(*b*) 1.7
21. The answer is a proof. **23.** 0.577 **25.** 0.097 m
27. 84 kg **29.** (*a*) +5.25 m/s (*b*) 1.41 m **31.** 1/3
33. (*a*) $v_{f1} = -0.400$ m/s, $v_{f2} = +1.60$ m/s
(*b*) $v_f = +0.800$ m/s
35. 0.19 s **37.** (*a*) 73.0° (*b*) 4.28 m/s
39. The 7.0 m/s ball has a final velocity of −4.0 m/s, opposite to its original direction. The 4.0 m/s ball has a final velocity of +7.0 m/s, opposite to its original direction.

41. 12 000 N **43.** 1.3×10^2 N **45.** 0.104 m
47. -8.7 kg·m/s **49.** (*a*) $+6.71$ m/s (*b*) 0.559
51. $+69$ N **53.** -2.6×10^3 N or 2.6×10^3 N, due west
55. (*a*) 2.13 m/s, upward (*b*) 0.231 m
57. 2.54×10^{-3} kg
59. (*a*) 5.56 m/s
(*b*) -2.83 m/s (1.50-kg ball), $+2.73$ m/s (4.60-kg ball)
(*c*) 0.409 m (1.50-kg ball), 0.380 m (4.60-kg ball)

CHAPTER 8

1. (*a*) 0.79 rad (*b*) 3.1 rad (*c*) 6.3 rad **3.** 4π rad
5. 13 rad/s **7.** (*a*) 9.4×10^{-4} s (*b*) 0.13 m
9. 157.3 rad/s **11.** 0.28 rad **13.** 2.00×10^{-2} s
15. 6.05 m **17.** (*a*) 4.00×10^1 rad (*b*) 15.0 rad/s
19. 4.50 s **21.** 1.6×10^3 rad/s^2
23. (*a*) 4.60×10^3 rad (*b*) 2.00×10^2 rad/s^2
25. 12.5 s **27.** 3.20×10^4 rad **29.** 7.37 s
31. (*a*) 0.332 rad/s (*b*) 128 m **33.** 466 m/s
35. 3.50×10^4 m/s **37.** (*a*) 1.3 m/s (*b*) 8.4 rev/s
39. 18.6° **41.** 380 m/s^2 **43.** 16
45. (*a*) 2.5 m/s^2 (*b*) 3.1 m/s^2 **47.** 1.88 m
49. 0.213 s **51.** 1450 rad
53. (*a*) 7.50 rad/s
(*b*) -1.73×10^{-3} rad/s^2, The angular velocity is decreasing.
55. 208 rad **57.** 11.8 rad **59.** 13.5 m/s
61. (*a*) 9.00 m/s^2
(*b*) The direction is radially inward, toward the center of the track.
63. (*a*) 15.0 rad/s (*b*) 2.55 m/s
65. (*a*) 54.0 rad/s (*b*) 486 rad **67.** 0.34 m
69. 28.0 rad/s **71.** $1/\sqrt{3}$ **73.** 1/2

CHAPTER 9

1. 0.25 m **3.** 1.5 N·m **5.** 1.3 **7.** 379 N
9. $-Fd$ **11.** 0.90 m **13.** 5.77×10^{-2} m
15. 1.20×10^3 N
17. $T = 56.4$ N (down), $F = 70.6$ N (up)
19. 3.40×10^3 N (front wheel), 2.28×10^3 N (rear wheel)
21. (*a*) 7.40×10^2 N (*b*) 0.851 m **23.** 228 N
25. 32° **27.** 17.5 N **29.** 1.1 N·m
31. 0.032 kg·m^2
33. (*a*) 0.131 kg·m^2 (*b*) 3.63×10^{-4} kg·m^2
(*c*) 0.149 kg·m^2
35. -3.32×10^{-2} N·m **37.** 2.4 rev **39.** 460 N
41. 0.34 N **43.** 1.33 m, 1.67 m **45.** 2.6×10^{29} J
47. 432 J **49.** 755 m **51.** 2/5 **53.** 1.26 m/s
55. 1.08×10^7 m **57.** 0.037 rad/s **59.** 0.34 rad/s
61. 8% increase **63.** 0.17 m **65.** 4.2 N·m
67. 1.83 rad/s
69. (*a*) $I_1 = 0.160$ kg·m^2, $I_2 = 2.50$ kg·m^2, $I_3 = 1.22$ kg·m^2 (*b*) 3.88 kg·m^2
(*c*) The smallest mass does not necessarily contribute the smallest amount to the total moment of inertia.
71. (*a*) $v_{T1} = 12.0$ m/s, $v_{T2} = 9.00$ m/s, $v_{T3} = 18.0$ m/s
(*b*) 1080 J (*c*) 60.0 kg·m^2 (*d*) 1080 J
73. (*a*) 0.500 rad/s, disk rotates opposite to person's motion
(*b*) 2.99 s
75. 2.12 s

CHAPTER 10

1. 3.7×10^{-5} m **3.** (*a*) 4.0×10^{-5} m (*b*) 1.0×10^{-5} m
5. 7.0×10^7 Pa
7. (*a*) 1.6×10^8 N/m^2 (*b*) 5.3×10^{-4}
(*c*) 3.0×10^{11} N/m^2
9. 1.6×10^5 N **11.** 1.2×10^3 N
13. (*a*) 6.3×10^{-2} m (*b*) 7.3×10^{-2} m
15. (*a*) 1.8×10^{-7} m (*b*) 1.0×10^{-6} m
17. 2.1×10^{-5} m
19. (*a*) 710 N/m^2 (*b*) 3.5×10^{-8} (*c*) 3.5×10^{-10} m
21. 3.6×10^{-4} m (tungsten), 6.4×10^{-4} m (steel)
23. (*a*) 7.44 N (*b*) 7.44 N **25.** 237 N **27.** 640 N/m
29. 0.012 m **31.** 0.311 m
33. (*a*) 1.00×10^3 N/m (*b*) 0.340
35. (*a*) 0.407 m (*b*) 397 N **37.** 3.5×10^4 N/m
39. 4.4 rad/s **41.** 0.069 m **43.** 9.93×10^{-3} m
45. 4.3 kg **47.** (*a*) 46.9 J (*b*) 55.9 m/s
49. 7.18×10^{-2} m **51.** 18 m/s
53.

h (meters)	KE	PE (gravity)	PE (elastic)	E
0	0	0	8.76 J	8.76 J
0.100	0.75 J	1.96 J	6.05 J	8.76 J
0.200	1.00 J	3.92 J	3.84 J	8.76 J
0.300	0.75 J	5.88 J	2.13 J	8.76 J
0.400	0	7.84 J	0.92 J	8.76 J

55. 16 m/s **57.** 1.3 m/s
59. 0.556 m/s (29.2-kg block), 1.11 m/s (14.6-kg block)
61. 2.08 m/s **63.** 0.99 m **65.** 6.0 m/s^2
67. 12.5 m **69.** 0.816
71. (*a*) 0.152 m (*b*) 1.85 m/s^2 **73.** 6.55 m/s
75. (*a*) 140 m/s (*b*) 1.7×10^{15} m/s^2 **77.** 260 m
79. (*a*) 3.59×10^{-2} m and 4.24 Hz
(*b*) 5.08×10^{-2} m and 4.24 Hz
81. 0.240 m **83.** 33.4 m/s **85.** 4.0×10^{-5} m
87. 66 Hz

CHAPTER 11

1. 6.8×10^9 N **3.** 3400 N **5.** 3.08×10^{-6} m^3
7. 0.13 m **9.** 63% **11.** 4.0×10^3 N

13. 4.33×10^6 Pa **15.** 29 bricks
17. (*a*) 0.097 m (*b*) 14 J **19.** 5.75×10^4 Pa
21. 1.21×10^5 Pa **23.** 10.3 m
25. (*a*) 9.98×10^4 Pa
(*b*) Worse, because atmospheric pressure is less, so there is less force to push the water up.
27. (*a*) 2.45×10^5 Pa (*b*) 1.73×10^5 Pa
29. 31.3 rad/s
31. 1.46 m from bottom on the left, and 0.54 m from bottom on right.
33. 2.1×10^4 N **35.** 3.8×10^5 N
37. 8.50×10^5 N·m **39.** 108 N **41.** 3.3×10^{-3} m^3
43. 2.2×10^3 kg **45.** 2.04×10^{-3} m^3
47. (*a*) 6.9 N (*b*) 7.0×10^{-4} m^3 **49.** 77.5%
51. 1120 N **53.** 4.5×10^{-5} kg/s **55.** 2.1 m/s
57. (*a*) 2.2×10^5 kg (*b*) 16 m/s
59. 46 Pa; air enters at B and exits at A
61. (*a*) 150 Pa
(*b*) The pressure inside the roof is greater than the pressure outside, so there is a net outward force.
63. (*a*) 22.2 m/s (*b*) 3.47×10^5 Pa
65. (*a*) 7.00 m/s (*b*) 1.21×10^5 Pa **67.** 38 m/s
69. 24 m/s **71.** 21 m/s **73.** 7.78 m/s **75.** 20 Pa
77. (*a*) 1.01×10^5 Pa (*b*) 1.19×10^5 Pa
79. 1.52×10^{-3} Pa·s
81. (*a*) 2.8×10^{-5} N (*b*) 1.0×10^1 m/s **83.** 20.6 m
85. 4.89 m **87.** 6.6×10^6 kg **89.** 1.91 m/s
91. (*a*) 7.5×10^{-2} N (*b*) 6.5 N
93. (*a*) 762 mm (*b*) 758 mm
95. (*a*) 1.6×10^{-4} m^3/s (*b*) 2.0×10^1 m/s
97. 0.11 m **99.** 1.57 kg **101.** 60.3%
103. (*a*) The answer is a proof. (*b*) The answer is a proof.

CHAPTER 12

1. (*a*) −12.2 °C and 41 °C (*b*) 261 K and 314 K
3. −196 °C **5.** (*a*) 270 F° (*b*) 1.50×10^2 K
7. 18 °C **9.** −40.0 °C **11.** 0.084 m **13.** 1500 m
15. 8.0×10^{-4} **17.** -2.82×10^{-4} m **19.** 38 °C
21. 1.001 38 s
23. (*a*) Since the ruler shrinks as the temperature decreases, a tension is needed to stretch the ruler.
(*b*) 9.6×10^7 N/m^2
25. 0.6 **27.** 0.23 m^3 **29.** 18 **31.** 1.2×10^{-5} m^3
33. 1 penny **35.** 8.9×10^{-8} m^3 **37.** 45 atm
39. 4.0×10^{-4} (C°)$^{-1}$ **41.** 6.9 **43.** 36.2 °C
45. 940 °C **47.** 27.05 °C **49.** $230
51. 2.3×10^5 N **53.** 3100 W **55.** 1.3×10^5 J
57. 3.9×10^5 J **59.** 9.3 kg **61.** 3.9×10^{-3} kg
63. (*a*) 1.49×10^{14} J (*b*) 993 **65.** 0.223
67. 1.0×10^{-3} kg **69.** 1 ton **71.** 0.237 kg
73. Vapor phase **75.** 123 °C **77.** 40% **79.** 76%
81. 28% **83.** 13 °C **85.** 9.49×10^{-3} kg **87.** 5.8 m
89. 44.0 °C **91.** 25% **93.** 44 h **95.** 930 W
97. 1.36×10^{-2} C° **99.** 996 N

CHAPTER 13

1. 12 J
3. 25 °C if heat is gained from the outside or 17 °C if heat is lost to the outside
5. 2.0×10^{-3} m **7.** 0.078 m **9.** 17 **11.** 283 °C
13. (*a*) 130 °C (*b*) 830 J (*c*) 237 °C
15. (*a*) 100.78 °C (*b*) 107.2 °C **17.** 0.74 **19.** 0.3
21. 320 K **23.** 1.19 **25.** 532 K
27. (*a*) 9×10^6 m (*b*) 7×10^8 kg/m^3
29. (*a*) 18 J (*b*) 0.22 °C **31.** 14 h **33.** 5.0×10^1 J/s
35. Aluminum, copper, or silver **37.** 1.67
39. 21 °C at the plasterboard-brick interface and 18 °C at the brick-wood interface
41. (*a*) 2.0 (*b*) 0.61

CHAPTER 14

1. (*a*) 16.043 u (*b*) 2.6641×10^{-26} kg
3. (*a*) 893.51 u (*b*) 2680 g **5.** 1.18×10^{25} **7.** 141 g
9. 2.6×10^{-10} m **11.** 304 K **13.** 1050 K
15. 67.0 m^3
17. (*a*) 8×10^{-18} mol/m^3 (*b*) 2×10^{-16} Pa
19. 0.140 m **21.** 1.02 **23.** 1.5×10^5 g **25.** 0.205
27. 5.61×10^5 Pa **29.** 590 K **31.** 5.1×10^{-16} kg
33. (*a*) 1.23 (*b*) 1.11 **35.** 1.01×10^4 K **37.** 399 J
39. 9400 m
41. (*a*) 120 N (*b*) 120 N (*c*) 4.0×10^5 Pa
43. 1.34×10^{-7} kg
45. (*a*) 5.00×10^{-13} kg/s (*b*) 5.8×10^{-3} kg/m^3
47. 2.3×10^6 s **49.** 0.55 kg **51.** 2.2 kg/m^3
53. 19 K **55.** 0.27 m **57.** 1.0×10^5
59. 4.0×10^1 Pa **61.** 307 K

CHAPTER 15

1. -4.3×10^5 J, heat is given off
3. −3700 J, heat flows out of the gas
5. −261 J, work is done on the system
7. −185 J, so that work is done on the system and the magnitude of the work is 185 J
9. 3.0×10^5 Pa **11.** 13 000 J **13.** 1.2×10^7 Pa
15. 2.9 J **17.** 1500 J **19.** 4.99×10^{-6}
21. 1.29×10^4 J **23.** 434 K **25.** 3.17 **27.** 0.59 J
29. 8.49×10^5 Pa **31.** $T_f = 327$ K, $V_f = 0.132$ m^3
33. 2400 J **35.** 310 J **37.** 27 K **39.** 12.5 s

41. (*a*) 60.0% (*b*) 40.0% **43.** 1.82×10^4 J
45. 0.631 **47.** 0.84 **49.** $e_1 + e_2 - e_1e_2$ **51.** 3000 J
53. (*a*) 1260 K (*b*) 1.74×10^4 J **55.** 1150 K
57. (*a*) 0.360 (*b*) 1.3×10^{13} J **59.** 2.5×10^3 J
61. 5.7 C° **63.** 11 **65.** 5.86×10^5 J
67. (*a*) 1050 J (*b*) 2.99 **69.** 1.4 **71.** 279 s
73. 1260 K **75.** 1.19×10^4 J/K
77. (*a*) Reversible (*b*) -125 J/K
79. (*a*) $+541$ J/K
(*b*) The entropy of the universe increases.
81. -210 K, a decrease in temperature
83. -2.0×10^{-3} m³, a decrease in volume
85. (*a*) $+3680$ J/K (*b*) $+18\,200$ J/K
(*c*) The conversion of a liquid to a vapor creates more disorder.
87. 9.03 **89.** -3100 J
91. (*a*) 0.50 (*b*) 1.5 (*c*) 0.67
93. $T_iV_i^{\gamma-1} = T_fV_f^{\gamma-1}$ **95.** 15.0 K
97. $e_A = 5/9$, $e_B = 1/3$

CHAPTER 16

1. 0.083 Hz **3.** (*a*) 0.909 Hz (*b*) 12.7 m/s
5. 0.25 m
7. (*a*) 0.022 s (*b*) 0.49 m
(*c*) The amplitude cannot be determined.
9. 78 cm **11.** 3.4 m/s² **13.** 600 m/s **15.** 1
17. 8.67×10^{-3} kg/m **19.** 1.2 m/s²
21. 3.3×10^{-4} kg/m **23.** 79 m/s
25. $y = (0.15 \text{ m}) \sin (160t - 22x)$
27. (*a*) $+x$ direction (*b*) -0.080 m **29.** 2.5 N
31. (*a*) 20 m and 0.02 m, respectively
(*b*) The statement is not correct.
33. 2.8×10^{-4} s **35.** 6.7×10^{-4} s **37.** 3.2×10^{-6} m
39. (*a*) The order of arrival is: steel wave first, water wave second, and air wave third.
(*b*) 0.059 s and 0.339 s
41. 690 rad/s **43.** 2 **45.** 61 m **47.** 2.55 m
49. 0.404 m **51.** 6.7×10^{-9} W **53.** 190 m
55. 3.9×10^{-3} W **57.** 1.98% **59.** 1.0×10^4 W/m²
61. 2.6 s **63.** 25 **65.** 1000 **67.** 120 dB
69. 56.2 W **71.** 6.3 **73.** 2.39 dB **75.** 786 Hz
77. 838 Hz **79.** 1.21 **81.** 1350 Hz **83.** 31 m/s
85. 209 m **87.** 6300 **89.** 28.9 K **91.** 5.9 m/s
93. 1.3 **95.** (*a*) 7.87×10^{-3} s (*b*) 4.12 **97.** 860 Hz
99. -6.0 dB **101.** $m_1 = 28.6$ kg, $m_2 = 14.3$ kg
103. 21.2 m/s **105.** 76.8 dB
107. (*a*) 99.5 m (*b*) 0.9%

CHAPTER 17

1. The answer is a series of drawings.
3. The answer is a series of drawings.
5. 107 Hz **7.** 107 Hz **9.** 3.89 m **11.** 9700 Hz
13. 0.37 m **15.** 3.4° **17.** (*a*) 50 kHz (*b*) 90 kHz
19. 4 Hz **21.** 8 Hz **23.** 14 Hz **25.** 88 m/s
27. 5.30 **29.** 171 N **31.** 4 **33.** 0.077 m
35. 0.49 **37.** 12 Hz **39.** 8.6 m and 8.6×10^{-3} m
41. 1.96 m **43.** 512 m/s **45.** 602 Hz
47. $f_n = n\left(\dfrac{v}{2L}\right)$, where $n = 1, 2, 3, \ldots$
49. 1.68×10^5 Pa **51.** 110 Hz
53. Destructive interference **55.** 680 m/s **57.** $n = 65$
59. 1.2 m/s **61.** (*a*) 0.0352 m (*b*) 0.0249 m

CHAPTER 18

1. 2.4×10^{13} **3.** 3.1×10^{13} **5.** (*a*) $+4q$ (*b*) $+4q$
7. 26 N
9. (*a*) 0.83 N
(*b*) Attractive, since the spheres have charges of different polarities.
11. 120 N
13. 0.38 N, at 49° below the $-x$ axis (third quadrant)
15. 17.3 N, at 38.7° south of east
17. (*a*) 4.56×10^{-8} C (*b*) 3.25×10^{-6} kg
19. 2.6×10^{12}
21. (*a*) They have the same polarity, either both positive or both negative.
(*b*) 8.4×10^{-6} C
23. 4.20×10^{-25}
25. (*a*) 8.2×10^{-8} C (*b*) 8.2×10^{-3} N
27. 1.1 N, due north **29.** The answer is a drawing.
31. 2.5×10^{-5} C, The charge is negative.
33. (*a*) 4.10×10^{12} N/C, directed radially outward
(*b*) 6.56×10^{-7} N, directed radially inward
(*c*) 4.37×10^6 m/s
35. 7.1×10^{-2} m²
37. (*a*) The excess charge is positive, because the electric field is directed upward.
(*b*) 3.46×10^7 protons
39. 2.8×10^5 N/C, in the $-x$ direction
41. $q_1 = 0.716\,q$, $q_2 = 0.0895q$ **43.** 0.364
45. (*a*) The charges have the same polarities (*b*) 27.0
47. 3.25×10^{-8} C **49.** (*a*) 230 N/C (*b*) 880 N/C
51. -5.0×10^5 C
53. (*a*) The electric flux through the face lying in the x, z plane at $y = 0$ is -6.0×10^1 N·m²/C. The electric flux through the face parallel to the x, z plane and located at $y = 0.20$ m is 6.0×10^1 N·m²/C. The electric flux through the remaining four faces is zero.
(*b*) Flux = 0
55. The answer is a proof. **57.** 0.14 N **59.** 18 N/C
61. 6.5×10^3 N/C, directed toward the ground

63. (*a*) 1.4 N, in the $-x$ direction
(*b*) 1.4 N, in the $+x$ direction
65. (*a*) The charges have the same polarity, either both positive or both negative.
(*b*) 1.7×10^{-16} C
67. 5.53×10^{-2} m **69.** -3.3×10^{-6} C
71. (*a*) 15.4° (*b*) 0.813 N

CHAPTER 19

1. 1.5×10^{-20} J **3.** $+3.2 \times 10^{-3}$ J **5.** 9.4×10^{7} m/s
7. 7.0×10^{1} hp **9.** 339 V **11.** 19 m/s
13. $+3.6 \times 10^{-9}$ C **15.** No net work is required.
17. 4.7×10^{-2} J
19. A distance $d/3$ to the left and a distance d to the right of the negative charge.
21. 0.0342 m **23.** -4.8×10^{-6} C
25. (*a*) 8.68×10^{-18} J (*b*) 54.3 eV
27. (*a*) $-2q/3$ (*b*) $-2q$ **29.** 18 000 V **31.** 1.1 m
33. 9.0×10^{3} V **35.** 1700 V/m, to the left
37. (*a*) 2.54 m and 3.54 m
(*b*) 3.54 V (for $r = 2.54$ m), 2.54 V (for $r = 3.54$ m)
39. 12 V **41.** 5.3 **43.** 7.0×10^{13} **45.** 8.0×10^{-5} C
47. (*a*) 41 J (*b*) 8200 W **49.** 3.98×10^{-15} C
51. 7.7 V decrease **53.** The answer is a proof.
55. 0.43 m **57.** 2×10^{-8} F **59.** 1.0×10^{-4} m
61. 0.707 **63.** (*a*) 159 V (*b*) 125 V (*c*) 135 V
65. 2.77×10^{6} m/s

CHAPTER 20

1. 3.5×10^{4} A **3.** 8.6 A **5.** 82 Ω
7. (*a*) 7.9×10^{5} C (*b*) 350 A
9. (*a*) 4.7×10^{13} (*b*) 17 C° **11.** 0.0050 $(\mathrm{C}°)^{-1}$
13. 2.2×10^{-2} Ω/m **15.** 11 m **17.** 189 Ω
19. 360 °C **21.** 39.5 °C **23.** 1.0×10^{3} W
25. 1.2 A **27.** (*a*) 13 V (*b*) 37 V **29.** 970 kg
31. 1.77 A **33.** 1.8 A
35. (*a*) 1300 W (*b*) 480 W (*c*) 1.4 **37.** 1.8 h
39. 32 A **41.** (*a*) 145 Ω (*b*) 74 V **43.** 21.7 V
45. 32 Ω **47.** 3 resistors **49.** 5.3 Ω
51. (*a*) 5.0 A (*b*) 540 W **53.** 64 resistors **55.** 160 Ω
57. (*a*) 3.6 Ω
(*b*) 33 A, The breaker will open.
59. 3.58×10^{-8} m² **61.** 1.0×10^{2} Ω **63.** 6.76 Ω
65. 4.6 Ω **67.** 3.6 W
69. 0.750 A (20.0-Ω resistor), 2.11 A (9.00-Ω resistor)
71. 10.9 V **73.** 540 A **75.** (*a*) 8.99 V (*b*) 2.5 Ω
77. (*a*) 0.38 A (*b*) 2.0×10^{1} V (*c*) *B* **79.** 33 A
81. 0.75 V, The left end is at the higher potential.
83. 1.43 V, Point *A* is at the higher potential.
85. 3430 Ω **87.** 0.667 Ω **89.** (*a*) 30.0 V (*b*) 28.1 V
91. (*a*) 12.0 μF (*b*) 3.0×10^{-4} C **93.** 4.68×10^{-6} C
95. 1.88 **97.** 55 V **99.** 1.80 V **101.** 0.15 s
103. 0.0089 s
105. (*a*) 3.60×10^{-4} C (*b*) 8.00×10^{-5} C
107. 0.025 A **109.** 0.21 A **111.** 25 V
113. 6.00 Ω, 3.67 Ω, 2.75 Ω, 2.20 Ω, 1.50 Ω, 1.33 Ω, 0.833 Ω, 0.545 Ω
115. $92 **117.** (*a*) 1.2 Ω (*b*) 110 V **119.** 50 m
121. C_0

CHAPTER 21

1. 8.1×10^{-5} T
3. (*a*) 0.11 T, upward away from the earth's surface
(*b*) 0.11 T, downward toward the earth's surface
5. 4.1×10^{-3} m/s **7.** 51° **9.** 4.1×10^{-12} N
11. (*a*) The answer is a drawing. (*b*) 2.6×10^{-2} T
(*c*) 2.7×10^{16} m/s²
13. (*a*) 7.2×10^{6} m/s (*b*) 3.5×10^{-13} N
15. 3.27×10^{-25} kg **17.** 1830 **19.** 1.6×10^{-2} m
21. 1.50×10^{7} C/kg **23.** 256
25. (*a*) 4.3×10^{-3} N (*b*) 1.6×10^{-4} C
27. 9.6×10^{4} m/s **29.** 8.1 N
31. 0.96 N (top and bottom sides), 0 (left and right sides)
33. 57.6° **35.** 0.325 kg
37. (*a*) Counterclockwise (*b*) 1.1×10^{-2} m
39. (*a*) 37° (*b*) 0.49 N **41.** 8.2×10^{-27} N·m
43. 0.023 N·m **45.** $\tau_{max} = IL^2B/(16\,N)$ **47.** 2.6 N
49. 1.2×10^{-5} A·m² **51.** 8.0×10^{-5} T
53. 1.3×10^{-2} T **55.** (*a*) Down (*b*) 3.1×10^{-4} T
57. 8.6 A **59.** 1.5×10^{-4} N **61.** 1.04×10^{-2} T
63. I_3 is directed out of the paper. $I_3/I = 2$
65. (*a*) 1.6×10^{-5} T (*b*) 4.0×10^{-6} T
67. (*a*) The answer is a proof. (*b*) The answer is a proof.
69. 75.1° or 105° **71.** (*a*) 24 A·m² (*b*) 4.8 N·m
73. 1.9×10^{-4} N·m **75.** 1.5×10^{-8} s
77. 140 V/m, directed toward the bottom of the page
79. 1.13 **81.** 1/3 **83.** 1.13×10^{-4} T, downward

CHAPTER 22

1. 690 m/s **3.** (*a*) The driver's side. (*b*) 2.0 m
5. (rod A) emf = 0, (rod B) emf = 1.6 V and end 2 is positive, (rod C) emf = 0
7. 42 A **9.** (*a*) 3.3 m/s (*b*) 4.6 N
11. (*a*) 5.6×10^{-3} Wb (*b*) 0 Wb
13. 0.031 Wb (*xy* plane), 0 Wb (*xz* plane)
15. (*a*) 7.3×10^{-4} Wb (*b*) 0 Wb (*c*) 4.7×10^{-3} Wb
17. 7.7×10^{-3} Wb **19.** 2.8×10^{-3} V
21. 8.6×10^{-5} T **23.** 4.8 s **25.** 0.459 T
27. 0.045 V **29.** (*a*) left-to-right (*b*) right-to-left

31. (*a*) Clockwise for *A* and counterclockwise for *B*.
(*b*) Counterclockwise for *A* and clockwise for *B*.

33. Clockwise at position 1 and clockwise at position 2.

35. (*a*) Retardation occurs because of an induced current that is counterclockwise in the ring while the magnet approaches from above and clockwise while the magnet moves away from the bottom.
(*b*) No induced current can exist in the cut ring.

37. The maximum emf is 180 V, and the period is 0.42 s.

39. 4.0×10^5 **41.** 0.150 m **43.** 230 V **45.** 96 V

47. 1.4 V **49.** 220 **51.** 0.32 A **53.** 1.5×10^9 J

55. $L = 2\pi\mu_0 n^2 AR$ **57.** 1.6×10^{-5} A

59. (*a*) step-up (*b*) 55 : 1 **61.** step-down, 1 : 12

63. (*a*) 6.0×10^2 V (*b*) step-up, 5 : 1

65. (*a*) 7.0×10^5 W (*b*) 7.0×10^1 W **67.** 0.15 T/s

69. right-to-left (part b), left-to-right (part c) **71.** 170 V

73. 36 V **75.** 3.6×10^9 N/C

77. (*a*) 3.6×10^{-3} V
(*b*) 2.0×10^{-3} m²/s, The area must shrink.

79. $M = \mu_0 \pi N_1 N_2 R_2^2/(2R_1)$

CHAPTER 23

1. 126 Hz **3.** (*a*) 13.0 μF (*b*) 0.61 A

5. 5.00×10^{-2} s **7.** 0.28 A, increase

9. 6.8×10^{-2} H **11.** 160 Ω **13.** 310 Hz

15. 0.17 V **17.** 84.0 V **19.** 0.819

21. (*a*) 0.925 A (*b*) 31.7°

23. (*a*) *R* and *C* (*b*) $R = 49.7\ \Omega$ and $X_C = 185\ \Omega$ **25.** 3

27. 1100 Hz **29.** 3250 Hz

31. (*a*) 0.50 A (*b*) 0.34 A (*c*) 0.704 H

33. 0.81 W **35.** 123 V **37.** 8

39. $X_L/R = 4/\sqrt{3}$, $X_C/R = 1/\sqrt{3}$ **41.** 3.4×10^{-6} F

43. (*a*) 2.57 A (*b*) 2.47 A (*c*) 0.10 A, decrease

45. 1.7 **47.** 113 Ω and +21°

49. (*a*) 2.94×10^{-3} H (*b*) 4.84 Ω (*c*) 0.164

51. 3.11×10^3 Hz and 7.50×10^3 Hz

CHAPTER 24

1. 70.9 min **3.** 6.8×10^{-11} F

5. (The answers are in graphical form.)

7. 1.4×10^{17} Hz **9.** 1.25 m

11. (*a*) 1090 kHz (*b*) AM radio wave **13.** 3.7×10^4

15. 1.2×10^3 m **17.** (*a*) 1.28 s (*b*) 1.9×10^2 s

19. 8.8 y **21.** 8.75×10^5 **23.** 3.8×10^2 W/m²

25. 7×10^{-2} N/C

27. (*a*) 6.81×10^5 N/C (*b*) 2.27×10^{-3} T

29. The answer is a proof.

31. (*a*) 27.3 N/C (*b*) 6.60×10^{-9} J/m³ (*c*) 1.98 W/m²

33. 1.2×10^3 W/m²

35. (*a*) 0.550 W/m² (*b*) 3.7×10^{-2} W/m² **37.** 71.6°

39. 14 W/m² **41.** 20 **43.** (*a*) 3 (*b*) 9

45. 5.00×10^2 s **47.** 602 W/m²

49. (*a*) 0.82 (*b*) 0.18

51. The angle must be increased by 9.3°.

53. (*a*) 2.7×10^{-5} N (*b*) 3.3×10^{-9} N

55. $E_{rms} = 5.30$ N/C, $B_{rms} = 1.77 \times 10^{-8}$ T

CHAPTER 25

1. 55° **3.** (*a*) 0.80 m (*b*) 0.75 m **5.** 10°

7. (*a*) 62.1° (*b*) 80.0° **9.** 1.73

11. Image distance is 3.0×10^1 cm behind the mirror. Image height is 5.0 cm, and the image is upright.

13. Image distance is 16.7 cm behind the mirror. Image height is 6.67 cm, and the image is upright.

15. 1.3 cm **17.** −8.7 cm

19. (*a*) −11.7 cm (*b*) −14.5 cm

21. (*a*) +62 cm (*b*) +0.35 (*c*) Upright (*d*) Smaller

23. (*a*) −1.8 m (*b*) Virtual (*c*) 0.19 m

25. (*a*) *R* (*b*) −1 (*c*) Inverted

27. (*a*) 6.0×10^1 cm (*b*) The answer is a drawing.

29. +42.0 cm

31. (*a*) The answer is a proof.
(*b*) The answer is a proof.

33. (*a*) $+2.0 \times 10^2$ cm (*b*) −6.3 cm
(*c*) Since h_i is negative, the image is inverted. Therefore, to produce a normal-appearing image, the slide should be oriented upside down.

35. (*a*) The answer is a drawing.
(*b*) The answer is a drawing.

37. +22 cm **39.** +46.7 cm

41. 80.0 cm, toward the lens **43.** 33.7°

CHAPTER 26

1. 2.00×10^8 m/s **3.** 0.5411 **5.** 2.0×10^{-11} s

7. 1.40 **9.** (*a*) 43° (*b*) 31° **11.** 40.8°

13. 1.92×10^8 m/s **15.** 2.05 cm **17.** 12.1 m

19. 2.1 cm **21.** 3.23 cm

23. (*a*) 11.3 cm (*b*) 33.8 cm (*c*) Larger **25.** 1.54

27. (*a*) Point *B* (*b*) Point *A* **29.** Floating on the surface.

31. 1.45 **33.** 1.52 **35.** (*a*) 53.12° (*b*) 52.62°

37. 33.67° **39.** The answer is a proof. **41.** 0.35°

43. 29.5° **45.** 21.5° **47.** 18 cm **49.** 61.1 mm

51. (*a*) −24 cm (*b*) 3.0 **53.** 0.23 cm

55. (*a*) −16 cm (*b*) 0.57 (*c*) Virtual (*d*) Upright
(*e*) Reduced

57. (*a*) 6.74×10^{-7} m² (*b*) 7.86×10^5 W/m²

59. (*a*) 4.52×10^{-4} m (*b*) 6.10×10^{-2} m **61.** 8.0 cm

63. (*a*) -0.7 m
(*b*) Virtual, because a plane mirror always produces a virtual image.
65. (*a*) 4.00 cm to the left of the diverging lens. (*b*) -0.167 (*c*) Virtual (*d*) Inverted (*e*) Smaller
67. (*a*) 10.4 cm to the left of the diverging lens, $m = 0.52$
(*b*) The answer is a confirmation.
69. (*a*) 18.1 cm (*b*) Real (*c*) Inverted **71.** -220 cm
73. 86.9 cm **75.** (*a*) Hyperopic (*b*) 2.8 diopters
77. 26.9 cm **79.** (*a*) 31.3 cm (*b*) 2.43 m **81.** 3.7
83. 0.13 m **85.** (*a*) 2.0×10^1 diopters (*b*) 5.6
87. 0.64 rad **89.** 3.0 cm **91.** 0.435 cm
93. 0.261 cm **95.** -2.0×10^2
97. (*a*) The 1.3-diopter lens. (*b*) 0.86 m (*c*) -8.5
99. (*a*) -194 (*b*) -7.8×10^{-5} m (*c*) 1.94×10^6 m
101. (*a*) -24 cm (*b*) 1.6 **103.** (*a*) 33° (*b*) 32°
105. 38.0 cm **107.** 0.600 **109.** 21.7°
111. (*a*) 0.775 m (*b*) 0.780 m
113. (*a*) Converging (*b*) Farsighted (*c*) 96.3 cm
115. (*a*) -10.0 cm (*b*) 0.500
117. (*a*) Converging (*b*) $d_i = 2f$ and $d_o = 2f$
119. 0.333 m **121.** -181
123. (*a*) 22.4 cm (*b*) 28.4 cm

CHAPTER 27

1. Constructive **3.** 660 nm **5.** 6.0×10^{-5} m
7. 0.43 cm **9.** 0.0309 m **11.** 207 nm **13.** Blue
15. 612 nm **17.** (*a*) 1.0×10^{-7} m (*b*) 2.1×10^{-7} m
19. 440 nm **21.** (*a*) 0.21° (*b*) 22°
23. (*a*) 4.2×10^{-5} degrees (*b*) 59° **25.** 490 nm
27. 0.012 m **29.** 0.447 **31.** (*a*) 2.59 m (*b*) 1.58 m
33. 14 000 m **35.** (*a*) 7.6 m (*b*) 56 m
37. 3.2×10^3 m **39.** 3.2 s **41.** (*a*) 37° (*b*) 22°
43. 5.90×10^{-7} m **45.** 4.0×10^{-6} m **47.** 24°
49. (*a*) 0.36 m (*b*) 1.1 m **51.** (*a*) 24° (*b*) 39°
53. $\theta_1 = 0.68°$, $\theta_2 = 1.4°$, $\theta_3 = 2.0°$
55. (*a*) 150 nm (*b*) 210 nm **57.** 0.52 m **59.** 3/4
61. (*a*) 7.9° (violet), 13° (red) (*b*) 16° (violet), 26° (red) (*c*) 24° (violet), 41° (red)
(*d*) The second and third orders overlap.
63. (*a*) 2 (*b*) $m_B = 4$, $m_A = 2$ and $m_B = 6$, $m_A = 3$

CHAPTER 28

1. 2.4×10^8 m/s **3.** 54.7 s **5.** 0.15 rad/s
7. 5.57 s **9.** 530 m **11.** 5.4×10^2 y
13. 3.0×10^6 m **15.** Length = 3400 m, width = 1500 m
17. 3.0 m × 1.3 m **19.** 3.0×10^9 kg·m/s
21. 1.80×10^8 m/s **23.** -2.0 m/s **25.** 3.3×10^5
27. 3.72×10^{-12} kg, the water **29.** (*a*) 1.0 (*b*) 6.6
31. (*a*) 3.8×10^{-12} J (*b*) 0.9998*c* **33.** *c* **35.** $-0.23c$
37. 0.94*c* **39.** 1.1 kg **41.** 2.60×10^8 m/s
43. 5.0×10^{-13} J **45.** $-0.406c$
47. (*a*) 4.3 *y* (*b*) The twin who travels at a speed of 0.500*c*.

CHAPTER 29

1. 310 nm **3.** 7.7×10^{29} photons/s **5.** 6.3 eV
7. 1.26 eV **9.** 73 **11.** 9.32×10^5 m/s
13. (*a*) 2.1×10^{24} (*b*) 32 **15.** 5.1×10^{-33} kg·m/s
17. 3.6×10^{-9} m **19.** 73° **21.** 9.50×10^{-17} m
23. 2.6×10^{-28} m **25.** 1.41×10^3 m/s
27. 2.5×10^3 m/s **29.** 1.2×10^{-36} m
31. 8.5×10^{-12} m **33.** 8.0×10^{-6} m/s
35. 5.3×10^{-20} kg·m/s **37.** 37 J·s
39. (*a*) 1.2×10^{-6} m (*b*) 1.2×10^{-9} m
41. (*a*) 4.50×10^{-36} m/s (*b*) 7.05×10^{27} years
43. (*a*) 1.97×10^{-24} kg·m/s (*b*) 1.95×10^{-24} kg·m/s
45. (*a*) 1.10×10^3 m/s (*b*) 5.93×10^5 m/s **47.** 42.8

CHAPTER 30

1. 3 km **3.** 6.88×10^{-15} m **5.** 7.3×10^{-14} m
7. 434.1 nm
9. (*a*) 7458 nm (*b*) 2279 nm (*c*) Infrared **11.** 5
13. 4.41×10^{-10} m **15.** (*a*) 16 (*b*) 1/4
17. KE = +4.90 eV, EPE = −9.80 eV **19.** $n_i = 6$, $n_f = 2$
21. 3920 lines/cm **23.** −3, −2, −1, 0, +1, +2, +3
25. 2, 3, 4, 5
27. (*a*) $L = h/(2\pi)$ (Bohr model), $L = 0$ (quantum mechanics)
(*b*) $L = 2h/(2\pi)$ (Bohr model)
$\left.\begin{array}{l}\text{for } \ell = 0,\ L = 0 \\ \text{for } \ell = 1,\ L = \sqrt{2}h/(2\pi)\end{array}\right\}$ (quantum mechanics)
29. 26.6° **31.** $1s^2\,2s^2\,2p^6\,3s^2\,3p^6\,4s^2\,3d^{10}\,4p^6\,5s^2\,4d^1$
33. (*a*) Not allowed (*b*) Allowed (*c*) Not allowed
(*d*) Allowed (*e*) Not allowed
35. 2.0×10^{-11} m **37.** 7.230×10^{-11} m
39. 1.11×10^{-10} m **41.** 21 600 V
43. A calculation is required to verify the laser wavelength of 633 nm.
45. 5.2×10^{17} **47.** $1s^2\,2s^2\,2p^6\,3s^2\,3p^6\,4s^2\,3d^{10}\,4p^3$
49. 54.4 eV **51.** 1.08×10^{-14} J or 6.75×10^4 eV
53. 3 **55.** 4 **57.** $6 \le n_i \le 19$ **59.** 30.39 nm

CHAPTER 31

1. (*a*) X = Pt, 117 neutrons (*b*) X = S, 16 neutrons
(*c*) X = Cu, 34 neutrons (*d*) X = B, 6 neutrons
(*e*) X = Pu, 145 neutrons
3. A calculation is required to verify an average nuclear mass of 35.45 u.
5. 35.2 **7.** Carbon-12 contains 3.9×10^{21} more atoms.

9. 1.14 **11.** (*a*) 0.555 357 u (*b*) 9.2217×10^{-28} kg
13. 39.2 MeV
15. (*a*) 1.858 968 u (*b*) 1732 MeV
(*c*) 7.66 MeV/nucleon
17. 2.017 330 u
19. (*a*) $^{235}_{92}U \rightarrow {}^{231}_{90}Th + {}^{4}_{2}He$ (*b*) $^{239}_{94}Pu \rightarrow {}^{235}_{92}U + {}^{4}_{2}He$
21. $^{60}_{27}Co \rightarrow {}^{60}_{28}Ni + {}_{-1}^{0}e$ **23.** 0.019 MeV
25. (*a*) $^{18}_{9}F \rightarrow {}^{18}_{8}O + {}_{+1}^{0}e$ (*b*) $^{15}_{8}O \rightarrow {}^{15}_{7}N + {}_{+1}^{0}e$
27. $^{212}_{84}Po$ **29.** 1.82 MeV **31.** 3.0 days
33. 8.00 days **35.** 379 years
37. 146 disintegrations/min **39.** 0.838 **41.** 7.23 days
43. 13 000 years
45. (*a*) 0.999 (*b*) 1.36×10^{-9} (*c*) 0.755
47. (*a*) 3.0×10^{4} years (*b*) 3.3×10^{4} years
49. 6900 years, maximum error is 900 years
51. (*a*) 8 protons, 10 neutrons (*b*) 50 protons, 70 neutrons
53. 19.9 **55.** 90.9% **57.** 4 782 969 electrons
59. 1.59×10^{7} m/s

CHAPTER 32

1. 0.26 R **3.** 50 rd **5.** 2.8×10^{-3} J
7. 2.4×10^{4} rem **9.** 3.01×10^{8} rd **11.** 30
13. (*a*) $^{27}_{13}Al$ (n, p) $^{27}_{12}Mg$ (*b*) $^{40}_{18}Ar$ (α, p) $^{43}_{19}K$
15. (*a*) Unknown = proton, $^{1}_{1}H$
(*b*) Unknown = alpha particle, $^{4}_{2}He$
(*c*) Unknown = lithium, $^{6}_{3}Li$ (*d*) Unknown = nitrogen, $^{14}_{7}N$
(*e*) Unknown = manganese, $^{56}_{25}Mn$
17. 13.6 MeV **19.** 2 neutrons **21.** 9×10^{-4}
23. 3.9×10^{18} fissions/s **25.** 2.7×10^{6} kg
27. 1.8×10^{4} kg/s **29.** 1200 kg **31.** 3.3 MeV
33. 1.0 gal
35. A calculation is required to verify that the proton-proton cycle generates about 25 MeV of energy.
37. The three possibilities are: (1) *u, d, s*; (2) *u, d, b*; (3) *u, s, b*
39. 140.9 MeV
41. (*a*) 1.9×10^{-20} kg·m/s (*b*) 3.6×10^{-14} m
43. 3.7×10^{-6} J **45.** Oxygen, $Z = 8$, $A = 16$
47. 9.6×10^{8} W **49.** 7.4×10^{9} kg **51.** 160 MeV

PHOTO CREDITS

TABLE OF CONTENTS Page vii: Greg Sams/Science Photo Library/Gamma Liaison. Page viii: Sygma. Page ix: Photo by Al Seib, courtesy of Cirque du Soleil. Page x: Geco UK/Science Photo Library/Photo Researchers. Page xi: Courtesy NASA. Page xii: Johnny Autery. Page xiii: Don Smetzer/Tony Stone Worldwide. Page xiv: Dave Reede/Westlight. Page xv: (top) Robin Moyer/Gamma Liaison; (bottom) E.R. Degginger/Animals Animals. Page xvi: Science Photo Library/Photo Researchers.

PART 1 *Opener:* Darrell Jones/AllStock, Inc.

CHAPTER 1 *Opener:* Photo by Murray Close; reproduced with permission of Universal City Studios and Amblin Entertainment. *Page 3:* William Sterne/William Sterne Photography. Figures 1.1 and 1.2: Courtesy Bureau International des Poids et Mesures, France. Figure 1.3: Courtesy National Institute of Standards and Technology. *Page 6:* UPI/Bettmann Newsphotos. *Page 7:* Pitkin Studio. *Page 8:* Sinclair Stammers/Photo Researchers *Page 9:* Gregory Sams/Science Photo Library/Photo Researchers. *Page 13:* Dave Black/Allsport. *Page 19:* Ron Sanford/f/STOP Pictures.

CHAPTER 2 *Opener:* Robert Holland/The Image Bank. *Page 31 (top):* Tony Duffy/Allsport. Figure 2.2: © Jeanne Drake. *Page 32:* Dale E. Boyer/Photo Researchers. *Page 34:* Sylvain Cazenave/Agence Vandystadt/Allsport. *Page 37:* Focus on Sports. *Page 38:* John Nicholson/Allsport. *Page 40:* Todd Buchanan/Matrix International, Inc. *Page 42:* Courtesy NASA. *Page 48:* Marc Romanelli/The Image Bank.

CHAPTER 3 *Opener:* Stephen Dalton/Animals Animals. *Page 67:* Renee Lynn/AllStock, Inc. *Page 69:* Agence Vandystadt/Photo Researchers. *Page 72:* William R. Sallaz/Duomo Photography, Inc. *Page 76:* Courtesy Jerry Ohlinger's Movie Material Store.

CHAPTER 4 *Opener:* David Madison/Duomo Photography, Inc. Figure 4.2: Focus on Sports. *Page 96:* Ronnie Kaufman/The Stock Market. *Page 98:* Mark Junak/Tony Stone Worldwide. *Page 99:* Courtesy Earth Satellite Corporation. *Page 102:* Rick Ridgeway/Adventure Photo. Figure 4.18a: Lester Sloan/Woodfin Camp & Associates. *Page 104 (bottom):* Marc Romanelli/The Image Bank. Figure 4.19: Rich Chisholm/The Stock Market. *Page 108:* Robert Noonan. *Page 109:* Kennan Ward/Adventure Photo. *Page 111:* Henri Bureau/Sygma. *Page 121:* Lawrence Migdale/Photo Researchers.

CHAPTER 5 *Opener:* Steven E. Sutton/Duomo Photography, Inc. Figure 5.6: Courtesy Ringling Brothers and Barnum & Bailey Circus. *Page 146:* Courtesy NASA. *Page 150:* Agence Vandystadt/Photo Researchers. *Page 151:* Mitchell Funk/The Image Bank.

CHAPTER 6 *Opener:* Tony Stone Worldwide. *Page 162:* Lew Long/The Stock Market. *Page 169:* David Woods/The Stock Market. *Page 174:* Allsport. Figure 6.17: Bruce Barthel/The Stock Market. Figure 6.19: Courtesy Daniel L. Feicht, The Amazement Park, Cedar Point. Figure 6.22a: Zimmerman/FPG International. Figure 6.24a: Ron Chapple/FPG International. *Page 195:* Louis Psihoyos/Matrix.

CHAPTER 7 *Opener:* Stuart McCall/Tony Stone Worldwide. *Page 198:* Photo Researchers. Figure 7.7: Photofest. *Page 206:* Brian Drake/Allsport. *Page 209:* Jane Faircloth/Transparencies, Inc. Figure 7.12a: Courtesy NASA. *Page 212:* Adam J. Stoltman/Duomo Photography, Inc.

CHAPTER 8 *Opener:* Stephen Marks/Stockphotos. Figure 8.5c: Roger Ressmeyer/Starlight. *Page 223:* Focus on Sports. Figure 8.7a: Courtesy Jerry Ohlinger's Movie Material Store. *Page 226:* Dominique Sarraute/The Image Bank. *Page 231:* Bob Daemmrich/The Image Works.

CHAPTER 9 *Opener:* Courtesy Ringling Brothers and Barnum & Bailey Circus. *Page 247:* Zefa Mathis/The Stock Market. *Page 251:* Al Seib Photograph, courtesy of Cirque du Soleil. Figure 9.13a: AP/Wide World Photos. Figure 9.18: Bernard Asset/Allsport. *Page 264:* Science Photo Library/Photo Researchers. *Page 268:* Tim Defrisco/Allsport. *Page 270:* Paul Miller.

CHAPTER 10 *Opener:* Simon Ward/Allsport. *Page 280:* Ben Rose/The Image Bank. Figure 10.4: Courtesy Ringling Brothers and Barnum & Bailey Circus. Page 295: Ken Levine/Allsport. Figure 10.30: Robert Mathena/Fundamental Photographs. Figure 10.34: Everett Johnson/Leo de Wys, Inc.

CHAPTER 11 *Opener:* D. Klein/Gamma Liaison. Figure 11.4: Michael Melford. Figure 11.5: Alan Carey/Photo Researchers. Figure 11.7a: Jim Richardson/Westlight. *Page 319 (center):* Norbert Wu/Peter Arnold. Figure 11.17: *(top left)* Phil Degginger; *(top right)* Aram Gesar/The Image Bank; *(center)* Phil Prosen/The Image Bank. Figure 11.23a: Rick Rickman/Duomo Photography, Inc. Figure 11.25: Philippe Blondel/Allsport. Figure 11.28b: Courtesy General Motors Corporation. *Page 333:* Larry Dale Gordon/The Image Bank. Figure 11.36b: Gary Gladstone/The Image Bank. Figure 11.37: David Madison/Duomo Photography, Inc. *Page 359b:* Frank Fisher/Gamma Liaison.

PART 2 *Opener:* Roger Ressmeyer/Starlight.

CHAPTER 12 *Opener:* Brian Bailey/Tony Stone Worldwide. *Figure 12.5b:* Courtesy Omega Engineering, Inc. Figure 12.6: Science Photo Library/Photo Researchers. Figure 12.7: Courtesy Dr. Sumio Uematsu, Johns Hopkins Hospital. Figure 12.8: Courtesy NASA. Figure 12.12a: AP/Wide World Photos. Figure 12.12b: Bob Daemmrich/The Image Works. *Page 376:* Tony Stone Worldwide. *Page 379:* Robert Semeniuk/The Stock Market. *Page 382:* Orion/Sipa Press. Figure 12.32: Mike Coltman/Planet Earth Pictures.

CHAPTER 13 *Opener:* Wayne Lynch/Masterfile. Figure 13.1: John David Fleck/Gamma Liaison. Figure 13.5: Gerard Fritz/Tony Stone Worldwide. *Page 402:* Christel Hesse/Tony Stone Worldwide. Figure 13.12: Wendell Metzen/Southern Stock Photos. Figure 13.16a: Gary Milburn/Tom Stack & Associates. Figure 13.16b: Frans Lanting/Minden Pictures, Inc. Figure 13.17: Courtesy NASA. *Page 421:* D. and S. Heaton/West Light.

CHAPTER 14 *Opener:* Stephen Frink/AllStock, Inc. *Page 424:* Ann Purcell/Photo Researchers. Figure 14.2: Courtesy Smithsonian Institution. Figure 14.4: Ken Karp. Figure 14.6: Courtesy Richard Zare, Stanford University. Figure 14.8a: Chris McLaughlin/The Stock Market. *Page 435:* Donovan Reese/Tony Stone Worldwide. Figure 14.13a: Al Satterwhite/The Image Bank. *Page 440:* Andy Washnik.

CHAPTER 15 *Opener:* Jim Nilsen/AllStock, Inc. Figure 15.1: Agence Vandystadt/Allsport. *Page 459:* Courtesy General Motors. *Page 470:* Laurence Hughes/The Image Bank. *Page 481A:* Ted Soqui/Sygma. *Page 454:* Geco UK/Photo Researchers.

PART 3 *Opener:* Murray & Associates/Tony Stone Worldwide.

CHAPTER 16 *Opener:* Van Nostrand/Photo Researchers. *Page 485:* Sylvain Cazenave/Allsport. Figure 16.15: Stephen Dalton/Animals Animals. *Page 492:* E.R. Degginger/Bruce Coleman, Inc. *Page 496:* Susan Copen Oken/Dot, Inc. *Page 499:* Ron Garrison photo, courtesy of The Zoological Society of San Diego. Figure 16.28: *(left)* G. Marche/FPG International; *(right)* Howard Sochurek/The Stock Market.

CHAPTER 17 *Opener:* Mark E. Gibson/The Stock Market. Figure 17.10: Andy Washnik. Figure 17.15: Courtesy Sony Corporation. Figure 17.18: From Physical Science Study Committee, PHYSICS, D.C. Heath, Lexington, Mass., 1960. *Page 539:* Sherry Rayne Barnett/Michael Ochs Archives, Venice, CA. *Page 547B:* Gary Gladstone/The Image Bank.

PART 4 *Opener:* Peter Menzel.

CHAPTER 18 *Opener:* Peter Menzel. *Page 553:* Leif Skoogfors/Woodfin Camp & Associates. *Page 556:* Courtesy BMW of North America, Inc. *Page 560:* Johnny Autery. *Page 571:* Image generated by Brian Rasnow, Christopher Assad and Mark Nelson, California Institute of Technology. Photographed by Mark Nelson/Peter Arnold.

CHAPTER 19 *Opener:* Jeff Smith/The Image Bank. *Page 587:* Shahn Kermani/Gamma Liaison. *Page 588:* Alex Kerstitch/Bruce Coleman, Inc. *Page 590:* Rick Blumberg/Stock, Boston. *Page 601:* Matthew McVay/SABA. *Page 603:* M. Hans/Vandystadt/Photo Researchers. *Page 613:* Marty Snyderman/Planet Earth Pictures.

CHAPTER 20 *Opener:* Toyohiro Yamada/FPG International. *Page 617:* Don Smetzer/Tony Stone Worldwide. Figure 20.5: Courtesy Apple Computers. *Page 619:* Mark Antman/The Image Works. *Page 622:* Courtesy Kinetic Corporation. *Page 624:* Dave Reede/West Light.

CHAPTER 21 *Opener:* Pekka Parviainen/Photo Researchers. Figure 21.4b: Andy Washnik. Figure 21.11a: Lawrence Berkeley Laboratory/Photo Researchers. Figure 21.19a: Dennis Budd Gray. Figure 21.37: Simon Fraser/Science Photo Library/Photo Researchers. Figure 21.38: Howard Sochurek/The Stock Market. *Page 689:* Courtesy Dr. Pepper Co. and Young and Rubicam. Figure 21.44a: Courtesy Deutsche Bundesbahn/Transrapid. *Page 703:* Don Kincaid.

CHAPTER 22 *Opener:* Fotex/Shooting Star. *Page 722:* Novosti Press Agency/Photo Researchers. *Page 732:* Will McIntyre/Photo Researchers.

CHAPTER 23 *Opener:* T. Tracy/FPG International. Figure 23.18: Matt Bradley/Bruce Coleman, Inc. *Page 758:* H. R. Bramaz/Peter Arnold. Figure 23.32: Photograph by Tom Way, courtesy IBM, *Page 767B:* Peter Menzel/Stock, Boston.

PART 5 *Opener:* Donovan Reese/Tony Stone Worldwide.

CHAPTER 24 *Opener:* Fritz Prenzel/Tony Stone Worldwide. *Page 771:* Jurgen Vogt/The Image Bank. *Page 773:* Courtesy Astronomical Society of the Pacific. Figure 24.9: Courtesy Anglo Australian Telescope Board. Figure 24.16: Diane Schiumo/Fundamental Photographs. *Page 786:* Hans Neleman/The Image Bank. *Page 792:* Georges Ollen/Sygma.

CHAPTER 25 *Opener:* Mitch Reardon/Tony Stone Worldwide. Figure 25.5b: Gary Bistram/The Image Bank. *Page 798:* Dan McCoy/Rainbow. Figure 25.15: Science Source/Photo Researchers. Figure 25.19b: Ken Karp. Figure 25.20a: Courtesy Pontiac Motor Division. Figure 25.21b: Paul Silverman/Fundamental Photographs. *Page 807:* Jim and Cathy Church/FPG International. *Page 809:* S. Achernar/The Image Bank. *Page 810:* Robin Moyer/Gamma Liaison. *Page 812:* Courtesy Sandia National Laboratories.

CHAPTER 26 *Opener:* Kuhn Studios. *Page 825:* Masterfile. Figure 26.12b: H.M.S. Images/The Image Bank. *Page 827 (top):* Courtesy Johns Hopkins University School of Medicine. *Page 828:* Ann States/SABA. Figure 26.14c: Science Source/Photo Researchers. *Page 833:* Daryl Benson/Masterfile. Figure 26.26: Any Washnik. *Page 845:* Courtesy Bausch & Lomb. Figure 26.40: Courtesy Tasco. Figure 26.41: Courtesy Celestron, Inc.

CHAPTER 27 *Opener:* Joe McDonald/Bruce Coleman, Inc. Figure 27.7: From *Atlas of Optical Phenomena,* Michael Cagnet, Springer-Verlag, Berlin. Figure 27.10: Andy Washnik. *Page 875:* E.R. Degginger/Animals Animals. Figure 27.13: Mickey Pfleger Photography. Figure 27.15 and 27.16: Courtesy Bausch & Lomb. Figure 27.19: Courtesy Education Development Center. Figures 27.24 and 27.29b: From *Atlas of Optical Phenomena,* Michael Cagnet, Springer-Verlag, Berlin. Figure 27.31: Courtesy The Art Institute of Chicago. Figure 27.39: Courtesy Edwin Jones, University of South Carolina. *Page 901b:* Sipa Press.

PART 6 *Opener.* Roger Ressmeyer/Starlight.

CHAPTER 28 *Opener: (top left and bottom right)* Courtesy AIP Neils Bohr Library; *(top and center)* Courtesy Lotte Jacobi; *(top right)* The New York Times News Service; *(bottom left)* UPI/Bettmann Archive; *(center)* Philippe Halsman Studio. *Page 907:* Mark Antman/The Image Works. *Page 909:* Sygma. *Page 910:* Courtesy David Malin/Anglo Australian Telescope Board. Figure 28.7: Bill Marsh/Photo Researchers.

CHAPTER 29 *Opener:* David Scharf/Peter Arnold. *Page 934:* Science Photo Library/Photo Researchers. Figure 29.8a: From "Physics Review" 73,527 (1948), by Wollan, Shull, and Marney. Figure 29.8b: Courtesy Edwin Jones, University of South Carolina. Figure 29.9: Courtesy Akira Tonomura, J. Endo, T. Matsuda and T. Kawasaki. *Page 942:* Courtesy Central Scientific Co.

CHAPTER 30 *Opener:* Roger Ressmeyer/Starlight. Figure 30.3: Courtesy Bausch & Lomb. *Page 951:* Arthur Hustwitt/Leo de Wys, Inc. *Page 968 (center):* ATC Productions/The Stock Market. Figure 30.19b: Hank Morgan/Photo Researchers. Figure 30.20a: Mehau Kulyk/Photo Researchers. Figure 30.20b: Photo Researchers. *Page 969 (bottom):* Alexander Tsiaras/Photo Researchers. Figure 30.28: Ronald R. Erickson/Media Interface Ltd.

CHAPTER 31 *Opener:* Dan Winters/Onyx Enterprises, Inc. *Page 995:* Ricki Rosen/SABA. *Page 996:* Louie Psihoyos/Matrix. Figure 31.11: Sygma. *Page 998:* William Rivelli/The Image Bank. Figure 31.15: Dan McCoy/Rainbow. *Page 1006:* George Rodger/Life Magazine.

CHAPTER 32 *Opener:* Courtesy NASA. Figure 32.9: Courtesy Lawrence Livermore Laboratory. Figure 32.10: Science Photo Library/Photo Researchers. *Page 1022:* Will & Deni McIntyre/Photo Researchers. *Page 1031B:* Courtesy of The Miami Project.

INDEX

Q

R

S